Atomic structure

Atomic Structure

E. U. CONDON

Formerly Professor Emeritus and Fellow Emeritus,
Joint Institute for Laboratory Astrophysics,
University of Colorado, Boulder, Colorado

HALİS ODABAŞI

Professor, Department of Physics
Boğaziçi University, Istanbul, Turkey

CAMBRIDGE UNIVERSITY PRESS

Cambridge
London New York New Rochelle
Melbourne Sydney

CAMBRIDGE UNIVERSITY PRESS
Cambridge, New York, Melbourne, Madrid, Cape Town, Singapore, São Paulo, Delhi, Dubai, Tokyo

Cambridge University Press
The Edinburgh Building, Cambridge CB2 8RU, UK

Published in the United States of America by Cambridge University Press, New York

www.cambridge.org
Information on this title: www.cambridge.org/9780521298933

First published 1980
Re-issued in this digitally printed version 2010

A catalogue record for this publication is available from the British Library

Library of Congress Cataloguing in Publication data

Condon, Edward Uhler, 1902–1974.

Atomic structure.

Includes bibliographical references and index.

1. Atomic structure.
I. Odabaşi, Halis, joint author.
II. Title.
QC173.C6115 1978 539′.14 77-87378

ISBN 978-0-521-21859-7 Hardback
ISBN 978-0-521-29893-3 Paperback

Contents

Tables

Foreword by Ugo Fano

The publication in 1935 of *The Theory of Atomic Spectra* by E. U. Condon and G. H. Shortley was a milestone for modern physics. It concluded the sweep achieved by quantum mechanics in a mere decade by accounting for the complexities of atomic spectra in unified fashion and in considerable detail. The clarity and completeness of the Condon-Shortley treatment, which ranged from theoretical foundations all the way to the analysis of countless experimental data, have remained unsurpassed for over forty years. The text, often and affectionately quoted by its familiar initials "TAS," still provides a framework and a main reference for all workers in its broad and burgeoning field.

By 1950 the subject of TAS had grown in depth and scope, and the original printing was exhausted. As I recall, Cambridge University Press approached Professor Condon at that time and proposed a new and expanded edition, but Condon was then directing large research organizations and thus was unavailable for the major task laid before him. Cambridge University Press then decided wisely to reprint the original work; it did so repeatedly and eventually in a compact paperback format, which has now spread far and wide.

In 1962, however, Condon moved to the University of Colorado as a professor in the Department of Physics and Astrophysics and fellow of the Joint Institute for Laboratory Astrophysics, thus becoming free of managerial responsibilities – at least he so hoped. He returned then to the idea of a new TAS. To this scholarly work he would devote much of his remaining activity.

The range of new material to be incorporated was of major scope. Racah's work in the 1940s took the framework of TAS as a springboard but expanded its concepts to such depth and range that it exerted a major influence on nuclear and particle physics. Judd, in the 1960s, complemented Racah's approach by reinterpreting fractional parentage through the concept of quasi-spin, thus reducing it to a direct application of angular-momentum theory. These theoretical advances made extensive use of group-theoretical concepts and techniques well beyond the scope of TAS. (More recent theoretical approaches, such as the linking of spectral and collision theories by quantum defect procedures or the application of many-body theories to atoms, came too late and remain too fluid for incorporation in a current text.) From another side, the advent of computers – for *ab-initio* calculations as well as for data analysis – and the further impulse to Hartree-Fock calculations generated by Roothaan's analytical approach have multiplied the scope of theoretical applications. And all along, the gathering and analyzing of spectral data have continued and have spurted again with the advent of laser techniques of ever-widening diversity. Achieving a new synthesis of all this progress thus constituted an encyclopedic task.

The scope of the task Condon set for himself was further expanded by some of his specific interests and goals. First, and foremost for its consequences, was a drive toward recapturing the essence of the history and development of quantum theory, of which atomic spectra have been the main proving ground. As a result, the opening chapters of this new text have been greatly expanded and deepened as compared with the old TAS. In fact, they may well constitute the most successful product of the new effort. Condon's second drive was to involve himself again in actual research on atomic spectra to renew his firsthand "feel" for the subject in its current context. He started to do this with young collaborators, first with R. N. Zare and then with H. Odabasi, who soon became a partner in the entire undertaking. The last of the chapters included in the present volume, and the one closest to detailed applications, bears the influence of this original research.

Condon's ebullient nature led to his involvement in other activities, but he did work on the new TAS rather steadily until his sudden demise in 1974. The momentum of the undertaking was maintained by Odabasi's activity in spadework and in actual drafting. Thus by 1974 roughly half of the contemplated work, designed to consist of two volumes, was laid down in various stages of refinement.

The material that is now presented in this volume includes nearly all of the general theory that was envisaged and very little discussion of its specific applications. What exists stands by itself because most of it was constructed, in layers of increasing penetration into the subject, so that only occasionally does it require from the reader an investment of effort for a long-delayed payoff. Much important material will be missed, but what we have will very likely prove important and instructive.

Following Condon's death a collective effort was organized to assist Odabasi in his editorial task of tightening and finalizing the text. Professor Roy Garstang, of the University of Colorado, was of major and immediate help. Later the NBS Optical Physics Division, particularly John W. Cooper, carried a main responsibility for this effort, and their roles and import are credited in Odabasi's own preface. Care was taken not to interfere with the authors' original intention and style, even where the process of assimilation of new material and of generation of a new synthesis had not been fully developed. The material that is presented should, however, afford the reader a dependable and unified guide.

Preface

I first met Professor E. U. Condon in an overcrowded classroom when I enrolled for a graduate-level atomic spectroscopy course in the spring of 1964. This was the first course he taught at the University of Colorado. A full year later, I was trying, like every second-year graduate student, to find a paying summer job. Encouraged by the fact that I had done well in his course, I asked Professor Condon for a job; he said he would think about it. A few weeks later Dr. Richard Zare, then a research associate working with Professor Condon, called to tell me that Professor Condon wanted to see me about my summer job. This is how my association with Professor E. U. Condon began. It did not start off at all well, because I was busy that summer with many other things, mainly with getting married. However, later in the summer I was productive enough to be forgiven – so much so that he asked me to stay on to work with him. Needless to say, I accepted and regarded the opportunity as an honor. I have made many decisions in my life that I later regretted, but this is not one of them. I consider myself very fortunate to have had the opportunity to work with Professor Condon. He was not only a great physicist, but also a great person – pleasant, friendly, concerned, and above all humorous. He opposed any kind of prejudice. He said "If we could raise a generation without prejudice we would solve almost all of the problems on earth."

After obtaining my Ph.D. in the spring of 1968 I stayed on at the University of Colorado as a research associate to finish some calculations. In the summer of 1969 Professor Condon asked me to help him complete work on this book, which he had begun, and I gladly agreed. At that time the first two chapters were almost in their present form. After two further years of work Chapters 3 through 7 were completed.

We planned to write Chapters 8 and 9 during 1971; these two chapters would have completed the first volume of the planned two-volume work. Unfortunately, Mrs. Condon underwent surgery and was seriously ill for the rest of her life. This greatly saddened Professor Condon, and in addition, he had a recurrence of heart trouble and later had to have major heart surgery. These events greatly reduced his ability to work. I continued to write, and during 1972 Chapters 8 and 9 were drafted and a chapter on radiative transitions was also begun; this was intended for the second volume.

After spending a few months in the Department of Chemistry at

Yale University, I returned to Turkey to take a teaching position in the Physics Department of Boğaziçi University. Professor Condon and I corresponded regularly and made plans to get together during the summer of 1974 to continue work on the book. His letters after his heart surgery were very encouraging. Unfortunately, these plans were never realized because Professor Condon died on March 26, 1974. A year later Mrs. Condon also passed away. Together they had survived many hardships, troubles, and injustices. An account of the Condons' troubled years may be found in *Topics in Modern Physics – A Tribute to E. U. Condon* (W. E. Brittin and H. Odabasi, eds., Boulder, Colorado Associated University Press, 1971).

After Professor Condon's death, Cambridge University Press, with whom we had a contract to publish the book, kindly agreed to publish the nine chapters completed under Professor Condon's guidance. These nine chapters describe the fundamentals of atomic structure and form this book.

At the outset, it must be stated that the present book is not a "new Condon and Shortley," for their superb classic cannot possibly be replaced. This volume is intended to be a text for undergraduate seniors or first-year graduate students in physics, astrophysics, and chemistry. It assumes that the reader has at least an undergraduate-level background in quantum mechanics. The first chapter reviews the historical developments in atomic physics that lead to the development of quantum mechanics. The second chapter is a condensed description of quantum mechanical methods, techniques, and concepts. It also includes a treatment of the WBKJ approximation. This might seem a little out of place, because the central-field approximation has not yet been introduced. We felt that it belonged in this chapter, because it clarifies the connection between the radial-wave equation and the original quantum condition of Sommerfeld-Wilson as well as some other important features, such as quantum mechanical tunneling. While reading this section, keep in mind that the central-field approximation on the radial-wave equation will be explained in Chapter 4. Chapter 3 deals with angular momentum in general and spin in particular. It is now common for *nj* symbols to be widely used in the formulation of the coupling of angular momenta. Although they are neither compact nor handy, we have used them throughout the book. As mentioned previously, Chapter 4 defines the central-field approximation and covers the so-called "zeroth-order scheme" for determining the configuration energies and the Russell-Saunders term energies. More elegant and powerful schemes, such as Racah methods, are introduced and discussed in Chapter 5.

Group theory now plays an important role in our understanding of atomic structures and in our calculations. Much as we would have liked to discuss applications of group theory to atomic physics without having to give an introduction to group theory, we decided after some debate that we should present the theory of groups as well. The facts influencing our decision are discussed at the beginning of Chapter 6, which contains the theory of groups in general terms. The reader already familiar with the principles of group theory may omit this chapter altogether. In Chapter 7 the application of group theory to atomic physics is presented, and, at the end of this chapter, we have tried to give some insights into a very promising development in atomic physics – the application of noncompact groups. This sheds new light on the understanding of certain degeneracies and selection rules that cannot be explained by the symmetries contained in compact algebras. The subject of self-consistent field theories is taken up in Chapter 8, where the conventional Hartree-Fock theory and some approximate methods are introduced. The contents of Chapter 9 include the observed regularities in atomic spectra and a discussion of the accuracy of theoretical Russell-Saunders term energies computed by conventional Hartree-Fock or other approximate methods.

We could have added another chapter in order to include spin-orbit interaction, level energies, and different coupling schemes, but I think that the natural sequence of treating the magnetic interactions in general, and spin-orbit interaction in particular, comes *after* the introduction of the Dirac theory, the Breit equation, and the Pauli approximation. Professor R. H. Garstang and I are now working on a volume that is intended to be a sequel to the present one. It will contain discussions of Dirac theory, relativistic and magnetic effects, alternative coupling schemes, configuration mixing, radiative transitions, X-ray spectra, Zeeman and Stark effects, hyperfine structure, and ligand fields.

References given in the present book are mostly to materials that appeared before 1972. However, some later references were added during 1976 when the chapters were reviewed.

The completion of a project of this size depends on the assistance of many people. First of all, I would like to say how much I appreciated Mrs. Condon's enthusiastic encouragement and inspiration to us both. Many thanks are due to the Joint Institute for Laboratory Astrophysics and to the University of Colorado for providing our bread during the years that Professor Condon and I were working on the book. Specifically I would like to thank Drs. L. M. Branscomb, P. L. Bender, and S. J. Smith, and Professor J. Cooper. I also thank Mrs. K. Shapley for her invaluable help in typing and editing most of the chapters and Mrs. L. H. Volsky for her work on the last two chapters.

Professor A. G. Shenstone, Professor E. Merzbacher, Professor U. Fano, Dr. J. W. Cooper, Professor L. Armstrong, Jr., Professor M. Hamermesh, Professor B. R. Judd, Dr. A. W. Weiss, and Dr. R. D. Cowan read Chapters 1, 2, 3, 4, 5, 6, 7, 8, and 9, respectively, and made helpful comments. Drs. K. G. Kessler and J. W. Cooper organized and arranged all this; I greatly appreciate their kindness in taking the time and trouble to help me.

I owe special thanks to Dr. J. W. Cooper for spending so much of his time in revising parts of Chapters 8 and 9 and for going over the recommendations received from the readers. His contribution is a considerable one, and it is only because of his modesty that he does not receive recognition for his efforts on the title page of this volume. Special thanks are also due to Professor R. H. Garstang for independently reading the first five chapters and recommending certain modifications. I also would like to thank Professors A. O. Barut and W. E. Brittin for their encouragement and for proofreading Chapter 6. I thank Professor C. Froese Fischer for calculating certain parameters when we needed them and for commenting on Chapter 8. My thanks also go to Professor U. Fano for writing the Foreword.

I would like to thank Professor Apdullah Kuran, president of the Boğaziçi University, and Professor Erdal İnönü, dean of Arts and Sciences, for allowing me to spend many months in the United States during the completion of this volume.

It has been a pleasure to work with the American branch of Cambridge University Press. I owe many thanks to Mr. K. I. Werner and to Ms. Rhona B. Johnson for their invaluable help and cooperation.

Last but not least, I thank my wife, Marian Jeanne, for putting up with all of this and for being so patient during the years – nearly a decade – that it took to complete this book.

Halis Odabaşı

Istanbul, May 1978

1. Pre-quantum mechanical developments

> But I must confess I am jealous of the term *atom:* for though it is very easy to talk of atoms, it is very difficult to form a clear idea of their nature, especially when compound bodies are under consideration.
>
> Michael Faraday, *Experimental Researches in Electricity,* §869 (December 31, 1833).

1¹. Atomic physics before the twentieth century

The idea that all matter consists of aggregates of large numbers of relatively few kinds of fundamental particles is an old one. Traces of it are found in Indian philosophy about twelve centuries before the Christian era. Anaxagoras, and also Aristotle, expounded the opposite view: that matter is continuous and so can be subdivided without limit.

Among the Greeks, Democritus and Epicurus were exponents of the *atomic* view. The basic units of matter were called atoms from Greek roots meaning uncuttable or unsubdividable. The atomic view is eloquently stated in classic literature in *De Rerum Natura* by the Roman poet Lucretius.

The Aristotelian view prevailed in the Middle Ages. With the birth of the modern period of experimental science about four hundred years ago, the atomic view was revived by Galileo, Descartes, Robert Boyle, and Newton.

Although mostly studied for its astronomical content, Newton's *Principia* states the program which was later followed by theoretical chemistry in these words: " . . . I am induced by many reasons to suspect that they [the phenomena of nature] all depend on certain forces by which the particles of bodies, by some causes hitherto unknown, are either mutually impelled towards each other and cohere in regular figures, or are repelled and recede from each other; which forces being unknown, philosophers have hitherto attempted the search of nature in vain" [1].

The eighteenth century saw little progress in the development of atomism. Atomic ideas remained qualitative until John Dalton (1766–1844) gave them a quantitative form in the period 1803–1810, his work culminating in his *New System of Chemical Philosophy* [2]. He developed the idea that the atoms of different chemical elements have different weights, or masses. This view provides a simple interpretation of the observed facts concerning the relative combining weights of the elements in various chemical compounds. It was thus possible to infer *relative* atomic weights from simple hypotheses about the atomic composition of molecules of compounds.

Detailed working out of a consistent scheme of relative atomic weights was a slow process which largely occupied the attention of chemists in the first half of the nineteenth century. It was greatly helped by the formulation in 1811 by Amadeo Avogadro (1776–1856) of what we know today as Avogadro's hypothesis: that all gases at low pressure contain the same number of molecules in unit volume at the same conditions of temperature and pressure. Here molecule means the units that fly around in the gas, giving rise to the pressure as the average impulse in unit area of their collisions with the walls in unit time. This rule, together with the idea that gases of many of the elements, such as hydrogen, oxygen, nitrogen, consist of diatomic molecules, gives an interpretation of Guy-Lussac's experiments concerning the relative volumes of gaseous reactants and products involved in reactions of gases. More than half a century passed before chemistry realized the full significance of Avogadro's ideas, after Stanislao Cannizzaro (1826–1910) drew attention to them in 1858 [3].

At first little thought was given to the question of a possible underlying unity of the atoms of different chemical elements. Then in 1815 William Prout (1785–1850), observing that the then quite inaccurately known relative atomic weights were nearly all quite close to being integral multiples of that of hydrogen, suggested that atoms of all the elements might be compounds of hydrogen atoms. This idea became known as *Prout's hypothesis.* It was later discredited when accurate measurements showed that many of the atomic weights do not have integral values. We now know that this departure is due to the fact that the observed weights are averages over several isotopes, and also to mass loss occurring in formation of the atoms from their constituents. Thus, Prout's hypothesis survives in the modern form which supposes that all atoms of a particular isotope consist of hydrogen atoms and neutrons.

The laws of electrolysis, which were discovered by Michael Faraday (1791–1867) in 1833–34, clearly indicate the intimate association of electric charge with the structure of atoms, and also point to the atomicity of electric charge. In Section 852 of his *Experimental Researches in Electricity* he wrote: "The atoms of matter are in some way endowed or associated with electrical powers, to which they owe their most striking qualities, and amongst them their mutual chemical affinity." In Section 869 he wrote: "Or if we adopt the atomic theory or phraseology, then the atoms of bodies which are equivalent to each other in their ordinary chemical action, have equal quantities of electricity naturally associated with them. . . ."

The mid-nineteenth century was a period in which electrical science was concerned with the discovery and formulation of macroscopic laws of the electromagnetic field, so for a time this early discovery of atomicity of charge went uncultivated. It was emphasized again in 1881 by Hermann von Helmholtz (1821–1894) in a lecture to the Chemical Society in London [4] in which he said: "If we accept the hypothesis that the elementary substances are composed of atoms, we cannot avoid concluding that electricity also, positive as well as negative, is divided into definite elementary portions which behave like atoms of electricity."

Paralleling these advances were those involving the interpretation of thermal phenomena in terms of random molecular motions. Heat energy was interpreted as residing in the kinetic and potential energy of atomic motions. The second law of thermodynamics thus found interpretation as a natural tendency for molecular motions to proceed from ordered to disordered forms [5].

A most important step on the chemical side was the discovery in 1869 by Dmitri Mendeleev (1834–1907) of the periodic classification of the elements. When the elements are arranged in order of increasing atomic weight, there is a roughly periodic recurrence of elements having similar chemical properties. Similarity of chemical properties must mean similarity of atomic structure. This discovery gave the first convincing evidence that there is a common pattern to the structure of atoms of different elements. However, this evidence alone does not provide specific indications about the nature of the pattern [6].

In arriving at these results Mendeleev built mainly on the concept of chemical valence that had been introduced by Edward Frankland (1825–1899) in 1852.

Modern views of atomic structure stem from a group of discoveries all of which were made in the last decade of the nineteenth century:

X rays were discovered by Wilhelm Röntgen (1845–1923) in 1895, but their nature as akin to light of extremely short wavelength was not established until 1912 with the discovery of X-ray diffraction by crystals by Max Von Laue [7].

Radioactivity was discovered in 1896 by Henri Becquerel [8].

The *electron,* named by G. Johnstone Stoney in 1874 as the unit of electric charge, was shown by J. J. Thomson in 1897 to be a common constituent of many kinds of matter. Electrons are negatively charged particles with a mass about 1/1837 of that of a hydrogen atom [9].

The *quantum* of radiant energy was introduced by Max Planck in 1900 in his theory of the energy-frequency dependence of the spectrum of black-body radiation. Albert Einstein in 1905 showed the fruitfulness of the light quantum idea by applying it to interpretation of Stokes's law in fluorescence (that the fluorescent light usually has a longer wavelength than the light that excites it) and the photoelectric effect that the maximum energy of the emitted photoelectrons increases linearly with the frequency of the light causing them to be emitted. The name *photon* for light quantum was coined by G. N. Lewis in 1926 [10].

2¹. The nuclear atom model

By 1900 the main facts about the composition of chemical compounds were clearly understood in terms of the idea that matter is made up of atoms of some ninety elements, of which hydrogen has the least mass. These atoms were known from the facts of electrolysis to be associated with integral multiples of a basic atomic electric charge. They were known to have some kind of common pattern of structure, from the facts of the periodic system. They were known to include electrons, negative particles of mass 1/1837 that of hydrogen, as a common constituent in the structure of all of them.

In 1911 Ernest Rutherford at Manchester [11] studied the scattering of α particles from radioactive elements by thin foils of various materials. Although the greatest number of the α particles were observed to be deflected through small angles, a few were found to be deflected or scattered through large angles varying up to π. By studying the relative frequency of scattering through various angles in its dependence on the foil thickness, he showed that this large-angle scattering is attributable to single scattering acts, not to the chance aggregation of a large angle of deflection from many small deflections. In order for such large deflections to be produced, one must suppose that very large electric fields exist inside the atom. He was thus led to postulate what we know as the *nuclear* atom model, according to which atoms consist of a massive central nucleus, carrying a positive electric charge, which is an integral multiple of the magnitude of the negative charge carried by an electron. This nucleus is normally surrounded by enough electrons to make the atom neutral. Because of the smallness of their mass, the electrons do not contribute appreciably to the scattering of the α particles.

The simplest nucleus is that of hydrogen. Its nucleus has unit positive charge and a mass about 1836 times that of an electron. The hydrogen nucleus is called the *proton,* after a suggestion of Rutherford in 1920.

In 1913 Frederick Soddy and Kasimir Fajans had inferred, from systematic study of the chemistry of the naturally occurring radioactive elements, that the same chemical element could exist in more than one form. Atoms of different forms were referred to as *isotopes* of the element, the word referring to the property of having the same place in the periodic system. In that year J. J. Thomson demonstrated the existence of isotopes of several chemical elements by measuring the charge-to-mass ratio of their positive ions in an instrument that was a forerunner of the present mass spectrograph [12].

It thus appeared that the chemical properties are wholly determined by the atomic number, Z, which is the number of electrons contained in the atom, and is also equal to the integral number of positive charges in the nucleus [13]. Table 1[1] gives the names of the known elements, their chemical symbols and the values of Z.

At this stage only two fundamental particles were known: the electron and the proton. The masses of the isotopic species were all rather close to integral multiples of the mass of one hydrogen atom. Thus, as in Prout's hypothesis, it was natural to assume that the nuclei, themselves, are made of protons and electrons in a particularly stable and compact state of combination. However, this view was never a satisfactory one in relation to other facts of atomic physics.

In 1932 the picture was clarified by the discovery, by James Chadwick, of another type of fundamental particle [14]. It has slightly more mass than a hydrogen atom, and is electrically neutral and so is called the *neutron*. Discovery of the neutron made it possible to assume, as first proposed by Heisenberg and by Iwanenko, that the atomic nucleus is a compact compound of protons and neutrons, and that all the electrons in the atom are outside the nucleus. This view has stood the test of time and is undoubtedly correct. The generic term *nucleon* is now used to mean proton or neutron.

The total number of nucleons in the nucleus is called the *nucleon number*, denoted by A, so that a nucleus consists of Z protons and $(A - Z)$ neutrons.

For a neutral atom the number of electrons outside the nucleus, N, is equal to Z. When $N = Z - p$ the ion is said to be p-fold ionized. Sometimes p is indicated in the chemical literature by writing p plus signs as a superscript to the chemical symbol, thus Al^{+++} indicates an Al ion ($Z = 13$) having $N = 10$. Another common notation is to follow the chemical symbol by $C = (Z - N + 1)$ written as a roman numeral. Thus Fe XIV indicates an ion of iron ($Z = 26$) from which 13 electrons have been removed. The cases $N > Z$ correspond to negative ions, that is, atoms to which one or more extra electrons are attached. Species having the same number of electrons N form an *isoelectronic sequence*.

It is customary [15] to attach A as a left superscript to the chemical symbol. Then $(Z - N)$ is indicated by plus signs as a right superscript. Thus ${}^{238}U^{+++}$ means a threefold ionized atom of the $A = 238$ isotope of uranium. This leaves the right subscript free to be given its traditional meaning: the number of atoms contained in a molecule of a compound.

Writing $M({}^{A}Z)$ for the mass of a neutral atom, where Z stands for the chemical symbol of the element having this atomic number, and

Table 1[1]. Atomic numbers and symbols of the elements

Name	Symbol	*Z*	Name	Symbol	*Z*
Actinium	Ac	89	Lanthanum	La	57
Aluminum	Al	13	Lawrencium	Lw	103
Americium	Am	95	Lead	Pb	82
Antimony	Sb	51	Lithium	Li	3
Argon	Ar	18	Lutetium	Lu	71
Arsenic	As	33	Magnesium	Mg	12
Astatine	At	85	Manganese	Mn	25
Barium	Ba	56	Mendelevium	Md	101
Berkelium	Bk	97	Mercury	Hg	80
Beryllium	Be	4	Molybdenum	Mo	42
Bismuth	Bi	83	Neodymium	Nd	60
Boron	B	5	Neon	Ne	10
Bromine	Br	35	Neptunium	Np	93
Cadmium	Cd	48	Nickel	Ni	28
Calcium	Ca	20	Niobium	Nb	41
Californium	Cf	98	Nitrogen	N	7
Carbon	C	6	Nobelium	No	102
Cerium	Ce	58	Osmium	Os	76
Cesium	Cs	55	Oxygen	O	8
Chlorine	Cl	17	Palladium	Pd	46
Chromium	Cr	24	Phosphorus	P	15
Cobalt	Co	27	Platinum	Pt	78
Copper	Cu	29	Plutonium	Pu	94
Curium	Cm	96	Polonium	Po	84
Dysprosium	Dy	66	Potassium	K	19
Einsteinium	Es	99	Praseodymium	Pr	59
Erbium	Er	68	Promethium	Pm	61
Europium	Eu	63	Protactinium	Pa	91
Fermium	Fm	100	Radium	Ra	88
Fluorine	F	9	Radon	Rn	86
Francium	Fr	87	Rhenium	Re	75
Gadolinium	Dg	64	Rhodium	Rh	45
Gallium	Ga	31	Rubidium	Rb	37
Germanium	Ge	32	Ruthenium	Ru	44
Gold	Au	79	Rutherfordium	Rf	104
Hafnium	Hf	72	Samarium	Sm	62
Hahnium	Ha	105	Scandium	Sc	21
Helium	He	2	Selenium	Se	34
Holmium	Ho	67	Silicon	Si	14
Hydrogen	H	1	Silver	Ag	47
Indium	In	49	Sodium	Na	11
Iodine	I	53	Strontium	Sr	38
Iridium	Ir	77	Sulfur	S	16
Iron	Fe	26	Tantalum	Ta	73
Krypton	Kr	36	Technetium	Te	43

Table 1[1] (cont.)

Name	Symbol	Z	Name	Symbol	Z
Tellurium	Te	52	Uranium	U	92
Terbium	Tb	65	Vanadium	V	23
Thallium	Tl	81	Xenon	Xe	54
Thorium	Th	90	Ytterbium	Yb	70
Thulium	Tm	69	Yttrium	Y	39
Tin	Sn	50	Zinc	Zn	30
Titanium	Ti	22	Zirconium	Zr	40
Tungsten	W	74			

regarding the atom as made up of Z hydrogen atoms, ^{1}H and $(A - Z)$ neutrons, the total mass of the constituents is

$$[ZM(^1\mathrm{H}) + (A - Z)M_n]$$

in which M_n is the mass of the neutron. Experiment shows that $M(^AZ)$ in all cases is less than this by somewhat less than 1 percent. This is known as the *mass defect* of the atom,

$$\Delta(^AZ) = ZM(^1\mathrm{H}) + (A - Z)M_n - M(^AZ) \tag{1}$$

It is interpreted as energy by Einstein's (1905) discovery of the equivalence of mass and energy [16] given by $E = Mc^2$, where c is the velocity of light. When the constituents come together to form a stable atom, the binding energy is released in various forms. The loss in energy shows itself as a loss in mass. There is presumably a similar reduction in mass when a stable chemical compound is formed from its constituent atoms. However, chemical bond energies are of the order of 10^{-6} of those involved in the formation of nuclei. Thus the mass defects due to molecule formation are of the order of 10^{-8} of the total mass, leading to an imperceptibly small violation of the familiar law of conservation of total mass in chemical reactions.

Determination of the atomic masses, $M(^AZ)$, has had a long history. Chemical methods based on gravimetric analysis of pure compounds give only *relative* values of $\overline{M}(Z)$, the mean of the $M(^AZ)$ over the isotopes of element Z, weighted by their actual relative abundances in the determination in question. These were long expressed in the *chemical* scale which arbitrarily adopts $\overline{M}(\mathrm{O}) = 16$ exactly for oxygen. Later this was replaced by the *oxygen-physical* scale, which arbitrarily adopts $M(^{16}\mathrm{O}) = 16$ exactly, because natural relative abundances were found not to be constant.

In 1960 the International Union of Pure and Applied Physics, and the International Union of Pure and Applied Chemistry adopted the *carbon-physical* scale, which arbitrarily adopts $M(^{12}\mathrm{C}) = 12$ exactly. The quantity $u = (1/12)M(^{12}\mathrm{C})$ is now called the *unified mass unit*. It is slightly larger than $(1/16)M(^{16}O)$, and therefore atomic masses on the carbon-physical scale are represented by slightly smaller numbers than the same masses on the oxygen-physical scale; this relation is

$$M_{\text{phys-oxygen}} = (1.000317917 \pm 0.000000017)M_{\text{phys-carbon}}$$

The charged particles, electrons and protons, interact according to Coulomb's law, suitably modified to take into account magnetic effects arising from their relative motion, and magnetic dipole interaction due to their possession of magnetic dipole moments. Although electrically neutral, the neutron also possesses a magnetic dipole moment and thereby interacts magnetically with other particles in an atom. In addition, the nucleons interact with each other by means of specifically nuclear forces, called *strong interaction,* which are not known to be directly related to the electromagnetic interactions. Strong interaction is short range, being negligible between particles separated by more than 10^{-12} cm, as contrasted with the long-range character of the Coulomb interaction energy which varies as the inverse first power of the distance between the interacting charged particles.

In the early study of radioactivity the radiations emitted by natural radioactive elements were recognized to be of three types, called α, β, and γ, at a time when their nature was not known. The α particle was later found to be a helium nucleus, $^4He^2$, ejected at high speed from the parent nucleus. The transformation is written as

$$^{A}Z \rightarrow {}^{(A-4)}(Z-2) + {}^{4}\text{He} \tag{2}$$

The symbols represent neutral atoms. The helium atom actually is doubly ionized, and also the daughter nucleus is multiply ionized, so that several of the Z electrons are free from the ions. These recombine later with the ions to form neutral atoms.

The γ radiation was recognized to be electromagnetic radiation such as that of light and X rays, but it originated in the nucleus.

The β- radiation was observed to consist of high-speed electrons emitted from the transforming nucleus, in a process by which the daughter nucleus had the same A as the parent, but one higher value of Z. The electrons are emitted with a statistical distribution of energies from parents that are in a definite initial energy state, and give rise to daughters in a definite final energy state. In this result there was an apparent violation of the principle of conservation of energy. This led to the recognition by Pauli that the process involves the emission of another particle, of zero-charge and zero-rest mass, which was not detected in the usual experiments. This particle is called the *neutrino.* In different instances of disintegration the constant total available energy is divided in different ways between the electron and the neutrino, to give a statistical distribution of the observed electron energy.

The process is thus representable by the relation

$$^{A}Z \rightarrow {}^{A}(Z+1)^{+} + e + \nu \tag{3}$$

where ν is written for the neutrino. Ultimately, an electron is captured by the daughter ion to form a neutral daughter atom,

$$e + {}^{A}(Z+1)^{+} \rightarrow {}^{A}(Z+1) \tag{4}$$

In 1932 the positive electron or *positron* was discovered by Carl Anderson. It is a particle having the same mass as an electron, but a positive charge of magnitude equal to that of the electron. We denote it by e^{+}. Such particles have only a transient existence, in view of various processes by which an e^{+} can combine with an e, both being annihilated, their energy usually appearing as electromagnetic radiation. In 1934 artificial radioactive elements were discovered of a type which emit positrons, and a different kind of neutrino, here written ν', according to the relation

$$^{A}Z \rightarrow {}^{A}(Z-1) + e + e^{+} + \nu' \tag{3'}$$

Here e^{+} is the emitted positron, and ν' the accompanying neutrino, and e is the extra electron set free from the daughter atom because of its reduced atomic number. Subsequently the positron annihilates with an electron to convert their energy into photons.

Both types of radioactivity, (3) and (3′) are called β decay. A unified theory for them has been developed in terms of another kind of interaction that is called *weak interaction*. The weak interaction is also sometimes called the four-fermion interaction, because basically four particles are involved, and these are of the type known as fermions. In the first type a neutron in the nucleus is transformed into a proton, an electron and a neutrino, the latter two being ejected. In the second type a proton in the nucleus is transformed into a neutron, a positron and a neutrino, the latter two being ejected [17].

To summarize, an atom consists of a nucleus which contains A nucleons, Z protons and $(A - Z)$ neutrons, held together by strong interaction forces, modified slightly by electromagnetic interactions of the nucleons, the nucleus being surrounded in the neutral atom by Z electrons which interact electromagnetically with each other and with the nucleons of the nucleus. In addition, the particles of the atom interact with the electromagnetic field to produce emission, absorption and scattering of light, the word light being used here to mean electromagnetic radiation of any frequency.

For a given A, only certain values of Z give nuclei that are stable. Values of Z that are too large for stability are reduced by β processes in which a positron is emitted, and those that are too small for stability have their Z increased by β processes in which an electron is emitted.

3[1]. Spectroscopy before the twentieth century

A great deal of what is known about atomic structure is derived from the study of the light that is emitted and absorbed by atoms, especially when they are in the gaseous state, which minimizes the effects of mutual disturbances.

The basic discovery that white light consists of a variety of colors, which can be separated as to direction of propagation because the index of refraction of glass (and other transparent materials) increases on going from red to violet, was made in 1666 by Isaac Newton [18].

Even though diffraction of light at a single hair was discovered by Grimaldi (1613–63) and the interference colors produced by thin films had been studied by Newton, it was not until 1801 that Thomas Young discovered the essential feature of optical interference, which is the combination of positive and negative quantities in light rays which are brought together after travelling from the same source by two different paths [19]. His key conclusion is contained in these words: "Whenever two portions of the same light arrive to the eye by different routes, either exactly or very nearly in the same direction, the light becomes most intense when the difference of the two routes is any multiple of a certain length, and least intense in the intermediate state of the interfering portions; and this length is different for light of different colours." This characteristic length is what we now call the wavelength.

In 1802 W. H. Wollaston repeated Newton's prism experiment of 1666 but let in the sunlight through a narrow slit, parallel to the prism axis, instead of using a round opening as Newton had, and discovered the solar spectrum to be crossed by seven dark lines. He also studied spectra of flames and electric sparks [20]. The discovery went unnoticed until made again by Joseph von Fraunhofer who, in 1814–15, presented to the Munich Academy a memoir mapping many dark lines in the solar spectrum which he named by letters of the alphabet. Some of his terminology is still in use, as when we speak of the D lines of sodium and the H and K lines of ionized calcium. These dark lines in the solar spectrum, of which he mapped 350, have ever since been called *Fraunhofer lines*. Fraunhofer also observed dark lines in the spectra of stars. Noting differences in the lines occurring in the stellar and solar spectra he inferred that they do not originate in the earth's atmosphere.

Discovery of the invisible extension of the spectrum beyond the red to the infrared, made by Sir William Herschel [21], and beyond the violet to the ultraviolet, made by J. W. Ritter [22], also occurred in the opening years of the nineteenth century.

A number of investigations of the spectral lines in emission and absorption was published in the first half of the nineteenth century

[23]. In these there is a more or less clear recognition of the correlation of particular lines with particular metal vapors in the flame that serves as the light source [24].

Brewster devoted attention to study of absorption spectra. J. D. Forbes expressed the view that coincidence of Fraunhofer lines in the solar spectrum with those in flame spectra proves the presence of certain substances in the sun, but his identifications were faulty.

D. Brewster and J. H. Gladstone [25] published a detailed drawing of the solar spectrum, showing more than 8000 lines. They had shown that some of the lines originate in the earth's atmosphere and then drew the erroneous conclusion that they all do.

The first photograph of the solar spectrum was published by E. Becquerel and independently by J. W. Draper [26].

In 1849 Léon Foucault observed that the *D* lines in the solar spectrum are made even darker by passing the sunlight through an electric arc, although the *D* lines appear bright in the spectrum of the arc by itself. He did not give an interpretation of these observations nor did he connect the *D* lines with sodium.

Numerous spectroscopic papers were published in the 1850's but perhaps the most important were those of J. Plücker [27], who began the study of spectra emitted by electric glow discharges through gas at low pressures (Geissler tubes). Under these conditions, he found the spectrum to be determined by the gas in the tube and not affected by the metal of which the electrodes were made.

A. J. Ångström [28] expressed the view that the dark absorption lines and the bright emission lines in the spectrum arise because of characteristic internal vibration frequencies of the atoms or molecules in the source. This fundamental result precedes the work of Balfour Stewart [29] who built on Prévost's law of exchanges, according to which bodies in thermal equilibrium are simultaneously absorbing and emitting radiant energy, that a body radiates in the same way whether other bodies are present or not, and that emission and absorption rates are equal. Stewart first enunciated the conclusion that for each wavelength the emission rate must balance the absorption rate.

These ideas concerning relation of emission rate to absorption rate were independently developed by Gustav Kirchhoff [30].

The final result is that for a given type of radiation, specified by wavelength, direction of propagation and state of polarization, the emissive power of any body at a given temperature, divided by its coefficient of absorption, is the same for all bodies, and therefore equal to the emissive power for an ideal black body whose coefficient of absorption is unity.

The first discovery of a numerical regularity in atomic spectra was made by W. N. Hartley [31]. In the spectrum of zinc are triplets, that is, groups of three closely spaced lines. When the lines are character-

Table 2[1]. The wavelengths of the first four Balmer lines and of the Balmer series limit*

	n	λ_n (10^{-8} cm)	λ_n^{-1} (cm^{-1})	$R = \lambda_n^{-1}\left(\frac{1}{4} - \frac{1}{n^2}\right)^{-1}$
H_α	3	6562.8	15238.	109713.
H_β	4	4861.3	20571.	109712.
H_γ	5	4340.5	23039.	109710.
H_δ	6	2101.7	24380.	109710.
lim	∞	3646.	27427.	109708.

*The calculated values of R are constant to the accuracy of these λ_n. The series limit, λ_∞ is calculated by assuming the value of R on the last line.

ized by the wave number (number of waves in unit length, reciprocal of vacuum wavelength) he found that the intervals between the lines in various triplets are the same. From that time (1883) forth, wave numbers should have been generally used to characterize the lines, but the practice of stating wavelengths was already so well entrenched that the change was never made.

The discovery that sets of lines in the spectra of alkalis and alkaline earths are related in series was made in 1879 by G. D. Liveing and James Dewar [32]. They found series of singlets, doublets and triplets, converging to limits at short wavelengths, some series characterized by having all members sharp, or all diffuse, with intensity diminishing and diffuseness increasing as the series limit is approached.

Six years later, in 1885, came the discovery by J. J. Balmer [33] of a simple formula relating the wavelengths of a series of lines in the visible spectrum of hydrogen. This is now famous as the *Balmer series.* It was first explained in terms of the nuclear model of the hydrogen atom by Niels Bohr in 1913. Although Balmer expressed it in terms of wavelengths it is written more significantly in wave numbers

$$\sigma_n = \lambda_n^{-1} = R\left(\frac{1}{2^2} - \frac{1}{n^2}\right) \tag{1}$$

in which $R = 109{,}677.58\ \text{cm}^{-1}$. The Balmer lines are usually designated as H_α, H_β, H_γ, . . . , although after the first seven or eight it is usual to drop the Greek letters and resort to numbering. The wavelengths of the first four lines, and of the series limit are given in Table 2[1].

Five years later the Balmer formula was generalized by J. R. Rydberg [34] to represent the series which Liveing and Dewar had found in the alkali spectra. The Rydberg formula, expressed in wave numbers, is a simple generalization of Balmer's:

$$\lambda_n^{-1} = \lambda_\infty^{-1} - \frac{R}{(n - \Delta)^2} \tag{2}$$

Here n takes on integral values, λ_∞^{-1} is the wave number of the series limit, and Δ, now called the quantum defect, is nearly constant in a particular series. The striking discovery was made that R has nearly the same value in all series and in all elements. It is now known as the *Rydberg constant.*

In 1890 H. Kayser and C. Runge [35] introduced the names: principal series, first (diffuse) subordinate series, and second (sharp) subordinate series, for the doublet series found in alkali spectra by Liveing and Dewar. The values of Δ are different in the three series, but it is found that for each element λ_∞^{-1} has the same value for the diffuse series and the sharp series. The two components of the same doublets are often represented by two slightly different values of Δ.

In 1900 Rydberg recognized that the wave numbers of all of the lines in the several series of the same element could be represented by taking appropriate differences between *terms* (expressed in wave numbers). The importance of this result was also stressed by W. Ritz so it is now commonly known as the *Rydberg-Ritz* combination principle [36].

In all of this early work it was found that the wavelength of a line remains quite constant whether in emission or absorption and through wide variations in the physical conditions in the source. In contrast, the relative intensities are extremely sensitive to source conditions.

This brief review of early spectroscopy closes with mention of the discovery in 1896 by P. Zeeman [37] of the effect on the spectral lines of placing the light source in a magnetic field, now called the *Zeeman effect.* He found the lines to split into patterns of close components, some polarized along the magnetic field, some perpendicular thereto. The perpendicular components show circular polarization, some right and some left, when observed in the direction of the magnetic field. At once H. A. Lorentz [38] developed a theory of the effect based on the idea that the light is emitted by an isotropic harmonic oscillator, made of a particle of charge e and mass m, whose motion is disturbed by the magnetic force $(e/c)\mathbf{v} \times \mathbf{B}$ acting on the moving charge. Lorentz's theory accounted for many of the observed facts of the Zeeman effect on assuming that the radiating charged particle in the atom has the same e/m as the cathode ray particles studied by J. J. Thomson in beams outside of atoms. This result gave a powerful impetus to the then new idea that electrons are a universal constituent of all atoms.

The Lorentz theory predicts the same behavior for all spectral lines. The light emitted in directions normal to the magnetic field consists of an undisplaced component (ν_0) with its linearly polarized electric vector along **B**, and two displaced components, having frequencies ($\nu_0 \pm eB/4\pi mc$), which are polarized with the electric vector perpendicular to the magnetic field. In directions along the magnetic field, the displaced components are found to have circular polarization in opposite senses.

These simple predictions were all verified on the lines first studied by Zeeman. But this happy state of affairs ended when Thomas Preston [39] found lines in the spectrum of zinc (4722) and cadmium (4800) which split into quadruplets instead of triplets in the magnetic field. Similar results were soon found by many others. All of these exceptional cases came to be called the *anomalous* Zeeman effect, although the exceptions soon proved to be much more numerous than the rule. The explanation of the anomalous effect had to wait until the hypothesis of electron spin was introduced in 1925.

4[1]. Introduction of Planck's constant

The most outstanding feature of quantum ideas involves the concept that many of the important quantities associated with the description of atomic phenomena can assume only a *discrete set of allowed values.* In contrast, on the classical theory analogous quantities take on a continuous range of values. Examples are the angular momentum of a particle moving in a central field of force, the values of the energy of an atom or molecule in closed states (in which none of the constituents move in an unlimited region of space).

This idea made its first appearance in physics in the derivation given by Max Planck [40] of his law for the relative energy density at different frequencies of electromagnetic radiation that is in thermodynamic equilibrium with matter. Such radiation is commonly called *black-body radiation,* because the relative intensity at different frequencies is that due to a perfect absorber or "black body." By general thermodynamic arguments, this equilibrium distribution is independent of the particular kind of emitting and absorbing atoms by which energy is exchanged between matter and the radiation field. He therefore chose to study the problem of equilibrium exchange of radiant energy between a set of harmonic oscillators and the radiation field, as being one which he expected to be simpler than others. In order to get agreement with the experimental observations, he had to assume that such an oscillator can only exist in states whose energy W is one of the discrete values,

$$W_n = nh\nu, \quad n = 0, 1, 2, 3, \ldots \tag{1}$$
$$(h = 6.26 \times 10^{-27} \text{ erg} \cdot \text{sec, App. 1})$$

where ν is the frequency (cycles/unit time) of the oscillator, h is a new universal constant, now called *Planck's constant,* and n is an integer. On this view, when an oscillator emits radiation it does so by making a transition to the next lower level ($n \rightarrow n - 1$) thereby emitting one quantum of light ($W = h\nu$) to the radiation field.

He assumed that absorption of radiant energy also takes place discontinuously, corresponding to the absorption from the radiation field of one quantum ($W = h\nu$) of radiant energy accompanied by a transition of the oscillator to the next higher level ($n \rightarrow n + 1$). This revolutionary view brought physics face-to-face with the wave-particle duality that troubled physicists for the next thirty years. All of the nineteenth-century successes of the electromagnetic theory of light, in explaining interference and diffraction effects, were based on the view that light energy is continuously distributed over the region of space occupied by an electromagnetic field. Efforts to interpret such phenomena on the view that light energy is propagated as a stream of discrete localized particles had all proved unsuccessful. Hence, by 1900 physicists had generally adopted the view that light energy is continuously distributed in space as an electromagnetic wave field.

This gives rise to a difficulty with the problem of absorption of light from weak beams. On ordinary views the absorbing oscillator is influenced only by the electromagnetic field in its immediate vicinity and if the beam is sufficiently weak the energy involved in such a region will be but a small fraction of one quantum. This led Planck for a time to suppose that the absorption process must be continuous, while only the emission process is discontinuous, but this view was later given up as unfruitful.

Contrary to the prevalent idea that all progress in physics is made by very young men, the major discovery by Planck, which led him to introduce this constant into physics, occurred when he was 42. Planck was born in 1858 and became a student at the University of Munich in 1875. He went to Berlin in 1877 where he studied under Kirchhoff and Helmholtz and received his doctorate at Munich in 1879 for a thesis on the second law of thermodynamics. In 1889 he returned to Berlin as the successor of Kirchhoff in the theoretical professorship there.

Thus 21 years elapsed between his doctorate and the publication of the work that completely altered the course of physics. These were years which he mainly devoted to studies of thermodynamics – what we regard as physical chemistry today – topics such as the thermodynamics of potential differences between dilute solutions of electrolytes. Classical thermodynamics remained a major interest all his life; he published a paper of major importance on the Braun-LeChâtelier principle in 1935 at the age of 77.

He began to be actively interested in the theory of the full radiation from hot bodies in 1896, stimulated by the experimental work on this subject at the Physikalisch-Technischen Reichsanstalt and the discovery in 1894 by Willy Wien of the Wien displacement law.

Although the study of such radiation goes back to the eighteenth century, theoretical application of thermodynamics to it dates from the work of Balfour Stewart and Kirchhoff in 1859. Boltzmann gave the thermodynamic derivation for the T^4 variation of total radiation in 1884, when Planck was only 26 years old.

In 1900, Lord Rayleigh published a law based on equipartition of energy in the degrees of freedom of the radiation field which agrees with experiment for long wavelength. James Jeans corrected an error of a factor of eight that Rayleigh had made in counting the number of degrees of freedom per unit volume of the electromagnetic field in a given frequency range. The law became the Rayleigh-Jeans law.

New measurements of the energy distribution over the spectrum were being made at the Physikalisch-Technischen Reichsanstalt. On 19 October 1900 Planck presented the radiation formula he had found on a semiempirical basis to the Physical Society of Berlin. That evening Heinrich Rubens carefully compared the best measurements of O. Lummer and E. Pringsheim with the Planck formula, found that they fit it extremely well, and called on Planck the next morning to tell him so; this result was communicated to the Prussian Academy on October 25.

Planck then set to work seriously to find a theoretical basis for the formula. He has spoken of this period as the most difficult of his whole life. On 14 December 1900 he gave a paper to the Physical Society of Berlin in which he took the decisive step. Applying Boltzmann's principle for the connection of entropy with probability, he worked out the spectral distribution of energy that would be in equilibrium with a system of electrical oscillators.

To obtain the desired result he had to suppose that the energy of each oscillator was built up of finite quanta of energy. In all of physics hitherto energy could apparently vary over a continuous range of values; Planck had now to assume that it varies discontinuously. To agree with the Wien displacement law he had to assume the finite size of the quanta to be proportional to the frequency, ν, so the energy quanta were $h\nu$. Finally there resulted the famous formula

$$\rho(\nu) = \frac{8\pi}{c^3}\frac{h\nu^3}{e^{h\nu/kT} - 1} \tag{2}$$

for the density of energy in unit volume, per unit frequency interval at ν. Here c is the speed of light, and k is what we now call the *Boltzmann constant*. The formula can be written more significantly as

$$\rho(\nu)d\nu = \frac{h\nu}{e^{h\nu/kT} - 1}\frac{8\pi\nu^2 d\nu}{c^3} \tag{3}$$

The first factor on the right is the mean energy at temperature T of an oscillator which can take on energies that are restricted to integral multiples of $h\nu$ and $8\pi\nu^2 d\nu/c^3$ is the number of equivalent oscillators in unit volume in this frequency range. The Rayleigh-Jeans law applies in the limit $\nu \rightarrow 0$ for which

$$\frac{h\nu}{e^{h\nu/kT} - 1} \rightarrow kT. \tag{4}$$

In the original form Planck's formula was written

$$\rho d\nu = \frac{c_1 \lambda^{-5} d\lambda}{e^{c_2/\lambda T} - 1} \tag{5}$$

where $\lambda = c/\nu$ and c_1 and c_2 are two uninterpreted constants to be fitted empirically to the observations. The derivation that Planck produced gave interpretations for c_1 and c_2, namely,

$$c_1 = 8\pi hc, \quad c_2 = \frac{hc}{k} \tag{6}$$

Regarding c_1 and c_2 as constants to be chosen to fit the radiation measurements, one can solve for the two constants h and k that were introduced by Planck in his quantum statistical derivation. With the values then available he found

$$\begin{aligned} h &= \frac{c_1}{8\pi c} = 6.55 \times 10^{-27} \text{ erg} \cdot \text{sec} \quad (6.6262) \\ k &= \frac{c_1}{8\pi c_2} = 1.346 \times 10^{-16} \text{ erg/deg} \quad (1.38062) \end{aligned} \tag{7}$$

This value for h is about 1 percent too low and for k about 2.5 percent too low, relative to the best modern values (as given in parentheses, Appendix 1), indicating the accuracy of the early Physikalisch-Technischen Reichsanstalt measurements. These measurements were later greatly improved by W. W. Coblentz at the National Bureau of Standards.

This first calculation of h in Planck's lecture of 14 December 1900 marks the real birth date of quantum physics; Planck's h was an entirely new natural constant whose meaning long remained obscure. More interest attached immediately to k, for

$$R = Nk$$

where R is the ideal gas constant per mole and N is Avogadro's number. Also

$$F = Ne$$

where F is the Faraday, which is defined as the charge carried in electrolysis by a mole of univalent ions, and e is the charge of the electron. At that time both R and F, being macroscopic molar quantities, were quite accurately known. The Planck theory of radiation, in giving an experimental value for k served at once to determine N and thus to determine the charge on the electron.

The electron had first been recognized by J. J. Thomson as an important entity only three years earlier, and at that time its charge was only very inaccurately known. Richarz had given $e = 1.29 \times 10^{-10}$ esu from F and an estimate of N based on the constant b in van der Waals's equation.

Thomson gave $e = 6.5 \times 10^{-10}$ esu from measurements on charged cloud droplets, which in 1902 he revised down to $e = 3.4 \times 10^{-10}$ esu.

Planck's 1900 value of k, combined with the then known value of the Faraday, led him to $e = 4.69 \times 10^{-10}$ esu, which was only 2.3 percent below today's recognized value.

Thus Planck had accomplished all of this in the fall of 1900: (1) He found an empirical formula that to this day gives an accurate representation of the spectral distribution of radiant energy; (2) he found a derivation of that formula which introduced the extraordinary idea of energy quantization into physics; and (3) he obtained an excellent value for the charge on the electron. One would expect all this to cause a great deal of excitement among physicists. But it did not; almost nothing was written about Planck's work in 1901–04.

The first real extension came with Einstein's famous paper of 1905. Nothing much happened in between. Even Planck wrote only one other paper on the subject [41]. There is one paper by S. H. Burbury on "Irreversible Processes and Planck's Theory" [42], and a paper by J. D. van der Waals, Jr., which tries to apply the methods of Gibbs' statistical mechanics to radiation [43]. Otherwise nothing; Planck was almost completely ignored.

In his brief *Scientific Autobiography* Planck tells [44] of his own attitude toward h:

> While the significance of the quantum of action for the interrelation between entropy and probability was thus conclusively established, the great part played by this new constant in the uniform regular occurrence of physical processes still remained an open question. I therefore tried immediately to weld the elementary quantum of action h somehow into the framework of classical theory. But in the face of all such attempts, the constant showed itself to be obdurate.
>
> So long as it could be regarded as infinitesimally small, i.e. when dealing with higher energies and longer periods of time, everything was in perfect order. But in the general case difficulties would arise at one point or another, difficulties which became more noticeable as higher frequencies were taken into consideration. The failure of every attempt to bridge this obstacle soon made it evident that the elementary quantum of action plays a fundamental part in atomic physics, and that its introduction opened up a new era in natural science. For it heralded the advent of something entirely unprece-

dented, and was destined to remodel basically the physical outlook and thinking of man which, ever since Leibnitz and Newton laid the groundwork for infinitesimal calculus, were founded on the assumption that all causal interactions are continuous.

My futile attempts to fit the elementary quantum of action somehow into the classical theory continued for a number of years, and they cost me a great deal of effort. Many of my colleagues saw in this something bordering on a tragedy. But I feel differently about it. For the thorough enlightenment I thus received was all the more valuable. I now knew for a fact that the elementary quantum of action played a far more significant part in physics than I had originally been inclined to suspect, and this recognition made me see clearly the need for the introduction of totally new methods of analysis and reasoning in the treatment of atomic problems.

In spite of Jeans' intimate association with the radiation problem, we find no mention of Planck's work in the first edition of his *Dynamical Theory of Gases,* published in 1904. In the Landolt-Bornstein Tabellen for 1905 we find quite discordant values for the number of molecules in 1 cm^3 under standard conditions: 2.1×10^{19} for air, 4.2×10^{19} for nitrogen, and 7.3×10^{19} for hydrogen – in direct contradiction to the well-established rule of Avogadro that this number is the *same* for all gases in the ideal gas limit. Planck found 2.76×10^{19} for this number.

Apparently the quantum hypothesis used by Planck in 1900 was so foreign to the modes of thought of physicists that they could not take seriously a value for the charge on the electron obtained in this way.

Josiah Willard Gibbs was America's first great theoretical physicist. He died in 1903 at the age of 64. There is no indication that he paid any attention to Planck's work, although he had puzzled over the problem of the specific heats of polyatomic gases which are too low to correspond to the classical value from equipartition of energy [45]. Apparently he found these difficulties somewhat depressing, as indicated for instance in this paragraph from the preface to his *Statistical Mechanics* [46]:

In the present state of science it seems hardly possible to frame a dynamic theory of molecular action which shall embrace the phenomena of thermodynamics, of radiation and of the electrical manifestations which accompany the union of atoms. Yet any theory is obviously inadequate which does not take account of all these phenomena. Even if we confine our attention to the phenomena distinctively thermodynamic we do not escape difficulties in as simple a matter as the number of degrees of freedom of a diatomic gas. It is well known that while theory would assign to the gas six degrees of freedom per molecule, in our experiments on specific heat we cannot account for more than five. *Certainly one is building on an insecure foundation, who rests his work on hypotheses concerning the constitution of matter.*

Lord Kelvin died in 1907 at the age of 83. His Baltimore lectures were delivered at Johns Hopkins in 1884 but were not published until

1904 after a good deal of revision. It includes as Appendix B his famous lecture, "Nineteenth Century Clouds over the Dynamical Theory of Heat and Light," delivered before the Royal Institution on 27 April 1900 and originally printed in the *Philosophical Magazine* for July 1901. It makes no reference to black-body radiation or to Planck's work, although "Cloud II" is the apparent failure of equipartition of energy as evidenced by specific heats of gases, the same problem that was troubling Gibbs.

Lord Rayleigh published what we now call the Rayleigh-Jeans law in 1900 [47] but did not return to the subject again until 1905 in several notes in *Nature*. In this he comments on Planck's determination of the Boltzmann k and remarks that he has "not succeeded in following Planck's reasoning."

Although he actively published papers through 1919, he seems to have had no more to say on black-body radiation in all of the rest of his published work. A search reveals only two more items relating to modern quantum physics.

In 1906 he comments [48] on the classical radiative properties of atom models resembling those of J. J. Thomson but treating the behavior of the electrons as if they form a continuous fluid. In a note added in 1911, he refers back to an old paper [49], which points out that nearly all models of vibrating systems give the *square* of the frequency as additive in contributions from different quantum numbers (as we would call them now), and thus are not in agreement with the Rydberg-Kayser-Runge constant frequency differences as found empirically in spectroscopy. Then he says "A partial escape from these difficulties might be found in regarding actual spectrum lines as due to *difference tones* from primaries of much higher pitch."

He devoted a paragraph to the problem of the sharpness of spectral lines despite the random character of the conditions of excitation and concludes with a paragraph that sounds very modern:

> It is impossible, however, that the conditions of stability or of exemption from radiation may after all demand this definiteness, notwithstanding that in the comparatively simple cases treated by Thomson the angular velocity is open to variation.
>
> *According to this view the frequencies observed in the spectrum may not be frequencies of disturbance or of oscillation in the ordinary sense at all, but rather form an essential part of the original constition of the atom as determined by conditions of stability.*

Even as late as 1911 we find him concerned about Kelvin's Cloud II, the specific-heat worry. In a letter of Prof. W. Nernst he expresses his concern this way:

> If we begin by supposing an elastic body to be rather stiff, the vibrations have their full share (of kinetic energy) and this share cannot be diminished by

increasing the stiffness . . . Perhaps this failure might be invoked in support of the views of Planck and his school that the laws of dynamics (as hitherto understood) cannot be applied to the smallest parts of bodies. But I must confess that I do not like this solution of the puzzle . . . I have a difficulty in accepting it as a picture of what actually takes place.

We do well I think to concentrate attention upon the diatomic gaseous molecule. Under the influence of collisions the molecule freely and rapidly acquires rotation. Why does it not also acquire vibration along the line joining the two atoms? If I rightly understand, the answer of Planck is that in consideration of the stiffness of the union the amount of energy that should be acquired at each collision falls below the minimum possible and that therefore none at all is acquired–an argument which certainly sounds paradoxical.

He goes on to discuss an idea of Jeans that perhaps the difficulty can be solved if there is an extremely low rate of transfer of energy to the vibrational mode so that equilibrium is not actually reached in the specific heat measurements. Although this particular use of the idea has not stood the test of time we know that it applies in other areas, such as the slow rate of interconversion of the ortho- and para- forms of molecular hydrogen.

5[1]. Einstein's paper on light quanta

The next major step was taken by A. Einstein [16] in 1905. Planck's problem had been a statistical one, involving calculation of statistical equilibrium between many oscillators and many light quanta. Therefore there might be a possibility of a loophole in the probability arguments which were involved. Einstein directed attention to the advantages of assuming the reality of light quanta in order to provide a picture for other optical effects. He avoided the difficulties about absorption in weak light beams by supposing that the energy of one quantum remains localized in a small region of unspecified size. When a diverging light beam becomes less intense, fewer quanta cross unit area in unit time, but the individual quanta themselves continue to have the same energy content.

Einstein's paper, "A Heuristic Viewpoint Concerning the Emission and Transformation of Light," was especially cited when he received the Nobel Prize in physics in 1921. In the introductory paragraphs he writes:

The energy of a ponderable body cannot be divided into indefinitely many indefinitely small parts, whereas the energy emitted by a point light source is regarded on the Maxwell theory (or more generally according to every wave theory) as continuously spread over a continuously increasing volume.

Such wave theories of light have given a good representation of purely optical phenomena and will surely not be replaced by any other theory. It is to be remembered that the optical observations refer to time mean values, not to instantaneous values, and it is quite conceivable that, in spite of complete success in dealing with diffraction, reflection, refraction, dispersion, etc. such a theory of continuous fields, could lead to contradictions with experience when applied to phenomena of light emission and absorption.

After a little more discussion there comes the key declaration:

According to the supposition here considered, the energy in the light propagated in rays from a point is not smeared out continuously over larger and larger volumes, but rather consists of a finite number of energy quanta localized at space-points, which move without breaking up, and which can be absorbed or emitted only as wholes.

The first six sections are taken up with the statistical treatment of blackbody radiation from this viewpoint. Then Section 7 interprets Stokes's rule for photoluminescence in terms of single-light quanta producing excitation of single molecules. Finally Section 8 interprets the photoelectric effect – the subject on which the paper is nearly always cited today.

It had long been known that when matter is excited to fluorescence, the emitted fluorescent light is nearly always of longer wavelength than the incident exciting light. Where this is not exactly true the wavelength emitted is not much shorter than that of the incident light. This is known as *Stokes's law* [50]. It was not explained by the wave theory of light. Einstein pointed out that if a molecule of the substance is excited to fluorescence by absorption of one quantum, $h\nu$, of radiant energy, then the energy of the quanta that it is able to emit must be less than this (that is, the fluorescent quanta correspond to lower frequencies and hence to longer wavelengths). The known exceptions to the rule could be explained on the basis that some of the molecules that absorb quanta already have a small amount of energy of thermal excitation and in fluorescing may make a transition to a lower energy state than that from which they were excited.

The photoelectric effect refers to the ability of light falling on a solid body to cause it to lose *negative* charge (Hertz, 1887). In the next few years this effect was extensively investigated by P. Lenard and others. It was found that there was a close relationship between the contact potentials for various metals and the frequency of light necessary to produce the effect. The negative charge lost by the body behaved in deflection experiments like a stream of particles having the same ratio of charge to mass as that of the particles that occur in cathode rays – the particles that G. Johnstone Stoney proposed be called *electrons*. Moreover, the ejected photoelectrons were found to have a maximum kinetic energy, which increases with the frequency of the incident light. This was determined by seeing how great a positive potential must be applied to the photo-emitter in order to reduce the rate of loss of charge to zero. Finally, the maximum energy of ejection was found to be independent of the intensity of the incident light. These experimental results on the photoelectric effect following Hertz's original discovery led Einstein to propose that a maximum energy could occur in the photoelectric effect given by the relation

$$E_{\max} = h\nu - h\nu_0 \tag{1}$$

where ν_0 is characteristic of the emitter.

It is an odd fact that nowhere in the paper does Einstein use h to represent the Planck constant. Instead he writes β for what we now call h/k, and writes R/N for what we now call k, so that the energy of one light quantum appears as $(R\beta\nu/N)$.

Einstein showed that this is what is to be expected on the hypothesis that light is absorbed in quanta of amount $h\nu$. He supposed that the energy of one light quantum is given directly by the absorption process to one electron. Thus the number of electrons affected is proportional to the number of quanta present in the incident beam. The energy given in this way to each affected electron is $h\nu$. The light penetrates a short distance into the metal. Some electrons are held more firmly to the metal than others. They are given momentum in a wide variety of directions, so that only some of those which absorb a quantum move to the surface, and different ones of these lose differing amounts of energy along the way. Hence, if $h\nu$ is the minimum amount of energy needed to extract one electron from the metal, some will be emitted with the energy given by (1), but most will be emitted with less than this, with a statistical distribution of individual energies E in the range $0 < E < (h\nu - h\nu_0)$.

The relation (1) is known as *Einstein's photoelectric equation*. It had an enormous influence on the early development of quantum theory. Modern work has shown that there is not an absolutely sharp cutoff to the maximum energy of emission of electrons as in (1), but that a very small number are emitted with slightly higher energies, just as some fluorescent light slightly violates Stokes's law. This is interpreted in the same way, as being a manifestation of the thermal distribution of energy of the electrons in a metal: those having thermal energy of motion toward the surface can be liberated by smaller light quanta. This detail was developed later to provide a valuable means of studying the distribution of thermal energy of electrons in metals and semiconductors.

Nowadays it is hard to realize how fragmentary and uncertain were the experimental results on the photoelectric effect that Einstein interpreted on the light quantum hypothesis [51]. It was two years *after* 1905 that the first measurements were made with monochromatic light by E. Ladenburg, and he represented his results on a basis equivalent to supposing that $E_{\max}$ is proportional to ν^2. But A. Joffé showed that his data were equally well consistent with Einstein's linear relation of E_{max} to ν. A. L. Hughes in 1912 found unambiguous evidence for the linear relation, and found an experimental value of h/e from the rate of increase of V_{max} with ν, where V_{max} is the retarding potential difference needed to prevent the most energetic electron from reaching the collecting electrode. Four years later, in 1916, R. A. Millikan made more elaborate measurements of this kind to get a more precise experimental value for h/e.

Two years before Einstein's 1905 paper, J. J. Thomson [52] had anticipated the idea of light quanta when considering the ionization of a gas by a beam of X rays passing through it. After reviewing the experimental fact that X rays have no effect at all except on an exceedingly minute fraction of all of the molecules they pass through, he says:

Other possible explanations of the small number of molecules dissociated by the rays are (1) that the rays are of such a kind that only a small fraction of the molecules are exposed to the full force of their influence: that if for example we consider a plane at right angles to the direction of propagation of the rays the energy is not directed uniformly over this plane, but that the distribution of energy has as it were a structure, although an exceedingly fine one, places where the energy is large alternating with places where it is small, like the mortar and bricks in a wall; thus if the places where the energy is intense enough to produce ionisation of a molecule occupied but a small fraction of the area of the plane at right angles to the rays, the rays would be able to pass through a gas and yet only a small fraction of the molecules would be exposed to their maximum influence, just as in the case when a beam of cathode rays passes through the gas; we shall return to this point when we consider the nature of the Röntgen rays. Another view which might be taken is that all the molecules of a gas, even though the gas may be like hydrogen, an element, are not of the same kind, and that mixed with the ordinary molecules we have a few which are of such a kind as to be very easily ionised, and that the number of molecules of this kind, which are practically molecules of a different gas, is not given by Maxwell's law of distribution. The idea that even a gas is not uniform in composition, but contains, as it were mixed with it, small quantities of other gases – not necessarily as impurities due to its method of preparation but as an essential constituent of it – may appear at first stating so opposed to the ordinary facts of chemistry as not to be worthy of discussion. We may however point out that the quantities of such gases, if we may take the ionisation as their measure, are so small as to be utterly beyond the power of chemical analysis to detect, so that it cannot be by chemical considerations that the truth or falsehood of this hypothesis can be decided."

He deals with the same question also [53] in his Silliman lectures of 1903 where he says:

Röntgen rays are able to pass very long distances through gases, and as they pass through the gas they ionise it: the number of molecules so split up is, however, an exceedingly small fraction, less than one-billionth, even for strong rays, of the number of molecules in the gas. Now, if the conditions in the front of the wave are uniform, all the molecules of the gas are exposed to the same conditions: how is it, then, that so small a proportion of them are split up? . . . The difficulty in explaining the small ionisation is removed if, instead of supposing the front of the Röntgen ray to be uniform, we suppose that it consists of specks of great intensity separated by considerable intervals where the intensity is very small.

6¹. Heat capacity of solids and gases

The next major step led off in a different direction – to the interpretation of the low-temperature heat capacity of simple solids. This program was initiated by Einstein [54] in December 1906.

In 1819 Dulong and Petit [55] had discovered an empirical regularity that is still known by their names: that all simple elementary solids have the same atomic heat capacity, defined as the product of the heat capacity per unit mass and the atomic weight of the element. The new universal constant involved here is about 6(cal/°C·gram)·(atomic weight). At this early date neither heat capacities nor atomic weights were accurately known. Heat capacities were measured at room temperature. The subject was studied thoroughly by H. F. Weber [56], who established that the elements carbon (both as diamond and as graphite) and boron and silicon, depart from the law in that (1) the atomic heat capacity at room temperature for these elements is much less than 6 cal/°C·gram-atom, and (2) the atomic heat capacity is strongly temperature dependent, rising toward this value at high temperatures. Lothar Meyer [57] enunciated the rule that elements of small atomic weight and small atomic volume depart most from the Dulong-Petit law.

The extension of the Dulong-Petit rule to compound substances is attributable to F. E. Neumann [58]. He showed that for substances that are chemically similar, the product of specific heat and molecular weight is constant. J. P. Joule [59] showed that the heat capacity of many compounds could be represented additively as contributions from each of the constituent atoms, as if each atom always has the same specific heat. This rule, subjected to careful critical testing by Kopp [60] led to the idea of ascribing atomic heat capacity to each constituent atom in a compound. This rule was found to be roughly applicable even to glass [61].

Work along these lines showed that for most of the elements the atomic heat capacity was 6.4 cal/°C·gram-atom, close to the Dulong and Petit value. However, very definitely smaller values had to be assigned to some of the elements:

Hydrogen	2.3	Fluorine	5.0
Boron	2.7	Silicon	3.8
Carbon	1.8	Phosphorus	5.4
Oxygen	4.0	Sulfur	5.4

The interpretation of the rule of Dulong and Petit was made by Boltzmann [62]. His idea was that each atom in the solid has three degrees of freedom and so has, at absolute temperature T, a mean kinetic energy equal to $(3/2)kT$. In addition, it oscillates as a spatial harmonic oscillator about its mean equilibrium position, so has also a mean potential energy of $(3/2)kT$. Hence a mole of such atoms will possess a mean thermal energy of $3NkT$ or $3RT$, where $R = Nk$ is the molal gas constant in $pV = RT$; thus the rate of increase of the mean energy with temperature has the constant value $3R$. From the equation of state of gases, and the mechanical equivalent of heat it is known that $R = 1.98$ cal/°C · mole. Boltzmann's view gave both a picture of the nature of the heat motions in solids, and also a quantitative interpretation of the Dulong and Petit constant.

Similar developments had occurred in connection with the specific heat of gases. Considering the heat capacity at constant volume, C_v, each atom of a monatomic gas has three translational degrees of freedom, so it should have for this reason a molal heat capacity of $(3/2)R$. This agrees with experiment, indicating that the mean potential energy of the weak intermolecular forces in a gas at low density makes a negligible contribution to the heat capacity. For diatomic gases, $C_v = (5/2)R$. Thus, although the number of degrees of freedom per mole is here twice what it is for a monatomic gas, the heat capacity only increases in the ratio 5:3 instead of 6:3 as expected on Boltzmann's views.

This was the nineteenth-century cloud of Kelvin, mentioned in Section 4[1], which also troubled Gibbs. It thus appeared that the Boltzmann value was a kind of upper limit, and that the experimental values *often fell decidedly below this,* in the case of gases as well as in the case of solids.

The departure from the Dulong and Petit values was emphasized by the work of Dewar [63] in the earliest cryogenic measurements of the heat capacity of diamond. These showed that in the range of liquid hydrogen to liquid air temperatures (~10 to 100°K) the heat capacity of diamond is less than 1 percent of the Dulong and Petit value.

Einstein's contribution was the recognition that these effects have their explanation in Planck's idea that an oscillator of frequency ν can only exist in discrete states of energy $W_n = nh\nu$, with $n = 0, 1, 2, 3, \ldots$ instead of in a continuous range of possible energy states as had been assumed by Boltzmann. According to a natural modification of the Boltzmann distribution law, the relative numbers of oscillators in a state of energy W_n is $\exp(-W_n/kT)$. Hence the mean energy is

$$W_{\mathrm{Av}} = \frac{\sum_n W_n \exp(-W_n/kT)}{\sum_n \exp(-W_n/kT)} \tag{1}$$

which with $W_n = nh\nu$ gives

$$W_{Av} = \frac{h\nu}{\exp(h\nu/kT) - 1} \tag{2}$$

which tends to the Boltzmann value for $h\nu << kT$.

In this first paper [54], Einstein assumed for simplicity that the natural frequency ν of vibration of each atom in a simple monatomic solid is the same. Each atom has three degrees of freedom so for one mole of such a solid the internal energy is

$$U = 3RT \frac{x}{e^x - 1} \tag{3}$$

where $x = h\nu/kT$ and the molal heat capacity becomes

$$C = \frac{\partial U}{\partial T} = 3R \frac{x^2 e^x}{(e^x - 1)^2} \tag{4}$$

At high temperatures, $x \rightarrow 0$, and $C \rightarrow 3R$, the classical or unquantized value. At a temperature such that $kT = h\nu$, the specific heat has already dropped to 0.92 of the classical value.

In the second [54] of the 1907 papers, Einstein pointed out that not all of the oscillators in a solid have the same frequency, but that there is a spectrum of frequencies such that the frequency ν_k is represented by f_k degrees of freedom in a mole, where $\Sigma f_k = 3N$, and that it is necessary to make a more detailed dynamical analysis of the crystal to get at the spectrum. The total energy is now

$$U = kT \sum_k f_k \frac{x_k}{\exp(x_k) - 1} \tag{5}$$

where $x_k = h\nu_k/kT$, so the main result, indicating how the quantization leads to low values of U at low temperatures is generally valid.

The later development of the theory of heat capacity has now a large literature. All of it proceeds along the line of providing models leading to the determination of the distribution of the spectral frequencies of the oscillators whose energy levels are quantized. In 1912 two distinct major steps were taken, one by Debye [64], who regarded the crystal for this purpose as an elastic continuum, and the other by Born and von Karman [65], who treated it as a dynamical lattice of elastically coupled mass-points.

From that time forward all interpretation of the heat capacity of solids proceeded on the basis of Einstein's 1907 analysis in terms of the quantization of energy levels. Modern development of these ideas is an important part of the quantum physics of solids.

Simultaneously with these developments the idea of quantized energy levels was applied to clear up the problem of the specific heat of gases made of diatomic and polyatomic molecules. This received a great impetus from the experimental work of A. Eucken [66], who in 1912 discovered that hydrogen (H_2) gas has $C_v = (3/2)R$ below 60°K, which increases to about $(5/2)R$ above about 300°K. Thus in this respect it behaves like a monatomic gas at low temperatures, and like a diatomic gas at high temperatures. Einstein [67a] and Stern and Ehrenfest [67b] pointed out that this change could be a result of quantization of the rotational motion of the diatomic molecules about an axis perpendicular to the line of centers, and this viewpoint was carried farther by Ehrenfest.

In these papers we encounter for the first time the quantization of a rotator by the assumption that the angular momentum must be an integral multiple of $h/2\pi$, and the idea that each quantum state is to be given the same weight in statistical considerations, because each is associated with the same volume in phase-space.

These considerations led Einstein and Stern to the formula

$$W_n = \frac{n^2\hbar^2}{2I} \tag{6}$$

for the discrete energy levels of a rigid rotator. Here $\hbar = h/2\pi$, a convenient notation introduced much later by Dirac, and I is the moment of inertia of the molecule. These levels were used in (1), where they were given unit statistical weight. At the end the result was multiplied by two, as a way of taking into account that there are two degrees of rotational freedom each with moment of inertia I. Later it was recognized that this is incorrect: the proper procedure is to modify (1) to

$$W_{\mathrm{Av}} = \frac{\sum_n g_n W_n \exp(-W_n/kT)}{\sum_n g_n \exp(-W_n/kT)} \tag{7}$$

where g_n is the statistical weight for a state of angular momentum $n\hbar$, associated with the fact that angular momentum is a vector capable of having different orientations.

Although the final solution of the problem of the rotational specific heat of H_2 gas was not worked out until 1927 by Dennison [68], the main idea that this is another consequence of the restriction of the rotating molecule to a discrete set of rotational energy levels had been clearly recognized in 1913.

7[1]. The Bohr model for hydrogen

The stage was set for a combination of the quantum ideas of Planck and Einstein, the Rydberg-Ritz combination principle, and the nuclear atom model of Rutherford, to provide a model of atomic structure. This was done for hydrogen by Niels Bohr [69] in 1913. He regarded the motion of the electron around the proton as a dynamical system obeying Newtonian mechanics, modified by certain quantum principles.

Foremost was the idea of discrete or *quantized* energy levels: that a closed dynamical system can only exist in states of certain allowed energies. This idea is basic to all later developments.

Next is the idea that emission and absorption of radiation is associated with transitions or quantum jumps between states in such a way that the change in energy from E_i to E_f (where i and f refer to "initial" and "final") produces a single quantum $h\nu$ of radiation, that is emitted if $E_i > E_f$ or absorbed if $E_i < E_f$, and such that

$$h\nu = |E_i - E_f| \tag{1}$$

This idea immediately gives an interpretation of the Rydberg-Ritz combination principle, found in empirical analysis of spectra. Dividing by hc we have

$$\sigma = \lambda^{-1} = \left|\frac{E_i}{hc} - \frac{E_f}{hc}\right| \tag{2}$$

So the *terms* of the combination principle, measured in wave numbers, give the observed values of the energy levels when multiplied by hc.

These two ideas represent a radical departure from the laws of mechanics and electrodynamics as they had been inferred from experimental study of larger systems. According to classical mechanics, the total energy of such a system is a constant of the motion, which can be assigned any value, rather than a discrete set of values as postulated by Bohr. According to classical electrodynamics the accelerated electrons in the atom should radiate continuously, so the total energy would slowly and continuously diminish. Except in the case of a harmonic oscillator, a change in the frequency of the radiation accompanies the continuous change in energy. Therefore, in a gas consisting of an ensemble of many such atoms, different atoms would possess a continuum of different energy values and the radiation would consist of a continuous spectrum of frequencies, rather than a line spectrum.

Since a line spectrum of discrete frequencies is observed it is evident that a drastic departure from classical theory is involved: It is merely a question of what kind of drastic change must be made.

In Bohr's original theory of the hydrogen atom he considered only the circular orbits of a single electron going around the nucleus. For these the centripetal force mv^2/r is provided by the Coulomb attraction e^2/r^2 between electron and nucleus so

$$\frac{mv^2}{r} = \frac{e^2}{r^2}$$

He also introduced as a quantum condition that only those circular orbits are allowed in which the orbital angular momentum, mvr, is an integral multiple, n, of $\hbar = h/2\pi$ which is assumed to be a natural quantum unit of angular momentum, that is

$$mvr = n\hbar$$

Solving these two relations for the radius of the nth orbit, a_n, and the speed of the electron in it, v_n, we find

$$a_n = \frac{n^2\hbar^2}{me^2}; \quad v_n = \frac{e^2}{n\hbar} \tag{3}$$

The total kinetic and potential energy of the electron in the nth orbit is thus

$$E_n = \frac{1}{2} mv_n^2 - \frac{e^2}{a_n} = -\frac{1}{n^2} \cdot \frac{me^4}{e\hbar^2} \tag{4}$$

The positive kinetic energy is numerically one half of the negative potential energy, so the negative total energy is one half of the negative potential energy. The quantity of energy E_1 for $n = 1$ is called one *Rydberg*.

Combining (4) with (2) this predicts for the spectral line wave number involved in a transition from $n \to n'$ (where $n > n'$)

$$\sigma(n, n') = \left(\frac{1}{n'^2} - \frac{1}{n^2}\right) \frac{me^4}{4\pi\hbar^3 c} \tag{5}$$

This agrees, in form, with the formula found in 1885 by Balmer for the Balmer series, if one takes $n' = 2$, and leads to the prediction that the Rydberg constant, R, hitherto only known empirically, should have the value

$$R = \frac{me^4}{4\pi\hbar^3 c} \tag{6}$$

When this theoretical value was calculated from the known values of m, e, h, c, good agreement with the empirical value was obtained. Thus Bohr's model not only predicted the form of the Balmer formula, but also, for the first time, related the Rydberg constant to other known physical constants.

By taking $n' = 3$, the relation (5) predicts another series of lines in the infrared. The prediction, for $n = 4$ and 5, corresponds with two lines which had been found by F. Paschen [70] in 1908. It is now called the *Paschen series*. By taking $n' = 1$, another series in the ultraviolet is predicted: This was discovered the next year by T. Lyman [71], and is now called the *Lyman series*. In 1922, in the far infrared, F. S. Brackett [72a] discovered two lines of the series for $n' = 4$, now called the *Brackett series*. Shortly afterward Pfund [72b] discovered the first member of the series with $n' = 5$. Humphreys [72c] discovered the first member of the series with $n' = 6$ and observed the third, fourth, and fifth members of the Brackett series and the second member of the Pfund series.

Further experimental verifications were obtained. The Balmer series is not obtained in absorption in the laboratory, because at ordinary temperatures there is a negligible population of hydrogen atoms in the $n' = 2$ state. In stellar atmospheres the temperature is high enough to provide enough atoms in the $n' = 2$ state to give the Balmer series as it is actually observed in absorption.

The relations as written, (3) to (6), are correct in the approximation that the proton is infinitely massive compared with the electron. More accurately in place of m in these relations one should use the reduced mass, μ, where

$$\mu = \frac{mM}{m + M} \tag{7}$$

and M is the mass of the nucleus. Ionized helium (He II) consists of one electron going around a doubly charged nucleus whose mass is about four times that of the proton. The double charge makes the force be Ze^2/r^2, where $Z = 2$, so in (3) and (4) e^2 is to be replaced by Ze^2 and m by μ.

It is now customary to write R_∞ for the R of (6) and R_{He}, and so on for the effective value $(\mu/m)R_\infty$ for the nuclear mass indicated by the subscript. Thus Bohr predicted, for lines in the He II spectrum ($Z = 2$),

$$\sigma_{\mathrm{He\ II}}(n,n') = 4\frac{\mu}{m}R_\infty\left(\frac{1}{n'^2} - \frac{1}{n^2}\right)$$

in which μ is evaluated using the mass of the helium nucleus for M. From this we see that He II has a series for $n' = 4$, in which the lines for *even* values of n correspond very closely (since $\mu/m \approx 1$) with the Balmer series of H I. The complete series in He II includes additional lines corresponding to odd values of n. E. C. Pickering [73a] in 1896 found a series of lines in the bright southern star ζ Puppis, which were not known at that time in any other celestial or laboratory source. This *Pickering series* was later found by him in many other hot stars; their modern wavelengths are 5411, 4542, 4200, 4026, 3923, 3858, and 3813Å. These lines alternated with the Balmer series and had the same series limit. The lines were produced by A. Fowler [73b] in a hydrogen and helium mixture, and later it was established by Fowler and by E. J. Evans [73c] that they arose from helium. Bohr [73d] attributed the lines to He II. The seven lines are attributable to $n' = 4$, $n = 7, 9, 11, 13, 15, 17$, and 19. The lines with even n would be almost coincident with the Balmer series; these lines were first seen separated from the Balmer series by H. H. Plaskett [73e] in the spectra of hot stars. Bohr accounted quantitatively for the slight displacement in terms of the two different reduced masses, μ, involved. The magnitude of the displacement is indicated by the wavelengths:

	Balmer line	*Corresponding He II line*
H_α	6562.8	6560.37
H_β	4861.3	4859.53
H_γ	4340.5	4338.86
H_δ	4101.7	4100.22

This small shift of the Balmer lines with nuclear mass was used as the means of detecting the presence of the $A = 2$ isotope of hydrogen, now called *deuterium,* in 1932 [74].

A line at λ4686 was seen by Pickering in some of the stars he studied; it had been observed visually by Sir William Huggins[73f] in the planetary nebula NGC 7662 in 1864, and was observed by A. Fowler and W. J. S. Lockyer on spectra of the solar chromosphere taken by them during Sir Norman Lockyer's [73g] 1898 eclipse expedition. The λ 4686 line was produced by Fowler in the laboratory, along with additional lines at 3203, 2733, 2511, 2385, 2306, and 2253Å. Bohr showed that the lines were also a result of He II, with $n' = 3$, $n = 4, 5, 6, 7, 8, 9$, and 10. Since that time other He II series have been seen. T. Lyman [73h] was the first to detect several lines in the series with $n' = 1$ and $n' = 2$. The line with $n' = 4$, $n = 5$ was seen by H. D. Babcock and C. E. Moore [73i] in the solar spectrum. A large number of transitions from $n' = 5$ were seen in stellar spectra ($n = 15$ to 21) by R. Wilson [73j] and in the planetary nebula NGC 7027 ($n = 9$ to 40) by L. H. Aller, I. S. Bowen and R. Minkowski [73k].

In 1915 the Bohr theory was extended, independently and simultaneously, by W. Wilson and A. Sommerfeld [75,76] to cover the case of quantized elliptic orbits. Classical orbit theory gives ellipses as well as circles under the inverse square law of electrical attraction. The idea of the orbital angular momentum being an integral multiple of $\hbar$ was retained in the form $\oint p_\theta \, d\theta = 2\pi k\hbar$. The integration is over one period of θ, that is, $0 \leqslant \theta \leqslant 2\pi$. The assumption was made that the radial momentum is to be quantized in the analogous way. Only those values of energy are allowed for which

$$\oint p_r \, dr = 2\pi(n - k)\hbar \tag{8}$$

Here the integral is extended over a complete cycle of r from $r_{\min}$ to $r_{\max}$ and return, and n as well as k are integers. Circular orbits are those for which $n = k$. The semimajor and semiminor axes, a_{nk} and b_{nk}, are given by formulas which are closely related to (3),

$$a_{nk} = \frac{n^2}{Z}\frac{\hbar^2}{me^2}; \quad b_{nk} = \frac{nk}{Z}\frac{\hbar^2}{me^2} \tag{9}$$

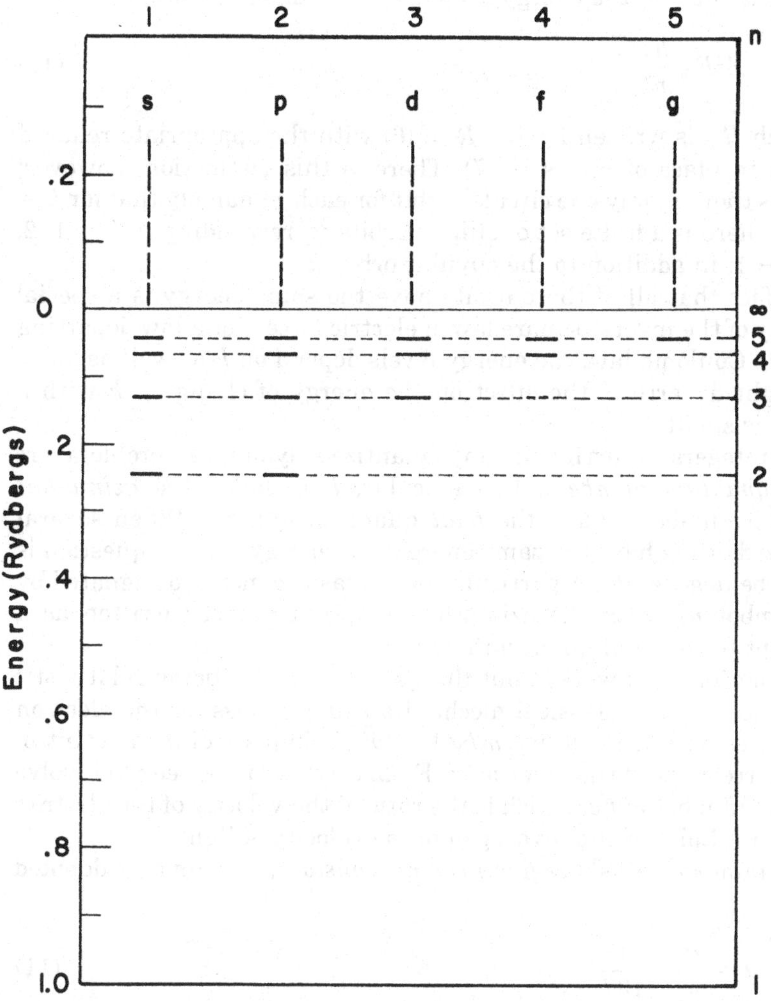

Figure 1[1]. Energy levels of the n_k orbits in the Bohr-Sommerfeld model of hydrogen. Columns give k values, with $n \geqslant k$.

The states of positive total energy correspond to hyperbolic orbits. For these it is still possible to quantize the angular momentum, but the radial quantum condition (8) is not applicable here because the motion in r is aperiodic, therefore all positive energies are allowed.

It is convenient to display the allowed energy levels in an energy level diagram in which the allowed energy levels are shown on a vertical scale, and are placed in different columns arranged by values of the orbital angular momentum. Figure 1[1] is such a diagram for hydrogen.

It turns out that the energy levels are the same as before,

$$-Z^2 R_M \frac{hc}{n^2} \tag{10}$$

in which R_M is written for the R of (6) with the appropriate reduced mass μ in place of m, as in (7). There is this distinction, however; whereas there is only one circular orbit for each n, namely that for $k = n$, now there is a finite set of elliptic orbits corresponding to $k = 1, 2, \ldots, n - 1$, in addition to the circular orbit, $k = n$.

The fact that all of these orbits have the same energy is a special property of the inverse square law of electric force. For a law departing from the Coulomb law, the energy levels depend on k as well as on n. For slight departure, the effect on the energy of changing k with a fixed n is small.

The integers occurring in any quantized dynamical problem are called *quantum numbers*. In particular, k is called the *azimuthal* quantum number, and n the *total* quantum number. When several quantized orbits have the same energy, the energy level in question is said to be *degenerate*. A particular orbit was commonly designated by the symbol n_k, with the azimuthal quantum number written as a subscript to the total quantum number.

Sommerfeld [77] worked out the quantized orbits using relativistic mechanics, that is, classical mechanics using a mass for the electron that varies with the velocity: $m/\sqrt{1 - (v^2/c^2)}$ in place of m. This introduces corrections of the order v^2/c^2. From (3) this can be seen to involve the pure number $e^2/\hbar c$, which is the ratio of the velocity of the electron in the first Bohr orbit of hydrogen to the velocity of light.

This number, called the *fine structure constant*, is commonly denoted by α,

$$\alpha = \frac{e^2}{\hbar c} \approx \frac{1}{137} \tag{11}$$

With the introduction of variable electron mass the quantized energy levels are found to depend slightly on k, with a fixed n. Thus a particular Balmer line $n \rightarrow (n' = 2)$ is the near superposition of a variety of lines corresponding to various k values, giving rise to fine structure.

The fine structure observed agrees remarkably well with that which was predicted by the Sommerfeld extension of the Bohr theory, so this was hailed as a great triumph. Later it was found that the real situation is actually considerably more complicated.

In the early development of the Bohr theory of hydrogen, calculations were also made successfully for the perturbing action of a magnetic field (Zeeman effect) and of an electric field (Stark effect).

Any one orbit in hydrogen occurs in a plane. At first it was believed that the plane could have any orientation in space. The orbital angular momentum of a particle at **r** with linear momentum **p** is the vector

$$\mathbf{L} = \mathbf{r} \times \mathbf{p} \tag{12}$$

So the quantization of angular momentum that Bohr postulated amounted to restricting the magnitude of **L** to the values $k\hbar$, with $k = 1, 2, 3, \ldots$ In the theory of the Zeeman effect it was also necessary to postulate quantization of the component of **L** in the direction of the magnetic field **B** to values $m\hbar$, in which m is an integer called the *magnetic* quantum number, and the range of m is $-k \leqslant m \leqslant k$. Direct experimental evidence for the reality of this space quantization, or directional quantization, was soon provided by the early molecular beam experiments of Stern and Gerlach [78].

8¹. Excitation of spectra

It was at once evident that (1) the idea of quantized energy levels and (2) the idea that emission and absorption of light are connected with transitions between energy levels through 7¹1 could be taken over to other systems than the hydrogenic atoms and ions. The Bohr model of hydrogen gave an immense stimulus to the study of atomic spectra. These two ideas were soon applied semiquantitatively to other atoms as well, and also to the vibrations and rotations of molecules, especially diatomic molecules, to provide interpretations of their spectra.

In the older work, based on the empirical Rydberg-Ritz combination principle, the spectra were analyzed into *terms, T,* usually measured in cm^{-1}, counting $T = 0$ at the series limit and counting the terms as positive numbers. As the energies in the hydrogen atom are negative (7¹4), that is, the energy of an atom whose electron is bound in a closed orbit is less than when that electron is free, the relation between an energy level and the corresponding term is

$$E = -hcT \tag{1}$$

when both are measured from the same origin. In most tabulations of atomic energy levels published since 1945 the magnitude of the energies are described by stating the values of $(T_0 - T)$ in cm^{-1}. This gives the value of E/hc for a particular level measured up from the lowest level, or largest term, T_0.

The lowest level is called the *normal level* or *ground state*. Higher levels are called *excited states*. The least energy E needed to remove a single electron from an atom or ion in its normal level is called the *ionization potential* $I(A)$, or IP(A), of the atom or ion, A. The word "potential" is used because it is often stated in volts, V, such that

$$E = e\left(\frac{V}{c \times 10^{-8}}\right) \tag{2}$$

The energy corresponding to V volts is said to be V electron volts, abbreviated (eV), and the factor $c \times 10^{-8}$ converts V in volts into electrostatic units. Here E is in ergs if e is in statcoulombs and c is in cm/sec.

The term value $T(\mathrm{cm}^{-1})$ that corresponds to an energy measured in electron volts, V (volts),

$$T = \frac{eV/(c \times 10^{-8})}{hc} \tag{3}$$

Atoms in an excited state, when left to themselves, spontaneously make transitions to lower levels, emitting radiation. Therefore, after a time all of the atoms will be in a normal state, after which no more radiation is emitted.

In order for a gas to continue to be luminous, processes of excitation must be occurring, which take atoms from the normal state, or other states of low energy by supplying energy to them, and which leave them in the excited states from which they can emit radiation by making radiative transitions to lower levels. The most common kinds of excitation processes are those in which excitation energy is supplied by absorption of a photon, or by *collision* of an electron, or of another atom or positive ion, with the atom in question.

By collision is here meant a coming near to each other of two or more atomic systems in such a way that there is appreciable interaction between the parts, which then separate to large distances after the collision. The collisions of interest here are those in which at least one of the partners has internal structure, which may be changed as a result of the interaction.

Collisions are called *elastic* if both partners are left in the same internal energy state as they had before interaction, otherwise *inelastic*.

Those inelastic collisions in which one or both partners are left in a higher state of internal energy are called collisions of the *first kind*. Those in which one or both partners are initially in excited states and are left after the collision with less excitation energy and a corresponding increase in kinetic energy of relative motion are called collisions of the *second kind*. They are also sometimes called *superelastic*.

When the partners, A and B, are composite systems, the collision may result in *rearrangement* whereby the products, C and D, which separate after interaction are different from A and B, subject to the restriction that C and D together contain the same number of protons, neutrons and electrons as did A and B together. At very high energies this restriction may be removed through the occurrence of processes of *creation* and *annihilation*.

With every collision $A + B \rightarrow C + D$ (including the case in which C is A and D is B, so no rearrangement) there is a possible *inverse collision* in which $C + D$ collide to produce $A + B$.

We consider first an electron–atom collision in which the atom A is in its normal state, and is at rest. Let the electron's initial kinetic energy be E. The system in question consists of A and the electron regarded as one system, so the center of mass is not at rest, and a small part of E is kinetic energy of the motion of the center of mass. The greater part of it, namely,

$$E' = \frac{E}{1 + (m/M_A)}$$

is energy of the motion of the electron relative to the atom. If this quantity is greater than the smallest energy of excitation of A, then it is possible for an electron to separate again to large distances, while leaving the atom in an excited state A^*,

$$e + A \rightarrow e + A^* \tag{4}$$

In this case the electron energy relative to the atom after the collision, E'', is less than E' by the amount by which the internal energy of A^* exceeds that of A. In such a collision there are two possible outcomes: The atom may be left unexcited, or it may be left in an excited state.

When E' exceeds the energy of the second excited state of A, there are three possible outcomes, (1) that A be left unexcited, (2) that A be left in the first excited state, and (3) that A be left in the second excited state. The situation is more complicated in that there is an angular distribution of directions of motion of the electron relative to its initial direction. This depends on E' and on which process actually occurred.

As E' is increased further more possible outcomes open up. Each possible outcome is called a collision *channel*. A channel is *open* if the energy is great enough to allow the electron to leave the atom in the excited state associated with that particular channel; otherwise it is *closed*.

Finally, when E' exceeds the ionization energy of A, a new set of channels opens in which A has an electron knocked out of it leaving the ion A^+ behind

$$e + A \rightarrow A^+ + e + e \tag{5}$$

(Here, if A is itself an ion, A^+ is to be construed as the next higher ion.) In this process the total kinetic energy available to the two outgoing electrons is $E' - I(A)$. This may be distributed between the two electrons in a way that is a characteristic of the atom A and the energy E'. If E' exceeds the sum of the ionization energies of A and A^+, then it is possible for two electrons to be removed in the collision

$$e + A \rightarrow A^{++} + e + e + e \tag{6}$$

where now the excess energy, $E' - I(A) - I(A^+)$ is divided among the three outgoing electrons in a way that is characteristic of the particular atom A and the energy E'.

The first experiments showing the quantized energy losses of energy of electrons in a gas were those of Franck and Hertz [79], published very soon after Bohr's first papers on the quantized energy level model of the hydrogen atom. These experiments opened up a vast field of experimental work which served to give full support to the picture of quantized energy levels. This kind of work, in which atoms were excited to emit light by impact of electrons of controlled and nearly homogeneous energy, proved to be an important tool in the analysis of spectra.

The great bulk of the excitation that produces luminosity in electric discharges through gases is caused by electron–atom collisions.

Atoms can also be excited by absorption of light. In such a process the photon disappears or is annihilated,

$$h\nu + A \rightarrow A^* \tag{7}$$

so that here there is no question of balancing the equation as to the number of photons on the two sides. Because of this, the process is sharply resonant [80]. The energy of the photon must be very nearly equal to the energy of excitation of the atom.

As a result, if the light available for excitation has a continuous spectrum, only a minute fraction of it is available for putting atoms in excited states. When $h\nu$ exceeds the ionization energy of A, the process of photoionization can occur:

$$h\nu + A \rightarrow A^+ + e \tag{8}$$

Such processes provide a much more efficient utilization of a continuous spectrum, since the photoelectron can have a wide continuous range of kinetic energies after ejection from the atom.

Experimental evidence [81] for the transfer of excitation energy from one atom to another in collision was obtained by mixing Hg vapor and Tl vapor. Thallium atoms do not absorb the λ2537 light of mercury. But when the mixed vapors are illuminated with a beam of λ2537 light, the resulting fluorescent light contains light emitted by the Tl spectrum. This indicates that excited mercury atoms can transfer their excitation energy to thallium atoms, by collisions of the second kind

$$\mathrm{Hg}^* + \mathrm{Tl} \rightarrow \mathrm{Hg} + \mathrm{Tl}^* + \text{kinetic energy} \tag{9}$$

In general the excitation energy of thallium does not equal that of mercury and the excess of $E(\mathrm{Hg}^*)$ over $E(\mathrm{Tl}^*)$ is taken up as kinetic energy of relative motion of the two atoms after collision. For λ2537 the excitation energy is 4.9 eV. Consistent with this view, the Tl lines that are excited are those requiring less that 4.9 eV for their excitation.

The yield, or rate of occurrence of collision processes of all kinds is stated in terms of an effective *collision cross section*. This is defined as follows: Let one of the partners, B, be at rest, and let a parallel beam of A, having an intensity of I_A atoms of A per unit area per unit time, go past the place where B is situated. Then the collision cross section $\sigma(A, B; C, D)$ is measured as an area such that $\sigma(A, B; C, D)\, I_A$ is the number of collisions occurring between the beam of A and one B atom in unit time, which result in production of C and D. The collision cross section depends on A and B, and on their relative kinetic energy, and on the specific products of the collision, including details such as the direction of motion of the products after the collision.

A large part of physics – molecular, atomic, nuclear, and particle – consists of devising experiments for measurement of effective collision cross sections and of theories for their calculation from first principles.

9^1. Alkali spectra

The alkali atoms and the associated ions in the corresponding *isoelectronic sequences* (shown in the table) have particularly simple level schemes. Some of these spectra had already been analyzed into series (Sec. 3^1) before the development of the Bohr model.

N	C: 1	2	3
3	Li	Be^+	B^{++} . . .
11	Na	Mg^+	Al^{++} . . .
19	K	Ca^+	Sc^{++} . . .
37	Rb	Sr^+	Y^{++} . . .
55	Cs	Ba^+	La^{++} . . .

The idea of a single electron moving in quantized states, in a central field, which differs from the Coulomb field, because of the average action of the other electrons in the *core* on the single outer *valence electron*, was quickly applied to give a semiquantitative account of the resemblance to and differences from the still simpler set of energy levels in the $N = 1$ isoelectronic sequence.

It is supposed that $(N - 1)$ of the electrons are tightly bound in a spherically symmetric core. When the single outer valence electron is outside the core its net Coulomb interaction with the nucleus and the core is represented by

$$V(r) = -\frac{Ze^2}{r} + \frac{(N-1)e^2}{r} = -C\frac{e^2}{r} \tag{1}$$

in which $C = (Z - N + 1)$.

When the valence electron is within the core, the net interaction becomes stronger, because the spherically symmetrical distribution of core charges at greater distances exerts no net force on the valence electron. It is thus convenient to postulate that the valence electron interacts with nucleus and core by the relation

$$V(r) = -\frac{Z(r)e^2}{r} \tag{2}$$

in which $Z(r)$ is an effective charge acting on the valence electron at various distances. As $r \to 0$ we have $Z(r) \to Z$ because the electron is then affected by the full field of the nucleus. But as r increases, the nucleus is screened more and more by the core electrons so $Z(r) \to C$.

The way in which $Z(r)$ varies between these limits depends through potential theory on the unknown radial charge distribution of the core electrons.

Because the field is spherically symmetric the orbits are still characterized by the quantum number k that gives the orbital angular momentum. The levels of a given k and changing n are said to form a *series*. In quantum mechanics the orbital angular momentum is given by $(k - 1)\hbar$ rather than $k\hbar$, and the letter l is used to denote the new value,

$$l = (k - 1) \qquad l = 0, 1, 2, 3, \ldots \tag{3}$$

Within a series the allowed energy values are determined from the radial quantum condition

$$\oint p_r \, dr = (n - k)h \tag{4}$$

so the lowest level corresponds to $n = k$. For this we must have $p_r = 0$, so this corresponds to a circular orbit. For the larger values of k the entire orbit lies outside the core where $Z(r)$ has the constant value C. Therefore, for these orbits the energy levels have (Sec. 7^1) the hydrogenic values

$$W(n_k) = -C^2 \frac{Rhc}{n^2} \tag{5}$$

For the series of smaller k the orbits penetrate inside the core, where the valence electron is subjected to stronger fields, so these energy levels are lower than the corresponding hydrogenic values. These are called *penetrating orbits*. It is sometimes convenient to describe the levels in terms of an *effective total quantum number*, ν, defined so that the actual energy level $W(n_k)$ is represented by (5); that is,

$$\nu = C\left[\frac{R}{T(n_k)}\right]^{1/2} \tag{6}$$

in which $T(n_k)$ is the term value of the level. The levels lie lower than the hydrogenic value, so $\nu < n$. The *quantum defect*

$$\Delta = n - \nu \tag{7}$$

is therefore a positive quantity. Terms that obey a Rydberg series formula (Sec. 3^1), are ones for which the quantum defect Δ is nearly constant along a series.

Although ν is defined by (6) in terms of known quantities, one does not have a unique way of choosing n. The earlier tendency was to choose n as the largest integer in ν so that Δ would be less than unity. However (Sec. 11^1), it proves to be more rational to assign the smaller values of n to the states occupied by the electrons in the core. This then uniquely determines the numbering for the valence electron outside the core.

All of these expectations are borne out by the observational data on the spectra of the alkali isoelectronic sequences [76,82]. We expect the one-electron central field model to be particularly applicable here because from the chemical evidence these are univalent atoms, and the single valence electron moves outside the unusually stable quasi-rigid core distribution that corresponds to a structure such as one of the noble gases: He, Ne, Ar, Kr, Xe.

Historically, the first instance of a comparison of this kind is due to A. Fowler [83a]. He showed the close relations existing between the levels of Mg II and Na I. Then Paschen [83b] extended the sequence to Al III and Fowler to Si IV. Such spectra of highly ionized atoms, or "stripped atoms" as R. A. Millikan called them, occur more deeply in the extreme ultraviolet as C increases. The discovery of regularities of this sort provided a great incentive to the experimental study of vacuum ultraviolet spectroscopy. A vigorous program was carried out by Millikan and Bowen [84], which quickly gave rather complete and detailed knowledge of the Li I, Be II, B III, C IV, . . . sequence and also of sequences having more complex spectra. The study of the highly ionized members of various isoelectronic sequences has contributed greatly to the knowledge of atomic structure.

The foregoing picture is incomplete in one important respect: It does not account for the *fine structure,* the structure by which there are two D lines in the Na I spectrum rather than one. This feature was clarified by Uhlenbeck and Goudsmit [85] in 1925 with the introduction of the hypothesis of *electron spin.* According to this, the electron has an intrinsic spin angular momentum $(1/2)\hbar$, and an associated magnetic moment $e\hbar/2mc$, which is called the *Bohr magneton* and which is antiparallel to the electron spin. The motion of the electron in the electric field of the nucleus and core is equivalent to its being in a magnetic field. The spin of the electron can orient itself either parallel or antiparallel to the orbital angular momentum, except in the case of the $l = 0$ orbits. Thus the $l = 0$ orbits remain single, but levels of all of the other series become double. The magnetic energy that results in this doubling is called *spin-orbit interaction.* The doublet intervals are quite small, because the magnetic effects are much weaker than the electric effects. Many doublet intervals are too small to be resolved.

In the alkalis the lowest level is that of the lowest level in the $l = 0$ series. The *principal series* arises from the combination of the doublet terms of the $l = 1$ series, with the lowest level of the $l = 0$ series. Because of this it became customary to call the terms of the $l = 1$ series, p terms. The doublet interval of the p terms rapidly decreases as one goes up the series until ultimately the principal series doublets cannot be resolved.

The *sharp series* arises from the combination of higher members of the single $l = 0$ series of terms with the lowest doublet level of the $l = 1$ series, that is, the lowest p level. Here the doubling arises from the common pair of final levels. It follows that the doublet interval is constant throughout the sharp series, and is equal to the doublet interval of the resonance line, or first line of the principal series. Because the $l = 0$ terms describe the sharp series, it is customary to call them s terms.

The *diffuse series* arises from the combination of the double $l = 2$ series of terms with the same lowest doublet level of the $l = 1$ series. For this reason the $l = 2$ terms are called d terms.

The *fundamental series* arises from combination of the $l = 3$ doublet levels with the lowest level pair of the $l = 2$ or d series. For this reason the $l = 3$ levels are called f terms.

In order to specify a level completely one must give the n, the l (or k), and the value of the vector resultant total angular momentum which is denoted by $j\hbar$. It is usual to state the value through the letter designation $s, p, d, f, \ldots$ and to indicate the j value as a subscript to the letter.

Thus we have

$$s_{1/2}, p_{1/2}, p_{3/2}, d_{3/2}, d_{5/2}, f_{5/2}, f_{7/2}, \ldots$$

as the basic structure of the l and j values in alkali spectra. The *normal* order of levels in a doublet pair is for the level of lower j to have lower energy. Some exceptions occur and such doublets are said to be *inverted*. Another common notation uses, for example, $^2S_{1/2}$ in place of $s_{1/2}$.

The principal features of the lower levels in the alkalis are shown in Figure 2[1]. The fine structure doubling is too small to be shown on this scale. The n numbering of the terms is determined by the considerations of Section 11[1], according to which lowest n values are assigned as needed to the core electrons. Figure 2[1] exemplifies the various points already mentioned: In Li the 2-electron core is so small that the $2p$ level is nearly hydrogenic. In Na the 10-electron core is larger but still small enough that $3d$ is quite hydrogenic, as it also is in K with an 18-electron core. The lowest s levels, the ground states, stay close to the same value, that is, ν a little less than 2, but n increases by one unit for

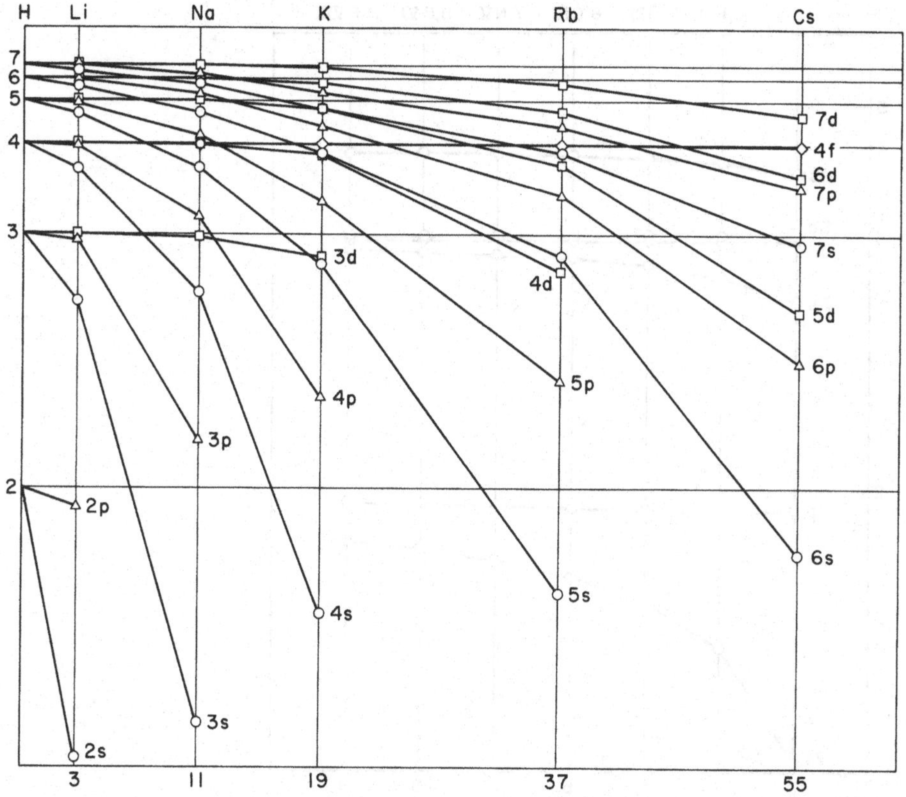

Figure 2[1]. Observed energy levels of one-electron states in hydrogen and the alkali metals plotted against Z. The hydrogenic values, indicated by n values on the left, are $e^2/2an^2$, measured from $n = \infty$ as zero point.

each successive alkali, so Δ, which is a better measure of the departure from hydrogen, increases greatly from Li to Cs.

The $4f$ levels, and all higher f levels, remain very close to hydrogenic values even up to Cs, showing that the 54-electron Cs core is smaller than the radius of the $4f$ circular orbit.

Figure 3[1]a and 3[1]b show the behavior of some of the levels along the $N = 3$(Li) and $N = 19$(K) isoelectronic sequences. Here the ordinates are the observed energy values divided by the C^2 occurring in (5).

In the Li sequence, the $2p$ level is nearly hydrogenic in lithium and remains so as one goes to higher degrees of ionization. $2s$ is quite nonhydrogenic in Li, but becomes more nearly so at higher stages of ionization. As Z increases, the core shrinks, but so also does the outer orbit, the two tendencies working in opposite directions. The figure shows clearly that $2s$ becomes proportionately less penetrating at higher ionization.

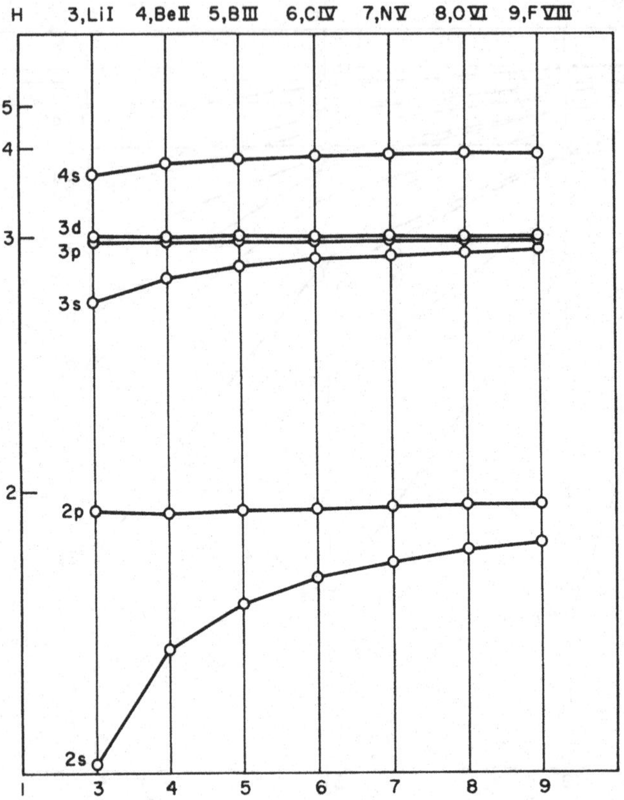

Figure 3[1]a. Low levels in the lithium ($N = 3$) isoelectronic sequence; ordinates $W(nl)/C^2$, abscissas C. Hydrogenic values for $n = 2, 3, 4, 5$, are shown at the left; the top boundary line corresponds to $n = \infty$.

The $3s$ levels in the Li sequence show similar behavior, as do the $4s$, $5s, \ldots$ and $4p, 5p, \ldots$ levels in the K sequence.

In the K sequence the behavior of the $3d$ and $4d$ levels is quite different from that of the s and p levels. Here the $3d$ level is quite closely hydrogenic in K I, but on going to Ca II it has a considerably increased quantum defect. This indicates that the $3d$ orbit in K is on the verge of being penetrating. The increase in Z on going to Ca II has shrunk the core, but has shrunk the $3d$ orbit more, so it is now penetrating, and so is exposed to a stronger field. This same tendency continues to Sc III, but at Ti IV the $3d$ level has started to increase again, becoming more hydrogenic, indicating that the core is shrinking more than the $3d$ orbit for each unit increase in Z.

The complete working out of the description of the alkali spectra in terms of Bohr orbit theory in the early 1920s gave an immense stimulus to the study of spectra and to the general acceptance of what a decade earlier was a most revolutionary development.

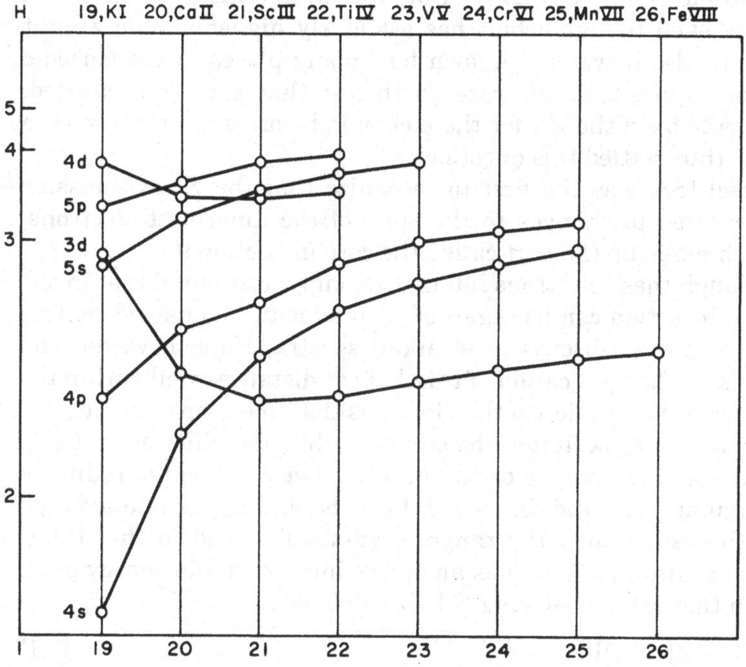

Figure 3¹b. Low levels in the potassium ($N = 19$) isoelectronic sequence; ordinates $W(nl)/C^2$, abscissas C. Hydrogenic values for $n = 2, 3, 4, 5$, are shown at the left; the top boundary line corresponds to $n = \infty$.

10¹. X-ray spectra

X-ray spectroscopy became possible with the discovery of X-ray diffraction by crystals by von Laue in 1912. The field was opened up at once by Moseley [86], working in Rutherford's laboratory in Manchester. His pioneering experiments covered the range $0.4 > \lambda > .08$ Å, using 38 different elements as targets in the X-ray sources.

For each element he found two close groups of lines which he called the K and L radiations, following a notation introduced considerably earlier by Barkla, who found that the radiation from a particular element consists of radiations having several different "hardnesses" or penetration powers [87].

Moseley found that the square roots of the wave numbers of the K and L radiations are approximately linear functions of the atomic number,

$$\sqrt{\sigma_k} = K(Z - S_k)$$

and similarly for the L lines.

At that time the Z values for the elements were not accurately known, except that in a general way Z increases with the atomic weight. Moseley found that Z for iron, cobalt, and nickel must be 26, 27, and 28, even though cobalt has a slightly higher atomic weight than nickel. Also it was not known how many places in the periodic table were required for the rare earths, at that time unseparated. Moseley's rule fixed the Z's for the elements before and after the rare earths and thus settled this question.

W. Kossel [88] was the first to recognize that the X-ray emission lines are related to changes in the state of the innermost electrons, those which make up the part called the core in Section 9^1.

Even though these electrons interact strongly, it is found that in the main the interaction can be regarded as producing a screened central field ($9^1 3$) as in the discussion of alkali spectra. Here, however, the emphasis is on the application of this $V(r)$ at distances well within the core to describe the motion of the electrons that make up the core.

It is easiest to think first of the circular orbits for which $n_k = 1_1, 2_2, 3_3, \ldots$ because these have a definite radius. Let $r(n_k)$ be the radius of such a circular orbit and for $n > k$ let it be an appropriate average value of the radius over the range of values involved in that orbit. Then it is natural to take $9^1 5$ as an approximation to the energy of an electron in that orbit, replacing C by Z^* defined as

$$Z^* = Z[r(n_k)] \tag{1}$$

This procedure requires fuller justification. Provisionally we accept it as giving an indication of the energies with which the different electrons are bound.

To this must be added the idea of a *maximum number* of electrons in each kind of orbit. Orbits of a given n_k are said to belong to the same *shell*. The filling of shells with increase of Z is the idea underlying the model's explanation of the periodic behavior of the chemical properties which was discovered by Mendeleev in 1869 (sec. 11^1).

On this view an atom in the middle range of Z in its normal state consists of a number of electrons in closed or filled shells, plus several outer or valence electrons. Excitation of the atom for X-ray emission consists in the removal of an electron from an inner filled shell, an X-ray photon being emitted when one of the less tightly bound electrons in an outer shell makes a transition inward to fill the inner shell in which a vacancy was made by the excitation process. As with visible spectra, this picture proved to be immensely valuable in correlating the large amount of X-ray spectroscopic data that was obtained in the decade following Moseley's first work in 1913.

Discovery of crystal diffraction of X rays also made possible development of absorption spectroscopy of X rays [89]. Here the pioneering was by M. de Broglie, who discovered the *absorption edges*. The absorption of X rays is found to be an additive property of the elements present in the absorber. The absorption coefficient, k, in $I = I_0 e^{-kz}$ for the fraction of incident intensity transmitted through a thickness z, depends on the wave number σ as

$$k(\sigma) = \sum_Z N_Z S_Z(\sigma) \tag{2}$$

in which N_Z is the number of atoms of atomic number Z in unit volume and $S_Z(\sigma)$ is an effective absorption cross section area characteristic of each element and nearly independent of its state of chemical composition.

The dependence of S_Z on σ is illustrated in Figure 4[1], which shows the observed results for Fe (26), Ag (47), and Pb (82). At a wave number just higher than that of the prominent K lines, there is an abrupt increase in the absorption coefficient, which then decreases as σ is further increased. A similar behavior is indicated for the L lines, except that here there are three distinct wave numbers at which abrupt increases occur in $S_Z(\sigma)$. These three edges are called the L_{III}, L_{II}, L_{I} in order of increasing σ. A similar behavior is observed for all elements and all edges, but there are five M edges, seven N edges, and so on.

For σ just less than the K edge, a photon does not have enough energy to lift an electron out of a K or 1_1 orbit. It cannot lift it to an L or M state because these states as well as N and O states for heavy atoms are already occupied to the maximum degree. If a photon raises an electron to an unoccupied or valence level the electron orbit will be unstable because of the presence of a vacancy in the shell with $k = 1$. Therefore, the photon can be absorbed only when it can remove the electron to a state in which it leaves the atom. The situation is to be contrasted with that involved in the absorption of light in the principal series of sodium. Here a long series of sharp absorption lines, involved in transitions from the 3_1 valence level to the n_2 levels occurs, followed by a continuous region of absorption due to transitions into the nonquantized levels beyond the series limit. In the X-ray case absorption will be continuous above the absorption edge.

The fact that there is but one K absorption edge indicates that the two 1_1 electrons are equally bound. But the fact that there are three L absorption edges indicates that there are *three* kinds of $n = 2$ electrons which differ in their tightness of binding. The simple orbit theory would give two types, 2_1 and 2_2. Thus there is additional structure to the inner levels. This extra complexity is closely related to the extra complexity occurring in the alkali spectra when the fine structure is taken into account. As with the alkali spectra it was found that the extra structure could be described by assigning a third quantum number, j, to label the X-ray levels. As in the alkali spectra, for a given n_k, j can have the two values $j = k \pm 1/2$, except that for $k = 1$, only the one value $j = 1/2$ occurs.

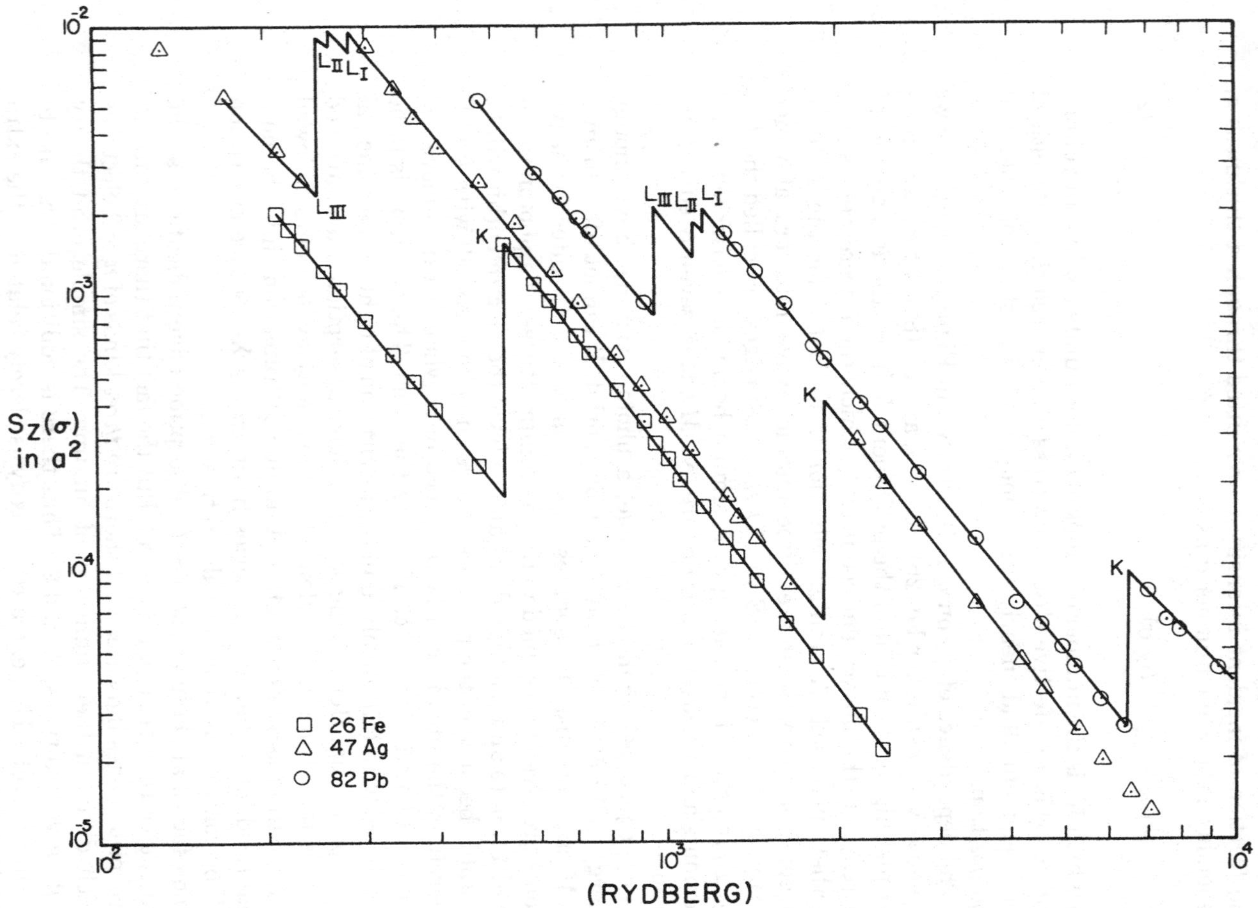

Figure 4[1]. X-ray absorption cross sections $S_Z(\sigma)$, for Fe, Ag, and Pd, with S in a^2 as unit, at photon energy, σ in Rydbergs. (Compiled by S. J. M. Allen, *Handbook of Chemistry and Physics*, 49th ed., Chemical Rubber Publishing Co., 1968–69, p. E133.)

Altogether the study of the regularities of both X-ray absorption and emission spectra came at an opportune time to be stimulated by the Bohr theory and to provide a wide range of experimental results, which contributed greatly to its acceptance.

11^1. The periodic system; Pauli exclusion principle

That the structures of different atoms must be regularly recurring variants of a basic pattern became clear long before the recognition of the nature of that pattern, through the discovery of the periodic system of the elements by Mendeleev [90] in 1869. Along with other contemporaries, he observed the more or less regular recurrence of elements with similar chemical properties when these are arranged in order of increasing atomic weight.

Although others also had noted some regularities of this kind, Mendeleev deserves the main credit because he carried their study to such a point that he was able to predict successfully the chemical properties of several elements which were not known at that time. His predictions proved to be remarkably accurate when the corresponding elements were later found. No doubt remained thereafter but that these empirical regularities must have profound significance with respect to the general pattern of atomic structure.

With the discovery of isotopes and atomic numbers, it became clear that Z, rather than the mean atomic weight, is the characteristic of an element giving a truly significant correlation. With the discovery of X-ray spectra and their interpretation in terms of shells containing maximum numbers of electrons, Bohr [91] was able to give a general qualitative picture of the building up of the complex atoms.

At this stage there was no basic understanding of a rationale for the idea of a maximum permissible number of electrons in an electron shell. An extensive literature developed as various physicists groped for the proper way to characterize individual electron states. The subject was confused by uncertainty as to the n values to be assigned to electrons in the outer orbits of atoms of the middle and higher values of Z.

When the situation was finally cleared up it appeared that the states of an individual electron in an atom could be characterized by *four* quantum numbers:

n the principal or total quantum number of earlier discussion.

l the orbital angular quantum number in units of $\hbar$, equal to $(k - 1)$ of the original Bohr theory and so taking on the values $l = 0, 1, 2, \ldots, (n - 1)$ for a given n; usually coded as $s, p, d, f, \ldots$, as in Section 9^1.

m the component of orbital angular momentum along a particular direction, in units of $\hbar$, and so taking on the $-l, -l+1, \ldots, l-1, l$, or a total of $(2l+1)$ values.

m_s the component of spin angular momentum along a particular direction, in units of $\hbar$, and always restricted to the values $\pm 1/2$.

Before the discovery of electron spin, the need for such a two-valued quantum number was recognized, but without a clear interpretation of its physical meaning.

Thus an individual quantum state requires (n, l, m, m_s) for its complete characterization. The (n, l) values characterize an *electron shell.* As m_s always has two possible values, and m has $2l + 1$ possible values, there are $2(2l + 1)$ distinct states associated with a given shell.

Pauli [92] showed in 1925 that the lengths of the periods of the periodic system of the elements fall neatly into place, as also do many other observed facts of the complex structure of atomic spectra if one postulates that in any quantum dynamical system, a given state can be either unoccupied or *occupied at most by one electron.* This is known as the *Pauli exclusion principle.* It is now known to play an important role in all electron containing systems, in the theory of chemical valence bonds in molecules, and in the theory of solid metals, semiconductors, and insulators. Later it was learned that the same principle applies to protons and neutrons. Because of the work of Fermi in applying the principle to a general study of statistical mechanics, particles that obey the principle are now called *fermions.* Electrons, protons, and neutrons are all fermions.

These general ideas served to clarify the periodic system. Table 3^1 shows the periodic table of the elements in one of its traditional forms and Table 4^1 shows it in a modern form in which the rows are labelled by n values of the valence electrons and the columns correspond to the l designations of the electrons outside closed shells. Table 1^1 gives an alphabetical index to Tables 3^1 and 4^1.

It is customary to indicate the number of electrons in a given nl state by $q(nl)$ written as an exponent on the nl symbol. Thus, the normal state of neutral nitrogen ($Z = 7$) is $1s^22s^22p^3$, and that of phosphorus ($Z = 15$) in the row below is $1s^22s^22p^63s^23p^3$. Thus the two structures agree in consisting of three p electrons outside of closed shells. This is the basis of classification by columns in Table 4^1.

The tightness of binding decreases with increasing l, for a fixed n, as mentioned in Section 9^1. The magnitude of this dependence on l increases with Z, because the penetrating orbits reach into regions of much stronger fields as Z increases. At first the departure from the main hydrogenic behavior is not great, so that shells are filled in the order $1s$, $2s$, $2p$, $3s$, $3p$, up to $_{18}$Ar. If this rule continued to be obeyed the next ten elements would then involve the successive addition of ten $3d$ electrons.

However, the increasing departure from hydrogenic values has the effect that at this point the $4s$ state is lower in energy than the $3d$ state. Therefore the next two elements add $4s$ and $4s^2$ to form their normal states, after which the filling of the $3d$ shell begins. As Table 4^1 shows, all s elements are *alkalis* (except hydrogen), and all s^2 elements are *alkaline earths* (except helium). It is customary to call the ten $3d$ shell elements the first *transition group*.

The extreme tightness of binding of the $1s$ state as compared with others is such as to put the elements $_1$H and $_2$He in a class by themselves. Admittedly hydrogen is not an alkali metal nor helium an alkaline earth, but it is better to place them in the s and s^2 columns, respectively, in order that the table serve as a correct indication of the electronic state occupancies in the normal state.

In the transition elements there is a relatively small energy difference between s^2d^q and s^1d^{q+1}, so which one of these happens to give rise to the lowest state in these atoms fluctuates a little from atom to atom. The first transition group ends with $_{30}$Zn after which the next six elements, to $_{36}$Kr, are involved in the filling of the $4p$ shell.

There are many details to be considered in a full discussion of this important subject. This brief account will suffice as a pre-quantum-mechanical historical introduction.

12^1. Bohr's correspondence principle

A conspicuous feature of any emission or absorption spectrum is the wide variation in the relative intensity of the various lines. So great is the range, in fact, that lines are not observed at all corresponding to transitions between some pairs of energy levels.

Such pairs are said not to combine, or that the combination is *forbidden*. Rules expressing which pairs of levels combine are called *selection rules*. For example, in alkali spectra, it is observed that s levels combine only with p levels, p only with s or d, d only with p or f, and so on. This may be summarized in the rule that $\Delta l = \pm 1$ in the allowed transitions.

When the fine structure is described by use of the j quantum number it is found to conform to the rule that $\Delta j = 0, \pm 1$ except $0 \to 0$. For example, the combination $^2D \to {}^2P$ gives three lines, not four, because $^2D_{5/2} \to {}^2P_{1/2}$ would violate this rule.

Among the allowed lines themselves there are wide variations of intensity. We now consider the early theoretical developments relating to attempts to correlate such observations. The first interpretation of the Δl rule was given by Rubinowicz [93], based on considerations of conservation of angular momentum in the radiation process. Efforts to interpret intensity questions led Bohr to the formulation of his *corre-*

Table 3[1]. Periodic table of the elements in traditional form

Period	0	I_a	I_b	II_a	II_b	III_a	III_b	IV_a	IV_b	V_a	V_b	VI_a	VI_b	VII_a	VII_b	VIII
0		${}_{1}$H														
1	${}_{2}$He	${}_{3}$Li		${}_{4}$Be		${}_{5}$B		${}_{6}$C		${}_{7}$N		${}_{8}$O		${}_{9}$F		
2	${}_{10}$Ne	${}_{11}$Na		${}_{12}$Mg		${}_{13}$Al		${}_{14}$Si		${}_{15}$P		${}_{16}$S		${}_{17}$Cl		
3	${}_{18}$Ar	${}_{19}$K		${}_{20}$Ca			${}_{21}$Sc		${}_{22}$Ti		${}_{23}$V		${}_{24}$Cr		${}_{25}$Mn	${}_{26}$Fe ${}_{27}$Co ${}_{28}$Ni
			${}_{29}$Cu		${}_{30}$Zn	${}_{31}$Ga		${}_{32}$Ge		${}_{33}$As		${}_{34}$Se		${}_{35}$Br		
4	${}_{36}$Kr	${}_{37}$Rb		${}_{38}$Sr			${}_{39}$Y		${}_{40}$Zr		${}_{41}$Nb		${}_{42}$Mo		${}_{43}$Tc	${}_{44}$Ru ${}_{45}$Rh ${}_{46}$Pd
			${}_{47}$Ag		${}_{48}$Cd	${}_{49}$In		${}_{50}$Sn		${}_{51}$Sb		${}_{52}$Te		${}_{53}$I		
5	${}_{54}$Xe	${}_{55}$Cs		${}_{56}$Ba			${}_{57}$La	rare earths	${}_{72}$Hf		${}_{73}$Ta		${}_{74}$W		${}_{75}$Re	${}_{76}$Os ${}_{77}$Ir ${}_{78}$Pt
			${}_{79}$Au		${}_{80}$Hg	${}_{81}$Tl		${}_{82}$Pb		${}_{83}$Bi		${}_{84}$Po		${}_{85}$At		
6	${}_{86}$Rn	${}_{87}$Fr		${}_{88}$Ra			${}_{89}$Ac									
4*f*, lanthanide series				${}_{58}$Ce	${}_{59}$Pr	${}_{60}$Nd	${}_{61}$Pm	${}_{62}$Sm	${}_{63}$Eu	${}_{64}$Gd	${}_{65}$Tb	${}_{66}$Dy	${}_{67}$Ho	${}_{68}$Er	${}_{69}$Tm	${}_{70}$Yb ${}_{71}$Lu
5*f*, actinide series				${}_{90}$Th	${}_{91}$Pa	${}_{92}$U	${}_{93}$Np	${}_{94}$Pu	${}_{95}$Am	${}_{96}$Cm	${}_{97}$Bk	${}_{98}$Cf	${}_{99}$Es	${}_{100}$Fm	${}_{101}$Md	${}_{102}$No ${}_{103}$Lw

Table 4[1]. Periodic table of the elements by electron configuration*

n	s	s^2	p	p^2	p^3	p^4	p^5	p^6	d	d^2	d^3	d^4	d^5	d^6	d^7	d^8	d^9	d^{10}
1	$_{1}$H	$_{2}$He																
2	$_{3}$Li	$_{4}$Be	$_{5}$B	$_{6}$C	$_{7}$N	$_{8}$O	$_{9}$F	$_{10}$NE										
3	$_{11}$Na	$_{12}$Mg	$_{13}$Al	$_{14}$Si	$_{15}$P	$_{16}$S	$_{17}$Cl	$_{18}$Ar										
$4s$, $3d$	$_{19}$K	$_{20}$Ca							$_{21}$Sc	$_{22}$Ti	$_{23}$V	$_{24}$Cr	$_{25}$Mn	$_{26}$Fe	$_{27}$Co	$_{28}$Ni	$_{29}$Cu	$_{30}$Zn
$4p$			$_{31}$Ga	$_{32}$Ge	$_{33}$As	$_{34}$Se	$_{35}$Br	$_{36}$Kr										
$5s$, $4d$	$_{37}$Rb	$_{38}$Sr							$_{39}$Y	$_{40}$Zr	$_{41}$Nb	$_{42}$Mo	$_{43}$Tc	$_{44}$Ru	$_{45}$Rh	$_{46}$Pd	$_{47}$Ag	$_{48}$Cd
$5p$			$_{49}$In	$_{50}$Sn	$_{51}$Sb	$_{52}$Te	$_{53}$I	$_{54}$Xe										
$6s$, $5d$, $4f$	$_{55}$Cs	$_{56}$Ba							$_{57}$La	(lanthanide series below)								
										$_{72}$Hf	$_{73}$Ta	$_{74}$W	$_{75}$Re	$_{76}$Os	$_{77}$Ir	$_{78}$Pt	$_{79}$Au	$_{80}$Hg
$6p$			$_{81}$Tl	$_{82}$Pb	$_{83}$Bi	$_{84}$Po	$_{85}$At	$_{86}$Rn										
$7s$, $6d$, $5f$	$_{87}$Fr	$_{88}$Ra								(actinide series below)								
lanthanides		df	$5d$ +	f	f^2	f^3	f^4	f^5	f^6	f^7	f^8	f^9	f^{10}	f^{11}	f^{12}	f^{13}	f^{14}	
				$_{58}$Ce	$_{59}$Pr	$_{60}$Nd	$_{61}$Pm	$_{62}$Sm	$_{63}$Eu	$_{64}$Gd	$_{65}$Tb	$_{66}$Dy	$_{67}$Ho	$_{68}$Er	$_{69}$Tm	$_{70}$Yb	$_{71}$Lu	
actinides			$6d$ +	f	f^2	f^3	f^4	f^5	f^6	f^7	f^8	f^9	f^{10}	f^{11}	f^{12}	f^{13}	f^{14}	
				$_{89}$Ac	$_{90}$Th	$_{91}$Pa	$_{92}$U	$_{93}$Np	$_{94}$Pu	$_{95}$Am	$_{96}$Cm	$_{97}$Bk	$_{98}$Cf	$_{99}$Es	$_{100}$Fm	$_{101}$Md	$_{102}$No	$_{103}$Lw

*Elements for $Z > 92$ are made artificially.

spondence principle, which clarified many points and pointed the way that led Heisenberg to the discovery of quantum mechanics in 1925.

In 1917 Einstein [94] gave a statistical discussion of the relations between rates of emission and absorption which must exist in order to preserve the thermal equilibrium between a gas of atoms in a Boltzmann distribution and the radiation in a Planck distribution. Because he was concerned only with such equilibrium, his original treatment did not concern itself with questions of the polarization or the angular distribution of the radiation emitted in a transition between a given pair of states.

Let n be the set of quantum numbers needed to specify the distinct levels, and $g(n)$ be the number of distinct quantum states belonging to the level $E(n)$. Then Einstein postulated, without giving any details of an electromagnetic basis, that the upper-level n' is characterized by a *spontaneous* transition rate $A(n', n'')$ for transitions to a lower level n'', with emission of a photon, $h\nu = E(n') - E(n'')$.

In addition, he postulated that radiation of frequency ν acts on atoms in the level n' inducing or *stimulating* them to make emissive transitions to n'' at a rate proportional to $\rho(\nu)$, the energy density of radiation of the same frequency. For this rate he wrote $B(n', n'')\rho(\nu)$. Similarly, the radiation of this frequency is absorbed by atoms in the state n'', to produce a rate of transition upward for the rate of which he wrote $B(n'', n')\rho(\nu)$.

Thus emissive and absorptive processes between the levels are governed by three rate coefficients. These are related, however, so that only one of these is needed, by use of the *principle of detailed balancing,* which says that the total upward rate and the downward rate of transitions between each pair of levels must be equal in thermal equilibrium, that is

$$N(n')[A(n', n'') + B(n', n'')\rho(\nu)] = N(n'')B(n'', n')\rho(\nu)$$

The ratio of the population of states in levels n' and n'' is given by the Boltzman distribution.

$$\frac{N(n')}{N(n'')} = \frac{g(n')}{g(n'')}\exp(-h\nu/kT)$$

and the energy density by the Planck distribution law (4^13)

$$\rho(\nu) = 8\pi h\,\frac{\nu^3}{c^3}\,\frac{1}{\exp(h\nu/kT)-1}$$

From this we find the Einstein relations

$$A(n', n'') = 8\pi h\,\frac{\nu^3}{c^3}\,B(n', n'') \qquad (1)$$

and

$$g(n')B(n', n'') = g(n'')B(n'', n') \tag{2}$$

by which all three rate coefficients may be expressed in terms of any one of them.

This argument does not provide an evaluation of the basic rate: It could be fast or slow, or even vanish and not upset thermal equilibrium, as long as (1) and (2) are satisfied. Bohr undertook to extend the theory so as to give actual values for the rates.

We consider now the classical view concerning radiation emission from a system consisting of a number of moving charges. For simplicity, we may deal with a case in which the charges remain within a region that is small compared with the wavelength of light that would be radiated. In this case the electric and magnetic fields, $\mathbf{E}(\mathbf{r},t)$ and $\mathbf{B}(\mathbf{r},t)$, at large distances because of the moving charges can be expressed in terms of the electric dipole moment

$$\mathbf{P} = \Sigma\, e\mathbf{r} \tag{3}$$

of the moving charges. This is given by use of the retarded potential solutions of Maxwell's equations for the electromagnetic field [95] as

$$\begin{aligned} \mathbf{E}(\mathbf{r}, t) &= (c^2r)^{-1}[\ddot{\mathbf{P}} - \hat{\mathbf{r}} \cdot \ddot{\mathbf{P}} \cdot \hat{\mathbf{r}}] = (c^2r)^{-1}\,\ddot{\mathbf{P}}_\perp \\ \mathbf{B}(\mathbf{r}, t) &= (c^2r)^{-1}\hat{\mathbf{r}} \times \ddot{\mathbf{P}}_\perp \end{aligned} \tag{4}$$

Here $\hat{\mathbf{r}}$ is the unit vector in the direction from the center of the atom to the field point and $\ddot{\mathbf{P}}_\perp = \ddot{\mathbf{P}} - (\hat{\mathbf{r}}\cdot\ddot{\mathbf{P}}\cdot\hat{\mathbf{r}})$ is the component of $\ddot{\mathbf{P}}$ perpendicular to $\mathbf{r}$, and $\ddot{\mathbf{P}}_\perp$ is to be evaluated at $(t - r/c)$.

This field represents an expanding spherical wave with a state of polarization and wave form given by the time variation of $\mathbf{P}$. Because of the occurrence of $\ddot{\mathbf{P}}_\perp$, the amplitude, and so the intensity, of the radiated wave depends on the direction of $\hat{\mathbf{r}}$. The intensity is given by the Poynting vector

$$\mathbf{S} = \frac{c}{4\pi}\mathbf{E} \times \mathbf{B} = [\ddot{\mathbf{P}} - (\hat{\mathbf{r}} \cdot \ddot{\mathbf{P}} \cdot \hat{\mathbf{r}})]^2 \frac{\hat{\mathbf{r}}}{4\pi c^3 r^2} \tag{5}$$

Integration of this over a spherical surface of radius r gives for the total rate of outward flow of energy in all directions

$$\frac{2}{3c^3}\,\ddot{\mathbf{P}}^2 \tag{6}$$

There is a sharp conflict between this picture and that of the quantum theory. On the classical view the atom is continuously radiating waves of various frequencies, which are gradually changing as the system loses energy. On the quantum view, when it is in an excited state, the atom radiates not at all until it makes a particular transition, *one* among a number of possible ones, in which case it emits a photon with a particular polarization in a particular direction, the different possible directions being statistically distributed in an ensemble of many transitions.

Despite the great differences, Bohr knew that the classical theory gives a correct description of the low frequency electric waves used in radio communication, and so the quantum description must be such that, asymptotically, in the limit of large quantum numbers and small changes in them, it would agree with the classical description. It is difficult to give a one-sentence definition of the correspondence principle. Perhaps one should say it is the ensemble of the consequences of following up this guiding idea, and they were many.

To carry the work further it is necessary to review briefly the Hamiltonian form of the classical mechanics of a multiply-periodic system [96]. Let its generalized coordinates be $q_1, \ldots, q_s$ and let $T(q, \dot{q})$ represent the kinetic energy, which is a function of all the q's and $\dot{q}$'s.

The conjugate momentum p_r is defined as

$$p_r = \frac{\partial T}{\partial \dot{q}_r} \tag{7}$$

which can be solved to express each $\dot{q}_r$ in terms of the q_r and p_r.

The Hamiltonian function is defined as

$$H(p, q) = \sum_r p_r \dot{q}_r - T + V \tag{8}$$

in which H is expressed as a function of p's and q's by using (7) to eliminate the $\dot{q}$'s. The Hamiltonian canonical equations of motion are

$$\dot{p}_r = -\frac{\partial H}{\partial q_r}, \quad \dot{q}_r = \frac{\partial H}{\partial p_r} \tag{9}$$

A transformation from the p's and q's to new variables P's and Q's will lead to a new Hamiltonian $\overline{H}(P, Q)$ obtained by expressing the p's and q's in $H(p, q)$ in terms of the P's and Q's. For an arbitrary change of variables the equations of motion no longer have the canonical form. But there is a restricted class of such transformations that do preserve the form (9): These are called *canonical* transformations.

Among the canonical transformations of a given system there is a particularly simple one that, if it can be found, leads directly to the general solution of the problem. This has the property that $\overline{H}(P, Q)$, which in general depends, as indicated, on the P's and Q's, actually depends only on the P's, so that $\partial \overline{H}/\partial Q_r = 0$.

In consequence from (9) the $\dot{P}_r = 0$ and so the P_r are constants, so $\dot{Q}_r = \partial H/\partial P_r = \nu_r$ depending on the values of the P_r so

$$Q_r = \nu_r t + \delta_r$$

where the δ_r are another set of integration constants. Altogether then a particular state of motion is defined by $2s$ integration constants (P_r and δ_r). The momenta and coordinates, P and Q, introduced in this way are called *action* and *angle* variables, respectively, and are usually denoted by J and w.

In this formulation the Bohr-Sommerfeld quantum conditions (Sec. 7¹) take the form

$$J_r = n_r h \tag{10}$$

where the n_r are a set of s quantum numbers defining a particular state of motion. At first the n_r were taken to be integers: What is essential is that the *differences* between allowed values of each n_r be integers, but this detail does not matter here.

For systems in which each Cartesian coordinate has a restricted range of variation, they can be expanded in multiple Fourier series having combination frequencies $\Sigma\tau_r\nu_r$ with coefficients that depend both on the n's and the integral τ's. Therefore, for the time variation of the electric dipole moment we have

$$\mathbf{P}(t) = \frac{1}{2}\sum_{\tau_r} \mathbf{P}(n_r, \tau_r) \exp[2\pi i(\Sigma\tau_r\nu_r)t] \tag{11}$$

Here the summation extends over all integer values of the τ's and, because $\mathbf{P}(t)$ is real,

$$\mathbf{P}(n_r, -\tau_r) = \mathbf{P}^*(n_r, \tau_r) \tag{12}$$

In classical mechanics, a state of the system is said to be *degenerate* if two or more of the ν_r are equal, or even if there are commensurable relations between some of them, so that several terms of (11) have equal frequency. This introduces special features that are analogous to those of degenerate systems in quantum mechanics but these do not need to be considered in this brief survey.

The frequencies, $\Sigma\tau_r\nu_r$, which appear in (11), do not agree with the quantum frequencies, but there is an important relation between them that was discovered by Kramers [97]. With the quantum transition $n_r' \rightarrow n_r''$ we associate the two terms in (11) for which the τ's are

$$\tau_r = \pm(n_r' - n_r'') \tag{13}$$

In classical mechanics the n's are not restricted to integer values so we may think of the n's as varying continuously from their initial to final values as

$$n_r = n_r' - \lambda(n_r' - n_r'') \tag{14}$$

where λ varies from 0 to 1. The quantum frequency,

$$\left|\frac{\overline{H}(n'_r h) - \overline{H}(n''_r h)}{h}\right|$$

is now easily seen to be exactly equal to the mean value of the frequency of the associated Fourier component over this set of intermediate states, because

$$\left|\overline{H}(n'_r h) - \overline{H}(n''_r h)\right| = \int_0^1 \sum_{\tau_r} \tau_r \frac{\partial H}{\partial n_r}\, d\lambda \qquad (15)$$

This is often called the correspondence *theorem* for frequencies, because it is an exact result.

In (15) for large n's and small τ's, the integrand does not vary much with λ, so in this limit the quantum frequency agrees with that of the associated Fourier component.

This correspondence strongly suggests that the ensemble of dipole moment Fourier components $\mathbf{P}(n_r, \tau_r)$ for the same set of intermediate states (14) between the initial and final, may govern, through an appropriate adaptation of classical formulas like (4), (5), and (6), the radiation emitted in the quantum jump from n'_r to n''_r. This is known as the correspondence *principle* for polarization and intensity. In particular, if $\mathbf{P}(n_r, \tau_r)$ vanishes for each λ in the range 0 to 1, one might expect that transition to be forbidden. When worked out in detail for a number of simple systems, this principle is found to give the empirical selection rules correctly.

In analogy with (15) efforts were made to discover some kind of average of $\mathbf{P}(n_r, \tau_r)$ with respect to λ to give the appropriate $\mathbf{P}$ to be used in the classical formulas for radiation, but such specific efforts were not fruitful.

In the limit of large n's and small τ's however, $\mathbf{P}(n_r, \tau_r)$ does not vary much with λ so the polarization was taken to be that given by using this $\mathbf{P}(n_r, \tau_r)$ in (4) and the transition probability $A(n'_r, n''_r)$ was assumed to be given by

$$A(n'_r, n''_r)h\nu = \frac{(2\pi\nu)^4}{3c^3}[\mathbf{P}(n_r, \tau_r)]^2 \qquad (16)$$

by equating the mean rate of radiation on the quantum view to that on the classical view.

By this time it was recognized [98] that the Fourier components of the classical motion are not the thing to be used in (16). But the formal relations were such that they suggested that the as-yet-undiscovered quantum mechanics of atomic systems would provide an appropriate $\mathbf{P}(n_r, \tau_r)$ dipole moment amplitude to use that would make (16) into an exact relation, for all pairs of sets of n's. Thus was the stage set for Heisenberg's discovery of the matrix formulation of quantum mechanics in 1925.

The correspondence principle was extended by Van Vleck [99] to the treatment of absorption and induced emission by considering the perturbation of a classical multiply-periodic system by the fields of a light wave. Absorption and induced emission occur when the frequency of the perturbing light wave agrees with one of the combination tones ($\Sigma \tau_r \nu_r$), so analogous alterations in the classical formulas are necessary to get approximate quantum-theoretic formulas for the Einstein B's.

At light frequencies away from resonance with a combination tone, the light wave produces forced oscillations of its own frequency, ν, and also $\nu \pm \Sigma \tau_r \nu_r$. Those of frequency ν have a coherent phase relation with the light wave, and give rise to a scattered wave which combines with the incident wave to alter the phase velocity of propagation *(dispersion)*. Those of frequency $\nu \pm \Sigma \tau_r \nu_r$ indicate that there would be scattered light of frequency modified by those of quantum jumps in the atom. These were discovered in 1928 and later became generally known as the *Raman effect* (Sec. 13[1]).

The classical theory of dispersion is based on the following connection between the refractive index, n, and the polarizability, α, of the molecules of which there are N per unit volume

$$\frac{n^2 - 1}{n^2 + 2} = \frac{4\pi}{3} N\alpha(\nu) \tag{17}$$

Here $\alpha(\nu)$ is the ratio of the induced coherent dipole moment $\mathbf{P}e^{2\pi i\nu t}$ to the amplitude $\mathbf{E}\,e^{2\pi i\nu t}$ of the light wave that induces it, a quantity known as the *polarizability* of the molecule. The simple dynamics of forced oscillation of an harmonic oscillator of natural frequency ν_0 gives

$$\alpha(\nu) = \frac{e^2}{m} \frac{1}{4\pi^2(\nu_0^2 - \nu^2)}$$

From early experimental work of R. W. Wood [100] it was known that sodium vapor shows the kind of dispersion described by this formula, provided ν_0 is equal to the quantum absorption frequency of the D lines of sodium, and provided also that one assigns empirical effective strength constants to each absorption line produced from the normal state. The appropriate form for α was found by Ladenburg [101] to be

$$\alpha(\nu) = \frac{e^2}{m} \sum_k \frac{f_k}{4\pi^2(\nu_k^2 - \nu^2)}$$

The quantities f_k have sometimes been called the number of dispersion electrons per atom, but more recently are simply called *oscillator strengths*.

The correspondence principle was applied to the problem by Kramers [102]. The most important and surprising result to come out was that for atoms which are in excited states i, there are terms like the classical ones but with *negative* values of f. That is, for atoms in the state i, one has a term $(e^2/4\pi^2 m)f_{ik}/(\nu_{ik}^2 - \nu^2)$ with $f_{ik} > 0$ when k refers to an energy level above i, but with $f_{ik} < 0$ when k refers to a level below i.

The final result is that the effective value of $N\alpha$ to be used in calculating the refractive index is

$$N\alpha = \frac{e^2}{m}\sum_{k,i}\frac{(N_i - N_k)f_{ik}}{4\pi^2(\nu_{ik}^2 - \nu^2)} \tag{18}$$

in which N_i is the number of atoms per unit volume in the ith state and the sum is over those k for each i for which $E_k > E_i$.

By correspondence principle arguments Ladenburg also arrived at a relation between the A_{ki} for spontaneous emission from k to i, and the oscillator strength, f_{ik}, appearing in this dispersion formula

$$\begin{aligned} A_{ki} &= \frac{2}{3}(2\pi\nu_{ik})^2\frac{e^2}{mc^3}f_{ik} \\ &= \frac{1}{6}\alpha^4\left(\frac{\sigma_{ik}}{R}\right)^2\frac{c}{a}f_{ik} \end{aligned} \tag{19}$$

in which $\alpha = e^2/\hbar c$ is the fine-structure constant, $7^1 11$, R is the Rydberg constant, and $a = \hbar^2/me^2$ is the radius of the first orbit in hydrogen ($7^1 3$). The physical reality of the negative dispersion terms in (18) was demonstrated experimentally by Ladenburg by measuring the dispersion of light by electrically excited gases [103].

During the late pre-quantum mechanical period the correspondence principle was used also to develop formulas for the relative intensities of the components of a line in the Zeeman effect, and also for the relative intensity of the component lines of a multiplet.

13^1. Waves and photons

Between 1905 and 1925 physicists were troubled by the contradiction that was inherent in the wave and corpuscle descriptions of light.

All phenomena connected with propagation, including the behavior of gratings and interferometers, the resolving power of microscopes and telescopes, were well described by regarding light waves as continuous electromagnetic wave solutions of Maxwell's equations. In all phenomena concerned with emission or absorption of light by matter, the corpuscular view of the photon picture seemed more appropriate. In some discussions, like those based on the correspondence principle, a curious vague mixture of the two was used to guess at approximate answers to important questions.

The experimentally demonstrated phenomenon of light pressure and corresponding developments based on the Maxwell equations showed that [104] a unidirectional light beam of intensity I erg/cm^2 · sec is also transferring momentum $\mathbf{I}/c$ (gr·cm/sec ÷ cm^2) where $\mathbf{I}$ is a vector in the direction of propagation.

This could be taken over into the photon view by supposing that each photon carries a momentum $(h\nu/c)$ in the direction of propagation, as well as energy $h\nu$. Thus the relations between energy momentum of a photon and frequency wave number of the associated wave were

$$E = h\nu = \hbar\omega, \quad \mathbf{p} = h\boldsymbol{\sigma} = \hbar\mathbf{k} \tag{1}$$

This is satisfactory from the viewpoint of the special theory of relativity because both $(E, \mathbf{p})$ and $(\nu, \boldsymbol{\sigma})$ can be regarded as the components of a 4-vector of zero length, so (1) is preserved in Lorentz transformations between different reference frames. For a particle of rest mass m the squared magnitude of the 4-vector $(E, \mathbf{p})$ is

$$\left(\frac{E}{c}\right)^2 - \mathbf{p}^2 = (mc)^2$$

so the fact that $(\nu, \boldsymbol{\sigma})$ is of zero magnitude indicates that the photon has to be regarded as having zero rest mass.

Thus when an atom emits a photon in the direction of $\boldsymbol{\sigma}$, the atom must recoil in the opposite direction with equal momentum. Thereby part of the energy of the transition $(E_i - E_f)$ is taken up in the kinetic energy of recoil of the atom. An atom of mass M takes up a recoil energy $(h\nu)^2/2Mc^2$. This makes a correction to the Bohr frequency condition for the photon, which now becomes

$$h\nu\left(1 + \frac{h\nu}{2Mc^2}\right) = E_i - E_f \tag{2}$$

For light in the visible part of the spectrum the correction is of the order 10^{-9}, so it is undetectable and is usually ignored.

Similarly, in the absorption of a photon $h\nu$, part of the incident photon's energy goes into the recoil energy of the atom, and so $h\nu$ must be larger than what is called for by the Bohr condition in order that the remainder be enough to produce the change $(E_i - E_f)$ in the atom's internal energy. Thus the relation for absorption is

$$h\nu\left(1 - \frac{h\nu}{2Mc^2}\right) = E_i - E_f \tag{3}$$

Therefore, if the Bohr condition has to be fulfilled *exactly*, an atom at rest cannot absorb in ($E_f \rightarrow E_i$) the radiation that another atom of like kind emits in an ($E_i \rightarrow E_f$) transition.

Many years later than the period under discussion this effect was demonstrated experimentally in gamma-ray spectra in the Mössbauer effect.

The question involves a basic difference between classical radiation and the emission of photons. On the classical view the atom radiates in continuous waves in all directions, and although the intensity depends on direction, the intensities in the directions $\hat{\mathbf{r}}$ and $-\hat{\mathbf{r}}$ are the same so there is no resultant recoil of the atom, so the Bohr condition is correct in its original form.

Early in the study of X rays, a classical wave theory of their scattering by free electrons was developed by J. J. Thomson [105]. The electric vector E cos ωt of the incident rays would set a free electron in forced oscillation with an amplitude of $(e^2/m)E$. Using this amplitude in the classical formula for dipole radiation, this predicts a scattered wave polarized like a dipole field with $\mathbf{E}$ as its axis and a total rate of radiation (from 12^16)

$$\frac{2}{3c^3}\overline{\ddot{\mathbf{P}}^2} = \frac{8\pi}{3}\left(\frac{e^2}{mc^2}\right)^2 I_0$$

where I_0 is the intensity of the incident beam. Here I_0 is multiplied by an area, which is known as the *Thomson cross section for scattering,*

$$\sigma_T = \frac{8\pi}{3}\left(\frac{e^2}{mc^2}\right)^2 \tag{4}$$

Translated into photon language, one supposes that each electron scatters photons that are unaltered in frequency and that have a statistical distribution of scattering angle given by the classical dipole field's distribution of radiated intensity.

This simple view gave a fairly good account of the experimental results for soft X rays, but the results for hard X rays and γ rays began to reveal serious discrepancies, concerning both total amount of scattering and angular distribution. Moreover, the scattered radiation was found by absorption measurements to be definitely softer than the incident radiation.

In 1922, A. H. Compton and Peter Debye [106], independently, made a radical application of the photon concept to this problem. Discarding the wave picture altogether, they regarded the scattering process as a collision between a photon and an electron with conservation of momentum and energy. To conserve both, the electron involved must recoil, thereby taking energy, so the energy $h\nu'$ of the scattered photon must be less than that of the incident photon, $h\nu$.

That resembles the processes considered in (2) and (3) but differs in being a scattering rather than an absorptive or emissive process. More important, the recoiling particle is an electron rather than an atom, and, because its mass is 10^{-4}, that of an atom, the effect is much larger.

Simple calculation shows that the wavelength λ' after scattering through angle θ is related to that before, λ, by

$$\lambda' = \lambda + 2\frac{h}{mc}\sin^2\frac{\theta}{2} \tag{5}$$

where $h/mc = 0.024$ Å approximately.

So far the picture is completely corpuscular but gives no estimate of the effective cross section or the distribution in angle. That requires a correspondence principle discussion of the process, so the two views were intertwined after all.

At first it was thought there might not be a correlation in time between the appearance of a scattered photon and the recoil electron, that is, no detailed conservation of momentum and energy in individual processes. But the time correlation was soon shown experimentally [107]. This work of Bothe and Geiger is also interesting in the history of physics as being the first use of time-coincidence counting circuits which now play a great role in all branches of physics.

As another example of wave particle duality, we may consider the action of a plane diffraction grating of infinite extent [108]. On Duane's view, such a grating of spacing d becomes identical with itself when displaced by d in its own plane in a direction perpendicular to its rulings. Therefore, its momentum is restricted by the Bohr-Sommerfeld conditions to values which are integral multiples of h/d.

When such a grating scatters a photon of momentum $h\sigma$ at normal incidence through an angle θ, the scattered photon acquires a transverse momentum component $h\sigma \sin\theta$, and the grating as a whole recoils by this amount. But because the momentum of the grating in this direction is quantized, the momentum change is restricted to nh/d where n is an integer. Therefore, the photon can only be scattered through angles such that $h\sigma \sin\nu = nh/d$; that is,

$$n\lambda = d\sin\theta$$

in agreement with the result usually obtained from the wave theory of the grating.

Of importance is an experiment by G. I. Taylor [109], who arranged a light source and a two-slit opening and a photographic plate inside a closed box to photograph the interference fringes in the usual way. Then the intensity of the source was reduced so much that on the average there was scarcely more than one photon in flight through the apparatus at a time. Nevertheless, he found that interference fringes were still there when sufficiently long exposures were taken. This was puzzling: A photon, as an undividable entity, must go through one slit or the other, so there is a slit through which it did not go. And still the experiment seemed to show that where it lands on the plate depends on there having been open a slit through which it did not go.

Finally, as a sample of the way physicists that thought about quanta at that time, it is interesting to consider an experiment of Lawrence and Beams [110]. They arranged a fast-acting Kerr cell in such a way that it could chop a light beam into pulses of a duration of 10^{-10} seconds. When these pulses fell on a sensitive surface, photoelectrons were observed as usual. They concluded, "If light quanta are of the commonly understood wave nature, they are less than 3 cm in length, and an electron absorbs a light quantum photoelectrically in less than 10^{-10} sec."

14^1. The discovery of quantum mechanics

By 1924 physicists realized quite generally that the fruitfulness of the ideas of quantized classical orbits and the correspondence principle, which had been very great indeed, was about played out. The ideas, at best, were beginning to give incorrect answers. Calculations were made by perturbation theory of the energy of the ground state of helium and for some of the excited states. The values obtained were in glaring disagreement with the facts [111].

Physicists were groping for some approach that would give a new formulation of the laws of *quantum mechanics* [112], which they expected would be quite different from what they had been working with but must agree with classical mechanics in the limit of large quantum numbers.

The new developments broke like a flood between 1924 and 1926, so completely altering the outlook of theoretical physics that the new developments constituted a major revolution. In the fifty years that have elapsed, a whole generation of physicists has grown up whose training in atomic physics usually begins with 1926, almost as if nothing had been accomplished before then.

Quantum mechanics arose from two independent streams of thought: (1) the introduction of the concept of matter waves that was initiated by Louis de Broglie [113] in the fall of 1923, which, however, first began to have a major impact after the publication of Schrödinger's series of papers on wave mechanics in the spring of 1926 [114], and (2) the development of matrix mechanics by Heisenberg, Born, and Jordan in the fall of 1925 as a more precise formulation of the idea of representing the radiative properties of an atom by an array of virtual oscillators, one for each *pair* of states [115].

By the summer of 1926 it was well recognized that the two formulations were two different mathematical approaches to the same physical content. New ways were available for finding the energy levels and transition probabilities of atomic systems that were proving successful as they were rapidly applied to a wide range of specific problems. But there was a great deal of uncertainty at first as to the deeper physical meaning of the new methods.

This situation began to clear up in the fall of 1926 and during 1927, with the introduction by Born of the probability interpretation of $\psi^*\psi$ where ψ is the amplitude of the de Broglie wave produced as a solution of the Schrödinger wave equation. In the paper [116] in which he laid down the fundamental basis for applying wave mechanics to collision problems, he also states the probability interpretation, which holds the field to this day.

This step was so important for the history of physics that it is interesting to read the essential paragraphs in translation. The paper is entitled *Quantenmechanik der Stossvorgänge* and begins:

Collision processes have provided not only the most convincing experimental proofs of the basic assumptions of the quantum theory, but also appear to be suited to give a clarification of the physical meaning of the formal laws of the so-called "quantum mechanics." These give, as it seems, always the correct term values of the stationary states, and the correct amplitudes of the emitted radiation associated with transitions, but opinions are divided about the physical interpretation of the formulas. The matrix form of quantum mechanics, as developed by Heisenberg, and extended by him with Jordan, and the author of this paper, starts from the idea that an exact description of the processes in space and time is not possible, and is therefore content with setting up relations between observable quantities, which can only be interpreted as properties of motions in the classical limit. Schrödinger however ascribed to the waves which, following de Broglie, he regards as the carriers of the atomic processes, a reality of the same kind as that of light waves; he tries to build wave groups which have in all directions relatively small dimensions, which will represent directly the moving corpuscles.

Neither of these two viewpoints seem satisfactory to me. I wish to try to give here a third interpretation and to investigate its usefulness in treating collision processes. For this I build on a remark of Einstein concerning the relation of the wave field and light quanta. He said in effect that the waves are only there to point the way to the corpuscular light quanta, and in this sense he spoke of a "ghost-field." This ghost-field determines the probability that a light quantum, as the carrier of energy and momentum, shall take a definite path. No energy or momentum belongs to the field itself.

To put these ideas into the framework of quantum mechanics had better be delayed until one has found how to build the electromagnetic field into the formalism. But in view of the complete analogy between light quanta and electrons, one may undertake to formulate the laws of electron motion in a similar way. In this way it is natural to regard the de Broglie-Schrödinger waves as the "ghost-field" or better the "guiding-field."

I wish therefore to follow up the view: that the guiding field, representated by a scalar function ψ of the coordinates of all particles in the system and the time, is propagated according to the Schrödinger differential equation. Momentum and energy however are transported as if corpuscles (electrons) are actually flying about. The paths of the particles are only determined to the extent that they are limited by the conservation of momentum and energy theorems. After that the choice of a particular path is determined by probability in accordance with the field of values of the function ψ. One can express this somewhat paradoxically: the motion of the particles follows probability laws, but the probability itself is propagated in accordance with a causal law. [This means that knowledge of the states in all points at one instant determines the distribution of the state at all later times.]

Considering the three stages of the development of quantum theory, one sees that the lowest, that of the periodic processes, is quite unsuited for a test of the usefulness of such an idea. The second stage, that of aperiodic stationary states gives somewhat more; we shall be concerned with these in the present paper. Actually decisive will be the third stage, that of nonstationary occurrences; in this we have to see whether the interference of damped "probability waves" will suffice to explain those phenomena which seem to point to a coupling outside of space and time.

The main body of the paper is devoted to a development of wave-mechanical formalism for collision processes. It then concludes:

On the basis of the preceding considerations, I am of the opinion that the quantum mechanics can formulate and solve the problem of the transition processes, as well as that of the stationary states. The Schrödinger formulation seems to be the easiest to handle, moreover it allows one to hold on to the usual ideas in space and time in which the events take place in a normal way. On the other hand the theory here does not lead to a causal determination of individual events. In my preliminary communication I have particularly emphasized this indeterminism since it seems to be in close agreement with the actual practice of the experimentalists. But naturally those who are not content with this may suppose that there exist further parameters which have not yet been introduced into the theory which would determine the course of individual events. These are the "phases" of the motion in classical mechanics, for example the coordinates of the particles at a definite instant. It seems improbable to me that one can introduce quantities into the new theory that correspond to the phases, but Herr Frenkel has told me that perhaps it can be done. However that may be, this possibility does not alter the practical indeterminism of collision processes, since one does not know the value of the phases, and so must end up with formulas like those of the "phaseless" theory here developed.

In 1927 Heisenberg [117] made plain the underlying reason for the statistical nature of quantum mechanics: that the conditions required for exact application of classical mechanics could not be fulfilled because of the reaction of the observing apparatus on the system being observed. He showed that this was of such a nature that the uncertainty Δx in the measurement of a coordinate x and the corresponding uncertainty Δp in the measurement of its conjugated momentum p are always such that

$$\Delta x \Delta p > \hbar \qquad (1)$$

This is also an attribute of the statistical uncertainties that are predicted by the new mathematical formalism. Heisenberg saw in this feature of the measurement process the fundamental physical reason why the correct formalism had to exhibit this feature. This general set of ideas became known as the *Heisenberg uncertainty principle.* This step represented the first clear elimination from fundamental physics of the exact causal determinism, a feature of Newtonian mechanics. Such a step was not accepted without a struggle, which to some extent is still going on. The issue was not so much whether the uncertainty principle represents a correct assessment of our present knowledge, but whether this is really a fundamental limitation, an inherent feature of the nature of things, or merely an inadequacy of the knowledge as reached so far, which we shall overcome some time in the future. Einstein adhered to the latter view for the rest of his life. He used to say, "*der liebe Gott würfelt nicht*"; that is, "the good Lord does not throw dice."

In 1928, Bohr [118] extended Heisenberg's ideas in the formulation of his *complementarity principle.*

In parallel with these philosophical considerations, the ideas of wave mechanics received stimulating support from the discovery of electron diffraction [119].

Basic to de Broglie's idea was that 13¹1 could be applied to the waves associated with moving material particles, with the difference that the 4-vector (E/c, $\mathbf{p}$) would have magnitude mc for a particle of rest-mass m. From 13¹1 we derive the convenient rule that an electron having V electron volts of energy has a de Broglie wavelength $\lambda = \sqrt{150/V}$ in Ångström units. Thus low-voltage electrons have wavelengths of the order of the lattice constants of crystals. Davisson and Kunsman had studied the angular distribution of electrons scattered by platinum. Elsasser pointed out that if electrons are somehow governed by de Broglie waves that these results might be interpreted as diffraction of the electrons by the crystal structure of the platinum. He suggested more decisive results could be obtained by studying electron scattering from a single crystal.

Elsasser also suggested that the abnormally long free paths of slow electrons in helium and argon might also be a consequence of their motion being governed by scattering of the de Broglie waves by inert gas atoms.

In 1927, Davisson and Germer found clear diffraction peaks on the back scattering of slow electrons from a single crystal of nickel. Later that year, G. P. Thomson found convincing evidence of electron diffraction in the scattering of fast electrons (~30 keV) in transmission through various thin metallic and oxide films. This was, for electrons, the analogue of the Debye-Scherrer method in X-ray diffraction by randomly oriented small crystals. With the lattice constants known from X-ray work, it was possible to infer the de Broglie wavelength of the electrons from the diffraction angles of their diffraction pattern, and to show that this agreed with de Broglie's hypothesis.

After this there could be no question of the reality of de Broglie waves and their role in determining the motion of electrons.

The tempo at which the development of quantum mechanics occurred, particularly in the period 1925–28, after the long period of preparation from 1900–25, was without doubt the most spectacular revolution on a wide variety of fronts that has ever occurred in the history of physical science. Important papers, still being cited and still worthy of careful study, were appearing in nearly every issue of leading physical journals.

Physicists were swamped in their efforts to assimilate and comprehend this outpouring of new results. Condon spent the fall of 1926 in Göttingen, working with Max Born, when Born was developing the probability amplitude interpretation of ψ in the treatment of collision processes, and the spring of 1927 in Munich, working with Arnold Sommerfeld, when he was initiating the modern theory of metals by applying Fermi statistics to the gas of quasi-free electrons in metals that had always been handled by Boltzmann statistics, with many consequent failures (these were largely cleared up by Sommerfeld's work). Returning to New York in the fall of 1927, at Bell Telephone Laboratories and at Columbia University, Condon found that his was not the only case of intellectual indigestion in American physics. Perhaps the mood was best summed up by Bergen Davis (1869–1958), distinguished X-ray physicist and professor at Columbia, who commented on quantum mechanics in the spring of 1928 that, "I don't think you young fellows understand it any better than I do, but you all stick together and say the same thing!" This has been called the conspiracy interpretation of quantum mechanics.

Today a complete treatment of quantum mechanical principles and their specific applications would fill many volumes. Many independent volumes have been written on various aspects of the subject. In this book we confine ourselves narrowly to the detailed treatment of the subject of atomic structure, the perturbation of atoms by static or low-frequency external electromagnetic fields, and the quantized interaction of individual atoms with the electromagnetic field. Thereby we omit from consideration vast and important areas of modern physics: the quantum mechanics of intermolecular forces, that of the chemical bonds between atoms in a molecule or in those very large molecules called crystals, the nature of the strong forces that bind nucleons into atomic nuclei and give them a level structure in many ways resembling that of the extranuclear electronic structure of atoms. Also omitted are the large topics of applications of collision theory to (1) the interpretation of chemical kinetics, the formation and dissociation of molecules; (2) the interpretation of nuclear kinetics, the formation and dissociation of atomic nuclei; and (3) the interpretation of particle physics, the creation and annihilation of mesons, leptons, and hyperons.

In the treatment of extranuclear electronic structure of isolated atoms that follows there is material enough for a large book. In the interest of brevity we shall, in ensuing chapters, devote much less space than in this first chapter to outlining the historical details of the growth. Nevertheless we shall attempt to give enough of such material that one need have no difficulty in supplementing the story by tracing out the main steps of the history through the references that are given.

Before leaving the explicit historical treatment of the subject with which this chapter has been concerned, perspective will be gained by summarizing the principal points of the major developments which were described in papers published in the five years 1925–29 inclusive:

1925

L. de Broglie's 1924 Paris thesis proposed the existence of waves associated with particle motion and that quantization of such particle motion is related to boundary-value problems of such waves.

W. Heisenberg's paper on formal changes in classical mechanics removed inaccuracies of the correspondence principle and led, with M. Born and P. Jordan, to the discovery of matrix mechanics late that year and in early 1926.

Pauli developed his exclusion principle and its application to the relation of the complex structure of the levels of an element to its place in the periodic system.

G. E. Uhlenbeck and S. Goudsmit discovered that a previously discovered double-valuedness in the states of an atom can be interpreted in terms of the electron's having a spin-angular momentum of $(1/2)\hbar$.

1926

W. Heisenberg discovered the special properties of quantum dynamical systems consisting of several identical particles and the identification of unexpectedly large spin-spin interaction energies in atoms with these properties of many-electron systems.

In the spring E. Schrödinger published his famous series of papers on wave mechanics and about his discovery of the relationship of his wave mechanics to the matrix mechanics of Heisenberg, Born, and Jordan.

Fermi applied the exclusion principle of Pauli, together with electron spin, to the formulation of Fermi-Dirac statistics. Born formulated the quantum mechanics of collision processes and probability interpretation of ψ.

1927

Diffraction of an electron beam by a nickel crystal was discovered by C. J. Davisson and L. H. Germer and later electron diffraction studied by G. P. Thomson in 1928.

P. A. M. Dirac developed the quantization of the electromagnetic field and unified interpretation of spontaneous as well as induced radiative emission processes.

A. Sommerfeld applied Fermi-Dirac statistics to initiate an improved version of electron theory of metals.

W. Heisenberg first recognized the uncertainty principle as arising from randomness of interaction between thing observed and the observing apparatus.

The theory of the chemical bond was initiated by (1) Burrau's solution of energy levels of H_2^+, (2) Condon's primitive extension of this to a molecular orbital theory for H_2, (3) Heitler and London's treatment by valence bond methods of the binding energy of H_2, (4) Born and Oppenheimer's general extension of perturbation theory to the treatment of molecular problems, (5) Condon's wave-mechanical interpretation of the features of vibrational changes in molecules that accompany electronic transitions, which later became known as the Franck-Condon principle, and (6) Hund's initial systematization of electronic structure of molecules as an extension of the use of the exclusion principle and electron spin.

Pauli developed Pauli matrices for the spin-angular momentum vector of the electron as the appropriate means for dealing with systems having angular momentum equal to an odd multiple of $(1/2)\hbar$.

H. Weyl and E. P. Wigner independently initiated work showing the usefulness of the theory of groups in the study of quantum mechanical systems.

1928

Dirac discovered a proper relativistic equation for dealing with quantum mechanics of the electron.

The statistical central-field model of the atom was formulated by L. H. Thomas and E. Fermi (independently).

Further work by F. Hund, which was extended independently by R.

S. Mulliken, accomplished the assignment of quantum numbers to electrons in molecules.

Particle tunnelling through potential barriers as a nonclassical property of wave mechanics was discovered by J. R. Oppenheimer to explain emission of electrons from cold metals at high negative potential. Its principal application was to the explanation of α-particle radioactivity of heavy elements, which was worked out independently by R. W. Gurney and Condon, and by George Gamow. This was the first major application of quantum mechanics to the not-yet-established field of nuclear physics.

Heisenberg interpreted ferromagnetism as arising from strong Coulombic apparent spin-spin interaction in metals related to the fact that the electrons are governed by the Pauli exclusion principle.

Hartree developed his initial form of quantum-mechanical self-consistent field theory.

1929

Slater developed a simplified way of application of the spin and exclusion principle to the calculation of energy levels of atoms having several electrons outside of closed shells.

Bethe initiated the ligand field theory, the perturbation of the levels of ions in crystals having incomplete d and f shells by electric fields caused by neighboring ions.

Dirac used creation and annihilation operators in the quantum mechanics of many-electron systems.

2. Principles of quantum mechanics

> In mathematics the greatest degree of self-evidence is usually not to be found quite at the beginning, but at some later point; hence the early deductions, until they reach this point, give reasons rather for believing the premises because true consequences follow from them, than for believing the consequences because they follow from the premises.
>
> Bertrand Russell, *Principia Mathematica,* Preface (Cambridge: Cambridge University Press, 1914).

Since 1926, study of quantum mechanics has become part of the training of all physicists. The Schrödinger wave-function methods are easier for most students to understand than the matrix methods, so these are usually emphasized in elementary instruction [1].

The books by Dirac [2] and von Neumann [3] have had a wide influence over the years. The application of the theory of groups to quantum mechanics has a specialized literature of its own [4]. The book by Born and Jordan [5] emphasizes matrix methods. Of the many books on quantum mechanics, a few that are particularly useful are listed [6].

Quantum mechanics describes observable properties of dynamical systems at the atomic level, where the finiteness of the quantum constant $\hbar$ plays an essential role. Its mathematical structure is patterned on and generalized from that of linear algebra [7], which provides the mathematical description of finite-dimensional vector spaces. These methods are inadequate for even the simplest quantum mechanical problems, so the mathematical structure needs to be extended to include linear analysis [8], which provides the mathematical description of infinite-dimensional vector spaces.

The first step in extending the algebra of Nth-order matrices to the analysis of infinite processes is the simple extension of N to infinity. This was the guiding idea of Hilbert's pioneering work on integral equations. Here the dimensionality of the space, although infinite, remains denumerable. Next one considers spaces in which several (M) indices, each having an infinite range of discrete values, are needed to label the linearly independent vectors. Such a space has N^M dimensions where N is infinite and M is finite. Next one has to consider spaces whose dimensionality requires a continuous variable for the labeling of its dimensions and then spaces that require several such continuous variables for the labeling of linearly independent vectors.

Finally, one has to consider spaces in which an infinite number of continuous variables is needed to label its linearly independent vectors. Such a space has dimensionality N^M in which both N and M have become nondenumerably infinite [e.g., the "Fock space" of quantum field theory describing infinitely many discrete and/or continuous states i each having an indefinite occupation number n_i (number of particles in the state i)].

In the rigorous mathematical framework of nonrelativistic quantum mechanics, which goes back to von Neumann, the state space is the standard (separable) Hilbert space, e.g., $L^2(\mathbf{R}^{3n})$, square integrable functions over the configuration space of n particles. Nevertheless, in the case of continuous spectrum, for example scattering states, the *generalized propervectors* such as plane waves that are not in Hilbert space are frequently used. This practice can be justified rigorously by going to a larger space, such as rigged Hilbert space, and has often practical advantages over the use of wave packets, which are in the Hilbert space.

1². Mathematical formalism

Quantum mechanics describes a dynamical system through the use of a many-dimensional *system-space*. A particular state of the system is represented by a vector in this system-space.

The dimensionality of the system-space, therefore, has to be great enough so that the vectors in it can span all possible states of the system. This already requires that the dimensionality be nondenumerably infinite, even for such a simple case as describing the motion of a single particle in one spatial dimension. It is clear that for a particle to be at x' is a different state than for it to be at x'', when $x'' \neq x'$, and so there are distinct states of such a system in one-to-one correspondence with the continuum of real values of x. Hence for a system containing N particles in space, the dimensionality of the system space must be that of the one-dimensional continuum to the $3N$th power.

In the system space the quantities that are the objects of physical measurement, called *observables,* are represented as linear operators, which act on a vector to convert it into another vector. Here linear means that the result of operation on the sum of two or more vectors is the same as the sum of the results of operation on these vectors separately.

The explicit expression of a vector in any space is accomplished through the introduction of a complete *basis* of linearly independent vectors. Linearly independent here means that no linear combination of them gives a null vector. Complete means that there are enough of them that any vector in the space can be represented as a linear combination of the vectors of the basis.

In a given space many different choices of the basis are possible. The specific form of representation of a vector, and of an operator, changes on making a change in the basis. Another kind of mathematical entity arises in this connection: a *transformation matrix,* which expresses the vectors of one basis in terms of those of another.

To achieve the generality needed for the purposes of atomic physics it is necessary to work with a system space in which the vectors have components that are complex numbers. Because the results that are to be compared with theory must be real, it is necessary to introduce a dual system-space in which there is a dual basis, and on which each vector has a component that is the complex conjugate of the component in the first system space. The theory is completely symmetrical between the dual system-spaces; that is, each system space is dual to the other.

Because the system space is used to describe actual physical systems in Euclidean 3-space, we also deal with quantities that are vectors and tensors in the ordinary sense in the 3-space. Thus the position vector **r** of a particle is a vector in 3-space, which has the usual implications about how its components transform on changing the reference frame in 3-space. In system space each of its 3-space components is an operator, defined by its mode of action on the vectors of system space. Such a set of three operators is called a *vector operator*. To differentiate between the 3-space in which physical events occur and the system space of quantum mechanics, we consistently speak of a set of coordinate axes in 3-space as a *frame,* and of a complete set of linearly independent vectors in system space as a basis. Thus the explicit form of **r** depends both on the choice of frame and on the choice of basis. Similar remarks apply to the 3-space tensors that arise in physics, such as the quadrupole moment of an electric charge distribution, which give rise to types of quantities which are called *tensor operators.*

Notation in physics and mathematics is exceedingly important. Many different notations are in use for the description of vectors. In 3-space the one that is more widely used by physicists is that introduced by Gibbs and Heaviside in which the vector is printed in boldface type, as was done for **r** in the preceding paragraph. A frame is specified by giving three orthogonal vectors of unit length, say $\hat{\mathbf{k}}_1$, $\hat{\mathbf{k}}_2$, and $\hat{\mathbf{k}}_3$. In terms of this frame any vector **A** is given by stating its components on these vectors so that

$$\mathbf{A} = \hat{\mathbf{k}}_1 A_1 + \hat{\mathbf{k}}_2 A_2 + \hat{\mathbf{k}}_3 A_3 = \hat{\mathbf{k}}_i A_i \tag{1}$$

The second form arises when the Einstein convention of summing over repeated indices is used.

Unit 3-vectors are also said to be normal, or normalized, and a set whose members are also orthogonal is called *orthonormal*. This means that the scalar products, $\hat{\mathbf{k}}_i \cdot \hat{\mathbf{k}}_j$, defined as the product of magnitudes by cosine of the included angle, are

$$\hat{\mathbf{k}}_i \cdot \hat{\mathbf{k}}_j = \delta_{ij} = \begin{cases} 1 & i = j \\ 0 & i \neq j \end{cases} \tag{2}$$

Therefore, the components of $\mathbf{A}$ on the $\mathbf{k}$-frame are

$$\hat{\mathbf{k}}_i \cdot \mathbf{A} = A_i \tag{3}$$

Inserting these values in (1) we may regard $\mathbf{A}$ on the left as the result of operation on $\mathbf{A}$ on the right with the identity operator, $\hat{\mathbf{k}}_i\hat{\mathbf{k}}_i$, giving

$$\mathbf{A} = \hat{\mathbf{k}}_i\hat{\mathbf{k}}_i \cdot \mathbf{A} \tag{4}$$

This notation will be used for vectors and tensors in 3-space.

In many textbooks of vector algebra, this notation is extended to n-space merely by extending the summation range on i to $1, 2, \ldots, n$, instead of 1, 2, 3. For several reasons a different notation is desirable for the vectors of the system space of quantum mechanics. In the first place, the notation ought to distinguish clearly between state vectors in system space, and 3-vectors in 3-space, because both kinds usually occur together in the same discussion. In the second place, the notation needs to provide for a clear distinction between the dual forms of the same state vector. Last, and perhaps most important, we need to label the axes of a basis in system space by continuous variables as well as by discrete indices, and we need to provide for a physical meaning to attach to these indices. The number of indices needed for a system of N electrons is $4N$ (three for positional coordinates of each electron and one two-valued coordinate for its spin), so the printing of such a large number of indices as subscripts (320 for atomic mercury!) becomes typographically awkward.

For these reasons, it is convenient to use the notation devised by Dirac [2]. The same state vector in its dual forms is designated by symbols of the type $\langle\,|$ and $|\,\rangle$, called a *bra* and a *ket,* respectively. Symbols of this type are the system-space analogue of the boldface type for a vector in 3-space, of the Gibbs-Heaviside notation. When a statement is quite general and refers to any state vector, these symbols are left quite empty, but when it is desired to refer to a particular state vector, suitable labels can be inserted on the line in the symbol. Thus a state vector that carries the labels $\alpha_1', \alpha_2', \ldots, \alpha_f'$ would be written $\langle\alpha_1', \alpha_2', \ldots, \alpha_f'|$ and $|\alpha_1', \alpha_2', \ldots, \alpha_f'\rangle$ in its bra and ket forms.

The values that can be assumed by these labels have to be independently specified. A particular label, such as α_i', can assume either a discrete set of values, or continuous ranges, or a mixture of both, as discussed more explicitly later. For brevity we may write A' for the ensemble of a specific set of values of $\alpha'_1 \ldots \alpha'_f$, also $\langle A'|$ and $|A'\rangle$ for the associated bra and ket state vectors.

Associated with any bra $\langle A'|$ and any ket $|B'\rangle$ where B' stands for any set of labels $\beta'_1 \ldots \beta'_f$, which may be other values of the α's or a partially or completely different set of labels, is a scalar product, $\langle A'|B'\rangle$, which is, in general, a complex number. Alternatively, the same pair of state vectors have the dual representation $\langle B'|$ and $|A'\rangle$ and a dual scalar product $\langle B'|A'\rangle$. The duality is such that these two complex numbers are complex conjugates, so that

$$\langle B'|A'\rangle = \langle A'|B'\rangle^* \tag{5}$$

for any pair of state vectors. No provision is made for scalar multiplication of two bras or two kets; the scalar product is always formed with a bra preceding a ket. Thus, open or incomplete bracket symbols are state vectors, and complete bracket symbols are scalar products that are complex numbers.

From (5) it follows that $\langle A'|A'\rangle$ is real for any state vector. This is called its *norm*. In the standard development of the mathematics of Hilbert space, restriction is usually made to state vectors of finite norm. However, this restriction is too severe for the needs of quantum mechanics, so we must be prepared to use a more general form of linear analysis in which this restriction is not made. It also follows from (5) that if $\langle B'|A'\rangle = 0$, then also $\langle A'|B'\rangle = 0$; the two state vectors are then said to be *orthogonal*.

If C is an ordinary number, the state vector $\langle A'|C^*$ or $C|A'\rangle$ is defined as one whose scalar products with all other state vectors are, respectively, C^* or C times those of $\langle A'|$ or $|A'\rangle$, respectively. Thus the norm of C times $\langle A'|$ is $C^*C\langle A'|A'\rangle$, so if $\langle A'|A'\rangle$ is finite, one can choose $|C|$ so that the norm of $C|A'\rangle$ is unity, in which case the resultant state vector is said to be *normalized*. This process does not determine the phase of normalization, concerning which an arbitrary choice must be made. Even when $\langle A'|A'\rangle$ is infinite, so the state vector cannot be normalized to unity, the physical interpretation of the theory is independent of C. This is sometimes phrased by saying that the state is described by a direction or *ray* in system space, rather than by a vector.

A set of state vectors given in either bra or ket form is said to be linearly independent if no linear relation exists between them,

$$a_1|A_1\rangle + a_2|A_2\rangle + a_3|A_3\rangle + \ldots = 0 \tag{6}$$

other than that in which $a_1 = a_2 = a_3 \ldots = 0$. Such a set of state vectors, in general, is said to *span* a *subspace* of the system space. This subspace is the ensemble of all state vectors that can be written as a linear form in the state vectors of this set

$$|\,\rangle = a_i|A_i\rangle \quad \text{and} \quad \langle\,| = \langle A_i|a^*{}_i \tag{7}$$

over the field of the a_i as independent complex numbers. If the set includes enough state vectors so that the entire system space is spanned by them, it is called a *complete* set and determines a *basis*. This definition does not require that the $|A_i\rangle$ be *orthonormal,* that is, mutually orthogonal and each of unit norm. However, the algebraic work is simpler when this is the case, so this property will be assumed for all sets used to span a subspace or as a basis for the whole system space unless the contrary is stated.

The orthonormal property for the basis is expressed by

$$\langle A'|A''\rangle = \delta(\alpha_1' - \alpha_1'')\delta(\alpha_2' - \alpha_2'') \ldots \delta(\alpha_f' - \alpha_f'') \tag{8}$$

where $\delta(\epsilon) = 1$ for $\epsilon = 0$, and $\delta(\epsilon) = 0$ for $\epsilon \neq 0$, in case the adjacent values of α'' have a finite difference between them, that is, the *spectrum* of ϵ is discrete. But if the range of $\epsilon = \epsilon' - \epsilon''$ is continuous in the neighborhood of $\epsilon = 0$, then $\delta(\epsilon)$ is zero for $\epsilon \neq 0$, but infinite at $\epsilon = 0$ in such a way that $\int\delta(\epsilon)\, d\epsilon = 1$ on integration from $\epsilon < 0$ to $\epsilon > 0$. This $\delta(\epsilon)$ is known as the *Dirac delta function.*

Suppose that an orthonormal basis consists of the $\langle A'|$ and $|A'\rangle$ for a specified domain of the labels $\alpha_1' \ldots \alpha_f'$ symbolized by A'. An arbitrary state, $|a\rangle$, can then be written in the dual forms

$$|a\rangle = \sum_{A'} x_{A'}|A'\rangle \qquad \langle a| = \sum_{A'} y_{A'}\langle A'|$$

Here $\sum_{A'}$ means summation over all *discrete* sets of values of the labels $\alpha_1' \ldots \alpha_f'$ plus an integration with respect to $d\alpha_1' \ldots d\alpha_f'$ over any *continuous* ranges of those labels that there may be. The coefficients $x_{A'}$ and $y_{A'}$ in the expansions may be found by multiplying the first expansion by $\langle A''|$ and the second expansion by $|A''\rangle$; then because of the orthogonality of the basis all the terms in the expansions except one in each drop out, and we are left with $x_{A''} = \langle A''|a\rangle$ and $y_{A''} = \langle a|A''\rangle$. The expansions then become

$$|a\rangle = \sum_{A'} |A'\rangle\langle A'|a\rangle \qquad \langle a| = \sum_{A'} \langle a|A'\rangle\langle A'| \tag{9}$$

As with ordinary vectors the state vectors satisfy the *Schwarz inequality*

$$|\langle a|b\rangle|^2 \leqslant \langle a|a\rangle\langle b|b\rangle \tag{10}$$

This is derived in the usual way from the fact that the norm of $|a\rangle + c|b\rangle$ must be real and positive and vanishes only in case $|a\rangle$ and $|b\rangle$ are linearly dependent so that there exists a c for which $|a\rangle + c|b\rangle = 0$, in which case the equal sign holds in (10).

We write γ for a general linear operator and $\gamma|a\rangle$ for the ket resulting from the action of γ on $|a\rangle$. This operator is fully characterized on the

A-basis by giving the components of $\gamma|A'\rangle$ for each A', that is, the coefficients $\langle A''|\gamma|A'\rangle$ in

$$\gamma|A'\rangle = \sum_{A''} |A''\rangle\langle A''|\gamma|A'\rangle \tag{11}$$

The bra form of this ket, as a series in the $\langle A''|$, has the complex conjugates of these same coefficients, namely

$$\langle A'|\gamma = \sum_{A''} \langle A''|\gamma|A'\rangle^* \langle A''|$$

This can be put in standard matrix form, where the sum is over adjacent indices by introducing $\gamma\dagger$, known as the *Hermitian adjoint* of γ so that the bra belonging to the ket $\gamma|A'\rangle$ is

$$\langle A'|\gamma\dagger = \sum_{A''} \langle A'|\gamma\dagger|A''\rangle\langle A''| \tag{12}$$

Comparing these two forms we see that the relation between the coefficients of γ and $\gamma\dagger$ is

$$\langle A'|\gamma\dagger|A''\rangle = \langle A''|\gamma|A'\rangle^* \tag{13}$$

The matrix of $\gamma\dagger$ is obtained from that of γ by transposing (interchange of rows and columns) and taking the complex conjugate of each element.

The ensemble of the coefficients $\langle A''|\gamma|A'\rangle$ is called the *matrix* of γ on the A basis. A particular coefficient is called a *matrix element*. Because A' and A'' stand for a multiplicity of labels, some or all of which may have continuous ranges, it is not possible to write out all of the elements in a square array as is usually done with finite matrices of n rows and n columns. Nevertheless, A'' is called the *row* index and A' the *column* index, and an element for which $A'' = A'$ is called a *diagonal* element. A matrix for which all the elements for which $A'' \neq A'$ vanish is called a diagonal matrix. A matrix for which some but not all of the $\alpha_i'' \neq \alpha_i'$ vanish is called a *step* matrix. It may be said to be diagonal in some of the labels but not in others.

Just as a state vector is a linear combination of the $\langle A'|$ or $|A'\rangle$ in its dual forms, a linear operator γ is a linear combination of *basis dyadics* $|A''\rangle\langle A'|$. These are the system-space analogues of the $\mathbf{k}_i\mathbf{k}_i$ dyadics of 3-space. This form of expression for γ and $\gamma\dagger$ is

$$\gamma = \sum_{A'',A'} |A''\rangle\langle A''|\gamma|A'\rangle\langle A'| \tag{14}$$

$$\gamma\dagger = \sum_{A'',A'} |A''\rangle\langle A'|\gamma|A''\rangle^*\langle A'| \tag{14'}$$

Evidently $(\gamma\dagger)\dagger = \gamma$; the Hermitian adjoint of $\gamma\dagger$ is γ.

Perhaps the simplest operator is the identity, I, which leaves any state-vector unaltered. It has only diagonal elements equal to δ functions, so

$$I = \sum_{A'',A'} |A''\rangle\delta(A'' - A')\langle A'| = \sum_{A'} |A'\rangle\langle A'| \tag{15}$$

Closely related to the identity are the *projection operators*. These have the same form as I in (15) except that the sum $|A'\rangle\langle A'|$ is over a specified subset of the A''s,

$$\sum_{A'}' |A'\rangle\langle A'| \tag{16}$$

Evidently the action of such an operator on $|a\rangle$ is to leave unaltered the part of $|a\rangle$ that is in the subspace spanned by the specified subset contained in the sum, and to annul the part outside the subspace. That is, each projection operator projects the state vector $|a\rangle$ onto its characteristic subspace.

The *product* of two operators α and β is written $\alpha\beta$ and is defined as the single operator which produces the same effect on arbitrary $|a\rangle$ as α acting on $\beta|a\rangle$. In general, such multiplication is not commutative, $\alpha\beta \neq \beta\alpha$. We have

$$\alpha\beta = \sum_{A'',A''',A'} |A''\rangle\langle A''|\alpha|A'''\rangle\langle A'''|\beta|A'\rangle\langle A'| \tag{17}$$

whereas

$$\beta\alpha = \sum_{A''\,A''',A'} |A''\rangle\langle A''|\beta|A'''\rangle\langle A'''|\alpha|A'\rangle\langle A'|$$

The A'' component of $\beta|a\rangle$ is

$$\langle A''|\beta|a\rangle = \sum_{A'} \langle A''|\beta|A'\rangle\langle A'|a\rangle \tag{18}$$

so it is obtained by multiplication of the "square" matrix representing β into the one-column matrix representing $|a\rangle$. Similarly, the matrix elements of $\alpha\beta$ are related to those of α and β by the ordinary rule for multiplying matrices

$$\langle A''|\alpha\beta|A'\rangle = \sum_{A'''} \langle A''|\alpha|A'''\rangle\langle A'''|\beta|A'\rangle \tag{19}$$

For any two operators α and β the operator

$$[\alpha, \beta] = \alpha\beta - \beta\alpha \tag{20}$$

is called their *commutator*. If the commutator vanishes, the operators are said to commute or to be commuting operators. Vanishing of an operator Ω means that $\Omega|a\rangle = 0$ for all $|a\rangle$. A great deal of the specific development of quantum mechanics consists in the discovery of specific operators to represent various physical quantities and the discovery of their commutation rules, that is, the value of their commutators.

As (14) shows, the direction of the vector $\gamma|a\rangle$ is in general different from that of $|a\rangle$. The operator γ, however, usually possesses particular state vectors, called the proper states of γ such that the operation of γ on them is a simple multiple of the state. We write

$$\gamma|\gamma'\rangle = \gamma'|\gamma'\rangle \tag{21}$$

for such a state, where γ' is the value of the multiplier. The various values of γ' for which (21) has solutions are called *proper values* of γ and the associated states are called its *proper states*. Proper states are also frequently called eigenstates and proper values are then called eigenvalues.

The normal thing is for a particular proper value γ' to be *degenerate,* which means that there are linearly independent states $|\gamma', \alpha'\rangle, |\gamma', \alpha''\rangle$ . . . each of which satisfy (21) with the same proper value, γ'. Because of the linearity of (21), it follows that any linear combination of these also satisfies (21). Thus the proper states associated with γ' span a subspace of the system space whose dimensionality is equal to the number of linearly independent solutions of (21). This number is called the *order of degeneracy* of γ' and is itself quite commonly infinite in simple examples.

Associated with γ and the state $|a\rangle$ is a number $\langle a|\gamma|a\rangle$, which is the scalar product of $\langle a|$ and $\gamma|a\rangle$ or

$$\langle a|\gamma|a\rangle = \sum_{A'',A'} \langle a|A''\rangle\langle A''|\gamma|A'\rangle\langle A'|a\rangle \tag{22}$$

The complex conjugate of this number is

$$\begin{aligned}\langle a|\gamma|a\rangle^* &= \sum_{A'',A'} \langle a|A'\rangle\langle A''|\gamma|A'\rangle^*\langle A''|a\rangle \\ &= \sum_{A'',A'} \langle a|A''\rangle\langle A''|\gamma^\dagger|A'\rangle\langle A'|a\rangle\end{aligned}$$

Therefore, $\langle a|\gamma|a\rangle$ is real provided that $\gamma^\dagger = \gamma$ (that is, provided that γ is equal to its own Hermitian adjoint). Operators having this property are called *Hermitian* or *self-adjoint.* A physical interpretation that is given to $\langle a|\gamma|a\rangle$ in Section 2^2 requires that real physical observables be represented by Hermitian operators.

In particular, if $|a\rangle$ is a proper state of γ with the proper value γ', the preceding argument shows that γ' is real if γ is Hermitian. The proper values of a Hermitian operator are real.

Similarly, an operator is called *anti-Hermitian,* or *skew-Hermitian,* if $\gamma^\dagger = -\gamma$. For such operators $\langle a|\gamma|a\rangle$ is purely imaginary. If γ is anti-Hermitian, then $i\gamma$ is Hermitian and vice versa. Also, the proper values of an anti-Hermitian operator are purely imaginary. Any γ can be written as the sum of a Hermitian operator and an anti-Hermitian operator

$$\gamma = \frac{1}{2}(\gamma + \gamma^\dagger) + \frac{1}{2}(\gamma - \gamma^\dagger) \tag{23}$$

which is the analogue of separating a complex number into real and imaginary parts.

From the definition of the relation of $\gamma^\dagger$ to γ it follows that

$$(\alpha\beta)^\dagger = \beta^\dagger\alpha^\dagger \tag{24}$$

the product of the Hermitian conjugates in reverse order. Therefore, the commutator of two Hermitian operators is anti-Hermitian and vice versa.

The proper states of a Hermitian operator γ that are associated with unequal proper values, $\gamma' \neq \gamma''$, are *orthogonal*. This theorem is simply proved. We have

$$\gamma|\gamma', a\rangle = \gamma'|\gamma', a\rangle$$

and

$$\gamma|\gamma'', b\rangle = \gamma''|\gamma'', b\rangle$$

in which a and b stand for all other labels than γ' and γ'' needed to label the states. Because γ is Hermitian, the second of these may also be written

$$\langle\gamma'', b|\gamma = \langle\gamma'', b|\gamma''$$

Multiplying this by $|\gamma', a\rangle$ on both sides and the first equation above by $\langle\gamma'', b|$ on both sides, and subtracting, we find

$$(\gamma' - \gamma'')\langle\gamma'', b|\gamma', a\rangle = 0 \tag{25}$$

Therefore, the states are orthogonal if $\gamma' \neq \gamma''$. Thus each Hermitian operator is characterized by its characteristic values, with each of which is associated a subspace, and the subspaces for unequal proper values are orthogonal; that is, any state vector in one subspace is orthogonal to any state vector in the other subspace.

Suppose now that β commutes with γ. We have

$$\gamma|\gamma', a\rangle = \gamma'|\gamma', a\rangle$$

and therefore

$$\beta\gamma|\gamma', a\rangle = \sum_{\gamma'', b} |\gamma'', b\rangle\langle\gamma'', b|\beta|\gamma', a\rangle\gamma'$$

Also

$$\beta|\gamma', a\rangle = \sum_{\gamma'', b} |\gamma'', b\rangle\langle\gamma'', b|\beta|\gamma', a\rangle$$

so that

$$\gamma\beta|\gamma', a\rangle = \sum_{\gamma'',b} |\gamma'', b\rangle\langle\gamma'', b|\beta|\gamma', a\rangle\gamma''$$

Subtracting, since $\beta\gamma - \gamma\beta = 0$, the coefficient of each $|\gamma'', b\rangle$ must vanish, and therefore

$$\langle\gamma'', b|\beta|\gamma', a\rangle(\gamma' - \gamma'') = 0 \tag{26}$$

Hence, if two Hermitian operators commute, the matrix elements of one of them are diagonal in the proper values of the other one. In other words, $\beta|\gamma', a\rangle$ lies entirely within the subspace associated with the proper value γ'.

Within each γ' subspace we can find proper values and proper states of β. These are simultaneous proper states of γ and β which may be denoted by $|\gamma', \beta', c\rangle$ in which c stands for all other labels needed to distinguish individual states within the (γ', β') subspace. Continuing this process, if α commutes with both β and γ then $\alpha|\gamma', \beta', c\rangle$ lies entirely within the (γ', β') subspace and so it is possible to find proper values α' whose proper states $|\alpha', \beta', \gamma', d\rangle$ are simultaneous proper states of α, β, and γ, with d standing for all other labels. This process can be continued until enough commuting Hermitian operators have been introduced that their simultaneous proper states provide a unique labeling of these states. Such a set of commuting operators is called a *complete* set.

When the A basis was introduced, no specific meaning was assigned to the labels, $\alpha_1', \alpha_2', \ldots, \alpha_f'$, by which the states $|A'\rangle$ were identified. We now see that the significant thing to do is to regard these labels as sets of proper values of commuting Hermitian operators $(\alpha_1, \alpha_2, \ldots, \alpha_f)$ for which the $|A'\rangle$ are simultaneous proper states so that

$$\alpha_i|A'\rangle = \alpha_i'|A'\rangle \qquad (i = 1, 2, \ldots f) \tag{27}$$

Let U represent a linear operator, none of whose proper values is equal to zero. Then there exists an operator, which we denote by U^{-1}, which has the property that

$$UU^{-1} = U^{-1}U = I \tag{28}$$

so that U^{-1} is the reciprocal of U in the usual sense.

We can use U to transform a set of operators by the relations

$$\alpha_U = U\alpha U^{-1}, \quad \beta_U = U\beta U^{-1}, \ldots \tag{29}$$

We call the passage from $\alpha, \beta, \gamma, \ldots$ to $\alpha_U, \beta_U, \gamma_U, \ldots$ a *canonical transformation*. It has the important property that algebraic relations that exist between the $\alpha, \beta, \gamma, \ldots$ also hold in *invariant* form between $\alpha_U, \beta_U, \gamma_U$. For example

$$U(\alpha + \beta)U^{-1} = U\alpha U^{-1} + U\beta U^{-1} = \alpha_U + \beta_U$$

and

$$U\alpha\beta U^{-1} = U\alpha U^{-1}U\beta U^{-1} = \alpha_U\beta_U$$

and similarly for more complicated expressions involving more of the operators.

Moreover, α_U has the same proper values as α and the proper states of α_U are simply related to those of α. From

$$\alpha|\alpha'\rangle = \alpha'|\alpha'\rangle$$

we have

$$U\alpha U^{-1}U|\alpha'\rangle = \alpha' U|\alpha'\rangle = \alpha_U U|\alpha'\rangle$$

and therefore $U|\alpha'\rangle$ is a proper state of α_U associated with the value α'.

It is convenient to impose on U an additional restriction such that $U|\alpha'\rangle$ is normalized if $|\alpha'\rangle$ is. The bra form of $U|\alpha'\rangle$ is $\langle\alpha'|U^\dagger$ and therefore the norm of $U|\alpha'\rangle$ is $\langle\alpha'|U^\dagger U|\alpha'\rangle$. In order that this be unity for complete sets of proper states of all of the operators α, β, γ, . . . , we must have

$$UU^\dagger = I \quad \text{or} \quad U^\dagger = U^{-1} \tag{30}$$

An operator having this property is called *unitary*. Thus canonical transformations (29) are made with unitary operators.

Suppose V is another unitary operator and we use it as in (29) to make a canonical transformation from the α_U to the α_{VU} so that

$$\alpha_{VU} = V\alpha_U V^{-1} = (VU)\alpha(VU)^{-1} \tag{31}$$

and similarly for the other operators. The combined effect of the canonical transformation by U followed by that of V is thus the same as that of a single transformation effected by the product VU, which is also a unitary operator because

$$VU(VU)^\dagger = VUU^\dagger V^\dagger = I$$

Because I itself is unitary, and because to each U there is also a unitary U^{-1}, it follows that the canonical transformations of operators have the *group* property. The same is true of the proper states: $VU|\alpha'\rangle$ is the proper state of $(VU)\alpha(VU)^\dagger$.

Of great importance are *infinitesimal* unitary transformations, those in which U differs from the identity by a quantity that is so small that squares of it are negligible. We write, for real ϵ,

$$U = I + i\epsilon F$$
$$U^\dagger = I - i\epsilon F^\dagger$$

so that

$$UU^\dagger = I = I + i\epsilon(F - F^\dagger). \tag{32}$$

To fulfill the unitary condition (30) we must have $F = F^\dagger$. Therefore, F must be Hermitian for U to be unitary.

For finite ϵ we can define

$$U = I + i\epsilon F + \frac{1}{2!}(i\epsilon F)^2 + \ldots = e^{i\epsilon F} \tag{33}$$

and

$$U^\dagger = e^{-i\epsilon F} \quad \text{so that} \quad UU^\dagger = I$$

provided that F is Hermitian. Thus, each Hermitian operator F defined in the system space generates in this way a one-parameter group of unitary transformations.

In the foregoing discussion the $|A'\rangle$ stand for any orthonormal basis in the system space. The formalism of transforming between different orthonormal bases greatly resembles that of canonical transformations of operators. This fact has a simple geometrical basis. The algebraic relations between various operators $\alpha, \beta, \gamma, \ldots$ are something intrinsic to themselves, and determine the relative orientation of the several sets of proper states $|\alpha'\rangle, |\beta'\rangle, |\gamma'\rangle, \ldots$ In a canonical transformation, the algebraic relations between the operators are invariant, and these relationships of proper states are preserved. What is changed by U is the relationship of all of these state vectors to the A basis. The change is analogous to a rigid rotation of all of these state vectors relative to the vectors of the A basis. The same sort of change may alternatively be described from a standpoint that regards the interrelated state vectors as unchanged but describes in terms of the B basis rotated from the A basis in the opposite sense to that involved in the transformation U.

Suppose that the B basis consists of the $|B'\rangle$ kets where B' stands for a complete set of labels $\beta_1', \beta_2', \ldots, \beta_f'$. Then all of the equations of the preceding discussion which are written for the A basis hold in the same form with the B basis.

The relation between the bases may be stated by giving each $|B'\rangle$ in terms of the $|A'\rangle$ or vice versa,

$$|B'\rangle = \sum_{A'} |A'\rangle\langle A'|B'\rangle \tag{34}$$

and

$$\langle B''| = \sum_{A''} \langle B''|A''\rangle\langle A''|$$

The ensemble of the coefficients for all A' and B' with $\langle B'|A'\rangle = \langle A'|B'\rangle^*$ in these relations make up the *transformation matrix* between the two bases. It is the analogue of the orthogonal matrix of direction cosines $\hat{\mathbf{k}}_i \cdot \hat{\mathbf{k}}_j'$ in a change from the $\hat{\mathbf{k}}$ basis to the $\hat{\mathbf{k}}'$ basis in 3-space.

The transformation matrix must satisfy a condition which ensures that the B basis is orthonormal as well as the A basis. We have

$$\begin{aligned}\langle B''|B'\rangle &= \sum_{A'',A'}\langle B''|A''\rangle\langle A''|A'\rangle\langle A'|B'\rangle \\ &= \sum_{A'}\langle B''|A'\rangle\langle A'|B'\rangle = \delta(B''-B')\end{aligned} \tag{35}$$

Therefore, the matrix must be such that the matrix of the $\langle B''|A'\rangle$ is the reciprocal of the $\langle A'|B'\rangle$. If the latter is denoted by U then the former is $U^\dagger$ and so the transformation matrix is unitary (30) and describes the same kind of transformation as is involved in a canonical transformation.

With the state vectors transforming in this way, it follows that the transformation of matrix elements is given by the rule

$$\langle B''|\gamma|B'\rangle = \sum_{A'',A'}\langle B''|A''\rangle\langle A''|\gamma|A'\rangle\langle A'|B'\rangle \tag{36}$$

The *trace* of a matrix is defined as the sum of its diagonal elements,

$$\mathrm{Tr}(\gamma) = \sum_{A'}\langle A'|\gamma|A'\rangle \tag{37}$$

In the B representation we have

$$\begin{aligned}\mathrm{Tr}(\gamma) &= \sum_{B'}\langle B'|\gamma|B'\rangle \\ &= \sum_{A'',A'}\langle A''|\gamma|A'\rangle\sum_{B'}\langle A'|B'\rangle\langle B'|A''\rangle \\ &= \sum_{A'}\langle A'|\gamma|A'\rangle\end{aligned} \tag{38}$$

This proves that the value of the trace is invariant to a unitary transformation of the basis.

This result is most frequently used in finite form. It finds wide application in atomic theory where it is called the *diagonal sum rule.* Suppose that the B basis is the same as the A basis except for a finite number of the $|A'\rangle$, which are changed into linear combinations of a finite number of the $|B'\rangle$. In such a case most of the $\langle B'|\gamma|B'\rangle$ are individually equal to corresponding individual $\langle A'|\gamma|A'\rangle$. The diagonal-sum rule then takes the finite form

$$\sum_{B'}{}'\langle B'|\gamma|B'\rangle = \sum_{A'}{}'\langle A'|\gamma|A'\rangle \tag{39}$$

in which the Σ' are over the states that are actually changed in the finite unitary transformation.

In a similar way we have

$$\begin{aligned}\sum_{B'',B'}|\langle B''|\gamma|B'\rangle|^2 = \sum &\langle B''|A''\rangle\langle A''|\gamma|A'\rangle\langle A'|B'\rangle\langle B'|A'''\rangle \\ &\times\langle A'''|\gamma|A^{\mathrm{iv}}\rangle\langle A^{\mathrm{iv}}|B''\rangle\end{aligned}$$

where the sum is over the repeated indices, A'', A', A''', and A^{iv} as well as over the B' and B''. The sum on B' produces a factor $\delta(A' - A''')$ and that over B'' produces a factor $\delta(A'' - A^{iv})$, and therefore

$$\sum_{B'',B'} |\langle B''|\gamma|B'\rangle|^2 = \sum_{A'',A'} |\langle A''|\gamma|A'\rangle|^2 \tag{40}$$

As before, this result finds most frequent application when the B basis differs from the A basis in only a finite number of states, in which case (40) holds for the finite sum over the indices of the states that are changed in the change of basis. This is the matrix-theoretic basis for a physical result known as the *principle of spectroscopic stability.*

Returning to (11), it is evident that the basis is used in two ways: first, as the source of the complete set of states $|A'\rangle$ on which γ acts, and second, as the basis in terms of which the $\gamma|A'\rangle$ are expressed. It is not necessary that the same basis be used for both purposes. This gives rise to the possibility of defining a mixed basis representative for γ as

$$\gamma|A'\rangle = \sum_{B'} |B'\rangle\langle B'|\gamma|A'\rangle \tag{41}$$

Or in terms of mixed dyadics, (14) becomes

$$\gamma = \sum_{B',A'}{}' |B'\rangle\langle B'|\gamma|A'\rangle\langle A'| \tag{42}$$

Such mixed representations are not used much, but occasionally they are convenient.

2². Physical postulates

The formalism of the preceding section is used by adopting Hermitian operators to represent real physical quantities.

The postulate is made that when any such quantity is measured the numerical value obtained will be one of the proper values of the operator γ that represents the quantity. If the state vector $|a\rangle$, representing the system, lies precisely in the γ' subspace of the quantity γ, then a measurement of γ gives the value γ' with certainty, that is, with unit probability. Likewise, after a measurement of γ has been made in which the result γ' was observed, the state vector $|a\rangle$ is known to be in the γ' subspace.

If the state vector $|a\rangle$ has components in several or all γ' subspaces, then a measurement of γ may give any of the γ' values that are represented in $|a\rangle$. The reaction of the observing equipment on the system observed is such that after the measurement $|a\rangle$ is projected into the particular γ' subspace that corresponds to the outcome of the measurement. This is where the idea of an undetermined, although statistically controlled, outcome of measurements on the state $|a\rangle$ enters the theory. It is postulated that if the system is somehow prepared repeatedly in the same state $|a\rangle$, and measurements of the same γ are made each time, there will be a statistical series of values $\gamma', \gamma'', \gamma''', \ldots$ observed such that the mean value of them is

$$\gamma_{\mathrm{Av}} = \langle a|\gamma|a\rangle \tag{1}$$

Writing $P(\gamma'; a)$ for the probability that the value γ' is observed for γ when the system is in the state $|a\rangle$, it is clear that γ_{Av} is also given by

$$\gamma_{\mathrm{Av}} = \sum_{\gamma'} P(\gamma', a)\gamma' \tag{2}$$

The probability is conveniently expressed in terms of the projection operators defined in $1^2$16. Choosing a basis in which the states are labeled by γ' and Γ', where Γ' stands for all others than γ of a complete set of commuting operators, then

$$\sum_{\Gamma'} |\gamma', \Gamma'\rangle\langle\gamma', \Gamma'| \tag{3}$$

is the operator that projects an arbitrary state $|a\rangle$ onto the γ' subspace. For a normalized state $\langle a|a\rangle = 1$ the sum of the norms of these projections on to various γ' subspaces is unity. Moreover, on this basis

$$\gamma_{\mathrm{Av}} = \sum_{\gamma'} \gamma' \sum_{\Gamma'} \langle a|\gamma', \Gamma'\rangle\langle\gamma', \Gamma'|a\rangle \tag{4}$$

from which it is clear, on comparing with (2), that

$$P(\gamma', a) = \sum_{\Gamma'} |\langle\gamma', \Gamma'|a\rangle|^2 \tag{5}$$

The probability of obtaining the value γ' in a measurement of γ on the system in the state $|a\rangle$ is the norm of the projection of $|a\rangle$ on the γ' subspace.

Because the sum of $P(\gamma', a)$ over all γ' is equal, on this hypothesis, to $\langle a|a\rangle$, the total probability is equal to unity. This same relation (5) may be used to give *relative* probabilities for cases in which $|a\rangle$ does not have a finite norm.

This result is easily generalized to the case of simultaneous measurement of two or more commuting physical quantities. The result can be stated in terms of the expressions written down by a slight extension of the meanings of the symbols. In Section 1^2 following $1^2$26 the idea of states that are simultaneous proper states of several commuting observables was introduced. We now interpret γ' as standing for a set of simultaneous proper values of the several commuting quantities and Γ' as standing for the operators of a complete set other than the several that are included in γ'.

With this interpretation (3) gives the projection operator onto the simultaneous proper space of these several γ''s. There is an expression like (4) for each of the γ_i that are being simultaneously measured, and finally (5) gives the probability of the particular set of simultaneous proper values of the γ_i that is represented by γ'.

It is clear, physically, that what determines $P(\gamma', a)$ is the norm of the projection of $|a\rangle$ onto the γ' subspace. In order to calculate this explicitly, one needs some complete basis to label the individual states within the γ' subspace. But the projection operator (3) ought to give the same result no matter what choice is made of orthonormal basis within this subspace. Expression (3) does have this necessary property. Suppose that Δ' stands for the values of some other set of commuting observables (commuting with themselves and with the γ_i' that are symbolized by γ') and then (3) takes the form

$$\sum_{\Delta'} |\gamma', \Delta'\rangle\langle\gamma', \Delta'| \tag{3'}$$

when expressed on this basis.

By 1²34 and 1²35 the $|\gamma', \Delta'\rangle$ are related to the $|\gamma', \Gamma'\rangle$ by a unitary transformation which is diagonal in γ', that is

$$|\gamma', \Delta'\rangle = \sum_{\Gamma'} |\gamma', \Gamma'\rangle\langle\gamma', \Gamma'|\gamma', \Delta'\rangle$$

and therefore the projection operator in (3′) is

$$\begin{aligned}\sum_{\Delta'} |\gamma', \Delta'\rangle\langle\gamma', \Delta'| &= \sum_{\Delta'\Gamma''\Gamma'} |\gamma', \Gamma'\rangle\langle\gamma', \Gamma'|\gamma', \Delta'\rangle\langle\gamma', \Delta'|\gamma', \Gamma''\rangle\langle\gamma', \Gamma''| \\ &= \sum_{\Gamma'} |\gamma', \Gamma'\rangle\langle\gamma', \Gamma'|\end{aligned}$$

so (3′) is the same as (3) because of the unitary nature of the change from the $\gamma'\Gamma'$ basis to the $\gamma'\Delta'$ basis. So the value of $P(\gamma', a)$ is independent of the choice of basis within the γ' subspace.

At first the operators representing familiar physical quantities were inferred from classical descriptions by use of the correspondence principle. Later the theory was extended to systems without classical analogues, by suitable choice of operators to give desired results.

The operators representing the Cartesian coordinates on any frame of the position of a particle are postulated to have a continuum of allowed values from $-\infty$ to $+\infty$. On the basis derived from proper states of these coordinates the state $|a\rangle$ for a single particle may be expressed in terms of an integral over the basis states $|x_1, x_2, x_3\rangle$, with the coefficients $\langle x_1, x_2, x_3|a\rangle$, given by the generalization of 1²9 to a continuous set of basis states. These coefficients, which have the dual form

$$\langle a|x_1, x_2, x_3\rangle = \langle x_1, x_2, x_3|a\rangle^* \tag{6}$$

may be taken as representing the state $|a\rangle$. Because (x_1, x_2, x_3) are the components of a 3-vector, $\mathbf{r}$, it is often convenient to write $\langle \mathbf{r}|a\rangle$ for $\langle x_1, x_2, x_3|a\rangle$.

This single-valued continuous function of the position of the particle is sometimes called a *probability amplitude*, because its squared magnitude gives the probability density of finding the particle in $dx_1\,dx_2\,dx_3$ at (x_1, x_2, x_3). It is the same function of position as is often written $\psi_a(x_1, x_2, x_3) = \psi_a(\mathbf{r})$ and called the *wave function* of the state in Schrödinger's formulation.

Similarly for N particles, the Cartesian coordinates of each of them x_{1i}, x_{2i}, x_{3i} with $i = 1, 2, \ldots, N$ are postulated to be commuting Hermitian operators with the same continuous range and the wave function becomes

$$\langle x_{11}, x_{21}, x_{31}, \ldots, x_{1N}, x_{2N}, x_{3N}|a\rangle \quad \text{or} \quad \langle \mathbf{r}_1, \mathbf{r}_2, \ldots, \mathbf{r}_N|a\rangle$$

a function of $3N$ arguments as in the Schrödinger form for the wave function of an N-particle system. The operator for $x_{\alpha i}$, the x_α coordinate of the ith particle, in a form for application to this state is simply multiplication of it by $x_{\alpha i}$.

If the particles are electrons having a two-valued spin variable it is necessary to introduce a fourth coordinate s for each electron, where s can take on only two values, usually taken to be $\pm 1/2$, so that for one particle we get $\langle x_1, x_2, x_3, s|a\rangle = \langle \mathbf{r}, s|a\rangle$ to represent the state. We may thus say that the system space has $2 \times \infty^3$ dimensions for each electron. Therefore, for N electrons, the system space has $2^N \times \infty^{3N}$ dimensions, as there is a two-valued spin coordinate for each of the N electrons.

In Heisenberg's original formulation of quantum mechanics he worked entirely with bases which put the matrices in discrete form, that is, without any continuous variables labeling rows or columns. In this connection he discovered the commutation rule between a Cartesian coordinate, $x_{\alpha i}$ and its conjugate momentum $p_{\alpha i}$, from a correspondence principle argument:

$$[x_{\alpha i}, p_{\beta j}] = x_{\alpha i}p_{\beta j} - p_{\beta j}x_{\alpha i} = i\hbar\delta_{\alpha\beta}\delta_{ij} \tag{7}$$

A particular coordinate $x_{\alpha i}$ commutes with all momenta except its own conjugate $p_{\alpha i}$. He also postulated that all momentum components commute with each other. Later when the spin variables s_i were introduced the spin of any electron was assumed to commute with the spin of any other electron as well as with the positions and linear and angular momenta of the same and other electrons.

Although Schrödinger's original approach was by way of giving precision to the ideas of Louis de Broglie about electron waves (Sec. 14[1]), his basic discovery, from the standpoint of the present formulation, was that (7) can be satisfied in the coordinate representation by using

$$p_{\beta j} \rightarrow -i\hbar\frac{\partial}{\partial x_{\beta j}} \tag{8}$$

as the operator representing the linear momentum of each particle.

Let $F(x_{\alpha i}\ldots)$ be a function of the coordinates only, then any coordinate, $x_{\alpha i}$, commutes with F, but for any momentum component $p_{\alpha i}$ we have

$$[p_{\alpha i}, F]\psi_a = -i\hbar \frac{\partial F}{\partial x_{\alpha i}} \psi_a$$

where $\psi_a = \langle \mathbf{r}_1, \mathbf{r}_2, \ldots, \mathbf{r}_N | a \rangle$. Hence the operators have the commutation relations

$$[x_{\alpha i}, F] = 0, \quad [p_{\alpha i}, F] = -i\hbar \frac{\partial F}{\partial x_{\alpha i}} \tag{8a}$$

Similarly, if $G(P_{\alpha i} \ldots)$ is a function of the momentum components only, we have

$$[p_{\alpha i}, G] = 0, \quad [x_{\alpha i}, G] = +i\hbar \frac{\partial G}{\partial p_{\alpha i}} \tag{8b}$$

which is the corresponding relation for the momentum representation (Sec. 5^2).

This operator representation for the momenta permits the formulation of a differential operator for the Hamiltonian, $H(p, x)$ for the total energy. In this way the proper-value equation for H becomes formulated as a boundary value problem for a differential equation, in the way that is thoroughly familiar from introductory presentations of quantum mechanics. For example, the classical nonrelativistic Hamiltonian for N electrons moving in the field of a fixed nucleus of charge Ze is

$$H = \sum_{\alpha} \left(\frac{1}{2m} \mathbf{p}_{\alpha}^2 - \frac{Ze^2}{r_{\alpha}} \right) + \sum_{\alpha<\beta} \frac{e^2}{r_{\alpha\beta}} \tag{9}$$

in which $\mathbf{p}_{\alpha} = \mathbf{k}_i\, p_{\alpha i}$, $r_{\alpha} = (x_{\alpha 1}^2 + x_{\alpha 2}^2 + x_{\alpha 3}^2)^{1/2}$ and $r_{\alpha\beta} = |\mathbf{r}_{\alpha} - \mathbf{r}_{\beta}|$. The terms in the $\mathbf{p}_{\alpha}^2$ represent the kinetic energy of the electrons, the $-Ze^2/r_{\alpha}$, represent the potential energy of electrostatic interaction of each electron with the nucleus, while the $e^2/r_{\alpha\beta}$ represent the potential energy of electrostatic interaction between all pairs of electrons. The sum on $\alpha < \beta$ therefore contains $(1/2)N(N - 1)$ terms.

The substitution (8) then gives for the allowed values of H, which is denoted as W instead of H', the nonrelativistic Schrödinger wave equation.

$$\left[\sum_{\alpha} \left(-\frac{\hbar^2}{2m} \nabla_{\alpha}^2 - \frac{Ze^2}{r_{\alpha}} \right) + \sum_{\alpha<\beta} \frac{e^2}{r_{\alpha\beta}} \right] \psi = W\psi \tag{10}$$

The allowed values of energy, W, are those for which (10) possesses solutions that are finite, continuous, and single-valued functions of the $3N$ position coordinates of the electrons.

When magnetic fields are present it is necessary to introduce a vector potential, **A**, as well as the scalar potential ϕ by the relations

$$\mathbf{E} = -\mathbf{grad}\,\phi - \frac{1}{c}\dot{\mathbf{A}}, \quad \mathbf{B} = \mathbf{curl}\,\mathbf{A}$$

When an electron is regarded simply as having a charge $-e$ and a mass m, the Hamiltonian for an N electron atom containing a central nucleus of charge Ze is

$$H = \sum_{i=1}^{N} \left\{ \frac{1}{2m} \left[\mathbf{p}_i + \frac{e}{c}\mathbf{A}(\mathbf{r}_i) \right]^2 - e\phi(\mathbf{r}_i) \right\} + \sum_{i<j} \frac{e^2}{r_{ij}} \tag{9'}$$

Here $\phi(\mathbf{r}_i)$ includes the term $+Ze/r_i$, which represents the Coulomb field of the nucleus plus additional terms to represent other static or time-varying electric fields that may be present.

The forms (9) or (9′) suffice for many purposes. But, in addition, it is necessary to add

$$\mu_B \boldsymbol{\sigma}_k \cdot \mathbf{B}(\mathbf{r}_k) \tag{9''}$$

to represent the direct interaction of the electron's intrinsic magnetic moment $\mu_B \boldsymbol{\sigma}_k$ with the magnetic field, where $\mu_B = e\hbar/2mc$ is the Bohr magnaton and $\boldsymbol{\sigma}_k$ is the vector defined in $2^36'$, originally introduced by Pauli to describe electron spin-angular momentum (Sec. 5^3). Other terms have to be added to represent *spin-orbit interaction,* which arises from the *motion* of each electron's magnetic moment in the electric field of the nucleus and the other electrons.

Dynamics deals with the change in the system with lapse of time. The physical postulate is that this can be expressed as a parametric dependence of the state $|a\rangle$ on time t expressed by the formal equation of motion

$$i\hbar \frac{\partial}{\partial t} |a\rangle = H|a\rangle \tag{11}$$

where H is the Hamiltonian operator. Accordingly, $|a\rangle$ at time $t + dt$ is obtained from $|a\rangle$ at time t by the infinitesimal transformation operator

$$T = I - i\,dt\frac{H}{\hbar} \tag{11'}$$

which by 1^232 defines a unitary transformation, because H is a Hermitian operator.

The interpretation of this equation of motion depends on the assumption made about the time dependence of the state vectors of the basis. Let its labels be W' and A' where A' stands for a set of commuting operators, which also commute with H and together with H make up a complete set. The basis kets are then $|W', A'\rangle$ where

$$H|W', A'\rangle = W'|W', A'\rangle$$

and

$$|a\rangle = \sum_{W'A'} |W', A'\rangle\langle W', A'|a\rangle$$

At this point a choice, in fact many choices, can be made. In the *Schrödinger description,* the $|W', A'\rangle$ are regarded as *not* depending on the time, and the time dependence is associated entirely with the time variation of the coefficients $\langle W', A'|a\rangle$. We may emphasize this choice by writing $|W', A'\rangle_0\langle W', A'|a\rangle_t$ for terms in the expansion of $|a\rangle$. Then (11) gives

$$\langle W', A'|a\rangle_t = \langle W', A'|a\rangle_0 \exp\frac{-iW't}{\hbar} \tag{12}$$

and so

$$|a\rangle_t = \sum_{W',A'} |W', A'\rangle_0\langle W', A'|a\rangle_0 \exp\frac{-iW't}{\hbar} \tag{12'}$$

and therefore the mean value of an operator γ in the state $|a\rangle_t$ is

$$\begin{aligned}\gamma_{\text{Av}}(t) &= \langle a|\gamma|a\rangle_t \\ &= \sum_{W',W''} \exp\frac{i(W'-W'')t}{\hbar} \sum_{A',A''}\langle a|W', A'\rangle_0 \\ &\quad\times \langle W', A'|\gamma|W'', A''\rangle_0\,\langle W'', A''|a\rangle_0\end{aligned} \tag{13}$$

In this expression there are constant terms ($W' = W''$) and terms which vary harmonically with the Bohr transition frequencies. In case the spectrum of allowed energy values is continuous there will be a continuous spectrum of values for which $(W' - W'') \to 0$. These give rise to slow or secular changes in $\gamma_{\text{Av}}(t)$.

Of particular importance are those operators that commute with the Hamiltonian. By $1^2 26$, the matrix elements vanish for such operators between two states of unequal energy, so if γ is such a quantity only constant terms are present in (13). Therefore, all quantities that commute with the Hamiltonian are constants and are analogous to the integrals of motion in classical dynamics.

In similar manner the Schödinger wave function for $|a\rangle$ is

$$\langle x_1, x_2, \ldots |a\rangle_t = \sum_{W'A'} \langle x_x, x_2, \ldots |W', A'\rangle_0 \exp\frac{-iW't}{\hbar}$$

so that the time dependence of the probability density of finding the particles in unit volume at (x_1, x_2, x_3) is

$$|\langle x_1, x_2, x_3|a\rangle_t|^2 = \sum_{W', W''} \exp \frac{i(W' - W'')t}{\hbar} \sum_{A', A''} \langle W', A'|x\rangle_0 \times \langle x|W'', A''\rangle_0 \tag{14}$$

An alternative choice for description of the time dependence is known as the *Heisenberg description*. In (12′) we can associate the exponential factor of each term with $|W', A'\rangle_0$ and regard this as defining a moving basis consisting of

$$|W', A'\rangle_t = |W', A'\rangle_0 \exp \frac{-iW't}{\hbar} \tag{15}$$

on which then $|a\rangle_t$ would be expressed with the constant coefficients, $\langle W', A'|a\rangle_0$. This makes no change in the overall expression of $|a\rangle_t$ but adopts a different viewpoint as to whether it is the basis kets or the components of $|a\rangle_t$ that are changing.

More generally, one can separate each W' in any arbitrary way into two parts, $W' = W_1' + W_2'$, and treat the $\exp(-iW_1't/\hbar)$ as showing the time dependence of the $\langle W', A'|a\rangle_t$ while treating the $\exp(-iW_2't/\hbar)$ as showing the time dependence of the basis kets in (15). All such schemes give the same overall results and so are equally valid.

The result (12′) for the time dependence of an arbitrary state is particularly simple because the representation is in terms of proper states of the Hamiltonian. For any other basis $|B'\rangle$ whose commuting operators do not include the Hamiltonian the corresponding expansion is

$$|a\rangle_t = \sum |B'\rangle\langle B'|W', A'\rangle \exp\left(\frac{-iW't}{\hbar}\right) \langle W', A'|B''\rangle\langle B''|a\rangle_0$$

which is summed on B', (W', A') and B''.

If H does not itself depend on the time, it is convenient to define a time dependent operator $T(t)$ by the relation

$$T(t) = \exp \frac{-iHt}{\hbar} \tag{16}$$

so that $T(0)$ is the operator for the identity. This is called the *evolution operator*, or the *time development operator*. Its matrix elements on the arbitrary basis $|B'\rangle$ are

$$\langle B'|T(t)|B''\rangle = \sum_{W', A'} \langle B'|W', A'\rangle \exp\left(\frac{-iW't}{\hbar}\right) \langle W', A'|B''\rangle \tag{17}$$

The time dependence of the general state is then given by

$$|a\rangle_t = \sum_{B',B''} |B'\rangle\langle B'|T(t)|B''\rangle\langle B''|a\rangle_0 \tag{18}$$

which can also be written to emphasize the departure of $|a\rangle_t$ from its initial value

$$|a\rangle_t = |a\rangle_0 + \sum_{B',B''} |B'\rangle\langle B'|T(t) - 1|B''\rangle\langle B''|a\rangle_0 \tag{18a}$$

The probability that the system, initially in state $|a\rangle_0$ will be found in state $|b\rangle$ at time t is

$$|\langle b|a\rangle_t|^2 = \Big|\sum_{B',B''} \langle b|B'\rangle\langle B'|T(t)|B''\rangle\langle B''|a\rangle_0\Big|^2 \geq 0 \tag{19}$$

This is, as it should be, independent of the arbitrary normalization phase of $|a\rangle_0$ or of $|b\rangle$. From its form it is obviously real and nonnegative. This probability takes a simpler form when $\langle b|$ and $|a\rangle_0$ are described in the $|W', A'\rangle$representation using (17),

$$|\langle b|a\rangle_t|^2 = \sum_{W',W''} C_{ba}(W'', W') \exp \frac{i(W'' - W')t}{\hbar} \tag{20}$$

in which

$$\begin{aligned} C_{ba}(W'', W') &= \sum_{A',A''} {}_0\langle a|W'', A''\rangle\langle W'', A''|b\rangle\langle b|W', A'\rangle\langle W', A'|a\rangle_0 \\ &= C^{(r)}_{ba}(W'', W') + iC^{(i)}_{ba}(W'', W') \end{aligned} \tag{21}$$

and

$$C_{ba}(W', W'') = C^*_{ba}(W'', W')$$

Hence the probability is the sum of a constant real term

$$\sum_{W'} C_{ba}(W', W') \tag{22}$$

and terms that vary harmonically with the time with the Bohr frequencies,

$$2 \sum_{W''>W'} C^{(r)}_{ba}(W'', W') \cos \frac{(W'' - W')t}{\hbar} - C^{(i)}_{ba}(W'', W') \sin \frac{(W'' - W')t}{\hbar} \tag{23}$$

For some purposes it is convenient to write $|\langle b|a\rangle_t|^2$ in terms of its departure from its initial value:

$$\begin{aligned} |\langle b|a\rangle_t|^2 = |\langle b|a\rangle_0|^2 - 2 \sum_{W''>W'} \Big\{ & C^{(r)}_{ba}(W'', W') \left[1 - \cos \frac{(W'' - W')t}{\hbar}\right] \\ & + C^{(i)}_{ba}(W'', W') \sin \frac{(W'' - W')t}{\hbar} \Big\} \end{aligned} \tag{24}$$

As a simple example, suppose the system has only two states. We write $|1\rangle$ and $|2\rangle$ for the two stationary states with energies W_1 and W_2. We suppose that it is initially in

$$|a\rangle_0 = |1\rangle\alpha + |2\rangle\beta$$

with $|\alpha|^2 + |\beta|^2 = 1$, then at time t,

$$|a\rangle_t = |1\rangle\alpha \exp\frac{-iW_1 t}{\hbar} + |2\rangle\beta \exp\frac{-iW_2 t}{\hbar}$$

The probability at time t of the system's being found in $|b\rangle$ where

$$|b\rangle = |1\rangle\gamma + |2\rangle\delta$$

is then

$$|\langle b|a\rangle_t|^2 = |\alpha|^2|\gamma|^2 + |\beta|^2|\delta|^2 + \gamma^*\delta\alpha\beta^* \exp\frac{i(W_2 - W_1)t}{\hbar} + \text{c.c.}$$

where c.c. means complex conjugate of the immediately preceding expression. Thus the probability of being found in state b fluctuates harmonically with the Bohr frequency of transition between the stationary states $|1\rangle$ and $|2\rangle$.

3^2. The uncertainty principle

The uncertainty principle was recognized by Heisenberg (Sec. 14^1) as a consequence of the incompatibility of precise measurements of a position coordinate and its conjugate momentum. In the mathematical formalism this comes about because the operators for x and p_x do not commute. A similar result holds [9] for any pair of noncommuting operators, say α and β.

Suppose that α', A' serve to label the simultaneous proper states of a set $|\alpha', A'\rangle$ that includes α, and β', B' a set $|\beta', B'\rangle$ that includes proper states of β. Because α and β do not commute this means that, in general, α has nonvanishing nondiagonal elements on the β', B' basis, and β has nonvanishing nondiagonal elements on the α', A' basis.

When a measurement of α is made, if the value α' is obtained the system is thereby put in a state $|\alpha', A'\rangle$ belonging to the α' subspace. In such a state there are finite probabilities of finding various proper-values of β, namely by 2^25, the norms of the projections of the $|\alpha', A'\rangle$ state on the β' subspaces for various values β'. In exceptional cases a particular $|\alpha', A'\rangle$ may lie wholly in some one β' subspace, but this cannot happen for an arbitrary $|\alpha', A'\rangle$; otherwise α and β would commute. The general rule is that precision in α implies a statistical distribution of β values and vice versa.

In an arbitrary state $|a\rangle$ we have

$$\alpha_{\text{Av}} = \langle a|\alpha|a\rangle, \quad \beta_{\text{Av}} = \langle a|\beta|a\rangle \tag{1}$$

The uncertainties, $\Delta\alpha$ and $\Delta\beta$ associated with repeated measurements on the state $|a\rangle$ may be taken as standard deviations,

$$(\Delta\alpha)^2 = \langle a|(\alpha - \alpha_{\text{Av}})^2|a\rangle = \langle a|(\alpha - \alpha_{\text{Av}})(\alpha - \alpha_{\text{Av}}|a\rangle$$

and similarly for $\Delta\beta$.

Because α and β are Hermitian, their commutator is $i\gamma$, where γ is Hermitian. The state vector

$$|b\rangle = [\lambda(\alpha - \alpha_{\text{Av}}) + i(\beta - \beta_{\text{Av}})]|a\rangle$$

has a norm $N(\lambda)$ that is greater than 0 for all λ except a value that makes the state vector vanish identically. Moreover,

$$N(\lambda) = \lambda^2(\Delta\alpha)^2 + \lambda\gamma_{\text{Av}} + (\Delta\beta)^2 \geqslant 0$$

The minimum of $N(\lambda)$ as a function of λ is $(\Delta\beta)^2 - \lambda_{\text{Av}}^2/4(\Delta\alpha)^2$, and therefore for an arbitrary state

$$\Delta\alpha \times \Delta\beta \geqslant \frac{1}{2}|\gamma_{\text{Av}}| \tag{2}$$

This is the general form of the Heisenberg uncertainty principle: the product of the uncertainties is not less than half the mean of i times their commutator.

In particular, for the most familiar case, that of position and conjugate momentum we have

$$\Delta x \Delta p \geqslant \frac{1}{2}\hbar \tag{3}$$

The state that leads to the minimum uncertainty product is that for which the state $|b\rangle$ vanishes. On using the Schrödinger representation for p_x we find

$$\left[\lambda(x - x_{\text{Av}}) + i\left(-i\hbar\frac{\partial}{\partial x} - p_{\text{Av}}\right)\right]\langle x|a\rangle = 0$$

The normalized solution of this first-order differential equation is

$$\langle x|a\rangle = \left(\frac{\lambda}{\pi\hbar}\right)^{1/4} \exp\frac{ip_{\text{Av}}x}{\hbar} \exp\frac{-\lambda(x - x_{\text{Av}})^2}{2\hbar} \tag{4}$$

From this we may calculate the uncertainties

$$\Delta x = \left(\frac{\hbar}{2\lambda}\right)^{1/2}, \quad \Delta p = \left(\frac{\hbar\lambda}{2}\right)^{1/2} \tag{5}$$

so that their product is verified to be the minimum value, $(1/2)\hbar$. The parameter λ may have any value. It is equal to the ratio $\Delta x/\Delta p$ showing that the same result holds whether the uncertainty is largely in Δx or largely in Δp.

4². The statistical operator (density matrix)

Situations arise in which there is incomplete knowledge of the state $|a\rangle$ of a system. Such partial knowledge may be described as that in which it is only known that $|a\rangle$ is in one of a set (finite or infinite in number) of orthogonal states with P_a being the probability of the state being $|a\rangle$. Here the P_a are nonnegative with $\sum_a P_a = 1$.

Such a situation may be visualized by considering an ensemble of a large number of independent noninteracting systems such that the number of them in each state $|a\rangle$ is proportional to P_a. An ensemble of this kind cannot be described by a single state. Instead, it is described by a *statistical operator* or *density matrix* [10].

If the system were in a particular state $|a\rangle$ then the mean value of γ would be $\langle a|\gamma|a\rangle$. Therefore, the ensemble average of γ is

$$\gamma_{\mathrm{Av}} = \sum_a P_a\langle a|\gamma|a\rangle \tag{1}$$

Written explicitly on the B basis this is

$$\gamma_{\mathrm{Av}} = \sum_{B'',B'} \langle B''|\gamma|B'\rangle\langle B'|\rho|B''\rangle \tag{2}$$

in which

$$\langle B'|\rho|B''\rangle = \sum_a \langle B'|a\rangle P_a\langle a|B''\rangle \tag{3}$$

Here ρ is the matrix of the statistical operator written on the B basis. Written for the arbitrary basis it is

$$\rho = \sum_a |a\rangle P_a\langle a| \tag{4}$$

With this definition of ρ, the average of γ can be written in a form that does not make explicit reference to any particular basis,

$$\gamma_{\mathrm{Av}} = \mathrm{Tr}(\gamma\rho) = \mathrm{Tr}(\rho\gamma) \tag{5}$$

From this definition of ρ it follows that the $|a\rangle$ are proper states of ρ and the proper values are the associated P_a. The statistical operator is therefore a Hermitian operator, the sum of whose proper values is equal to unity.

In the special case that (4) contains only a single term, that is, $P_a =$

1 for one state $|a\rangle$ and is zero for all others, the statistical matrix describes an ensemble in which all of the systems are in this state. Such a ρ satisfies the equation

$$\rho^2 - \rho = 0 \tag{6}$$

showing that it has one proper value unity and that the value $\rho' = 0$ is associated with all other axes.

In some ways the statistical matrix $|a\rangle\langle a|$ is a more satisfactory way of describing the definite state $|a\rangle$ than is $|a\rangle$ itself. Because of the juxtaposition here of $|a\rangle$ and $\langle a|$, the arbitrary phase of normalization in $|a\rangle$ does not appear in $|a\rangle\langle a|$. But the habit of thinking in terms of the vector $|a\rangle$ rather than the dyad $|a\rangle\langle a|$ is probably too deeply ingrained for such a change to be made.

It is important to be clear on the distinction between a system's being in a definite state of superposition of two states, with equal likelihood of being in $|a\rangle$ or $|b\rangle$ (where these are orthogonal), and that of being in an ensemble in which $P_a = 1/2$ and $P_b = 1/2$.

In the former case the state $|c\rangle$ is a definite or "pure" one, represented by

$$|c\rangle = 2^{-1/2}(|a\rangle + e^{i\delta}|b\rangle)$$

where δ is a definite real number. For such a state γ_{Av} is

$$\gamma_{\text{Av}} = \langle c|\gamma|c\rangle = \frac{1}{2}[\langle a|\gamma|a\rangle + \langle b|\gamma|b\rangle] + \text{Re}(\langle a|\gamma|b\rangle e^{-i\delta})$$

so that γ_{Av} depends explicitly on δ, the relative phase of superposition.

In the second case, the ensemble is represented by

$$\rho = \frac{1}{2}[|a\rangle\langle a| + |b\rangle\langle b|]$$

Thus the ensemble average is equal to the state average when that is averaged over all values of δ, these being considered equally likely.

More generally, a definite superposition of a number of states is

$$|c\rangle = \sum_a C_a|a\rangle$$

giving

$$\gamma_{\text{Av}} = \sum_a |C_a|^2\langle a|\gamma|a\rangle + \sum_{a=b} C^*{}_a C_b\langle a|\gamma|b\rangle$$

Again, if all relative phases are equally likely, the nondiagonal elements drop out, giving a result that agrees with that of an ensemble average with $P_a = |C_a|^2$.

In statistical quantum mechanics, the entropy, S, of an ensemble of N systems, in units of the Boltzmann constant, k, is represented by

$$S = -N\,\mathrm{Tr}(\rho \log \rho) = N \sum_{a} P_a \log P_a^{-1} \tag{7}$$

This is equal to zero if all members of the ensemble are in the same state, for then only one $P_a = 1$ and the others vanish. When several do not vanish, all are less than unity so (7) gives a positive entropy.

The *canonical* ensemble of Gibbs is that in which the ensemble is in thermal equilibrium with a heat reservoir at temperature T. It is most simply expressed in terms of the states $|W', A'\rangle$ that are labeled by values of the energy W' and values of other commuting operators A'. The *statistical weight*, $g(W')$ of the energy W' is the number of states having this energy, that is, the number of distinct values of A' which give a state $|W', A'\rangle$.

The partition function is defined as

$$Z(T) = \sum_{W'} g(W') \exp \frac{-W'}{kT} \tag{8}$$

and the statistical operator for the canonical ensemble is

$$\rho = Z^{-1} \sum_{W', A'} |W', A'\rangle \exp\left(\frac{-W'}{kT}\right) \langle W', A'| \tag{9}$$

The mean energy per system over the ensemble is

$$W_{\mathrm{Av}} = \frac{\partial \log Z}{\partial(1/kT)} \tag{10}$$

and the entropy of the ensemble of N systems is

$$S = N\left[\log Z + \frac{W_{\mathrm{Av}}}{kT}\right] \tag{11}$$

There is a close tie here with information theory [11].

5². The momentum representation

Because of the symmetry of the commutation relations between the position vector, $\mathbf{r}$, and the conjugate momentum vector, $\mathbf{p}$, of a particle, it is possible to set up operators, analogous to Schrödinger's, for action on $\langle \mathbf{p}|a\rangle$, the momentum representative of the one-particle state $|a\rangle$.

Just as $|\langle \mathbf{r}|a\rangle|^2$ gives the probability density for finding the particle in unit volume at $\mathbf{r}$ when in the state $|a\rangle$, so $|\langle \mathbf{p}|a\rangle|^2\, dp_x\, dp_y\, dp_z$ gives the probability of finding the linear momentum of the particle in $dp_x\, dp_y\, dp_z$ at $\mathbf{p}$. The momentum representation is useful in the study of collision problems.

The basic commutation rules of 2²7 for a single particle can be written

$$[\mathbf{r}, \mathbf{p}] = i\hbar\mathbf{I} \tag{1}$$

or

$$[\mathbf{r}, \mathbf{k}] = i\mathbf{I}$$

on writing

$$\mathbf{p} = \mathbf{k}\hbar$$

where $\mathbf{k}$ is called the wave vector. Here $\mathbf{I}$ is the dyad representing the identity, $(\hat{\mathbf{k}}_1\hat{\mathbf{k}}_1 + \hat{\mathbf{k}}_2\hat{\mathbf{k}}_2 + \hat{\mathbf{k}}_3\hat{\mathbf{k}}_3)$, and it acts as an identity operator in system space.

In the Schrödinger scheme for $\langle \mathbf{r}|a\rangle$, $\mathbf{r}$ becomes multiplication by $\mathbf{r}$ and $\mathbf{k}$ becomes $-i\nabla$. Alternatively, the commutation rules (1) for $\langle \mathbf{k}|a\rangle$ can be satisfied by taking $\mathbf{k}$ to be multiplication by $\mathbf{k}$ and $\mathbf{r}$ to be represented by $+i\nabla_k$ in which ∇_k means the gradient operator in the 3-space of $\mathbf{k}$ vectors,

$$\nabla_k = \hat{\mathbf{k}}_1 \frac{\partial}{\partial k_1} + \hat{\mathbf{k}}_2 \frac{\partial}{\partial k_2} + \hat{\mathbf{k}}_3 \frac{\partial}{\partial k_3} \tag{2}$$

The notation $\dfrac{\partial}{\partial \mathbf{k}}$ is sometimes used for ∇_k.

The amplitude $\langle \mathbf{k}|a\rangle$ is the Fourier transform of $\langle \mathbf{r}|a\rangle$. In the Schrödinger scheme, the amplitude $\langle \mathbf{r}|\mathbf{k}'\rangle$ for a precise value $\mathbf{k}'$ of $\mathbf{k}$ is found from

$$i\nabla\langle \mathbf{r}|\mathbf{k}'\rangle = \mathbf{k}'\langle \mathbf{r}|\mathbf{k}'\rangle$$

to be

$$\langle \mathbf{r}|\mathbf{k}'\rangle = A \exp(i\mathbf{k}' \cdot \mathbf{r}) \tag{3}$$

This is finite for all real values of $\mathbf{k}'$, so all $\mathbf{k}'$ are proper values of $\mathbf{k}$, just as all $\mathbf{r}'$ are proper values of $\mathbf{r}$.

The normalization constant A is defined so that

$$|A|^2 \iiint \exp[i(\mathbf{k}' - \mathbf{k}'')\cdot\mathbf{r}]\, dx_1\, dx_2\, dx_3 = \delta(\mathbf{k}' - \mathbf{k}'') \tag{4}$$

Its value is known from the standard theory of Fourier integrals [12] to be $|A| = (2\pi)^{-3/2}$ in accordance with the symmetrical one-dimensional relations,

$$\begin{aligned} f(x) &= (2\pi)^{-1/2} \int_{-\infty}^{+\infty} e^{ikx} g(k)\, dk \\ g(k) &= (2\pi)^{-1/2} \int_{-\infty}^{+\infty} e^{-ikx} f(x)\, dx \end{aligned} \tag{5}$$

Making an arbitrary phase choice, we adopt the real factor $(2\pi)^{-1/2}$ for each coordinate-momentum pair in (3), so we have

$$\langle \mathbf{k}|a\rangle = (2\pi)^{-3/2} \iiint \exp(-i\mathbf{k}\cdot\mathbf{r})\langle \mathbf{r}|a\rangle\, dx_1\, dx_2\, dx_3$$

and

$$\langle \mathbf{r}|a\rangle = (2\pi)^{-3/2} \iiint \exp(+i\mathbf{k}\cdot\mathbf{r})\langle \mathbf{k}|a\rangle\, dk_1\, dk_2\, dk_3 \tag{6}$$

In one dimension the state that corresponds to k_1 having the precise value k_1' from (3) is $\langle x_1|k_1'\rangle = (2\pi)^{-1/2}\exp(ik_1'x_1)$ so that the probability of finding the particle between x_1 and $x_1 + dx_1$ is independent of x_1. This corresponds to an extreme situation of the uncertainty relation; precise knowledge of k_1 implies total uncertainty about position. At the other extreme, $\langle x|x'\rangle = \delta(x - x')$ corresponds to a state $|x'\rangle$in which the position is known to be precisely x'. The corresponding $\langle k_1|x'\rangle$ is $(2\pi)^{-1/2}\exp(-ik_1x')$, which implies complete uncertainty about the value of the momentum.

6^2. Symmetry group of the Hamiltonian

The Hamiltonian operator for an N-electron atom is a function of the position vectors $\mathbf{r}_\alpha$ $(\alpha = 1, \ldots, N)$ and the conjugate momenta $\mathbf{p}_\alpha$ and of the spin coordinates. The approximate form $2^2 9$ does not explicitly depend on the spin coordinates, s_α, but these appear in better approximations.

If new variables $\mathbf{r}_1' \ldots \mathbf{r}_N', \mathbf{p}_1' \ldots \mathbf{p}_N'$ are introduced as linear combinations of the $\mathbf{r}_1 \ldots \mathbf{r}_N$ and $\mathbf{p}_1 \ldots \mathbf{p}_N$ in such a way that the commutation rules ($2^2 7$) hold in the primed variables, the Hamiltonian, when expressed in terms of the new variables is, in general, a different function H' of the new variables than H was of the old.

There exist special transformations such that H' is the same function of the new variables that H was of the old. Any such transformation, which leaves the Hamiltonian *invariant,* is called *symmetry* of H.

Clearly every H is invariant to the *identity,* that trivial transformation in which each new variable is the same as the corresponding old one. Also if H is invariant to a transformation it is invariant to the inverse transformation. Also if it is invariant to two transformations, it is invariant to the combined transformation obtained by carrying them out in succession in either order. Therefore the transformations that leave H invariant form a *group* [4]. This is called the *symmetry group* of the Hamiltonian.

For example, the Hamiltonian $2^2 9$ is invariant to any permutation of the electron labels $\alpha = 1, 2, \ldots, N$ appearing in it. The permutation that is written as $P(p_1, p_2, \ldots, p_N)$, means that 1 is to be replaced by p_1 and so on. This is a linear transformation,

$$\mathbf{r}'_{p_\alpha} = \mathbf{r}_\alpha, \quad \mathbf{p}'_{p_\alpha} = \mathbf{p}_\alpha, \quad s'_{p_\alpha} = s_\alpha \tag{1}$$

which leaves the commutation rules satisfied.

The Hamiltonian for a free atom is also left unaltered by any rigid rotation of the reference frame,

$$\begin{aligned} x'_{\alpha i} &= \sum_j R_{ij} x_{\alpha j} \\ p'_{\alpha i} &= \sum_j R_{ij} p_{\alpha j} \end{aligned} \tag{2}$$

Here $\mathbf{x}'_{\alpha i}$ gives the coordinates of the αth particle in the rotated frame. The property of being an orthogonal substitution means that the form of the scalar product of two vectors is unchanged,

$$\mathbf{a} \cdot \mathbf{b} = \sum_i a'_i b'_i = \sum_{ijk} R_{ij} a_j R_{ik} b_k = \sum_j a_j b_j$$

which requires that

$$\sum_i R_{ij} R_{ik} = \sum_i R_{ji} \tilde{R}_{ik} = \delta_{jk} \tag{2'}$$

Pure or *proper rotations* are those for which $\det(R_{ij}) = +1$. An orthogonal substitution for which $\det(R_{ij}) = -1$ is called an *improper rotation*.

The simplest case of an improper rotation is the inversion, $\mathcal{I}$, in which R_{ij} has the form

$$\mathcal{I}_{ij} = -\delta_{ij} \tag{3}$$

This is equivalent to reversing the sense of each of the basic unit vectors of the frame. Every improper rotation may be expressed as a proper rotation preceded or followed by the inversion.

The viewpoint regarding the effect of R as that of the description of the same system in a new frame is often called the *passive* interpretation of the foregoing relations. The *active* interpretation regards the frame as remaining fixed, but the physical system itself as rotated to a new orientation, or inverted (transformed from right-handed to left-handed form).

We shall usually use the active interpretation by which the physical system is rotated relative to a fixed frame.

If the atom of $2^2 9$ is placed in an external electric field such that the electrostatic potential at $\mathbf{r}$ is $\phi(\mathbf{r})$, then a term

$$-e \sum_\alpha \phi(\mathbf{r}_\alpha) \tag{4}$$

has to be added to the Hamiltonian to take account of the interaction of each electron with this field. This additional term is also invariant to the permutations of the electron labels, because e is the same for all electrons. But, unless the field happens to be spherically symmetric, the modified Hamiltonian is no longer invariant under rotations of the atom.

Such a term is said to *break* the rotation symmetry of the Hamiltonian. Even though the Hamiltonian is no longer invariant to the full rotation group, it may still be invariant under a subgroup of the full rotation group. For example, if $\phi(\mathbf{r})$ has the form $-\epsilon x_3$, representing a uniform electric field along the third axis, the Hamiltonian is still invariant with regard to any rotation of the form,

$$R_{ij} = \begin{pmatrix} R_{11} & R_{12} & 0 \\ R_{21} & R_{22} & 0 \\ 0 & 0 & 1 \end{pmatrix} \tag{5}$$

which is a rotation around the third axis. This kind of electric field produces the *Stark effect,* the effect on atomic spectra arising from their being perturbed by such a uniform electric field.

As another example, $\phi(\mathbf{r})$ may arise from a charge distribution external to the atom which has tetrahedral symmetry. This means that it is unchanged by the finite group of rotations which carry a regular tetrahedron into itself. Finite symmetries of this kind are important in ligand field theory.

The importance of the symmetries of the Hamiltonian lies in the fact that they provide a partial determination of the proper states of H. Let $|W', A'\rangle$ correspond to the energy W' and the values A' of such other labels as are needed to label the state, so that

$$H|W', A'\rangle = W'|W', A'\rangle$$

or

$$H\langle \mathbf{r}_1 \ldots \mathbf{r}_N|W', A'\rangle = W'\langle \mathbf{r}_1 \ldots \mathbf{r}_N|W', A'\rangle$$

for the relation in terms of the Schrödinger representative of the state.

Let R be defined as the operator which represents the effect on such a representative of a particular transformation giving

$$RH\langle \mathbf{r}_1 \ldots \mathbf{r}_N|W', A'\rangle = W'R\langle \mathbf{r}_1 \ldots \mathbf{r}_N|W', A'\rangle$$
$$RHR^{-1}R\langle \mathbf{r}_1 \ldots \mathbf{r}_N|W', A'\rangle = W'R\langle \mathbf{r}_1 \ldots \mathbf{r}_N|W', A'\rangle$$

In the particular case that R leaves H invariant, the effect is the same as if $RH = HR$, and therefore $R\,\langle \mathbf{r}_1 \ldots \mathbf{r}_N|W', A'\rangle$ is also a solution of the wave equation for the energy W'. It can therefore be expanded in terms of the states $|W', A''\rangle$ that belong to this same proper value, W'.

We write

$$R|W',A'\rangle = \sum_{A''} |W',A''\rangle\langle A''|D(R)|A'\rangle \tag{6}$$

for the expansion of the transformed $R|W',A'\rangle$ in terms of the untransformed states $|W',A''\rangle$.

In this way each R has associated with it a square matrix of order equal to that of the degeneracy $g(W')$ of the W' energy level. Such an association of a matrix with each element of a group is called a *representation* of the group. The letter D in (6) comes from the word *Darstellung,* which means representation in German.

The complex conjugate functions transform by equations involving Hermitian adjoint matrices $D^\dagger(R)$

$$\langle A'|D^\dagger(R)|A''\rangle = \langle A''|D(R)|A'\rangle^*$$

so that

$$\langle W',A''|R^\dagger = \sum_{A'} \langle A''|D^\dagger(R)|A'\rangle\langle W',A'| \tag{6'}$$

The operations R carry an orthonormal basis into an orthonormal basis and therefore

$$\begin{aligned}\langle W',A''|R^\dagger R|W',A'\rangle &= \delta(A'',A') \\ &= \sum_{A'''} \langle A''|D^\dagger(R)|A'''\rangle\langle A'''|D(R)|A'\rangle\end{aligned}$$

showing that the representation is by unitary matrices,

$$D^\dagger(R)D(R) = 1 \tag{7}$$

The choice of basis in the W' subspace is arbitrary, so a different set of orthonormal states $|W',B'\rangle$ could be used, which are related to the $|W',A'\rangle$ by a unitary transformation as in $1^2 34$ and $1^2 35$. We therefore have

$$\begin{aligned}R|W',B'\rangle &= \sum_{A''} R|W',A''\rangle\langle A''|B'\rangle \\ &= \sum_{A'',A'''} |W',A'''\rangle\langle A'''|D(R)|A''\rangle\langle A''|B'\rangle \\ &= \sum_{B''} |W',B''\rangle \sum_{A'',A'''} \langle B''|A'''\rangle\langle A'''|D(R)|A''\rangle\langle A''|B'\rangle\end{aligned}$$

Thus the basis states $|W',B'\rangle$ transform by matrices that are obtained from those for $|W',A'\rangle$ by a unitary transformation

$$\langle B''|D(R)|B'\rangle = \sum_{A''',A''} \langle B''|A'''\rangle\langle A'''|D(R)|A''\rangle\langle A''|B'\rangle \tag{8}$$

These matrices for the ensemble of the R's in the group form a *different* representation than the $\langle A''|D(R)|A'\rangle$ but one that is considered equivalent.

It may happen that, by a suitable choice of the basis $|W', B'\rangle$, the matrices in (8) for *every* R are diagonal in one or more of the labels β. We now write β for those labels in which each of the matrices is diagonal and B for all others of the complete set B. Thus each matrix for every R takes the form

$$\langle \beta'', B''|D(R)|\beta', B'\rangle = \delta(\beta''; \beta')\langle \beta', B''|D(R)|\beta', B'\rangle \tag{9}$$

In this case the representation $D(R)$, even in the equivalent form $\langle A''|D(R)|A'\rangle$ is called *reducible*. If no basis exists to put the $D(R)$ matrices for every R in the form of step matrices, the representation is called *irreducible*.

When all matrices of a reducible representation are put in the step-matrix form (9) to the maximum extent, so that the matrices on the right are themselves irreducible, the representation may, in a symbolic sense, be regarded as the *sum* of the irreducible representations that appear in the steps. Distinguishing different irreducible representations by an index λ, writing $D^{(\lambda)}(R)$, we would write

$$D(R) = \sum_{\lambda} D^{(\lambda)}(R) \tag{10}$$

in which the sum is over those values of λ which actually occur in (9) for the various values of β'. Here the same value of λ may occur several times over. We write $g(\lambda)$ for the order of the matrices in $D^{(\lambda)}(R)$, so the total dimensionality of $D(R)$ is

$$g = \sum_{\lambda} g(\lambda)$$

States that transform by the same irreducible representation have the same energy. Those that transform by different irreducible representations need not have the same energy, and usually they do not. When their energies are equal, this is called an *accidental* degeneracy.

The important thing is that the irreducible representations are a characteristic of the particular group to which they belong. All atoms include the rotation-inversion group in their symmetries, so that properties associated with the representations of this group appear again and again in different atoms.

7^2. Indistinguishable particles

The fact that the Hamiltonian of an atom is invariant to permutation of the electron labels $i = 1, 2, \ldots, N$ has important consequences for atomic structure [13] and provides the basis for the Pauli exclusion principle (Sec. 11^1).

The relations are particularly simple for $N = 2$. Let x_1 and x_2 stand for $\mathbf{r}_1$, s_1 and $\mathbf{r}_2$, s_2, the position and spin coordinates of the first and second electron and write $\psi(x_1, x_2)$ for any state function $\langle x_1, x_2 | \ \rangle$. The probability density $|\psi(x_1, x_2)|^2$ must be a symmetric function of x_1 and x_2 because of the indistinguishability of the two electrons. This can occur in either of two ways

$$\psi(x_2, x_1) = \pm\psi(x_1, x_2) \tag{1}$$

These are called *symmetric* (+) and *antisymmetric* (−) states, respectively.

With $N = 2$ the permutation group consists of $2! = 2$ elements that are E, the identity, and P, which interchanges the labels 1 and 2. Evidently $P^2 = E$, because repetition of the interchange restores the original state. Hence the proper values of P are ± 1, the symmetric states being proper states of P belonging to the value of $+1$, and the antisymmetric states being those belonging to the value -1. Because P commutes with H, the proper states of H can be labeled by proper values of P, that is, classified as *symmetric* or *antisymmetric*.

Any operator γ that refers to the two electrons will be unaffected by P, that is, it commutes with P. Otherwise, the electrons could be distinguished by the difference of their response in any measurement of γ. In consequence, by $1^2 26$, the matrix elements of γ between unequal values of P' vanish. Matrix elements of any symmetric operator vanish between states of opposite symmetry.

In particular, this is true for the matrix elements of H. Because P commutes with H, P is a constant of the motion (Sec. 2^2). Therefore, if the system is initially in a state of either symmetry it continues in a state of the same symmetry. This is true no matter how the system is affected by external perturbations, provided only that these do not distinguish between the two electrons and so commute with P.

Similar results hold for N indistinguishable particles, for which there are $N!$ permutations including the identity, E. $P(p_1, p_2, \ldots, p_N)$ is written for a particular one of these as in Sec. 6^2. The order of listing is evidently without significance. There are N possible choices for p_1, then $(N - 1)$ for p_2, and so on, giving the stated result that altogether there are $N!$ permutations.

If Q is another permutation, QP means that Q is to be applied after P to whatever stands on the right, then QP is equivalent to one of the permutations. Likewise, every permutation has an inverse, and E is one of them, so they form a finite group of $N!$ elements.

In general, the permutations do not commute with each other, although each individual P commutes with the N-particle Hamiltonian and with all operators that involve the N particles in an indistinguishable way.

Because the P's do not commute, it is not possible for a given $\psi(x_1, \ldots, x_N)$ to be a simultaneous proper state of all of them. It is necessary to find suitable commuting functions of the P's with which to label the proper states of the Hamiltonian.

Two such functions that commute with any operator that is invariant to the P's are called the symmetrizer, $\mathcal{S}$, and the antisymmetrizer, $\mathcal{A}$. After the identity, the simplest P's are those which are called transposition; these leave all but two of the labels unaffected and these are interchanged. A convenient notation for these is P_{ab} for the one which interchanges a and b while leaving the other $(N - 2)$ labels unchanged. Evidently there are $(1/2)N(N - 1)$ of these, as there is one for each pair, $a \neq b$. For each transposition we have $P_{ab}^2 = 1$, but in general these transpositions do not commute with each other.

Each P can be expressed as a product of transpositions. Although such an expression is not unique, the parity (oddness or evenness) of the number of transpositions in the expression is unique. Thus, the P's can be separated into two classes, called odd and even, according to the parity, p, of the number of transpositions in such an expression for P.

The symmetrizer and antisymmetrizer are then defined by

$$\begin{aligned} \mathcal{S} &= (N!)^{-1/2} \sum_{p} P \\ \mathcal{A} &= (N!)^{-1/2} \sum_{p} (-1)^p P \end{aligned} \tag{2}$$

the sum being over the $N!$ permutations and p being the parity of P.

Writing p^0 for the parity of P^0 where this is some particular permutation, it follows that

$$P^0 \mathcal{S} = \mathcal{S} \tag{3}$$

because P^0P is another one of the P's so, because P runs over all of the permutations in the sum for $\mathcal{S}$, so also will P^0P, giving all of them merely listed in a different order, which does not affect the sum.

Similarly,

$$P^0 \mathcal{A} = (-1)^{p^0} \mathcal{A}$$

and

$$P_{ab} \mathcal{A} = -\mathcal{A} \tag{4}$$

where P_{ab} is any one of the transpositions.

Also

$$\begin{aligned} \mathcal{S}^2 &= (N!)^{1/2} \mathcal{S} \\ \mathcal{A}^2 &= (N!)^{1/2} \mathcal{A} \\ \mathcal{A}\mathcal{S} &= \mathcal{S}\mathcal{A} = 0 \end{aligned} \tag{5}$$

because there are equal numbers of odd and even permutations, so

$$\sum_{P} (-1)^p = 0$$

From these results it follows that the proper values of $\mathcal{S}$ and $\mathcal{A}$ are $(N!)^{1/2}$ and 0 for each of them, and that a proper state of $\mathcal{S}$ with proper value $(N!)^{1/2}$ is a proper state of $\mathcal{A}$ with proper value 0 and vice versa.

Although the P's do not commute, a proper state of $\mathcal{S}$ for the value $(N!)^{1/2}$ is also a proper state for each P with the proper value $+1$, as is clear from (3).

Similarly, (4) shows that a proper state for $\mathcal{A}$ with value $(N!)^{1/2}$ is a proper state for each P with the value $+1$ for the even P's and -1 for the odd P's.

A proper state for $\mathcal{S}$ with the value $(N!)^{1/2}$ is called a *symmetric* state, one corresponding to this value for $\mathcal{A}$ is called an *antisymmetric* state.

Starting with an arbitrary $\psi(x_1, x_2, \ldots, x_N)$ it is seen from (5) that, provided they do not vanish identically $\mathcal{S}\psi$ is a symmetric state and $\mathcal{A}\psi$ is an antisymmetric state. Thus $\mathcal{A}$ and $\mathcal{S}$ behave like projection operators so that, in general, $\mathcal{S}\psi$ and $\mathcal{A}\psi$ are not normalized even if ψ was.

Any operator that is symmetric in the labels $1, 2, \ldots, N$ commutes with $\mathcal{S}$ and with $\mathcal{A}$ and therefore, by 1²26, its matrix elements between states of opposite symmetry vanish. In particular, this is true for the Hamiltonian, even when symmetric external perturbations are involved, and so the symmetry of ψ is a constant of the motion.

This result does not prove that ψ must be purely symmetric or purely antisymmetric. If it has components of both symmetry initially, it continues to have them in the same proportions. If initially it is purely symmetric or purely antisymmetric it continues to remain so. It appears to be true empirically that the states occurring in nature do have pure symmetry or pure antisymmetry. This is an attribute of each type of particle that has to be determined experimentally.

Particles whose state functions are symmetric are called *bosons,* and are said to obey Einstein-Bose statistics. Those whose state functions are antisymmetric are called *fermions* and are said to obey Fermi-Dirac statistics. The three kinds of particles involved in atomic structure: protons, neutrons, and electrons, are all fermions. For all states involved in atomic structure the state function is antisymmetric with regard to transpositions among each of these three types of indistinguishable particles.

Much of the theory of atoms, molecules and nuclei is based on the *independent particle model.* In this, a set of approximate basis states for a system of N interacting identical particles is set up by assuming that specific pair-wise interactions of the particles can be replaced by an effective field $V(\mathbf{r})$ in which each particle moves. Thus, for an N-electron atom or ion the Hamiltonian would be

$$H' = \sum_k H_k \quad \text{with} \quad H_k = \mathbf{p}_k^2/2m + V(\mathbf{r}_k) \quad \text{and} \quad H_k\phi(a, k) = W_a\phi(a, k) \tag{6}$$

Then the energy levels and approximate wave functions for the Hamiltonian 2²9 are calculated by treating $(H - H')$ as a perturbation (Sec. 9²).

A state of the one-electron wave functions may be labeled by an *individual set* of quantum numbers denoted for short by $a, b, c, \ldots$. More explicitly, in a central field problem, a stands for (n_a, l_a, m_a, m_{sa}). The energy level of this quantum state in the one-particle Hamiltonian is denoted by $W(a)$.

To specify a state of the N-electron problem, it is necessary to give N individual sets of quantum numbers, one for each electron, written as $(a, b, c, \ldots, n)$.

The simple product wave function is

$$\phi(a; 1)\phi(b; 2)\ldots\phi(n; N) = \langle \mathbf{r}_1, s_1|a\rangle\langle \mathbf{r}_2, s_2|b\rangle\ldots\langle \mathbf{r}_N, s_N|n\rangle \tag{7}$$

The corresponding zero-order energy is

$$W(a, b, c, \ldots, n) = W(a) + W(b) + \ldots + W(n) \tag{8}$$

The zero-order approximation to the N-electron wave function results from the action of the antisymmetrizer on the product wave function (7). This gives a determinantal wave function, often called a *Slater determinant.* For these a variety of notations are in use. We write

$$\begin{aligned}
\Phi(a, b, c, &\ldots, n; 1, 2, \ldots, N) \\
&= \mathcal{A}\,\phi(a; 1)\phi(b; 2)\ldots\phi(n; N) \\
&= \langle \mathbf{r}_1, s_1; \ldots \mathbf{r}_N, s_N|a, b, \ldots, n\rangle \\
&= \{a, b, \ldots n\}_N \\
&= \{n_a, l_a, m_a, m_{sa}; n_b, l_b, m_b, m_{sb}; \ldots\} \\
&= (N!)^{-1/2}\begin{vmatrix} \phi(a; 1) & \phi(a; 2) & \ldots & \phi(a; N) \\ \phi(b; 1) & \phi(b; 2) & \ldots & \phi(b; N) \\ & \ldots & & \ldots \\ \phi(n; 1) & \phi(n; 2) & \ldots & \phi(n; N) \end{vmatrix} \\
&= (N!)^{-1/2}\begin{vmatrix} \langle \mathbf{r}_1, s_1|a\rangle & \langle \mathbf{r}_2, s_2|a\rangle & \ldots & \langle \mathbf{r}_N, s_N|a\rangle \\ \langle \mathbf{r}_1, s_1|b\rangle & \langle \mathbf{r}_2, s_2|b\rangle & \ldots & \langle \mathbf{r}_N, s_N|b\rangle \\ & \ldots & & \ldots \\ \langle \mathbf{r}_1, s_1|n\rangle & \langle \mathbf{r}_2, s_2|n\rangle & \ldots & \langle \mathbf{r}_N, s_N|n\rangle \end{vmatrix}
\end{aligned} \tag{9}$$

Sometimes it is convenient to write

$$(a, b, c, \ldots, n) = \langle a, b, c, \ldots, n|H|a, b, c, \ldots, n\rangle \tag{10}$$

that is, the diagonal matrix element of the Hamiltonian.

A determinant changes sign on interchanging any two of its rows (or columns). Hence (9) vanishes identically if any two of the individual sets are alike. This property gives expression to the Pauli exclusion principle, according to which no two fermions can be in the same quantum state.

Likewise (9) vanishes if the coordinates, including spin, of any two electrons are the same, for in that case the determinant has two equal columns. Thus, in states described by an antisymmetric state, there is a low probability of two electrons of the same spin being near each other. This is over and above the correlation in their motion, which arises from the fact that they repel each other. This property of an antisymmetric state is sometimes called the *Fermi hole:* Around each electron there is a small region where no other electrons (of the same spin) are likely to be found. Although presented here as a property of a Slater determinant on the independent particle model, it also holds for any antisymmetric function of the coordinates of the N electrons.

Basis states of the independent-particle model, for either bosons or fermions, are conveniently described in the *occupation number representation*

$$|n_a, n_b, \ldots\rangle \tag{11}$$

Here n_a is the number of particles occupying the ath state of the one-particle Hamiltonian (6). As there are generally an infinite number of such states, and a finite total number of particles $N = \Sigma n_a$, most of the n_a are equal to zero.

In the boson case each n_a can have any nonnegative integral value. The N-particle wave function for the state (11) is

$$\langle x_1, \ldots, x_N|n_a, n_b \ldots\rangle = \mathcal{S}\phi(a, 1)\phi(a, 2) \ldots \phi(n, N) \tag{12}$$

In the product following on the right, each factor $\phi(a, i)$ occurs n_a times. The symmetrizer involves summation over $N!$ permutations. Permutation of the same particle indices within the n_a factors $\phi(a, i)$ gives the same term, so in (12) there are

$$\frac{N!}{n_a!n_b! \ldots} \tag{13}$$

different terms, each repeated $n_a!n_b! \ldots$ times in the full sum. Because (12) is symmetric, the wave function remains the same for any order of listing of the one-particle states.

In the fermion case, each n_a is restricted to the values 0 and 1. Hence we have

$$\langle x_1 \ldots x_N|n_a, n_b \ldots\rangle = \mathcal{A}\phi(a, 1)\phi(b, 2) \ldots \phi(n, N) \tag{14}$$

in which each of the occupied states $(a, b, \ldots, n)$ is different, otherwise (14) would represent a determinant with two or more identical columns, vanishing identically. For the state in which $a, b, \ldots, n$ are occupied and all other states are unoccupied, the shorter notation $\{a, b, \ldots, n\}_N$ is also useful.

Because of the antisymmetry of (14) the sign of this wave function depends on the order of listing of the one-particle states. Any order can be adopted as standard, but it is usual to choose one in which increasing a means increasing $W(a)$.

It is convenient to introduce *destruction* and *creation operators*, η_a and $\eta^\dagger{}_a$ for each of the one-particle states [14]. These take different forms in the boson and fermion cases.

For the boson case, the destruction operator is defined by

$$\eta_a | \ldots n_a \ldots \rangle = \sqrt{n_a} | \ldots n_a - 1 \ldots \rangle \tag{15}$$

Thus, it acts on an N-particle state to give an $(N - 1)$-particle state in which n_a is reduced by unity, hence the name destruction operator. The least value of n_a is zero, and from (15) it follows that there are no states corresponding to negative values of n_a. The nonvanishing matrix elements of η_a are, accordingly,

$$\langle \ldots n'_a \ldots | \eta_a | \ldots n_a \ldots \rangle = \sqrt{n_a} \tag{16}$$

in which $n'_a = n_a - 1$, and all other $n'_b = n_b$, for $b \neq a$. Therefore, the nonvanishing matrix elements for the Hermitian conjugate operator $\eta^\dagger{}_a$ are

$$\langle \ldots n'_a \ldots | \eta^\dagger{}_a | \ldots n_a \ldots \rangle = \sqrt{n'_a} \tag{17}$$

and $n'_b = n_b$ for $b \neq a$. This gives the operator form of $\eta^\dagger{}_a$ analogous to (15),

$$\eta^\dagger{}_a | \ldots n_a \ldots \rangle = \sqrt{n_a + 1} | \ldots n_a + 1 \ldots \rangle \tag{15'}$$

Thus $\eta^\dagger{}_a$ acts on an N-particle state to give an $(N + 1)$-particle state in which n_a is increased to $n_a + 1$ and all other n's are left unchanged. Hence the name, creation operator.

The same state results from $\eta_a \eta_b$ or $\eta_b \eta_a$. These are $(N - 2)$-particle states in which n_a and n_b are each decreased by unity and the other n's are the same. Similarly $\eta^\dagger{}_a \eta^\dagger{}_b$ and $\eta^\dagger{}_b \eta^\dagger{}_a$ are the same $(N + 2)$-particle states in which n_a and n_b are each increased by unity and the other n's are the same. Likewise for $b \neq a$, $\eta_a \eta^\dagger{}_b$ and $\eta^\dagger{}_b \eta_a$ are the same N-particle state with $n'_b = n_b + 1$ and $n'_a = n_a - 1$ and all other n's are the same. These combinations are therefore related to a transition of a single particle from state a to state b. For $b = a$ we have, from (15) and (15′)

$$\eta^\dagger{}_a \eta_a = n_a \quad \text{and} \quad \eta_a \eta^\dagger{}_a = (n_a + 1)$$

in both cases times the N-particle state with all n's unchanged. Accordingly the n's are the proper values of the operators $\eta^\dagger{}_a \eta_a$.

All of these results are summarized in the statement that the η's obey the following commutation rules:

$$\begin{aligned} \eta_a \eta_b - \eta_b \eta_a &= 0 \\ \eta^\dagger{}_a \eta^\dagger{}_b - \eta^\dagger{}_b \eta^\dagger{}_a &= 0 \\ \eta_a \eta^\dagger{}_b - \eta^\dagger{}_b \eta_a &= \delta_{ab} \end{aligned} \tag{18}$$

We may write $|0\rangle$ for the "vacuum" state in which all of the n's are zero. Then from (15′) it follows that a general N-particle state can be written as

$$|n_a, n_b, \ldots\rangle = \left[\frac{(\eta\dagger_a)^{n_a}}{\sqrt{n_a!}} \times \frac{(\eta\dagger_b)^{n_b}}{\sqrt{n_b!}} \cdots\right]|0\rangle \tag{19}$$

in which, because of the second of (18) the operator factors can be put in any convenient order.

Analogous destruction and creation operators exist for the fermion case. The operators have different properties because now the allowed values are 0 and 1 instead of all nonnegative integers, and because the state is antisymmetric instead of symmetric. Analogous to (15), the destruction operator η_a, acting on $|\ldots n_a \ldots\rangle$ gives 0 if $n_a = 0$ and a multiple of $|\ldots n_a - 1 \ldots\rangle$ if $n_a = 1$. The multiplying factor has to be chosen to be consistent with the antisymmetry of the state.

This may be done as follows. Using the notation $\{a, b, \ldots, n\}_N$ for the N-particle state in which these N states are occupied, we arbitrarily choose the relation

$$\eta_a\{a, b, \ldots, n\}_N = \{b, \ldots, n\}_{N-1} \tag{20}$$

Thus η_a annihilates the occupied state that is first in the list. For another state x, supposing it is the ν_xth from the left, this state can first be moved to first place by ν_x transpositions with its neighbors to the left, giving a factor $(-1)^{\nu_x}$. Thus the general definition of η_x is

$$\eta_x\{a, b, \ldots, x, \ldots, n\}_N = (-1)^{\nu_x}\{a, b, \ldots, n\}_{N-1} \tag{20′}$$

Applied to an N-particle state in which x is initially an unoccupied state, η_x gives a zero result.

The analogous definition for the creation operator, $\eta\dagger_x$ is that it gives zero if the state x is already occupied and gives

$$\eta\dagger_x\{a, b, \ldots, n\}_N = (-1)^{\nu_x}\{a, b, \ldots, x, \ldots, n\}_{N+1} \tag{21}$$

in which ν_x is the number of places to the left at which the added state x appears in the standard order of listing.

From these definitions we readily find, in contrast with the result for the boson case, that the operators anticommute:

$$\begin{aligned} \eta_a\eta_b + \eta_b\eta_a &= 0 \\ \eta\dagger_a\eta\dagger_b + \eta\dagger_b\eta\dagger_a &= 0 \\ \eta_a\eta\dagger_b + \eta\dagger_b\eta_a &= \delta_{ab} \end{aligned} \tag{22}$$

The simplest type of symmetric operator for N particles is a sum from 1 to N of the *same* one-particle operator $f(i)$ for each particle

$$F^{(1)} = \sum_{i=1}^{N} f(i) \tag{23}$$

Next in simplicity is a type consisting of over all pairs of different particles, each counted once, of a symmetric two-particle operator, $f(i,j) = f(j,i)$,

$$F^{(2)} = \sum_{i<j=1}^{N} f(i,j) \tag{24}$$

These are the first two members of a hierarchy of which the general member is

$$F^{(\kappa)} = \sum_{i<j<\ldots 1}^{N} f(i,j,k,\ldots) \tag{25}$$

in which $f(i,j,k,\ldots)$ is a symmetric function of κ particle indices. Such multiparticle interactions have not been found in nature so we confine attention to $F^{(1)}$ and $F^{(2)}$.

For $\kappa = 1$ the general matrix element is

$$\begin{aligned}\langle n_1', n_2', \ldots |F^{(1)}| n_1, n_2, \ldots\rangle \\ = (N!)^{-1} \sum_{P'P} \sum_{i} \int \ldots \int \phi^*(1, P'1)\, \phi^*(1, P'2) \ldots f(i)\phi(1, P1) \\ \times \phi(1, P2) \ldots dx_1\, dx_2 \ldots dx_N\end{aligned}$$

for the boson case. Here there are N factors involving $\phi^*(\ ,\ )$ the factor $\phi^*(1,\ )$ appearing n_1' times, and so on. Likewise there are N factors involving $\phi(\ ,\ )$, with the factor $\phi(1,\ )$ occurring n_1 times. A simple calculation shows that the diagonal element is

$$\langle n_1, n_2, \ldots |F^{(1)}| n_1, n_2, \ldots\rangle = \sum_a n_a \langle a|f|a\rangle \tag{26}$$

All nondiagonal elements vanish except those in which one of the n's increases by unity, another decreases by unity and the others are unchanged, with the result

$$\langle \ldots n_a \ldots n_b - 1 \ldots |F^{(1)}| \ldots n_a - 1 \ldots n_b \ldots\rangle = \sqrt{n_a n_b}\,\langle a|f|b\rangle \tag{26$'$}$$

in which the one-particle matrix elements are

$$\langle a|f|b\rangle = \int \phi^*(a;x) f(x) \phi(b;x)\, dx \tag{27}$$

These results have a simple expression in terms of creation and destruction operators. The matrix elements of $F^{(1)}$ in (26) and (26$'$) are the same as those of

$$F^{(1)} = \sum_{a,b} \eta^\dagger_a \langle a|f|b\rangle \eta_b \tag{28}$$

This result is easily generalized to apply to the κ-particle case. For $\kappa = 2$ the result is

$$F^{(2)} = \frac{1}{2} \sum_{a,b,c,d} \eta^\dagger_a \eta^\dagger_b \langle a, b|f^{(2)}|c, d\rangle \eta_d \eta_c \tag{29}$$

in which

$$\langle a, b|f^{(2)}|c, d\rangle = \iint \phi^*(a; x_1)\phi^*(b; x_2)f(x_1, x_2)\phi(c; x_1)\phi(d; x_2)\, dx_1\, dx_2 \tag{30}$$

The same expressions apply to the fermion case when the fermion η's are used in place of the boson η's.

In the fermion case the expression analogous to (19) takes a simpler form, because all of the n's are 0 or 1,

$$\{a, b, \ldots, n\}_N = \eta^\dagger{}_n \ldots \eta^\dagger{}_b\eta^\dagger{}_a|0\rangle \tag{31}$$

but here it is necessary to apply the creation operators in standard order, or to make proper allowance for the sign change involved if they are applied in an order which is an odd permutation of the standard order.

The normal states of boson and fermion systems are quite different. For a boson system, the lowest state is that in which all of the particles are in the one-particle state of lowest energy. For a fermion system, the N states of lowest energy are occupied, one particle per state from the lowest up to a highest, up to an energy W_F, such that there are N one-particle states of lower energy. States of energy higher than W_F are unoccupied. The set of filled states is often called the *Fermi sea* and the energy W_F is called the *Fermi energy*.

Excitation of a fermion system consists in taking one or more particles from the Fermi sea and putting them in higher states that are normally unoccupied. Each such excitation puts a particle in a normally unoccupied state and generates a vacancy or *hole* within the sea of normally occupied states.

It is sometimes convenient to reckon the excited states from the normal state rather than from the vacuum state. When this is done the destruction operators, η_a, for the normally occupied states are regarded as creation operators for holes, and the creation operators, $\eta^\dagger{}_a$, for such states are regarded as destruction operators for holes.

8². Perturbation theory: variation method

There are not many problems for which the proper energy values can be found by direct analytical solution of

$$H|W'\rangle = W'|W'\rangle \tag{1}$$

Most discussion of atomic systems is therefore based on approximation procedures, known as perturbation theories.

One of the most widely useful of these is called the Rayleigh-Ritz variation method [15a,b]. For any state vector $|a\rangle$, $\langle a|H|a\rangle$ is the mean value of the energy for that state. If the solutions of (1) were known, $|a\rangle$ could be expanded in its proper states

$$|a\rangle = \sum_{W',A'} |W', A'\rangle\langle W', A'|a\rangle \tag{2}$$

giving

$$\langle a|H|a\rangle = \sum_{W',A'} W'|\langle W', A'|a\rangle|^2 \tag{3}$$

From this it is clear that $\langle a|H|a\rangle$, regarded as a function of its expansion coefficients, has an absolute minimum, equal to the lowest energy level, W_0, when all other expansion coefficients vanish, and this particular one is equal to unity.

This statement is true for work on any basis. Therefore, working on the Schrödinger basis as a special example, one can use a trial function $\langle \mathbf{r}_1 \ldots \mathbf{r}_N|a\rangle$ depending on a finite number of free parameters, λ, μ, ν, . . . , in which case $\langle a|H|a\rangle$ becomes a function of these parameters. By choosing the parameters to minimize $\langle a|H|a\rangle$ an upper bound is found for the value of the unknown lowest energy level. If the family of trial functions happens to include the exact unknown $\langle \mathbf{r}_1 \ldots \mathbf{r}_N|W_0\rangle$, then the $\langle a|H|a\rangle$, when minimized, gives W_0 exactly, and the corresponding values of the parameters make $\langle \mathbf{r}_1 \ldots \mathbf{r}_N|a\rangle$exactly equal to the $\langle \mathbf{r}_1 \ldots \mathbf{r}_N|W_0\rangle$.

Even when the family of trial solutions does not include the exact function, an upper bound is obtained for W_0 and the corresponding $\langle \mathbf{r}_1 \ldots \mathbf{r}_N|a\rangle$ is an approximation to $\langle \mathbf{r}_1 \ldots \mathbf{r}_N|W_0\rangle$ which may be useful. In practice, one can introduce trial solutions containing more and more parameters and use modern computers to carry out the minimizing process. Calculations of high accuracy have been made in this way on the normal states of two-electron atoms [15c].

As additional parameters are introduced, if the minimum $\langle a|H|a\rangle$ seems to approach a limit, there is a strong temptation to suppose that this limiting value is W_0. This is not a valid inference. Despite the extra flexibility gained by introducing more parameters, it could be that all of the trial functions differ from the true one in some essential way. In that case the procedure could produce a limiting $\langle a|H|a\rangle_{\min}$, which remains definitely higher than W_0.

However, if γ stands for an operator, or several operators, which commute with H and whose proper values and proper states are exactly known, these can be used as a partial guide to the form of the trial functions. These can be prepared in a form such that they are accurately orthogonal to each other when they relate to unequal values of γ'. Then when $\langle a|H|a\rangle_{\min}$ is calculated it gives an upper bound to the least energy of each of the series of states for various values of γ'. In this way the variation method has been usefully extended to the calculation of some excited energy levels.

The Rayleigh-Ritz method also provides a procedure for obtaining excited energy levels corresponding to equal values of γ. Trial functions, which correspond to a given value of γ and which are accurately orthogonal to each other, can easily be formed. Then, for a given value of γ, one of these trial functions, $|a_\gamma\rangle$, will provide the lowest value $\langle a_\gamma|H|a_\gamma\rangle = W_{\gamma 0}$. Thus, $W_{\gamma 0}$ is a lowest bounds for energy levels of a given γ, that is, of the same symmetry. Trial functions $|b_\gamma\rangle$ that are orthogonal to $|a_\gamma\rangle$, that is, $\langle b_\gamma|a_\gamma\rangle = 0$, may then be used to obtain upper bounds of excited states of the same symmetry $\langle b_\gamma|H|b_\gamma\rangle = W_{\gamma 1}$ [15d].

9². Perturbation theory: discrete energy levels

Although it is natural to suppose that if the Hamiltonian depends on a parameter λ, its proper values and proper states depend continuously on λ, this is not generally true. The study of this problem presents many mathematical difficulties which have not been completely elucidated. Although a formal perturbation procedure can be written down for an expansion of the proper states and proper values in power of λ, the method becomes too complicated for practical use except at the first one or two levels of approximation.

The method [16] was originally developed by Lord Rayleigh for acoustical problems, in a much more specialized context, and was adapted to quantum mechanics by Schrödinger. It has a wide range of practical usefulness in atomic theory.

The Hamiltonian is written

$$H = H_0 + \lambda H_1 \tag{1}$$

where the proper values and states of H_0 are known, as a rigorously solved problem, and H_1 is called the perturbation.

It is convenient to divide the commuting operators which are constants of the motion for H_0 into three groups denoted as α, β, γ. Let W be an allowed value of H and assume for its ψ an expansion in terms of the ψ_0's of H_0. Here we depart from the bra and ket notation to conform to more common usage

$$\psi = \sum_{\alpha'\beta'\gamma'} \psi_0(\alpha', \beta', \gamma')C(\alpha', \beta', \gamma') \tag{2}$$

The C's and W are found by approximate treatment of an infinite system of linear equations which result from using (2) in $H\psi = W\psi$, then using Equation (1²11), and finally equating the coefficient of each $\psi_0(\alpha', \beta', \gamma')$ to zero:

$$[W_0(\alpha', \beta', \gamma') - W]C(\alpha', \beta', \gamma') + \sum_{\alpha''\beta''\gamma''} \langle \alpha', \beta', \gamma'|\lambda H_1|\alpha'', \beta'', \gamma''\rangle C(\alpha'', \beta'', \gamma'') = 0 \tag{3}$$

The grouping of quantum numbers is as follows: α stands for all of the constants of the motion which commute with H_1 as well as H_0. In consequence the matrix elements of H_1 in (3) are diagonal in the α's, so the sum on α'' gives but one term, that for $\alpha'' = \alpha'$. In consequence system (3) breaks up into a series of uncoupled sets of equations, one set for each value of α'. These can, therefore, be discussed separately, one system at a time. To simplify the notation, the particular value of α' in the set under consideration will not be explicitly written.

The operators α are *exact* quantum numbers for the perturbed states as well as the unperturbed.

In (3) the diagonal matrix elements of H_1 occur additively to $W_0(\alpha', \beta', \gamma')$. Therefore they may be taken into account exactly by writing

$$W_1(\beta', \gamma') = W_0(\beta', \gamma') + \langle\beta', \gamma'|\lambda H_1|\beta', \gamma'\rangle \tag{4}$$

and writing (3) in the form

$$[W_1(\beta', \gamma') - W]C(\beta', \gamma') + \sum_{\beta''\gamma''}{}' \langle\beta', \gamma'|\lambda H_1|\beta'', \gamma''\rangle C(\beta'', \gamma'') = 0 \tag{3'}$$

where prime on the summation indicates that the diagonal term

$$\langle\beta', \gamma'|\lambda H_1|\beta', \gamma'\rangle$$

is excluded.

The distinction between the β and the γ labels is that $W_1(\beta', \gamma')$ changes considerably for a change in β' but changes only a small amount or not at all with a change in γ'. This amounts to saying that the $W_1(\beta', \gamma')$ values occur in *clusters* that are labeled by different β' values, with individual states within a cluster labeled by γ' values.

More precisely, the relevant quantity is the ratio

$$\frac{\langle\beta', \gamma'|\lambda H_1|\beta'', \gamma''\rangle}{W_1(\beta', \gamma') - W_1(\beta'', \gamma'')} \tag{5}$$

The classification by β and γ is that this ratio is small compared with unity for $\beta'' \neq \beta'$ and for all γ'' and γ', but is comparable with or even large compared with unity for $\gamma'' \neq \gamma'$ when $\beta'' = \beta'$. The effectiveness of a matrix element of λH_1 in producing a change in the states of $H_0 + \lambda H_1$ is measured by the ratio in (5). A nondiagonal element is more effective when it connects two states of nearly equal energy than when it connects states having widely different energies.

We may picture the perturbation as being turned on by increasing λ from zero. At $\lambda = 0$, the levels in a given β cluster (β'') are given by the unperturbed values $W_0(\beta''', \gamma''')$. The first-order effects in λ are given by using the $W_1(\beta''', \gamma''')$. At relatively small values of λ, the perturbed states become linear combinations of $\psi_0(\beta''', \gamma''')$ for various γ''' belonging to states in the β''' cluster, with coefficients $C(\beta''', \gamma''')$ of the same order of magnitude. This means that, relatively speaking, the γ's are *not good* quantum numbers for the perturbed states of the cluster. The γ's cease to be good quantum numbers when λ has reached such a value that the within-cluster matrix elements of λH_1 are comparable with the within-cluster first-order energy differences $[W_1(\beta''', \gamma') - W_1(\beta''', \gamma''')]$.

In the extreme case that the β''' cluster is partially or wholly degenerate, that is, $W_1(\beta''', \gamma''')$ independent of γ''', then even at the outset, with λ very small, the perturbed states involve a linear transformation away from the $\psi_0(\beta''', \gamma''')$ to such an extent that they need different labels, say ϵ''' instead of γ'''.

In a state belonging to the β''' cluster, the C's with $\beta' \neq \beta'''$ are quite small, but not vanishingly so, so the perturbed $\psi(\beta''')$ is no longer exactly labeled by the β''' value, but is to some extent *contaminated* by the presence of other β values in its expansion in terms of the $\psi_0(\beta', \gamma')$. But because all of the C's for $\beta' \neq \beta'''$ are rather small, we say that β''' is a *good* (rather than an *exact*) quantum number.

This concept plays an important role in atomic theory. In many cases atomic states are labeled by good quantum numbers rather than exact ones, recognizing that the state really involves a variety of values of the good labels which are good only in the sense that these values predominate in the expansion of ψ in terms of the ψ_0's.

To summarize, the α's are exact for H as well as H_0, the β's are exact for H_0 but only good for H, and the γ's are not good for H except perhaps for very small values of λ.

The analysis supporting this qualitative discussion will now be given.

For simplicity, consider first the case in which the β''' cluster consists of but one state, so no γ'''label is necessary. Its perturbed energy W is close to $W_1(\beta''')$. Therefore, in those equations of the set (3′) for which $\beta' \neq \beta'''$, the energy differences $|W_1(\beta', \gamma') - W|$ are large and the $C(\beta', \gamma')$ small. Hence, in these equations each of the terms in the Σ' are small of second order and can be neglected, except the particular one for which $\beta'' = \beta'''$. This gives the approximate values (for $\beta' \neq \beta'''$)

$$C(\beta', \gamma') = \frac{\langle \beta', \gamma' | \lambda H_1 | \beta''' \rangle C(\beta''')}{W_1(\beta''') - W_1(\beta', \gamma')} \tag{6}$$

By hypothesis each of these is small compared with unity. In this approximation, the perturbed ψ can be labeled by $\beta'''$and is

$$\psi_1(\beta''') = C(\beta''') \left[\psi_0(\beta''') + \sum_{\beta'\gamma'}{}' \frac{\psi_0(\beta', \gamma') \langle \beta', \gamma' | \lambda H_1 | \beta''' \rangle}{[W_1(\beta''') - W_1(\beta', \gamma')]} \right] \tag{7}$$

The values of $|C(\beta''')|$ are found by normalization of $\psi_1(\beta''')$, and an arbitrary phase choice is made. Because of the orthonormality of the $\psi_0(\beta', \gamma')$,

$$1 = |C(\beta''')|^2 \left[1 + \sum_{\beta'\gamma'}{}' \frac{|\langle \beta', \gamma' | \lambda H_1 | \beta''' \rangle|^2}{[W_1(\beta''') - W_1(\beta', \gamma')]^2} \right] \tag{8}$$

is the equation determining $|C(\beta''')|$.

The perturbed energy is found from the $\beta' = \beta'''$member of the set (3), using in it the values of the C's from (6) and choosing W so that it is satisfied with $C(\beta''') \neq 0$. The result is

$$W_2(\beta''') = W_1(\beta''') + \sum_{\beta'\gamma'}{}' \frac{|\langle \beta', \gamma' | \lambda H_1 | \beta''' \rangle|^2}{W_1(\beta''') - W_1(\beta', \gamma')} \tag{9}$$

This expression formally contains terms to order λ^2 but its actual dependence on λ is not exactly quadratic because of the appearance of λ in the energy difference denominators. In special cases this may produce large effects, if the denominator becomes small or vanishes, even though for $\lambda = 0$ there was a large difference in the $W_0(\beta''') - W_0(\beta', \lambda')$. Effects of this kind go by the name of *level-crossing phenomena*.

The form of (9) shows that $W_2(\beta''')$ is increased by interaction with the states whose $W_1(\beta', \gamma')$ is lower than $W_1(\beta''')$, and is decreased by interaction with those above. This feature may be described by saying the perturbation causes the levels to *repel* each other on the energy scale.

Suppose that the β' cluster consists of $g(\beta')$ states for this number of distinct values of γ', so $g(\beta')$ is the statistical weight of the cluster. By hypothesis each of the $W_1(\beta', \gamma')$ is close to the mean,

$$W_1(\beta', \text{Av}) = [g(\beta')]^{-1} \sum_{\gamma'} W_1(\beta', \gamma')$$

and therefore in (9) it is a good approximation to replace each $W_1(\beta', \gamma')$ by the average $W_1(\beta', \text{Av})$. The numerator of the terms for the β' cluster is then

$$\sum_{\gamma'} |\langle \beta', \gamma' | \lambda H_1 | \beta''' \rangle|^2$$

By $1^2$40 this sum is invariant to any unitary transformation of the states in the β' cluster that is occasioned by the action of λH_1 on it.

We therefore have the important result that $W_2(\beta''')$ is not affected even though λH_1 is strong enough to produce a drastic change in the wave functions of the states within the β' cluster. This is particularly important because if we happen to be interested only in the calculation of $W_2(\beta''')$, we do not need to go to the trouble of finding the strongly perturbed states within a cluster that is well separated from $W_1(\beta''')$.

This concludes the discussion of the perturbation of a "cluster" that consists of one state. The calculation will now be extended to the case in which the β''' cluster consists of $g(\beta''')$ states, labeled by this number of distinct values of γ'''.

As above, the $C(\beta'', \gamma''')$ for $\beta'' \neq \beta'''$ are small, whereas the $C(\beta''', \gamma''')$ for each γ''' are of the order unity. Using the equation (3′) we now get, as the analogue of (6)

$$C(\beta', \gamma') = \sum_{\gamma''} \frac{\langle \beta', \gamma' | \lambda H_1 | \beta''', \gamma'' \rangle C(\beta''', \gamma'')}{W_1(\beta''', \gamma'') - W_1(\beta', \gamma')} \tag{6'}$$

When these values are used in each of the $g(\beta''')$ equations of the set (3′) for which $(\beta', \gamma') = (\beta''', \gamma')$ there results the following set of $g(\beta''')$ simultaneous linear equations for the $C(\beta''', \gamma')$:

$$[W_1(\beta''', \gamma') - W]C(\beta''', \gamma') + \sum_{\gamma'''}{}' [\beta''', \gamma' | \lambda H_1 | \beta''', \gamma'''] C(\beta''', \gamma''') = 0 \tag{10}$$

in which the brackets [ ] stand for the perturbed-matrix elements,

$$[\beta''', \gamma' | \lambda H_1 | \beta''', \gamma'''] = \langle \beta''', \gamma' | \lambda H_1 | \beta''', \gamma''' \rangle + \sum_{\beta''\gamma''\ (\beta'' \neq \beta''')}{}' \frac{\langle \beta''', \gamma' | \lambda H_1 | \beta'', \gamma'' \rangle \langle \beta'', \gamma'' | \lambda H_1 | \beta''', \gamma''' \rangle}{W_1(\beta''', \gamma''') - W_1(\beta'', \gamma'')} \tag{11}$$

For the set (10) to have a nontrivial solution W must satisfy the *secular equation* formed by equating to zero the determinant of the coefficients in (10). The matrix of coefficients is Hermitian, so the roots are real.

Let them be denoted by $W_2(\beta''', \epsilon')$ where ϵ' takes on $g(\beta''')$ distinct values. For each ϵ' the set (10) can be solved for a set of ratios of the C's, which we denote by $C(\beta''', \gamma'; \epsilon')$. These can be normalized so that

$$\sum_{\gamma'} C^*(\beta''', \gamma'; \epsilon') C(\beta''', \gamma'; \epsilon'') = \delta(\epsilon', \epsilon'') \tag{12}$$

with one arbitrary phase choice for each ϵ'.

It may happen that the secular equation has multiple roots. In that case the β''' cluster of $g(\beta''')$ states is partially degenerate with statistical weights $g(\beta''', \epsilon')$ whose sum equals $g(\beta''')$. Within each degenerate subspace the choice of basis is arbitrary.

After the $C(\beta''', \gamma'; \epsilon')$ have been determined, the corresponding perturbed states are (from 6′)

$$\psi_1(\beta''', \epsilon') = \sum_{\gamma''} C(\beta''', \gamma''; \epsilon') \left[\psi_0(\beta''', \gamma'') + \sum_{\beta'\gamma'} \frac{\psi_0(\beta', \gamma') \langle \beta', \gamma' | \lambda H_1 | \beta''', \gamma'' \rangle}{W_1(\beta''', \gamma'') - W_1(\beta', \gamma')} \right] \tag{13}$$

Here β''' is a good quantum number, but γ' has lost its validity and has had to be replaced by ϵ'.

As with the discussion of (9), in (11) we can replace the individual $W_1(\beta'', \gamma'')$ in the denominator by their average $W_1(\beta'', \text{Av})$ in which case the sum over γ'' is over quantities in the numerator only,

$$\frac{\sum_{\gamma''}(\beta''', \gamma'|\lambda H_1|\beta'', \gamma'')(\beta'', \gamma''|\lambda H_1|\beta''', \gamma''')}{W_1(\beta''', \gamma''') - W_1(\beta'', \text{Av})}$$

By 1^240 this sum is invariant to choice of basis states in the β'' cluster and so can be calculated from the original β'', γ' states even though these are strongly perturbed.

The values of $W_2(\beta''', \epsilon')$ are obtained by solution of the secular equation, and therefore it is inconvenient to give explicit formulas for them [and impossible for $g(\beta''') > 4$]. Nevertheless, it is often useful to have an explicit formula for the mean of these roots, $W_2(\beta''', \text{Av})$. This can be done because of the invariance of the trace of a matrix to a unitary transformation of its basis as shown in 1^239. Hence, the mean of the roots is the mean of the diagonal elements of the matrix of the coefficients in (11). Therefore,

$$W_2(\beta''', \text{Av}) = [g(\beta''')]^{-1}\left[\sum_{\gamma'} W_0(\beta''', \gamma') + \lambda\sum_{\gamma'}(\beta''', \gamma'|\lambda H_1|\beta''', \gamma') + \lambda^2\sum_{\gamma'}\frac{\sum_{\beta''\gamma''}|(\beta''', \gamma'|\lambda H_1|\beta'', \gamma'')|^2}{W_1(\beta''', \text{Av}) - W_1(\beta'', \text{Av})}\right] \tag{14}$$

This is closely analogous to (9) for the one-state cluster.

10^2. Perturbation theory: change of coupling

Continuing the discussion of the preceding section, certain aspects of the within-cluster behavior of states and energy levels deserve more explicit discussion because they play an important role in atomic physics.

Attention may be confined to a particular β'''cluster of $g(\beta''')$ states, which therefore define a β''' subspace of the system-space. By hypothesis each of the $W_0(\beta''', \gamma')$ has a small deviation from their mean, which is denoted by $W_0(\beta''', \text{Av})$.

It is convenient to separate H_0 into two parts, $H_0 = H_{00} + H_{01}$ such that H_{00} is degenerate in γ' with the proper value $W_0(\beta''', \text{Av})$ having a $g(\beta''')$ order of degeneracy. Then H_{01} is defined as an operator in this subspace whose proper values are $W_0(\beta''', \gamma') - W_0(\beta''', \text{Av})$, and whose proper states are the $|\beta''', \gamma')$ of H_0.

If the Hamiltonian had been H_{00}, the states within the β''' subspace would be quite indeterminate, or labile, or uncoupled. Any choice of a basis within the subspace would be as good as any other. The part H_{01}

provides a *coupling* of the states in the β''' subspace. But because the energy differences $W_0(\beta''', \gamma') - W_0(\beta''', \text{Av})$ are small we say that the coupling is weak. Situations can also arise in which H_{01} does not completely remove the degeneracy, so that (β''', γ') has a degeneracy of order $g(\beta''', \gamma')$ where $\Sigma_{\gamma'} g(\beta''', \gamma') = g(\beta''')$. In that case H_{01} does not provide a complete coupling of the states but leaves them indertermi-nate within the (β''', γ') subspaces.

That part H_1 that is made up of within-cluster matrix elements, the part that is diagonal in β''', consisting of $g(\beta''')$ rows and columns,

$$(\beta''', \gamma'|\lambda H_1|\beta''', \gamma'') \tag{1}$$

defines a coupling within the β''' subspace that is different from that provided by H_{01}. A system whose Hamiltonian is $H_{00} + \lambda H_1$, with omission of H_{01}, would have energy values given by the proper-values of the matrix (1), and proper states within the $g(\beta''')$ dimensional sub-space defined as the proper states of the matrix (1). These may be labeled by ϵ', so that the proper values of (1) are $\lambda W_1(\beta''', \epsilon')$, accurately linear in λ. Therefore, the proper values of $H_{00} + \lambda H_1$ are

$$W_0(\beta''', \text{Av}) + \lambda W_1(\beta''', \epsilon') \tag{2}$$

and the associated proper states are $|\beta''', \epsilon'\rangle$.

Thus H_{01} sets up one kind of coupling in the β''' subspace and H_1 sets up a different kind, when either one acts alone. H_{01} defines the states $|\beta''', \gamma'\rangle$ with levels

$$W_0(\beta''', \text{Av}) + [W_0(\beta''', \gamma') - W_0(\beta''', \text{Av})]$$

and proper states $|\beta''', \gamma'\rangle$, while H_1 defines the states with the levels (2).

When both H_{01} and λH_1 act in the Hamiltonian with comparable strengths the proper states and energy levels are those defined as the proper states of

$$(\beta''', \gamma'|H_{01} + \lambda H_1|\beta''', \gamma'') \tag{3}$$

For each λ this defines a set of states of *intermediate coupling*. For $\lambda \to 0$ the states are the proper states of H_{01} and for large λ the states are those defined by H_1. As λ is increased parametrically from zero to a large value the proper states gradually turn in the β''' subspace from the $|\beta''', \gamma'\rangle$ set toward the $|\beta''', \epsilon'\rangle$ set.

This may be regarded as a kind of competition between H_{01} and λH_1, each of which is "trying" to establish its own coupling scheme. What actually results is an intermediate scheme that depends on the relative strengths of these two parts of the Hamiltonian.

This kind of competition between coupling schemes is encountered again and again in atomic physics. For example, the proper states of the Hamiltonian, $2^2 9$, define a kind of coupling known as *Russell-Saunders coupling.* A more accurate Hamiltonian includes magnetic interaction terms. When these are weak it produces a small separation of levels in the degenerate terms in Russell-Saunders coupling, and defines a particular choice of states within the subspace of such a term. But when the magnetic interaction is of greater strength the state vectors are changed over to a condition of intermediate coupling.

Another example of a change of coupling is that which occurs in the Zeeman effect, the perturbation of levels and states produced by an external magnetic field. For weak fields the perturbation defines a particular choice of states, and the perturbed levels are linear in the magnetic field. As the field is increased, there is a gradual change to a different coupling scheme. The coupling scheme that is valid for large fields is called the *Paschen-Back* effect, and the perturbed levels again become linear in the magnetic field.

Still another example is found in the action on an ion in a crystal due to the electric fields of its neighbors, commonly called ligand fields. When these are weak the coupling is the same as that in a free ion, not perturbed by external fields, but as the ligand field becomes stronger, there is a change of coupling to states whose nature is largely determined by the crystal symmetry of the ligand field.

11^2. Perturbation theory: time dependent

When the Hamiltonian of a system depends explicitly on the time, the state of the system makes transitions between what would otherwise be stationary states [17]. We suppose that

$$H = H_0 + H_t \tag{1}$$

where H_0 represents the unperturbed part whose states $|a\rangle$, $|b\rangle$, ... and energy levels W_a, W_b, ... satisfying

$$H_0|a\rangle = W_a|a\rangle \tag{2}$$

are supposed known. The time-dependent perturbing part can then be described by its matrix elements on the $|a\rangle$ basis,

$$H_t|a\rangle = \sum_b |b\rangle\langle b|H_t|a\rangle \tag{3}$$

The Schrödinger description (Sec. 2^2) is being used so the $|a\rangle$, $|b\rangle$ ... are not time dependent, but the matrix elements in (3) depend on the time because of the time dependence of H_t.

The problem is to find solutions of the dynamical equation

$$i\hbar\frac{\partial}{\partial t}|\ \rangle_t = (H_0 + H_t)|\ \rangle_t \tag{4}$$

for a special solution that corresponds to the system's being certainly in the state $|a\rangle_0$ at $t = 0$. Such a solution can be described by time-varying components on the fixed vectors of the $|a\rangle$ basis

$$|a\rangle_t = \sum_b |b\rangle\langle b|a\rangle_t \tag{5}$$

with the initial conditions

$$\langle b|a\rangle_0 = \delta(b, a) \tag{5'}$$

Substitution of (5) in (4) and equating coefficients of $|b\rangle$ on both sides gives a set of simultaneous linear first-order differential equations for the $\langle b|a\rangle_t$,

$$i\hbar \frac{\partial}{\partial t}\langle b|a\rangle_t = [W_b + \langle b|H_t|b\rangle]\langle b|a\rangle_t + \sum_c{}' \langle b|H_t|c\rangle\langle c|a\rangle_t \tag{6}$$

in which Σ' means that the term $c = b$ is omitted, because it has been explicitly written on the first line.

The first term on the right gives rise to the usual time dependence of the expansion coefficients, modified by the diagonal elements of the perturbation. We write

$$W_{bt} = W_b t + \int_0^t \langle b|H_t|b\rangle\, dt \tag{7}$$

Also

$$\langle b|a\rangle_t = [b|a]_t \exp\frac{-iW_{bt}}{\hbar} \quad \text{and} \quad \langle c|a\rangle_t = [c|a]_t \exp\frac{-iW_{ct}}{\hbar}$$

and

$$[b|H_t|c] = \langle b|H_t|c\rangle \exp\frac{i(W_{bt} - W_{ct})}{\hbar}$$

or

$$\langle b|H_t|c\rangle = [b|H_t|c] \exp\frac{-i(W_{bt} - W_{ct})}{\hbar}$$

and find that the equations for the $[b|a]_t$ are

$$i\hbar\frac{\partial}{\partial t}[b|a]_t = \sum_c{}' [b|H_t|c][c|a]_t \tag{8}$$

These equations show that all time variation of the $[b|a]_t$ arises from the nondiagonal elements $[b|H_t|c]$, which include the time variation arising from the unperturbed levels and the time-dependent diagonal matrix elements of H_t.

If we find a solution of (8) satisfying the initial conditions (5′) then the mean value at t of γ, an operator representing and observable, is

$$\begin{aligned}\gamma_{Av}(t) &= \sum_{b,c} \langle a|b\rangle\langle b|\gamma|c\rangle\langle c|a\rangle \\ &= \sum_{b,c} [a|b][b|\gamma|c][c|a]\end{aligned} \tag{9}$$

In particular, if γ is defined as having the value 1 for $|b\rangle$ and 0 for all other states so $\langle b|\gamma|c\rangle = \delta(b, c)$ then for this γ, $\gamma_{Av}(t)$ is the probability that the system is in the state b at time t, having been certainly in a at $t = 0$. Writing $P(a \to b)$ for this induced transition probability we have

$$P(a \to b) = |\langle b|a\rangle|^2 = |[b|a]|^2 \tag{9′}$$

Time-dependent perturbation theory undertakes to provide an approximate solution of (8) on the supposition that H_t is small. If the $[b|H_t|c]$ are sufficiently small then the rate of change of $[b|a]_t$ is small, so in (8) we can substitute the initial values on the right. This gives, approximately,

$$[b|a]_t = -\frac{i}{\hbar}\int_0^t [b|H_t|a]\, dt \tag{10}$$

Therefore, in this approximation

$$P(a \to b) = h^{-2}\left|\left|\int_0^t [b|H_t|a]\, dt\right|\right|^2 \tag{11}$$

Clearly this approximation is valid only if it gives

$$\sum_b{}' P(a \to b) << 1 \tag{12}$$

Otherwise it would imply a large decrease in $[a|a]_t$ from its initial value 1 and a considerable buildup in the other $[b|a]_t$ that were initially zero, contradicting the assumption leading to (10).

An important property of (11) is its symmetry between initial and final states. Interchange of b and a gives $P(b \to a)$, the probability of making the transition to a if initially in b. Because of the Hermitian character of H_t we have from (11)

$$P(b \to a) = P(a \to b) \tag{13}$$

Thus, if $\langle b|H_t|a\rangle$ vanishes, then $P(a \to b)$ vanishes, in this first approximation. Let c stand now for all of those states for which both

$$\langle c|H_t|a\rangle \quad \text{and} \quad \langle b|H_t|c\rangle$$

have nonvanishing values. Then values of $[c|a]_t$ are given in first approximation by (10) with c in place of b. Using these values in (8) we get a second approximation for $[b|a]_t$

$$[b|a]_t = -i\hbar^{-2} \int_0^t \sum_c [b|H_t|c] \int_0^{u=t} [c|H|a]\, du\, dt \tag{10'}$$

which gives a nonvanishing $P(a \to b)$ when used in (9′). It is a second approximation in the sense that (10′) depends quadratically on the perturbation matrix elements instead of linearly as in (10).

Sometimes in the literature the result (10′) is described by saying that the system arrives at $|b\rangle$ by making a *virtual* transition from $a \to c$ followed by a second virtual transition from $c \to b$. This mode of description is to be avoided unless there is clear understanding that this is just a way of describing the structure of (10′) as quadratic in the matrix elements, in short, it must be clearly recognized that a virtual transition is not a real transition. The sum over c in (10′) indicates that there are many virtual paths $a \to c \to b$ that contribute to the second-order transition probability.

If there are no states c for which both $\langle c|H_t|a\rangle$ and $\langle b|H_t|c\rangle$ are nonvanishing then $[b|a]_t = 0$ in (10′) and the transition probability $P(a \to b)$ vanishes in the second approximation. But if there are states d and c such that

$$\langle c|H_t|a\rangle, \langle d|H_t|c\rangle \quad \text{and} \quad \langle b|H_t|d\rangle$$

all have nonvanishing values, then a third-order expression analogous to (10′) can be set up to give a third-order approximation to $P(a \to b)$, and so on for higher orders.

Returning now to (10), suppose that the time dependence of H_t is through a factor $\cos(\omega t + \delta)$ in the interval $0 < t < T$ and that $H_t = 0$ outside this interval. For brevity we write

$$\omega_{ba} = \frac{W_{bt} - W_{at}}{\hbar}$$

Then the time integral in (10) gives for $t < T$,

$$\int_0^T \frac{1}{2}\{\exp[i(\omega_{ba} + \omega)t]e^{i\delta} + \exp[i(\omega_{ba} - \omega)t]e^{-i\delta}\}\, dt$$
$$= \frac{1}{2}\left\{\frac{\exp[i(\omega_{ba} + \omega)T] - 1}{\omega_{ba} + \omega} e^{i\delta} + \frac{\exp[i(\omega_{ba} - \omega)T] - 1}{\omega_{ba} - \omega} e^{-i\delta}\right\}$$

Writing

$$\theta_1 = \frac{1}{2}(\omega_{ba} + \omega)T, \quad \theta_2 = \frac{1}{2}(\omega_{ba} - \omega)T$$

we get

$$[b|a]_T = \frac{\langle b|H_t|a\rangle T}{2i\hbar}\left[\frac{\sin\theta_1}{\theta_1}\exp i(\theta_1+\delta) + \frac{\sin\theta_2}{\theta_2}\exp i(\theta_2-\delta)\right]$$

from which

$$P(a\to b) = \hbar^{-2}|\langle b|H_t|a\rangle|^2\left(\frac{T}{2\hbar}\right)^2\left[\left(\frac{\sin\theta_1}{\theta_1}\right)^2 + \left(\frac{\sin\theta_2}{\theta_2}\right)^2 + 2\frac{\sin\theta_1}{\theta_1}\frac{\sin\theta_2}{\theta_2}\cos(\theta_1-\theta_2+2\delta)\right] \tag{14}$$

The $(\sin\theta)/\theta$ factors for large T are sharply peaked on a frequency scale at $\theta = 0$, that is, near $\nu_{ba} = -\nu$ ($\omega_{ba} = -\omega$) for the one involving θ_1 and near $\nu_{ba} = +\nu$ ($\omega_{ba} = \omega$) for the other one. From maxima of unity they drop to zero at $|\nu_{ba}| = \nu \pm T^{-1}$ ($\omega_{ba} = \omega \pm 2\pi T^{-1}$). Hence, in practice, when ν_{ba} is near one of these values the other $(\sin\theta)/\theta$ is negligibly small, so the cross-product term is negligible. In any case it vanishes when averaged over δ.

Historically the Bohr rule

$$h\nu = |h\nu_{ba}| = |(W_b - W_a)|$$

with $W_b > W_a$ for absorption or $W_b < W_a$ for induced emission, was a postulate (Sec. 7¹). The preceding results amount to a derivation of this rule from the time-dependent perturbation theory, in that to get an appreciable value of $P(a\to b)$ the value of W_b must make either $(\sin\theta_1)/\theta_1$ for induced emission or $(\sin\theta_2)/\theta_2$ for absorption take on a value near its maximum. The derivation also shows that the Bohr rule need not be exactly fulfilled, but that transitions occur for a discrepancy between ν and $|\nu_{ba}|$ of the order of T^{-1}. This takes care of the peculiarity by which $P(a\to b)$ increases as T^2. If the perturbation, acting in the interval $0 < t < T$ has *exactly* the correct frequency then P does increase as T^2. But if the perturbation consists of a continuous spectrum of frequency components near the Bohr frequency $|\nu_{ba}|$, then as T increases, the band of effective frequencies narrows down as T^{-1}, giving a $P(a\to b)$ for their joint action that increases as T instead of T^2.

It is instructive to reach this same result in another way. Suppose the time dependence of H_t is through a factor $f(t)$ where $f(t) = 0$ outside of the active interval, $0 < t < T$. The slight extra generality that different parts of H_t have different time dependences, $H_t = \sum_\alpha H_{t\alpha}f_\alpha(t)$, can be included in what follows without essential change in the final results.

Then from (10) we have

$$[b|a]_T = -i\hbar^{-1}\langle b|H_t|a\rangle\int_0^T f(t)\exp(i\omega_{ba}t)\,dt$$

where the integration limits can be extended to $-\infty$ to $+\infty$, because $f(t)$ vanishes outside of $0 < t < T$. In that case we can write

$$[b|a]_T = -i\hbar^{-1}\langle b|H_t|a\rangle\sqrt{2\pi}g(\omega_{ba}) \tag{15}$$

where

$$g(\omega) = (2\pi)^{-1/2}\int_{-\infty}^{+\infty} f(t)\exp(i\omega t)\,dt$$

is the Fourier integral component of $f(t)$ for the radial frequency ω_{ba} given by the Bohr rule for the transition in question. Here

$$f(t) = (2\pi)^{-1/2}\int_{-\infty}^{+\infty} g(\omega)\exp(-i\omega t)\,d\omega$$

From its definition $g(\omega)$ is a time-interval so $[b|a]_T$ is a dimensionless number. Thus the part of H_t that is effective in producing the transition $a \to b$ is the Fourier integral component for the frequency $\omega_{ba} > 0$ for absorption or $\omega_{ba} < 0$ for induced emission. Hence, finally

$$P(a \to b) = 2\pi\left|\frac{\langle b|H_t|a\rangle}{\hbar}\right|^2 \cdot |g(2\pi\nu_{ba})|^2 \tag{16}$$

shows that the transition probability depends both on a nonvanishing $\langle b|H_t|a\rangle$ and on a frequency dependence that fulfills the Bohr rule.

An important special case is that in which $H_t = 0$ for $t < 0$ and

$$H_t = H_{t0}\exp\left(\frac{-\gamma t}{2}\right)\cos(\omega t + \delta) \tag{17}$$

for $t > 0$. In this case (10) gives

$$[b|a]_t = i(2\hbar)^{-1}\langle b|H_{t0}|a\rangle\left[\frac{e^{i\delta}}{i(\omega_{ba}+\omega)-(\gamma/2)} + \frac{e^{-i\delta}}{i(\omega_{ba}-\omega)-(\gamma/2)}\right]$$

for $t \to \infty$. Hence from (11), averaging over δ

$$P(a \to b) = (2\hbar)^{-2}|\langle b|H_{t0}|a\rangle|^2\left[\frac{1}{(\omega_{ba}+\omega)^2+(\gamma/2)^2} + \frac{1}{(\omega_{ba}-\omega)^2+(\gamma/2)^2}\right] \tag{18}$$

This resembles (14), with a different line shape factor, the first of the two terms in brackets gives absorption, the second induced emission. The line shape, dependence of P on ω that appears in this case, is known as the *Lorentz profile* [18].

In many cases the perturbation H_t in (1) does not depend on the time. The time development of the system can then be given in terms of the proper values of the unperturbed Hamiltonian H_0. Because, by (8),

$$i\hbar\frac{\partial}{\partial t}[b|a]_t = \langle b|H_t|a\rangle\exp(i\omega_{ba}t) \qquad b \neq a$$

one can easily perform the integration obtaining

$$[b|a]_t = \frac{\langle b|H_t|a\rangle[1 - \exp(i\omega_{ba}t)]}{W_b^{(0)} - W_a^{(0)}} \tag{19}$$

where the initial time is chosen to be zero, $t_0 = 0$, and $[b|b]_0 = 0$, $[a|a]_0 = 1$. The probability that the system will be in the state b at a time t becomes

$$|[b|a]_t|^2 = \frac{2|\langle b|H_t|a\rangle|^2(1 - \cos\omega_{ba}t)}{[W_b^{(0)} - W_a^{(0)}]^2} \tag{20}$$

which is an oscillating function of time.

If $|\langle b|H_t|a\rangle|/[W_b^{(0)} - W_a^{(0)}]$ is small, the proper states of the unperturbed Hamiltonian H_0 are nearly proper states of the total Hamiltonian H. Therefore, for small perturbations, if the energy difference between states a and b is large, the system will probably remain in the initial state. But if $W_b \approx W_a$, assuming that $|\omega_{ba}|t << 1$, we have

$$|[b|a]|^2 \approx \hbar^{-2}|\langle b|H_t|a\rangle|^2 t^2 \tag{21}$$

That is, the probability increases quadratically with the time. If there are many final states b such that $W_b \approx W_a$, as when the b states form a continuum, the total probability $\Sigma_b|[b|a]|^2$ approaches to unity rapidly. This situation is responsible for the *radiationless transitions* to the continuum, such as *Auger transitions,* predissociation, and *Coster-Kronig transitions.*

In order to replace the sums by integrals we introduce an energy density $\rho(W) = \rho(\hbar\omega)$, the number of states per unit energy interval. Then the total transition probability becomes

$$\sum_b |[b|a]|^2 = 2\int\left|\frac{\langle b|H_t|a\rangle}{W_b^{(0)} - W_a^{(0)}}\right|^2 (1 - \cos\omega_{ba}t)\rho[W_b^{(0)}]\,dW_b^{(0)} \tag{22}$$

The time rate of change of the above total probability is

$$\frac{\partial}{\partial t}\sum_b |[b|a]|^2 = \frac{2}{\hbar^2}\int |\langle b|H_t|a\rangle|^2 \frac{\sin\omega_{ba}t}{\omega_{ba}}\rho[W_b^{(0)}]\,dW_b^{(0)} \tag{23}$$

Usually the quantities $\langle b|H_t|a\rangle$ and $\rho[W_b^{(0)}]$ are fairly constant over the energy range ΔW in the neighborhood of the final energy W_b. Because $\Delta\omega = \Delta W/\hbar$, for times, t, much larger than $\hbar/\Delta W$ the term $(\sin\omega_{ba}t)/\omega_{ba}$ oscillates very rapidly. Assuming that $\Delta W \approx W_a$, the condition $t >> \hbar/\Delta W$ is easily satisfied. The term $(\sin\omega_{ba}t)/\omega_{ba}$ has a pronounced maximum at $W_b^{(0)} = W_a^{(0)}$, so (23) may be approximated as

$$\begin{aligned}\frac{\partial}{\partial t}\sum_b |[b|a]|^2 &= \frac{2}{\hbar}|\langle b|H_t|a\rangle|^2\rho[W^{(0)}]\int_{-\infty}^{\infty}\frac{\sin\omega_{ba}t}{\omega_{ba}}\,d\omega_{ba} \\ &= \frac{2\pi}{\hbar}|\langle b|H_t|a\rangle|^2\rho[W_a^{(0)}] \qquad t > 0\end{aligned} \tag{24}$$

where the limits of the integral are justified in view of the condition $t >> \hbar/W$.

Equation (24) is a compromise between the relation (21), which gives the probability for transitions where the unperturbed energy is strictly conserved ($W_a^{(0)} = W_b^{(0)}$) and changes quadratically with time, and the relation (20), which is for the transitions where the unperturbed energy is not conserved ($W_a^{(0)} \neq W_b^{(0)}$) and oscillates with time. The time rate of change of such a compromised transition probability is constant. Equation (24) is sometimes called the *golden rule of the time dependent perturbation.*

12^2. Wentzel-Brillouin-Kramers-Jeffreys approximation

Solution of the equation for radial factors of central-field wave functions is always done now with digital electronic computers. Nevertheless, the WBKJ method [19] of studying the radial functions remains of historical importance because it clarifies the connection of the radial wave equation and the original quantum condition of Sommerfeld-Wilson, 7^18, and also some other important features such as quantum mechanical tunneling, or barrier leakage. This way of treating one-dimensional wave problems also finds application in other branches of mathematical physics and is fully treated in a book by Heading [20] in which the critical historical account (pp. 1–24) is most illuminating.

Here we write $V(r)$ for the total acting radial potential energy, including the centrifugal barrier (Sec. 1^4), if any. The equation to be discussed is

$$-\frac{\hbar^2}{2m}P''(r) + V(r)P(r) = WP(r) \tag{1}$$

where $P(0) = 0$ and $V(r) \to 0$ as $r \to \infty$. Various special cases need to be considered for various applications. Here attention is confined to $V(r) < 0$ in most of the finite range of r, where $V(r) \to -\infty$ for s states, or to $V(r) \to +\infty$ for $l \neq 0$, at the origin.

This form of $V(r)$ is also applicable to the vibration-rotation states of diatomic molecules. For the barrier-leakage theory of α decay of nuclei the $V(r) \to +2Ze^2/r$ for large r, corresponding to repulsion of the α particle by the daughter nucleus (Z); accordingly some details of the following discussion need to be changed.

In atomic applications there are sometimes two "valleys" of $V(r)$, that is, ranges in which $W > V(r)$. In classical mechanics the motion remains entirely in one such range; if there are several ranges, initial conditions determine which one. But in quantum mechanics for a given allowed W there is a single $P(r)$ defined for the whole range, $0 < r < \infty$. This has the effect that motions in the separate valleys are linked together by the specifically quantum mechanical effect of barrier leakage.

In the WBKJ method we find approximate solutions of an oscillatory character in the ranges for which $W > V(r)$, and of a quasi-exponential character where $W < V(r)$. Both of these forms are inapplicable in the immediate neighborhood of roots of $W = V(r)$, and so the forms for $W > V(r)$ and $W < V(r)$ have to be joined together by bridging solutions involving Airy functions, which are closely related to Bessel functions of order 1/3.

In a range for which $W > V(r)$ we write

$$k^2(r) = \frac{2m}{\hbar^2}[W - V(r)]$$

so (1) becomes

$$P''(r) + k^2(r)P(r) = 0 \tag{2}$$

We seek a solution of the form

$$P(r) = \exp[i(S - iT)] \tag{3}$$

Substitution in (2) gives for $S(r)$ and $T(r)$,

$$(-S'^2 + k^2) + i(2S'T' + S'') + (T'^2 + T'') = 0$$

This can be approximately satisfied by neglecting T'^2 and T'', and choosing

$$S'(r) = \pm k(r), \quad S(r) = \pm \int_{r_2}^{r} k(r)\, dr \tag{4}$$

Here an arbitrary choice of integration constant is made: r_2 is the left boundary of the region in which $k(r)$ is real. Then setting $(2S'T' + S'')$ equal to zero we find an approximation for $T(r)$,

$$T'(r) = -\frac{1}{2}\left(\frac{k'}{k}\right) \quad \text{so} \quad T(r) = \ln \frac{A}{\sqrt{k(r)}} \tag{5}$$

where A is an integration constant. Thus in any range in which $W > V(r)$ the solution of (1) is approximated by

$$\begin{aligned} P(r) &= k^{-1/2}[A\, e^{iS(r)} + A^*\, e^{-iS(r)}] \\ &= k^{-1/2}|A| \cos [S(r) - \delta] \end{aligned} \tag{6}$$

This diverges where $k(r) \to 0$, and, moreover, the neglected quantities

$$T'^2 + T'' = \frac{3}{4}\left(\frac{k'}{k}\right)^2 - \frac{1}{2}\frac{k''}{k} \tag{7}$$

become infinite there, so (6) is not applicable near where $k(r) = 0$.

Similarly in a range where $W < V(r)$ we write

$$\kappa^2(r) = \frac{2m}{\hbar^2}[V(r) - W]$$

and

$$P''(r) - \kappa^2(r)P(r) = 0$$

and seek a solution of the form

$$P(r) = e^{(S+T)}$$

By a procedure similar to that just used, we find a quasi-exponential approximate solution

$$P(r) = \kappa^{-1/2}[Be^{S(r)} + Ce^{-S(r)}] \tag{8}$$

in which

$$S(r) = \int_{r_1}^{r} \kappa(r)\, dr \tag{9}$$

where r_1 is the left boundary of the region in which $W < V(r)$. This approximate solution also diverges near $\kappa(r) = 0$ and is not applicable near such a turning point of the classical motion.

To deal with the turning point regions, we consider first $r \approx r_2$ where $V'(r) < 0$ and assume that a linear approximation is adequate,

$$V(r) = V(r_2) - (r - r_2)f_2 \tag{10}$$

with $f_2 > 0$. Similarly near $r = r_1$ we have

$$V(r) = V(r_1) + (r - r_1)f_1 \tag{10'}$$

with $f_1 > 0$. The corresponding approximate equation for $P(r)$ is

$$P''(r) + \frac{2mf_2}{\hbar^2}(r - r_2)P(r) = 0 \tag{11}$$

Introducing some abbreviations

$$\alpha = \left(\frac{\hbar^2}{2mf_2}\right)^{1/3}, \qquad s = \frac{r - r_2}{\alpha} \tag{12}$$

this becomes

$$\frac{d^2P}{ds^2} + sP = 0 \tag{13}$$

When f is negative, corresponding substitutions lead to

$$\frac{d^2P}{ds^2} - sP = 0 \tag{13'}$$

If $P(s)$ satisfies (13), then $P(-s)$ satisfies (13′), so one solution covers both cases.

Solutions of (13) or (13′) are known as *Airy functions* [20] and were first studied in 1838 by G. B. Airy in connection with diffraction of light at a caustic. Later G. G. Stokes discovered an extraordinary property of asymptotic expansions in the complex plane, now known as *Stokes's phenomenon,* while studying these function. An extensive compilation of properties of Airy functions may be found in [21e].

Different choices of particular solutions are found in the literature. Jeffreys and Jeffreys work with (13′) and call their standard solutions $Ai(s)$ and $Bi(s)$. Smirnov works with (13) and calls his basic functions $U_1(s)$ and $U_2(s)$. As auxiliary functions, Jeffreys and Jeffreys write

$$y_1(s) = 1 + \sum_{n=1}^{\infty} \frac{1 \cdot 4 \cdot 7 \ldots (3n-2)}{(3n)!} s^{3n}$$

$$y_2(s) = s + \sum_{n=1}^{\infty} \frac{2 \cdot 5 \cdot 8 \ldots (3n-1)}{(3n+1)!} s^{3n+1} \tag{14}$$

In terms of these, $Ai(s)$ is defined as

$$\begin{aligned} Ai(s) &= \left[3^{2/3}\Gamma\left(\frac{2}{3}\right)\right]^{-1} y_1(s) - \left[3^{1/3}\Gamma\left(\frac{1}{3}\right)\right]^{-1} y_2(s) \\ &= 0.355 y_1(s) - 0.259 y_2(s) \end{aligned}$$

This particular combination of $y_1(s)$ and $y_2(s)$ is characterized by the property that it decreases exponentially with increasing positive s, and the asymptotic formula is

$$Ai(s) \approx (2\pi^{1/2})^{-1} s^{-1/4} \exp\left(-\frac{2}{3}s^{3/2}\right)\left(1 - \frac{1 \cdot 5}{1!48}s^{-3/2} + \frac{1 \cdot 5 \cdot 7 \cdot 11}{2!48^2}s^{-3} \ldots\right) \tag{15}$$

The asymptotic formula for $Ai(s)$ for $s < 0$ shows that in this region the function is oscillatory with a determined phase,

$$\begin{aligned} Ai(s) \approx \pi^{-1/2}(-s)^{-1/4} \Bigg\{ & P(-s)\cos\left[\frac{2}{3}(-s)^{3/2} - \frac{\pi}{4}\right] \\ & + Q(-s)\sin\left[\frac{2}{3}(-s)^{3/2} - \frac{\pi}{4}\right]\Bigg\} \end{aligned} \tag{15′}$$

in which

$$\begin{aligned} P(\zeta) &= 1 - \frac{1 \cdot 5 \cdot 7 \cdot 11}{2!48^2}\zeta^{-3} \\ &\quad + \frac{(1 \cdot 7 \cdot 13 \cdot 19)(5 \cdot 11 \cdot 17 \cdot 23)}{4!48^4}\zeta^{-6} \ldots \\ Q(\zeta) &= \frac{1 \cdot 5}{1!48}\zeta^{-3/2} - \frac{(1 \cdot 7 \cdot 13)(5 \cdot 11 \cdot 17)}{3!48^3}\zeta^{-9/2} + \ldots \end{aligned}$$

The second solution is $Bi(s)$,

$$\begin{aligned} Bi(s) &= \sqrt{3}\left[3^{2/3}\Gamma\left(\frac{2}{3}\right)\right]^{-1} y_1(s) + \left[3^{1/3}\Gamma\left(\frac{1}{3}\right)\right]^{-1} y_2(s) \\ &= 0.615 y_1(s) + 0.448 y_2(s) \end{aligned} \tag{16}$$

With P and Q as before, its asymptotic formulas are

$$
\begin{aligned}
s > 0 \qquad & Bi(s) \approx \pi^{-1/2} s^{-1/4} \exp\left(+ \frac{2}{3} s^{3/2} \right) \\
& \times \left(1 + \frac{1 \cdot 5}{1!48} s^{-3/2} + \frac{1 \cdot 5 \cdot 7 \cdot 11}{2!48^2} s^{-3} + \ldots \right) \\
s < 0 \qquad & Bi(s) \approx \pi^{-1/2} (-s)^{-1/4} \left\{ P(-s) \sin\left[\frac{2}{3} (-s)^{3/2} - \frac{\pi}{4} \right] \right. \\
& \left. + Q(-s) \cos\left[\frac{2}{3} (-s)^{3/2} - \frac{\pi}{4} \right] \right\}
\end{aligned}
\tag{16'}
$$

The combination $(2/3)|s|^{3/2}$ is recognized as that given by the WBKJ method since here k or κ equals $|s|$.

From (14) we see that

$$
\begin{aligned}
y_1(0) &= 1 \qquad & y_1'(0) &= 0 \\
y_2(0) &= 0 \qquad & y_2'(0) &= 1
\end{aligned}
$$

By comparison of series, Smirnov's $U_1(s)$ and $U_2(s)$ are

$$
U_1(s) = y_1(-s) \quad \text{and} \quad U_2(s) = y_2(-s) \tag{17}
$$

which are needed when Smirnov's tables are used.

Clearly the leading terms in the asymptotic expansions for $Ai(s)$ and $Bi(s)$ are those given by the WBKJ approximation. What is added by the specific relations quoted is the exact connections between the expressions valid for $s > 0$ and for $s < 0$. The region in which the WBKJ approximation is badly off for $Ai(s)$ near $s = 0$ is quite narrow as is shown in Figure 1^2. Here (a) is a plot of $Ai(s)$ and (b) is a plot of the leading term in its asymptotic expansion, corresponding to the WBKJ approximation to $Ai(s)$. In figure (c) is a plot of the $\pi^{-1/2}/|s|^{1/4}$ factor.

The simplest application of these results is that of finding the Sommerfeld-Wilson quantum condition for energy levels of oscillation around a single minimum of $V(r)$. Here we suppose that $[V(r) - W]$ becomes and remains large for $r < r_2$ and for $r > r_1$. We need a $P(r)$ that goes to zero quasi-exponentially in both regions. This condition will be satisfied by using $Ai(-s)$ near r_2. Then the phase in $\cos[S(r) - \delta]$ in (6) must be chosen equal to $\pi/4$ for (6) to join on to $Ai(-s)$ in the region $r > r_2$. Similarly, near $r = r_1$ the phase must be $\cos[S(r_1) - \pi/4]$ to join smoothly to $Ai(+s)$, where this s is measured back from r_1. This phase must equal $(n + 1/4)\pi$ with $n = 0, 1, 2, \ldots$ for smooth joining to the modification of (6) applicable near $r = r_1$. Hence the energy W must be such that

$$
S(r_1) = \int_{r_2}^{r_1} k(r)\, dr = \left(n + \frac{1}{2} \right) \pi \tag{18}
$$

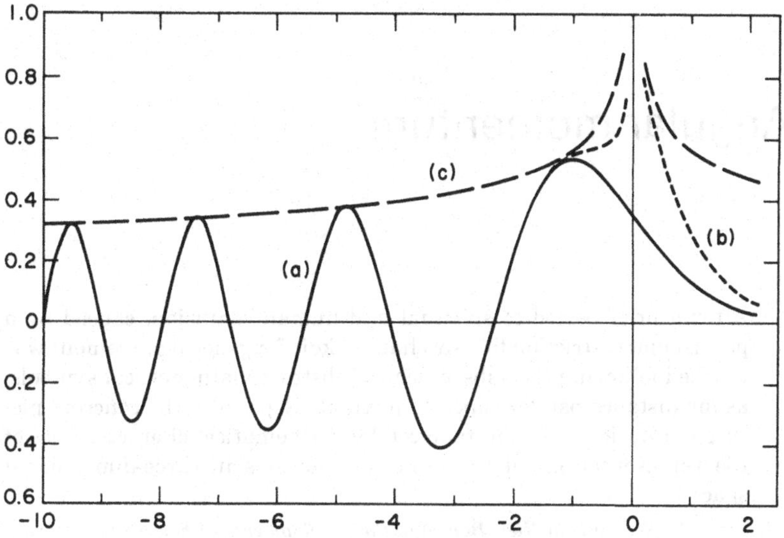

Figure 1^2. The Airy function, $Ai(s)$, and its WBKJ approximation: (a) $Ai(s)$; (b) WBKJ approximation to $Ai(s)$, which on this scale is indistinguishable from (a) for $s < -1$; (c) the amplitude factor $\pi^{-1/2}|s|^{-1/4}$.

which is equivalent to the Sommerfeld-Wilson condition, $7^1 8$, with half-integral quantum numbers

$$\oint p(r)\, dr = \left(n + \frac{1}{2}\right)h \tag{18'}$$

Here n is the number of nodes of $P(r)$ in the range of classical motion from r_2 to r_1.

Some limitations of (18) need to be mentioned. It is implicit in the foregoing that r_2 is large enough, in units of α_2, that the exponential tail is fully developed in the range $0 < r < r_2$, in order to satisfy $P(0) = 0$. If r_2/α_2 is not large enough that $Ai\,[-(r - r_2)/\alpha_2] = 0$ at $r = 0$, some $Bi(-s)$ has to be introduced to satisfy $P(0) = 0$. This has the effect of introducing a phase near $r = r_2$ somewhat different from $\pi/4$, with a corresponding change in (18).

Likewise, for $r > r_1$, it may be that $V(r)$ does not go on rising linearly as postulated in use of the Airy function $Ai(s)$ in that region. In that case it may be necessary to patch the solution onto the asymptotic form (8) with $B = 0$ for $r > r_1$, which also makes a change in (18). Thus at each end the quantum condition is somewhat affected by the detailed shape of the $V(r)$ curve in the ranges for which $W < V(r)$, which are inaccessible to classical motion.

3. Angular momentum

> After a brief period of spiritual and human confusion, caused by a provisional restriction to "Anschaulichkeit," a general agreement was reached following the substitution of abstract mathematical symbols, as for instance psi, for concrete pictures. Especially the concrete picture of rotation has been replaced by mathematical characteristics of the representations of the group of rotations in three-dimensional space."
>
> W. Pauli in *Niels Bohr and the Development of Physics* (New York: McGraw-Hill, 1955), p. 30.

1^3. Commutation rules and allowed values

Angular momentum [1] is an important integral of motion in the classical mechanics of a particle in a central field. It also plays a role of great importance in quantum mechanics, because of the invariance of the Hamiltonian for free atoms to the group of rotations in 3-space (Sec. 6^2). Many of the results can be derived by algebraic methods from the commutation rules. This approach was followed exclusively in TAS. Application of the theory of the rotation group stems from the work of Wigner and of Weyl. From 1942 on, great contributions to the theory were made in an important series of papers by G. Racah [2].

Writing $|a\rangle$ for any state of the atomic system that has been prepared or selected with apparatus set up in a certain first orientation, we may write $|a'\rangle$ for the state which in every respect shows the same properties relative to a second set of preparing or observing apparatus of the same kind, but turned relative to the first set through an angle ω about an axis $\mathbf{n}$, which rotation we denote by $R \equiv R(\omega, \mathbf{n})$.

We write R for the unitary operator that changes $|a\rangle$ into $|a'\rangle$ so that

$$|a'\rangle = R|a\rangle, \quad \langle a'| = \langle a|R^\dagger \tag{1}$$

The explicit form assumed by R depends on the particular form of description used for the states $|a\rangle$ and $|a'\rangle$.

Let γ stand for any observable in a form to operate on the $|a\rangle$, which describes the behavior of apparatus in the first orientation, and γ_R stand for that observable in a form to operate on the $|a'\rangle$, which describes the behavior of the same kind of apparatus, but in the second orientation. When we say that the apparatus represented by γ and γ_R

differ only with respect to orientation, we mean that every relation of the form

$$\gamma|a\rangle = \Sigma|b\rangle\langle b|\gamma|a\rangle$$

implies that

$$\gamma_R|a'\rangle = \Sigma|b'\rangle\langle b'|\gamma_R|a'\rangle$$

in which

$$\langle b'|\gamma_R|a'\rangle = \langle b|\gamma|a\rangle \tag{2}$$

for all corresponding states and all corresponding operators. Because the states are related as in (1) we must have

$$\langle b'|\gamma_R|a'\rangle = \langle b|R^\dagger\gamma_R R|a\rangle$$

Therefore, combining this relation with relation (2), which holds for all $|a\rangle$ and $|b\rangle$, we have this law of transformation of the operator for rotation of the apparatus it represents,

$$\gamma_R = R\gamma R^\dagger \tag{3}$$

Strictly speaking, more needs to be said about (2). For a particular *R*, all that we have a right to require is that the absolute value on the left equals the absolute value on the right, and this would permit a relation in which the element on the left is equal to the complex conjugate of the element on the right. However, equality as in (2) must hold in case *R* is the identity (no rotation) and continuity requires equality, because the *R* form a continuous group in which there are rotations that differ infinitesimally from the identity.

Suppose that the system under observation contains a single (spinless) particle so that $|a\rangle$ is described by a Schrödinger function $\langle x_1, x_2, x_3|a\rangle$, where (x_1, x_2, x_3) are the particle's position relative to the first orientation of the apparatus. Then the state $|a'\rangle$ will be represented by $\langle x_1', x_2', x_3'|a'\rangle$, which is a function with the same values as $\langle x_1, x_2, x_3|a\rangle$ in which the point (x_1', x_2', x_3') in the second orientation corresponds to the point (x_1, x_2, x_3) in the first orientation, that is, as in Section 6^2, they are related by

$$x_i' = \sum_j R_{ij}x_j; \qquad x_i = \sum_j R_{ij}^{-1}x_j' \tag{4}$$

in which R_{ij} is the matrix that shows the relation of corresponding points in the two orientations. Thus the functional form of $\langle x_1', x_2', x_3'|a'\rangle$ must be different from that of $\langle x_1, x_2, x_3|a\rangle$ in such a way that

$$\langle x_1, x_2, x_3|a'\rangle = \langle R_{1j}^{-1}x_j, R_{2j}^{-1}x_j, R_{3j}^{-1}x_j|a\rangle \tag{5}$$

in which the primes have been dropped. For the sake of simplicity the summation signs in (5) are omitted.

In particular, suppose that the rotation is through γ about the x_3 axis. Then

$$R_{ij}(3,\gamma)=\begin{pmatrix}\cos\gamma & -\sin\gamma & 0\\ \sin\gamma & \cos\gamma & 0\\ 0 & 0 & 1\end{pmatrix};\quad R_{ij}^{-1}(3,\gamma)=\begin{pmatrix}\cos\gamma & \sin\gamma & 0\\ -\sin\gamma & \cos\gamma & 0\\ 0 & 0 & 1\end{pmatrix}$$

so the operation on $|a\rangle$ represented in (5) becomes

$$\langle x_1\cos\gamma + x_2\sin\gamma,\ -x_1\sin\gamma + x_2\cos\gamma,\ x_3|a\rangle$$

For $\gamma \to 0$ we have $\sin\gamma \to \gamma$ and $\cos\gamma \to 1$ to the second order in γ. Therefore this is, for an infinitesimal $\delta\gamma$

$$\left[1 - \delta\gamma\left(x_1\frac{\partial}{\partial x_2} - x_2\frac{\partial}{\partial x_1}\right)\right]\langle x_1, x_2, x_3|a\rangle$$

From the discussion of infinitesimal unitary transformations in Section 1[2] we know that the operator can be written $(1 - i\delta\gamma L_3)$ where L_3 is Hermitian and equal to

$$L_3 = -i\left(x_1\frac{\partial}{\partial x_2} - x_2\frac{\partial}{\partial x_1}\right) \tag{6}$$

By repeating the infinitesimal transformation we find for finite γ

$$R = e^{-i\gamma L_3} \tag{7}$$

as the form applicable to the Schrödinger function $\langle x_1, x_2, x_3|a\rangle$ for rotation of the apparatus about the x_3 axis through the finite γ.

Similarly, we find the operators

$$L_1 = -i\left(x_2\frac{\partial}{\partial x_3} - x_3\frac{\partial}{\partial x_2}\right),\quad L_2 = -i\left(x_3\frac{\partial}{\partial x_1} - x_1\frac{\partial}{\partial x_3}\right)$$

to represent the operators for infinitesimal rotation about the x_1 and x_2 axes giving

$$R = (1 - i\omega\mathbf{n}\cdot\mathbf{L}) \tag{8}$$

for infinitesimal ω about the $\mathbf{n}$ axis where $\omega\mathbf{n} = \alpha\hat{\mathbf{i}} + \beta\hat{\mathbf{j}} + \gamma\hat{\mathbf{k}}$, and $\mathbf{L}$ is written for the vector

$$\mathbf{L} = \hat{\mathbf{i}}L_1 + \hat{\mathbf{j}}L_2 + \hat{\mathbf{k}}L_3 \tag{9}$$

By direct calculation from the forms given we find that the L_i do not commute. Instead we have

$$[L_i, L_j] = iL_k \quad \text{(cyclic)} \tag{10}$$

The Hermitian operators obtained in this way by consideration of the effects on $\langle x_1, x_2, x_3|a\rangle$ of a changed orientation of the apparatus are those that represent the orbital angular momentum of a spinless particle, measured in units of $\hbar$, and defined by

$$\mathbf{L}\hbar = \mathbf{r} \times \mathbf{p} \tag{11}$$

leading to the commutation relations (10).

In a system that involves several particles, an orbital angular momentum $\mathbf{L}_\alpha$ for each particle is defined by (11) as $\hbar^{-1}\mathbf{r}_\alpha \times \mathbf{p}_\alpha$. Because coordinates and momenta commute when they refer to different particles, (10) can be extended to

$$[L_{\alpha i}, L_{\beta j}] = iL_{\alpha k}\delta(\alpha,\beta) \tag{10$'$}$$

The vector resultant orbital angular momentum

$$\mathbf{L} = \sum_\alpha \mathbf{L}_\alpha \tag{12}$$

then obeys the commutation rules (10), the same as for a single particle.

In addition to the orbital angular momentum, elementary particles have a spin-angular momentum $\mathbf{S}$, which we consider in Section 5³. Anticipating, the spin operators for a system of particles will be found to obey the commutation rules (10′), as does

$$\mathbf{S} = \sum_\alpha \mathbf{S}_\alpha \tag{12$'$}$$

the resultant spin. In addition a resultant angular momentum

$$\mathbf{J}_\alpha = \mathbf{L}_\alpha + \mathbf{S}_\alpha \tag{13}$$

may be defined for each particle and a resultant for all the particles

$$\mathbf{J} = \mathbf{L} + \mathbf{S} = \sum_\alpha \mathbf{L}_\alpha + \sum_\alpha \mathbf{S}_\alpha \tag{14}$$

which also obeys the basic commutation rules (10).

We turn now to the determination of the allowed values of any angular momentum vector, which can be derived from a study of the commutation relations.

Each component of $\mathbf{J}$ commutes with

$$\mathbf{J}^2 = J_x^2 + J_y^2 + J_z^2 \tag{15}$$

so there exist states that are simultaneous proper states of $\mathbf{J}^2$ and any one component, say, J_z. We write

$$\mathbf{J}^2 = j(j+1) \quad \text{and} \quad J_z = m \tag{15$'$}$$

for the simultaneous proper values, where the allowed values of (j, m) will now be found.

It proves convenient to work with two non-Hermitian operators, which are called *shift operators;* these are defined as

$$J_\pm = J_x \pm iJ_y \tag{16}$$

In terms of these (10) readily give

$$[J_\pm, J_\mp] = 2J_z \quad \text{and} \quad [J_z, J_\pm] = \pm J_\pm \tag{16'}$$

We write $|\alpha, j, m\rangle$ for a simultaneous proper state of $\mathbf{J}^2$ and J_z and such other commuting observables α as are necessary, so that

$$\mathbf{J}^2|\alpha, j, m\rangle = |\alpha, j, m\rangle j(j+1)$$
$$J_z|\alpha, j, m\rangle = |\alpha, j, m\rangle m$$

and, therefore,

$$(J_x^2 + J_y^2)|\alpha, j, m\rangle = |\alpha, j, m\rangle[j(j+1) - m^2]$$

As $J_x^2 + J_y^2$ is essentially positive, there is a largest value of m^2 associated with a given j. Let m_1 be the largest and m_2 the smallest m, so that a single string of states are labeled by $m_1, m_1 - 1, m_1 - 2, \ldots, m_2$.

From the second of (16′) we find

$$J_zJ_\pm|\alpha, j, m\rangle = J_\pm|\alpha, j, m\rangle(m \pm 1) \tag{17}$$

Hence $J_\pm|\alpha, j, m\rangle$ is proportional to a state in which J_z has the value $(m \pm 1)$. This is the reason for calling $J_\pm$ shift operators: J_+ applied to $|\alpha, j, m\rangle$ gives a state in which J_z is increased by one unit and J_- gives a state in which J_z is decreased by one unit.

But J_+ applied to $|\alpha, j, m_1\rangle$ must give zero; otherwise it would give a state corresponding to $(m_1 + 1)$, contradicting the hypothesis that m_1 is the largest. Operate on

$$J_+|\alpha, j, m_1\rangle = 0$$

with J_- to find

$$J_-J_+|\alpha, j, m_1\rangle = 0 = |\alpha, j, m_1\rangle[j(j+1) - m_1(m_1+1)]$$

and therefore $m_1 = j$. Similarly from $J_-|\alpha, j, m_2\rangle = 0$, by operating with J_+ we find that $m_2 = -j$. Therefore, $m_1 - m_2 = 2j$, which is an integer, so the allowed values of j are

$$j = 0, \frac{1}{2}, 1, \frac{3}{2}, 2, \ldots \tag{18}$$

and the values of m associated with each j are the $(2j + 1)$ values, which have integral difference and extend over the range

$$-j \leqslant m \leqslant +j \tag{19}$$

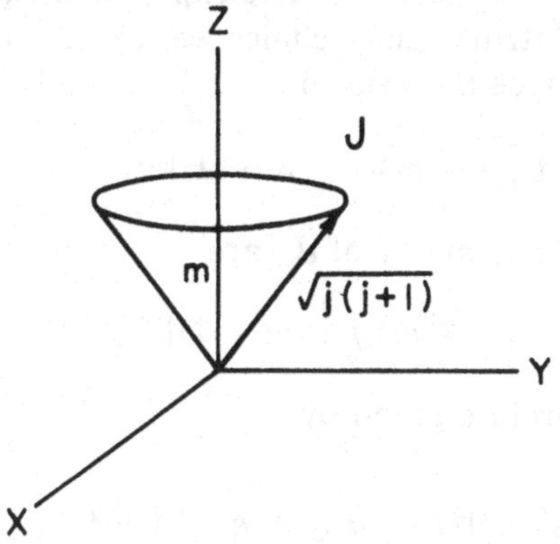

Figure 1³. Distribution of **J** in the state $|\alpha, j, m>$.

This concludes the determination of the allowed values of $\mathbf{J}^2$ and J_z. We shall see (Sec. 4³) that the integral values of j are appropriate for the description of orbital angular momentum, and (Sec. 5³) the value $j = 1/2$ is appropriate to the description of electron spin.

Except for the state $|\alpha, 0, 0\rangle$, the lack of commutation of components of **J** implies that all three components cannot be precisely known at the same time. It is therefore incorrect to picture **J** as a directed line segment in the way that is usually done for vectors. Instead we can picture the distribution of values of **J** in the state $|\alpha, j, m\rangle$ as lying on a half-cone, as in Figure 1³. This has an altitude m and a slant height $\sqrt{j(j+1)}$, all azimuths of $J_x^2 + J_y^2$ being equally likely. In stationary states the situation is static, so **J** does not precess, but the direction of **J** on this cone is indeterminate. For large j and $m = j$, the opening of the cone becomes small, indicating that in this limit the **J** vector can be specified quite accurately as having very nearly the direction of the **k** axis.

2³. Matrices for angular momentum

From 1³17 we infer that

$$|\alpha, j, m \pm 1\rangle = N_\pm J_\pm |\alpha, j, m\rangle \quad (1)$$

where $N_\pm$ is a normalizing factor. Because $J\dagger_\pm = J_\mp$ we have

$$\langle \alpha, j, m \pm 1| = \langle \alpha, j, m| \mathrm{J}_\mp \mathrm{N}_\pm^*$$

giving

$$\begin{aligned}\langle \alpha, j, m \pm 1|\alpha, j, m \pm 1\rangle &= |N_\pm|^2 \langle \alpha, j, m|J_\mp J_\pm|\alpha, j, m\rangle \\ &= |N_\pm|^2[j(j+1) - m(m \mp 1)]\end{aligned}$$

where $|N_\pm|$ must be determined to make the left expression equal to unity. Making a convenient arbitrary phase choice we take $N_\pm$ as real and positive, $N_\pm = |N_\pm|$. This gives the relation

$$J_\pm|\alpha, j, m\rangle = |\alpha, j, m \pm 1\rangle[(j \mp m)(j \pm m + 1)]^{1/2} \tag{2}$$

Hence the nonvanishing matrix elements of $J_\pm$ are

$$\langle \alpha, j, m \pm 1|J_\pm|\alpha, j, m\rangle = [(j \mp m)(j \pm m + 1)]^{1/2} \tag{3}$$

The vector operator $\mathbf{J}$ is therefore given by

$$\begin{aligned}\mathbf{J}|\alpha, j, m\rangle = {} & |\alpha, j, m + 1\rangle(\hat{\mathbf{i}} - i\hat{\mathbf{j}})\frac{1}{\sqrt{2}}[(j - m)(j + m + 1)]^{1/2} \\ & + |\alpha, j, m - 1\rangle(\hat{\mathbf{i}} + i\hat{\mathbf{j}})\frac{1}{\sqrt{2}}[(j + m)(j - m + 1)]^{1/2} \\ & + |\alpha, j, m\rangle\hat{\mathbf{k}}\, m\end{aligned} \tag{4}$$

This completes the determination of the matrix elements of angular momentum.

Another way of writing this result is often used:

$$\begin{aligned}\langle \alpha', j', m'|J_x|\alpha, j, m\rangle &= \delta(\alpha', \alpha)\delta(j', j)\frac{1}{2}[j(j + 1) - mm']^{1/2} \\ &\qquad \times [\delta(m', m - 1) + \delta(m', m + 1)] \\ \langle \alpha', j', m'|J_y|\alpha, j, m\rangle &= -i(m' - m)\langle \alpha', j', m'|J_x|\alpha, j, m\rangle \\ \langle \alpha', j', m'|J_z|\alpha, j, m\rangle &= \delta(\alpha', \alpha)\delta(j', j)\delta(m', m)m\end{aligned} \tag{5}$$

The second of these is a direct consequence of $J_y = -i(J_zJ_x - J_xJ_z)$.

An arbitrary state $|a\rangle$ has the coefficients $\langle \alpha, j, m|a\rangle$ when expanded in terms of the $|\alpha, j, m\rangle$. The set of these for the $(2j + 1)$ values of m for a given j in 1³18 may be regarded as a one-column $(2j + 1)$-row matrix. Then the matrix for $\mathbf{J}$ consists of $(2j + 1) \times (2j + 1)$ steps that are diagonal in α and j; this matrix multiplies the one-column matrix $\langle \alpha, j, m|a\rangle$ to give the column matrix for $\mathbf{J}|a\rangle$. It is conventional to write this matrix with the m values in *decreasing* order. Starting at the upper left corner this matrix, $\langle \alpha', j', m'|\mathbf{J}|\alpha, j, m\rangle$, has the form

$$[\delta(\alpha', \alpha)\delta(j', j)]\begin{bmatrix} \hat{\mathbf{k}}j & (\hat{\mathbf{i}} - i\hat{\mathbf{j}})\frac{1}{2}[1(2j)]^{1/2} & 0 & \cdots \\ (\hat{\mathbf{i}} + i\hat{\mathbf{j}})\frac{1}{2}[1(2j)]^{1/2} & \hat{\mathbf{k}}(j - 1) & (\hat{\mathbf{i}} - i\hat{\mathbf{j}})\frac{1}{2}[2(2j - 1)]^{1/2} & \cdots \\ 0 & (\hat{\mathbf{i}} + i\hat{\mathbf{j}})\frac{1}{2}[2(2j - 1)]^{1/2} & \hat{\mathbf{k}}(j - 2) & \cdots \\ \cdots & & \cdots & \end{bmatrix} \tag{6}$$

Therefore the special forms for $j = 1/2$ and $j = 1$ are

$$\frac{1}{2}\begin{bmatrix} \hat{\mathbf{k}} & \hat{\mathbf{i}} - i\hat{\mathbf{j}} \\ \hat{\mathbf{i}} + i\hat{\mathbf{j}} & -\hat{\mathbf{k}} \end{bmatrix} \quad \text{and} \quad \begin{bmatrix} \hat{\mathbf{k}} & \dfrac{\hat{\mathbf{i}} - i\hat{\mathbf{j}}}{\sqrt{2}} & 0 \\ \dfrac{\hat{\mathbf{i}} + i\hat{\mathbf{j}}}{\sqrt{2}} & 0 & \dfrac{\hat{\mathbf{i}} - i\hat{\mathbf{j}}}{\sqrt{2}} \\ 0 & \dfrac{\hat{\mathbf{i}} + i\hat{\mathbf{j}}}{\sqrt{2}} & -\hat{\mathbf{k}} \end{bmatrix}$$

The matrix for $j = 1/2$ are often written in terms of the *Pauli matrices,*

$$\sigma_1 = \begin{pmatrix} 0 & 1 \\ 1 & 0 \end{pmatrix}, \quad \sigma_2 = \begin{pmatrix} 0 & -i \\ i & 0 \end{pmatrix}, \quad \sigma_3 = \begin{pmatrix} 1 & 0 \\ 0 & -1 \end{pmatrix} \tag{6'}$$

as $(1/2)[\hat{\mathbf{i}}\sigma_1 + \hat{\mathbf{j}}\sigma_2 + \hat{\mathbf{k}}\sigma_3]$.

Repeated application of (2) gives a relation by which any one of the $|\alpha, j, m\rangle$ can be expressed in terms of the others of the same j:

$$|\alpha, j, m \pm k\rangle = (J_\pm)^k|\alpha, j, m\rangle \left[\frac{(j \pm m)!\,(j \mp m - k)!}{(j \mp m)!\,(j \pm m + k)!}\right]^{1/2} \tag{7}$$

In particular each one is expressible in terms of the $|\alpha, j, j\rangle$,

$$|\alpha, j, m\rangle = (J_-)^{j-m}|\alpha, j, j\rangle \left[\frac{(j + m)!}{(2j)!(j - m)!}\right]^{1/2} \tag{8}$$

The diagonal elements of J_x and J_y vanish, so this is also true of those of $(J_x \cos\phi + J_y \sin\phi)$ for any ϕ. Hence the mean value of the projection of **J** on the equatorial plane in any azimuth is zero, so that **J** is equally distributed in azimuth as was assumed in Figure 1[3].

Another derivation of the matrix for **J** is instructive. For convenience we suppress the α, j labels in which **J** is diagonal. From the second of (5) we have

$$(m' - m)\langle m'|J_\pm|m\rangle = \pm\langle m'|J_\pm|m\rangle$$

Therefore, $\langle m'|\mathrm{J}_\pm|m\rangle$ vanishes unless $m' = m \pm 1$; that is, $\langle m'|\mathrm{J}_+|m\rangle = 0$ unless $m' = m + 1$, and $\langle m'|\mathrm{J}_-|m\rangle = 0$ unless $m' = m - 1$. For the diagonal element of the first of (5) we have

$$\langle m|J_\pm|m \mp 1\rangle\langle m \mp 1|J_\mp|m\rangle = j(j + 1) - m(m \mp 1)$$

Because $J\dagger_\mp = J_\pm$, this gives

$$|\langle m \mp 1|J_\mp|m\rangle|^2 = j(j + 1) - m(m \mp 1)$$

Using the lower sign, where $m = m_1$ its largest value, the matrix element must vanish, so $m_1 = j$. Likewise, using the upper sign, when $m = m_2$, its least value, the matrix element must vanish, so $m_2 = -j$.

Therefore $m_1 - m_2 = 2j$, an integer, the same result as was found in Section 1^3. The same arbitrary phase choice as before now gives

$$\langle m \pm 1|J_\pm|m\rangle = [j(j+1) - m(m \pm 1)]^{1/2}$$

agreeing with (3).

Because the commutation rules $1^3 10$ are the same for any two orientations of apparatus as considered in Section 1^3, it follows that the allowed values of $\mathbf{J}^2$ and J_z, and also the matrices for $\mathbf{J}$, have the same form in the basis states of the two orientations. We write m for a proper value of J_z and μ for one of J'_z, the states of the second orientation.

The states $|\alpha, j, \mu\rangle$ of the second orientation are then expressible in terms of those of the first orientation,

$$|\alpha, j, \mu\rangle = \sum_m |\alpha, j, m\rangle D^{(j)}_{m\mu}(R) \tag{9}$$

where the coefficients $D^{(j)}_{m\mu}(R)$ for the various R constite a set of $(2j + 1) \times (2j + 1)$ unitary matrices that provide an irreducible representation of the three-dimensional rotation group, $SO(3)$.

Many important results follow from the mere existence of such matrices, as will be seen in Section 3^3. Explicit determination of them is made in Section 6^3.

3^3. Addition of two angular momenta

In atomic systems there are several angular momentum vectors that refer to different parts of the system, such as the orbital motion of each electron, the spin of each electron, and the resultant angular momentum of the nucleons in the nucleus. In particular, if there are two angular moments, $\mathbf{J}_1$ and $\mathbf{J}_2$, the components of each satisfy $1^3 10$, and each component of $\mathbf{J}_1$ commutes with each component of $\mathbf{J}_2$. Therefore, states $|\alpha, j_1, j_2; m_1, m_2\rangle$ exist for which both of the $\mathbf{J}^2$ and J_z have precise values.

For a given (j_1, j_2) there are $(2j_1 + 1)(2j_2 + 1)$ such states, obtained by varying (m_1, m_2) independently over their allowed values $1^3 19$. In these states $\mathbf{J}_1$ and $\mathbf{J}_2$ are said to be *uncoupled.*

Alternatively, there are *coupled* states $|\alpha, j_1, j_2; j, m)$, which are labeled by precise values of $(\mathbf{J}_1 + \mathbf{J}_2)^2 = j(j+1)$ and $(J_{1z} + J_{2z}) = m$. These states are mutually orthonormal. The coupled and uncoupled states are therefore connected by a unitary transformation

$$|\alpha, j_1, j_2; j, m) = \sum_{m_1 + m_2 = m} |\alpha, j_1, j_2; m_1, m_2\rangle\langle m_1, m_2|j, m) \tag{1}$$

The coefficients $\langle m_1, m_2|j, m)$ depend also on j_1 and j_2. When necessary to avoid ambiguity we write the coefficients as $\langle m_1, m_2|(j_1, j_2)j, m)$. The states on the left side of (1) are an orthonormal set. Multiplying (1) by its Hermitian conjugate we obtain the orthogonality relation

$$|\alpha, j_1, j_2; m_1, m_2\rangle = \sum_{j,m} |\alpha, j_1, j_2; j, m)(j, m|m_1, m_2\rangle \tag{2}$$

Similarly, from the orthonormality of (2) we obtain

$$\sum_{m_1,m_2} (j', m'|m_1, m_2\rangle\langle m_1, m_2|j, m) = \delta(j, j')\delta(m', m) \tag{3}$$

and also

$$\sum_{j,m} \langle m_1', m_2'|j, m)(j, m|m_1, m_2\rangle = \delta(m_1', m_1)\delta(m_2', m_2) \tag{3'}$$

The notation for the coefficients could become ambiguous when numerical values are substituted, so a round bracket is used for the coupled states and an angular bracket for the uncoupled states. Because of unitarity $(j, m|m_1, m_2\rangle$ is the complex conjugate of $\langle m_1, m_2|j, m)$; actually we shall see that these coefficients can be made real, so they are equal.

The coefficients in (1) and (2) are of great importance in atomic physics. They are variously called coupling coefficients, Clebsch-Gordan coefficients, or Wigner coefficients. They depend on six parameters: (j_1, j_2, j) and (m_1, m_2, m), but only five of them are independent because of the relation $m = m_1 + m_2$.

Wigner [3] has shown that the properties of the coefficients can be nicely exhibited by relating them to a 3j *symbol.*

$$\langle m_1, m_2|j, m) = (j, m|m_1, m_2\rangle = (-1)^{j_2-j_1-m}\sqrt{2j+1}\begin{pmatrix} j_1 & j_2 & j \\ m_1 & m_2 & -m \end{pmatrix} \tag{4}$$

The $3j$ symbol vanishes unless $m_1 + m_2 = m$, that is, unless $m_1 + m_2 - m = 0$. The relation

$$\langle m_2, m_1|(j_2, j_1)j, m) = (-1)^{j-j_1-j_2}\langle m_1, m_2|(j_1, j_2)j, m)$$

follows from (4) and from the symmetry properties of the $3j$ symbols stated after (15).

Many other notations appear in the literature. Rose [1a] writes

$$C(j_1, j_2, j; m_1, m_2) = \langle m_1, m_2|j, m) \tag{5}$$

Schwinger [4] writes

$$X(j_1, j_2, j_3; m_1, m_2, m_3) = \begin{pmatrix} j_1 & j_2 & j_3 \\ m_1 & m_2 & m_3 \end{pmatrix} \tag{6}$$

Racah [2] writes

$$V(j_1, j_2, j_3; m_1, m_2, m_3) = (-1)^{j_1-j_2-j_3}\begin{pmatrix} j_1 & j_2 & j_3 \\ m_1 & m_2 & m_3 \end{pmatrix} \tag{7}$$

Landau and Lifshitz [5] write

$$S_{j_1,m_1;\,j_2,m_2;\,j_3,m_3} = (-1)^{j_1-j_2+j_3}\begin{pmatrix} j_1 & j_2 & j_3 \\ m_1 & m_2 & m_3 \end{pmatrix} \tag{8}$$

The largest value of m is $j_1 + j_2$, with $m_1 = j_1$ and $m_2 = j_2$, corresponds to a single uncoupled state, $|\alpha, j_1, j_2; j_1, j_2\rangle$. Therefore the largest value of j is $j_1 + j_2$. With a proper choice of normalization phase we can equate this state to the coupled state $|\alpha, j_1, j_2; j_1 + j_2, j_1 + j_2\rangle$.

Barring the trivial case in which j_1 or j_2 is zero, the value $m = j_1 + j_2 - 1$ can be realized in two ways,

$$(m_1, m_2) = (j_1 - 1, j_2) \quad \text{or} \quad (j_1, j_2 - 1)$$

In the 2-space spanned by those two uncoupled states, there will be one coupled state $|\alpha, j_1, j_2; j_1 + j_2, j_1 + j_2 - 1\rangle$. To span the space there must also be a coupled state $|\alpha, j_1, j_2; j_1 + j_2 - 1, j_1 + j_2 - 1\rangle$. Hence $j = j_1 + j_2 - 1$ is an allowed value. For this value of m the coupling coefficients form a 2×2 matrix.

Continuing in this way, the value $m = j_1 + j_2 - k$ can be realized in $(k + 1)$ ways, provided that k is less than or equal to the lesser of j_1 and j_2:

$$(m_1, m_2) = (j_1 - k, j_2), (j_1 - k + 1, j_2 - 1) \ldots (j_1, j_2 - k)$$

Thus each decrease of m by unity brings in an additional allowed value for j. The allowed values of the vector sum of $(\mathbf{J}_1 + \mathbf{J}_2)^2 = j(j + 1)$ are therefore given by

$$j = (j_1 + j_2), (j_1 + j_2 - 1), \ldots |j_1 - j_2| \tag{9}$$

The original matrix of the coupling coefficient, with $(2j_1 + 1)(2j_2 + 1)$ rows and columns, breaks up into a set of step matrices, labeled by values of m from $(j_1 + j_2)$ down to $-(j_1 + j_2)$. The step for $m = j_1 + j_2 - k$ has $(k + 1)$ rows and columns for $k < \min(2j_1, 2j_2)$. The matrix order is then constant until m is reduced to $-|j_1 - j_2|$. Further unit decreases of m reduce the order by unity, ending with $m = -(j_1 + j_2)$ for which m there is but a single state, as with $m = +(j_1 + j_2)$.

The allowed values of j in (9) are symmetric in (j_1, j_2, j), forming a triangle whose perimeter, $j_1 + j_2 + j$, is an integer. Also $j_1 + j_2 - j$, $j_2 + j - j_1$ and $j + j_1 - j_2$ must be nonnegative integers. It is convenient to define a triangular delta function,

$$\Delta(j_1, j_2, j) = \begin{matrix} 1 & \text{for a triangle of integral perimeter} \\ 0 & \text{otherwise} \end{matrix} \tag{10}$$

The coupling coefficients in (1) and (2) and the $3j$ symbols in (4) may be regarded as defined for all arguments by including $\Delta(j_1, j_2, j)$ and $\delta(m, m_1 + m_2)$ as factors.

Evaluation of the coupling coefficients proceeds in two major steps:

1. The relation $J_+|\alpha, j, j\rangle = 0$ is used to find the $\langle m_1, m_2|j, j\rangle$ for each value of j, with $m = j$. A phase normalization must be fixed by convention.

2. Then the $\langle m_1, m_2|j, m\rangle$ are found from the $\langle m_1, m_2|j, j\rangle$ by application of $2^3 7$.

For brevity we omit the (α, j_1, j_2) from the symbols, writing

$$|j, j\rangle = \sum_{m_1+m_2=j} |m_1, m_2\rangle\langle m_1, m_2|j, j\rangle$$

Application of J_+ to this expression gives zero. The result is

$$\begin{aligned} 0 &= \sum\{|m_1 + 1, m_2\rangle[(j_1 - m_1)(j_1 + m_1 + 1)]^{1/2} \\ &\qquad + |m_1, m_2 + 1\rangle[(j_2 - m_2)(j_2 + m_2 + 1)]^{1/2}\}\langle m_1, m_2|j, j\rangle \\ &= \sum |m_1, m_2 + 1\rangle\{\sqrt{(j_1 - m_1 + 1)(j_1 + m_1)}\langle m_1 - 1, m_2 + 1|j, j\rangle \\ &\qquad + \sqrt{(j_2 + m_2 + 1)(j_2 - m_2)}\langle m_1, m_2|j, j\rangle\} \end{aligned}$$

A recursion relation is obtained by equating to zero the coefficient of each $|m_1, m_2 + 1\rangle$,

$$\frac{\langle m_1, m_2|j, j\rangle}{\langle m_1 - 1, m_2 + 1|j, j\rangle} = -\left[\frac{(j_1 - m_1 + 1)(j_1 + m_1)}{(j_2 - m_2)(j_2 + m_2 + 1)}\right]^{1/2}$$

This can be iterated to give

$$\frac{\langle m_1, m_2|j, j\rangle}{\langle m_1 - \lambda, m_2 + \lambda|j, j\rangle} = (-1)^{\lambda}\left[\frac{(j_1 - m_1 + \lambda)!(j_1 + m_1)!(j_2 - m_2 - \lambda)!(j_2 + m_2)!}{(j_1 - m_1)!(j_1 + m_1 - \lambda)!(j_2 - m_2)!(j_2 + m_2 + \lambda)!}\right]^{1/2}$$

Rearrangement shows that

$$\langle m_1, m_2|j, j\rangle = C_j(-1)^{j_1 - m_1}\left[\binom{j_1 + m_1}{j_2 - m_2}\binom{j_2 + m_2}{j_1 - m_1}\right]^{1/2} \quad (m_1 + m_2 = j) \tag{11}$$

where C_j is a normalizing factor, and $\binom{a}{b}$ means the binomial coefficient $a!/b!(a - b)!$. The coefficient C can be evaluated as follows. Because

$$(1 + \chi)^{-a-b} = (1 + \chi)^{-a}(1 + \chi)^{-b}$$

it follows that

$$\sum_n \binom{a + b + n - 1}{n}(-\chi)^n = \left[\sum_\alpha \binom{a + \alpha - 1}{\alpha}(-\chi)^\alpha\right]\left[\sum_\beta \binom{b + \beta - 1}{\beta}(-\chi)^\beta\right]$$

Equating coefficients of χ^n,

$$\sum_{\alpha+\beta=n} \binom{a+\alpha-1}{\alpha}\binom{b+\beta-1}{\beta} = \binom{a+b+n-1}{n}$$

The C_j must be such that

$$\sum_{m_1+m_2=j} |\langle m_1, m_2|j, j)|^2 = 1$$

Requiring, by convention, C_j to be real and positive, we find

$$C_j = \binom{j_1+j_2+j+1}{j_1+j_2-j}^{-1/2} \tag{11'}$$

Use of this value in (11) completes the first step, the determination of $|\alpha, j, j)$ in terms of the $|\alpha, j_1 j_2; m_1, m_2\rangle$ for $m_1 + m_2 = j$, where j is one of the allowed values in (9).

The coefficients for $m < j$ are now found by using $2^3 8$,

$$|j, m) = \left[\frac{(j+m)!}{(2j)!(j-m)!}\right]^{1/2} \sum_{\kappa=0}^{j-m} \binom{j-m}{\kappa} J_{1-}^{j-m-\kappa} J_{2-}^{\kappa}|j, j)$$

The values of $|j, j)$ from (11) are substituted and $2^3 7$ is used to evaluate the action of J_{1-} and J_{2-} on $|m_1, m_2\rangle$. The final result is

$$\frac{\langle m_1, m_2|j, m)}{\sqrt{2j+1}}$$

$$= \left[\frac{(j_1-m_1)!(j_2-m_2)!(j_1+j_2-j)!(j-m)!(j+m)!}{(j_1+m_1)!(j_2+m_2)!(j_1-j_2+j)!(j_2-j_1+j)!(j_1+j_2+j+1)!}\right]^{1/2}$$

$$\times \sum_{\kappa} \frac{(-1)^{j_1-m_1+\kappa}(j_1+m_1+\kappa)!(j_2+j-m_1-\kappa)!}{(j_1-m_1-\kappa)!(j_2-j+m_1+\kappa)!\kappa!(j-m-\kappa)!} \tag{12}$$

The sum on κ runs over all nonnegative integers that do not give rise to factorials of negative integers. Because all the coefficients (12) are real,

$$(j, m|m_1, m_2\rangle = \langle m_1, m_2|j, m)$$

as already stated.

The states obtained by adding $\mathbf{J}_1$ to $\mathbf{J}_2$ are physically the same as those obtained by adding $\mathbf{J}_2$ to $\mathbf{J}_1$, but the phase convention adopted leads to different signs. The rule is

$$|j_b, j_a; j, m) = (-1)^{j_a+j_b-j}|j_a, j_b; j, m) \tag{13}$$

Racah [2(a)II, p. 440; *QTAM*, p. 148] has transformed (12) to a more symmetrical form. Stated in terms of the $3j$ symbol defined in (4) with

Table 1³. $\langle m - m_2, m_2 | (j_1, 1/2) j, m \rangle$

$j =$	$m_2 = \pm 1/2$
$j_1 + 1/2$	$\left(\frac{j_1 \pm m + 1/2}{2j_1 + 1}\right)^{1/2}$
$j_1 - 1/2$	$\mp \left(\frac{j_1 \mp m + 1/2}{2j_1 + 1}\right)^{1/2}$

Table 2³. $\langle m - m_2, m_2 | (j_1, 1) j, m \rangle$

$j =$	$m_2 = \pm 1$	$m_2 = 0$
$j_1 + 1$	$\left[\frac{(j_1 \pm m)(j_1 \pm m + 1)}{(2j_1 + 1)(2j_1 + 2)}\right]^{1/2}$	$\left[\frac{(j_1 - m + 1)(j_1 + m + 1)}{(2j_1 + 1)(j_1 + 1)}\right]^{1/2}$
j_1	$\mp\left[\frac{(j_1 \pm m)(j_1 \mp m + 1)}{2j_1(j_1 + 1)}\right]^{1/2}$	$\frac{m}{[j_1(j_1 + 1)]^{1/2}}$
$j_1 - 1$	$\left[\frac{(j_1 \mp m)(j_1 \mp m + 1)}{2j_1(2j_1 + 1)}\right]^{1/2}$	$-\left[\frac{(j_1 - m)(j_1 + m)}{j_1(2j_1 + 1)}\right]^{1/2}$

$m_1 + m_2 + m_3 = 0$, this result is

$$\begin{pmatrix} j_1 j_2 j_3 \\ m_1 m_2 m_3 \end{pmatrix} = (-1)^{j_1 - j_2 - m_3} \left[\frac{(j_1 + j_2 - j_3)!(j_2 + j_3 - j_1)!(j_3 + j_1 - j_2)!}{(j_1 + j_2 + j_3 + 1)!}\right]^{1/2}$$
$$\times \sum_\kappa (-1)^\kappa \left[\frac{[(j_1 + m_1)!(j_1 - m_1)!(j_2 + m_2)!]^{1/2}}{\kappa!(j_1 + j_2 - j_3 - \kappa)!(j_1 - m_1 - \kappa)!(j_2 + m_2 - \kappa)!}\right.$$
$$\left. \times \frac{[(j_2 - m_2)!(j_3 + m_3)!(j_3 - m_3)!]^{1/2}}{(j_3 - j_2 + m_1 + \kappa)!(j_3 - j_1 - m_2 + \kappa)!}\right] \tag{14}$$

The $3j$ symbol possesses symmetry properties that can be derived from this expression. Regge [6] expresses these by defining a 3×3 symbol:

$$\begin{pmatrix} j_1 j_2 j_3 \\ m_1 m_2 m_3 \end{pmatrix} = \begin{bmatrix} j_3 + j_2 - j_1 & j_1 + j_3 - j_2 & j_2 + j_1 - j_3 \\ j_1 - m_1 & j_2 - m_2 & j_3 - m_3 \\ j_1 + m_1 & j_2 + m_2 & j_3 + m_3 \end{bmatrix} \tag{15}$$

The symmetries are: (1) An even permutation of the columns makes no change, whereas an odd permutation multiplies by $(-1)^{j_1+j_2+j_3}$; (2) the same statement applies to permutations of the rows; (3) the value of the 3×3 symbol is unaltered by transposing its rows and columns.

For practical applications, it is essential to have tables of the cou-

Table 3[3]. $\langle m - m_2, m_2 | (j_1, 3/2) j, m \rangle$

$j =$	$m_2 = \pm 3/2$	$m_2 = \pm 1/2$
$j_1 + 3/2$	$\left[\frac{(j_1 \pm m - 1/2)(j_1 + m + 1/2)(j_1 \pm m + 3/2)}{(2j_1 + 1)(2j_1 + 2)(2j_1 + 3)}\right]^{1/2}$	$\left[\frac{3(j_1 \pm m + 1/2)(j_1 \pm m + 3/2)(j_1 \mp m + 3/2)}{(2j_1 + 1)(2j_1 + 2)(2j_1 + 3)}\right]^{1/2}$
$j_1 + 1/2$	$\mp\left[\frac{3(j_1 \pm m - 1/2)(j_1 \pm m + 1/2)(j_1 \mp m + 3/2)}{2j_1(2j_1 + 1)(2j_1 + 3)}\right]^{1/2}$	$\mp(j_1 \mp 3m + 3/2)\left[\frac{j_1 \pm m + 1/2}{2j_1(2j_1 + 1)(2j_1 + 3)}\right]^{1/2}$
$j_1 - 1/2$	$\left[\frac{3(j_1 \pm m - 1/2)(j_1 \mp m + 1/2)(j_1 \mp m + 3/2)}{(2j_1 - 1)(2j_1 + 1)(2j_1 + 2)}\right]^{1/2}$	$-(j_1 \pm 3m - 1/2)\left[\frac{j_1 \mp m + 1/2}{(2j_1 - 1)(2j_1 + 1)(2j_1 + 2)}\right]^{1/2}$
$j_1 - 3/2$	$\mp\left[\frac{(j_1 \mp m - 1/2)(j_1 \mp m + 1/2)(j_1 \mp m + 3/2)}{2j_1(2j_1 - 1)(2j_1 + 1)}\right]^{1/2}$	$\pm\left[\frac{3(j_1 + m - 1/2)(j_1 - m - 1/2)(j_1 \mp m + 1/2)}{2j_1(2j_1 - 1)(2j_1 + 1)}\right]^{1/2}$

Table 4[3]. $\langle m - m_2, m_2 | (j_1, 2) j, m \rangle$

$j =$	$m_2 = \pm 2$	$m_2 = \pm 1$	$m_2 = 0$
$j_1 + 2$	$\left[\frac{(j_1 \pm m - 1)(j_1 \pm 1)(j_1 \pm 1 + 1)(j_1 \pm m + 2)}{(2j_1 + 1)(2j_1 + 2)(2j_1 + 3)(2j_1 + 4)}\right]^{1/2}$	$\left[\frac{(j_1 - m + 2)(j_1 \pm m + 1)(j_1 \pm m)(j_1 + m + 2)}{(2j_1 + 1)(j_1 + 1)(2j_1 + 3)(j_1 + 2)}\right]^{1/2}$	$\left[\frac{3(j_1 - m + 2)(j_1 - m + 1)(j_1 + m + 2)(j_1 + m + 1)}{(2j_1 + 1)(2j_1 + 2)(2j_1 + 3)(j_1 + 2)}\right]^{1/2}$
$j_1 + 1$	$\mp\left[\frac{(j_1 \pm m - 1)(j_1 \pm 1)(j_1 \pm m + 1)(j_1 \mp m + 2)}{2j_1(j_1 + 1)(j_1 + 2)(2j_1 + 1)}\right]^{1/2}$	$\mp(j_1 \mp 2m + 2)\left[\frac{(j_1 \pm m + 1)(j_1 \pm m)}{2j_1(2j_1 + 1)(j_1 + 1)(j_1 + 2)}\right]^{1/2}$	$m\left[\frac{3(j_1 - m + 1)(j_1 + m + 1)}{j_1(2j_1 + 1)(j_1 + 1)(j_1 + 2)}\right]^{1/2}$
j_1	$\left[\frac{3(j_1 \pm m - 1)(j_1 \pm m)(j_1 \mp m + 1)(j_1 \mp m + 2)}{(2j_1 - 1)2j_1(j_1 + 1)(2j_1 + 3)}\right]^{1/2}$	$(1 \mp 2m)\left[\frac{3(j_1 \pm m)(j_1 \mp m + 1)}{(2j_1 - 1)2j_1(j_1 + 1)(2j_1 + 3)}\right]^{1/2}$	$\frac{3m^2 - j_1(j_1 + 1)}{[(2j_1 - 1)j_1(j_1 + 1)(2j_1 + 3)]^{1/2}}$
$j_1 - 1$	$\mp\left[\frac{(j_1 \pm m - 1)(j_1 \mp m)(j_1 \mp m + 1)(j_1 \mp m + 2)}{(j_1 - 1)2j_1(j_1 + 1)(2j_1 + 1)}\right]^{1/2}$	$\pm(j_1 \pm 2m - 1)\left[\frac{(j_1 \mp m + 1)(j_1 \mp m)}{(j_1 - 1)2j_1(j_1 + 1)(2j_1 + 1)}\right]^{1/2}$	$-m\left[\frac{3(j_1 - m)(j_1 + m)}{(j_1 - 1)j_1(j_1 + 1)(2j_1 + 1)}\right]^{1/2}$
$j_1 - 2$	$\left[\frac{(j_1 \mp m - 1)(j_1 \mp m)(j_1 \mp m + 1)(j_1 \mp m + 2)}{(2j_1 - 2)(2j_1 - 1)2j_1(2j_1 + 1)}\right]^{1/2}$	$-\left[\frac{(j_1 \mp m + 1)(j_1 \mp m)(j_1 - m - 1)(j_1 + m - 1)}{(j_1 - 1)(2j_1 - 1)j_1(2j_1 + 1)}\right]^{1/2}$	$\left[\frac{3(j_1 - m)(j_1 - m - 1)(j_1 + m)(j_1 + m - 1)}{(2j_1 - 2)(2j_1 - 1)j_1(2j_1 + 1)}\right]^{1/2}$

pling coefficients. In one form of these [7] the values of $\langle m_1, m_2|j, m)$ are tabulated as function of j, and m, the columns being labeled by values of m_2, and the rows by values of j. A separate table is made for each value of j_2.

Tables 1^3, 2^3, 3^3, and 4^3 give the $\langle m - m_2|(j_1, j_2)j, m)$ in this way for $j_2 = 1/2, 1, 3/2, 2$. Various numerical tables are available [8, 9].

As an explicit example, from Table 1^3 and (1) we can express the result of adding $j_2 = 1/2$ to any j_1,

$$\begin{aligned}\left|\alpha, j_1 + \frac{1}{2}, m\right) &= \left|\alpha, j_1, \frac{1}{2}, m - \frac{1}{2}, \frac{1}{2}\right) \left[\frac{j_1 + m + 1/2}{2j_1 + 1}\right]^{1/2} \\ &\quad + \left|\alpha, j_1, \frac{1}{2}, m + \frac{1}{2}, \frac{1}{2}\right) \left[\frac{j_1 - m + 1/2}{2j_1 + 1}\right]^{1/2} \\ \left|\alpha, j_1 - \frac{1}{2}, m\right) &= - \left|\alpha, j_1, \frac{1}{2}, m - \frac{1}{2}, \frac{1}{2}\right) \left[\frac{j_1 - m + 1/2}{2j_1 + 1}\right]^{1/2} \\ &\quad + \left|\alpha, j_1, \frac{1}{2}, m + \frac{1}{2}, -\frac{1}{2}\right) \left[\frac{j_1 + m + 1/2}{2j_1 + 1}\right]^{1/2}\end{aligned} \tag{16}$$

By considering the coupling process in two frames related so that R is the rotation that carries the unprimed frame over to the primed frame, we can find important relations between the coupling coefficients and the rotation matrices. In group theory these relations are expressed in terms of a series of matrices. The coefficients of these form the *Clebsch-Gordan series*. Applications to atomic theory were first made by Wigner.

An uncoupled state $|\alpha, j_1, j_2; m_1, m_2\rangle$ may be regarded as a formal product of $|\alpha, j_1, m_1\rangle$ and $|\alpha, j_2, m_2\rangle$. Therefore, such states transform, on rotation of the frame, according to $2^3 9$, that is, by

$$|\alpha, j_1, j_2; \mu_1, \mu_2\rangle = \sum_{m_1, m_2} |\alpha, j_1, j_2; m_1, m_2\rangle D^{(j_1)}_{m_1,\mu_1} D^{(j_2)}_{m_2,\mu_2} \tag{17}$$

The coupled states also transform via a single D matrix,

$$\begin{aligned}|\alpha, j_1, j_2; j, \mu) &= \sum_m |\alpha, j_1, j_2; j, m) D^{(j)}_{m,\mu} \\ &= \sum_m \sum_{m_1, m_2} |\alpha, j_1, j_2; m_1, m_2\rangle\langle m_1, m_2|j, m) D^{(j)}_{m,\mu}\end{aligned} \tag{18}$$

Alternatively, we have

$$\begin{aligned}|\alpha, j_1, j_2; \mu_1, \mu_2\rangle &= \sum_j |\alpha, j_1, j_2; j, \mu)(j, \mu|\mu_1, \mu_2\rangle \\ &= \sum_{\substack{j \\ m_1+m_2=m \\ \mu_1+\mu_2=\mu}} |\alpha, j_1, j_2, m_1, m_2\rangle\langle m_1, m_2|j, m) D^{(j)}_{m,\mu}(j, \mu|\mu_1, \mu_2\rangle\end{aligned} \tag{19}$$

Here the sum on j is over the values given by the vector-addition rule in (10) and the sums on (m_1, m_2) and (μ_1, μ_2) are subject to $m_1 + m_2 =$

m and $\mu_1 + \mu_2 = \mu$. Equating coefficients of $|\alpha, j_1, j_2; \mu_1, \mu_2\rangle$ in (17) and (19) we obtain

$$D^{(j_1)}_{m_1,\mu_1} D^{(j_2)}_{m_2,\mu_2} = \sum_{j,m,\mu} \langle m_1, m_2|j, m\rangle D^{(j)}_{m,\mu} (j, \mu|\mu_1, \mu_2\rangle \tag{20}$$

In terms of the $3j$ symbols this is

$$D^{(j_1)}_{m_1,\mu_1} D^{(j_2)}_{m_2,\mu_2} = \sum_j (2j+1) \begin{pmatrix} j_1 & j_2 & j \\ m_1 & m_2 & -m \end{pmatrix} D^{(j)}_{m,\mu} \begin{pmatrix} j_1 & j_2 & j \\ \mu_1 & \mu_2 & -\mu \end{pmatrix} \tag{20'}$$

Because the coupling coefficients describe a unitary transformation, we can multiply (20) on the left by $(j', m|m_1, m_2\rangle$ and on the right by $\langle \mu_1, \mu_2|j', \mu)$ and sum over $m_1 + m_2 = m$ and $\mu_1 + \mu_2 = \mu$. Using the orthogonality properties

$$\sum_{m_1+m_2=m} (j, m|m_1, m_2\rangle\langle m_1, m_2|j', m) = \delta(j', j)$$
$$\sum_{\mu_1+\mu_2=\mu} (j, \mu|\mu_1, \mu_2\rangle\langle \mu_1, \mu_2|j', \mu) = \delta(j, j')$$

the right side reduces to a single term, the result being

$$\begin{aligned} D^{(j)}_{m,\mu} &= \sum_{\substack{m_1+m_2=m \\ \mu_1+\mu_2=\mu}} (j, m|m_1, m_2\rangle D^{(j_1)}_{m_1,\mu_1} D^{(j_2)}_{m_2,\mu_2} \langle \mu_1, \mu_2|j, \mu) \\ &= \sum (2j+1) \begin{pmatrix} j_1 & j_2 & j \\ m_1 & m_2 & -m \end{pmatrix} D^{(j_1)}_{m_1,\mu_1} D^{(j_2)}_{m_2,\mu_2} \begin{pmatrix} j_1 & j_2 & j \\ \mu_1 & \mu_2 & -\mu \end{pmatrix} \end{aligned} \tag{21}$$

This relation provides a way of building up the $D^{(j)}_{m,\mu}$ from those of lower j_1 and j_2. Hence $D^{(j)}_{m,\mu}$ for any j may be constructed by starting from the $D^{(1/2)}_{m,\mu}$, which we shall determine in Section 5^3.

An alternative method [9g] for determining the coefficients $\langle m_1, m_2|j, m)$ consists of obtaining the left side of (1) by applying the operator $2\mathbf{J}_1 \cdot \mathbf{J}_2$, and its equivalent form $2J_{1z}J_{2z} + J_{1+}J_{2-} + J_{1-}J_{2+}$, to expression on the right side and then equating coefficients of each state vector. This procedure produces as many equations as there are coupling coefficients. The proper values of $2\mathbf{J}_1 \cdot \mathbf{J}_2$ are $j(j+1) - j_1(j_1 + 1) - j_2(j_2 + 1)$ and the proper values of the operator $2J_{1z}J_{2z} + J_{1+}J_{2-} + J_{1-}J_{2+}$ can be easily found by using the relation $J_z|\alpha, j, m\rangle = |\alpha, j, m\rangle m$ and $2^3 2$. At the end the coupling coefficients are normalized with the assumption that the phases can be chosen so that the coefficients are real.

For example, the expansion (1) of the spectroscopic state $|S = 1; L = 4; J = 3, M = 2\rangle$ where j_1, j_2, j, m are S, L, J, M, respectively, gives

$$\begin{aligned} |1, 4; 3, 2\rangle &= |1, 4; 1, 1\rangle\langle 1, 1|3, 2) + |1, 4; 0, 2\rangle\langle 0, 2|3, 2) \\ &\quad + |1, 4; -1, 3\rangle\langle -1, 3|3, 2) \\ &= a|1, 4; 1, 1\rangle + b|1, 4; 0, 2\rangle + c|1, 4; -1, 3\rangle \end{aligned}$$

The proper value of $2\mathbf{S} \cdot \mathbf{L}$, for the given state vectors, is -10. By acting on the right side of the above equality with the operator $2S_zL_z + S_+L_- + S_-L_+$ and by equating coefficients of each state vector we obtain:

$$\begin{aligned} -10a &= 2a + 6b \\ -10b &= 6a + 2\sqrt{7}c \\ -10c &= 2\sqrt{7}b - 6c \end{aligned}$$

Therefore,

$$b = -2a \quad \text{and} \quad c = \sqrt{7}a$$

The normalization condition

$$a^2 + b^2 + c^2 = a^2 + 4a^2 + 7a^2 = 1 \quad \text{gives} \quad a = \langle 1, 1|3, 2) = \pm \frac{1}{2\sqrt{3}}$$

To agree with the conventional phase we choose the positive value for $\langle 1, 1|3, 2)$ so that

$$b = \langle 0, 2|3, 2) = -\frac{1}{\sqrt{3}}$$

$$c = \langle -1, 3|3, 2) = \frac{1}{2}\left(\frac{7}{3}\right)^{1/2}$$

4^3. Orbital angular momentum

The Schrödinger operators for the orbital angular momentum of a particle having position $\mathbf{r}$ and momentum $\mathbf{p}$ are denoted by the vector $\mathbf{L}$, defined by Equation $1^3 11$ and hence obeying the commutation rules $1^3 10$.

Introducing spherical polar coordinates $(r, \theta, \phi,)$ in the usual way we have

$$\begin{aligned} \mathbf{r} &= r(\hat{\mathbf{i}} \sin\theta \cos\phi + \hat{\mathbf{j}} \sin\theta \sin\phi + \hat{\mathbf{k}} \cos\theta) \\ &= r\left[\frac{1}{2}(\hat{\mathbf{i}} + i\hat{\mathbf{j}}) \sin\theta\, e^{-i\phi} + \frac{1}{2}(\hat{\mathbf{i}} - i\hat{\mathbf{j}}) \sin\theta\, e^{i\phi} + \hat{\mathbf{k}} \cos\theta\right] \end{aligned} \tag{1}$$

and the operators suitable for acting on $\langle \mathbf{r}|\ \rangle$ are easily found to be

$$\begin{aligned} L_\pm &= e^{\pm i\phi}\left(\pm\frac{\partial}{\partial\theta} + i\cot\theta\frac{\partial}{\partial\phi}\right) \\ L_z &= -i\frac{\partial}{\partial\phi} \end{aligned} \tag{2}$$

from which we find

$$\mathbf{L}^2 = -\frac{1}{\sin\theta}\frac{\partial}{\partial\theta}\left(\sin\theta\frac{\partial}{\partial\theta}\right) - \frac{1}{\sin^2\theta}\frac{\partial^2}{\partial\phi^2} \tag{3}$$

The state $\langle \mathbf{r}|\alpha,l,m\rangle$ in which $\mathbf{L}^2 = l(l+1)$ and $L_z = m$ have precise values has the form of a product of an arbitrary function of r, multiplied by a function of θ and ϕ, which is customarily denoted by $Y(l, m; \theta, \phi) = \Theta_{lm}(\theta)\Phi_m(\phi)$. These functions are known as *spherical harmonics* [10]. They are well known in classical mathematical physics, having been first studied by P. S. Laplace and A. M. Legendre, in the calculation of the gravitational potential of nearly spherical mass distributions. Choices of the normalizing phases other than that used here are found in the literature.

The l values are usually indicated in atomic and nuclear physics by the letter code whose historical origins were given in Section 9[1]:

$$\begin{array}{l} l = 0\ 1\ 2\ 3\ 4\ 5\ 6\ 7\ \ldots \\ \text{code: } s\ p\ d\ f\ g\ h\ i\ k\ \ldots \end{array}$$

After 7 the letters follow alphabetically with omission of p and s, but such high l values do not occur in practice.

The dependence on ϕ is given by

$$L_z Y_{lm} = m\, Y_{lm}$$

and is therefore expressed by the factor $\Phi_m(\phi)$,

$$\Phi_m(\phi) = \frac{e^{im\phi}}{(2\pi)^{1/2}} \tag{4}$$

Here m and therefore l are restricted to integral values in order that $\langle \mathbf{r}|\alpha, l, m\rangle$ be single valued [11].

For the state $m = l$ the relation $L_+\langle \mathbf{r}|\alpha, l, l\rangle = 0$ gives a simple differential equation for $\Theta_{ll}(\theta)$,

$$\Theta_{ll}' - (l \cot \theta)\Theta_{ll} = 0,$$

which is easily found to have the normalized solution

$$\Theta_{ll}(\theta) = (-1)^l \left[\frac{1}{2}(2l+1)!\right]^{1/2} \frac{\sin^l \theta}{2^l l!} \tag{5}$$

where the arbitrary phase $(-1)^l$ is introduced for consistency with the choice made in $2^3 2$. Writing $R(\alpha l; r)$ for the radial factor and using $2^3 8$ we find

$$\langle \mathbf{r}|\alpha, l, 0\rangle = R(\alpha l; r)\left(\frac{2l+1}{4\pi}\right)^{1/2} \frac{(-1)^l}{2^l l!}(L_-)^l (\sin^l \theta\, e^{il\phi})$$

The following easily derived relation is useful

$$\begin{aligned} &(L_\pm)^k[f(\theta)\, e^{in\phi}] \\ &\quad = (\mp 1)^k \exp[i(n \pm k)]\phi \sin^{k\pm n}\theta \frac{d^k}{(d\cos\theta)^k}[\sin^{\pm n}\theta f(\theta)] \end{aligned} \tag{6}$$

We apply this with the lower sign, setting $f(\theta) = \sin^l \theta$ and $k = n = l$ to find

$$Y(l, 0; \theta, \phi) = \left(\frac{2l+1}{4\pi}\right)^{1/2} P_l(\cos\theta) \tag{7}$$

in which $P_l(\mu)$ is the lth *Legendre polynomial* in $\mu = \cos\theta$ as given by Rodrigues's formula,

$$P_l(\mu) = \frac{1}{2^l l!}\frac{d^l}{d\mu^l}(\mu^2 - 1)^l \tag{8}$$

To find the $Y(l, m; \theta, \phi)$ for $m \neq 0$ we use 2^37 and (6). For m > 0 we use the upper sign in 2^37 setting $m = 0$ and $k = m$ to get,

$$\langle \mathbf{r}|\alpha, l, m\rangle = \left[\frac{(l-m)!}{(l+m)!}\right]^{1/2} (L_+)^m \langle r|\alpha, l, 0\rangle$$

We now use (6) with the upper sign, $n = 0$ and $f(\theta) = P_l(\cos\theta)$ and get

$$(m > 0) \qquad Y(l, m; \theta, \phi) = (-1)^m e^{im\phi}\left[\frac{2l+1}{4\pi}\frac{(l-m)!}{(l+m)!}\right]^{1/2} \sin^m\theta \frac{d^m}{d\mu^m}P_l(\mu) \tag{9}$$

For $m < 0$ we use the lower sign in 2^37, setting $m = 0$ and $k = |m|$ to find

$$\langle \mathbf{r}|\alpha,l,m\rangle = \left[\frac{(l-|m|)!}{(l+|m|)!}\right]^{1/2} (L_-)^{|m|}\langle \mathbf{r}|\alpha, l, 0\rangle$$

We now use (6) with the lower sign, $n = 0$ and $f(\theta) = P_l(\cos\theta)$ to find

$$(m < 0) \qquad Y(l, m; \theta, \phi) = e^{im\phi}\left[\frac{2l+1}{4\pi}\frac{(l-|m|)!}{(l+|m|)!}\right]^{1/2} \sin^{|m|}\theta \frac{d^{|m|}}{d\mu^{|m|}}P_l(\mu) \tag{9'}$$

Thus the natural choice of phase made in Section 2^3 leads to a rather curious occurrence of $(-1)^m$ for $m > 0$ but not for $m < 0$. The dependence of probability distribution on θ is therefore the same for $\pm m$.

The dependence on θ of the $Y(l, m; \theta, \phi)$ involves the *associated Legendre functions*. Many slightly different notations are found in the literature. We write ($m \geqslant 0$),

$$P_l^m(\mu) = \frac{1}{2^l l!}(1-\mu^2)^{m/2}\frac{d^{l+m}}{d\mu^{l+m}}(\mu^2-1)^l = (1-\mu^2)^{m/2}\frac{d^m}{d\mu^m}P_l(\mu) \tag{10}$$

so that (9) and (9′) can be combined as

$$Y(l, m; \theta, \phi) = \left[\frac{2l+1}{4\pi}\frac{(l-|m|)!}{(l+|m|)!}\right]^{1/2} e^{im\phi}(-1)^{(m+|m|)/2}P_l^{|m|}(\mu) \tag{11}$$

Table 5³. Legendre polynomials, $P_k(z)$

k	Code	$P_k(z)$
0	s	1
1	p	z
2	d	$(3/2)\ z^2 - 1/2$
3	f	$(5/2)\ z^3 - (3/2)\ z$
4	g	$(35/8)\ z^4 - (15/4)\ z^2 + 3/8$

$$P_k(z) = 2^{-k}\ (k!)^{-1} \sum_{v=0}^{k} (-1)^{k-v} \binom{k}{v} \frac{(2v)!}{(2v-k)!} z^{2v-k}$$

for either sign of m. Also

$$Y^*(l, m; \theta, \phi) = (-1)^m Y(l, -m; \theta, \phi) \tag{11'}$$

At the pole ($\theta = 0$, $\mu = 1$) the Legendre polynomials are equal to unity. From (8) we have

$$\frac{d^l}{d\mu^l}(\mu^2 - 1)^l = \frac{d^l}{d\mu^l}(\mu - 1)^l(\mu + 1)^l = l!(\mu + 1)^l$$
$$+ \text{ terms that vanish when } \mu = 1$$

For $\mu = 1$ the first term is equal to $2^l l!$; hence $P_l(1) = 1$.

The spherical harmonics have an *inversion symmetry* property of importance. The direction opposite to (θ, ϕ) is $(\pi - \theta, \phi + \pi)$, so $\cos(\pi - \theta) = -\cos\theta$. The $(1 - \mu^2)^{|m|/2}$ factor is unaltered. The polynomial factor in μ is of degree $(l - |m|)$ and so is multiplied by $(-1)^{l-|m|}$. The $e^{im\phi}$ factor is multiplied by $(-1)^m$; therefore,

$$Y(l, m; \pi - \theta, \phi + \pi) = (-1)^l Y(l, m; \theta, \phi) \tag{12}$$

We say that the spherical harmonic has *positive parity* for even l, and *negative parity* for odd l.

The Legendre polynomials are sometimes defined by a *generating function*,

$$|\chi| < 1: \qquad \frac{1}{[1 - 2\mu\chi + \chi^2]^{1/2}} = \sum_{n=0}^{\infty} P_n(\mu)\chi^n \tag{13}$$

This result is of importance in calculating the matrix elements of Coulomb interaction between electrons.

Table 5³ gives some of the Legendre polynomials and Tables 6³ and 7³ give some of the spherical harmonics.

The quantity $Y^*(l, m)Y(l, m)\ d\Omega$ gives the probability that the particle's direction from the origin lies in the solid angle $d\Omega$ at (θ, ϕ).

Table 6[3]. Spherical harmonics, $Y(l, m; \theta, \phi)$

s	$Y(0, 0; \theta, \phi)$	$= (4\pi)^{-1/2}$
p	$Y(1, \pm 1; \theta, \phi)$	$= \mp(3/8\pi)^{1/2} \sin\theta\, e^{\pm i\phi}$
	$Y(1, 0; \theta, \phi)$	$= (3/4\pi)^{1/2} \cos\theta$
d	$Y(2, \pm 2; \theta, \phi)$	$= (15/32\pi)^{1/2} \sin^2\theta\, e^{\pm 2i\phi}$
	$Y(2, \pm 1; \theta, \phi)$	$= \mp(15/8\pi)^{1/2} \sin\theta \cos\theta\, e^{\pm i\phi}$
	$Y(2, 0; \theta, \phi)$	$= (5/4\pi)^{1/2}[(3/2)\cos^2\theta - 1/2]$
f	$Y(3, \pm 3; \theta, \phi)$	$= \mp (35/64\pi)^{1/2} \sin^3\theta\, e^{\pm 3i\phi}$
	$Y(3, \pm 2; \theta, \phi)$	$= (105/32\pi)^{1/2} \sin^2\theta \cos\theta\, e^{\pm 2i\phi}$
	$Y(3, \pm 1; \theta, \phi)$	$= \mp(21/64\pi)^{1/2} \sin\theta(5\cos^2\theta - 1)\, e^{\pm i\phi}$
	$Y(3, 0; \theta, \phi)$	$= (7/4\pi)^{1/2}[(5/2)\cos^3\theta - (3/2)\cos\theta]$
g	$Y(4, \pm 4; \theta, \phi)$	$= (3/8)(35/8\pi)^{1/2} \sin^4\theta\, e^{\pm 4i\phi}$
	$Y(4, \pm 3; \theta, \phi)$	$= \mp(3/4)(35/4\pi)^{1/2} \sin^3\theta \cos\theta\, e^{\pm 3i\phi}$
	$Y(4, \pm 2; \theta, \phi)$	$= (3/4)(5/8\pi)^{1/2} \sin^2\theta(7\cos^2\theta - 1)e^{\pm 2i\phi}$
	$Y(4, \pm 1; \theta, \phi)$	$= \mp(3/4)(5/4\pi)^{1/2} \sin\theta(7\cos^3\theta - 3\cos\theta)e^{\pm i\phi}$
	$Y(4, 0; \theta, \phi)$	$= (9/4\pi)^{1/2}[(35/8)\cos^4\theta - (15/4)\cos^2\theta + (3/8)]$

Note: $Y^*(l, m) = (-1)^m Y(l, -m)$

Table 7[3]. Squared values of spherical harmonics at directions of maximum

s	$\|Y(0, 0)\|^2 = 0.07958$	in all directions
p	$\|Y(1, 0)\|^2 = 0.23873$	at $\theta = 0$
d	$\|Y(2, 0)\|^2 = 0.39788$	at $\theta = 0$
f	$\|Y(3, 0)\|^2 = 0.55704$	at $\theta = 0$
g	$\|Y(4, 0)\|^2 = 0.71619$	at $\theta = 0$
p	$\|Y(1, \pm 1)\|^2 = 0.11937$	at $\theta = 90°$
d	$\|Y(2, \pm 1)\|^2 = 0.14921$	at $\theta = 45°$
f	$\|Y(3, \pm 1)\|^2 = 0.19806$	at $\theta \cong 31.1°$
g	$\|Y(4, \pm 1)\|^2 = 0.24959$	at $\theta \cong 23.9°$
d	$\|Y(2, \pm 2)\|^2 = 0.14921$	at $\theta = 90°$
f	$\|Y(3, \pm 2)\|^2 = 0.15474$	at $\theta \cong 54.7°$
g	$\|Y(4, \pm 2)\|^2 = 0.18498$	at $\theta \cong 40.9°$
f	$\|Y(3, \pm 3)\|^2 = 0.17407$	at $\theta = 90°$
g	$\|Y(4, \pm 3)\|^2 = 0.16523$	at $\theta = 60°$
g	$\|Y(4, \pm 4)\|^2 = 0.19583$	at $\theta = 90°$

The state $\langle \mathbf{r}|\alpha, 0, 0\rangle$ is spherically symmetric, thus representing a large departure from the original Bohr picture of motion in a plane orbit. It may be regarded as describing a situation in which every orientation of the orbit plane has equal probability.

These distributions are all independent of ϕ, in accord with the statement in Section 1^3 that all azimuths of **L** are equally likely.

The distribution in θ for $m = l$ varies as $\sin^{2l+1}\theta\, d\theta$. Its maximum is therefore in the equatorial plane ($\theta = \pi/2$). With increasing l the distribution is increasingly concentrated near this plane [12]. For $|m| < l$, but both large, the distribution is less concentrated near $\theta = \pi/2$, corresponding to a librational motion in θ over the range of angles expected from the orientation of **L** shown in Figure 1^3. Figure 2^3 shows the polar diagrams of some of these probability distributions in θ for (l, m) values that actually occur in atoms.

The spherical harmonic *addition theorem,* which expresses the Legendre polynomials of the angle ω between two specified directions (θ, ϕ) and (θ', ϕ') in terms of the spherical harmonics of (θ, ϕ) and (θ', ϕ'), has many applications. Let the direction of $\mathbf{r}$ be specified by (θ, ϕ) and (θ', ϕ') in two different frames. Then the function $Y(l, \mu; \theta, \phi)$ is expressible in terms of the $Y(l, m; \theta', \phi')$ of the other frame through a unitary matrix $D^{(l)}_{m\mu}$,

$$Y(l, \mu; \theta, \phi) = \sum_m Y(l, m; \theta', \phi')D^{(l)}_{m\mu}$$

The matrix $D^{(l)}_{m\mu}$ is unitary because each of the sets is orthonormal. The elements of $D^{(l)}_{m\mu}$ depend parametrically on the angles that specify the relations of the two frames. For two directions $\mathbf{r}_1$ and $\mathbf{r}_2$ we have

$$\begin{aligned}\sum_\mu & Y^*(l, \mu; \theta_1, \phi_1)Y(l, \mu; \theta_2, \phi_2)\\ &= \sum_{n,m} Y^*(l, n; \theta_1', \phi_1')\left(\sum_\mu D^{(l)*}_{n\mu}D^{(l)}_{m\mu}\right) Y(l, m; \theta_2', \phi_2') \qquad (14)\\ &= \sum_m Y^*(l, m; \theta_1', \phi_1')Y(l, m; \theta_2', \phi_2')\end{aligned}$$

since $\sum_\mu D^*_{n\mu}\, D_{m\mu} = \delta_{nm}$, because the D matrix is unitary. Hence the bilinear form in (14) is invariant to choice of axes. Now choose the primed system in such a way that its pole lies along the direction of $\mathbf{r}_2$, and its x-axis is in the plane of $\mathbf{r}_1$ and $\mathbf{r}_2$. Then all of the $Y(l, m; \theta_2', \phi_2')$ vanish for $m \neq 0$, and the right side of (14) reduces to $Y^*(l, 0; \omega, 0)Y(l, 0; 0, 0)$ in which ω is the angle between $\mathbf{r}_1$ and $\mathbf{r}_2$. From (7) this is $[(2l + 1)/4\pi]\, P_l(\cos\omega)$ giving the addition theorem for any pair of directions $\mathbf{r}_1$ and $\mathbf{r}_2$ relative to any coordinate frame,

$$\frac{4\pi}{2l + 1}\sum_\mu Y^*(l, \mu; \theta_1, \phi_1)Y(l, \mu; \theta_2, \phi_2) = P_l(\cos\omega) \qquad (15)$$

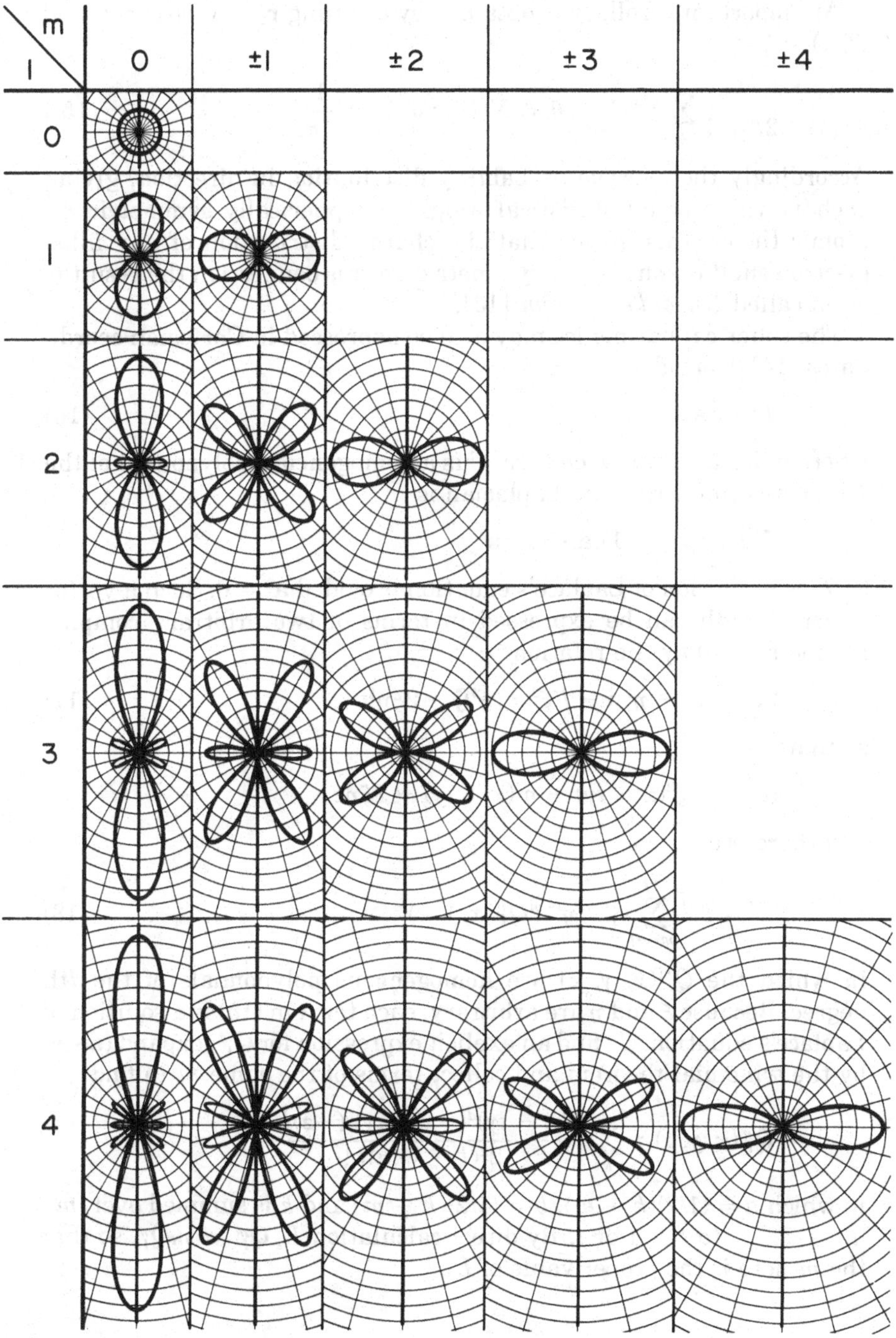

Figure 2[3]. Polar diagrams of spherical harmonics (squared). (Those for $m \neq 0$ are multiplied by 2.)

An important corollary is obtained by choosing $\mathbf{r}_1 = \mathbf{r}_2$ so $\omega = 0$ and $P_l(1) = 1$,

$$\frac{1}{2l+1}\sum_{\mu} Y^*(l, \mu; \theta, \phi)Y(l, \mu; \theta, \phi) = \frac{1}{4\pi} \tag{15'}$$

Accordingly the average probability distribution in direction, giving each m value equal statistical weight, is spherically symmetric. In atomic theory this means that the charge distribution of any filled electron shell is spherically symmetric. In this connection the result is often called *Unsöld's theorem* [13].

The spherical harmonics may also be expressed in Cartesian coordinates [14]. Consider

$$T^l = (\mathbf{a} \cdot \mathbf{r})^l \tag{16}$$

where $\mathbf{a}$ is a constant vector, so T^l is a homogeneous polynomial of the lth degree in (x, y, z). Its Laplacian is

$$\nabla^2 T^l = l(l-1)(\mathbf{a} \cdot \mathbf{r})^{l-2}\mathbf{a}^2$$

so T^l is a solution of Laplace's equation provided $\mathbf{a}^2 = 0$. Such a vector of zero length can be expressed in terms of two arbitrary complex numbers (ξ, η) by the relation

$$2\mathbf{a} = (\xi^2 - \eta^2)\hat{\mathbf{i}} - i(\xi^2 + \eta^2)\hat{\mathbf{j}} - 2\xi\eta\hat{\mathbf{k}} \tag{17}$$

so that

$$\mathbf{a} \cdot \mathbf{r} = \xi^2(x - iy) - \eta^2(x + iy) - 2\xi\eta z$$

and therefore

$$T^l = 2^{-l}\sum_{m=-l}^{+l} \xi^{l-m}\eta^{l+m}Q_l^m(x, y, z) \tag{18}$$

in which the $Q_l^m(x, y, z)$ are homogeneous polynomials of the lth degree. Because ξ and η are arbitrary, each term in (18) is a solution of Laplace's equation. To find an explicit expression for Q_l^m, expand $(\mathbf{a} \cdot \mathbf{r})$ by the trinomial theorem and collect terms of the same m to find

$$Q_l^m(x, y, z) = \sum_k \frac{l!(x - iy)^a(-x - iy)^b(-z)^{l-k}}{2^k(l-k)!a!b!}$$

in which $a = (1/2)(k - m)$, $b = (1/2)(k + m)$ and k is summed over $|m|$, $|m| + 2, \ldots$ to $l - 1$ or l. By direct calculation $L_z Q_l^m = mQ_l^m$ so that the m here is the proper value of L_z.

5³. Spin-angular momentum

In addition to the position vector, $\mathbf{r}$, an electron or nucleon requires an additional coordinate, s, for the description of its internal spin (Sec.

9^1). Usually s is taken to be equal to an allowed value of $S_z = \mathbf{k} \cdot \mathbf{S}$ in some particular frame. Here $\mathbf{S}$ is the vector operator describing spin angular momentum whose components satisfy $1^3 10$.

A particle is said to have spin S if the domain of values of s is S to $-S$ by integer steps. For electrons and nucleons the domain is $\pm 1/2$, which we abbreviate to $\pm$, S is always equal to 1/2 and so does not need to be explicitly written.

Pauli and C. G. Darwin independently [15], showed the applicability of the matrices $2^3 6$ with $j = 1/2$ to the description of a particle of spin 1/2. Thus the Schrödinger function for an electron or nucleon in state $|\alpha\rangle$ is given by $\langle \mathbf{r}, \mathrm{s}|\alpha\rangle$. Such a two-component quantity is called a *spinor,* the word having been coined by Paul Ehrenfest [16]. An extensive literature on spinors now exists [17].

In accordance with general principles, $|\langle \mathbf{r}, + |\alpha\rangle|^2$ and $|\langle \mathbf{r}, - |\alpha\rangle|^2$ give, respectively, the probabilities of finding the electron in unit volume at $\mathbf{r}$ with its S_z equal to $+1/2$ or $-1/2$.

Writing $u_\pm(\mathbf{r})$ for the functions $\langle \mathbf{r}, \pm |\alpha\rangle$, we now wish to find the spinor

$$\begin{bmatrix} u'_+(\mathbf{r}') \\ u'_-(\mathbf{r}') \end{bmatrix}$$

which describes the same state with reference to the rotated frame whose basis is $(\hat{\mathbf{i}}', \hat{\mathbf{j}}', \hat{\mathbf{k}}')$. We know $(1^2 29)$ that it transforms by a unitary 2×2 matrix, U,

$$\begin{bmatrix} u'_+ \\ u'_- \end{bmatrix} = U \begin{bmatrix} u_+ \\ u_- \end{bmatrix} = \begin{bmatrix} \xi & \eta \\ \lambda & \mu \end{bmatrix} \begin{bmatrix} u_+ \\ u_- \end{bmatrix} \tag{1}$$

and that $\mathbf{S}'$ in the primed frame has the same form $2^3 6$ as $\mathbf{S}$ has in the unprimed frame where

$$\mathbf{S}' = U\mathbf{S}U^\dagger \tag{2}$$

First we find the most general unitary 2×2 matrix, and then find which rotation of the frame corresponds to it. Unitarity requires

$$UU^\dagger = \begin{bmatrix} 1 & 0 \\ 0 & 1 \end{bmatrix} = \begin{bmatrix} \xi^*\xi + \eta^*\eta & \lambda^*\xi + \mu^*\eta \\ \xi^*\lambda + \eta^*\mu & \lambda^*\lambda + \mu^*\mu \end{bmatrix} \tag{3}$$

Vanishing of the nondiagonal elements gives

$$\mu = -\frac{\xi^*}{\eta^*\,\lambda}$$

and this with the requirement that the diagonal elements equal unity gives

$$\lambda^*\lambda = \eta^*\eta \quad \text{and} \quad \mu^*\mu = \xi^*\xi$$

These are satisfied, for example, by choosing

$$\lambda = -\eta^* \quad \text{and} \quad \mu = \xi^*$$

where ξ and η are any two complex numbers for which

$$\xi^*\xi + \eta^*\eta = 1 \tag{4}$$

Thus the 2 × 2 unitary matrices have the form

$$U = \begin{bmatrix} \xi & \eta \\ -\eta^* & \xi^* \end{bmatrix} \qquad U\dagger = \begin{bmatrix} \xi^* & -\eta \\ \eta^* & \xi \end{bmatrix} \tag{5}$$

The product of any two U's with parameters (ξ, η) and (ξ', η') is a U with parameters $(\xi\xi' - \eta\eta'^*, \xi\eta' + \xi'^*\eta)$, which satisfy (4). Hence the U matrices are the elements of a group. This group is called the two-dimensional special unitary group and is designated $SU(2)$. It is closely related to $O(3)$, the orthogonal group of rotations in three dimensions, but the two groups are not isomorphic (see Chap. 6).

From (3) we now find that the rotation described by U is

$$\begin{aligned} \hat{\mathbf{i}}' + i\hat{\mathbf{j}}' &= \xi^{*2}(\hat{\mathbf{i}} + i\hat{\mathbf{j}}) - \eta^{*2}(\hat{\mathbf{i}} - i\hat{\mathbf{j}}) - 2\xi^*\eta^*\hat{\mathbf{k}} \\ \hat{\mathbf{i}}' - i\hat{\mathbf{j}}' &= -\eta^2(\hat{\mathbf{i}} + i\hat{\mathbf{j}}) + \xi^2(\hat{\mathbf{i}} - i\hat{\mathbf{j}}) - 2\xi\eta\hat{\mathbf{k}} \\ \hat{\mathbf{k}}' &= \xi^*\eta(\hat{\mathbf{i}} + i\hat{\mathbf{j}}) + \xi\eta^*(\hat{\mathbf{i}} - i\hat{\mathbf{j}}) + (\xi^*\xi - \eta^*\eta)\hat{\mathbf{k}} \end{aligned} \tag{6}$$

Because the coefficients here are quadratic in ξ and η, it follows that $(-\xi, -\eta)$ represents the same rotation as (ξ, η). This mode of description of a rigid rotation has long been known in classical kinematics, where (ξ, η) are called the Cayley-Klein parameters [18].

A general rotation may be described by three Euler angles (α, β, γ). In the literature there is a confusing variety of definitions of them. We use the definition exhibited in Figure 3[3]:

α is a rotation about the $\hat{\mathbf{k}} = \hat{\mathbf{k}}'$ axis
β is a rotation about the $\hat{\mathbf{j}}' = \hat{\mathbf{j}}''$ axis
γ is a rotation about the $\hat{\mathbf{k}}'' = \hat{\mathbf{k}}'''$ axis

For the α rotation, because $\hat{\mathbf{k}} = \hat{\mathbf{k}}'$ we have $\eta = 0$, so $\xi^*\xi = 1$. Also because $(\hat{\mathbf{i}}' + i\hat{\mathbf{j}}') = e^{-i\alpha}(\hat{\mathbf{i}} + i\hat{\mathbf{j}})$, we find for the U_α,

$$U_\alpha = \pm \begin{bmatrix} e^{i\alpha/2} & 0 \\ 0 & e^{-i\alpha/2} \end{bmatrix} \tag{7}$$

This same form is also applicable to the γ rotation, because it is also about a $\hat{\mathbf{k}}$ axis.

Let V_β stand for the matrix for the β rotation, represented by

$$\begin{aligned} \hat{\mathbf{i}}'' &= \hat{\mathbf{i}}' \cos\beta - \hat{\mathbf{k}}' \sin\beta \\ \hat{\mathbf{k}}'' &= \hat{\mathbf{i}}' \sin\beta + \hat{\mathbf{k}}' \cos\beta \end{aligned}$$

From $\hat{\mathbf{k}}'' \cdot \hat{\mathbf{k}}' = \xi^*\xi - \eta^*\eta = \cos\beta$ and $\xi^*\xi + \eta^*\eta = 1$ we find

$$\xi^*\xi = \cos^2\frac{\beta}{2} \quad \text{and} \quad \eta^*\eta = \sin^2\frac{\beta}{2}$$

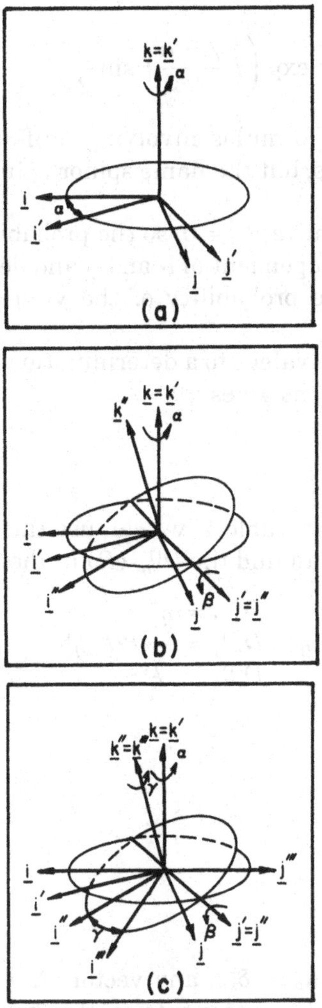

Figure 3³. Specification of rotation of reference frames by Euler's angles: α about $\mathbf{k} = \mathbf{k}'$, then β about $\mathbf{j}' = \mathbf{j}''$, then γ about $\mathbf{k}'' = \mathbf{k}'''$.

From $(\hat{\imath}'' \cdot \hat{\mathbf{k}}' - \hat{\mathbf{k}}'' \cdot \hat{\imath}')$ we find that $(\xi^* - \xi)(\eta^* - \eta) = 0$, so either one or both are real. From $\hat{\imath}'' \cdot \hat{\mathbf{j}}' = 0$ we find $(\xi^* - \xi)(\xi^* + \xi) + (\eta^* - \eta)(\eta^* + \eta) = 0$, so if either is real, the other one is also real. Therefore, both are real and

$$V_\beta = \pm \begin{bmatrix} \cos \beta/2 & \sin \beta/2 \\ -\sin \beta/2 & \cos \beta/2 \end{bmatrix} \qquad (8)$$

The U representing the product of these three rotations is

$$U = U_\gamma V_\beta U_\alpha = \pm \begin{bmatrix} \xi & \eta \\ -\eta^* & \xi^* \end{bmatrix}$$

with

$$\xi = \exp\left(i\frac{\gamma+\alpha}{2}\right)\cos\frac{\beta}{2}, \quad \eta = \exp\left(i\frac{\gamma-\alpha}{2}\right)\sin\frac{\beta}{2} \tag{9}$$

Because the spinors transform with formulas involving half-angles, they are sometimes called semivectors, but the name spinor is in more general use.

In a state for which $S_k = +1/2$ we have $u_- = 0$, so the probability of $S_{k'''}$ equaling $+1/2$ is $\xi^*\xi$, which is independent of α and γ and depends on β through $\cos^2 \beta/2$. Similarly the probability of the value $-1/2$ equals $\sin^2 \beta/2$.

The determination of U in (9) is equivalent to a determination of the $D^{(1/2)}_{m\mu}$ of $2^3 9$. Comparison of the notations gives

$$\begin{bmatrix} D_{++} & D_{+-} \\ D_{-+} & D_{--} \end{bmatrix} = \begin{bmatrix} \xi & -\eta^* \\ \eta & \xi^* \end{bmatrix} \tag{10}$$

As an exercise in the use of $3^3 21$ and Table 1^3 we can use the above expression to verify that $D^{(0)}_{00} = 1$ and to find the $D^{(1)}_{m\mu}(R)$ in the form

$$\begin{array}{lll} D^{(1)}_{11} = \xi^2 & D^{(1)}_{10} = -\sqrt{2}\xi\eta^* & D^{(1)}_{1-1} = \eta^{*2} \\ D^{(1)}_{01} = \sqrt{2}\xi\eta & D^{(1)}_{00} = \xi^*\xi - \eta^*\eta & D^{(1)}_{0-1} = -\sqrt{2}\xi^*\eta^* \\ D^{(1)}_{-11} = \eta^2 & D^{(1)}_{-10} = \sqrt{2}\xi^*\eta & D^{(1)}_{-1-1} = \xi^{*2} \end{array} \tag{11}$$

If we introduce the vector basis

$$\begin{aligned} \hat{\mathbf{a}}_1 &= -\frac{\hat{\mathbf{i}} + i\hat{\mathbf{j}}}{\sqrt{2}} \\ \hat{\mathbf{a}}_0 &= \hat{\mathbf{k}} \\ \hat{\mathbf{a}}_{-1} &= \frac{\hat{\mathbf{i}} - i\hat{\mathbf{j}}}{\sqrt{2}} \end{aligned} \tag{12}$$

which is orthonormal because $\hat{\mathbf{a}}^*_\lambda \cdot \hat{\mathbf{a}}_\mu = \delta_{\lambda\mu}$, any vector, $\mathbf{A}$, can be expressed as $\sum_\lambda (\hat{\mathbf{a}}_\lambda \cdot \hat{\mathbf{a}}^*_\lambda) \cdot \mathbf{A}$, so that its components on this basis are $\hat{\mathbf{a}}^*_\lambda \cdot \mathbf{A}$. The transformation (6) in terms of these basis vectors is

$$\hat{\mathbf{a}}'_\mu = \sum_\lambda \hat{\mathbf{a}}_\lambda D^{(1)*}_{\lambda\mu} \tag{13}$$

Therefore, these components of $\mathbf{A}$ transform by $D^{(1)}_{\mu\lambda}$

$$\begin{aligned} A'_\mu = \hat{\mathbf{a}}'^*_\mu \cdot \mathbf{A} &= \sum_\lambda \hat{\mathbf{a}}^*_\lambda \cdot \mathbf{A} D^{(1)}_{\lambda\mu} \\ &= \sum_\lambda A_\lambda D^{(1)}_{\lambda\mu} \end{aligned} \tag{14}$$

This set of components of an arbitrary vector transforms by the use of the $D^{(1)}_{\mu\lambda}$ matrix. Because of this relation to the irreducible representations of the rotation group formulas involving vectors are more simply expressed in terms of the $\hat{\mathbf{a}}_\mu$ basis than in terms of the more familiar $(\hat{\mathbf{i}}, \hat{\mathbf{j}}, \hat{\mathbf{k}})$ basis.

6³. Transformation of the $|\alpha, j, m\rangle$ under rotations

In the preceding section we found the law of transformation of a spinor under rotation of the coordinate frame. It is

$$\begin{aligned} \alpha' &= \alpha D^{(1/2)}_{++} + \beta D^{(1/2)}_{-+} \\ \beta' &= \alpha D^{(1/2)}_{+-} + \beta D^{(1/2)}_{--} \end{aligned} \tag{1}$$

in which the matrix for $D^{(1/2)}$ is

$$D^{(1/2)}_{m\mu} = \begin{bmatrix} \xi & -\eta^* \\ \eta & \xi^* \end{bmatrix} \tag{2}$$

as in 5³10 with (ξ, η) given by 5³9 in terms of the Euler angles defined in Figure 3³.

By use of 3³21 and Table 1³ we found the matrix 5³11 for $D^{(1)}_{m\mu}$ by regarding $j = 1$ as the sum of $j_1 = 1/2$ and $j_2 = 1/2$. We could find the $D^{(j)}_{m\mu}$ for arbitrary j by a simple continuation of this process, regarding j as arising from the addition of $j_2 = 1/2$ to $j_1 = j - 1/2$. Alternatively we can get the $D^{(j)}_{m\mu}$ for integral values of j by using Table 2³ to add $j_2 = 1$ to $j_1 = (j - 1)$ starting with the values in 5³11 for $D^{(1)}_{m\mu}$.

This procedure, while straightforward in principle, rapidly becomes laborious, not lending itself to the discovery of a general result. Another approach is more satisfactory.

By direct calculation

$$(\alpha' \beta') \begin{pmatrix} \alpha' \\ \beta' \end{pmatrix}^* = \alpha'^*\alpha' + \beta'^*\beta' = \alpha^*\alpha + \beta^*\beta$$

Hence $(\alpha^*\alpha + \beta^*\beta)$ is invariant under rotations of the frame. This is also true of $(\alpha^*\alpha + \beta^*\beta)^{2j}$. Therefore the $(2j + 1)$ quantities defined by

$$u_m = \frac{\alpha^{j+m}\beta^{j-m}}{\sqrt{(j+m)!(j-m)!}} \qquad (m = j, \ldots, -j) \tag{3}$$

have the property that

$$\sum_m u_m{}^* u_m = \frac{(\alpha^*\alpha + \beta^*\beta)^{2j}}{(2j)!}$$

is invariant. It follows that the quantities u_m transform by a unitary transformation. Writing u'_m for the quantities (3) in terms of α' and β' we have

$$\begin{aligned} \sqrt{(j+\mu)!(j-\mu)!}\; u'_\mu &= (\alpha\xi + \beta\eta)^{j+\mu}(-\alpha\eta^* + \beta\xi^*)^{j-\mu} \\ &= \sum_{\nu,\lambda} \binom{j+\mu}{\nu}\binom{j-\mu}{\lambda} \\ &\quad \times \alpha^{2j-\nu-\lambda}\beta^{\nu+\lambda}\xi^{j+\mu-\nu}\xi^{*\lambda}\eta^{\nu}(-\eta^*)^{j-\mu-\lambda} \end{aligned} \tag{4}$$

By defining m by the relations

$$j + m = 2j - (\nu + \lambda) \quad \text{and} \quad j - m = (\nu + \lambda)$$

the products of α and β occurring here can be replaced by u_m. The result is the explicit formula for transformation of the u's:

$$u'_\mu = \sum_m u_m D^{(j)}_{m\mu}$$

in which

$$D^{(j)}_{m\mu} = \left[\frac{(j+m)!(j-m)!}{(j+\mu)!(j-\mu)!}\right]^{1/2} \sum_\lambda \binom{j+\mu}{j-m-\lambda}\binom{j-\mu}{\lambda} \times \xi^{\mu+m+\lambda}(\xi^*)^\lambda \eta^{j-m-\lambda}(-\eta^*)^{j-\mu-\lambda} \tag{5}$$

In (4) the sum over ν runs from 0 to $j + \mu$ and that over λ from 0 to $j - \mu$, so in (5) the sum on λ runs from the larger of 0 and $-(\mu + m)$ up to the smaller of $(j - \mu)$ and $(j - m)$.

By using 5³9 the matrix elements $D_{m\mu}$ can be expressed in terms of the Euler angles of R. The result is

$$D^{(j)}_{m\mu}(R) = e^{im\alpha}\, d^{(j)}_{m\mu}(\beta) e^{i\mu\gamma} \tag{6a}$$

in which

$$d^{(j)}_{m\mu}(\beta) = \left[\frac{(j+m)!(j-m)!}{(j+\mu)!(j-\mu)!}\right]^{1/2} \left(\cos\frac{\beta}{2}\right)^{\mu+m} \left(\sin\frac{\beta}{2}\right)^{\mu-m} \times \sum_\lambda (-1)^{j-\mu-\lambda} \binom{j+\mu}{j-m-\lambda}\binom{j-\mu}{\lambda} \left(\cos\frac{\beta}{2}\right)^{2\lambda} \left(\sin\frac{\beta}{2}\right)^{2(j-\mu-\lambda)} \tag{6b}$$

Because $\cos^2 \beta/2 = (1/2)(1 + \cos\beta)$ and $\sin^2 \beta/2 = (1/2)(1 - \cos\beta)$ the sum over λ is a polynomial in $\cos\beta$.

These polynomials are known as *Jacobi polynomials.* Using the notation of Szegö [19], they may be defined as

$$P^{(a,b)}_n(x) = \frac{(-1)^n}{2^n n!}(1-x)^{-a}(1+x)^{-b} \frac{d^n}{dx^n}\left[(1-x)^{a+n}(1+x)^{b+n}\right]$$
$$= 2^{-n} \sum_{\nu=0}^{n} \binom{a+n}{\nu}\binom{b+n}{n-\nu}(x-1)^{n-\nu}(x+1)^\nu \tag{7}$$

These polynomials have the property

$$P^{(a,b)}_n(-x) = (-1)^n P^{(b,a)}_n(x)$$

In terms of them (6) becomes

$$d^{(j)}_{m\mu}(\beta) = \left[\frac{(j+m)!(j-m)!}{(j+\mu)!(j-\mu)!}\right]^{1/2} \left(\cos\frac{\beta}{2}\right)^{m+\mu} \left(\sin\frac{\beta}{2}\right)^{m-\mu} P^{(m-\mu,m+\mu)}_{j-m}(\cos\beta) \tag{8}$$

Comparing these results with the definitions of spherical harmonics in Section 4³ we find these relations (for integer l)

$$D^{(l)}_{m0} = (-1)^m \left(\frac{4\pi}{2l+1}\right)^{1/2} Y(l, m; \beta, \alpha) = (-1)^m D^{(l)*}_{m0} \tag{9}$$

$$D^{(l)}_{0\mu} = \left(\frac{4\pi}{2l+1}\right)^{1/2} Y(l, \mu; \beta, \gamma) \tag{10}$$

$$D^{(l)}_{00} = P_l(\cos\beta) \tag{11}$$

The state in which $\mathbf{J}$ has its maximum component, $\mu = j$, along the k' direction is one for which the Jacobi polynomial is of zeroth degree, and (6) reduces to

$$d^{(j)}_{mj}(\beta) = (-1)^{j-m}\left[\frac{(2j)!}{(j+m)!(j-m)!}\right]^{1/2}\left(\cos\frac{\beta}{2}\right)^{j+m}\left(\sin\frac{\beta}{2}\right)^{j-m} \tag{12}$$

For $j = 1/2$ this reduces to results already derived in Section 5³.

The Jacobi polynomials are orthogonal on the interval $-1 < x < +1$. The normalization integral can be calculated from (7) by repeated integrations by parts. The result is

$$\int_{-1}^{+1} (1-x)^a(1+x)^b P_n^{(a,b)}(x) P_m^{(a,b)}(x)\, dx = \delta(n, m)\frac{2^{a+b+1}}{2n+a+b+1}\frac{(a+n)!(b+n)!}{n!(a+b+n)!} \tag{13}$$

The Jacobi polynomials are related to the hypergeometric function,

$$F(\alpha, \beta, \gamma, z) = 1 + \frac{\alpha\beta}{\gamma} z + \frac{\alpha(\alpha+1)\beta(\beta+1)}{2!\gamma(\gamma+1)} z^2 \ldots \tag{14}$$

The relation is

$$P_n^{(a,b)}(x) = \frac{(-1)^n(b+n)!}{b!n!} \times F[a+b+n+1, -n, b+1, (1+x)/2] \tag{15}$$

The Jacobi polynomials may also be defined in terms of their rather complicated generating function,

$$\frac{(1+x)^{-a}(1-x)^{-b}[t+1-(1-2xt+t^2)^{1/2}]^a}{t^{a+b}} \times \frac{[t-1+(1-2xt+t^2)^{1/2}]^b}{(1-2xt+t^2)^{1/2}} = \sum_{n=0} t^n P_n^{(b,a)}(x) \tag{16}$$

The $D^{(j)}_{m\mu}$ (α, β, γ), regarded as functions of the Euler angles by (6) and (9) form an orthogonal function set over the range of variables: 0

to 2π for α and γ and -1 to $+1$ for $\cos\beta$. The overall result is

$$\int_0^{2\pi}\int_0^{\pi}\int_0^{2\pi} D^{(j_1)*}_{m_1\mu_1} D^{(j_2)}_{m_2\mu_2}\, d\gamma \sin\beta\, d\beta\, d\alpha = \frac{8\pi^2}{2j+1}\delta(j_1, j_2)\delta(m_1, m_2)\delta(\mu_1, \mu_2) \tag{17}$$

This completes the determination of the explicit matrices for transformation of the u_m under rotation of the frame. The Jacobi polynomials also occur in the theory of rotation of molecules with two equal moments of inertia (symmetric top) [20].

Whenever $j_2 = j_1$, a possible value for j is $j = 0$ for which $D^{(0)}_{00} = 1$. Applying $3^3 21$ gives

$$\sum_{m,\mu} (0, 0|m, -m\rangle D^{(j)}_{m,-\mu} D^{(j)}_{-m,\mu} \langle -\mu, \mu|0, 0) = 1 \tag{18}$$

in which the sum over both m and μ is from j to $-j$.

The series $3^3 20$ in combination with (17) permits the calculation of the integral over the Euler angles of the product of three D's:

$$\frac{1}{8\pi^2}\int_0^{2\pi}\int_0^{\pi}\int_0^{2\pi} D^{(j_1)}_{m_1\mu_1} D^{(j_2)}_{m_2\mu_2} D^{(j)}_{m\mu}\, d\gamma \sin\beta\, d\beta\, d\alpha = \begin{pmatrix} j_1 & j_2 & j \\ m_1 & m_2 & -m \end{pmatrix}\begin{pmatrix} j_1 & j_2 & j \\ \mu_1 & \mu_2 & -\mu \end{pmatrix} \tag{19}$$

To obtain this one uses $3^3 20$ to express the product of any two D's in a series of simple D's and then uses (17) to evaluate the resulting integrals.

Using (9), (10), and (11) we can find in this way the value of an integral over all directions of the product of three spherical harmonics:

$$\int_0^{2\pi}\int_0^{\pi} Y(l_1m_1)Y(l_2m_2)Y(l_3m_3)\sin\theta\, d\theta\, d\phi = \left[\frac{(2l_1+1)(2l_2+1)(2l_3+1)}{4\pi}\right]^{1/2}\begin{pmatrix} l_1 & l_2 & l_3 \\ 0 & 0 & 0 \end{pmatrix}\begin{pmatrix} l_1 & l_2 & l_3 \\ m_1 & m_2 & m_3 \end{pmatrix} \tag{20}$$

The integral vanishes (by parity) unless $(l_1 + l_2 + l_3)$ is an even integer, and also unless $m_1 + m_2 + m_3 = 0$. Using $4^3 11'$ this gives integrals in which one or more of the Y's are replaced by Y^*'s. In particular a commonly used set of coefficients is defined by

$$c^k(l, m; l'm') = \left(\frac{4\pi}{2k+1}\right)^{1/2}\int_0^{2\pi}\int_0^{\pi} Y(k, m-m') \times Y^*(l, m)Y(l', m')\sin\theta\, d\theta\, d\phi \tag{20'}$$

which is related to the $3j$ symbols in this way:

$$c^k(l, m; l'm') = (-1)^m\sqrt{(2l+1)(2l'+1)} \times \begin{pmatrix} l & l' & k \\ 0 & 0 & 0 \end{pmatrix}\begin{pmatrix} l & l' & k \\ -m & m' & m-m' \end{pmatrix} \tag{21}$$

These are used in calculation of matrix elements of electrostatic interaction (Sec. 7^4) and are tabulated in Table 6^4 of that section.

With the explicit formula for the rotation matrices it is easy to discuss their symmetry properties. By using the fact that $j + \mu$ is integral (for any j and μ) the symmetry properties of $d_{m\mu}^{(j)}$, which is introduced in (6a) and defined in (6b), can be given as

$$d_{m\mu}^{(j)}(-\beta) = (-1)^{m-\mu} d_{m\mu}^{(j)}(\beta) \tag{22a}$$
$$d_{m\mu}^{(j)}(\beta) = d_{-\mu,-m}^{(j)}(\beta) \tag{22b}$$
$$d_{\mu m}^{(j)}(\beta) = d_{m\mu}^{(j)}(-\beta) \tag{22c}$$

A further relation can be obtained by putting $\pi - \beta$ for β in the complete formula. It is

$$d_{m\mu}^{(j)}(\pi - \beta) = (-1)^{j-\mu} d_{-m,\mu}^{(j)}(\beta) \tag{23}$$

If equations (22a) and (22c) are combined,

$$d_{\mu m}^{(j)}(\beta) = (-1)^{m-\mu} d_{m\mu}^{(j)}(\beta) \tag{24}$$

From the explicit expression given in (6b) and (8) it follows that

$$d_{m\mu}^{(j)}(\beta + 2\pi) = (-1)^{2j} d_{m\mu}^{(j)}(\beta) \tag{25}$$

That is, if j is integral, the matrix elements $d_{m\mu}^{(j)}$ are unchanged when β is increased by 2π. However, if j is half-integral (1/2, 3/2, 5/2, ...), the matrix elements $d_{m\mu}^{(j)}$ change sign when β is increased by 2π, and an increase by 4π is needed to make a complete period. Therefore, the spinors change sign under a rotation 2π.

From the above relations the symmetry properties of the complete $D_{m\mu}^{(j)}$ matrix elements may be obtained. By using equations (6a) and (12c) and the reality of the $d_{m\mu}^{(j)}$ matrix elements,

$$D_{m\mu}^{(j)}(-\gamma, -\beta, -\alpha) = [D_{\mu m}^{(j)}(\alpha, \beta, \gamma)]^* \tag{26}$$

Similarly

$$[D_{m\mu}^{(j)}(\alpha, \beta, \gamma)]^* = (-1)^{m-\mu} D_{-m,-\mu}^{(j)}(\alpha, \beta, \gamma) \tag{27}$$

and

$$D_{m\mu}^{(j)}(\pi + \alpha, \pi - \beta, -\gamma) = e^{i\pi j} D_{m,-\mu}^{(j)}(\alpha, \beta, \gamma) \tag{28}$$

In the important case when j is an integer the factor in equation (28) becomes $(-1)^j$. Equation (28) is a useful relation when considering the effect of inverting all the coordinates in the origin ($x_i \rightarrow -x_i$, $y_i \rightarrow -y_i$, $z_i \rightarrow -z_i$ for all the particles in the system).

7^3. Irreducible tensors

Any scalar A is invariant under rotations of the frame, and therefore commutes with $\mathbf{J}$ and hence with $\mathbf{J}^2$. Its matrix is therefore diagonal in j and m.

From the fact that $J_\pm A = AJ_\pm$ we have

$$J_\pm A|\alpha, j, m\rangle = \sum_\alpha |\alpha', j, m\pm 1\rangle[(j \mp m)(j \pm m + 1)]^{1/2}\langle\alpha', j, m|A|\alpha, j, m\rangle$$

and also

$$AJ_\pm|\alpha, j, m\rangle = \sum_\alpha |\alpha', j, m \pm 1\rangle\langle\alpha', j, m \pm 1|A|\alpha, j, m \pm 1\rangle \times [(j \mp m)(j \pm m + 1)]^{1/2}$$

and therefore

$$\langle\alpha', j, m \pm 1|A|\alpha, j, m \pm 1\rangle = \langle\alpha', j, m|A|\alpha, j, m\rangle \tag{1}$$

Thus the matrix elements of a scalar are not only diagonal in m, but the actual value is independent of m. Applied to the Hamiltonian this shows that the energy levels of an atomic system (whose Hamiltonian commutes with $\mathbf{J}$) are degenerate in m.

The commutation rules for a vector $\mathbf{A}$ with $\mathbf{J}$ can be found by application of the definition $1^3 11$ for the cases $\mathbf{A} = \mathbf{r}$ or $\mathbf{p}$ to be

$$\begin{aligned} [J_i, A_j] &= iA_k \quad \text{(cyclic)} \\ [J_i, A_i] &= 0 \end{aligned} \tag{2}$$

In terms of $J_\pm = J_x \pm iJ_y$ and $A_\pm = A_x \pm iA_y$ the commutation relations are

$$\begin{aligned} &[J_\pm, A_\mp] = \pm 2A_z \\ &[J_\pm, A_\pm] = 0 \qquad [J_z, A_z] = 0 \\ &[J_\pm, A_z] = A_\pm \qquad [J_z, A_\pm] = \pm A_\pm \end{aligned} \tag{3}$$

By the use of fairly elaborate algebra, one can find the selection rules on (j, m) for the matrix elements of $\mathbf{A}$, and also the dependence on m' and m of the nonvanishing matrix elements. The results are that $\langle\alpha', j', m'|\mathbf{A}|\alpha, j, m\rangle$ vanishes unless

$$j' = j - 1, j, j + 1 \quad \text{and} \quad m' = m - 1, m, m + 1 \tag{4}$$

The m selection rule is easily found. From $[J_z, A_z] = 0$ we find that

$$(m' - m)\langle\alpha', j', m'|A_z|\alpha, j, m\rangle = 0,$$

and therefore this matrix element vanishes unless $m' = m$. Similarly, from $[J_z, A_\pm] = \pm A_\pm$ we find

$$[m' - (m \pm 1)]\langle\alpha', j', m'|A_\pm|\alpha, j, m\rangle = 0$$

Hence the matrix element of A_+ vanishes unless $m' = m + 1$, and that of A_- vanishes unless $m' = m - 1$.

Comparison of these results with the selection rules $j' = j$ and $m' = m$ for a scalar suggests that the scalar and vector may be the first two members of a hierarchy of $(2k + 1)$-component quantities. This is in fact the case, so we proceed to explore this idea instead of carrying out the special calculations for the vector **A**.

When we write a vector in invariant form in terms of its components A_i and of frame vectors $\hat{\mathbf{e}}_i$

$$\mathbf{A} = \sum_j \hat{\mathbf{e}}_j A_j = \sum_i \hat{\mathbf{e}}'_i A'_i$$

for two bases that are connected by the rotation R (Sec. 6^2), we have

$$A'_i = R_{ij}\mathrm{A}_j \tag{5}$$

and

$$\mathbf{A} = \sum_j \hat{\mathbf{e}}_j \mathrm{A}_j = \sum_{i,j} \hat{\mathbf{e}}'_i R_{ij} A_j$$

and therefore

$$\hat{\mathbf{e}}_j = \sum_i \hat{\mathbf{e}}'_i R_{ij}$$

showing that the frame vectors transform contragradiently to the components.

The traditional generalization to Cartesian tensors of higher order involves introduction of quantities having 3^k components, distinguished as $A_{lmn}\ldots$ with k indices each varying over 1, 2, 3, which transform on rotation of the frame by

$$A'_{lmn\ldots} = \sum [R_{li}R_{mj}R_{nk}\ldots]A_{ijk\ldots} \tag{6}$$

with k indices on the A' and A and k factors involving components of the R matrix.

This particular generalization is inconvenient for description of the kinds of directed quantities that occur in atomic physics. This is because the tensors built on the pattern of (6) do not transform by irreducible rotations of the rotation group in general (except for $k = 0$, 1).

Suppose that we have a $(2k + 1) \times (2k + 1)$ representation of the rotation group, $E^{(k)}_{ij}(R)$. Then we can define as *tensors*, sets of $(2k + 1)$ quantities, or operators, which transform by the representation

$$A^{(k)\prime}_i = E^{(k)}_{ij}(R)A^{(k)}_j \tag{7}$$

Here i and j range over the values $k, k - 1, \ldots -k$. $\mathbf{A}^{(k)}$ for the tensor of order k whose components are the $A^{(k)}_j$ of (7).

Suppose now that a unitary matrix T can be found such that, for every R

$$TE^{(k)}(R)T^{-1} = D^{(k)}(R) \tag{8}$$

in which $D^{(k)}(R)$ is one of the irreducible representations of the rotation group as defined in Section 6^2 and found explicitly in Section 6^3.

Then from (7) we have

$$\sum_i A_i^{(k)'} T_{il}^{-1} = \sum_m \left(\sum_j A_j^{(k)} T_{jm}^{-1} \right) \left(\sum_{n,p} T_{mn} E_{np}^{(k)}(R) T_{pl}^{-1} \right)$$
$$= \sum_m \left(\sum_j A_j^{(k)} T_{jm}^{-1} \right) \mathrm{D}_{ml}^{(k)}(R) \qquad (9)$$

Thus the set of $(2k + 1)$ quantities

$$\sum_j A_j^{(k)} T_{jm}^{-1}$$

can be adopted as an alternative set for the description of the tensor $\mathbf{A}^{(k)}$. These components are so chosen that they transform, according to (9) by means of the irreducible representation matrices $D_{ml}^{(k)}(R)$, as found in Section 6[3].

A tensor that can be described in this way, by components that transform by a single $D^{(k)}(R)$ matrix, is called an *irreducible tensor.* Although the components need not be given in this form, in fact we shall always work with components that do transform in this way. The components that transform in this way are called the *spherical components* of the tensor.

The case $k = 1$ for an ordinary vector provides an important example. The Cartesian components (A_x, A_y, A_z) transform by the R matrix as in (5). But if $\mathbf{A}$ is described by means of its spherical components, which transform by $D_{ml}^{(1)}$ with m and l taking the values ± 1 and 0, we find that these are related to the Cartesian components by

$$A_{\pm 1}^{(1)} = \mp \frac{1}{\sqrt{2}} (A_x \pm iA_y), \quad A_0^{(1)} = A_z \qquad (10)$$

Returning to (7), if no T can be found that transforms $E^{(k)}(R)$ for every R to a single $D^{(k)}(R)$, then, instead, a T can be found to transform every $E^{(k)}(R)$ to a step matrix consisting of

$$D^{(k_1)}(R), D^{(k_2)}(R), \ldots$$

along the principal diagonal, where

$$k_1 + k_2 + \ldots = k$$

and particular values of the k_a may be repeated. In this case we say that $\mathbf{A}^{(k)}$ is a reducible tensor that contains the $\mathbf{A}^{(k_1)}, \mathbf{A}^{(k_2)}, \ldots$ as irreducible constituents.

The Cartesian tensor of the second order, having 3^2 components, $A_{xx}, A_{xy}, \ldots, A_{zz}$ provides another illustrative example. It is reducible to a scalar $\mathbf{A}^{(0)}$, a vector $\mathbf{A}^{(1)}$, and a tensor $\mathbf{A}^{(2)}$. The distribution by components is $1 + 3 + 5 = 9$. The relation of these irreducible constituents to the Cartesian components is easily found to be

$$\begin{aligned}
A_0^{(0)} &= A_{xx} + A_{yy} + A_{zz} \\
A_{\pm 1}^{(1)} &= \mp\frac{1}{2\sqrt{2}}[(A_{yz} - A_{zy}) \pm i(A_{zx} - A_{xz})] \\
A_0^{(1)} &= \frac{1}{2}[A_{xy} - A_{yx}] \qquad (11)\\
A_{\pm 2}^{(2)} &= [3/2]^{1/2}(S_{11} - S_{22} \pm 2iS_{12}) \\
A_{\pm 1}^{(2)} &= \mp\sqrt{6}(S_{13} \pm iS_{23}) \\
A_0^{(2)} &= 2S_{33} - S_{11} - S_{22}
\end{aligned}$$

where

$$S_{ij} = \frac{1}{2}(A_{ij} + A_{ji}) - \frac{1}{3}\delta_{ij}A_0^{(0)}$$

The spherical harmonics, regarded as functions of the direction of a vector **r**, may be written $Y(l, m; \mathbf{r})$. These transform by the D matrices (Sec. 4 ³) and hence $P(r)Y(l, m; \mathbf{r})$ (where $P(r)$ is an arbitrary function of r) are the $(2l + 1)$ components of a tensor of order l. By 4³11′, an irreducible tensor operator is Hermitian if

$$T_q^{(k)\dagger} = (-1)^q T_q^{(k)} \qquad (11')$$

We turn now to the investigation of the (j, m) dependence of the matrix elements of an irreducible tensor of order k. The general results were discovered independently by Eckart and by Wigner [21] and are now generally called the *Wigner-Eckart theorem.*

According to these results, the matrix elements for the various components $A_\kappa^{(k)}$ depend on j', m' and j, m through a vector coupling coefficient (Sec. 3³) that is the same for all tensors of order k:

$$\begin{aligned}
&\langle \alpha', j', m' | A_\kappa^{(k)} | \alpha, j, m\rangle \\
&\quad = (-1)^{j'-m'} \begin{pmatrix} j' & k & j \\ -m' & \kappa & m \end{pmatrix} \langle \alpha', j' \| A^{(k)} \| \alpha, j\rangle \qquad (12)
\end{aligned}$$

Here the $\langle \alpha', j' \| A^{(k)} \| \alpha, j\rangle$ are called *reduced-matrix elements.* The dependence on κ, m, and m' is entirely expressed by the vector coupling coefficient. This definition of the reduced-matrix elements is different from the one given by 9³11 of TAS. The above reduced-matrix elements are not Hermitian but satisfy the following relations:

$$\langle \alpha', j' \| A^{(k)} \| \alpha, j\rangle = (-1)^{j'-j} \langle \alpha, j \| A^{(k)\dagger} \| \alpha', j'\rangle^* \qquad (12')$$

for any tensor operator. These relations differ from a similar expression on page 63 of TAS, by a phase factor $(-1)^{j'-j}$ (see also Racah [2aII, p. 443]). From the results of Section 3³ we know that the coupling coefficients vanish unless

$$m' = m + \kappa$$

so this is the generalization of the selection rule on m'. They also vanish unless

$$j' = (j + k), (j + k - 1), \ldots, |j - k|$$

so this is the generalization of the selection rule on j' already stated for scalars and vectors.

Proof of the result (12) requires that we find the commutation rules of $A_\kappa^{(k)}$ with $\mathbf{J}$. For any rotation we have, from $1^3 3$,

$$A_\kappa^{(k)\prime} D = D A_\kappa^{(k)}$$

For an infinitesimal rotation through ω about the axis in the direction of $\boldsymbol{\omega}$ we have

$$D_{i\kappa}^{(k)}(R) = \delta_{i\kappa} - i\omega \langle k, i|J_\omega|k, \kappa\rangle$$

so that

$$J_\omega A_\kappa^{(k)} - A_\kappa^{(k)} J_\omega = \sum_i A_i^{(k)} \langle k, i|J_\omega|k, \kappa\rangle$$

Using the matrix elements of J_ω from Section 2^3 we find

$$[J_\pm, A_\kappa^{(k)}] = \mathrm{A}_{\kappa\pm1}^{(k)} \, [(k \mp \kappa)(k \pm \kappa + 1)]^{1/2} \tag{13a}$$

$$[J_z, A_\kappa^{(k)}] = A_\kappa^{(k)} \kappa \tag{13b}$$

From (13b) we find

$$[m' - (m + \kappa)] \langle \alpha', j', m'|A_\kappa^{(k)}|\alpha, j, m\rangle = 0$$

in agreement with the selection rule on m for the coupling coefficients. From (13a) we have

$$\langle \alpha', j', m'|J_\pm A_\kappa^{(k)}|\alpha, j, m\rangle - \langle \alpha', j', m'|A_\kappa^{(k)} J_\pm|\alpha, j, m\rangle$$
$$= [(k \mp \kappa)(k \pm \kappa + 1)]^{1/2} \langle \alpha', j', m'|A_{\kappa\pm1}^{(k)}|\alpha, j, m\rangle$$

In the first element, we replace $\langle \alpha', j', m'|J_\pm$ by $J_\mp|\alpha', j', m'\rangle$ and obtain the result

$$\langle \alpha', j', m' \mp 1|A_\kappa^{(k)}|\alpha, j, m\rangle [(j' \pm m')(j' \mp m' + 1)]^{1/2}$$
$$- \langle \alpha', j', m'|A_\kappa^{(k)}|\alpha, j, m \pm 1\rangle [(j \mp m)(j \pm m + 1)]^{1/2}$$
$$= \langle \alpha', j', m'|A_{\kappa\pm1}^{(k)}|\alpha, j, m\rangle [(k \mp \kappa)(k \pm \kappa + 1)]^{1/2}$$

Each term here vanishes unless $m' = m \pm 1 + \kappa$; accordingly we assign this value to m' and find

$$\langle \alpha', j', (m + \kappa)|A_\kappa^{(k)}|\alpha, j, m\rangle \{[j' \mp (m + \kappa)][j' \pm (m + \kappa) + 1]\}^{1/2}$$
$$= \langle \alpha', j', (m \pm 1) + \kappa|A_\kappa^{(k)}|\alpha, j, m \pm 1\rangle [(j \mp m)(j \pm m + 1)]^{1/2}$$
$$+ \langle \alpha', j', (m + \kappa) \pm 1|A_{\kappa\pm1}^{(k)}|\alpha, j, m\rangle [(k \mp \kappa)(k \pm \kappa + 1)]^{1/2} \tag{14}$$

We now consider the relations involved in coupling angular momentum vectors $\mathbf{j}$ and $\mathbf{k}$. From $3^3 1$,

$$|\alpha, j, k; j', m') = \sum_{m+\kappa=m'} |\alpha, j, k; m, \kappa\rangle \langle m, \kappa|j', m')$$

Applying the operator $J_\mp + K_\pm$ to both sides we find

$$|\alpha, j, k; j', m' \mp 1)[(j' \pm m')(j' \mp m' + 1)]^{1/2}$$
$$= \sum_{m+\kappa=m'} |\alpha, j, k; m \mp 1, \kappa\rangle\langle m, \kappa|j', m')[(j \pm m)(j \mp m + 1)]^{1/2}$$
$$+ |\alpha, j, k; m, \kappa \pm 1\rangle\langle m, \kappa|j', m')[(k \mp \kappa)(k \pm \kappa + 1)]^{1/2}$$

On the left we replace the coupled state by the series of uncoupled states $|\alpha, j, k; \mu, \lambda\rangle$. On the right we replace $m \mp 1$ with μ, and κ with λ on the first line, and m with μ and $\kappa \mp 1$ with λ on the second. Then equating coefficients of $|\alpha, j, k; \mu, \lambda\rangle$ on both sides, we find that the vector-coupling coefficients are connected by the same set of equations (14) as are satisfied by the matrix elements of $A_\kappa^{(k)}$. Therefore, each nonvanishing matrix element of $A_\kappa^{(k)}$ is proportional to the corresponding coupling coefficient, as stated in (12).

8³. Tensor products of irreducible tensors

A product of two tensors $\mathbf{A}^{(k)}$ and $\mathbf{B}^{(l)}$ is a quantity built out of the ensemble of $(2k + 1)(2l + 1)$ bilinear products $A_\kappa^{(k)}B_\lambda^{(l)}$ for the full range of values of κ and λ. A tensor is said to vanish if each of its components vanishes in any frame (if they all vanish in one frame, they vanish in every frame). Hence any tensor product having these bilinear monomials as components will have the familiar property that the product vanishes if either $\mathbf{A}^{(k)} = 0$ or $\mathbf{B}^{(l)} = 0$.

The tensor built of all of these components is reducible. This is an extension of the result about the product of two vectors $\mathbf{A}^{(1)}\mathbf{B}^{(1)}$, which has nine components. These nine components can be regarded as three distinct irreducible tensor products: One component belongs to the usual scalar product, $\mathbf{A} \cdot \mathbf{B}$; three more belong to the usual vector product, $\mathbf{A} \times \mathbf{B}$; and the other five belong to the symmetrical tensor product $(1/2)(\mathbf{AB} + \mathbf{BA}) - (1/3)\mathbf{A} \cdot \mathbf{B}\mathbf{I}$ in which $\mathbf{I}$ is the unit second-order tensor whose Cartesian expression is $(\hat{\mathbf{i}}\hat{\mathbf{i}} + \hat{\mathbf{j}}\hat{\mathbf{j}} + \hat{\mathbf{k}}\hat{\mathbf{k}})$.

In a similar way the product elements $A_\kappa^{(k)}B_\lambda^{(l)}$ may be transformed to a set of irreducible tensor products of orders, n, given by the usual vector-coupling values

$$n = k + l, k + l - 1, \ldots, |k - l| \tag{1}$$

We adopt the notation

$$[\mathbf{A}^{(k)} \times \mathbf{B}^{(l)}]^{(n)}$$

for the irreducible tensor of order n, which is one of the products of $\mathbf{A}^{(k)}$ and $\mathbf{B}^{(l)}$.

The rule of transformation of the basic components on making a rotation R of the coordinate frame is (Sec. 7³),

$$A_\kappa^{(k)\prime}\, B_\lambda^{(l)\prime} = \sum_{\kappa', \lambda'} A_{\kappa'}^{(k)} B_{\lambda'}^{(l)} D_{\kappa'\kappa}^{(k)}(R) D_{\lambda'\lambda}^{(l)}(R)$$

By using $3^3 20$ or $3^3 20'$ the products of the D's can be replaced by a series of single D's, giving

$$A_\kappa^{(k)'} B_\lambda^{(l)'} = \sum_{\kappa'+\lambda'=\nu} A_{\kappa'}^{(k)} B_{\lambda'}^{(l)} \sum_\nu \langle \kappa', \lambda' | (k, l) j, \nu \rangle \sum_\mu D_{\nu\mu}^{(j)} \langle j, \mu (k, l) | \kappa, \lambda \rangle$$

Multiplying by $\langle \kappa, \lambda | (k, l) n, \mu \rangle$, summing over (κ, λ), and using the orthogonality property $3^3 3$ of the coupling coefficients, we find

$$\sum_{\kappa+\lambda=\mu} A_\kappa^{(k)'} B_\lambda^{(l)'} \langle \kappa, \lambda | (k, l) n, \mu \rangle = \sum_{\kappa'+\lambda'=\nu} A_{\kappa'}^{(k)} B_{\lambda'}^{(l)} \langle \kappa', \lambda' | (k, l) n, \nu \rangle D_{\nu\mu}^{(n)} \tag{2}$$

This shows that the quantities on the left transform as the components of an irreducible tensor of order n, where n is one of the values in (1) for which the coupling coefficients exist. Therefore the nth-order irreducible tensor product has these components.

$$[\mathbf{A}^{(k)} \times \mathbf{B}^{(l)}]_\mu^{(n)} = \sum_{\kappa+\lambda=\mu} A_\kappa^{(k)} B_\lambda^{(l)} \langle \kappa, \lambda | (k, l) n, \mu \rangle \tag{3}$$

in which $\mu = n, n-1, \ldots, -n$.

This general result shows that a scalar product $(n = 0)$ exists only for two tensors of equal order, $k = l$. From $3^3 11$ we have

$$\langle \kappa, \lambda | (k, k)\, 0, 0 \rangle = \frac{(-1)^{k-\kappa}}{(2k+1)^{1/2}} \delta(\lambda, -\kappa)$$

and therefore the scalar product is

$$[\mathbf{A}^{(k)} \times \mathbf{B}^{(k)}]_0^{(0)} = \frac{(-1)^k}{(2k+1)^{1/2}} \sum_\kappa A_\kappa^{(k)} (-1)^\kappa B_{-\kappa}^{(k)} \tag{4}$$

In the special case of ordinary vectors $(k = 1)$ this reduces to $-\mathbf{A} \cdot \mathbf{B}/\sqrt{3}$, the usual scalar product, except that here it has an unusual normalization. Similarly, the components of the tensor product of two vectors are

$$\begin{aligned}
[\mathbf{A}^{(1)} \times \mathbf{B}^{(1)}]_{+1}^{(1)} &= \frac{1}{\sqrt{2}} (A_{+1} B_0 - A_0 B_{+1}) \\
[\mathbf{A}^{(1)} \times \mathbf{B}^{(1)}]_0^{(1)} &= \frac{1}{\sqrt{2}} (A_{+1} B_{-1} - A_{-1} B_{+1}) \\
[\mathbf{A}^{(1)} \times \mathbf{B}^{(1)}]_{-1}^{(1)} &= \frac{1}{\sqrt{2}} (A_0 B_{-1} - A_{-1} B_0) \\
[\mathbf{A}^{(1)} \times \mathbf{B}^{(1)}]_{\pm 2}^{(2)} &= A_{\pm 1} B_{\pm 1} \\
[\mathbf{A}^{(1)} \times \mathbf{B}^{(1)}]_{\pm 1}^{(2)} &= \frac{A_0 B_{\pm 1} + A_{\pm 1} \mathrm{B}_0}{\sqrt{2}}
\end{aligned} \tag{5}$$

$$[\mathbf{A}^{(1)} \times \mathbf{B}^{(1)}]_0^{(2)} = \frac{3A_0B_0 - \mathbf{A}\cdot\mathbf{B}}{\sqrt{6}}$$
$$= \frac{1}{\sqrt{6}}(A_{+1}B_{-1} + 2A_0B_0 + A_{-1}B_{+1}) \tag{5}$$

As a further example, the scalar product of the spherical harmonic tensors $Y(k,\kappa;\, \theta, \phi)$ and $Y(k,\kappa;\, \theta', \phi')$, by using $4^3 11'$, is

$$[\mathbf{Y}(k;\, \theta, \phi)\times\mathbf{Y}(k;\, \theta', \phi')]_0^{(0)} = \frac{(-1)^k}{(2k+1)^{1/2}}\sum_\kappa Y^*(k,\kappa;\, \theta, \phi)Y(k,\kappa;\, \theta', \phi')$$

Here we write $\mathbf{Y}(k;\, \theta, \phi)$ for the kth-order irreducible tensor whose components are the $Y(k, \kappa;\, \theta, \phi)$. Therefore, from $4^3 15$,

$$P_k(\cos\omega) = \frac{4\pi(-1)^k}{(2k+1)^{1/2}}[\mathbf{Y}(k;\, \theta, \phi) \times \mathbf{Y}(k;\, \theta', \phi')]_0^{(0)} \tag{6}$$

giving another form of the spherical harmonic addition theorem. Here ω is the angle between (θ, ϕ) and (θ', ϕ').

The Coulomb interaction energy between two electrons at $\mathbf{r}$ and $\mathbf{r}'$ may be expanded in terms of these scalar products of spherical harmonic tensors as

$$\frac{e^2}{|\mathbf{r} - \mathbf{r}'|} = 4\pi e^2 \sum_{k=0}^{\infty} \frac{r_<^k}{r_>^{k+1}} \frac{(-1)^k}{(2k+1)^{1/2}}[\mathbf{Y}(k;\, \theta, \phi) \times \mathbf{Y}(k;\, \theta', \phi')]_0^{(0)} \tag{7}$$

in which $r_<$ is the lesser and $r_>$ the greater of r and r'.

9^3. Coupling of more than two angular momenta

Situations arise in which it is necessary to transform from uncoupled to coupled states involving more than two angular momenta. This requires an extension of the discussion given for two angular momenta (Sec. 3^3).

First we consider three angular momenta, $\mathbf{J}_1$, $\mathbf{J}_2$, and $\mathbf{J}_3$, which refer to different parts of the system and hence commute with each other. The uncoupled states may be written

$$|j_1, m_1; j_2, m_2; j_3, m_3\rangle \tag{1}$$

of which there are $(2j_1 + 1)(2j_2 + 1)(2j_3 + 1)$ as the m's range over their possible values for given values of j_1, j_2, and j_3.

Introducing

$$\mathbf{J}_{12} = \mathbf{J}_1 + \mathbf{J}_2 \tag{2a}$$

and

$$\mathbf{J} = \mathbf{J}_{12} + \mathbf{J}_3 \tag{2b}$$

we have coupled states that are labeled by values of the partial sum $\mathbf{J}_{12}$ as well as of $\mathbf{J}$ and J_z. These are expressible in terms of the uncoupled states (1) by the use of generalized coupling coefficients,

$$|j_1, j_2, j_3; j_{12}, j, m) = \sum_{m_1+m_2+m_3=m} |j_1, m_1; j_2, m_2; j_3, m_3\rangle\langle m_1, m_2, m_3|j_{12}, j, m) \tag{3}$$

Here, as in the corresponding relation 3³1 for two angular-momentum vectors, the dependence of the coupling coefficients on j_1, j_2, j_3 has not been explicitly indicated.

The coefficients in (3) are easily expressible in terms of the coupling coefficients for two angular momenta in that one set serves to couple $\mathbf{J}_1$ and $\mathbf{J}_2$ to give $\mathbf{J}_{12}$, followed by another set to couple $\mathbf{J}_{12}$ and $\mathbf{J}_3$ to give $\mathbf{J}$. The relation is

$$\langle m_1, m_2, m_3|j_{12}, j, m) = \sum_{m_{12}} \langle m_1, m_2,|j_{12}, m_{12})(m_{12}, m_3|j, m) \tag{4}$$

in which m_{12} is restricted to the value $(m_1 + m_2)$, and we have omitted the arguments j_1 and j_2 in the first coefficient and j_{12} and j_3 in the second.

With three angular-momentum vectors, there are three distinct coupling possibilities. Instead of labeling states by values of $\mathbf{J}_{12} = \mathbf{J}_1 + \mathbf{J}_2$, different sets of states can be obtained in which the states are labeled by values of $\mathbf{J}_{23} = \mathbf{J}_2 + \mathbf{J}_3$ or by $\mathbf{J}_{13} = \mathbf{J}_1 + \mathbf{J}_3$.

Situations arise in which we need to be able to express the states of one coupling scheme in terms of another. This problem was first solved by Racah, who introduced his W coefficients for this purpose. These are closely related to the $6j$ symbols of Wigner. The relation between the two notations is

$$W(j_1, j_2, l_2, l_1; j_3, l_3) = (-1)^{j_1+j_2+l_1+l_2} \begin{Bmatrix} j_1 & j_2 & j_3 \\ l_1 & l_2 & l_3 \end{Bmatrix} \tag{5}$$

The relation of recoupling is expressed as

$$|j_1, j_2, j_3; j_{23}, j, m) = \sum_{j_{12}} |j_1, j_2, j_3; j_{12}, j, m)(j_1, j_2, j_3; j_{12}, j|j_1, j_2, j_3; j_{23}, j) \tag{6}$$

in which the coefficients are related to the $6j$ symbols by

$$(j_1, j_2, j_3; j_{12}, j|j_1, j_2, j_3; j_{23}, j) = (-1)^{j_1+j_2+j_3+j}[(2j_{12} + 1)(2j_{23} + 1)]^{1/2} \begin{Bmatrix} j & j_3 & j_{12} \\ j_2 & j_1 & j_{23} \end{Bmatrix} \tag{7}$$

Similarly,

$$|j_1, j_3, j_2; j_{13}, j, m) = \sum_{j_{12}} |j_1, j_2, j_3; j_{12}, j, m)(j_1, j_2, j_3; j_{12}, j|j_1, j_3, j_2; j_{13}, j) \tag{6'}$$

and

$$(j_1, j_2, j_3; j_{12}, j | j_1, j_3, j_2; j_{13}, j) = (-1)^{j_2+j_3+j_{12}+j_{13}}[(2j_{12}+1)(2j_{13}+1)]^{1/2} \begin{Bmatrix} j & j_3 & j_{12} \\ j_1 & j_2 & j_{13} \end{Bmatrix} \tag{7'}$$

The $6j$ symbols can be expressed [22] as a sum over $3j$ symbols:

$$\begin{Bmatrix} j_1 & j_2 & j_3 \\ l_1 & l_2 & l_3 \end{Bmatrix} \begin{pmatrix} j_1 & j_2 & j_3 \\ m_1 & m_2 & m_3 \end{pmatrix} = \sum_{\mu_1\mu_2\mu_3} (-1)^{l_1+l_2+l_3+\mu_1+\mu_2+\mu_3} \times \begin{pmatrix} j_1 & l_2 & l_3 \\ m_1 & \mu_2 & -\mu_3 \end{pmatrix} \begin{pmatrix} l_1 & j_2 & l_3 \\ -\mu_1 & m_2 & \mu_3 \end{pmatrix} \begin{pmatrix} l_1 & l_2 & j_3 \\ \mu_1 & -\mu_2 & m_3 \end{pmatrix} \tag{8}$$

It is possible to reduce (8) to a sum over a single variable. Then a $6j$ symbol is given as

$$\begin{Bmatrix} j_1 & j_2 & j_3 \\ l_1 & l_2 & l_3 \end{Bmatrix} = \Delta(j_1, j_2, j_3)\Delta(j_1, l_2, l_3)\Delta(l_1, j_2, l_3)\Delta(l_1, l_2, j_3) \times \sum_Z \frac{(-1)^Z (Z+1)!}{AB} \tag{9}$$

where

$$A = (Z - j_1 - j_2 - j_3)!(Z - j_1 - l_2 - l_3)!(Z - l_1 - j_2 - l_3)!(Z - l_1 - l_2 - j_3)!$$
$$B = (j_1 + j_2 + l_1 + l_2 - Z)!(j_2 + j_3 + l_2 + l_3 - Z)!(j_3 + j_1 + l_3 + l_1 - Z)!$$

and

$$\Delta(a, b, c) = \left[\frac{(a+b-c)!(a-b+c)!(b+c-a)!}{(a+b+c+1)!}\right]^{1/2}$$

In the sum over Z terms are retained if all the arguments of the factorials are nonnegative. The four triangular conditions for nonvanishing of a $6j$ symbol are contained in the four $\Delta(a, b, c)$'s in (9), the arguments of the factorials in all the $\Delta(a, b, c)$'s must be nonnegative.

The symmetries for $6j$ symbols are: (1) Any permutation of the columns of arguments makes no change; for example,

$$\begin{Bmatrix} j_1 & j_2 & j_3 \\ l_1 & l_2 & l_3 \end{Bmatrix} = \begin{Bmatrix} j_1 & j_3 & j_2 \\ l_1 & l_3 & l_2 \end{Bmatrix} \tag{10a}$$

(2) An interchange of upper and lower arguments in pairs makes no change; for example,

$$\begin{Bmatrix} j_1 & j_2 & j_3 \\ l_1 & l_2 & l_3 \end{Bmatrix} = \begin{Bmatrix} l_1 & l_2 & j_3 \\ j_1 & j_2 & l_3 \end{Bmatrix} \tag{10b}$$

The sum in (9) reduces to relatively simple equations for cases where the smallest argument is 0, 1/2, 1, 3/2, or 2. These relations are given in the literature [1c, 9a, 22]. Though the numerical values for the $3j$ and $6j$ symbols whose arguments do not exceed 8 are presented in the tables of Rotenberg et al. [9a], there are now fast and accurate computer programs to calculate the $3j$, $6j$, and $9j$ symbols for a wide range of arguments [9f].

The $3j$ and $6j$ symbols have orthogonality properties that follow from their definitions 3³4 and (7), respectively:

$$\sum_{j_3 m_3} (2j_3+1)\begin{pmatrix} j_1 & j_2 & j_3 \\ m_1 & m_2 & m_3 \end{pmatrix}\begin{pmatrix} j_1 & j_2 & j_3 \\ m_1' & m_2' & m_3 \end{pmatrix} = \delta(m_1, m_1')\delta(m_2, m_2') \tag{11a}$$

$$\sum_{m_1 m_2} \begin{pmatrix} j_1 & j_2 & j_3 \\ m_1 & m_2 & m_3 \end{pmatrix}\begin{pmatrix} j_1 & j_2 & j_3' \\ m_1 & m_2 & m_3' \end{pmatrix} = \frac{\delta(j_3, j_3')\delta(m_3, m_3')}{2j_3+1} \tag{11b}$$

$$\sum_{j_3} (2j_3+1)(2l_3+1)\begin{Bmatrix} j_1 & j_2 & j_3 \\ l_1 & l_2 & l_3 \end{Bmatrix}\begin{Bmatrix} j_1 & j_2 & j_3 \\ l_1 & l_2 & l_3' \end{Bmatrix} = \delta(l_3, l_3') \tag{12}$$

The two most useful summation properties of $6j$ symbols are

$$\sum_{l_3} (-1)^{j'_3+j_3+l_3}(2l_3+1)\begin{Bmatrix} j_1 & j_2 & j_3 \\ l_1 & l_2 & l_3 \end{Bmatrix}\begin{Bmatrix} j_1 & l_1 & j_3' \\ j_2 & l_2 & l_3 \end{Bmatrix} = \begin{Bmatrix} j_1 & j_2 & j_3 \\ l_2 & l_1 & j_3' \end{Bmatrix} \tag{13a}$$

and

$$\sum_{k} (-1)^{S+k}(2k+1)\begin{Bmatrix} l_1 & j_2 & l_3 \\ l_3' & l_2' & k \end{Bmatrix}\begin{Bmatrix} j_2 & j_3 & j_1 \\ l_1' & l_3' & k \end{Bmatrix}\begin{Bmatrix} l_1 & j_3 & l_2 \\ l_1' & l_2' & k \end{Bmatrix} = \begin{Bmatrix} j_1 & j_2 & j_3 \\ l_1 & l_2 & l_3 \end{Bmatrix}\begin{Bmatrix} l_3 & j_1 & l_2 \\ l_1' & l_2' & l_3' \end{Bmatrix} \tag{13b}$$

where

$$S = j_1 + j_2 + j_3 + l_1 + l_2 + l_3 + l_1' + l_2' + l_3'$$

Similar results are applicable for the addition of four angular momenta:

$$\mathbf{J} = \mathbf{J}_1 + \mathbf{J}_2 + \mathbf{J}_3 + \mathbf{J}_4 \tag{14}$$

Here we can label coupled states by the values of

$$\mathbf{J}_{12} = \mathbf{J}_1 + \mathbf{J}_2 \quad \text{and} \quad \mathbf{J}_{34} = \mathbf{J}_3 + \mathbf{J}_4 \tag{15}$$

where

$$|j_1, j_2, j_3, j_4; j_{12}, j_{34}, j, m) = \sum_{\substack{m_1+m_2=m_{12} \\ m_3+m_4=m_{34} \\ m_{12}, m_{34}}} |j_1, m_1; j_2, m_2; j_3, m_3; j_4, m_4\rangle\langle m_1, m_2|j_{12}, m_{12})\langle m_3, m_4|j_{34}, m_{34}) \tag{16}$$

The coefficients occurring here are a sum of products of the coupling coefficients for pairs of vectors. The case of three vectors, covered in (3), may be regarded as a special case of (16) in which $j_4 = 0$; $m_4 = 0$. Then j_{34} becomes the same as j_3.

Alternatively, the four vectors can be coupled by the scheme

$$\mathbf{J}_{13} = \mathbf{J}_1 + \mathbf{J}_3 \quad \text{and} \quad \mathbf{J}_{24} = \mathbf{J}_2 + \mathbf{J}_4 \tag{17}$$

to give a set of states labeled by j_{13}, j_{24}, j, m, analogous to those in (16) that are labeled by j_{12}, j_{34}, j, m, with a similar expression to that in (16) for their coupling coefficients.

As these states for a given (j, m) also form a complete basis, it follows that a unitary transformation exists between the two sets of states:

$$\begin{aligned} &|j_1, j_3, j_2, j_4; j_{13}, j_{24}, j, m) \\ &\quad = \sum_{j_{12}, j_{34}} |j_1, j_2, j_3, j_4; j_{12}, j_{34}, j, m)(j_{12}, j_{34}|j_{13}, j_{24}) \end{aligned} \tag{18}$$

in which the transformation matrix elements are independent of m. On substituting the expressions for the coupling coefficients from (16) we readily find that

$$\begin{aligned} (j_{12}, j_{34}|j_{13}, j_{24}) &= [(2j_{12}+1)(2j_{34}+1)(2j_{13}+1)(2j_{24}+1)]^{1/2} \\ &\times \sum_{j'} (-1)^{2j'}(2j'+1) \begin{Bmatrix} j_1 & j_2 & j_{12} \\ j_{34} & j & j' \end{Bmatrix} \begin{Bmatrix} j_3 & j_4 & j_{34} \\ j_2 & j' & j_{24} \end{Bmatrix} \begin{Bmatrix} j_{13} & j_{24} & j \\ j' & j_1 & j_3 \end{Bmatrix} \end{aligned} \tag{19}$$

The sum in this equation has symmetry properties that are manifested by identifying it as a $9j$ symbol; that is,

$$(j_{12}, j_{34}|j_{13}, j_{24}) = \sqrt{(2j_{12}+1)(2j_{34}+1)(2j_{13}+1)(2j_{24}+1)} \begin{Bmatrix} j_1 & j_2 & j_{12} \\ j_3 & j_4 & j_{34} \\ j_{13} & j_{24} & j \end{Bmatrix} \tag{20}$$

The following relation between $6j$ and $9j$ symbols is obvious:

$$\begin{aligned} \begin{Bmatrix} j_{11} & j_{12} & j_{13} \\ j_{21} & j_{22} & j_{23} \\ j_{31} & j_{32} & j_{33} \end{Bmatrix} &= \sum_j (-1)^{2j}(2j+1) \begin{Bmatrix} j_{11} & j_{21} & j_{31} \\ j_{32} & j_{33} & j \end{Bmatrix} \\ &\times \begin{Bmatrix} j_{12} & j_{22} & j_{32} \\ j_{21} & j & j_{23} \end{Bmatrix} \begin{Bmatrix} j_{13} & j_{23} & j_{33} \\ j & j_{11} & j_{12} \end{Bmatrix} \end{aligned} \tag{21}$$

The $9j$ symbol can also be defined in terms of a sum over products of six $3j$ symbols:

$$\begin{aligned} \begin{Bmatrix} j_{11} & j_{12} & j_{13} \\ j_{21} & j_{22} & j_{23} \\ j_{31} & j_{32} & j_{33} \end{Bmatrix} &= \sum_{\text{all } m\text{'s}} \begin{pmatrix} j_{11} & j_{12} & j_{13} \\ m_{11} & m_{12} & m_{13} \end{pmatrix} \begin{pmatrix} j_{21} & j_{22} & j_{23} \\ m_{21} & m_{22} & m_{23} \end{pmatrix} \\ &\times \begin{pmatrix} j_{31} & j_{32} & j_{33} \\ m_{31} & m_{32} & m_{33} \end{pmatrix} \begin{pmatrix} j_{11} & j_{21} & j_{31} \\ m_{11} & m_{21} & m_{31} \end{pmatrix} \begin{pmatrix} j_{12} & j_{22} & j_{32} \\ m_{12} & m_{22} & m_{32} \end{pmatrix} \begin{pmatrix} j_{13} & j_{23} & j_{33} \\ m_{13} & m_{23} & m_{33} \end{pmatrix} \end{aligned} \tag{22}$$

where m_{ij} takes values from $-j_{ij}$ to $+j_{ij}$ in unit steps.

From this definition it follows that the $9j$ symbol is unaltered by any even permutation of its rows and columns. Interchange of any two rows or columns multiplies the symbol by $(-1)^{j_{11}+j_{12}+j_{13}+j_{21}+j_{22}+j_{23}+j_{31}+j_{32}+j_{33}}$. Interchange of rows and columns leaves the symbol unaltered.

The orthonormality of the coupling coefficient factors leads to the equation

$$\sum_{j_{13},j_{23}} (2j_{13}+1)(2j_{23}+1) \begin{Bmatrix} j_{11} & j_{12} & j_{13} \\ j_{21} & j_{22} & j_{23} \\ j_{31} & j_{32} & j_{33} \end{Bmatrix} \begin{Bmatrix} j_{11} & j_{12} & j_{13} \\ j_{21} & j_{22} & j_{23} \\ j'_{31} & j'_{32} & j_{33} \end{Bmatrix} = \frac{\delta(j_{31}, j'_{31})\delta(j_{32}, j'_{32})}{(2j_{31}+1)(2j_{32}+1)} \tag{23}$$

The following sum rule is often useful

$$\sum_{\mu} (2\mu+1) \begin{Bmatrix} j_{11} & j_{12} & \mu \\ j_{21} & j_{22} & j_{23} \\ j_{31} & j_{32} & j_{33} \end{Bmatrix} \begin{Bmatrix} j_{11} & j_{12} & \mu \\ j_{23} & j_{33} & \lambda \end{Bmatrix} = (-1)^{2\lambda} \begin{Bmatrix} j_{21} & j_{22} & j_{23} \\ j_{12} & \lambda & j_{32} \end{Bmatrix} \begin{Bmatrix} j_{31} & j_{32} & j_{33} \\ \lambda & j_{11} & j_{21} \end{Bmatrix} \tag{24}$$

In the same manner as before the $12j$ symbol can be defined in connection with the coupling of five angular momenta and as a sum of four $6j$ symbols by

$$\begin{Bmatrix} c & & b & & h & & e & \\ & p & & s & & r & & q \\ d & & a & & g & & f & \end{Bmatrix} = (-1)^S \sum_x (2x+1)(-1)^{-x} \begin{Bmatrix} a & b & x \\ c & d & p \end{Bmatrix} \times \begin{Bmatrix} c & d & x \\ e & f & q \end{Bmatrix} \begin{Bmatrix} e & f & x \\ g & h & r \end{Bmatrix} \begin{Bmatrix} g & h & x \\ b & a & s \end{Bmatrix} \tag{25a}$$

where

$$S = a+b+c+d+e+f+g+h+p+q+r+s$$

The $12j$ symbol can also be given in terms of a sum of products of two $6j$ and one $9j$ symbols:

$$\begin{Bmatrix} c & & b & & h & & e & \\ & p & & s & & r & & q \\ d & & a & & g & & f & \end{Bmatrix} = \sum_x (2x+1)(-1)^{a+b+e+f+2x} \begin{Bmatrix} b & f & x \\ q & p & c \end{Bmatrix} \times \begin{Bmatrix} a & e & x \\ q & p & d \end{Bmatrix} \begin{Bmatrix} s & h & b \\ g & r & f \\ a & e & x \end{Bmatrix} \tag{25b}$$

The $6j$ and $9j$ symbols also find application in the problem of expressing the reduced-matrix elements of irreducible tensor products of two irreducible tensors in terms of the reduced-matrix elements of the factors.

We use 8³3 to express $[\mathbf{A}^{(k)} \times \mathbf{B}^{(l)}]^{(n)}_{\mu}$ in terms of products of $A^{(k)}_{\kappa} B^{(l)}_{\lambda}$ with coupling coefficients, which by 3³4 are expressed in terms of $3j$ symbols: (1) If $\mathbf{A}^{(k)}$ and $\mathbf{B}^{(l)}$ act on the same coordinates we have

$$\begin{aligned}\langle \alpha', j' \| [\mathbf{A}^{(k)} \times \mathbf{B}^{(l)}]^{(n)} \| \alpha, j \rangle &= (-1)^{j+j'+n} \sqrt{(2n+1)} \sum_{\alpha'' j''} \langle \alpha', j' \| A^{(k)} \| \alpha'', j'' \rangle \\ &\quad \times \langle \alpha'', j'' \| B^{(l)} \| \alpha, j \rangle \begin{Bmatrix} k & l & n \\ j & j' & j'' \end{Bmatrix} \qquad (26)\end{aligned}$$

(2) If $\mathbf{A}^{(k)}$ operates on a set of coordinates 1, and $\mathbf{B}^{(l)}$ operates on a set of coordinates 2, then we suppose that $|\alpha, \mathrm{j}, \mathrm{m}\rangle$ are the states resulting from the coupling of angular momenta j_1 and j_2.

The final result for the reduced-matrix elements in this case is

$$\begin{aligned}\langle \alpha_1', \alpha_2', j_1', j_2'; j' \| [\mathbf{A}^{(k)} \times \mathbf{B}^{(l)}]^{(n)} \| \alpha_1, \alpha_2, j_1, j_2; j \rangle &= \sqrt{(2j'+1)(2n+1)(2j+1)} \begin{Bmatrix} j_1' & j_1 & k \\ j_2' & j_2 & l \\ j' & j & n \end{Bmatrix} \\ &\quad \times \langle \alpha_1', j_1' \| A^{(k)} \| \alpha_1', j_1 \rangle \langle \alpha_2, j_2' \| B^{(l)} \| \alpha_2, j_2 \rangle \qquad (27)\end{aligned}$$

Equations (26) and (27), as well as the matrix elements for scalar products, are derived and fully discussed in Chapter 5.

4. Central-field approximation

A new phase of my scientific life began when I met Niels Bohr for the first time. This was in 1922, when he gave a series of guest lectures at Göttingen, in which he reported on his theoretical investigations on the periodic system of elements. I shall recall only briefly that the essential progress made by Bohr's considerations at that time was the explaining by means of the spherically symmetric atomic model the formation of the intermediate shells of the atom and the general properties of the rare earths. The question as to why all electrons for an atom in its ground state were not bound in the innermost shell had already been emphasized by Bohr in his earlier works. In his Göttingen lectures he treated particularly the closing of the innermost K shell in the helium atom and its essential connection with the two noncombining spectra of helium, the ortho- and para helium spectra. However, no convincing explanation for this phenomenon could be given on the basis of classical mechanics. It made a strong impression on me that Bohr at that time was looking for a *general* explanation which should hold for the closing of *every* electron shell and in which the number 2 was considered as essential as 8, in contrast to Sommerfeld's approach.

W. Pauli in Nobel Prize lecture, Stockholm, 13 December 1946, in *Nobel Lectures – Physics (1942–1962)* (New York: Elsevier, 1964), pp. 27–43.

As previously mentioned in Sections 9^1–11^1, the theory of N-electron atoms is based on the central field approximation in which each of the electrons is regarded as moving in the same central field, $V(r)$, as in $9^1 2$. In this chapter and in Chapters 5, 6, and 7, the application of perturbation theory to the nonrelativistic model is developed in a way that is largely independent of the particular choice made for $V(r)$. Then in Chapter 8 the means of making a definite choice of $V(r)$ are considered.

Study of the special problems of many fermion systems was initiated by Heisenberg [1] and carried farther by Dirac [2]. This led to the discovery of *exchange interaction* of electrons, whereby the non-spin-dependent Hamiltonian implies, through the Pauli exclusion principle (Sec. 7^2), an apparent strong spin dependence of the energy levels. This extraordinary result finds application in many other areas of physics, as in chemical bond theory and the origin of ferromagnetism.

This chapter treats the problem of exchange interaction by elementary use of perturbation theory (Sec. 9^2) following in the main the method originally developed by Slater [3]. The nonrelativistic Hamiltonian (2^29) commutes with the inversion operator (6^23). This permits accurate labeling of its states with $I = +1$ (even) or $I = -1$ (odd), which is called the *parity* of the state. This Hamiltonian also commutes with **S** and **L**, the vector sum of the N spins and the N orbital angular momenta, respectively. Thus the states can be accurately labeled by $\mathbf{S}^2 = S(S+1)$ and $\mathbf{L}^2 = L(L+1)$. States labeled in this way are said to be in *Russell-Saunders coupling* [4].

The basis set for use of perturbation theory is that in which each of the N electrons moves in the field $V(r)$ so that the states are labeled by a complete set of quantum numbers, which consists of N individual sets, (n, l, m, m_s). The one-electron levels, $W_0(n, l)$, are degenerate in m and m_s so all states having the same N sets of (n, l) values have the same energy in the zeroth approximation. The set of (n, l) values serves to label a *configuration*. For light elements, the dependence of $W_0(n, l)$ on l is sometimes negligible. The larger set of states characterized by the same set of N n's and the same parity, irrespective of the l's, is then called a *complex*. The concept of complex is introduced by Layzér [5].

Application of perturbation theory calls for calculation of the matrix elements of the Coulomb interaction,

$$\sum_{i<j=1}^{N} \frac{e^2}{r_{ij}}$$

of the electrons. The matrix elements of the Coulomb interaction have nonzero values only between states of the same parity. In Russell-Saunders coupling the matrix elements are also diagonal in **L** and **S**. Most important are the matrix elements between states belonging to the same configuration. These have the effect of splitting the degenerate energy of that configuration into Russell-Saunders *terms*. Next in importance are the matrix elements between states of different configurations of the same complex. Next in importance after these are matrix elements between different complexes and the same parity.

The effect of these last two kinds of matrix elements of the Coulomb interaction is to break down the accuracy of the assignment of configurations to a given term, a phenomenon known as *configuration mixing* or *configuration interaction*. This phenomenon is of great qualitative importance in describing some features of the energy level structure. It is usually rather small quantitatively. In the experimental literature a definite configuration is usually assigned to each term. This has to be understood as approximate in the sense that (if it is correctly given) the single configuration assigned to a term is the configuration of greatest importance in an expansion of the term's wave function, in a series of configurational wave functions.

Choice of $V(r)$ is usually made by the self-consistent field method of Hartree [6] as extended by Fock [7].

Despite the vast amount of work that has been done since 1900 in observing and analyzing atomic spectra, knowledge of the energy levels of all (Z, N) structures is still far from complete. An indispensable reference to this large amount of material is the critical compilation of C. E. Moore [8]. Other useful compilations exist [9].

Fairly complete analyses are available for the structures in which there are few electrons outside closed shells. Only for $Z < 13$ and for $Z = 15, 20, 22, 23, 26$ are some levels known in all stages ($N = 1, 2, 3, \ldots, Z$) of ionization [9i]. Analyses for the transition groups (partially filled d shells) are still quite incomplete. Analyses for the lanthanides ($4f$ shell) and actinides ($5f$ shell) are much less complete because of the great complexity of their structure.

1⁴. Hydrogenic atoms; discrete energy levels

The wave equation for one electron of mass m in the Coulomb field of a nucleus of charge $+Ze$ and mass M is (and here we use φ for the one-electron wave function instead of the usual ϕ, to prevent any confusion with the azimuthal angle ϕ)

$$H\varphi(r, \theta, \phi, s) = \left[-\frac{\hbar^2}{2\mu}\nabla^2 - \frac{Ze^2}{r}\right]\varphi = W\varphi \tag{1}$$

in which $\mu = mM/(m + M)$ is the electronic reduced mass (Sec. 7^1). Here s is written for the two-valued spin coordinate (Sec. 5^3). Because this Hamiltonian does not depend on $\mathbf{S}$, the wave function may be assumed to contain a factor $\delta(s, m_s)$, where $m_s = \pm 1/2$, is the precise value of S_z for the state. Thus each allowed W is twofold degenerate in m_s.

The Hamiltonian (1) is invariant under rotation of axes. It therefore commutes with the orbital angular momentum operators, $\mathbf{L}^2$ and L_z (Sec. 4^3) so the energy levels are labeled by (l, m), where $l(l + 1)$ is the proper value of $\mathbf{L}^2$ and m is that of L_z. This fixes the dependence of φ on (θ, ϕ) as being through the spherical harmonic $Y(l, m; \theta, \phi)$.

The Laplace operator is

$$\nabla^2 = \frac{1}{r^2}\frac{\partial}{\partial r}\left(r^2\frac{\partial}{\partial r}\right) - \frac{1}{r^2}\mathbf{L}^2 \tag{2a}$$

The equation (1) is separable, and the solution is a product of a radial function by a spherical harmonic by a delta function for the spin dependence.

$$\varphi = \frac{1}{r} P(n, l; r) Y(l, m; \theta, \phi) \delta(m_s, s) \tag{2b}$$

where $P(n, l; 0) = 0$ in order that $\varphi(0)$ be finite. A solution for the one-electron problem without the spin factor is called an *orbital*, and one that includes the spin factor is called a *spin orbital*.

The equation for the radial function is

$$-\frac{\hbar^2}{2\mu} P''(r) + V_l(r) P = WP \tag{3}$$

in which

$$V_l(r) = -\frac{Ze^2}{r} + \frac{\hbar^2}{2\mu} \frac{l(l+1)}{r^2} \tag{4}$$

The second term on the right represents the kinetic energy of the angular motion. It is often called the *centrifugal barrier* because in (3) it acts like a repulsive potential energy tending to keep the electron away from the nucleus. Curves for several $V_l(r)$ are shown in Figure 1^4. The minima occur at

$$r_l = l(l+1) \frac{a}{Z} \tag{5}$$

and the minimum values are

$$V_l(r_l) = -\frac{Z^2(e^2/a)}{2l(l+1)} \tag{6}$$

Here a is a modified atomic length unit

$$a = \frac{\hbar^2}{\mu e^2} \tag{7}$$

slightly larger than the a of ($7^1 3$).

It is convenient to write

$$\rho = \frac{2}{n} \frac{Zr}{a}, \qquad W = -n^{-2} \frac{Z^2 e^2}{2a} \tag{8}$$

The equation for $P(r)$ is then

$$\frac{d^2P}{d\rho^2} + \left[-\frac{1}{4} + \frac{n}{\rho} - \frac{l(l+1)}{\rho^2} \right] P = 0 \tag{9}$$

Because ρ approaches ∞, the solutions behave asymptotically like $e^{\pm\rho/2}$. The one that behaves like $e^{-\rho/2}$ is chosen, because it is finite in this limit.

The substitution

$$P = e^{-\rho/2} f(\rho)$$

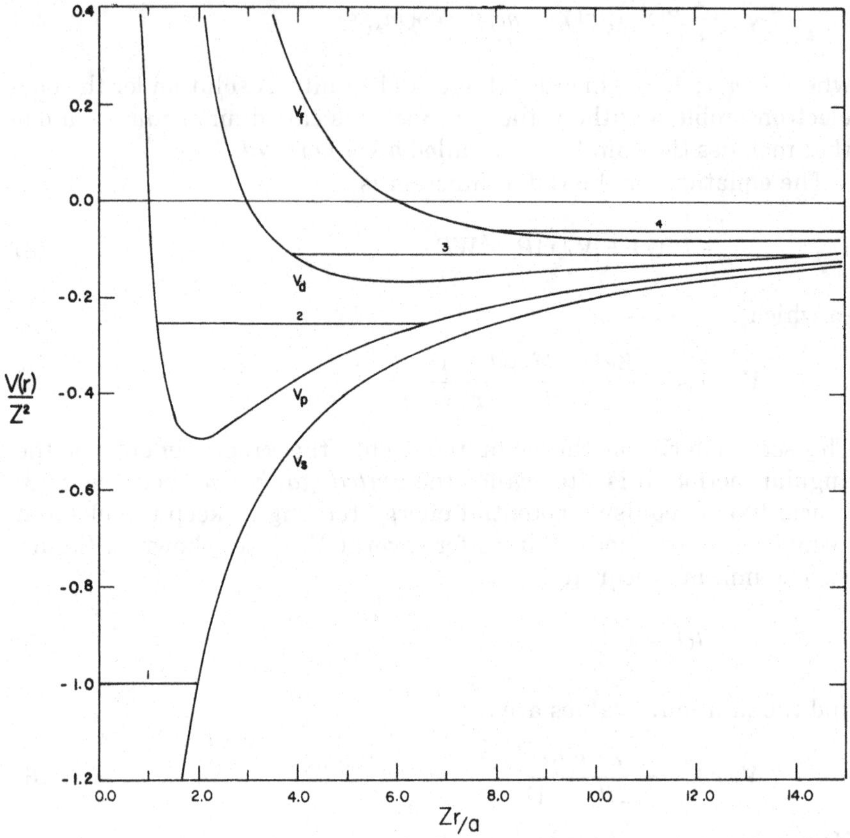

Figure 1[4]. Coulomb field potentials for s, p, d, and f states. Lowest energy levels ($n = 1, 2, 3, 4$) are shown as horizontal lines.

gives the equation for $f(\rho)$,

$$f'' - f' + [n\rho^{-1} - l(l+1)\rho^{-2}]f = 0 \tag{10}$$

Substituting the power series

$$f(\rho) = \sum_{\nu=0}^{\infty} a_\nu \rho^{\nu+\lambda} \qquad (a_0 \neq 0) \tag{11}$$

leads to a two-term recursion relation between the a_ν. The coefficient of a_0 with this $f(\rho)$ in (10) indicates that $\lambda = (l+1)$ or $-l$, but only the former satisfies the requirement that $f(0) = 0$, in which case the recursion relation is

$$a_{\nu+1}[(\nu + l + 1)(\nu + l + 2) - l(l+1)] + a_\nu[n - l - 1 - \nu] = 0 \tag{12}$$

When n is an integer, the series terminates as a polynominal of degree $(n - l - 1)$, multiplied by ρ^{l+1}. If n is not an integer, the series for $f(\rho)$ is infinite, defining a function that behaves like e^{ρ} as ρ approaches ∞, thus giving an unacceptable P. Thus the only allowed values of n are the integers $n \geqslant l + 1$. These give in (8) allowed energies agreeing with the Bohr theory and hence with the observed levels.

2⁴. Radial functions for hydrogenic atoms

The radial function for the state with quantum numbers (n, l) in a hydrogenic atom is

$$P(n, l; r) = -N_{nl}e^{-\rho/2}\rho^{l+1}L_{n+l}^{2l+1}(\rho) \tag{1}$$

in which $L_{n+l}^{2l+1}(\rho)$ is a polynomial of degree $(n - l - 1)$. The minus sign is inserted to conform to traditional usage, which makes $P(n, l; r) > 0$ for very small r and for all n and l. Here ρ is given by 1⁴8.

The polynomials occurring here are known as *associated Laguerre polynomials* [10]. These are defined as

$$L_\lambda^\mu(\rho) = \frac{d^\mu}{d\rho^\mu} L_\lambda(\rho); \qquad L_\lambda(\rho) = e^\rho \frac{d^\lambda}{d\rho^\lambda}(\rho^\lambda e^{-\rho}) \tag{2}$$

or, alternatively,

$$L_\lambda^\mu(\rho) = \frac{1}{\lambda!}\rho^{-\mu}e^\rho \frac{d^\lambda}{d\rho^\lambda}(\rho^{\lambda+\mu}e^{-\rho}) \tag{3}$$

From these definitions

$$L_\lambda^\mu(\rho) = (-1)^\mu(\lambda!)^2 \sum_{\nu=0}^{\lambda-\mu} \frac{(-\rho)^\nu}{\nu!(\mu+\nu)!(\lambda-\mu-\nu)!} \tag{4}$$

The associated Laguerre polynomials are related to the *confluent hypergeometric function* [11]

$$F(\alpha, \beta, \chi) = 1 + \frac{\alpha}{\beta \cdot 1!}\chi + \frac{\alpha(\alpha+1)}{\beta(\beta+1)2!}\chi^2 + \ldots \tag{5}$$

as follows

$$L_\lambda^\mu(\rho) = \frac{(-1)^\mu(\lambda!)^2}{\mu!(\lambda-\mu)!} F(-\lambda+\mu, \mu+1, \rho) \tag{6}$$

To evaluate N_{nl} it is convenient to calculate [12] a slightly more general integral

$$J(\lambda, \mu, \sigma) = \frac{1}{(\lambda!)^2}\int_0^\infty e^{-\rho}\rho^{\mu+\sigma}[L_\lambda^\mu(\rho)]^2\, d\rho \tag{7}$$

Table 1^4. Normalized hydrogenic radial functions [r in modified atomic unit (1^47)]

$$P(1s;\,r) = 2\sqrt{Z}(Zr)e^{-Zr}$$

$$P(2s;\,r) = \sqrt{2Z}(Zr/2)(1 - Zr/2)e^{-Zr/2}$$
$$P(2p;\,r) = \sqrt{2Z/3}(Zr/2)^2e^{-Zr/2}$$

$$P(3s;\,r) = \sqrt{4Z/3}(Zr/3)[1 - 2(Zr/3) + (2/3)(Zr/3)^2]e^{-Zr/3}$$
$$P(3p;\,r) = (8/3)\sqrt{Z/6}(Zr/3)^2\,[1 - (1/2)(Zr/3)]e^{-Zr/3}$$
$$P(3d;\,r) = (4/3)\sqrt{Z/30}(Zr/3)^3e^{-Zr/3}$$

$$P(4s;\,r) = \sqrt{Z}(Zr/4)[1 - 3(Zr/4) + 2(Zr/4)^2 - (1/3)(Zr/4)^3]e^{-Zr/4}$$
$$P(4p;\,r) = \sqrt{5Z/3}(Zr/4)^2\,[1 - (Zr/4) + (1/5)(Zr/4)^2]e^{-Zr/4}$$
$$P(4d;\,r) = \sqrt{Z/5}(Zr/4)^3\,[1 - (1/3)(Zr/4)]e^{-Zr/4}$$
$$P(4f;\,r) = (1/3)\sqrt{Z/35}(Zr/4)^4e^{-Zr/4}$$

$$P(5s;\,r) = 2\sqrt{Z/5}(Zr/5)[1 - 4(Zr/5) + 4(Zr/5)^2 - (4/3)(Zr/5)^3 + (2/15)(Zr/5)^4]e^{-Zr/5}$$
$$P(5p;\,r) = 8\sqrt{Z/30}(Zr/5)^2\,[1 - (3/2)(Zr/5) + (3/5)(Zr/5)^2 - (1/15)(Zr/5)^3]e^{-Zr/5}$$
$$P(5d;\,r) = (28/5)\sqrt{Z/70}(Zr/5)^3\,[1 - (2/3)(Zr/5) + (2/21)(Zr/5)^2]e^{-Zr/5}$$
$$P(5f;\,r) = (16/15)\sqrt{Z/70}(Zr/5)^4\,[1 - (1/4)(Zr/5)]e^{-Zr/5}$$
$$P(5g;\,r) = (4/45)\sqrt{Z/70}(Zr/5)^5e^{-Zr/5}$$

$$P(6s;\,r) = 2\sqrt{Z/6}(Zr/6)\,[1 - 5(Zr/6) + (20/3)(Zr/6)^2 - (10/3)(Zr/6)^3 + (2/3)(Zr/6)^4 - (54/1215)(Zr/6)^5]e^{-Zr/6}$$
$$P(6p;\,r) = (14/3)\sqrt{5Z/42}(Zr/6)^2\,[1 - 2(Zr/6) + (6/5)(Zr/6)^2 - (4/15)(Zr/6)^3 + (2/105)(Zr/6)^4]e^{-Zr/6}$$
$$P(6d;\,r) = (28/3)\sqrt{Z/105}(Zr/6)^3\,[1 - (Zr/6) + (2/7)(Zr/6)^2 - (1/42)(Zr/6)^3]e^{-Zr/6}$$
$$P(6f;\,r) = (72/54)\sqrt{Z/35}(Zr/6)^4\,[1 - (1/2)(Zr/6) + (3/54)(Zr/6)^2]e^{-Zr/6}$$
$$P(6g;\,r) = (2/27)\sqrt{Z/7}(Zr/6)^5\,[1 - (1/5)(Zr/6)]e^{-Zr/6}$$
$$P(6h;\,r) = (2/135)\sqrt{Z/77}(Zr/6)^6e^{-Zr/6}$$

$$P(7s;\,r) = 2\sqrt{Z/7}(Zr/7)\,[1 - 6(Zr/7) + 10(Zr/7)^2 - (20/3)(Zr/7)^3 + 2(Zr/7)^4 - (4/15)(Zr/7)^5 + (4/315)(Zr/7)^6]e^{-Zr/7}$$
$$P(7p;\,r) = 8\sqrt{Z/21}(Zr/7)^2\,[1 - (5/2)(Zr/7) + 2(Zr/7)^2 - (2/3)(Zr/7)^3 + (2/21)(Zr/7)^4 - (1/210)(Zr/7)^5]e^{-Zr/7}$$
$$P(7d;\,r) = 12\sqrt{Z/105}(Zr/7)^3\,[1 - (4/3)(Zr/7) + (4/7)(Zr/7)^2 - (2/21)(Zr/7)^3 + (1/189)(Zr/7)^4]e^{-Zr/7}$$
$$P(7f;\,r) = (16/7)\sqrt{Z/42}(Zr/7)^4\,[1 - (3/4)(Zr/7) + (1/6)(Zr/7)^2 - (1/90)(Zr/7)^3]e^{-Zr/7}$$
$$P(7g;\,r) = (44/63)\sqrt{Z/154}(Zr/7)^5\,[1 - (2/5)(Zr/7) + (2/55)(Zr/7)^2]e^{-Zr/7}$$
$$P(7h;\,r) = (8/105)\sqrt{Z/231}(Zr/7)^6\,[1 - (1/6)(Zr/7)]e^{-Zr/7}$$

$$P(8s;\,r) = \sqrt{Z/2}(Zr/8)\,[1 - 7(Zr/8) + 14(Zr/8)^2 - (35/3)(Zr/8)^3 + (14/3)(Zr/8)^4 - (14/15)(Zr/8)^5 + (4/45)(Zr/8)^6 - (1/315)(Zr/8)^7]e^{-Zr/8}$$
$$P(8p;\,r) = 7\sqrt{Z/14}(Zr/8)^2\,[1 - 3(Zr/8) + 3(Zr/8)^2 - (4/3)(Zr/8)^3 + (2/7)(Zr/8)^4 - (1/35)(Zr/8)^5 + (1/945)(Zr/8)^6]e^{-Zr/8}$$

Table 1[4]. (cont.)

$$P(8d;\, r) = 21\sqrt{Z/210}(Zr/8)^3\,[1 - (5/3)(Zr/8) + (20/21)(Zr/8)^2 - (5/21)(Zr/8)^3 + (5/189)(Zr/8)^4 - (1/945)(Zr/8)^5]e^{-Zr/8}$$

$$P(8f;\, r) = 11\sqrt{Z/462}(Zr/8)^4\,[1 - (Zr/8) + (1/3)(Zr/8)^2 - (2/45)(Zr/8)^3 + (1/495)(Zr/8)^4]e^{-Zr/8}$$

$$P(8g;\, r) = (11/9)\sqrt{Z/154}(Zr/8)^5\,[1 - (3/5)(Zr/8) + (6/55)(Zr/8)^2 - (1/165)(Zr/8)^3]e^{-Zr/8}$$

$$P(8h;\, r) = (13/15)\sqrt{Z/6006}(Zr/8)^6\,[1 - (1/3)(Zr/8) + (1/39)(Zr/8)^2]e^{-Zr/8}$$

in terms of which

$$(N_{nl})^2 \frac{na}{2Z}[(n + l)!]^2 J(n + l, 2l + 1, 1) = 1$$

We arbitrarily choose N_{nl} real and positive. Evaluation gives (for $\sigma > 0$),

$$J(\lambda, \mu, \sigma) = (-1)^\sigma \frac{\lambda!\sigma!}{(\lambda - \mu)!} \sum_{\beta=0}^{\sigma} (-1)^\beta \binom{\sigma}{\beta} \binom{\lambda + \beta}{\sigma} \binom{-\mu + \beta + \lambda}{\sigma} \tag{8}$$

Therefore, the normalized radial functions are

$$P(n, l;\, r) = -\left\{\frac{Z}{na}\frac{(n - l - 1)!}{n[(n + l)!]^3}\right\}^{1/2} e^{-\rho/2}\rho^{l+1}L_{n+l}^{2l+1}(\rho) \tag{9}$$

These are orthogonal on n,

$$\int_0^\infty P(n, l;\, r)P(n', l;\, r)\, dr = \delta(n, n') \tag{10}$$

Each P has n nodes at finite values of r.

Table 1[4] gives explicit expressions for some of the $P(n, l;\, r)$. Several $P^2(n, l;\, r)$ are plotted in Figure 2[4].

The value of $J(\lambda, \mu, \sigma)$ for $\sigma < 0$ is

$$J(\lambda, \mu, \sigma) = \frac{\lambda!}{(\lambda - \mu)!(s + 1)!} \sum_{\gamma=0}^{s} (-1)^{s-\gamma} \frac{\binom{s}{\gamma}\binom{\lambda - \mu + \gamma}{s}}{\binom{\mu + s - \gamma}{s + 1}} \tag{11}$$

in which $s = -(\sigma + 1)$. From (8) and (11) expressions can be obtained for the mean value of r^k in the (n, l) state:

$$\langle n, l|r^k|n, l\rangle = \left(\frac{na}{2Z}\right)^k N_{nl}^2[(n + l)!]^2 J(n + l, 2l + 1, k + 1) \tag{12}$$

Table 2[4] gives explicit expressions and values for several values of k.

Each $P(n, l;\, r)$ is oscillatory in the range of r for which the coefficient

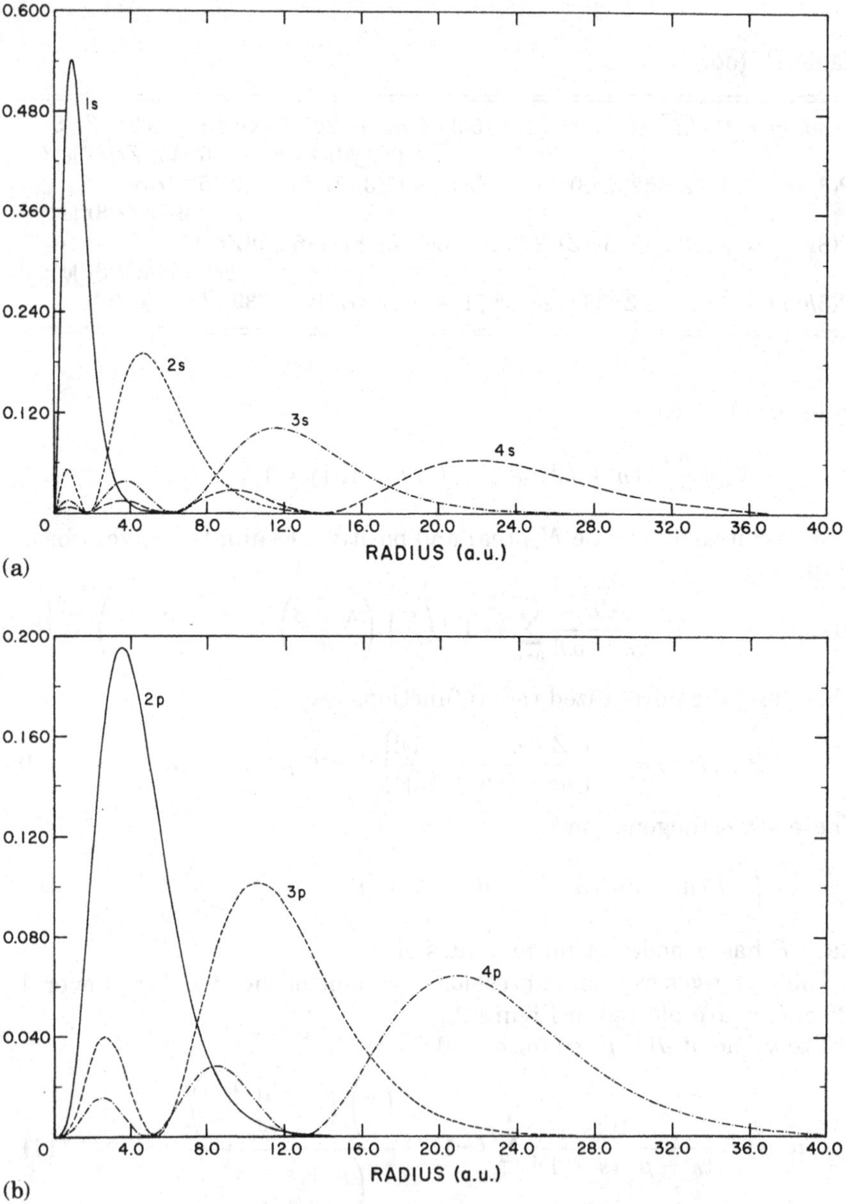

Figure 2[4]. Normalized squared hydrogenic radial functions, P^2 $(nl;\ r)$. (a) *s* states; (b) *p* states.

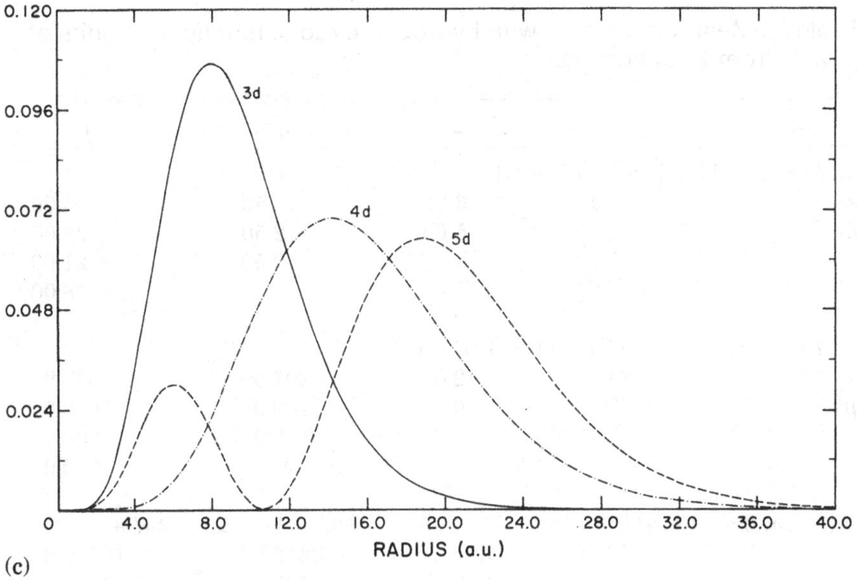

(c)

Figure 2⁴ (*cont.*) (c) *d* states.

of P in 1⁴3 is positive and goes rapidly to zero outside this range, the limits of which are

$$\frac{a}{Z}\, n^2 \left\{1 \pm \left[1 - \frac{l(l+1)}{n^2}\right]^{1/2}\right\}$$

These ranges are plotted in Figure 3⁴ for several states, as also are the kth roots of means of r^k for these states. These show that the main trend is for the radius of orbitals to increase as n^2 and to decrease as Z^{-1}.

The momentum representation (Sec. 5²) state functions for the hydrogen atom were obtained by Podolsky and Pauling [13] by direct calculation of the Fourier transform of the $\varphi(r, \theta, \phi, s)$ of 1⁴2b. The momentum function may be written

$$\chi(\mathbf{p}, s; n, l, m, m_s) = F(n, l; p)Y(l, m; \theta, \phi)\, \delta(m_s, s) \tag{13}$$

in which θ, ϕ specify the direction of the electron momentum, $\mathbf{p}$, instead of the direction of the electron position, and p is the magnitude of $\mathbf{p}$. With p in units of $(Z\hbar/a)$, the $F(n, l; p)$, normalized to

$$\int_0^\infty F^2(n, l; p)p^2\, dp = 1$$

Table 2[4]. Mean values of r^k with hydrogenic radial functions in units of $(a/Z)^k$ [from Equation (12)]

n:	1	2	3	4
$\langle n, l \lvert r \rvert n, l \rangle = (1/2)\,[3n^2 - l(l+1)]$				
s	1.50	6.00	13.50	24.00
p		5.00	12.50	23.00
d			10.50	21.00
f				18.00
$\langle n, l \lvert r^2 \rvert n, l \rangle = (1/2)n^2\,[5n^2 + 1 - 3l(l+1)]$				
s	3.0	42.0	207.0	648.0
p		30.0	180.0	600.0
d			126.0	504.0
f				360.0
$\langle n, l \lvert r^3 \rvert n, l \rangle = (1/8)n^2\,[35n^2(n^2-1) - 30n^2(l+2)(l-1) + 3(l+2)(l+1)l(l-1)]$				
s	7.5	330.0	3442.5	18720.0
p		210.0	2835.0	16800.0
d			1701.0	13104.0
f				7920.0
$\langle n, l \lvert r^{-1} \rvert n, l \rangle = \dfrac{1}{n^2}$				
$s, p, d, f \ldots$	1	1/4	1/9	1/16
$\langle n, l \lvert r^{-2} \rvert n, l \rangle = \dfrac{1}{n^3(l+1/2)}$				
s	2.0	1/4	2/37	1/32
p		1/12	2/81	1/96
d			2/135	1/160
f				1/224
$\langle n, l \lvert r^{-3} \rvert n, l \rangle = \dfrac{1}{n^3 l(l+1/2)(l+1)}$				
s	∞	∞	∞	∞
p		1/24	1/81	1/192
d			1/405	1/960
f				1/2688
$\langle n, l \lvert r^{-4} \rvert n, l \rangle = \dfrac{3n^2 - l(l+1)}{2n^5(l-1/2)l(l+1/2)(l+1)(l+3/2)}$				
s	∞	∞	∞	∞
p		1/24	10/729	23/3840
d			2/3645	1/3840
f				1/26880

Table 2⁴. (cont.)

n:	1	2	3	4
$\langle n, l\vert r^{-5}\vert n, l\rangle = \dfrac{5n^2 - 3l^2 - 3l + 1}{2n^5(l-1)(l-1/2)l(l+1/2)(l+1)(l+3/2)(l+2)}$				
s	∞	∞	∞	∞
p		∞	∞	∞
d			2/10935	1/10240
f				1/215040
$\langle n, l\vert r^{-6}\vert n, l\rangle = \dfrac{35n^4 - 5n^2(6l^2 + 6l - 5) + 3(l+1)l(l+1)(l+2)}{8n^7(l-3/2)(l-1)(l-1/2)(l+1)(l+3/2)(l+2)(l+5/2)}$				
s	∞	∞	∞	∞
p		∞	∞	∞
d			4/32805	13/184320
f				1/290240

is

$$F(n, l; p) = \left[\frac{2}{\pi}\frac{(n-l-1)!}{(n+l)!}\right]^{1/2} n^2 2^{2(l+1)} l! \frac{(np)^l}{(n^2p^2+1)^{l+2}} C_{n-l-1}^{l+1}\left(\frac{n^2p^2-1}{n^2p^2+1}\right) \tag{14}$$

in which C_μ^λ is a Gegenbauer polynomial, defined as the coefficient of a^μ in the expansion

$$(1 - 2ax + a^2)^{-\lambda} = \sum_{n=0}^{\infty} C_n^\lambda(x) a^n \tag{15}$$

Alternatively, the momentum state functions may be found by direct solution of the Schrödinger equation in momentum space. This takes the form [14]

$$\left(\frac{\mathbf{p}^2}{2\mu} - W\right)\chi(\mathbf{p}; W) + \int V'(\mathbf{p} - \mathbf{p}')\,\chi(\mathbf{p}'; W)\, d^3\mathbf{p} = 0 \tag{16}$$

in which $V'(\mathbf{p} - \mathbf{p}')$ is the Fourier transform of the potential energy, $V(\mathbf{r})$,

$$V'(\mathbf{p} - \mathbf{p}') = h^{-3}\int \exp\left[\frac{i}{\hbar}\mathbf{r}\cdot(\mathbf{p} - \mathbf{p}')\right] V(\mathbf{r})\, d^3\mathbf{r} \tag{17}$$

For the hydrogenic Coulomb case, in which $V(r) = -Ze^2/r$,

$$V'(\mathbf{p} - \mathbf{p}') = -\frac{Ze^2}{2\pi^2\hbar}|\mathbf{p} - \mathbf{p}'|^{-2} \tag{18}$$

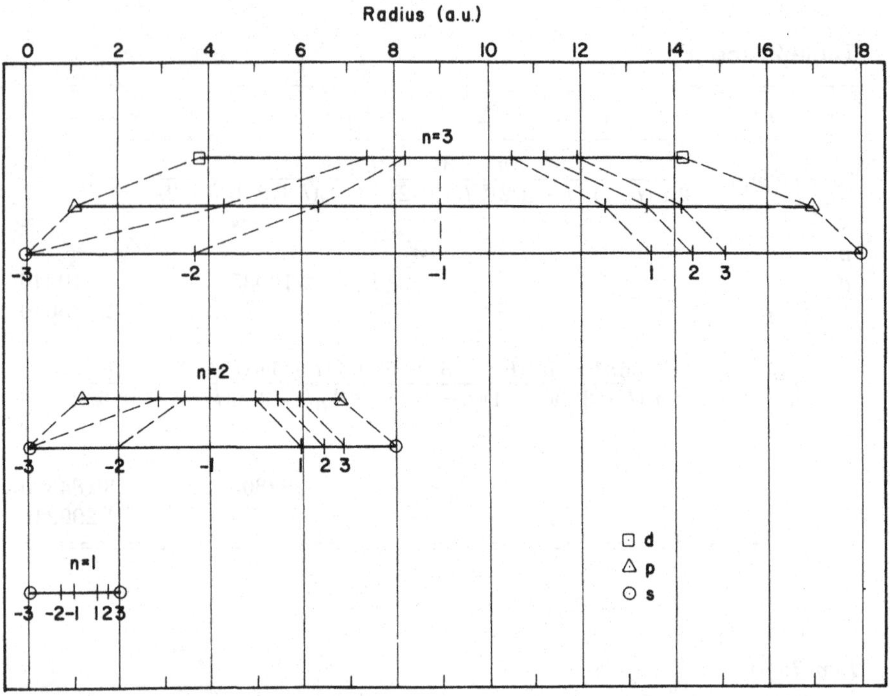

Figure 3[4]. Ranges of r in which hydrogenic radial functions $P(nl; r)$ are oscillatory and values of kth root of $<r^k>$ for various states.

For bound states, $W < 0$, it is convenient to write $p_4^2/2\mu = -W$, and to measure $\mathbf{p}$ and p_4 in units of $Z\hbar/a$, giving, finally, for the integral equation for $\chi(\mathbf{p})$,

$$(\mathbf{p}^2 + p_4^2)\chi(\mathbf{p}) = \pi^{-2}\int |\mathbf{p} - \mathbf{p}'|^{-2}\chi(\mathbf{p}')\, d^3\mathbf{p} \tag{19}$$

3[4]. General trends of $W_0(n, l)$ and $P(n, l; r)$

The Hamiltonian 2[2]9 is separated into a zero-order part, H_0, which defines the basis functions, and a first-order part, H_1, whose matrix elements define the perturbed levels and perturbed wave functions of H by the methods of Section 9[2]. We write

$$\begin{aligned} H_0 &= \sum_{i=1}^{N}\left[\frac{p_i^2}{2m} + V(r_i)\right] \\ H_1 &= \sum_{i=1} - V(r_i) + \sum_{i<j}\frac{e^2}{r_{ij}} \end{aligned} \tag{1}$$

in which

$$V(r) = -\frac{Z(r)e^2}{r} \tag{9^12}$$

with $Z(0) = Z$ and $Z(\infty) = C = Z - N + 1$. The calculations up to Chapter 8 are largely independent of the specific choice made for $V(r)$ or $Z(r)$, which is treated in Chapter 8.

The approach of $Z(r)$ to C turns out to be extremely rapid so that in practice this limiting value is usually reached with high accuracy for $r > 5a$.

Because of the form of (1), the one-electron wave functions of H_0 have the form $1^4$2b with the radial functions for each electron determined by $1^4$3 and $1^4$4 with $V(r)$ replacing $-Ze^2/r$. We write $W_0(n, l)$ for the resulting one-electron energy levels and $P(n, l; r)$ for the associated radial functions.

Because $V(r)$ has a Coulomb tail, the $P(n, l; r)$ at large r are determined by this feature of the one-electron wave equation. This calls for study of the asymptotic nature of the solutions for large r, for values $W_0(n, l)$ other than the hydrogenic values. At large r the radial function is dominated by an exponential factor

$$P(n, l; r) \approx \exp\left(-\frac{r}{r_{nl}}\right) \tag{2}$$

in which

$$r_{nl} = \frac{a}{C}\left[\frac{W_0(n, l)}{Rhc}\right]^{-1/2}$$

Near $r = 0$, the effective $V_l(r)$ is dominated by the centrifugal barrier, as in the hydrogen problem. This has the effect that in this limit

$$P(n, l; r) = A_{nl}r^{l+1} \tag{3}$$

as in hydrogen.

In the range of the classical motion, where $W_0(n, l) > V_l(r)$, the $P(n, l; r)$ is oscillatory, with $(n - l - 1)$ nodes. The exact form of $P(n, l; r)$ in this connecting range and the value of $W_0(n, l)$ are obtained by integrating $1^4$3 outward, starting with (3) in which A_{nl} is not yet determined, out to some value of r at which $V(r)$ has reached its Coulomb tail. The value of $W_0(n, l)$ is determined by the requirement that the function defined by outward integration must join smoothly to the form (2), defined by the wave equation in that outer range. Then the value of A_{nl} is fixed by normalization of $P(n, l; r)$.

For $n > l + 1$, the radial function has several loops. The amplitudes of these loops increase with increasing r. This corresponds to the fact that in the classical motion the radial velocity of the electron is least at large radii, as can be seen from the WBKJ approximation.

In the normal configurations of the atoms, most of the electrons are tightly bound in closed shells. These *inner* electrons are assumed not to change in the radiative transitions involved in optical spectra. A few *valence* electrons are in open shells in the normal configuration. These are raised to *outer* states in the excitation processes that generate optical spectra.

It is found convenient to represent the $W_0(n, l)$ by different types of empirical formula according to whether (n, l) in the structure under consideration refers to an inner or to a valence electron. For inner electrons we write

$$W_0(n, l) = -(Z - S_{nl})^2 \frac{Rhc}{n^2} \tag{4}$$

as in Section 10[1]. Here S_{nl} is called a *screening constant.* It is defined by (4) where $W_0(n, l)$ represents observed or calculated energy values.

For valence electrons the more appropriate type of empirical formula is

$$W_0(n, l) = -e^2 \frac{Rhc}{\nu_{nl}^2} \tag{5}$$

with

$$\nu_{nl} = n - \Delta_{nl}$$

Here ν_{nl}, often designated by n^* in the literature, is called the *effective* quantum number, and Δ_{nl} is called the *quantum defect,* as in Section 9[1]. Applied to levels outside the inner core, it is found that Δ_{nl} is nearly independent of n. It is also a rapidly decreasing function of l, in consequence of the fact that the centrifugal barrier increases rapidly with l, causing the $P(n, l; r)$ to be concentrated at large r where $Z(r)$ is close to C, so the field is Coulombic.

Relative to the hydrogenic solutions for $Z = C$, the $P(n, l; r)$ are in general distorted by contraction to smaller values of r. Rough calculations are sometimes made by representing the radial functions by the hydrogenic functions, 2[4]1, using an appropriately screened value of Z as in (4) in the relation $\rho = (2/n)[(Z - S_{nl})r/a]$. These are called *screened hydrogenic radial functions.* This assumes that the principal actual distortion of $P(n, l; r)$ is simply a change of scale factor on r. The actual distortions of $P(n, l; r)$ are given only a rough representation by this assumption. Moreover, these screened functions are no longer orthogonal (2[4]10) for $S_{n'l} \neq S_{nl}$.

4⁴. Term structure of configurations

We proceed now to calculation of the levels of a (Z, N) structure by applying perturbation theory to the N-electron states specified by N sets of individual quantum numbers $(a_1, a_2, \ldots, a_N)$, called A for brevity. It should be pointed out that all developments in this section and in the following sections are in Russell-Saunders coupling.

The zero order energy is simply

$$W_0(A) = \sum_{k=1}^{N} W_0(n_{a_k}, l_{a_k}) \tag{1}$$

In the state $(a_1, a_2, \ldots, a_N)$ that is represented by the Slater determinant (7^29) no two of the individual sets can be alike. The same state is given (except for phase) by the same N sets for any order of listing. To determine a definite phase a conventional order of listing of the individual sets is needed. We adopt this convention:

Individual sets are listed in order of increasing n, and for a given n in order of increasing l. Finally they are listed in order of decreasing m, and if both $m_s = \pm 1/2$ occur with a given (n, l, m), $m_s = +1/2$ is listed before $m_s = -1/2$.

For brevity, a negative m is indicated with a bar over its value, and the m_s is attached as an exponent on the value of m. Thus $3d\bar{2}^+$ refers to a set for which $n = 3$, $l = 2$, $m = -2$, and $m_s = +1/2$.

All sets with the same (n, l), of which there are $Q_l = 2(2l + 1)$, are said to belong to the same *shell*. We write q_{nl} for the actual number of occupied states in the (n, l) shell, so that $\Sigma q_{nl} = N$. Although it is impossible to associate a particular set of quantum numbers with a definite electron, we shall refer to q_{nl} as the number of electrons in the nl shell. A shell for which $q_{nl} = 0$ is called *empty*, one for which $0 < q_{nl} < Q_l$ is called *open*, and one for which $q_{nl} = Q_l$ is called *filled* or *closed*.

All states having the same set of q_{nl} are said to belong to the same *configuration*. The *normal configuration* of a structure is defined as that which gives rise to the lowest, or normal, energy level. Usually this is the same as the configuration that gives the lowest $W_0(A)$ in (1). But this is not always the case, especially where the spread of levels belonging to two configurations A and B is comparable with the difference of their $W_0(A)$ and $W_0(B)$.

The order of degeneracy of a configuration containing one open l shell with q electrons in it is $\binom{Q_l}{q_{nl}}$; this binomial coefficient is the number of ways that q_{nl} distinguishable counters can be drawn from Q_l without regard to order. The order of degeneracy of a closed shell is therefore unity. The order of degeneracy of a configuration with several open shells is the product of the factors $\binom{Q_l}{q_{nl}}$ for each open shell.

The classification of states by parity depends on the behavior under inversion of the spherical harmonic factors $Y(l, m; \theta, \phi)$. This is given by $4^3 12$. Because the parity does not depend on m or m_s, it is the same for all the states of a configuration, so configurations are even or odd. For N electrons the parity is $(-1)^{\Sigma l}$ where the sum is over the N values of l in the complete set. Because Σl is even for all closed shells, this reduces to a sum over the open shells. The s and d shells are even, so it follows that the parity of a configuration is that of the total number of p or f (or other odd l) electrons in the configuration.

The usual notation [4b] for a Russell-Saunders *term* is that in which the multiplicity $(2S + 1)$ is written as a left superscript on the uppercase letter in the S, P, D, F code that designates L. Thus ^5F (read "quintet eff") denotes the 35 states of a term with $S = 2$ and $L = 3$. Individual states, when **S** and **L** are uncoupled, are designated by giving their (M_S, M_L) in parentheses after the term symbol. Here $S(S + 1)$ and $L(L + 1)$ refer to proper-values of squares of the total spin and total orbital angular-momentum operators for the N-electron system. M_s and M_L are the corresponding proper values of the z-components of total spin and total orbital angular-momentum operators.

When **S** and **L** are coupled to give **J**, the value of J is attached as a subscript to the term symbol, thus $^5F_2$, to designate a *level*. Individual states of a level are designated by following the level symbol with (M_J) in parentheses.

We are now prepared to undertake the problem of finding the term structure of a configuration, that is, the kinds of Russell-Saunders terms to which it gives rise.

The general procedure is to develop a listing of all of the configuration's complete sets, classified by their (M_S, M_L) values. In doing this it is unnecessary to write out the individual sets belonging to closed shells, because closed shells do not contribute to the complexity of a configuration.

The simplest configurations are those composed entirely of closed shells. For these there is only one complete set. It is one for which $(M_S, M_L) = 0$, so it gives a ^1S term consisting only of a $^1S_0$ level.

Next in simplicity are the configurations with closed shells plus a single electron in one open shell. Here the (M_S, M_L) are simply the (m, m_s) of this single electron, as in the case of the alkali metals (Sec. 9^1). Therefore, these give a single ^2L term with levels $^2L_{l-1/2}$ and $^2L_{l+1/2}$ for $l > 0$ and simply $^2S_{1/2}$ for $l = 0$.

Of almost equal simplicity are the configurations consisting of closed shells plus $(np)^5$. These are characteristic of the halogens. In tabulating the possible (m_s, m) sets for p^5 we can use the same listing as for the p configuration, with the interpretation that (m_s, m) now means the individual set that is *missing* from an otherwise filled p shell. Then $(M_S, M_L) = (-m_s, -m)$. The range of values is the same as with p, so p^5 gives a ^2P term. Similarly, d^9 gives a ^2D and f^{13} a ^2F term.

The next case in respect of simplicity is that of two inequivalent electrons outside closed shells. They are called inequivalent if $(n, l) \neq (n', l')$, but equivalent if $n' = n$ and $l' = l$. For inequivalent electrons the exclusion principle does not operate to restrict assignments of the $(m_s, m; m_s', m')$, so there are a total of $2^2(2l + 1)(2l' + 1)$ possibilities.

The spins alone give four possibilities:

$$\text{with} \quad \begin{array}{lcccc} (m_s, m_s') = & (+,+) & (+,-) & (-,+) & (-,-) \\ M_s \quad = & 1 & 0 & 0 & -1 \end{array}$$

Hence the possible values of S are 0, 1, giving singlet and triplet terms.

The orbital (m, m') give $(2l + 1)(2l' + 1)$ possibilities:

M_L	(m, m')
$l + l'$	(l, l')
$l + l' - 1$	$(l - 1, l'), (l, l' - 1)$
$l + l' - 2$	$(l - 2, l'), (l - 1, l' - 1), (l, l' - 2)$
...	... ...

The number of possibilities increases by one with each unit decrease of M_L down to $M_L = |l - l'|$. After this the number of possibilities is constant down to $M_L = -|l - l'|$, after which each unit decrease of M_L decreases the possibilities by one down to one possibility, $(-l, -l')$, for $M_L = -(l + l')$. This situation was encountered in Section 9³, where it was found that the vector sum of **l** and **l**′ can take on the values by unit steps, $l + l'$ down to $|l - l'|$.

Therefore, the configurations involving two inequivalent electrons, $(n, l; n', l')$, give ^1L and ^3L terms with $(l + l') \leq L \leq |l - l'|$. In the particularly simple case in which one electron is in an S state so $l = 0$, L has the value l', giving a ^1L and a ^3L term. These configurations account for most of the excited levels of $_2$He and of the alkaline earths: $_4$Be, $_{12}$Mg, $_{20}$Ca, $_{38}$Sr, and $_{56}$Ba.

A slightly more complicated case is that in which the two electrons outside closed shells are equivalent. The exclusion principle then greatly reduces the number of complete sets from $2^2(2l + 1)(2l' + 1)$ to $\binom{2(2l+1)}{2}$, so the term structure is much simpler.

The largest M_L is $2l$, for which $(m, m') = (l, l)$. This makes two (n, l, m) sets alike, so the spins must be opposite $(+, -)$. Thus there is a ^1L with $L = 2l$, but the ^3L that occurs with inequivalent electrons has been excluded.

The next $M_L = 2l - 1$ can be realized in two ways:

$$(m, m') = (l - 1, l) \quad \text{and} \quad (l, l - 1)$$

Because these are unlike, all four (m_s, m_s') combinations can be put with $(l - 1, l)$. No additional states are given by associating the (m_s, m_s') with $(l, l - 1)$; this leads to the same complete sets in a different order of listing.

Of these four states one, with $M_S = 0$, is needed as a component of the ^1L already found. This leaves three states for $M_L = (2l - 1)$ with $M_S = 1, 0, -1$, indicating the occurrence of a $^3(L - 1)$ term but not a $^1(L - 1)$.

The next $M_L = (2l - 2)$ can be realized in three ways:

$$(m, m') = (l - 2, l), (l - 1, l - 1), (l, l - 2)$$

The four spin combinations can be associated with the first of these, but the spin combinations put with the third of these give the same four states in a permuted order of listing. The second of these are alike, so only the $(+, -)$ spin combination can be put with that. Altogether, then, the $M_L = (2l - 2)$ line has one state each for $M_S = 1$ and -1 and three for $M_S = 0$. All but one of these are needed as components of the ^1L and $^3(L - 1)$ already found, thus indicating the presence of a $^1(L - 2)$ without a $^3(L - 2)$.

Continuing in this way singlets and triplets alternate in occurrence from ^1L with $L = 2l$ down to ^1S. For example, $(nd, n'd)$ gives singlets and triplets from G down to S, but $(nd)^2$ gives ^1G, ^3F, ^1D, ^3P, ^1S.

In the $N = 4$ structures the normal configuration is $1s^22s^2$, and most of the excited levels are the singlets and triplets built on $1s^22snl$. The levels ^1D, ^3P, and ^1S are also known in which both $2s$ electrons are excited to the $2p$ shell, resulting in the configuration $1s^22p^2$.

In the $N = 6$ structures the normal configuration is $1s^22s^22p^2$ and it is found that the ^3P term is the normal term.

The $(3d)^2$ configuration is fully known as the normal configuration of $_{22}$Ti where ^3F is the normal term. It is known again in the (22, 20) structure of Ti^{++}, where the normal configuration is the same as that of $_{22}$Ti with $4s^2$ removed. In this ion the $3d4d$ configuration is completely known, including the ^3G, ^1F, ^3D, ^1P, and ^3S terms, which do not occur in the $3d^2$ configuration.

5⁴. Parentage of terms

With more than two electrons in open shells the term structure rapidly becomes much more complex. With three we encounter for the first time a case in which several $^{(2S+1)}L$ terms arise having the same (S, L). We are thus confronted with the problem of finding meaningful labels with which to distinguish them. Group theory finds application for this in more complicated cases (Chaps. 6 and 7), but simple examples can be discussed by elementary methods.

We start with the case (nl, $n'l'$, $n''l''$) of three nonequivalent electrons, for which the exclusion principle sets no restrictions on the sets of three (m_s, m) values. The order of degeneracy is therefore $2^3(2l + 1)(2l' + 1)(2l'' + 1)$.

The 2^3 spin possibilities for (m_s, m'_s, m''_s) are

M_S	(m_s, m'_s, m''_s)
$\frac{3}{2}$	$(+ + +)$
$\frac{1}{2}$	$(+ + -)\,(+ - +)\,(- + +)$
$-\frac{1}{2}$	$(- - +)\,(- + -)\,(+ - -)$
$-\frac{3}{2}$	$(- - -)$

From this we infer that with each allowed L there can be a ^4L and two ^2L terms. Thus this configuration confronts us with an abundance of like (S, L) terms, and some means of labeling them distinctively must be found.

The allowed L values are found by considering the vector coupling of three angular momenta (Sec. 9³), in this instance $\mathbf{l}$, $\mathbf{l}'$, and $\mathbf{l}''$. Three angular momenta can be coupled in three ways, by first coupling together any two of them, say l and l', to arrive at values of L_{12}, followed by addition of the third.

In this way the resulting terms are labeled, not only by the final L for all three electrons, but also by the value of L_{12} (or L_{23} or L_{13}) of the two that are coupled before the third is added. These intermediate resultants are called *parents* of the term.

In formal coupling theory, each mode of coupling provides a different scheme of parents for the terms, so the parents of a term are not unique. Each of the three schemes, labeled by L_{12}, L_{23}, or L_{13}, provides a complete basis for the terms of the configuration.

When the matrix of H is calculated, it is found that terms of like (S, L), but different parentage, are in general connected with nonvanishing matrix elements. The process of diagonalizing H then determines the actual (S, L) terms as linear combinations of those having different parents. This phenomenon is known as *parental mixing*. The end result is the same set of terms having mixed parentage, no matter which of the three intermediate L_{ab} are used. In practice it often happens that one of the electrons, say, ($n''l''$), is loosely coupled to the other two. In that case the scheme based on values of L_{12} gives an H

matrix that is more nearly diagonal than the other schemes, so that in this basis parental mixing is relatively small, and L_{12} as a parental label is meaningful.

The situation may be illustrated by choosing $l = l' = l'' = 1$, but with unequal n's, the case of three inequivalent p electrons. When the first two are coupled, the possible values of S_{12} and L_{12} give rise to singlet and triplet S, P, and D terms as parents. When the third electron is added, the singlet parents give doublets, and the triplet parents give doublets and quartets. Thus the two doublets for each L are distinguished by whether the parent is a singlet or triplet. The L values are those expected from coupling $l'' = 1$ on to the L_{12} values.

The usual notation for parents is to put the parental term symbol in parentheses preceding the final term symbol.

Altogether, then, the term structure for $pp'p''$ is:

(pp')	$(pp')p''$	(pp')	$(pp')p''$	
$(^1S_{12})$	^2P	$(^3S_{12})$	^2P	^4P
$(^1P_{12})$	$^2S\ ^2P\ ^2D$	$(^3P_{12})$	$^2S\ ^2P\ ^2D$	$^4S\ ^4P\ ^4D$
$(^1D_{12})$	$^2P\ ^2D\ ^2F$	$(^3D_{12})$	$^2P\ ^2D\ ^2F$	$^4P\ ^4D\ ^4F$

The order of degeneracy here is $6^3 = 216$, which is the sum of $(2S + 1)(2L + 1)$ for these terms, as it should be. Here the subscript 12 on the parents means the proper values of the $(\mathbf{p} + \mathbf{p}')$ orbital angular momenta. The two alternative schemes using subscript 23 for $(\mathbf{p}' + \mathbf{p}'')$, or 13 for $(\mathbf{p} + \mathbf{p}'')$, have the same term structure.

The complexities of parental mixing are considerable here, because the configuration gives rise to two ^2S, six ^2P, four ^2D, two ^2F, three ^4P, and two ^4D terms.

This complexity is greatly reduced by the exclusion principle if two of the n's are equal, giving p^2p', further still if all three are equal, giving p^3.

For p^2p' the order of degeneracy is $\binom{6}{2}\binom{6}{1} = 90$, reduced from 216. In p^2 the parents are reduced to ^1S, ^3P, ^1D so the resulting terms are:

p^2	p^2p'	p^2	p^2p'
(^1S)	^2P		
		(^3P)	$^2S\ ^2P\ ^2D\ ^4S\ ^4P\ ^4D$
(^1D)	$^2P\ ^2D\ ^2F$		

The $\Sigma(2S + 1)(2L + 1) = 90$. The total number of terms and the possibilities of parental mixing are greatly reduced. The p^3 configuration is considered in the next section. The terms are ^4S, ^2P, ^2D.

From the examples of this and the preceding section it is clear that the term structure of a configuration rapidly becomes exceedingly complicated as the number of electrons in open shells increases, and also when these include d and f electrons as contrasted with the relatively simple structures arising from s and p electrons.

6^4. Term energies by the diagonal sum rule

Slater [3a] showed how first-order perturbation theory can be simply expressed in terms of diagonal matrix elements of the electrostatic interaction in a large number of cases, by use of the diagonal sum rule, $1^2 39$.

The procedure avoids the need to solve secular equations, except when the configuration gives rise to more than one term of the same (S, L). Even in this case the procedure gives the mean value of the perturbed energy of all such like terms in terms of diagonal matrix elements.

Writing $(a_1, a_2, \ldots)$ for the diagonal matrix element of H for the complete set A, and writing ${}^{(2S+1)}L$ for the energy of the term with this term symbol, the diagonal sum rule gives one relation,

$$\sum {}^{(2S+1)}L = \sum (a_1, a_2, \ldots) \tag{1}$$

for each of the (M_S, M_L) values that occur for the configuration. The sum on the left is over all the (S, L) terms which appear for a given (M_S, M_L) set, that is, for which

$$|M_S| \leqslant S \quad \text{and} \quad M_L \leqslant L \tag{2}$$

The sum on the right is over all of the complete sets in the same cell, that is, in the same block in Table 3^4.

The symmetry in M_S and M_L is such that the same relations are contained in the four cells $(\pm M_S, \pm M_L)$. Accordingly the table of complete sets classified by (M_S, M_L) needs only to be prepared for $M_S \geqslant 0$ and $M_L \geqslant 0$.

The procedure may be illustrated on specific examples. We write $[M_S, M_L]$ for the sum on the right of (1).

Table 3^4 shows the analysis for pp' and p^2 whose term structure was found in Section 4^4. The first line in each cell gives the complete sets occurring in it; the second line gives the term symbols of the terms having components in that cell. For pp' the diagonal sum rule gives

$$\begin{aligned} {}^3D &= [1, 2] & {}^3D + {}^1D &= [0, 2] \\ {}^3D + {}^3P &= [1, 1] & {}^3D + {}^1D + {}^3P + {}^1P &= [0, 1] \\ {}^3D + {}^3P + {}^3S &= [1, 0] & {}^3D + {}^1D + {}^3P + {}^1P + {}^3S + {}^1S &= [0, 0] \end{aligned}$$

Table 3^4. (M_S, M_L) analysis for pp' and p^2 configurations, showing relation of term energies to the diagonal elements of H in the scheme of individual sets

pp'		M_S	
		1	0
M_L	2	$(1^+,1^+)$ ^3D	$(1^+,1^-),(1^-,1^+)$ ^3D, ^1D
	1	$(1^+,0^+),(0^+,1^+)$ ^3D, ^3P	$(1^+,0^-),(0^+,1^-),(1^-,0^+),(0^-,1^+)$ ^3D, ^3P, ^1D, ^1P
	0	$(1^+,\bar{1}^+),(\bar{1}^+,1^+),(0^+,0^+)$ ^3D, ^3P, ^3S	$(1^+,\bar{1}^-),(\bar{1}^+,1^-),(1^-,\bar{1}^+),(\bar{1}^-,1^+),(0^+,0^-),(0^-,0^+)$ ^3D, ^3P, ^3S, ^1D, ^1P, ^1S

$^3D = (1+,1+)$

$^3D + {}^1D = (1+,1-) + (1-,1+)$ $\quad$ $^1D = (1+,1-) + (1-,1+) - (1+,1+)$

$^3D + {}^3P = (1+,0+) + (0+,1+)$ $\quad$ $^3P = (1+,0+) + (0+,1+) - (1+,1+)$

$^3D + {}^3P + {}^1D + {}^1P = (1+,0-) + (0+,1-) + (1-,0+) + (0-,1+)$

$$^1P = (1+,0-) + (0+,1-) + (1-,0+) + (0-,1+) + (1+,1+) - (1+,0+) - (0+,1+) - (1+,1-) - (1-,1+)$$

$^3D + {}^3P + {}^3S = (1+,\bar{1}+) + (\bar{1}+,1+) + (0+,0+)$

$$^3S = (1+,\bar{1}+) + (\bar{1}+,1+) + (0+,0+) - (1+,0+) - (0+,1+)$$

$^3D + {}^3P + {}^3S + {}^1D + {}^1P + {}^1S = (1+,\bar{1}-) + (\bar{1}+,1-) + (1-,\bar{1}+) + (\bar{1}-,1+) + (0+,0-) + (0-,0+)$

$$^1S = (1+,\bar{1}-) + (\bar{1}+,1-) + (1-,\bar{1}+) + (\bar{1}-,1+) + (0+,0-) + (0-,0+) - (1+,0-) - (0+,1-) - (1-,0+) - (0-,1+) - (1+,\bar{1}+) - (\bar{1}+,1+) - (0+,0+) + (1+,0+) + (0+,1+)$$

p^2		M_S	
		1	0
M_L	2		$(1^+,1^-)$ ^1D
	1	$(1^+,0^+)$ ^3P	$(1^+,0^-),(1^-,0^+)$ ^3P, ^1D
	0	$(1^+,\bar{1}^+)$ ^3P	$(1^+,\bar{1}^-),(1^-,\bar{1}^+),(0^+,0^-)$ ^3P, ^1D, ^1S

$^1D = (1+,1-)$

$^3P = (1+,0+)$

$^3P + {}^1D + {}^1S = (1+,\bar{1}-) + (1-,\bar{1}+) + (0+,0-)$

$$^1S = (1+,\bar{1}-) + (1-,\bar{1}+) + (0+,0-) - (1+,0+) - (1+,1-)$$

From these we find

$$
\begin{aligned}
^3D &= [1, 2] \\
^3P &= [1, 1] - [1, 2] \\
^3S &= [1, 0] - [1, 1] \\
^1D &= [0, 2] - [1, 2] \\
^1P &= [0, 1] - [0, 2] - [1, 1] + [1, 2] \\
^1S &= [0, 0] - [0, 1] - [1, 0] + [1, 1]
\end{aligned}
$$

The procedure is similar but much simpler for p^2, whose (M_S, M_L) cells are shown on the right in Table 3^4. Here we find

$$
\begin{aligned}
^1D &= [0, 2] \\
^3P &= [1, 1] = [1, 0] \\
^1S &= [0, 0] - [0, 2] - [1, 1]
\end{aligned}
$$

Suppose that now we are dealing with a configuration in which a particular term $^{(2S+1)}L$ appears more than once. This situation is signalized in the (M_S, M_L) table by an increase in the *number* of complete sets in the (S, L) cell over the *number* accounted for by terms of greater S or L by the multiplicity of occurrence of this particular ${}^{(2S+1)}L$ term. The multiple *occurrence* of these components is represented in all of the cells for $M_S \leqslant S$ and $M_L \leqslant L$. The repeated term appears in each of these cells, so application of the outlined procedure gives the sum of the repeated term energies. Other calculations are needed to find each of them separately.

Table 4⁴ gives the (M_S, M_L) analysis for p^2p' whose term structure was worked out in Section 4⁴. For quartet terms we omit the parent designation as there are no repeated terms. From the $M_s = 3/2$ cells we find

$$
\begin{aligned}
{}^4D &= \left[\tfrac{3}{2}, 2\right] \\
{}^4P &= \left[\tfrac{3}{2}, 1\right] - \left[\tfrac{3}{2}, 2\right] \\
{}^4S &= \left[\tfrac{3}{2}, 0\right] - \left[\tfrac{3}{2}, 1\right]
\end{aligned}
$$

The doublet term energies may now be found in terms of the $[1/2, M_L] - [3/2, M_L]$. The final results are:

$$
\begin{aligned}
{}^2F &= \left[\tfrac{1}{2}, 3\right] \\
({}^3P)^2D + ({}^1D)^2D &= \left[\tfrac{1}{2}, 2\right] - \left[\tfrac{1}{2}, 3\right] - \left[\tfrac{3}{2}, 2\right] \\
({}^3P)^2P + ({}^1D)^2P + ({}^1S)^2P &= \left[\tfrac{1}{2}, 1\right] - \left[\tfrac{1}{2}, 2\right] - \left[\tfrac{3}{2}, 1\right] + \left[\tfrac{3}{2}, 2\right] \\
{}^2S &= \left[\tfrac{1}{2}, 0\right] - \left[\tfrac{1}{2}, 1\right] - \left[\tfrac{3}{2}, 0\right] + \left[\tfrac{3}{2}, 1\right]
\end{aligned}
$$

When the (M_S, M_L) table is worked out for the p^3 configuration the following term energies are obtained:

$$
\begin{aligned}
{}^2D &= \left[\tfrac{1}{2}, 2\right] \\
{}^2P &= \left[\tfrac{1}{2}, 1\right] - \left[\tfrac{1}{2}, 2\right] \\
{}^4S &= \left[\tfrac{3}{2}, 0\right]
\end{aligned}
$$

For configurations $(nl)^q$ in which all electrons other than those in closed shells are equivalent, the order of degeneracy of the configuration $(nl)^{Q-q}$ is the same as that for $(nl)^q$. These two associated configu-

Table 4[4]. (M_S, M_L) analysis for p^2p' configuration, showing relation of term energies to the diagonal elements of H in the scheme of individual sets

p^2p'		$M_S = \frac{3}{2}$	$M_S = \frac{1}{2}$
M_L	3		$(1^+,1^-;1^+)$ $(^1D)\,^2F$
	2	$(1^+,0^+;1^+)$ $(^3P)\,^4D$	$(1^+,0^-;1^+)$, $(1^-,0^+;1^+)$, $(1^+,0^+;1^-)$, $(1^+,1^-;0^+)$ $(^3P)\,^4D$, $(^1D)\,^2F$, $(^3P)\,^2D$, $(^1D)\,^2D$
	1	$(1^+,\bar{1}^+;1^+)$, $(1^+,0^+;0^+)$ $(^3P)\,^4D$, $(^3P)\,^4P$	$(1^+,1^-;\bar{1}^+)$, $(1^+,0^+;0^-)$, $(1^+,0^-;0^+)$, $(1^+,\bar{1}^-;1^+)$, $(1^-,0^+;0^+)$ $(1^+,\bar{1}^+;1^-)$, $(1^-,\bar{1}^+;1^+)$, $(0^+,0^-;1^+)$ $(^3P)\,^4D$, $(^3P)\,^4P$, $(^1D)\,^2F$, $(^1D)\,^2D$, $(^3P)\,^2D$, $(^3P)\,^2P$, $(^1D)\,^2P$, $(^1S)\,^2P$
	0	$(1^+,\bar{1}^+;0^+)$, $(1^+,0^+;\bar{1}^+)$, $(\bar{1}^+,0^+;1^+)$ $(^3P)\,^4D$, $(^3P)\,^4P$, $(^3P)\,^4S$	$(1^+,0^+;\bar{1}^-)$, $(1^+,0^-;\bar{1}^+)$, $(1^-,0^+;\bar{1}^+)$, $(1^+,\bar{1}^+;0^-)$, $(1^+,\bar{1}^-;0^+)$, $(1^-,\bar{1}^+;0^+)$, $(0^+,\bar{1}^+;1^-)$, $(0^+,\bar{1}^-;1^+)$, $(0^-,\bar{1}^+;1^+)$, $(0^+,0^-;0^+)$ $(^3P)\,^4D$, $(^3P)\,^4P$, $(^1D)\,^2F$, $(^1D)\,^2D$, $(^3P)\,^2D$, $(^3P)\,^2P$, $(^1D)\,^2P$, $(^1S)\,^2P$, $(^3P)\,^4S$, $(^1S)\,^2S$

$$(^1D)\,^2F = [3;1/2] = (1^+,1^-;1^+)$$

$$(^3P)\,^4D = [2;3/2] = (1^+,0^+;1^+)$$

$$(^3P)\,^2D + (^1D)\,^2D = [2;1/2] - [3;1/2] - [2;3/2]$$
$$= (1^+,0^-;1^+) + (1^-,0^+;1^+) + (1^+,0^+;1^-) + (1^+,1^-;0^+) - (1^+,1^-;1^+) - (1^+,0^+;1^+)$$

$$(^3P)\,^4P = [1;3/2] - [2;3/2] = (1^+,\bar{1}^+;1^+) + (1^+,0^+;0^+) - (1^+,0^+;1^+)$$

$$(^3P)\,^2P + (^1D)\,^2P + (^1S)\,^2P = [1;1/2] - [2;1/2] - [1;3/2] + [2;3/2]$$
$$= (1^+,1^-;\bar{1}^+) + (1^+,0^+;0^-) + (1^+,0^-;0^+) + (1^+,\bar{1}^-;1^+) + (1^-,0^+;0^+) + (1^+,\bar{1}^+;1^-) + (1^-,\bar{1}^+;1^+)$$
$$+ (0^+,0^-;1^+) - (1^+,0^-;1^+) - (1^-,0^+;1^+) - (1^+,0^+;1^-) - (1^+,1^-;0^+) - (1^+,\bar{1}^+;1^+) - (1^+,0^+;0^+)$$
$$+ (1^+,0^+;1^+)$$

$$(^3P)\,^4S = [0;3/2] - [1;3/2] = (1^+,\bar{1}^+;0^+) + (1^+,0^+;\bar{1}^+) + (\bar{1}^+,0^+;1^+) - (1^+,\bar{1}^+;1^+) - (1^+,0^+;0^+)$$

$$(^1S)\,^2S = [0;1/2] - [1;1/2] - [0;3/2] + [1;3/2]$$

rations, moreover, give rise to the same terms, as was mentioned with respect to p and p^5. To see that the terms are the same, suppose that an (M_S, M_L) analysis of the complete sets in $(nl)^q$ for $q \leqslant (2l+1)$ has been made. This same analysis can also be interpreted as applying to $(nl)^{Q-q}$ for its $(-M_S, -M_L)$ cell by interpreting the complete sets listed for $(nl)^q$ as being the ones that are missing from $(nl)^Q$ in $(nl)^{Q-q}$. Because the number of complete sets in each $(-M_S, -M_L)$ cell is the same as those in the (M_S, M_L) cell, the procedure reveals the same term structure for these two configurations.

Table 5⁴ lists the kinds of terms and the number of times that they are repeated for all of the $(nl)^q$ configurations that actually occur in the periodic table.

7⁴. Evaluation of N-electron matrix elements

Calculation [3] of the matrix elements of H and of other significant physical operators is simplified by classifying the operators by types. The Hamiltonian of 2²9 consists of a sum of the same operator $f(i)$ for each electron,

$$F = \sum_i f(i) \qquad f(i) = \frac{p_i^2}{2m} - \frac{Ze^2}{r_i} \tag{1}$$

and also of a symmetric operator that consists of the sum over all (i, j) pairs of the same symmetric operator $g(i, j)$ for each pair,

$$G = \sum_{i<j} g(i,j) \qquad g(i,j) = \frac{e^2}{r_{ij}} \tag{2}$$

Another example of an operator of this type arises in dealing with the magnetic interactions of pairs of electrons.

Clearly other symmetric operators could be based on an $h(i, j, k)$, which is symmetric in these indices summed over all triples of unequal indices, and so on for quadruples and higher (7²25). So far there is no indication of anything but single-particle terms as in F and pair-wise terms in G. This results in a great simplification of the calculations.

We write (by returning to our usual notation ϕ for the one-electron wave function):

$$\phi(A) = \phi(a_1; 1)\phi(a_2; 2) \ldots \tag{3}$$

for the product of one-electron wave functions of the complete set A and

$$P\phi(A) = \phi(Pa_1; 1)\phi(Pa_2; 2) \ldots \tag{4}$$

in which $Pa_k = a_{p_k}$ denotes a permutation of the a_i's (see Sec. 7²).

A general matrix element for F between N particle states B and A is

$$\langle B|F|A\rangle = (N!)^{-1} \sum_i \sum_{P,Q} (-1)^{p+q} \delta'(QB, PA)\langle Qb_i|f(i)|Pa_i\rangle \tag{5}$$

in which

$$\delta_i'(QB, PA) = \prod_{j \neq i} \delta(b_{q_j}, a_{p_j}) \tag{6}$$

Table 5^4. Russell-Saunders terms for $(nl)^q$ configurations*

s	^2S			
p, p^5	^2P			
p^2, p^4	^1SD	^3P		
p^3	^2PD	^4S		
d, d^9	^2D			
d^2, d^8	^1SDG	^3PF		
d^3, d^7	^2PDFGH 2	^4PF		
d^4, d^6	^1SDFGI 2 2 2	^3PDFGH 2 2	^5D	
d^5	$^2SPDFGHI$ 3 2 2	^4PDFG	^6S	
f, f^{13}	^2F			
f^2, f^{12}	^1SDGI	^3PFH		
f^3, f^{11}	$^2PDFGHIKL$ 2 2 2 2	^4SDFGI		
f^4, f^{10}	$^1SDFGHIKLN$ 2 4 4 2 3 2	$^3PDFGHIKLM$ 3 2 4 3 4 2 2	^5SDFGI	
f^5, f^9	$^2PDFGHIKLMNO$ 4 5 7 6 7 5 5 3 2	$^4SPDFGHIKLM$ 2 3 4 4 3 3 2	^6PFH	
f^6, f^8	$^1SPDFGHIKLMNQ$ 4 6 4 8 4 7 3 4 2 2	$^3PDFGHIKLMNO$ 6 5 9 7 9 6 6 3 3	$^5SPDFGHIKL$ 3 2 3 2 2	^7F
f^7	$^2SPDFGHIKLMNOQ$ 2 5 7 10 10 9 9 7 5 4 2	$^4SPDFGHIKLMN$ 2 2 6 5 7 5 5 3 3	^6PDFGHI	^8S

*The number of terms of the given type that occur in the configuration is written under the letters denoting L value.

and

$$\langle Qb_i|f(i)|Pa_i\rangle = \int \phi^*(Qb_i)f(\ )(Pa_i)\,dv \tag{7}$$

in which the coordinate labels are omitted, because the value of the integral is the same for each particle index, and Q denotes a permutation of the b_i's.

To arrive at a nonvanishing δ' in (6), each of the b_{q_j}; must be the same as a corresponding a_{p_j}. This can be satisfied if B differs from A in one individual set, or if B is the same as A in all N individual sets, that is, (5) is a diagonal element.

In the former case let a_k be the individual set in A that is not matched by the same set in B. Let R be the permutation $RB = B'$, and r the parity of the transposition R, for which the individual sets in B are matched to the order of A:

$$\begin{aligned} A&: a_1\, a_2 \ldots a_k \ldots \\ RB = B'&: a_1\, a_2 \ldots b_k \ldots \end{aligned}$$

In (5) the sum over QB can now be replaced by QB', because the sum is over all permutations.

In order to have a nonvanishing δ' factor, P and Q must now be restricted so that for the $(N-1)$ sets $j \neq i$

$$b_{q_j} = a_{p_j} \qquad j \neq i$$

This restricts P to one of the $(N-1)!$ permutations for which

$$Pa_i = a_k$$

and restricts Q to being the same as P. That is, the sum on Q involves but one term in which $Q = P$, so that they have the same parity, leaving $(-1)^{p+q+r} = (-1)^r$. Summing on P gives a factor $(N-1)!$. Summing on i gives a factor N.

The final result is that the only nonvanishing nondiagonal element for F is

$$\langle B|F|A\rangle = (-1)^r \langle b_k|f|a_k\rangle \tag{8}$$

so the N-electron element is reduced to a one-electron element connecting the single individual sets that are unlike in B and A.

For the diagonal element, $B = A$, so their conventional order will be the same. For each P in (5) we must have $Q = P$, so $(-1)^{p+q} = 1$. For a fixed Pa_i, there are $(N-1)!$ P's, each of which gives the same $\langle Pa_i|f(i)|Pa_i\rangle$. The value of this is the same for all i so the sum on i gives a factor N. The final result for the diagonal element of F is

$$\langle A|F|A\rangle = \sum_{i=1}^{N} \langle a_i|f|a_i\rangle \tag{9}$$

so the diagonal N-electron element is the sum of the N-diagonal one-electron elements in the complete set A.

The analogous calculation for two-electron operators, here called type G, is almost as simple. In place of (5) we have

$$\langle B|G|A\rangle = (N!)^{-1} \sum_{i<j} (-1)^{p+q} \delta''(QB, PA) \times \langle Qb_i, Qb_j|g(i,j)|Pa_i, Pa_j\rangle \tag{10}$$

where $\delta''_{ij}(QB, PA)$ is analogous to (6) in being a product of $(N-2)$ one-electron δ's with the omission of i and j.

In this case, to avoid vanishing of δ'', the set B can differ from A in at most two individual sets, because QB must agree with PA in respect of all electron indices other than i and j.

Suppose that B differs from A in respect to the two sets a_k and a_t. Then a permutation R exists that orders B' as in A. Another permutation of opposite parity to R exists that gives B'' with b_k and b_t interchanged:

$$\begin{aligned} A&: a_1\, a_2 \ldots a_k \ldots a_t \ldots \\ B'&: a_1\, a_2 \ldots b_k \ldots b_t \ldots \\ B''&: a_1\, a_2 \ldots b_t \ldots b_k \ldots \end{aligned}$$

Either B'' or B' with A gives in this case a nonvanishing δ'' factor. By arguments such as those leading to (8) we readily find

$$\langle B|G|A\rangle = (-1)^r[\langle b_k, b_t|g|a_k, a_t\rangle - \langle b_t, b_k|g|a_k, a_t\rangle] \tag{11}$$

Thus the N-electron nondiagonal element is reduced to a *pair* of two-electron integrals with opposite signs.

Similarly, if B differs from A by one individual set, let it be $b_k \neq a_k$. Let R be the permutation of B, which puts this in the place opposite B'. The nondiagonal matrix element in this case is

$$\langle B|G|A\rangle = (-1)^r \sum [\langle b_k, a_t|g|a_k, a_t\rangle - \langle a_t, b_k|g|a_k, a_t\rangle] \tag{12}$$

In the sum a_t runs over the $(N-1)$ sets common to A and B.

Finally, by similar calculations, the diagonal element for G is

$$\langle A|G|A\rangle = \sum_{k,t=1}^{N} [\langle a_k, a_t|g|a_k, a_t\rangle - \langle a_t, a_k|g|a_k, a_t\rangle] \tag{13}$$

In (12) and (13) the two-electron integrals with positive sign are called *direct;* those with negative sign are called *exchange integrals.*

Occurrence of integrals with opposite signs in (11), (12), and (13) is a characteristic feature of antisymmetric wave functions. When applied to matrix elements of H the integrals of negative sign contribute to the total energy of an atomic system that is labeled as *exchange energy.*

8[4]. One-electron central-field integrals

Calculation of $\langle b|f|a\rangle$ with f given by 7^41 and states a and b represented by 7^29 reduces to a radial integral only, because of the form of the central-field wave functions.

For brevity, from here on we write $P(a, 1)$ for $P(n_a, l_a; r_1)$ and $Y(a, 1)$ for $Y(l_a, m_a; \theta_1, \phi_1)$.

The radial integral in $\langle b|f|a\rangle$ is

$$\int_0^\infty P^*(b, 1)\left[-\frac{\hbar^2}{2m}\frac{d^2}{dr_1^2} + \frac{\hbar^2}{2m}\frac{l(l+1)}{r_1^2} - \frac{Ze^2}{r_1}\right]P(a, 1)\,dr_1 \tag{1}$$

This is multiplied by the integral over directions

$$\int_0^{2\pi}\int_0^{\pi} Y^*(b, 1)Y(a, 1)\sin\theta\,d\theta\,d\phi = \delta(l_b, l_a)\delta(m_b, m_a) \tag{2}$$

and the sum over spin factors

$$\sum_s \delta(m_{sb}, s)\delta(m_{sa}, s) = \delta(m_{sb}, m_{sa}) \tag{3}$$

Hence $\langle b|f|a\rangle$ is diagonal in l, m, m_s, and its value is independent of (m, m_s). The radial function $P(a, 1)$ satisfies

$$\left[-\frac{\hbar^2}{2m}\frac{d^2}{dr_1^2} + \frac{\hbar^2}{2m}\frac{l(l+1)}{r_1^2} + V(r_1)\right]P(a, 1) = W_0(a)P(a, 1) \tag{4}$$

Therefore, the radial integral in $\langle b|f|a\rangle$ is

$$W_0(a)\delta(n_b, n_a) - \int_0^\infty P(b, 1)\left[V(r_1) + \frac{Ze^2}{r_1}\right]P(a, 1)\,dr_1.$$

For the diagonal element, $b = a$, this integral is often denoted by $I(n_a, l_a)$ so

$$I(n_a, l_a) = W_0(a) - \int P^2(a, 1)\left[V(r_1) + \frac{Ze^2}{r_1}\right]dr_1 \tag{5}$$

Therefore from 7^49 the diagonal element of F in 7^41 is

$$\langle A|F|A\rangle = \sum_{\text{shells}} q_{nl}I(n, l) \tag{6}$$

having the same value for all states in a configuration.

9[4]. Coulomb interaction matrix elements

The two-electron integrals that appear in 7^410 through 7^413 can be further reduced to double radial integrals over r_1 and r_2 multiplied by coefficients that are integrals of products of three spherical harmonics, as evaluated in $6^3 20.$ and $6^3 20'$. These radial integrals are independent of the m's and m_s's and so are the same for all complete sets in a configuration.

In the calculation of $\langle a, b|e^2/r_{12}|c, d\rangle$ the sums on s_1 and s_2 give the factors

$$\delta(m_{sa}, m_{sc})\,\delta(m_{sb}, m_{sd}) \tag{1}$$

The matrix element is diagonal separately in the spins, as follows from the fact that e^2/r_{12} commutes with $\mathbf{S}_1$ and $\mathbf{S}_2$.

The integral over spatial coordinates is that of a classical Coulomb interaction energy between two volume densities of charge, which we designate as $\rho(ac; 1)$ and $\rho(bd; 2)$:

$$\int\int \rho(ac; 1)\frac{1}{|\mathbf{r}_1 - \mathbf{r}_2|}\rho(bd; 2)\, d\tau_1\, d\tau_2 \tag{2}$$

in which the densities are

$$\rho(ac; 1) = \frac{e}{r_1^2} P(a; 1)P(c; 1)Y^*(a; 1)Y(c; 1)$$

$$\rho(bd; 2) = \frac{e}{r_2^2} P(b; 2)P(d; 2)Y^*(b, 2)Y(d;(2)$$

It is useful to evaluate (2) in two stages, first calculating the electrostatic potential $\phi(ac; 2)$ at $\mathbf{r}_2$ because of charge distribution $\rho(ac; 1)$. This is

$$\phi(ac; 2) = \int \frac{\rho(ac; 1)}{|\mathbf{r}_1 - \mathbf{r}_2|}\, d\tau_1 \tag{3}$$

after which

$$\left\langle a, b\left|\frac{e^2}{r_{12}}\right|c, d\right\rangle = \delta(m_{sa}, m_{sc})\delta(m_{sb}, m_{sd}) \int \phi(ac, 2)\rho(bd, 2)\, d\tau_2 \tag{4}$$

The potentials that appear in (3) are useful in the study of self-consistent fields (Chap. 8).

To evaluate (3) we use $4^3 13$:

$$|\mathbf{r}_1 - \mathbf{r}_2|^{-1} = \sum_{k=0}^{\infty} \frac{r_<^k}{r_>^{k+1}} P_k(\cos\omega) \tag{5}$$

in which $r_<$ and $r_>$ are the lesser and greater of r_1 and r_2, and ω is the angle between $\mathbf{r}_1$ and $\mathbf{r}_2$.

Expanding $P_k(\cos\omega)$ by $4^3 15$ we find that $\phi(ac; 2)$ is expressed as a series in $Y(k, m; \theta_2, \phi_2)$. The coefficients contain integrals over products of three spherical harmonics as defined in $6^3 20'$,

$$\left(\frac{2k+1}{4\pi}\right)^{1/2} c^k(a; c) = \left(\frac{2k+1}{4\pi}\right)^{1/2} c^k(l_a, m_a; l_c, m_c)$$
$$= \int_0^{2\pi}\int_0^{\pi} Y(k, m; 1)Y^*(a; 1)Y(c; 1) \sin\theta_1 \, d\theta_1 \, d\phi_1$$

(which vanishes unless $m = m_a - m_c$). Values of c^k are given in Table 6^4. The coefficient also contain the radial integrals

$$R^k(ac; 2) = \int_0^\infty \frac{r_<^k}{r_>^{k+1}} P(a; 1)P(c; 1)\, dr_1 \tag{6}$$

The final result is

$$\phi(ac; 2) = 4\pi e \sum_k Y^*(k, m_a - m_c; 2)R^k(ac; 2)\left(\frac{2k+1}{4\pi}\right)^{1/2} c^k(a; c) \tag{7}$$

From Section 6^3 we know that the c's vanish unless $(k + l_a + l_c)$ is an even integer and $(l_a + l_c) \geq k \geq |l_a - l_c|$. In consequence the sum on k in (7) contains very few terms in usual cases.

The result gives the potential, as a series of spherical harmonics, caused by a charge density, which is itself given as a product of two spherical harmonics. A special case of importance arises in the calculation of diagonal matrix elements, in which case $c = a$. If, in addition, we are interested in the potential attributable to the full shell (7) has to be summed over m_{sa} (by multiplying by 2) and summed over m_a from l_a to $-l_a$. For the full shell a, the potential is

$$\phi(aa; 2) = 2\int |\mathbf{r}_1 - \mathbf{r}_2|^{-1} \frac{e}{r_1^2} P^2(a; 1)\left[\sum_{m_a} Y^*(l_a, m_a; 1)Y(l_a, m_a; 1)\right] d\tau_1$$

By $4^3 15'$ the sum on m_a is $(2l_a + 1)/4\pi$; therefore, the potential is spherically symmetric. The 4π is canceled on integrating over all directions, so the potential attributable to the full shell of $2(2l_a + 1)$ electrons is

$$\begin{aligned}\phi(aa; 2) &= 2(2l_a + 1)e\int_0^\infty \frac{1}{\mathrm{r}_>} P^2(a; 1)\, dr_1 \\ &= 2(2l_a + 1)e\left[\frac{1}{r_2}\int_0^{r_2} P^2(a; 1)\, dr_1 + \int_{r_2}^\infty \frac{1}{r_1} P^2(a; 1)\, dr_1\right]\end{aligned} \tag{8}$$

Now, substituting (7) in (4), we find that the integration over directions of $\mathbf{r}_2$ again introduces an integral of the product of three spherical harmonics. This vanishes unless $m = m_d - m_b$, when it has the value $\sqrt{(2k+1)/4\pi}\, c^k(d; b)$. The order (d, b) instead of (b, d) occurs because

Table 6[4]. $c^k(l_a, m_a; l_b, m_b)$ (defined in 6[3]20′). From the definition $c^k(l_b, m_b; l_a, m_a) = (-1)^{m_a - m_b} c^k(l_a, m_a; l_b, m_b)$, so this table can be used for $l_a > l_b$. Writing $c^k = y/x$, the value of x^2 is given in parentheses at the top of each column. Tabulated are values of y^2 and the sign of y. Values of $(c^k)^2$ are those of b^k, defined in 9[4]13′.

l_a, l_b	m_a, m_b	k = 0	k = 2	k = 4	k = 6
s,s		(1)			
	0 0	+1			
s,d			(5)		
	0 ±2		+ 1		
	0 ±1		− 1		
	0 0		+ 1		
p,p		(1)	(25)		
	±1 ±1	+1	− 1		
	±1 0	0	+ 3		
	0 0	+1	+ 4		
	±1 ∓1	0	− 6		
p,f			(175)	(189)	
	±1 ±3		+45	− 1	
	±1 ±2		−30	+ 3	
	±1 ±1		+18	− 6	
	±1 0		− 9	+10	
	0 ±3		0	− 7	
	0 ±2		+15	+12	
	0 ±1		−24	−15	
	0 0		+27	+16	
	±1 ∓3		0	−28	
	±1 ∓2		0	+21	
	±1 ∓1		+ 3	−15	
d,d		(1)	(49)	(441)	
	±2 ±2	+1	− 4	+ 1	
	±2 ±1	0	+ 6	− 5	
	±2 0	0	− 4	+15	
	±1 ±1	+1	+ 1	−16	
	±1 0	0	+ 1	+30	
	0 0	+1	+ 4	+36	
	±2 ∓2	0	0	+70	
	±2 ∓1	0	0	−35	
	±1 ∓1	0	− 6	−40	
f,f		(1)	(225)	(1089)	(184041)
	±3 ±3	+1	−25	+ 9	− 25
	±3 ±2	0	+25	−30	+ 175
	±3 ±1	0	−10	+54	− 700
	±3 0	0	0	−63	+ 2100
	±2 ±2	+1	0	−49	+ 900
	±2 ±1	0	+15	+32	− 2625
	±2 0	0	−20	− 3	+ 5600
	±1 ±1	+1	+ 9	+ 1	− 5625
	±1 0	0	+ 2	+15	+ 8750
	0 0	+1	+16	+36	+10000
	±3 ∓3	0	0	0	−23100
	±3 ∓2	0	0	0	+11550
	±3 ∓1	0	0	+42	− 5250
	±2 ∓2	0	0	+70	+12600
	±2 ∓1	0	0	−14	− 9450
	±1 ∓1	0	−24	−40	−10500

l_a, l_b	m_a, m_b	k = 1	k = 3	k = 5
s,p		(3)		
	0 ±1	− 1		
	0 0	+ 1		
s,f			(7)	
	0 ±3		− 1	
	0 ±2		+ 1	
	0 ±1		− 1	
	0 0		+ 1	
p,d		(15)	(245)	
	±1 ±2	− 6	+ 3	
	±1 ±1	+ 3	− 9	
	±1 0	− 1	+18	
	0 ±2	0	+15	
	0 ±1	− 3	−24	
	0 0	+ 4	+27	
	±1 ∓2	0	+45	
	±1 ∓1	0	−30	
d,f		(35)	(315)	(7623)
	±2 ±3	−15	+10	− 5
	±2 ±2	+ 5	−20	+ 25
	±2 ±1	− 1	+24	− 75
	±2 0	0	−20	+ 175
	±1 ±3	0	+25	− 35
	±1 ±2	−10	−15	+ 120
	±1 ±1	+ 8	+ 2	− 250
	±1 0	− 3	+ 2	+ 400
	0 ±3	0	+25	− 140
	0 ±2	0	0	+ 315
	0 ±1	− 6	− 9	− 450
	0 0	+ 9	+16	+ 500
	±2 ∓3	0	0	−1050
	±2 ∓2	0	0	+ 630
	±2 ∓1	0	+10	− 350
	±1 ∓3	0	0	− 420
	±1 ∓2	0	+25	+ 560
	±1 ∓1	0	−15	− 525

of a different occurrence of complex conjugates in the spherical harmonics. There also appears a new radial integral,

$$R^k(a, b; c, d) = \int_0^\infty R^k(ac; 2)P(b; 2)P(d; 2)\, dr_2$$
$$= \int_0^\infty \int_0^\infty \frac{r_<^k}{r_>^{k+1}} P(a; 1)P(c; 1)P(b; 2)P(d; 2)\, dr_1\, dr_2 \qquad (9)$$

The final result is

$$\left\langle a, b \left| \frac{e^2}{r_{12}} \right| c, d \right\rangle = \delta(m_{sa}, m_{sc})\delta(m_{sb}, m_{sd})\delta(m_a + m_b, m_c + m_d) \times e^2 \sum_k c^k(a; c)c^k(d; b)R^k(a, b; c, d) \qquad (10)$$

The matrix elements are diagonal in $L_z(1) + L_z(2)$ as expressed by the third delta function factor. The matrix elements are not diagonal separately in $L_z(1)$ and $L_z(2)$, expressing the physical fact that the interacting electrons may interchange orbital angular momentum.

The entire dependence of the matrix elements on the m's in (10) is contained in the c's. These are related to the $3j$ symbols and so the matrix element can be expressed in terms of them,

$$\left\langle a, b \left| \frac{e^2}{r_{12}} \right| c, d \right\rangle = \delta(m_{sa}, m_{sc})\delta(m_{sb}, m_{sd})\, e^2(-1)^{m_a+m_d}$$
$$\times [(2l_a + 1)(2l_b + 1)(2l_c + 1)(2l_d + 1)]^{1/2}$$
$$\times \sum_k \begin{pmatrix} l_a & l_c & k \\ 0 & 0 & 0 \end{pmatrix} \begin{pmatrix} l_d & l_b & k \\ 0 & 0 & 0 \end{pmatrix} \begin{pmatrix} l_a & l_c & k \\ -m_a & m_c & m \end{pmatrix} \qquad (11)$$
$$\times \begin{pmatrix} l_d & l_b & k \\ -m_d & m_b & m \end{pmatrix} R^k(a, b; c, d)$$

in which $m = m_a - m_c = m_d - m_b$.

Special attention is now given to the matrix elements involved in the calculation of diagonal matrix elements as in 7⁴13. There are the direct (J), and exchange (K) contributions defined as

$$J(a, b) = \left\langle a, b \left| \frac{e^2}{r_{12}} \right| a, b \right\rangle = \sum_k a^k(a, b)F^k(a, b) \qquad (12)$$

in which

$$a^k(a, b) = c^k(a, a)c^k(b, b)$$
$$= (-1)^{m_a+m_b}(2l_a + 1)(2l_b + 1) \begin{pmatrix} l_a & l_a & k \\ 0 & 0 & 0 \end{pmatrix} \begin{pmatrix} l_b & l_b & k \\ 0 & 0 & 0 \end{pmatrix}$$
$$\times \begin{pmatrix} l_a & l_a & k \\ -m_a & m_a & 0 \end{pmatrix} \begin{pmatrix} l_b & l_b & k \\ -m_b & m_b & 0 \end{pmatrix} \qquad (12')$$

$$F^k(a, b) = e^2 R^k(a, b; a, b) \qquad (12'')$$

and

$$K(a, b) = \left\langle a, b \left| \frac{e^2}{r_{12}} \right| b, a \right\rangle = \delta(m_{sa}, m_{sb}) \sum_k b^k(a, b) G^k(a, b) \tag{13}$$

in which

$$b^k(a, b) = [c^k(a, b)]^2 = (2l_a + 1)(2l_b + 1) \begin{pmatrix} l_a & l_b & k \\ 0 & 0 & 0 \end{pmatrix}^2 \begin{pmatrix} l_a & l_b & k \\ -m_a & m_b & m \end{pmatrix}^2 \tag{13'}$$

where $m = m_a - m_b$ and

$$G^k(a, b) = e^2 R^k(a, b; b, a) \tag{13''}$$

From its definition the F^k integral is positive and a decreasing function of k. Although from its definition such a statement cannot be made for the G^k integrals, Racah [2³aII] showed that they also are positive and $(2k + 1)G^k$ is a decreasing function of k.

The complete diagonal matrix element $\langle A|G|A\rangle$ is

$$\langle A|G|A\rangle = \sum_{\text{pairs}} [J(a, b) - K(a, b)] = \sum_{\text{pairs}} (m_a^{m_{sa}}, m_b^{m_{sb}}) \tag{14}$$

The sum runs over all pairs of individual sets ($a \neq b$), each pair counted once, so there $(1/2)N(N - 1)$ terms in it. Because of the peculiarity that $K(a, a) = J(a, a)$ the summand vanishes for $a = b$ so one can include the $a = b$ terms formally if that is convenient. Here $(m_a^{m_{sa}}, m_b^{m_{sb}})$ stands for the energy of a term, ${}^{2S+1}L$, or its component (see Equation 6⁴1) appearing in the (M_S, M_L) analysis of a pair in a given configuration, as shown in Table 3⁴ and as explained in Section 6⁴ (see also the example that will be given in Sec. 10⁴).

The expression can be further simplified by recognizing that in practice many of the individual sets refer to full shells. The sums can be carried out in such a way as to give expressions for interactions between all pairs of full shells, with each pair of shells counted once. We consider first the contribution from the direct integrals for an electron in the (n_b, l_b) shell interacting with the full (n_a, l_a) shell. It is

$$\begin{aligned} \sum_{m_{sa}, m_a} J(a, b) &= \sum_{m_{sa}, m_a} \sum_k a^k(a, b) F^k(a, b) \\ &= (-1)^{m_b}(2l_a + 1)(2l_b + 1) \\ &\quad \times \sum_k \begin{pmatrix} l_a & k & l_a \\ 0 & 0 & 0 \end{pmatrix} \begin{pmatrix} l_b & k & l_b \\ 0 & 0 & 0 \end{pmatrix} \begin{pmatrix} l_b & k & l_b \\ m_b & 0 & -m_b \end{pmatrix} \\ &\quad \times \sum_{m_{sa}, m_a} (-1)^{m_a} \begin{pmatrix} l_a & k & l_a \\ m_a & 0 & -m_a \end{pmatrix} F^k(a, b) \end{aligned} \tag{15}$$

The sum on m_a can be evaluated by using the following relation, which comes from coupling of two angular momenta:

$$\sum_m \langle lm; kq|lm\rangle = [2l+1]^{1/2} \sum_m (-1)^{k-l-m} \begin{pmatrix} l & k & l \\ m & q & -m \end{pmatrix} = (2l+1)\delta(k,0)\delta(q,0) \tag{16}$$

This leads to

$$\sum_m (-1)^m \begin{pmatrix} l & k & l \\ m & q & -m \end{pmatrix} = (-1)^l\,[2l+1]^{1/2}\delta(k,0)\delta(q,0) \tag{17}$$

The sum on m_{sa} produces a factor 2, so the contribution from direct integrals is

$$\sum_{m_{sa},\, m_a} J(a,b) = (-)^{l+m_b} 2(2l_a+1)(2l_b+1)[2l_a+1]^{1/2} \times \begin{pmatrix} l_a & 0 & l_a \\ 0 & 0 & 0 \end{pmatrix} \begin{pmatrix} l_b & 0 & l_b \\ 0 & 0 & 0 \end{pmatrix} \begin{pmatrix} l_b & 0 & l_b \\ m_b & 0 & -m_b \end{pmatrix} F^0(a,b) \tag{18}$$

where the $3j$ symbols can be evaluated from the simple relation

$$\begin{pmatrix} l & 0 & l \\ m & 0 & -m \end{pmatrix} = (-1)^{l-m} \frac{1}{[2l+1]^{1/2}} \tag{19}$$

By substituting (19) in (18) for $m = m_b$; and for $m = 0$, we finally get

$$\sum_{m_{sa}, m_a} J(a,b) = 2(2l_a+1)F^0(a,b)$$

for the direct interaction of one (n_b, l_b) electron with the $2(2l_a + 1)$ electrons of the full (n_a, l_a) shell. This is independent of (m_b, m_{bs}) and therefore the direct interaction with a full (n_b, l_b) shell is

$$(4l_a+2)(4l_b+2)F^0(n_a, l_a; n_b, l_b) \tag{15'}$$

Thus the *average direct* interaction per pair of electrons, one in (n_a, l_a), the other in (n_b, l_b) is $F^0(n_a, l_a; n_b, l_b)$.

The exchange interaction for a full (n_a, l_a) shell with an electron (n_b, l_b) is

$$\sum_{m_{sa}, m_a} K(a,b) = \delta(m_{sa}, m_{sb}) \sum_{m_{sa}, m_a} \sum_k b^k(a,b) G^k(a,b) = \sum_{m_a} \sum_k (2l_a+1) \times (2l_b+1) \begin{pmatrix} l_a & l_b & k \\ 0 & 0 & 0 \end{pmatrix}^2 \begin{pmatrix} l_a & l_b & k \\ -m_a & m_b & m_a - m_b \end{pmatrix}^2 G^k(a,b)$$

Unlike the contribution from direct integrals, this does depend on m_b. The exchange interaction for a full (n_a, l_a) shell with a full (n_b, l_b)

shell can be found by summing over m_{sb} and m_b. The result is

$$\sum_{m_{sb},m_b}\sum_{m_{sa},m_a} K(a,b) = 2(2l_a+1)(2l_b+1)\sum_k \begin{pmatrix} l_a & l_b & k \\ 0 & 0 & 0 \end{pmatrix}^2 G^k(a,b)$$
$$\times \sum_{m_a,m_b} \begin{pmatrix} l_a & l_b & k \\ -m_a & m_b & m_a-m_b \end{pmatrix}^2$$

From the orthogonality properties of the $3j$ symbols ($3^3 3$ and $3^3 4$),

$$\sum_{m_a,m_b} \begin{pmatrix} l_a & l_b & k \\ -m_a & m_b & m_a-m_b \end{pmatrix}^2 = 1$$

so the total exchange contribution becomes

$$\sum_{m_{sb},m_b}\sum_{m_{sa},m_a} K(a,b) = 2(2l_a+1)(2l_b+1)\sum_k \begin{pmatrix} l_a & l_b & k \\ 0 & 0 & 0 \end{pmatrix}^2 G^k(a,b)$$
$$= 2[(2l_a+1)(2l_b+1)]^{1/2}\sum_k c^k(l_a,0;l_b,0)G^k(a,b)$$

The average exchange interaction per pair of electrons can now be found by dividing the above result by the number of pairs, which is $4(2l_a+1)(2l_b+1)$. The result is

$$\frac{1}{4(2l_a+1)(2l_b+1)}\sum_{m_{sb},m_b}\sum_{m_{sa},m_a} K(a,b)$$
$$= \frac{1}{2}\sum_k \begin{pmatrix} l_a & l_b & k \\ 0 & 0 & 0 \end{pmatrix}^2 G^k(a,b)$$
$$= \frac{1}{2}\sum_k \frac{c^k(l_a,0;l_b,0)}{[(2l_a+1)(2l_b+1)]^{1/2}} G^k(a,b) \qquad (20)$$

Hence the *average interaction per electron pair* involving the (n_a, l_a) shell and the (n_b, l_b) shell is

$$\mathrm{Av}(a,b) = F^0(a;b) - \frac{1}{2}\sum_k \begin{pmatrix} l_a & l_b & k \\ 0 & 0 & 0 \end{pmatrix}^2 G^k(a,b)$$
$$= F^0(a;b) - \frac{1}{2}\sum_k \frac{c^k(l_a,0;l_b,0)}{[(2l_a+1)(2l_b+1)]^{1/2}} G^k(a,b) \qquad (21)$$

The total contribution of the pairs of electrons in the same (n_a, l_a) closed shell can be obtained by multiplying (21) for $(n_a, l_a) = (n_b, l_b)$ by the square of the number $2^2(2l_a+1)^2$ of electrons in the shell, replacing $G^k(a,a)$ by $F^k(a,a)$, and dividing by two because each pair of electrons is to be counted only once. This is

$$2(2l_a+1)^2\left\{F^0(a,a) - \frac{1}{2}\sum_k \begin{pmatrix} l_a & l_a & k \\ 0 & 0 & 0 \end{pmatrix}^2 F^k(a,a)\right\}$$

Because

$$\begin{pmatrix} l_a & l_a & 0 \\ 0 & 0 & 0 \end{pmatrix}^2 = \frac{1}{2l_a + 1}$$

the $k = 0$ term in the sum combines with the first term to give

$$(2l_a + 1)(4l_a + 1)F^0(a, a) - (2l_a + 1)^2 \sum_{k>0} \begin{pmatrix} l_a & l_a & k \\ 0 & 0 & 0 \end{pmatrix}^2 F^k(a, a) \qquad (22)$$

The *average interaction energy* for each of the $(2l_a + 1)(4l_a + 1)$ electron pairs in the same (n_a, l_a) shell, then, becomes

$$\begin{aligned} \mathrm{Av}(a, a) &= F^0(a, a) - \frac{2l_a + 1}{4l_a + 1} \sum_{k>0} \begin{pmatrix} l_a & l_a & k \\ 0 & 0 & 0 \end{pmatrix}^2 F^k(a, a) \\ &= F^0(a, a) - \frac{1}{4l_a + 1} \sum_{k>0} c^k(l_a, 0; l_a, 0) F^k(a, a) \qquad (23) \end{aligned}$$

Table 7[4] gives the average interaction energy per electron pair in different shells, and in the same shell, from (21) and (23), using the values of the c^k's from Table 6[4].

Finally, we consider those pairs in which one or both of the electrons, (n_a, l_a) and (n_b, l_b) are in open shells. The interaction of such a pair depends on the m_a and m_b, given by $J(a, b) - K(a, b)$ as defined in (12) and (13). However, it is often more convenient, as first recognized by Shortley and emphasized by Slater [15], to express this as the average value per pair plus a deviation from the average:

$$\begin{aligned} &J(a, b) - K(a, b) \\ &\qquad = \mathrm{Av}(a, b) + E(n_a, l_a, m_a, m_{sa}; n_b, l_b, m_b, m_{sb}) \qquad (24) \end{aligned}$$

The advantage lies in the fact that the average of the deviations, E, over all values of m and m_s is zero. Therefore, $\langle A|H|A\rangle_{\mathrm{Av}}$, the *average* of all of the $\langle A|H|A\rangle$ over the states of a configuration, is equal to the sum of the average interaction energies per pair multiplied by the number of pairs of each kind. By the diagonal sum rule 1[2]39, the mean energy of the terms [weighted by $(2S + 1)(2L + 1)$] of a configuration is the same as the mean of the $\langle A|H|A\rangle$ for the states of that configuration. Therefore, the average interaction energy represents the part that shows where the configuration as a whole lies on the energy scale, and the deviation interaction energy represents the part that gives the location of the terms with respect to the configuration average. Table 8[4] gives the deviation of $(J - K)$ from $\mathrm{Av}(a, b)$ as defined in (24) for the cases of most frequent use.

In writing the expression for W_{Av}, the one-electron integrals, 8[4]5, are the same for all states of the configuration so their sum contributes to the configuration average. Pairs of electrons in the same or different

Table 7[4]. Average-pair interaction energy $\mathrm{Av}(a, b) = [J(a, b) - K(a, b)]_{\mathrm{Av}} = \sum_k (f_k F^k - g_k G^k)$ where all $f_0 = 1$. [Also note that $g_k = 0$ (for all k) for equivalent orbitals and $f_k = 0$ ($k > 0$) for nonequivalent orbitals.]

(a, a)	f_2	f_4	f_6
s, s	0	0	0
p, p	−2/25	0	0
d, d	−2/63	−2/63	0
f, f	−4/195	−2/143	−100/5577

Nonequivalent orbitals

(a, b)	g_0	g_1	g_2	g_3	g_4	g_5	g_6
s, s'	1/2						
s, p	0	1/6					
s, d	0	0	1/10				
s, f	0	0	0	1/14			
p, p'	1/6	0	1/15				
p, d	0	1/15	0	3/70			
p, f	0	0	3/70	0	2/63		
d, d'	1/10	0	1/35	0	1/35		
d, f	0	3/70	0	2/105	0	5/231	
f, f'	1/14	0	2/105	0	1/77	0	50/3003

Table 8[4]. Deviation interaction energies, $E(n_a, l_a, m_a, m_{sa}; n_b, l_b, m_b, m_{sb})$ as defined in 9[4]24. On (m,m') take upper or lower signs together. Listed are coefficients of the integrals at the top of the column. × indicates states not allowed by exclusion principle.

$(ns,n's)$ m, m'	$m_s = m_s'$ $G^0/2$	$m_s \neq m_s'$ $G^0/2$
0, 0	−1	1

$(ns, n'p)$ m, m'	$m_s = m_s'$ $G^1/6$	$m_s \neq m_s'$ $G^1/6$
0, ±1	−1	1
0, 0	−1	1

$ns, n'd$ m, m'	$m_s = m_s'$ $G^2/10$	$m_s \neq m_s'$ $G^2/10$
0, ±2	−1	1
0, ±1	−1	1
0, 0	−1	1

$(np, n'p)$	$m_s = m_s'$			$m_s \neq m_s'$		
m, m'	$F^2/25$	$G^0/6$	$G^2/75$	$F^2/25$	$G^0/6$	$G^2/75$
±1, ±1	1	−5	2	1	1	5
±1, 0	−2	1	−4	−2	1	5
±1, ∓1	1	1	−13	1	1	5
0, 0	4	−5	−7	4	1	5

Table 8[4] (cont.)

(np, np) m, m'	$m_s = m'_s$ $F^2/25$	$m_s \neq m'_s$ $F^2/25$
$\pm1, \pm1$	$\times$	3
$\pm1, 0$	−3	0
$\pm1, \mp1$	−3	3
$0, 0$	$\times$	6

$(np, n'd)$	$m_s = m'_s$			$m_s \neq m'_s$		
m, m'	$F^2/35$	$G^1/15$	$G^3/490$	$F^2/35$	$G^1/15$	$G^3/490$
$\pm1, \pm2$	2	−5	15	2	1	21
$\pm1, \pm1$	−1	−2	3	−1	1	21
$\pm1, 0$	−2	0	−15	−2	1	21
$\pm1, \mp1$	−1	1	−39	−1	1	21
$\pm1, \mp2$	2	1	−69	2	1	21
$0, \pm2$	−4	1	−9	−4	1	21
$0, \pm1$	2	−2	−27	2	1	21
$0, 0$	4	−3	−33	4	1	21

(nd, nd)	$m_s = m_s'$		$m_s \neq m_s'$	
m, m'	$F^2/441$	$F^4/441$	$F^2/441$	$F^4/441$
±2, ±2	×	×	50	15
±2, ±1	−58	5	−4	10
±2, 0	−58	5	−22	20
±2, ∓1	−4	−25	−4	10
±2, ∓2	50	−55	50	15
±1, ±1	×	×	23	30
±1, 0	23	−40	32	−10
±1, ∓1	−31	−10	23	30
0, 0	×	×	50	50

$(nd, n'd)$	$m_s = m_s'$					$m_s \neq m_s'$				
m, m'	$F^2/49$	$F^4/441$	$G^0/10$	$G^2/245$	$G^4/2205$	$F^2/49$	$F^4/441$	$G^0/10$	$G^2/245$	$G^4/2205$
±2, ±2	4	1	−9	−13	58	4	1	1	7	63
±2, ±1	−2	−4	1	−23	38	−2	−4	1	7	63
±2, 0	−4	6	1	−13	−12	−4	6	1	7	63
±2, ∓1	−2	−4	1	7	−112	−2	−4	1	7	63
±2, ∓2	4	1	1	7	−287	4	1	1	7	63
±1, ±1	1	16	−9	2	−17	1	16	1	7	63
±1, 0	2	−24	1	2	−87	2	−24	1	7	63
±1, ∓1	1	16	1	−23	−137	1	16	1	7	63
0, 0	4	36	−9	−13	−117	4	36	1	7	63

Table 9[4]. Coulomb interaction integrals: hydrogenic ($Z = 1$) radial functions (Rydbergs; to be multiplied by Z)

Average interaction energies, Av(a, b)

	$1s$	$2s$	$2p$	$3s$	$3p$	$3d$	$4s$	$4p$
$1s$	1.25000	.39781	.46853	.19321	.21312	.22180	.11293	.12119
$2s$		.30078	.29492	.16075	.17719	.19985	.10031	.10700
$2p$			.34922	.17037	.18308	.20581	.10372	.10945
$3s$				.13281	.12348	.14171	.08603	.09160
$3p$					.13798	.14724	.08973	.09280
$3d$						.08366	.09350	.09722
$4s$							.03728	.03396
$4p$								.03734

Direct integrals, $F^0(a, b)$

	$1s$	$2s$	$2p$	$3s$	$3p$	$3d$	$4s$	$4p$
$1s$	1.25000	.41975	.48560	.19898	.21765	.22205	.11527	.12303
$2s$		.30078	.32422	.16823	.18002	.20716	.10270	.10799
$2p$			.36328	.17351	.18789	.21266	.10480	.11104
$3s$				.13281	.13759	.14627	.08980	.09302
$3p$					.14374	.15386	.09139	.09499
$3d$						.08605	.09433	.09825
$4s$							.03728	.03804
$4p$								.03894

Direct integrals, $F^k(a, b)$

$k = 2$:	$(2p, 2p)$	.17578	$(3p, 3p)$	.07198	$(4p, 4p)$	.01992
			$(3d, 3d)$	.04542		
$k = 4$:			$(3d, 3d)$	.02962		

Exchange integrals, $G^k(a, b)$ ($k = l_a + l_b$ except where two values are indicated. Then the value of k is written as a prefix.)

	$1s$	$2s$	$2p$	$3s$	$3p$	$3d$	$4s$	$4p$
$1s$		.04390	.10242	.01154	.02719	.00247	.00468	.01107
$2s$			.17578	.01495	.01699	.07304	.00479	.00587
$2p$				.01885	$^{(0)}$.01982 $^{(2)}$.02265	$^{(1)}$.07474 $^{(3)}$.04360	.00648	$^{(0)}$.00632 $^{(2)}$.00801
$3s$					.08464	.04557	.00755	.00854
$3p$						$^{(1)}$.06836 $^{(3)}$.04811	.00995	$^{(0)}$.00886 $^{(2)}$.01061
$3d$							.00835	$^{(1)}$.00925 $^{(3)}$.00960
$4s$								.02446

closed shells contribute to W_{Av} as given by Table 7^4. Pairs of electrons, one of which is in a closed shell, the other in an open shell, and finally both of which refer to open shells, also contribute to W_{Av} as given in Table 7^4.

To summarize, the configuration average $W_{\text{Av}}(\text{config})$ for a configuration in which there are $q(a), q(b), \ldots$ electrons in the $a, b, \ldots$ shells can be expressed as a sum over shells, and pairs of shells,

$$W_{\text{Av}}(\text{config}) = \sum_a \left[q(a)I(a) + \frac{1}{2}\, q(a)[q(a)-1]\text{Av}(a, a) + \sum_{a<b} q(a)q(b)\text{Av}(a, b) \right] \tag{25}$$

Here $I(a)$ is given by $8^4 5$, $\text{Av}(a, b)$ is given by (21) and $\text{Av}(a, a)$ by (23), and both are explicitly tabulated in Table 7^4.

As an example, $1s^2 2s^2 2p^3$ is the normal configuration for nitrogen ($N = 7$). For this the configuration average is

$$\begin{aligned} W_{\text{Av}}(1s^2 2s^2 2p^3) = {} & 2I(1s) + 2I(2s) + 3I(2p) + F^0(1s, 1s) + F^0(2s, 2s) \\ & + 3\left[F^0(2p, 2p) - \frac{2}{25} F^2(2p, 2p) \right] \\ & + 4\left[F^0(1s, 2s) - \frac{1}{2} G^0(1s, 2s) \right] \\ & + 6\left[F^0(1s, 2p) - \frac{1}{6} G^1(1s, 2p) \right] \\ & + 6\left[F^0(2s, 2p) - \frac{1}{6} G^1(2s, 2p) \right] \end{aligned} \tag{26}$$

The sum of the coefficients of pair-wise interactions here is $21 = 7(6/2)$, as it should be. The corresponding p^3 configuration in the next row of the periodic table is $1s^2 2s^2 2p^6 3s^2 3p^3$ so its W_{Av} contains 15 one-electron terms and $105 = 15(14/2)$pair-wise interaction terms.

Finally, Table 9^4 gives some values of the direct and exchange integrals of Coulomb interaction computed with hydrogenic radial functions.

10^4. Russell-Saunders term energies

Explicit formulas for the Russell-Saunders term energies in terms of the F and G double radial integrals defined in $9^4 12''$ and $9^4 13''$ can now be obtained.

Here, and in the following, we write $W(^{(2S+1)}L)$ for a term energy measured from an arbitrary origin, and $E(^{(2S+1)}L)$ for the same term energy measured from the configuration average, W_{Av}.

This means that the configuration average has to be subtracted from all diagonal elements of H that enter into the discussion by use of the diagonal sum rule as given in Section 6[4]. Combining 8[4]6 and 9[4]14, the diagonal element for the set $(a, b, c, \ldots, n)$ is

$$(a, b, c, \ldots, n) = \sum_a I(a) + \sum_{\text{pairs}} [J(a, b) - K(a, b)] \tag{1}$$

The $\sum_a I(a)$ is the same as appears in 9[4]25 for W_{Av}, so this goes out on subtracting W_{Av}. We use 9[4]24 to express each $[J(a, b) - K(a, b)]$ as $\text{Av}(a, b) + E(a, b)$.

If a or b refers to a full shell, then the sum over pairs of the $E(a, b)$ vanishes. Hence the result

$$(a, b, c, \ldots, n) = W_{\text{Av}} + \sum_{\text{pairs}}{}' \, E(a, b) \tag{2}$$

in which Σ' means summation over pairs counting only the pairs in which a or b refer to open shells.

This result, combined with the fact that the coefficients in Table 8[4] do not depend on the values of (n_a, n_b), shows that the formulas for the relative term energies of a configuration do not depend on which closed shells are involved in the configuration. Thus the formulas for (closed shells + $2p^3$) are the same as for (closed shells + $3p^3$), except for the appearance of $3p$ instead of $2p$ in the arguments of the F and G integrals which appear.

For two electrons outside closed shells $(n_a, l_a; n_b, l_b)$ with $(n_b, l_b) \neq (n_a, l_a)$ we found in Section 4[4] that the terms are ^1L and ^3L where L takes on the values from $(l_a + l_b)$ down to $|l_a - l_b|$. The explicit formulas depend on the actual values of l_a and l_b. We write them from Table 3[4] and Table 8[4] for the case of two nonequivalent p electrons, so $l_a = l_b = 1$, but $n_b \neq n_a$. The term energies relative to the configuration average are

$$\left.\begin{array}{l} ^3D, {}^1D \\[2ex] ^3P, {}^1P \\[2ex] ^3S, {}^1S \end{array}\right\} = \frac{1}{6}\left(G^0 + \frac{2}{5}G^2\right) + \left\{\begin{array}{l} +\dfrac{1}{25}F^2 \mp \left(G^0 + \dfrac{1}{25}G^2\right) \\[2ex] -\dfrac{5}{25}F^2 \pm \left(G^0 - \dfrac{5}{25}G^2\right) \\[2ex] +\dfrac{10}{25}F^2 \mp \left(G^0 + \dfrac{10}{25}G^2\right) \end{array}\right. \tag{3}$$

where the upper signs are for the triplets.

Similarly, from Table 3[4] and Table 8[4] we find the relations for a pair of equivalent p electrons $(n_b = n_a)$ to be

$$\begin{aligned} ^1D &= +\frac{3}{25}F^2 \\ ^3P &= -\frac{3}{25}F^2 \\ ^1S &= +\frac{12}{25}F^2 \end{aligned} \tag{4}$$

In applying the method to a three or more electron configuration, as for example the p^2p' configuration of Table 4[4] it is necessary to recall that the diagonal element for three electrons consists of the sum over three pairs of pair-wise interactions. Thus the diagonal element for $(1^+, 0^+; 1^+)$ appearing in the $(M_S, M_L) = (3/2, 2)$ cell of Table 4[4] is the sum of pair-wise interactions:

$$(1^+, 0^+; 1^+) = (1^+, 0^+) + (1^+; 1^+) + (0^+; 1^+)$$

With this understanding the formulas for term energies can be worked out as already explained to give

$$({}^3P)^4D = -\frac{3}{25}F^2(p,p) - \frac{1}{25}F^2(p,p') - \frac{2}{3}G^0(p,p') - \frac{2}{75}G^2(p,p')$$

$$({}^3P)^4P = -\frac{3}{25}F^2(p,p) + \frac{1}{5}F^2(p,p') - \frac{2}{3}G^0(p,p') - \frac{4}{15}G^2(p,p')$$

$$({}^3P)^4S = -\frac{3}{25}F^2(p,p) - \frac{2}{5}F^2(p,p') + \frac{7}{3}G^0(p,p') - \frac{4}{15}G^2(p,p')$$

$$({}^3P)^2S = -\frac{3}{25}F^2(p,p) - \frac{2}{5}F^2(p,p') - \frac{2}{3}G^2(p,p') + \frac{1}{3}G^2(p,p')$$

$$({}^1D)^2F = +\frac{3}{25}F^2(p,p) + \frac{2}{25}F^2(p,p') - \frac{2}{3}G^0(p,p') + \frac{7}{75}G^2(p,p')$$

$$({}^3P)^2D + ({}^1D)^2D = -\frac{8}{25}F^2(p,p') + \frac{5}{3}G^0(p,p') + \frac{14}{75}G^2(p,p')$$

$$({}^3P)^2P + ({}^1D)^2P + ({}^1S)^2P = \frac{12}{25}F^2(p,p) + \frac{12}{25}F^2(p,p') + G^0(p,p') + \frac{4}{25}G^2(p,p')$$

As previously explained the diagonal sum rule is unable to give more than the sum of the energies of the two ${}^2D$ or the three ${}^2P$ terms arising in this configuration. A general method due to Racah for finding the separate energies is given in Section 1[5].

Appendix 3 summarizes the formulas for the Russell-Saunders terms, relative to the configuration average for a large number of configurations. Many of these could be obtained by the use of the diagonal sum method, but, as explained in Section 4[4], more elaborate methods are needed when the configuration gives several terms of the same kind.

11⁴. Russell-Saunders wave functions

We now take up the problem of finding the wave functions of Russell-Saunders terms as linear combinations of zero-order Slater determinants. For example, for a system consisting of filled $1s$ and $2s$ shells but with an unfilled $2p$ shell containing three electrons, in states with labels $m_s = 1/2$, $m_l = 0$; $m_s = 1/2$, $m_l = -1$; $m_s = -1/2$, $m_l = -1$, we write:

$$\{1s, 0^+, 0^-; 2s, 0^+, 0^-; 2p, 0^+, \bar{1}^+, \bar{1}^-\} = \{0^+, \bar{1}^+, \bar{1}^-\} \tag{1}$$

for the Nth (here $N = 7$) order determinant, $7^2 9$, associated with the complete set of quantum numbers which is denoted by the same symbols in parentheses.

We write $\Psi(\alpha, S, L, M_S, M_L)$ for the (M_S, M_L) state of the Russell-Saunders term ${}^{(2S+1)}L$.

When the complete sets for a configuration are classified by their (M_S, M_L) values as in Section 6^4, the $\Psi(\alpha, S, L, M_S, M_L)$ is a linear combination of the zero-order determinants that appear in that cell. The problem is the determination of the coefficients in the unitary transformation from the zero-order (nlm) scheme to the (LSM) Russell-Saunders scheme. A method will be explained using the p^2+ closed shells configuration of Table 3^4 as an example. Eight of the eleven (M_S, M_L) cells are singly occupied, so the zero-order scheme for them is directly related to the Russell-Saunders scheme. This gives eight simple relations of which

$$\Psi(p^2, {}^1D, 0, 2) = \{1^+, 1^-\} \tag{2}$$

from the (0, 2) cell is an example.

The (0, 1) cell is doubly occupied by $\{1^+, 0^-\}$ and $\{1^-, 0^+\}$. Two orthonormal linear combinations of these are needed to represent $\Psi(p^2, {}^1D, 0, 1)$ and $\Psi(p^2, {}^3P, 0, 1)$. In this simple instance we know (Sec. 4^4) that the triplet is symmetric and the singlet antisymmetric in the spins. Therefore, the correct linear combinations are

$$2^{-1/2}[\{1^+, 0^-\} \pm \{1^-, 0^+\}] \tag{3}$$

the upper sign applying for ${}^3P$ and the lower for ${}^1D$.

Gray and Wills [16] showed how the angular momentum shift operators, $S_\pm$ and $L_\pm$, defined in $2^3 2$, can be used to determine the wave functions. These operators are of type F, a sum of one-electron angular-momentum operators. $L_\pm$, acting on a Russell-Saunders state, increases or decreases M_L by unity (Equation $2^3 2$):

$$\begin{aligned} L_\pm\Psi(\alpha, S, L, M_S, M_L) \\ = [(L \mp M_L)(L \pm M_L + 1)]^{1/2}\Psi(\alpha, S, L, M_S, M_L \pm 1) \end{aligned} \tag{4}$$

The same operator acting on a zero-order determinant gives

$$\begin{aligned} L_\pm\{m_1, m_2, m_3, \ldots, m_N\} = \sum_{k=1}^{N} [(l_k \mp m_k)(l_k \pm m_k + 1)]^{1/2} \\ \times \{m_1, \ldots, m_k \pm 1, \ldots, m_N\} \end{aligned} \tag{5}$$

Here the m_s symbols are omitted for clarity, because they are not affected by $L_\pm$. Similarly, $S_\pm$ increases or decreases each M_S and each m_s by unity. In place of (4) and (5) we have

$$S_\pm\Psi(\alpha, S, L, M_S, M_L) = [(S \mp M_S)(S \pm M_S + 1)]^{1/2}\Psi(\alpha, S, L, M_S \pm 1, M_L) \quad (4')$$

$$S_\pm\{m_{s1}, m_{s2}, \ldots, m_{sN}\} = \sum_{k=1}^{N} [(3/4) - m_{sk}(m_{sk} \pm 1)]^{1/2}\{m_{s1} \ldots m_{sk} \pm 1 \ldots m_{sN}\} \quad (5')$$

Here the M_S and m_s are acted upon. For each electron $s_k(s_k + 1) = 3/4$, as written in (5′).

Vanishing coefficients prevent the introduction of meaningless terms, such as $|M_L| > L$ or $|m_k| > l_k$. In some cases application of (5) or (5′) to a zero-order determinant gives rise to a determinant that vanishes because it violates the exclusion principle. Hence these do not enter the final result.

An important example is the application of $L_\pm$ or $S_\pm$ to zero-order determinants that involve closed shells. If the nl shell is full, the value nl is repeated $2(2l + 1)$ times with $l \geqslant m \geqslant -l$ and $m_s = \pm$ in the individual sets making up a complete set. When, for example, L_+ is applied to $\{m_1, m_2, \ldots, m_N\}$, the two terms in which m equals l have vanishing coefficients. A lower value of $m^\pm$ is converted into $(m^\pm + 1)$, which is already represented among the individual sets so such a Slater determinant vanishes. Hence closed shells do not contribute in (5) or (5′), and the sum is actually only over the individual sets in open shells. Even here some of the determinants vanish.

Applied to the (0, 2) cell of a p^2 configuration the shift operator L_- gives

$$L_-\Psi(p^2, {}^1D, 0, 2) = \sqrt{4 \cdot 1}\Psi(p^2, {}^1D, 0, 1) = \sqrt{2 \cdot 1}\{0^+, 1^-\} + \sqrt{2 \cdot 1}\{1^+, 0^-\}$$

This agrees with the relation in (3), because $\{0^+, 1^-\} = -\{1^-, 0^+\}$. We can apply L_- again to arrive at

$$\sqrt{6}\Psi(p^2, {}^1D, 0, 0) = \{1^+, \bar{1}^-\} - \{1^-, \bar{1}^+\} + 2\{0^+, 0^-\}$$

By applying L_- two more times, wave functions for all five of the ^1D states are found.

In a similar way the operator S_- can be used on $\Psi(p^2, {}^3P, 1, 1) = \{1^+, 0^+\}$ to find $\Psi(p^2, {}^3P, 0, 1)$ and again to find $\Psi(p^2, {}^3P, \bar{1}, 1)$. Then L_- can be applied twice to each of these three to complete the determination of all nine of the states for ^3P, some of which in this case are already known as belonging to singly occupied (M_S, M_L) cells.

To find $\Psi(p^2, {}^1S, 0, 0)$ we write it as a linear combination of the Slater determinants of the (0, 0) cell,

$$\Psi(p^2, {}^1S, 0, 0) = a\{1^+, \bar{1}^-\} + b\{1^-, \bar{1}^+\} + c\{0^+, 0^-\}$$

Relations for the coefficients are found by requiring this to be orthogonal to the other Russell-Saunders states $\Psi(p^2, {}^3P, 0, 0)$ and $\Psi(p^2, {}^1D, 0, 0)$ which also have states in that cell. Another relation from normalization completes determination of the coefficients except for the usual arbitrary phase.

In the (M_S, M_L) table of complete sets for a configuration, the occurrence of a term ${}^{(2S+1)}L$ is signalized by the fact that the (S, L) cell contains one more complete set than the larger of those in the $(S + 1, L)$ cell and the $(S, L + 1)$ cell. Assuming that the wave functions for the components of Russell-Saunders terms for larger S or larger L in the (S, L) cell have already been determined by the method described, the $\Psi(\alpha, S, L, M_S, M_L)$ can now be written as a linear combination of the zero-order wave functions in that cell, and the ratios of the coefficients found by requiring that this function be orthogonal to each of those previously found.

In this way use of $L_\pm$ and $S_\pm$ permits a complete determination of the Russell-Saunders wave functions in terms of the determinantal states of the zero-order scheme, provided that each kind of ${}^{(2S+1)}L$ term appears only once. Multiple occurrence of the ${}^{(2S+1)}L$ term, say K-fold, is signalized by the fact that the (S, L) cell in the (M_S, M_L) table of complete sets contains K more complete sets than the larger of those in the $(S + 1, L)$ and the $(S, L + 1)$ cells. In this case, the requirement that the linear combination of the zero-order states of the cell be orthogonal to the other Russell-Saunders terms in it does not suffice to determine its coefficients completely. It does determine a K-dimensional space in which the K Russell-Saunders terms ${}^{(2S+1)}L$ lie. In that K space the final determination of the wave functions is accomplished by diagonalization of the Hamiltonian.

As an example, we outline the analysis for the configuration d^3 in Table 10^4, which gives rise to two ${}^2D$ terms, as given by Ufford and Shortley [17]. For brevity the complete sets are designated by letters. The (1/2, 5) component of ${}^2H$ is A. Successive applications of L_- gives the components of ${}^2H$,

$$\Psi\left(d^3, {}^2H, \frac{1}{2}, 4\right) = 10^{-1/2}[6^{1/2}B - 2C]$$

$$\Psi\left(d^3, {}^2H, \frac{1}{2}, 3\right) = 30^{-1/2}[6^{1/2}E - 2F + 4G - 2H]$$

$$\Psi\left(d^3, {}^2H, \frac{1}{2}, 2\right) = 30^{-1/2}[J - K + 3L - 2M - 3N + 6^{1/2}O]$$

Table 10^4. The analysis for the configuration d^3.

M_L	$M_S = 3/2$	$M_S = 1/2$		
5		$A(2^+, 2^-, 1^+)$		^2H
4		$B(2^+, 2^-, 0^+)$	$C(2^+, 1^+, 1^-)$	^2H, ^2G
3	$D(2^+, 1^+, 0^+)$ ^4F	$E(2^+, 2^-, \bar{1}^+)$	$F(2^+, 1^+, 0^-)$	^2H, ^2G
		$G(2^+, 1^-, 0^+)$	$H(2^-, 1^+, 0^+)$	^4F, ^2F
2	$I(2^+, 1^+, \bar{1}^+)$ ^4F	$J(2^+, 2^-, \bar{2}^+)$	$K(2^+, 1^+, \bar{1}^-)$	^2H, ^2G
		$L(2^+, 1^-, \bar{1}^+)$	$M(2^-, 1^+, \bar{1}^+)$	^4F, ^2F
		$N(2^+, 0^+, 0^-)$	$O(1^+, 1^-, 0^+)$	^2D, ^2D

Then $\Psi(d^3, {}^2G, 1/2, 4)$ is a combination of B and C orthogonal to the $^2H(1/2, 4)$ with arbitrary phase, after which M_L is lowered by L_-,

$$\Psi\left(d^3, {}^2G, \frac{1}{2}, 4\right) = 10^{-1/2}[2B + 6^{1/2}C]$$

$$\Psi\left(d^3, {}^2G, \frac{1}{2}, 3\right) = 20^{-1/2}[6^{1/2}E + 3F - G - 2H]$$

$$\Psi\left(d^3, {}^2G, \frac{1}{2}, 2\right) = \left(\frac{3}{140}\right)^{1/2}\left[2J + 3K + L - 4M + 4N + \left(\frac{2}{3}\right)^{1/2} O\right]$$

The components for ^4F are

$$\Psi\left(d^3, {}^4F, \frac{3}{2}, 3\right) = D$$

$$\Psi\left(d^3, {}^4F, \frac{1}{2}, 3\right) = 3^{-1/2}[F + G + H]$$

$$\Psi\left(d^3, {}^4F, \frac{1}{2}, 2\right) = 3^{-1/2}[K + L + M]$$

Orthogonality to other Russell-Saunders terms determines $^2F(1/2, 3)$ and L_- gives $^2F(1/2, 2)$:

$$\Psi\left(d^3, {}^2F, \frac{1}{2}, 3\right) = 12^{-1/2}[-6^{1/2}E + F + G - 2H]$$

$$\Psi\left(d^3, {}^2F, \frac{1}{2}, 2\right) = 12^{-1/2}[-2J + K - L + 6^{1/2}O]$$

Thus explicit functions have been found for four of the six states in the (1/2, 2) cell. These represent two ^2D's, which are here arbitrarily designated ^2_aD and ^2_bD. The two ^2D's are distinguished in Section 8^5 by introduction of Racah's seniority number. For ^2_aD, Ufford and Shortley chose arbitrarily a state orthogonal to the ^2H, ^2G, ^4F, and ^2F components, and then for ${}^2_bD$, a state orthogonal to these and to the ${}^2_aD$. Their choice was

$$\Psi\left(d^3, {}^2_aD, \frac{1}{2}, 2\right) = \frac{1}{2}[-J - K + L + N]$$
$$\Psi\left(d^3, {}^2_bD, \frac{1}{2}, 2\right) = 84^{-1/2}[-5J + 3K + L - 4M - 3N - 24^{1/2}O]$$

Additional application of the L_- operator gives expressions for all of the remaining Ψ's for lower values of M_L down to $M_L = -5$.

The energies of all of the terms except ${}^2_aD$ and ${}^2_bD$ are found in the usual way by expressing them in terms of diagonal elements of the various zero-order states. To find the energies of the two ${}^2D$ terms it is necessary to calculate the nondiagonal element of H connecting them as well as the diagonal elements. With the choice made by Ufford and Shortley, we find

$$\begin{aligned} H_{aa} &= 3F^0(d,d) + \frac{7}{49}F^2(d,d) + \frac{63}{441}F^4(d,d) \\ H_{ab} &= H_{ba} = 3(21)^{1/2}\left[\frac{1}{49}F^2(d,d) - \frac{5}{441}F^4(d,d)\right] \\ H_{bb} &= 3F^0(d,d) + \frac{3}{49}F^2(d,d) - \frac{57}{441}F^4(d,d) \end{aligned} \tag{6}$$

The proper values are the roots of

$$(H_{aa} - \epsilon)(H_{bb} - \epsilon) - H_{ab}^2 = 0$$

giving

$$\epsilon = \frac{1}{2}(H_{aa} + H_{bb}) \pm \left[\frac{1}{4}(H_{aa} - H_{bb})^2 + H_{ab}^2\right]^{1/2} \tag{7}$$

for the energy levels and

$$\Psi = \alpha\Psi(d^3, {}^2_aD, M_S, M_L) + \beta\Psi(d^3, {}^2_bD, M_S, M_L)$$

with

$$\begin{aligned} \alpha &= \left[1 + \left(\frac{H_{aa} - \epsilon}{H_{ab}}\right)^2\right]^{-1/2} \\ \beta &= -\left(\frac{H_{aa} - \epsilon}{H_{ab}}\right)\alpha \end{aligned} \tag{8}$$

for the wave functions of the two ${}^2D$ terms. Substitution of the two values of ϵ from (7) in (8) gives the coefficients for the two ${}^2D$'s.

5. Racah methods

> The applications of the quantum theory of angular momentum, in particular the exploitation of the Racah-Wigner techniques, have by now become so extensive that it is hardly possible for us to attempt even a synoptic survey. Besides the field of angular correlations already noted, one may single out the two fields of atomic and nuclear spectroscopy where these new methods have had the greatest impact. One of the basic problems if not *the* basic problem in spectroscopy, both atomic and nuclear, is the construction of antisymmetric n-particle wave functions from the (degenerate) states of a given energy shell. Following Slater the standard procedure employed determinantal states to achieve antisymmetry, but the resulting wave functions are not necessarily eigenstates of the angular momentum. One technique, introduced by Gray and Wills, employed projection operators to project out of the determinantal state the desired angular momentum eigenstate. The disadvantage of such a procedure, as Racah pointed out, lies in the fact that it foregoes use of tensor operator methods, and is in consequence often unnecessarily cumbersome.
>
> L. C. Biedenharn and H. Van Dam, *Quantum Theory of Angular Momentum* (New York; Academic Press, 1965), p. 10.

We presented in Chapter 3 (Secs. 3^3, 7^3, 8^3, and 9^3) much of the mathematical formalism attributable to G. Racah in his epoch-making series of papers [1, also 2^3a]. These involved mainly: (1) the development of the concept of irreducible tensors (Sec. 7^3) as more suitable objects for description of quantum states than the Cartesian tensors that dominate classical physics, (2) the development of the concept of tensor products of irreducible tensors (Sec. 8^3), and (3) the development of transformations between alternative methods of coupling more than two angular momentum vectors. But there we were concerned only with the development of a mathematical apparatus to be used later.

In Chapter 4 we presented the basic ideas of the central-field approximation following methods largely initiated by Slater and have seen how they lead to direct and simple methods for calculation of the energy levels of the simpler configurations, namely, those consisting of closed shells plus open shells containing not more than two or three electrons, or lacking not more than two or three electrons from being closed. Thus these methods are adequate for dealing directly with configurations arising from s and p shells and even the d^2 and f^2 configurations. But these methods become extremely cumbersome for the more complicated d^n and f^n configurations and also those with several p electrons added to d^n and f^n. More powerful methods are needed to deal with those.

In this chapter we introduce the methods developed by Racah for dealing with more complicated configurations in the Russell-Saunders approximation. Then the account is interrupted in Chapter 6 to present more of the theory of groups than was covered in Chapter 2. The application to atomic spectra is resumed in Chapter 7.

1⁵. Tensor operators and their matrix elements

The important concepts presented by Racah [1] have greatly affected atomic and nuclear physics. These are the tensor operators, fractional parentage, and the application of the theory of continuous groups to the problem of finding the term energies of the f^n configurations. In what follows we make use of the first two of the above concepts in reformulating the relations for the energies and wave functions of the Russell-Saunders terms.

Most of the needed relations and definitions about irreducible tensors have been introduced in Sections 7³ and 8³. Let $X_Q^{(K)}$ stand for the product

$$[\mathbf{T}^{(k_1)} \times \mathbf{U}^{(k_2)}]_Q^{(K)},$$

as defined by 8³3. We need the matrix elements, in *LSM* quantization, of the mixed tensor operator $X_Q^{(K)}$. It is convenient to suppose that the system can be separated into two independent parts: part 1 and part 2. If it is a two-particle system, **T** could be a function of position and spin coordinates of the first particle and **U** of the second. For systems containing more than two particles, **T** can be taken as dependent on the spin alone and **U** on the space coordinates alone. Then the M dependence of the desired matrix elements, of the mixed tensor operator $X_Q^{(K)}$, can be separated by use of the Wigner-Eckart theorem (7³12), giving

$$\begin{aligned}(\alpha j_1 j_2; JM|X_Q^{(K)}|\alpha' j_1' j_2'; J'M')\\ = (-1)^{J-M}\begin{pmatrix} J & K & J' \\ -M & Q & M' \end{pmatrix}(\alpha j_1 j_2; J\|X^{(K)}\|\alpha' j_1' j_2'; J')\end{aligned} \tag{1}$$

By 8³3, because $X_Q^{(K)}$ is defined as a tensor product of two irreducible tensor operators each acting on the separate parts of the system, it may be written as

$$\begin{aligned}X_Q^{(K)} = \sum_{q_1,q_2} (-1)^{k_1-k_2+Q}\sqrt{2K+1}\\ \times \begin{pmatrix} k_1 & k_2 & K \\ q_1 & q_2 & -Q \end{pmatrix} T_{q_1}^{(k_1)} U_{q_2}^{(k_2)}\end{aligned} \tag{2}$$

Equation 1 then is

$$(\alpha j_1 j_2; JM|X_Q^{(K)}|\alpha' j_1' j_2'; J'M') = (\alpha j_1 j_2; JM| \sum_{q_1,q_2} (-1)^{k_1-k_2+Q} \times [2K+1]^{1/2} \begin{pmatrix} k_1 & k_2 & K \\ q_1 & q_2 & -Q \end{pmatrix} T_{q_1}^{(k_1)} U_{q_2}^{(k_2)} |\alpha' j_1' j_2'; J'M')$$

By substituting

$$|\alpha j_1 j_2; JM) = \sum_{m_1,m_2} (-1)^{j_1-j_2+M} [2J+1]^{1/2} \times \begin{pmatrix} j_1 & j_2 & J \\ m_1 & m_2 & -M \end{pmatrix} |j_1 m_1, j_2 m_2\rangle$$

and using the Wigner-Eckart theorem individually for $T_{q_1}^{(k_1)}$ and $U_{q_2}^{(k_2)}$ we obtain:

$$\begin{aligned}
&(-1)^{J-M} \begin{pmatrix} J & K & J' \\ -M & Q & M' \end{pmatrix} (\alpha j_1 j_2; J\|X^{(K)}\|\alpha' j_1' j_2'; J') \\
&= \sum_{m_1,m_2} \sum_{m'_1,m'_2} \sum_{q_1,q_2,\alpha''} (-1)^{j_1-j_2-M+j_1'-j_2'+M'+k_1-k_2+Q+j_1-m_1+j_2-m_2} [2K+1]^{1/2} \\
&\times [(2J+1)(2J'+1)]^{1/2} \begin{pmatrix} j_1 & j_2 & J \\ m_1 & m_2 & -M \end{pmatrix} \begin{pmatrix} j_1' & j_2' & J' \\ m_1' & m_2' & -M' \end{pmatrix} \\
&\times \begin{pmatrix} k_1 & k_2 & K \\ q_1 & q_2 & -Q \end{pmatrix} \begin{pmatrix} j_1 & k_1 & j_1' \\ -m_1 & q_1 & m_1' \end{pmatrix} \begin{pmatrix} j_2 & k_2 & j_2' \\ -m_2 & q_2 & m_2' \end{pmatrix} \\
&\times \langle \alpha j_1 \| T^{(k_1)} \| \alpha'' j_1' \rangle \langle \alpha'' j_2 \| U^{(k_2)} \| \alpha' j_2' \rangle
\end{aligned} \qquad (3)$$

When both sides are multiplied by $(-1)^{J-M} \begin{pmatrix} J & K & J' \\ -M & Q & M' \end{pmatrix}$ and summed over M, M', Q, the left side, by the orthonormality property of $3j$ symbols, becomes simply $(\alpha j_1 j_2; J\|X^{(K)}\|\alpha' j_1' j_2'; J')$. On the right side, the sum of six $3j$ symbols gives a $9j$ symbol, by 9[3]22. The phase introduced in the process of obtaining the proper alignment of the rows and columns and the proper signs cancels the phase factor in (3). The result is

$$\begin{aligned}
&(\alpha j_1 j_2; J\|X^{(K)}\|\alpha' j_1' j_2'; J') \\
&\quad = [(2J'+1)(2J+1)(2K+1)]^{1/2} \begin{Bmatrix} j_1 & j_2 & J \\ j_1' & j_2' & J' \\ k_1 & k_2 & K \end{Bmatrix} \\
&\quad \times \sum_{\alpha''} \langle \alpha j_1 \| T^{(k_1)} \| \alpha'' j_1' \rangle \langle \alpha'' j_2 \| U^{(k_2)} \| \alpha' j_2' \rangle
\end{aligned} \qquad (3')$$

The scalar product of two irreducible vector operators $\mathbf{T}^{(k)}$ and $\mathbf{U}^{(k)}$ is defined as

$$(\mathbf{T}^{(K)} \cdot \mathbf{U}^{(k)}) = \sum_q (-1)^q T_q^{(k)} U_{-q}^{(k)}$$

This definition differs from the definition of Fano-Racah [2³b, p. 84] by a phase factor $(-1)^k$. In terms of the tensor product:

$$(\mathbf{T}^{(k)} \cdot \mathbf{U}^{(k)}) = (-1)^k[2k+1]^{1/2}[\mathbf{T}^{(k)} \times \mathbf{U}^{(k)}]^{(0)}_0$$

The matrix elements are obtained by combining (1) and (3′):

$$\begin{aligned}(\alpha j_1 j_2; JM|(\mathbf{T}^{(k)} \cdot \mathbf{U}^{(k)})|\alpha' j_1' j_2'; J'M') &= (-1)^{k+J-M}[2k+1]^{1/2}\begin{pmatrix} J & 0 & J' \\ -M & 0 & M' \end{pmatrix}(\alpha j_1 j_2; J\|X^{(0)}\|\alpha' j_1' j_2'; J') \\ &= (-1)^{J-M}[(2k+1)(2K+1)(2J+1)(2J'+1)]^{1/2}\begin{pmatrix} J & 0 & J' \\ -M & 0 & M' \end{pmatrix} \\ &\times \begin{Bmatrix} j_1 & j_2 & J \\ j_1' & j_2' & J' \\ k & k & 0 \end{Bmatrix} \sum_{\alpha''} \langle \alpha j_1 \| T^{(k)} \| \alpha'' j_1' \rangle \langle \alpha'' j_2 \| U^{(k)} \| \alpha' j_2' \rangle\end{aligned}$$

Because

$$\begin{pmatrix} J & 0 & J' \\ -M & 0 & M' \end{pmatrix} = \delta(JJ')\delta(MM')\frac{(-1)^{J-M}}{[(2J+1)]^{1/2}} \tag{4}$$

and

$$\begin{Bmatrix} j_1 & j_2 & J \\ j_1' & j_2' & J \\ k & k & 0 \end{Bmatrix} = \frac{(-1)^{j'_1+j_2+J+k}}{[(2J+1)(2k+1)]^{1/2}}\begin{Bmatrix} j_1 & j_2 & J \\ j_2' & j_1' & k \end{Bmatrix} \tag{5}$$

the above relation reduces to

$$\begin{aligned}(\alpha j_1 j_2; JM|(\mathbf{T}^{(k)} \cdot \mathbf{U}^{(k)})|\alpha' j_1' j_2'; J'M') &= (-1)^{j'_1+j_2+J}\,\delta(JJ')\,\delta(MM')\begin{Bmatrix} j_1 & j_2 & J \\ j_2' & j_1' & k \end{Bmatrix} \\ &\times \sum_{\alpha''} \langle \alpha j_1 \| T^{(k)} \| \alpha'' j_1' \rangle \langle \alpha'' j_2 \| U^{(k)} \| \alpha' j_2' \rangle\end{aligned} \tag{6}$$

Two more useful relations, usually called Racah's T and U equations, can be obtained from (3′). By setting $k_2 = 0$, $\mathbf{X}^{(k)}$ reduces to $\mathbf{T}^{(k_1)}$ so that

$$\begin{aligned}(\alpha j_1 j_2; J\|T^{(k_1)}\|\alpha' j_1' j_2'; J') &= [(2J+1)(2J'+1)(2k_1+1)]^{1/2} \\ &\times \begin{Bmatrix} j_1 & j_2 & J \\ j_1' & j_2' & J' \\ k_1 & 0 & k_1 \end{Bmatrix} \sum_{\alpha''} \langle \alpha j_1 \| T^{(k_1)} \| \alpha'' j_1' \rangle \langle \alpha'' j_2 \| 1 \| \alpha' j_2' \rangle\end{aligned}$$

Because [see (10)]

$$\langle \alpha'' j_2 \| 1 \| \alpha' j_2' \rangle = \delta(j_2, j_2')\delta(\alpha', \alpha'')[(2j_2+1)]^{1/2}$$

and, by (5), the $9j$ symbol reduces to a constant times a $6j$ symbol.

Therefore, for a tensor operator acting on part 1 of the total system, we obtain

$$(\alpha j_1 j_2; J\|T^{(k_1)}\|\alpha' j_1' j_2'; J')$$
$$= (-1)^{j_1+j_2+J'+k_1}[(2J+1)(2J'+1)]^{1/2}\begin{Bmatrix} j_1 & k_1 & j_1' \\ J' & j_2 & J \end{Bmatrix}$$
$$\times \langle \alpha j_1\|T^{(k_1)}\|\alpha' j_1'\rangle \delta(j_2, j_2') \tag{7}$$

By setting $k_1 = 0$, and proceeding as before, we arrive at the relation for a tensor operator, which acts on part 2 of the total system:

$$(\alpha j_1 j_2; J\| U^{(k_2)}\|\alpha' j_1' j_2'; J')$$
$$= (-1)^{j_1+j'_2+J+k_2}[(2J+1)(2J'+1)]^{1/2}$$
$$\times \begin{Bmatrix} j_2 & k_2 & j_2' \\ J' & j_1 & J \end{Bmatrix} \langle \alpha j_2\| U^{(k_2)}\|\alpha' j_2'\rangle \delta(j_1, j_1') \tag{8}$$

It is not always possible to assume that each tensor operator acts on separate parts of a given system. We need the matrix elements of a mixed tensor operator $X_Q^{(K)}$ for the case where the tensor operators $\mathbf{T}^{(k_1)}$ and $\mathbf{U}^{(k_2)}$ do not act on separate parts of the system. From (1) and (2) we have

$$(\alpha JM|X_Q^{(K)}|\alpha' J'M') = \sum_{q_1,q_2} (-1)^{k_1-k_2+Q}[2K+1]^{1/2}\begin{pmatrix} k_1 & k_2 & K \\ q_1 & q_2 & -Q \end{pmatrix}$$
$$\times (\alpha JM|T_{q_1}^{(k_1)} U_{q_2}^{(k_2)}|\alpha' J'M')$$
$$= \sum_{q_1,q_2}\sum_{J''M''\alpha''} (-1)^{k_1-k_2+Q}[2K+1]^{1/2}\begin{pmatrix} k_1 & k_2 & K \\ q_1 & q_2 & -Q \end{pmatrix}$$
$$\times (\alpha JM|T_{q_1}^{(k_1)}|\alpha'' J''M'')(\alpha'' J''M''|U_{q_2}^{(k_2)}|\alpha' J'M')$$

Applying the Wigner-Eckart theorem to separate the M dependence of the matrix elements for the tensor operators $\mathbf{T}^{(k_1)}$ and $\mathbf{U}^{(k_2)}$ produces

$$(\alpha JM|X_Q^{(k)}|\alpha' J'M') = \sum_{q_1,q_2}\sum_{J''M''\alpha''} (-1)^{k_1-k_2+Q+J-M+J''-M''}[2K+1]^{1/2}$$
$$\times \begin{pmatrix} k_1 & k_2 & K \\ q_1 & q_2 & -Q \end{pmatrix}\begin{pmatrix} J & k_1 & J'' \\ -M & q_1 & M'' \end{pmatrix}\begin{pmatrix} J'' & k_2 & J' \\ -M'' & q_2 & M' \end{pmatrix}$$
$$\times (\alpha J\|T^{(k_1)}\|\alpha'' J'')\,(\alpha'' J''\|U^{(k_2)}\|\alpha' J')$$

By 9[3]8 the three $3j$ symbols summed over q_1, q_2, M'' reduce to a product of one $6j$ and one $3j$ symbol, giving

$$(\alpha JM|X_Q^{(K)}|\alpha' J'M') = (-1)^{J-M}\begin{pmatrix} J & K & J' \\ -M & Q & M' \end{pmatrix}(\alpha J\|X^{(K)}\|\alpha' J')$$
$$= \sum_{J'',\alpha''} (-1)^{J+J'+K+J-M}\begin{pmatrix} J & K & J' \\ -M & Q & M' \end{pmatrix}\begin{Bmatrix} J & K & J' \\ k_2 & J'' & k_1 \end{Bmatrix}$$
$$\times [2K+1]^{1/2}(\alpha J\|T^{(k_1)}\|\alpha'' J'')(\alpha'' J''\|U^{(k_2)}\|\alpha' J')$$

Finally, comparing these results, one obtains

$$(\alpha J\|X^{(K)}\|\alpha' J') = (-1)^{J+J'+K}[2K+1]^{1/2} \sum_{J'',\alpha''} \begin{Bmatrix} k_1 & k_2 & K \\ J' & J & J'' \end{Bmatrix} \times (\alpha J\|T^{(k_1)}\|\alpha'' J'')\,(\alpha'' J''\|U^{(k_2)}\|\alpha' J') \tag{9}$$

where the tensor operators $\mathbf{T}^{(k_1)}$ and $\mathbf{U}^{(k_2)}$ act on the same coordinates.

Next we consider the evaluation of the reduced-matrix elements for a few simple cases. We start with a scalar, the tensor of order zero, $T_0^{(0)}$, taken as equal to unity. Then

$$\begin{aligned}\langle \alpha jm|T_0^{(0)}|\alpha' j' m'\rangle &= \langle \alpha jm|1|\alpha' j' m'\rangle \\ &= (-1)^{j-m}\begin{pmatrix} j & 0 & j' \\ -m & 0 & m' \end{pmatrix}\langle \alpha j\|1\|\alpha' j'\rangle\end{aligned}$$

From the orthogonality properties of the wave functions

$$\langle \alpha jm|1|\alpha' j' m'\rangle = \delta(\alpha,\alpha')\delta(j,j')\delta(m,m')$$

and finally, by (4),

$$\langle \alpha j\|1\|\alpha' j'\rangle = \delta(\alpha,\alpha')\delta(j,j')\sqrt{2j+1} \tag{10}$$

A vector, such as the angular momentum operator **L** or the position vector **r**, is a tensor of rank (order) one, $k = 1$. For this we have

$$\langle \alpha jm|J_q^{(1)}|\alpha' j' m'\rangle = (-1)^{j-m}\begin{pmatrix} j & 1 & j' \\ -m & q & m' \end{pmatrix}\langle \alpha j\|J^{(1)}\|\alpha' j'\rangle$$

The three allowed values of q $(-1, 0, +1)$ give the three components of the vector. The left side can be calculated easily for the z-component of a vector, which corresponds to $q = 0$:

$$\langle \alpha jm|J_0^{(1)}|\alpha' j' m'\rangle = \langle \alpha jm|J_z|\alpha' j' m'\rangle = m'\delta(\alpha,\alpha')\delta(j,j')\delta(m,m')$$

Evaluating the $3j$ symbol explicitly we have

$$\begin{pmatrix} j & 1 & j' \\ -m & 0 & m \end{pmatrix} = \delta(j,j')\delta(m,m')(-1)^{j-m}\frac{m}{[j(j+1)(2j+1)]^{1/2}} \tag{11}$$

and therefore,

$$\langle \alpha j\|J\|\alpha' j'\rangle = \delta(\alpha,\alpha')\delta(j,j')[j(j+1)(2j+1)]^{1/2} \tag{12}$$

2[5]. Russell-Saunders term energies for two-electron systems

The spherical tensor $C_q^{(k)}$ is defined, in terms of the spherical harmonics of Section 4[3], by

$$C_q^{(k)} = \left(\frac{4\pi}{2K+1}\right)^{1/2} Y(k,q;\theta,\phi)$$

The inverse of the interelectron distance, $1/r_{ij}$, can be expressed as

$$\frac{1}{r_{ij}} = \frac{1}{|r_i - r_j|} = \sum_k \frac{r_<^k}{r_>^{k+1}} \mathbf{C}^{(k)}(i) \cdot \mathbf{C}^{(k)}(j) \tag{1}$$

where $\mathbf{C}^k(i)$ and $\mathbf{C}^k(j)$ are operators acting in coordinates of particles i and j, respectively. The reduced-matrix element for the tensor operator $\mathbf{C}^{(k)}$ is obtained by considering the defining equation

$$\langle \alpha l m | C_q^{(k)} | \alpha' l' m' \rangle = (-1)^{l-m} \begin{pmatrix} l & k & l' \\ -m & q & m' \end{pmatrix} \langle \alpha l \| C^{(k)} \| \alpha' l' \rangle$$

This reduces for $m = m' = 0$ to

$$\langle \alpha l 0 | C_q^{(k)} | \alpha' l' 0 \rangle = (-1)^l \begin{pmatrix} l & k & l' \\ 0 & q & 0 \end{pmatrix} \langle \alpha l \| C^{(k)} \| \alpha' l' \rangle \delta(q, 0)$$

Because $\mathbf{C}^{(k)}$ acts only on the angular part contained in $|\alpha l m\rangle$, and

$$|\alpha l 0\rangle = f(\alpha) Y(l, 0; \theta, \phi) = f^*(\alpha) Y^*(l, 0; \theta, \phi)$$

then

$$\begin{aligned} &\langle \alpha l 0 | C_q^{(k)} | \alpha' l' 0 \rangle \\ &\quad = \delta(\alpha, \alpha') \left(\frac{4\pi}{2k+1} \right)^{1/2} \int_0^{2\pi} \int_0^{\pi} Y(l, 0; \theta, \phi) Y(k, 0; \theta, \phi) \\ &\qquad \times Y(l', 0; \theta, \phi) \sin\theta \, d\theta \, d\phi \\ &\quad = \left(\frac{4\pi}{2k+1} \right)^{1/2} \left[\frac{(2l+1)(2l'+1)(2k+1)}{4\pi} \right]^{1/2} \begin{pmatrix} l & k & l' \\ 0 & 0 & 0 \end{pmatrix}^2 \end{aligned}$$

using the integral over three spherical harmonics from 6[3]20. By comparing results, we find

$$[(2l+1)(2l'+1)]^{1/2} \begin{pmatrix} l & k & l' \\ 0 & 0 & 0 \end{pmatrix}^2 \delta(\alpha, \alpha') = (-1)^l \begin{pmatrix} l & k & l' \\ 0 & 0 & 0 \end{pmatrix} \langle \alpha l \| C^{(k)} \| \alpha' l' \rangle$$

or, finally,

$$\langle \alpha l \| C^{(k)} \| \alpha' l' \rangle = (-1)^l [(2l+1)(2l'+1)]^{1/2} \begin{pmatrix} l & k & l' \\ 0 & 0 & 0 \end{pmatrix} \delta(\alpha, \alpha') \tag{2}$$

The wave function in the *LSM* quantization, in Russell-Saunders coupling, is obtained by coupling the angular and spin parts of the one-electron spin orbitals independently (Sec. 3[3]):

$$\begin{aligned} &|ab; S_{ab}, M_{S_{ab}}, L_{ab}, M_{ab}\rangle \\ &\quad = N^{-1/2} \sum_{m_a, m_b} \sum_{m_{sa}, m_{sb}} (-1)^{l_a - l_b + M_{ab} + M_{S_{ab}}} [(2L_{ab}+1)(2S_{ab}+1)]^{1/2} \\ &\quad \times \begin{pmatrix} 1/2 & 1/2 & S_{ab} \\ m_{sa} & m_{sb} & -M_{S_{ab}} \end{pmatrix} \begin{pmatrix} l_a & l_b & L_{ab} \\ m_a & m_b & -M_{ab} \end{pmatrix} \Phi(a, b; 1, 2) \end{aligned} \tag{3}$$

where the normalization constant $N = 1$ if $a \neq b$, and $N = 2$ if $a = b$, and $\Phi(a, b; 1, 2)$ is a Slater determinant as defined in 7²9. Note that $\Phi(a, b; 1, 2)$ depends explicitly on m_a, m_b, m_{sa}, m_{sb}. To find the Russell-Saunders term energies for two-electron systems, let $W(a, b; S, L)$ stand for the energy arising from the Coulomb interaction; then

$$W(a, b; S, L) = \left(a, b; SM_S, LM \left| \frac{e^2}{r_{12}} \right| a, b; S'M'_S, L'M' \right)$$
$$= e^2\left(a, b(1, 2); SM_S, LM \left| \sum_k \frac{r_<^k}{r_>^{k+1}} \mathbf{C}^{(k)}(1) \cdot \mathbf{C}^{(k)}(2) \right| a, b(1, 2); S'M'_S, L'M' \right)$$
$$- e^2\left(a, b(1, 2); SM_S, LM \left| \sum_k \frac{r_<^k}{r_>^{k+1}} \mathbf{C}^{(k)}(1) \cdot \mathbf{C}^{(k)}(2) \right| a, b(2, 1); S'M'_S, L'M' \right)$$

The first term on the right side is the "direct" part, which by 1⁵6 can be reduced to a product of two reduced-matrix elements. In the second term, which is the "exchange" part, the coordinates are mixed, and as a result 1⁵6 cannot readily be used. To align the coordinates properly we should exchange the quantum labels a and b. This corresponds to interchanging two columns of two $3j$ symbols in (3) and introduces a phase factor $(-1)^{1+l_a+l_b+L'+S'}$ so that

$$W(a, b; S, L)$$
$$= e^2\left(a, b(1, 2); SM_S, LM \left| \sum_k \frac{r_<^k}{r_>^{k+1}} \mathbf{C}^{(k)}(1) \cdot \mathbf{C}^{(k)}(2) \right| a, b(1, 2); S'M'_S, L'M' \right)$$
$$+ (-1)^{l_a+l_b+L'+S'} e^2$$
$$\times \left(a, b(1, 2); SM_S, LM \left| \sum_k \frac{r_<^k}{r_>^{k+1}} \mathbf{C}^{(k)}(1) \cdot \mathbf{C}^{(k)}(2) \right| b, a(1, 2); S'M'_S, L'M' \right)$$

Because $\mathbf{C}^{(k)}(1)$ and $\mathbf{C}^{(k)}(2)$ act only on the space coordinates, the spin part can be separated to give

$$(SM_S|1|S'M'_S) = (-1)^{S-M_S} \begin{pmatrix} S' & S & 0 \\ M'_S & -M_S & 0 \end{pmatrix} \times (S\|1\|S) = \delta(S, S')\delta(M_S, M'_S) \qquad (4)$$

for the direct as well as the exchange part. Therefore, by applying 1⁵6 to the space-dependent part of the Coulomb interaction, the expression for the energy, because of the coulomb interaction, may be further reduced to

$$W(a, b; S, L) = \delta(S, S')\delta(M_S, M_S')\delta(L, L')\delta(M, M')$$
$$\times \left[(-1)^{l_a+l_b+L}\sum_k \begin{Bmatrix} l_a & l_b & L \\ l_b & l_a & k \end{Bmatrix} \langle l_a \| C^{(k)} \| l_a\rangle\langle l_b \| C^{(k)} \| l_b\rangle F^k(a, b)\right.$$
$$\left. + (-1)^{l_a+l_b+S}\sum_k \begin{Bmatrix} l_a & l_b & L \\ l_a & l_b & k \end{Bmatrix} \langle l_a \| C^{(k)} \| l_b\rangle\langle l_b \| C^{(k)} \| l_a\rangle G^k(a, b)\right]$$

where $F^k(a, b)$ and $G^k(a, b)$ are the radial integrals defined by 9[4]12″ and 9[4]13″, respectively. Finally, by (2),

$$W(a, b; S, L) = \delta(S, S')\delta(M_S, M_S')\delta(L, L')\delta(M, M')(2l_a+1)(2l_b+1)$$
$$\times \left[(-1)^L \sum_k \begin{Bmatrix} l_a & l_b & L \\ l_b & l_a & k \end{Bmatrix} \begin{pmatrix} l_a & l_a & k \\ 0 & 0 & 0 \end{pmatrix} \begin{pmatrix} l_b & l_b & k \\ 0 & 0 & 0 \end{pmatrix} F^k(a, b)\right.$$
$$\left. + (-1)^S \sum_k \begin{Bmatrix} l_a & l_b & L \\ l_a & l_b & k \end{Bmatrix} \begin{pmatrix} l_a & l_b & k \\ 0 & 0 & 0 \end{pmatrix}^2 G^k(a, b)\right] \quad (5)$$

To express the term energies relative to the configuration average, it is necessary to subtract the average energy of an electron pair from (5). Here the distinction must be made between equivalent and nonequivalent electron orbitals, so that using 9[4]21 and 9[4]23 for $a \neq b$:

$$E(a, b; S, L) = W(a, b; S, L) - F^0(a, b) + \frac{1}{2}\sum_k \begin{pmatrix} l_a & l_b & k \\ 0 & 0 & 0 \end{pmatrix}^2 G^k(a, b) \quad (6)$$

and for $a = b$:

$$E(a, a; S, L) = W(a, a; S, L) - F^0(a, a) + \frac{2l_a + 1}{4l_a + 1}\sum_{k>0} \begin{pmatrix} l_a & l_a & k \\ 0 & 0 & 0 \end{pmatrix}^2 F^k(a, a) \quad (7)$$

where $W(a, a; S, L)$ is

$$W(a, a; S, L) = -(1/2)\,[1 + (-1)^{L+S}] \times (2l_a + 1)^2 \sum_k \begin{pmatrix} l_a & l_a & k \\ 0 & 0 & 0 \end{pmatrix}^2 \begin{Bmatrix} l_a & l_a & L \\ l_a & l_a & k \end{Bmatrix} F^k(a, a) \quad (5')$$

Equations (6) and (7) give the same results for the term energies, relative to the configuration average, as were calculated before in the zero-order scheme described in Chapter 4. For example, we consider the pd configuration. For $W(p, d; {}^3F) = W({}^3F)$ we have $l_a = 1$, $l_b = 2$, $L = 3$, and $S = 1$ so that

$$W({}^3F) = -15 \begin{pmatrix} 1 & 1 & 0 \\ 0 & 0 & 0 \end{pmatrix} \begin{pmatrix} 2 & 2 & 0 \\ 0 & 0 & 0 \end{pmatrix} \begin{Bmatrix} 1 & 1 & 0 \\ 2 & 2 & 3 \end{Bmatrix} F^0(p, d)$$
$$-15 \begin{pmatrix} 1 & 1 & 2 \\ 0 & 0 & 0 \end{pmatrix} \begin{pmatrix} 2 & 2 & 2 \\ 0 & 0 & 0 \end{pmatrix} \begin{Bmatrix} 1 & 1 & 2 \\ 2 & 2 & 3 \end{Bmatrix} F^2(p, d)$$

$$-15\begin{pmatrix}1 & 2 & 1\\ 0 & 0 & 0\end{pmatrix}^2\begin{Bmatrix}1 & 2 & 1\\ 1 & 2 & 3\end{Bmatrix}G^1(p,d)$$
$$-15\begin{pmatrix}1 & 2 & 3\\ 0 & 0 & 0\end{pmatrix}^2\begin{Bmatrix}1 & 2 & 3\\ 1 & 2 & 3\end{Bmatrix}G^3(p,d)$$
$$= F^0(p,d) + \frac{2}{35}F^2(p,d) - \frac{2}{5}G^1(p,d) - \frac{3}{245}G^3(p,d)$$

and

$$E(^3F) = {}^3F = W(^3F) - F^0(p,d) + \frac{1}{15}G^1(p,d) + \frac{3}{70}G^3(p,d)$$
$$= \frac{2}{35}F^2(p,d) - \frac{1}{3}G^1(p,d) + \frac{3}{98}G^3(p,d)$$

Proceeding in the same manner we find

$$^3D = -\frac{1}{5}F^2(p,d) + \frac{4}{15}G^1(p,d) - \frac{3}{70}G^3(p,d)$$
$$^3P = \frac{1}{5}F^2(p,d) - \frac{3}{14}G^3(p,d)$$
$$^1F = \frac{2}{35}F^2(p,d) + \frac{7}{15}G^1(p,d) + \frac{27}{490}G^3(p,d)$$
$$^1D = -\frac{1}{5}F^2(p,d) - \frac{2}{15}G^1(p,d) + \frac{9}{70}G^3(p,d)$$
$$^1P = \frac{1}{5}F^2(p,d) + \frac{2}{15}G^1(p,d) + \frac{3}{10}G^3(p,d)$$

As another example we take the d^2 configuration. From the rule that $L + S$ must be even for two equivalent electrons, (5′), the allowed terms are: ^1G, ^3F, ^1D, ^3P, ^1S. Then for $W(^1G)$ where $L = 4$, $S = 0$, $l_a = l_b = 2$ we have:

$$W(^1G) = 25\begin{pmatrix}2 & 2 & 0\\ 0 & 0 & 0\end{pmatrix}^2\begin{Bmatrix}2 & 2 & 4\\ 2 & 2 & 0\end{Bmatrix}F^0(d,d) + 25\begin{pmatrix}2 & 2 & 2\\ 0 & 0 & 0\end{pmatrix}^2$$
$$\times\begin{Bmatrix}2 & 2 & 4\\ 2 & 2 & 2\end{Bmatrix}F^2(d,d) + 25\begin{pmatrix}2 & 2 & 4\\ 0 & 0 & 0\end{pmatrix}^2\begin{Bmatrix}2 & 2 & 4\\ 2 & 2 & 4\end{Bmatrix}F^4(d,d)$$
$$= F^0(d,d) + \frac{4}{49}F^2(d,d) + \frac{1}{441}F^4(d,d)$$

and

$$^1G = W(^1G) - F^0(d,d) + \frac{2}{63}F^2(d,d) + \frac{2}{63}F^4(d,d)$$
$$= \frac{50}{441}F^2(d,d) + \frac{15}{441}F^4(d,d)$$

$$^3F = \quad -\frac{58}{441}F^2(d,d) + \frac{5}{441}F^4(d,d)$$

$$^1D = \quad -\frac{13}{441}F^2(d,d) + \frac{50}{441}F^4(d,d)$$

$$^3P = \quad \frac{11}{63}F^2(d,d) - \frac{10}{63}F^4(d,d)$$

$$^1S = \quad \frac{20}{63}F^2(d,d) + \frac{20}{63}F^4(d,d)$$

3⁵. Projection operators, wave functions for two-electron systems

The wave function for a two-electron system, 2⁵3, can be rewritten as

$$|a, b; SM_S, LM)$$

$$= \sum_{m_a, m_b} \sum_{m_{sa}, m_{sb}} (m_{sa}, m_{sb}|SM_S)(m_a, m_b|LM)\{m_a^{m_{sa}}, m_b^{m_{sb}}\} \tag{1}$$

By 3³2, the zero-order scheme wave functions are given in terms of LSM-quantized ones as

$$\{m_a^{m_{sa}}, m_b^{m_{sb}}\} = \sum_{S,L} (SM_S|m_{sa}, m_{sb}\rangle(LM|m_a, m_b\rangle|a, b; SM_S, LM) \tag{2}$$

Because the specification of m_a, m_{sa}, m_b, m_{sb} fixes the values for M and M_S, in (2), there is no summation over M and M_S.

When the projection operator, defined by

$$\mathbf{P}_{Q,K} = \prod_{\substack{S \neq Q \\ L \neq K}} [\mathbf{S}^2 - S(S+1)][\mathbf{L}^2 - L(L+1)] \tag{3}$$

acts on the right side of (2) all terms but $L = K$ and $S = Q$, that is, $|a, b; QM_S, KM)$, are zero, because $|a, b; SM_S, LM)$ is the proper function for the operators $\mathbf{L}^2$, $\mathbf{S}^2$, $\mathbf{L}_Z$, and $\mathbf{S}_Z$, with proper values $L(L+1)$, $S(S+1)$, M, and M_S, respectively. This operator, acting on the left side of (2), gives

$$\mathbf{P}_{Q,K}\{m_a^{m_{sa}}, m_b^{m_{sb}}\} = \prod_{\substack{S \neq Q \\ L \neq K}} [\mathbf{S}^2 - S(S+1)][\mathbf{L}^2 - L(L+1)]\,|1/2, m_{sa}\rangle|l_a, m_a\rangle|1/2, m_{sb}\rangle|l_b, m_b\rangle$$

where $\mathbf{S}^2 = \mathbf{S}_+\mathbf{S}_- + \mathbf{S}_Z^2 - \mathbf{S}_Z$ and $\mathbf{L}^2 = \mathbf{L}_+\mathbf{L}_- + \mathbf{L}_Z^2 - \mathbf{L}_Z$. Because, by 2³2,

$$\mathbf{L}_\pm|l_a, m_a\rangle|l_b, m_b\rangle = [(l_a \mp m_a)(l_a \pm m_a + 1)]^{1/2}|l_a, m_a \pm 1\rangle|l_b, m_b\rangle + [(l_b \mp m_b)(l_b \pm m_b + 1)]^{1/2}|l_a, m_a\rangle|l_b, m_b \pm 1\rangle$$

and

$$\mathbf{L}_Z|l_a, m_a\rangle|l_b, m_b\rangle = (m_a + m_b)|l_a, m_a\rangle|l_b, m_b\rangle$$

and, likewise, for the spin operators (Sec. 5[3]),

$$\begin{aligned}
&\mathbf{S}_\pm|1/2, \pm 1/2\rangle|1/2, \pm 1/2\rangle = 0\\
&\mathbf{S}_\pm|1/2, \mp 1/2\rangle|1/2, \pm 1/2\rangle = |1/2, \pm 1/2\rangle|1/2, \pm 1/2\rangle\\
&\mathbf{S}_\pm|1/2, \mp 1/2\rangle|1/2, \mp 1/2\rangle = |1/2, \pm 1/2\rangle|1/2, \mp 1/2\rangle\\
&\qquad\qquad\qquad\qquad\qquad\qquad + |1/2, \mp 1/2\rangle|1/2, \pm 1/2\rangle\\
&\mathbf{S}_z|1/2, \pm 1/2\rangle|1/2, \pm 1/2\rangle = \pm|1/2, \pm 1/2\rangle|1/2, \pm 1/2\rangle\\
&\mathbf{S}_z|1/2, \pm 1/2\rangle|1/2, \mp 1/2\rangle = 0
\end{aligned}$$

the left side becomes a sum of zero-order wave functions. Thus the *LSM* wave functions are expressed in terms of *nlm* wave functions.

The wave functions for simple cases, such as two-electron systems, can be found directly from 2[5]3 without use of the projection operators. For example, the pp' configuration for which the wave functions for $M = M_S = 0$ are

$$\begin{aligned}
|pp'; S0, L0) &= [(2S + 1)(2L + 1)]^{1/2}\\
&\times \sum_{m_a, m_b} \sum_{m_{sa}, m_{sb}} \begin{pmatrix} 1/2 & 1/2 & S \\ m_{sa} & m_{sb} & 0 \end{pmatrix} \begin{pmatrix} 1 & 1 & L \\ m_a & m_b & 0 \end{pmatrix} \{m_a^{m_{sa}}, m_b^{m_{sb}}\}
\end{aligned}$$

From the properties of $3j$ symbols (Sec. 3[3]), the right side vanishes unless $m_{sa} = -m_{sb}$ and $m_a = -m_b$ or $m_a = m_b = 0$, so that

$$\begin{aligned}
|pp'; S0, L0) &= [(2S + 1)(2L + 1)]^{1/2}\\
&\times \left[\begin{pmatrix} 1/2 & 1/2 & S \\ 1/2 & -1/2 & 0 \end{pmatrix} \begin{pmatrix} 1 & 1 & L \\ 1 & -1 & 0 \end{pmatrix} \{1^+, \bar{1}^-\} \right.\\
&+ \begin{pmatrix} 1/2 & 1/2 & S \\ -1/2 & 1/2 & 0 \end{pmatrix} \begin{pmatrix} 1 & 1 & L \\ 1 & -1 & 0 \end{pmatrix} \{1^-, \bar{1}^+\}\\
&+ \begin{pmatrix} 1/2 & 1/2 & S \\ 1/2 & -1/2 & 0 \end{pmatrix} \begin{pmatrix} 1 & 1 & L \\ 0 & 0 & 0 \end{pmatrix} \{0^+, 0^-\}\\
&+ \begin{pmatrix} 1/2 & 1/2 & S \\ -1/2 & 1/2 & 0 \end{pmatrix} \begin{pmatrix} 1 & 1 & L \\ 0 & 0 & 0 \end{pmatrix} \{0^-, 0^+\}\\
&+ \begin{pmatrix} 1/2 & 1/2 & S \\ 1/2 & -1/2 & 0 \end{pmatrix} \begin{pmatrix} 1 & 1 & L \\ -1 & 1 & 0 \end{pmatrix} \{\bar{1}^+, 1^-\}\\
&+ \left. \begin{pmatrix} 1/2 & 1/2 & S \\ -1/2 & 1/2 & 0 \end{pmatrix} \begin{pmatrix} 1 & 1 & L \\ -1 & 1 & 0 \end{pmatrix} \{\bar{1}^-, 1^+\} \right]
\end{aligned}$$

This is a general expression for any allowed S and L value. By substituting specific values for S and L we obtain

$$\begin{aligned}
{}^3D\colon |pp'; 10, 20) = \frac{1}{\sqrt{12}} [&\{1^+, \bar{1}^-\} + \{1^-, \bar{1}^+\} + 2\{0^+, 0^-\} + 2\{0^-, 0^+\}\\
&+ \{\bar{1}^+, 1^-\} + \{\bar{1}^-, 1^+\}]
\end{aligned}$$

$$^1D\colon |pp';\,00,\,20) = \frac{1}{\sqrt{12}}[\{1^+,\bar{1}^-\} - \{1^-,\bar{1}^+\} + 2\{0^+,0^-\} - 2\{0^-,0^+\} + \{\bar{1}^+,1^-\} - \{\bar{1}^-,1^+\}]$$

$$^3P\colon |pp';\,10,\,10) = \frac{1}{2}[\{1^+,\bar{1}^-\} + \{1^-,\bar{1}^+\} - \{\bar{1}^+,1^-\} - \{\bar{1}^-,1^+\}]$$

$$^1P\colon |pp';\,00,\,10) = \frac{1}{2}[\{1^+,\bar{1}^-\} - \{1^-,\bar{1}^+\} - \{\bar{1}^+,1^-\} + \{\bar{1}^-,1^+\}]$$

$$^3S\colon |pp';\,10,\,00) = \frac{1}{\sqrt{6}}[\{1^+,\bar{1}^-\} + \{1^-,\bar{1}^+\} - \{0^+,0^-\} - \{0^-,0^+\} + \{\bar{1}^+,1^-\} + \{\bar{1}^-,1^+\}]$$

$$^1S\colon |pp';\,00,\,00) = \frac{1}{\sqrt{6}}[\{1^+,\bar{1}^-\} - \{1^-,\bar{1}^+\} - \{0^+,0^-\} + \{0^-,0^+\} + \{\bar{1}^+,1^-\} - \{\bar{1}^-,1^+\}]$$

The electron orbitals for the p^2 configuration are indistinguishable, so

$$\{1^+,\bar{1}^-\} = -\{\bar{1}^-,1^+\};\{1^-,\bar{1}^+\} = -\{\bar{1}^+,1^-\};\{0^+,0^-\} = -\{0^-,0^+\}$$

In $2^5 3$ the normalization constant is $N = 2$, so that for $M = M_S = 0$ the wave functions are

$$^1D\colon |p^2;\,00,\,20) = \frac{1}{[6]^{1/2}}[\{1^+,\bar{1}^-\} - \{1^-,\bar{1}^+\} + 2\{0^+,0^-\}]$$

$$^3P\colon |p^2;\,10,\,10) = \frac{1}{[2]^{1/2}}[\{1^+,\bar{1}^-\} + \{1^-,\bar{1}^+\}]$$

$$^1S\colon |p^2;\,00,\,00) = \frac{1}{[3]^{1/2}}|\{1^+,\bar{1}^-\} - \{1^-,\bar{1}^+\} - \{0^+,0^-\}]$$

Those belonging to other values of M and M_S are found by use of the shift operators $\mathbf{S}_\pm$, $\mathbf{L}_\pm$ (Sec. 1^3).

4^5. Three-electron systems

The coupling order is not unique for three-electron systems (Sec. 9^3) and necessitates use of the recoupling relations, $9^3 7$ and $9^3 7'$. If the three-electron orbitals are equivalent, the wave functions for the Russell-Saunders terms of a kind are linear combinations of the individual functions, specified according to their parentage, each with a proper coefficient. These are *coefficients of fractional parentage* (Sec. 5^5). The case in which all three electrons are not equivalent is considered here.

By repeated application of the coupling procedure, (Sec. 3^3), the *LSM* wave function for the three-electron systems with a specified coupling order is found to be:

$$|abc;(S_{ab},L_{ab})SM_S,LM) = (-1)^{1/2+l_a+l_b+l_c-S_{ab}-L_{ab}-M_S-M} \times \frac{[(2S+1)(2L+1)(2S_{ab}+1)(2L_{ab}+1)]^{1/2}}{N_{ab}} \times \sum_{M_{ab},m_c}\sum_{M_{S_{ab}},m_{sc}} (-1)^{M_{ab}+M_{S_{ab}}} \begin{pmatrix} L_{ab} & l_c & L \\ M_{ab} & m_c & -M \end{pmatrix} \times \begin{pmatrix} S_{ab} & 1/2 & S \\ M_{S_{ab}} & m_{sc} & -M_S \end{pmatrix} \sum_{m_a,m_b}\sum_{m_{sa},m_{sb}} \begin{pmatrix} l_a & l_b & L_{ab} \\ m_a & m_b & -M_{ab} \end{pmatrix} \times \begin{pmatrix} 1/2 & 1/2 & S_{ab} \\ m_{sa} & m_{sb} & -M_{S_{ab}} \end{pmatrix} \Phi(a,b,c;1,2,3) \quad (1)$$

where $N = 1$ if $a \neq b$ and $N_{ab} = 2$ if $a = b$, and $\Phi(a, b, c; 1, 2, 3)$ is a Slater determinant.

The Russell-Saunders term energies, relative to the configuration average, for the general case $(ab)c$ will now be found. The parentheses around ab indicate that these electrons are coupled first. Letting $|SM_S, LM)$ stand for $|abc; (S_{ab}, L_{ab})SM_S, LM)$, the elements of the energy matrix for Coulomb interaction become

$$W[abc;(S_{ab},L_{ab})SL:(S'_{ab},L'_{ab})S'L'] = \left(SM_S,LM\left|\sum_{i<j}\frac{e^2}{r_{ij}}\right|S'M'_S,L'M'\right) = \left(SM_S,LM\left|\frac{e^2}{r_{12}}\right|S'M'_S,L'M'\right) + \left(SM_S,LM\left|\frac{e^2}{r_{13}}\right|S'M'_S,L'M'\right) + \left(SM_S,LM\left|\frac{e^2}{r_{23}}\right|S'M'_S,L'M'\right)$$

and, by $2^5 1$,

$$W[abc;(S_{ab},L_{ab})SL:(S'_{ab},L'_{ab})S'L'] = e^2\left(SM_S,LM\left|\sum_k\frac{r_<^k}{r_>^{k+1}}\mathbf{C}^{(k)}(1)\cdot\mathbf{C}^{(k)}(2)\right|S'M'_S,L'M'\right) + e^2\left(SM_S,LM\left|\sum_k\frac{r_<^k}{r_>^{k+1}}\mathbf{C}^{(k)}(1)\cdot\mathbf{C}^{(k)}(3)\right|S'M'_S,L'M'\right) + e^2\left(SM_S,LM\left|\sum_k\frac{r_<^k}{r_>^{k+1}}\mathbf{C}^{(k)}(2)\cdot\mathbf{C}^{(k)}(3)\right|S'M'_S,L'M'\right) \quad (2)$$

The first term on the right side gives

$$e^2\left(SM_S,LM\left|\sum_k\frac{r_<^k}{r_>^{k+1}}\mathbf{C}^{(k)}(1)\cdot\mathbf{C}^{(k)}(2)\right|S'M'_S,L'M'\right) = \delta(L,L')\delta(M,M')\delta(S,S')\delta(M_S,M'_S)e^2 \times\left(S_{ab}M_{S_{ab}}L_{ab}M_{ab}\left|\sum_k\frac{r_<^k}{r_>^{k+1}}\mathbf{C}^{(k)}(1)\cdot\mathbf{C}^{(k)}(2)\right|S'_{ab}M'_{S_{ab}}L'_{ab}M'_{ab}\right)$$

$$= \delta(L, L')\delta(M, M')\delta(S, S')\delta(M_S, M'_S)\delta(L_{ab}, L'_{ab})\delta(M_{ab}, M'_{ab})$$
$$\times\ \delta(S_{ab}, S'_{ab})\delta(M_{S_{ab}}, M'_{S_{ab}})(2l_a + 1)(2l_b + 1)$$
$$\times \Bigg[(-1)^{L_{ab}} \sum_k \begin{pmatrix} l_a & l_a & k \\ 0 & 0 & 0 \end{pmatrix} \begin{pmatrix} l_b & l_b & k \\ 0 & 0 & 0 \end{pmatrix} \begin{Bmatrix} l_a & l_b & L_{ab} \\ l_b & l_a & k \end{Bmatrix} \times F^k(a, b)$$
$$+ (-1)^{S_{ab}} \sum_k \begin{pmatrix} l_a & l_b & k \\ 0 & 0 & 0 \end{pmatrix}^2 \begin{Bmatrix} l_a & l_b & L_{ab} \\ l_a & l_b & k \end{Bmatrix} G^k(a, b) \Bigg] \qquad (3a)$$

The last step uses the results 2[5]4 and 2[5]5.

In order to evaluate the second and third terms of (2) it is necessary to recouple the state $|(ab)c)$ in such a way that the quantum labels a, b, c are associated with coordinate labels 1, 2, 3, respectively. The two electrons that are coupled first are indicated by the arguments of the spherical tensor operators forming the scalar products.

Taking the second term in (2) and recoupling the state $|(ab)c)$, by 9[3]7′ as $|(ac)b)$ we obtain

$$e^2\left(SM_S, LM \,\middle|\, \sum_k \frac{r_<^k}{r_>^{k+1}} \mathbf{C}^{(k)}(1) \cdot \mathbf{C}^{(k)}(3) \,\middle|\, S'M'_S, L'M' \right)$$
$$= \delta(L, L')\delta(M, M')\delta(S, S')\delta(M_S, M'_S)(-1)^{L_{ab}+L'_{ab}+2S+1}$$
$$\times\ e^2[(2L_{ab} + 1)(2L'_{ab} + 1)(2S_{ab} + 1)(2S'_{ab} + 1)]^{1/2}$$
$$\times \sum_{L_{ac}, L'_{ac}} \sum_{S_{ac}, S'_{ac}} (-1)^{L_{ac}+L'_{ac}} [(2L_{ac} + 1)(2L'_{ac} + 1)(2S_{ac} + 1)(2S'_{ac} + 1)]^{1/2}$$
$$\times \begin{Bmatrix} L_{ab} & l_c & L \\ L_{ac} & l_b & l_a \end{Bmatrix} \begin{Bmatrix} L'_{ab} & l_c & L \\ L'_{ac} & l_b & l_a \end{Bmatrix} \begin{Bmatrix} S_{ab} & 1/2 & S \\ S_{ac} & 1/2 & 1/2 \end{Bmatrix} \begin{Bmatrix} S'_{ab} & 1/2 & S \\ S'_{ac} & 1/2 & 1/2 \end{Bmatrix}$$
$$\times \left(S_{ac}M_{S_{ac}}L_{ac}M_{ac} \,\middle|\, \sum_k \frac{r_<^k}{r_>^{k+1}} \mathbf{C}^{(k)}(1) \cdot \mathbf{C}^{(k)}(3) \,\middle|\, S'_{ac}M'_{S_{ac}}L'_{ac}M'_{ac} \right)$$
$$= \delta(L, L')\delta(M, M')\delta(S, S')\delta(M_S, M'_S)\delta(L_{ac}, L'_{ac})\delta(M_{ac}, M'_{ac})\delta(S_{ac}, S'_{ac})$$
$$\times\ \delta(M_{S_{ac}}, M'_{S_{ac}})(-1)^{L_{ab}+L'_{ab}}(2l_a + 1)(2l_c + 1)$$
$$\times\ [(2L_{ab} + 1)(2L'_{ab} + 1)(2S_{ab} + 1)(2S'_{ab} + 1)]^{1/2}$$
$$\times \sum_{L_{ac}} \sum_{S_{ac}} \begin{Bmatrix} L_{ab} & l_c & L \\ L_{ac} & l_b & l_a \end{Bmatrix} \begin{Bmatrix} L'_{ab} & l_c & L \\ L_{ac} & l_b & l_a \end{Bmatrix} \begin{Bmatrix} S_{ab} & 1/2 & S \\ S_{ac} & 1/2 & 1/2 \end{Bmatrix} \begin{Bmatrix} S'_{ab} & 1/2 & S \\ S_{ac} & 1/2 & 1/2 \end{Bmatrix}$$
$$\times \Bigg[(-1)^{L_{ac}} \sum_k \begin{pmatrix} l_a & l_a & k \\ 0 & 0 & 0 \end{pmatrix} \begin{pmatrix} l_c & l_c & k \\ 0 & 0 & 0 \end{pmatrix} \begin{Bmatrix} l_a & l_c & L_{ac} \\ l_c & l_a & k \end{Bmatrix} F^k(a, c)$$
$$+ (-1)^{S_{ac}} \sum_k \begin{pmatrix} l_a & l_c & k \\ 0 & 0 & 0 \end{pmatrix}^2 \begin{Bmatrix} l_a & l_c & L_{ac} \\ l_a & l_c & k \end{Bmatrix} G^k(a, c) \Bigg]$$

where we have used the fact that $(2S + 1)$ is even for three-electron systems, so that $(-1)^{2S+1} = 1$. The summation over the spin quantum label S_{ac} can be carried out by use of 9[3]12 and 9[3]13a for the direct and exchange parts, respectively. By 9[3]13b and 9[3]21, the summation over L_{ac} simplifies to a product of two $6j$ symbols for the direct part and to a $9j$ symbol for the exchange; thus giving

$$e^2\left(SM_S, LM\left|\sum_k \frac{r_<^k}{r_>^{k+1}}\mathbf{C}^{(k)}(1)\cdot\mathbf{C}^{(k)}(3)\right|S'M'_S, L'M'\right)$$
$$= \delta(L,L')\delta(M,M')\delta(S,S')\delta(M_S,M'_S)\delta(L_{ac},L'_{ac})\delta(M_{ac},M'_{ac})$$
$$\times\,\delta(S_{ac},S'_{ac})\delta(M_{S_{ac}},M'_{S_{ac}})(2l_a+1)(2l_c+1)[(2L_{ab}+1)(2L'_{ab}+1)]^{1/2}$$
$$\times\left[(-1)^{L+l_b}\delta(S_{ab},S'_{ab})\sum_k\begin{Bmatrix} l_c & l_c & k\\ L_{ab} & L'_{ab} & L\end{Bmatrix}\begin{Bmatrix} L_{ab} & L'_{ab} & k\\ l_a & l_a & l_b\end{Bmatrix}\right.$$
$$\times\begin{pmatrix} l_a & l_a & k\\ 0 & 0 & 0\end{pmatrix}\begin{pmatrix} l_c & l_c & k\\ 0 & 0 & 0\end{pmatrix}F^k(a,c)$$
$$+(-1)^{L_{ab}+L'_{ab}}[(2S_{ab}+1)(2S'_{ab}+1)]^{1/2}\begin{Bmatrix} S & 1/2 & S_{ab}\\ 1/2 & 1/2 & S'_{ab}\end{Bmatrix}$$
$$\left.\times\sum_k\begin{Bmatrix} l_c & l_a & k\\ L & L'_{ab} & l_c\\ L_{ab} & l_b & l_a\end{Bmatrix}\begin{pmatrix} l_a & l_c & k\\ 0 & 0 & 0\end{pmatrix}^2 G^k(a,c)\right] \qquad (3b)$$

The third term on the right side of (2) is evaluated in the same way, by using 9³7 to recouple the state $|(ab)c\rangle$ as $|a(cb)\rangle$,

$$e^2\left(SM_S, LM\left|\sum_k \frac{r_<^k}{r_>^{k+1}}\mathbf{C}^{(k)}(2)\cdot\mathbf{C}^{(k)}(3)\right|S'M'_S, L'M'\right)$$
$$= \delta(L,L')\delta(M,M')\delta(S,S')\,\delta(M_S,M'_S)\delta(L_{bc},L'_{bc})\delta(M_{bc},M'_{bc})$$
$$\times\,\delta(S_{bc},S'_{bc})\delta(M_{S_{bc}},M'_{S_{bc}})(2l_b+1)(2l_c+1)\,[(2L_{ab}+1)(2L'_{ab}+1)]^{1/2}$$
$$\times\left[(-1)^{l_a+L_{ab}+L'_{ab}+L}\delta(S_{ab},S'_{ab})\sum_k\begin{Bmatrix} l_b & l_b & k\\ L_{ab} & L'_{ab} & l_a\end{Bmatrix}\begin{Bmatrix} L_{ab} & L'_{ab} & k\\ l_c & l_c & L\end{Bmatrix}\right.$$
$$\times\begin{pmatrix} l_b & l_b & k\\ 0 & 0 & 0\end{pmatrix}\begin{pmatrix} l_c & l_c & k\\ 0 & 0 & 0\end{pmatrix}F^k(b,c)+(-1)^{S_{ab}+S'_{ab}}[(2S_{ab}+1)(2S'_{ab}+1)]^{1/2}$$
$$\left.\times\begin{Bmatrix} 1/2 & 1/2 & S'_{ab}\\ 1/2 & S & S_{ab}\end{Bmatrix}\sum_k\begin{Bmatrix} l_b & l_c & k\\ l_a & L'_{ab} & l_b\\ L_{ab} & L & l_c\end{Bmatrix}\begin{pmatrix} l_b & l_c & k\\ 0 & 0 & 0\end{pmatrix}^2 G^k(b,c)\right] \qquad (3c)$$

The diagonal elements of the energy matrix in the Russell-Saunders scheme relative to the configuration average are now found by subtracting the configuration average. This is

$$E[abc;(S_{ab},L_{ab})SL] = W[abc;(S_{ab},L_{ab})SL]$$
$$-\sum_{\text{pairs}}\left[F^0(i,j)-\frac{1}{2}\sum_k\begin{pmatrix} l_i & l_j & k\\ 0 & 0 & 0\end{pmatrix}^2 G^k(i,j)\right] \qquad (4)$$

where $W[abc;(S_{ab},L_{ab})SL]$ stands for $W[abc;(S_{ab},L_{ab})SL\colon(S_{ab},L_{ab})SL]$ and is the sum of (3a,) (3b,) and (3c.) The indices i and j stand for the electron quantum labels a, b, and c. For the off-diagonal elements of the energy matrix

$$E[abc;(S_{ab},L_{ab})SL\colon(S'_{ab},L'_{ab})SL]$$
$$= W[abc;(S_{ab},L_{ab})SL\colon(S'_{ab},L'_{ab})SL] \qquad (4')$$

5⁵. Energy matrices for the $(sp)p'$, $s(pp')$, and d^2p configurations

The configuration $(sp)p'$, from 4⁵3a, 4⁵3b, 4⁵3c, 4⁵4, and for $l_a = 0$, $l_b = 1$, $l_c = 1$ gives

$$E[(^3P)^4D] = \frac{1}{25} F^2(pp') - \frac{1}{6} G^1(sp) - \frac{1}{6} G^1(sp') - \frac{5}{6} G^0(pp')$$
$$+ \frac{2}{75} G^2(pp')$$

$$E[(^3P)^4P] = -\frac{1}{5} F^2(pp') - \frac{1}{6} G^1(sp) - \frac{1}{6} G^1(sp') + \frac{7}{6} G^0(pp')$$
$$- \frac{2}{15} G^2(pp')$$

$$E[(^3P)^4S] = \frac{2}{5} F^2(pp') - \frac{1}{6} G^1(sp) - \frac{1}{6} G^1(sp') - \frac{5}{6} G^0(pp')$$
$$- \frac{1}{3} G^2(pp')$$

$$E[(^3P)^2D] = \frac{1}{25} F^2(pp') - \frac{1}{6} G^1(sp) + \frac{1}{3} G^1(sp') + \frac{2}{3} G^0(pp')$$
$$+ \frac{13}{150} G^2(pp')$$

$$E[(^1P)^2D] = \frac{1}{25} F^2(pp') + \frac{1}{2} G^1(sp) - \frac{1}{3} G^0(pp') + \frac{7}{150} G^2(pp')$$

$$E[(^3P)^2D: (^1P)^2D] = \frac{\sqrt{3}}{2}\left[\frac{1}{3} G^1(sp') - G^0(pp') - \frac{1}{25} G^2(pp') \right]$$

$$E[(^3P)^2P] = -\frac{1}{5} F^2(pp') - \frac{1}{6} G^1(sp) + \frac{1}{3} G^1(sp') - \frac{1}{3} G^0(pp')$$
$$+ \frac{1}{6} G^2(pp')$$

$$E[(^1P)^2P] = -\frac{1}{5} F^2(pp') + \frac{1}{2} G^1(sp) + \frac{2}{3} G^0(pp') - \frac{1}{30} G^2(pp')$$

$$E[(^3P)^2P: (^1P)^2P] = \frac{\sqrt{3}}{2}\left[\frac{1}{3} G^1(sp') + G^0(pp') - \frac{1}{5} G^2(pp') \right]$$

$$E[(^3P)^2S] = \frac{2}{5} F^2(pp') - \frac{1}{6} G^1(sp) + \frac{1}{3} G^1(sp') + \frac{2}{3} G^0(pp')$$
$$+ \frac{4}{15} G^2(pp')$$

$$E[(^1P)^2S] = \frac{2}{5} F^2(pp') + \frac{1}{2} G^1(sp) - \frac{1}{3} G^0(pp') - \frac{2}{15} G^2(pp')$$

$$E[(^3P)^2S: (^1P)^2S] = \frac{\sqrt{3}}{2}\left[\frac{1}{3} G^1(sp') - G^0(pp') - \frac{2}{5} G^2(pp') \right]$$

Repeating the calculation with the assumption that the electrons p and p' are coupled first, we obtain

$$E[(^3D)^4D] = \frac{1}{25} F^2(pp') - \frac{1}{6} G^1(sp) - \frac{1}{6} G^1(sp') - \frac{5}{6} G^0(pp') + \frac{2}{75} G^2(pp')$$

$$E[(^3P)^4P] = -\frac{1}{5} F^2(pp') - \frac{1}{6} G^1(sp) - \frac{1}{6} G^1(sp') + \frac{7}{6} G^0(pp') - \frac{2}{15} G^2(pp')$$

$$E[(^3S)^4S] = \frac{2}{5} F^2(pp') - \frac{1}{6} G^1(sp) - \frac{1}{6} G^1(sp') - \frac{5}{6} G^0(pp') - \frac{1}{3} G^2(pp')$$

$$E[(^3D)^2D] = \frac{1}{25} F^2(pp') + \frac{1}{3} G^1(sp) + \frac{1}{3} G^1(sp') - \frac{5}{6} G^0(pp') + \frac{2}{75} G^2(pp')$$

$$E[(^1D)^2D] = \frac{1}{25} F^2(pp') + \frac{7}{6} G^0(pp') + \frac{8}{75} G^2(pp')$$

$$E[(^3D)^2D\colon (^1D)^2D] = \frac{\sqrt{3}}{6}[G^1(sp) - G^1(sp')]$$

$$E[(^3P)^2P] = \frac{1}{5} F^2(pp') + \frac{1}{3} G^1(sp) + \frac{1}{3} G^1(sp') + \frac{7}{6} G^0(pp') - \frac{2}{15} G^2(pp')$$

$$E[(^1P)^2P] = -\frac{1}{5} F^2(pp') - \frac{5}{6} G^0(pp') + \frac{4}{15} G^2(pp')$$

$$E[(^3P)^2P\colon (^1P)^2P] = \frac{\sqrt{3}}{6}[G^1(sp) - G^1(sp')]$$

$$E[(^3S)^2S] = \frac{2}{5} F^2(pp') + \frac{1}{3} G^1(sp) + \frac{1}{3} G^1(sp') - \frac{5}{6} G^0(pp') - \frac{1}{3} G^2(pp')$$

$$E[(^1S)^2S] = \frac{2}{5} F^2(pp') + \frac{7}{6} G^0(pp') + \frac{7}{15} G^2(pp')$$

$$E[(^3S)^2S\colon (^1S)^2S] = \frac{\sqrt{3}}{6}[G^1(sp) - G^1(sp')]$$

These results show that the same energies are obtained regardless of the order of coupling for the Russell-Saunders terms that can be distinctly labeled by L and S quantum numbers, that is, for all quartets. However, for the terms with the same S and L labels, but different parents, the energy matrix is, as expected, not diagonal; and the diagonal as well as the off-diagonal elements are different in the two different coupling orders. When the energy matrices are diagonalized, the term energies are the same in both cases as they should be.

The elements of the energy matrix for the a^2c configuration can be deduced from the previous results for the general three-electron configuration $(ab)c$. In the evaluation of the first term on the right side, 4^52, the result $2^55'$ is used instead of 2^55. The last two terms become equal, which can be verified by changing all b's to a's, using the previously defined symmetry properties of nj symbols and keeping in mind that now $S_{aa} + L_{aa}$ and $S'_{aa} + L'_{aa}$ are even in 4^53b and 4^53c. The result is

$$\begin{aligned}
&W[a^2c; (S_{aa}, L_{aa})SL: (S'_{aa}, L'_{aa})S'L'] \\
&\quad = \delta(L, L')\delta(M, M')\delta(S, S')\delta(M_S, M'_S) \\
&\quad \times \Bigg\{ \delta(L_{aa}, L'_{aa})\delta(M_{aa}, M'_{aa})\delta(S_{aa}, S'_{aa})\delta(M_{S_{aa}}, M'_{S_{aa}}) \frac{(2l_a+1)^2}{2} \\
&\quad \times [1 + (-1)^{L_{aa}+S_{aa}}](-1)^{L_{aa}} \sum_k \begin{Bmatrix} l_a & l_a & L_{aa} \\ l_a & l_a & k \end{Bmatrix} \begin{pmatrix} l_a & l_a & k \\ 0 & 0 & 0 \end{pmatrix}^2 F^k(a, a) \\
&\quad + \delta(L_{ac}, L'_{ac})\delta(M_{ac}, M'_{ac})\delta(S_{ac}, S'_{ac})\,\delta(M_{S_{ac}}, M'_{S_{ac}})2(2l_a+1)(2l_c+1) \\
&\quad \times [(2L_{aa}+1)(2L'_{aa}+1)]^{1/2} \Bigg[\delta(S_{aa}, S'_{aa})\,(-1)^{L+l_a} \sum_k \begin{Bmatrix} L_{aa} & L'_{aa} & k \\ l_a & l_a & l_a \end{Bmatrix} \\
&\quad \times \begin{Bmatrix} L_{aa} & L'_{aa} & k \\ l_c & l_c & L \end{Bmatrix} \begin{pmatrix} l_a & l_a & k \\ 0 & 0 & 0 \end{pmatrix} \begin{pmatrix} l_c & l_c & k \\ 0 & 0 & 0 \end{pmatrix} F^k(a, c) \\
&\quad + (-1)^{L_{aa}+L'_{aa}}[(2S_{aa}+1)(2S'_{aa}+1)]^{1/2} \begin{Bmatrix} 1/2 & 1/2 & S'_{aa} \\ 1/2 & S & S_{aa} \end{Bmatrix} \\
&\quad \times \sum_k \begin{Bmatrix} l_a & l_c & k \\ l_a & L'_{aa} & l_a \\ L_{aa} & L & l_c \end{Bmatrix} \begin{pmatrix} l_a & l_c & k \\ 0 & 0 & 0 \end{pmatrix}^2 G^k(a, c)] \Bigg] \Bigg\} \qquad (1)
\end{aligned}$$

The diagonal elements of the energy matrix relative to the configuration average are

$$\begin{aligned}
E[a^2c; (S_{aa}, L_{aa})SL] = {} & W[a^2c; (S_{aa}, L_{aa})SL] - F^0(a, a) \\
& + \frac{2l_a+1}{4l_a+1} \sum_{k>0} \begin{pmatrix} l_a & l_a & k \\ 0 & 0 & 0 \end{pmatrix}^2 F^k(a, a) - 2F^0(a, c) \\
& + \sum_k \begin{pmatrix} l_a & l_c & k \\ 0 & 0 & 0 \end{pmatrix}^2 G^k(a, c) \qquad (2)
\end{aligned}$$

where, as before, $W[a^2c; (S_{aa}, L_{aa})SL]$ stands for $W[a^2c; (S_{aa}, L_{aa})SL: (S_{aa}, L_{aa})SL]$. The off-diagonal elements are

$$\begin{aligned}
&E[a^2c; (S_{aa}, L_{aa})SL: (S'_{aa}, L'_{aa})SL] \\
&\qquad = W[a^2c; (S_{aa}, L_{aa})SL: (S'_{aa}, L'_{aa})SL] \qquad (2')
\end{aligned}$$

The d^2p configuration demonstrates the use of (1) and (2). The off-diagonal element $E[d^2p; (^1D)^2P: (^1S)^2P]$ is given as an example. For $l_a = 2$, $l_c = 1$, $L_{aa} = 2$, $S_{aa} = 0$, $L'_{aa} = 0$, $S'_{aa} = 0$, $L = 1$, $S = 1/2$ we have

$$E[d^2p; (^1D)^2P: (^1S)^2P] = 2 \times 5 \times 3 \times \sqrt{5} \times \Bigg[- \begin{Bmatrix} 2 & 0 & 2 \\ 2 & 2 & 2 \end{Bmatrix} \begin{Bmatrix} 2 & 0 & 2 \\ 1 & 1 & 1 \end{Bmatrix}$$

$$\times \begin{pmatrix} 2 & 2 & 2 \\ 0 & 0 & 0 \end{pmatrix} \begin{pmatrix} 1 & 1 & 2 \\ 0 & 0 & 0 \end{pmatrix} F^2(d,p)$$

$$+ \begin{Bmatrix} 1/2 & 1/2 & 0 \\ 1/2 & 1/2 & 0 \end{Bmatrix} \begin{Bmatrix} 2 & 1 & 1 \\ 2 & 0 & 2 \\ 2 & 1 & 1 \end{Bmatrix} \begin{pmatrix} 2 & 1 & 1 \\ 0 & 0 & 0 \end{pmatrix}^2 G^1(d,p)$$

$$\times \begin{Bmatrix} 1/2 & 1/2 & 0 \\ 1/2 & 1/2 & 0 \end{Bmatrix} \begin{Bmatrix} 2 & 1 & 3 \\ 2 & 0 & 2 \\ 2 & 1 & 1 \end{Bmatrix} \begin{pmatrix} 2 & 1 & 3 \\ 0 & 0 & 0 \end{pmatrix}^2 G^3(d,p) \Bigg]$$

$$= \frac{4}{5\sqrt{7}} F^2(d,p) - \frac{\sqrt{7}}{15} G^1(d,p) - \frac{3}{35\sqrt{7}} G^3(d,p)$$

The remaining energy matrix elements can be worked out in similar fashion and are given in Appendix 3.

From these applications in generalizing the formalism from two-electron to three-electron systems, it is obvious how to go to systems with higher numbers of electrons. The procedure is straightforward, but it is laborious to evaluate the many *nj* symbols produced by coupling of the angular momentum of a large number of electrons. This problem may be avoided by starting with Racah formalism and going back to the zero-order scheme by use of the projection operators defined in $3^5 3$.

6^5. Equivalent-electron orbitals; coefficients of fractional parentage

The wave functions for the terms with the same L and S labels but different parents are linear combinations of the pure parentage wave functions, each multiplied by a factor that is found by diagonalizing the energy matrix (Sec. 5^4). For example, the three ^2P terms of the p^2p' configuration have wave functions of the type

$$\begin{aligned} \Psi(1, {}^2P) &= a_{11}\psi[(^3P)^2P] + a_{12}\psi[(^1D)^2P] + a_{13}\psi[(^1S)^2P] \\ \Psi(2, {}^2P) &= a_{21}\psi[(^3P)^2P] + a_{22}\psi[(^1D)^2P] + a_{23}\psi[(^1S)^2P] \\ \Psi(3, {}^2P) &= a_{31}\psi[(^3P)^2P] + a_{32}\psi[(^1D)^2P] + a_{33}\psi[(^1S)^2P] \end{aligned}$$

where the ψ's are the pure parentage *LSM*-quantized functions. The a_{ij}'s are the elements of a unitary $N \times N$ matrix that diagonalizes the energy matrix where N is the number of different terms with the same L, S labels. These are called *parental mixing coefficients*. From the properties of unitary transformations all Ψ's are orthogonal, provided that the pure parentage wave functions are orthogonal. The problem of parentage occurs for configurations with three of more electrons in open shells. When the *LSM*-quantized wave functions, $4^5 1$, for three-electron systems are examined to see under what conditions they are orthogonal, by using the summation properties $9^3 11$b and $9^3 13$a of $3j$ and $6j$ symbols, and considering all possible permutations of equivalent-electron orbitals, we obtain

$$(abc;(S_{ab}L_{ab})SL|abc;(S'_{ab}L'_{ab})SL) = \delta(L_{ab},L'_{ab})\delta(S_{ab},S'_{ab}) \quad (1a)$$

$$(a^2c;(S_{aa}L_{aa})SL|a^2c;(S'_{aa}L'_{aa})SL) = \frac{1}{2}\delta(L_{aa},L'_{aa})\delta(S_{aa},S'_{aa})[1+(-1)^{L_{aa}+S_{aa}}] \quad (1b)$$

$$\begin{aligned}(ab^2;(S_{ab}L_{ab})SL|ab^2;(S'_{ab}L'_{ab})SL) &= \delta(L_{ab},L'_{ab})\delta(S_{ab},S'_{ab}) \\ &+ (-1)^{L_{ab}+S_{ab}+L'_{ab}+S'_{ab}}[(2L_{ab}+1)(2S_{ab}+1)(2L'_{ab}+1)(2S'_{ab}+1)]^{1/2} \\ &\times \begin{Bmatrix}1/2 & 1/2 & S'_{ab}\\ 1/2 & S & S_{ab}\end{Bmatrix}\begin{Bmatrix}l_a & l_b & L'_{ab}\\ L & l_b & L_{ab}\end{Bmatrix}\end{aligned} \quad (1c)$$

$$\begin{aligned}(a^3;(S_{aa}L_{aa})SL|a^3;(S'_{aa}L'_{aa})SL) &= \delta(L_{aa},L'_{aa})\delta(S_{aa},S'_{aa}) \\ &+ 2[(2L_{aa}+1)(2S_{aa}+1)(2L'_{aa}+1)(2S'_{aa}+1)]^{1/2} \\ &\times \begin{Bmatrix}1/2 & 1/2 & S'_{aa}\\ 1/2 & S & S_{aa}\end{Bmatrix}\begin{Bmatrix}l_a & l_a & L'_{aa}\\ L & l_a & L_{aa}\end{Bmatrix}\end{aligned} \quad (1d)$$

Thus the *LSM*-quantized wave functions are orthogonal for the abc and a^2c configurations. In the case of the ab^2 configuration, where the coupling order remains as (ab), the wave functions are easily orthogonalized by recoupling them as (bb) and introducing a normalization constant $N_{\bar{b}\bar{b}}^{-1/2} = 1/\sqrt{2}$. When the summations over the parentage labels L_{ab}, S_{ab}, L'_{ab}, and S'_{ab}, which occur in recoupling, are carried out using the properties of $6j$ symbols given by $9^3 13$a and $9^3 12$, the relation (1c) reduces to (1b).

For three equivalent-electron orbitals the situation is more complex, and we cannot make them orthogonal by recoupling because the different coupling orders are no longer distinguishable. By treating the third-electron orbital as if it is coupled last to the first two, the wave function for the three equivalent orbitals can be written as a linear combination of the states $|a^2a;(S'L')SL)$ as

$$|a^3;(S_0L_0)\alpha SL) = \sum_{L'S'} A(S'L')|a^2a;(S'L')SL) \quad (2)$$

where the summation is over all allowed values of the parentage labels L' and S'. Though it is possible to use the *LSM*-quantized wave function $|a^3;(S_{aa}L_{aa})SL)$ in the computation without modification, the above expansion, first introduced by Racah [1c], is more convenient and better known. In (2) the expansion coefficients, $A(S'L')$, are called *coefficients of fractional parentage* defined in the principal parent scheme. To a certain degree these coefficients are analogous to the parental mixing coefficients A_{ij}. But it should be realized that here, in the case of equivalent-electron orbitals, the parents are not distinguishable. The quantum labels L_0 and S_0 identify a *principal parent,* which will be explained in detail by taking two examples in the next section, and α stands for any other label that may be necessary, in addition to L and S, for specifying a Russell-Saunders term. Following Racah the notation $(a^3\alpha SL\{|a^2S'L', a)$ is used instead of $A(S'L')$ for the coefficients of fractional parentage. Both sides of (2) are multiplied by $(a^2a; (S''L'')SL|$ and the results (1b) and (1d) are used to obtain

$$\begin{aligned}\sqrt{N}(a^3\alpha SL\{|a^2S'L', a) &= \delta(L_0,L')\delta(S_0,S') \\ &+ 2[(2L_0+1)(2S_0+1)(2L'+1)(2S'+1)]^{1/2} \\ &\times \begin{Bmatrix} 1/2 & 1/2 & S' \\ 1/2 & S & S_0 \end{Bmatrix} \begin{Bmatrix} l_a & l_a & L' \\ L & l_a & L_0 \end{Bmatrix}\end{aligned} \tag{3}$$

where N is the normalization constant and is given by

$$\begin{aligned}N = \sum_{L'S'} \Bigg[& \delta(L_0,L')\delta(S_0,S') + 2[(2L_0+1)(2S_0+1)(2L'+1) \\ & \times [(2S'+1)]^{1/2} \begin{Bmatrix} 1/2 & 1/2 & S' \\ 1/2 & S & S_0 \end{Bmatrix} \begin{Bmatrix} l_a & l_a & L' \\ L & l_a & L_0 \end{Bmatrix}\Bigg]^2 \Bigg]^{1/2}\end{aligned} \tag{4}$$

This procedure for three equivalent-electron orbitals can be generalized to express the wave functions for n equivalent orbitals in terms of the functions of the $a^{n-1}a$ configuration as

$$\begin{aligned}&|a^n; (S_0L_0)\alpha SL) \\ &\quad = \sum_{\alpha'L'S'} (a^n\alpha SL\{|a^{n-1}\alpha'S'L', a)|a^{n-1}a; (S'L')SL)\end{aligned} \tag{5}$$

where $n < 4l_a + 2$. Actually, (5) only partially defines the coefficient of fractional parentage, in the principal parent scheme. Specification of the phase is also needed. Furthermore, other schemes are needed to classify, or label, the Russell-Saunders terms of like kind, such as the *seniority scheme,* which will be discussed later. For a closed shell, that is, $n = 4l_a + 2$, $L = S = M = M_s = 0$, so the wave function becomes a single determinant whose elements correspond to m and m_s quantum numbers

$$\{m^+, m^-, (m-1)^+, (m-1)^-, \ldots, -(m-1)^+, -(m-1)^-, -m^+, -m^-\}$$

Equation (5) then reduces to

$$\Psi(a^{4l_a+2}; SL) = |a^{4l_a+1}a; 00) \tag{6}$$

7⁵. Coefficients of fractional parentage for the p^3 and d^3 configurations

The p^3 configuration may be taken as an example of the determination of the coefficients of fractional parentage. Because there are three parent terms, ^3P, ^1D, and ^1S, any one may be chosen as a "trial" principal parent to determine whether such a choice produces nonzero and independent coefficients of fractional parentage for all of the Russell-Saunders terms of the p^3 configuration. Taking the ^3P as the principal parent first, $L_0 = S_0 = 1$ so that by 6⁵3,

$$\sqrt{N_1}(p^3\,{}^4S\{|p^2\,{}^1S, p) = 2 \times 3 \begin{Bmatrix} 1/2 & 1/2 & 0 \\ 1/2 & 3/2 & 1 \end{Bmatrix} \begin{Bmatrix} 1 & 1 & 0 \\ 0 & 1 & 1 \end{Bmatrix} = 0$$

$$\sqrt{N_1}(p^3\,{}^4S\{|p^2\,{}^3P, p) = 1 + 2 \times 3 \times 3 \times \begin{Bmatrix} 1/2 & 1/2 & 1 \\ 1/2 & 3/2 & 1 \end{Bmatrix} \times \begin{Bmatrix} 1 & 1 & 1 \\ 0 & 1 & 1 \end{Bmatrix} = 3$$

$$\sqrt{N_1}(p^3\,{}^4S\{|p^2\,{}^1D, p) = 2 \times 3 \times \sqrt{5} \begin{Bmatrix} 1/2 & 1/2 & 0 \\ 1/2 & 3/2 & 1 \end{Bmatrix} \begin{Bmatrix} 1 & 1 & 2 \\ 0 & 1 & 1 \end{Bmatrix} = 0$$

and by 6⁵4, $\sqrt{N_1} = 3$. Therefore, the orthogonal wave function for the ^4S term is, by 6⁵2

$$|p^3; (^3P)^4S) = |p^2p; (^3P)^4S)$$

Proceeding in the same way we find

$$\sqrt{N_2}(p^3\,{}^2P\{|p^2\,{}^1S, p) = -1$$

$$\sqrt{N_2}(p^3\,{}^2P\{|p^2\,{}^3P, p) = \frac{3}{2}$$

$$\sqrt{N_2}(p^3\,{}^2P\{|p^2\,{}^1D, p) = \frac{\sqrt{5}}{2}$$

$$N_2 = 1 + \frac{9}{4} + \frac{5}{4} = \frac{18}{4}; \sqrt{N_2} = \frac{\sqrt{18}}{2}$$

$$|p^3; (^3P)^2P) = -\frac{1}{\sqrt{18}}[2|p^2p; (^1S)^2P) - 3|p^2p; (^3P)^2P) - \sqrt{5}|p^2p; (^1D)^2P)]$$

The above state has the opposite phase compared with that of Racah [1c]. Moreover,

$$\sqrt{N_3}(p^3\,{}^2D\{|p^2\,{}^1S, p) = 0$$

$$\sqrt{N_3}(p^3\,{}^2D\{|p^2\,{}^3P, p) = \frac{3}{2}$$

$$\sqrt{N_3}(p^3\,{}^2D\{|p^2\,{}^1D, p) = -\frac{3}{2}$$

$$\sqrt{N_3} = \frac{3}{2}\sqrt{2}$$

$$|p^3; ({}^3P){}^2D) = \frac{1}{(2)^{1/2}}[|p^2p; ({}^3P){}^2D) - |p^2p; ({}^1D){}^2D)]$$

The choice of either the ^1S or the ^1D term as principal parent gives zero for the coefficients of fractional parentage for the $|p^3\,{}^4S)$ LSM-quantized state, indicating the ^3P term as the principal parent for the three equivalent p-electron orbitals.

The d^3 configuration is another example. By taking the ^3P term as principal parent, $L_0 = S_0 = 1$ in 6⁵3, we have

$$\sqrt{N_1}(d^3\,{}^2P\{|d^2\,{}^1S, d) = 2 \times 3 \times \begin{Bmatrix} 1/2 & 1/2 & 0 \\ 1/2 & 1/2 & 1 \end{Bmatrix}\begin{Bmatrix} 2 & 2 & 0 \\ 1 & 2 & 1 \end{Bmatrix}$$
$$= 0$$

$$\sqrt{N_1}(d^3\,{}^2P\{|d^2\,{}^3P, d) = 1 + 2 \times 3 \times 3 \times \begin{Bmatrix} 1/2 & 1/2 & 1 \\ 1/2 & 1/2 & 1 \end{Bmatrix}\begin{Bmatrix} 2 & 2 & 1 \\ 1 & 2 & 1 \end{Bmatrix}$$
$$= \frac{7}{10}$$

$$\sqrt{N_1}(d^3\,{}^2P\{|d^2\,{}^1D, d) = 6 \times \sqrt{5} \times \begin{Bmatrix} 1/2 & 1/2 & 0 \\ 1/2 & 1/2 & 1 \end{Bmatrix}\begin{Bmatrix} 2 & 2 & 2 \\ 1 & 2 & 1 \end{Bmatrix}$$
$$= 7\frac{\sqrt{15}}{10}$$

$$\sqrt{N_1}(d^3\,{}^2P\{|d^2\,{}^3F, d) = 6 \times \sqrt{21} \times \begin{Bmatrix} 1/2 & 1/2 & 1 \\ 1/2 & 1/2 & 1 \end{Bmatrix}\begin{Bmatrix} 2 & 2 & 3 \\ 1 & 2 & 1 \end{Bmatrix}$$
$$= -\sqrt{7}\frac{\sqrt{8}}{10}$$

$$\sqrt{N_1}(d^3\,{}^2P\{|d^2\,{}^1G, d) = 6 \times 3 \times \begin{Bmatrix} 1/2 & 1/2 & 0 \\ 1/2 & 1/2 & 1 \end{Bmatrix}\begin{Bmatrix} 2 & 2 & 4 \\ 1 & 2 & 1 \end{Bmatrix}$$
$$= 0$$

$$N_1 = \frac{49}{100} + \frac{105}{100} + \frac{56}{100} = \frac{21}{10}; \sqrt{N_1} = \left(\frac{21}{10}\right)^{1/2}$$

$$|d^3; ({}^3P){}^2P) = \frac{1}{[30]^{1/2}}[\sqrt{7}|d^2d; ({}^3P){}^2P) + \sqrt{15}|d^2d; ({}^1D){}^2P)$$
$$- \sqrt{8}|d^2d; ({}^3F){}^2P]$$

For the other Russell-Saunders terms we arrive at

$$\sqrt{N_2}(d^3\,{}^4P\{|d^2\,{}^1S, d) = 0$$

$$\sqrt{N_2}(d^3\,{}^4P\{|d^2\,{}^3P, d) = \frac{8}{5}$$

$$\sqrt{N_2}(d^3\,{}^4P\{|d^2\,{}^1D, d) = 0$$

$$\sqrt{N}_2(d^3\,{}^4P\{|d^2\,{}^3F, d) = \sqrt{7}\,\frac{\sqrt{8}}{5}$$
$$\sqrt{N}_2(d^3\,{}^4P\{|d^2\,{}^1G, d) = 0$$
$$\sqrt{N}_2 = \left(\frac{24}{5}\right)^{1/2}$$
$$|d^3; ({}^3P)^4P) = \frac{1}{\sqrt{15}}[\sqrt{8}|d^2d; ({}^3P)^4P + \sqrt{7}|d^2d; ({}^3F)^4P]$$

$$\sqrt{N}_3(d^3\,{}^2D\{|d^2\,{}^1S, d) = -\frac{3}{5}$$
$$\sqrt{N}_3(d^3\,{}^2D\{|d^2\,{}^3P, d) = \frac{3}{2}$$
$$\sqrt{N}_3(d^3\,{}^2D\{|d^2\,{}^1D, d) = -\sqrt{5}\,\frac{3}{10}$$
$$\sqrt{N}_3(d^3\,{}^2D\{|d^2\,{}^3F, d) = 0$$
$$\sqrt{N}_3(d^3\,{}^2D\{|d^2\,{}^1G, d) = \frac{6}{5}$$
$$\sqrt{N}_3 = \left(\frac{45}{10}\right)^{1/2}$$
$$|d^3; ({}^3P)^2D) = \frac{1}{\sqrt{50}}[-2|d^2d; ({}^1S)^2D) + 5|d^2d; ({}^3P)^2D) - \sqrt{5}|d^2d; ({}^1D)^2D) + 4|d^2d; ({}^1G)^2D)]$$

$$\sqrt{N}_4(d^3\,{}^2F\{|d^2\,{}^1S, d) = 0$$
$$\sqrt{N}_4(d^3\,{}^2F\{|d^2\,{}^3P, d) = \frac{6}{5}$$
$$\sqrt{N}_4(d^3\,{}^2F\{|d^2\,{}^1D, d) = -3\,\frac{\sqrt{2}}{\sqrt{35}}$$
$$\sqrt{N}_4(d^3\,{}^2F\{|d^2\,{}^3F, d) = \frac{3}{5}$$
$$\sqrt{N}_4(d^3\,{}^2F\{|d^2\,{}^1G, d) = -\frac{3}{\sqrt{7}}$$
$$\sqrt{N}_4 = \left(\frac{2}{5}\right)^{1/2}$$
$$|d^3; ({}^3P)^2F) = \frac{1}{\sqrt{70}}[2\sqrt{7}|d^2d; ({}^3P)^2F) - \sqrt{10}|d^2d; ({}^1D)^2F) + \sqrt{7}|d^2d; ({}^3F)^2F) - 5|d^2d; ({}^1G)^2F)]$$

$$\sqrt{N}_5(d^3\,{}^4F\{|d^2\,{}^1S, d) = 0$$
$$\sqrt{N}_5(d^3\,{}^4F\{|d^2\,{}^3P, d) = \frac{3}{5}$$
$$\sqrt{N}_5(d^3\,{}^4F\{|d^2\,{}^1D, d) = 0$$

$$\sqrt{N_5}(d^3\,{}^4F\{|d^2\,{}^3F, d) = -\frac{6}{5}$$
$$\sqrt{N_5}(d^3\,{}^4F\{|d^2\,{}^1G, d) = 0$$
$$\sqrt{N_5} = \frac{3}{\sqrt{5}}$$
$$|d^3; ({}^3P)^4F) = \frac{1}{\sqrt{5}}[|d^2d; ({}^3P)^4F) - 2|d^2d; ({}^3F)^4F)]$$

For the ^2G and ^2H terms, choice of the ^3P as principal parent gives the coefficients of fractional parentage all equal to zero. This indicates that some other term must be chosen as principal parent. Trying first the ^1S term, the coefficients of fractional parentage for all the terms except the ^2D are zero. The nonzero values are

$$\sqrt{N_1}(d^3\,{}^2D\{|d^2\,{}^1S, d) = \frac{4}{5}$$
$$\sqrt{N_1}(d^3\,{}^2D\{|d^2\,{}^3P, d) = -\frac{3}{5}$$
$$\sqrt{N_1}(d^3\,{}^2D\{|d^2\,{}^1D, d) = -\frac{\sqrt{5}}{5}$$
$$\sqrt{N_1}(d^3\,{}^2D\{|d^2\,{}^3F, d) = -\frac{\sqrt{21}}{5}$$
$$\sqrt{N_1}(d^3\,{}^2D\{|d^2\,{}^1G, d) = -\frac{3}{5}$$
$$\sqrt{N_1} = 2\left(\frac{3}{5}\right)^{1/2}$$
$$|d^3; ({}^1S)^2D) = \frac{1}{\sqrt{60}}[4|d^2d; ({}^1S)^2D) - 3|d^2d; ({}^3P)^2D) - \sqrt{5}|d^2d; ({}^1D)^2D) - \sqrt{21}|d^2d; ({}^3F)^2D) - 3|d^2d; ({}^1G)^2D)]$$

Thus the two linearly independent wave functions for the ^2D terms can be classified by their principal parentage. However, this method gives no assurance that these two linearly independent functions are orthogonal. In fact,

$$(d^3; ({}^1S)^2D|d^3; ({}^3P)^2D) = -\left(\frac{3}{10}\right)^{1/2}$$

showing that they are not. Elementary vector algebra shows that if two vectors **A** and **B** are not orthogonal they can be orthogonalized by fixing one of them and rotating the other. For example, the vector $\mathbf{B}' = \mathbf{B} - (\mathbf{A}\cdot\mathbf{B})\mathbf{A}$ is orthogonal to the vector **A**. In this way

$$|d^3; (^3P)^2D)' = |d^3; (^3P)^2D) + \left(\frac{3}{10}\right)^{1/2} |d^3; (^1D)^2D)$$

$$= \frac{1}{10\sqrt{2}} [7|d^2d; (^3P)^2D) - 3\sqrt{5}|d^2d; (^1D)^2D) - \sqrt{21}|d^2d; (^3F)^2D) + 5|d^2d; (^1G)^2D)]$$

and after normalization

$$|d^3; (^3P)^2D)'' = \frac{1}{\sqrt{140}} [7|d^2d; (^3P)^2D) - 3\sqrt{5}|d^2d; (^1D)^2D) - \sqrt{21}|d^2d; (^3F)^2D) + 5|d^2d; (^1G)^2D)]$$

The ^1D term, as principal parent, gives

$$\sqrt{N_1}(d^3\ ^2P\{|d^2\ ^1S, d) = 0$$

$$\sqrt{N_1}(d^3\ ^2P\{|d^2\ ^3P, d) = \frac{\sqrt{105}}{10}$$

$$\sqrt{N_1}(d^3\ ^2P\{|d^2\ ^1D, d) = \frac{3}{2}$$

$$\sqrt{N_1}(d^3\ ^2P\{|d^2\ ^3F, d) = -\frac{\sqrt{30}}{5}$$

$$\sqrt{N_1}(d^3\ ^2P\{|d^2\ ^1G, d) = 0$$

$$N_1 = \frac{15}{4}$$

$$|d^3; (^1D)^2P) = \frac{1}{\sqrt{30}} [\sqrt{7}|d^2d; (^3P)^2P) + \sqrt{15}|d^2d; (^1D)^2P) - \sqrt{8}|d^2d; (^3F)^2P)]$$

$$= |d^3; (^3P)^2P)$$

This choice, therefore, does not produce a new linearly independent state for the ^2P term. In addition, the coefficients of fractional parentage are all zero for the quartets.

Using the ^3F term as principal parent,

$$|d^3; (^3F)^2P) = -\frac{1}{\sqrt{30}} [\sqrt{7}|d^2d; (^3P)^2P) + \sqrt{15}|d^2d; (^1D)^2P) - \sqrt{8}|d^2d(^3F)^2P)]$$

$$= -|d^3; (^3P)^2P)$$

$$|d^3; (^3F)^4P) = \frac{1}{\sqrt{15}} [\sqrt{8}|d^2d; (^3P)^4P) + \sqrt{7}|d^2d; (^3F)^4P)]$$

$$= |d^3; (^3P)^4P)$$

Table 1[5]. $|(p^3;\,\alpha SL\{|p^2\alpha' S' L', p)|^2$ as defined in $6^5 5$. The signs of the square roots are indicated. N is the normalization constant as presented in $6^5 3$.

p^3 \ p^2	N	^1S	^3P	^1D
^4S	1	0	1	0
^2P	18	4	−9	−5
^2D	2	0	1	−1

$$|d^3;\,(^3F)^2D) = -\frac{\sqrt{2}}{30}\,[2\sqrt{21}\,|d^2d;\,(^1S)^2D) - \frac{40\sqrt{3}}{\sqrt{35}}\,|d^2d;\,(^1D)^2D)$$
$$-\,15\,|d^2d;\,(^3F)^2D) - \frac{3\sqrt{3}}{\sqrt{7}}\,|d^2d;\,(^1G)^2D)]$$

$$|d^3;\,(^3F)^2F) = \frac{1}{\sqrt{70}}\,[2\sqrt{7}\,|d^2d;\,(^3P)^2F - \sqrt{10}\,|d^2d;\,(^1D)^2F)$$
$$+\,\sqrt{7}\,|d^2d;\,(^3F)^2F) - 5\,|d^2d;\,(^1G)^2F)]$$
$$= |d^3;\,(^3P)^2F)$$

$$|d^3;\,(^3F)^4F) = -\frac{1}{\sqrt{5}}\,[|d^2d;\,(^3P)^4F) - 2\,|d^2d;\,(^3F)^4F)]$$
$$= |d^3;\,(^3P)^4F)$$

$$|d^3;\,(^3F)^2G) = \frac{1}{\sqrt{42}}\,[-\sqrt{10}\,|d^2d;\,(^1D)^2G) + \sqrt{21}\,|d^2d;\,(^3F)^2G)$$
$$+\,\sqrt{11}\,|d^2d;\,(^1G)^2G)]$$

$$|d^3;\,(^3F)^2H) = \frac{1}{\sqrt{2}}\,[|d^2d;\,(^3F)^2H) - |d^2d;\,(^1G)^2H)]$$

since $(d^3;\,(^1S)^2D|d^3;\,(^3F)^2D) = -7/\sqrt{70}$ these two states need to be orthogonalized. Proceeding as before, we find

$$|d^3;\,(^3F)^2D)'' = -\frac{1}{\sqrt{140}}\,[7\,|d^2d;\,(^3P)^2D) - 3\sqrt{5}\,|d^2d;\,(^1D)^2D)$$
$$-\,\sqrt{21}\,|d^2d;\,(^3F)^2D) + 5\,|d^2d;\,(^1G)^2D)]$$
$$= -|d^3;\,(^3P)^2D)''$$

From the above calculations we deduce that the ^3F term is the principal parent for the d^3 configuration. Further, the two linearly independent ^2D states, classified by their principal parents, $6^5 3$, are not orthogonal. Finally, the phase depends on the choice of "principal" parent. For the p^3 and d^3 configurations, where the ^3P and ^3F terms are principal parents, respectively, the phases reduce to Racah's [1c] phase convention if each state is multiplied by $(-1)^L$ for $|L - L_0| \neq 0$ and $(-1)^{l_a}$ for $(L - L_0) = 0$.

Table 2[5]a. $|(d^3;\alpha SL\{|d^2\alpha' S'L', d)|^2$ as defined in 6[5]5. The sign of the square roots are indicated. N is the normalization constant as presented in 6[5]3.

d^3 \ d^2	N	${}^{1}_{0}S$	${}^{3}_{2}P$	${}^{1}_{2}D$	${}^{3}_{2}F$	${}^{1}_{2}G$
${}^{2}_{3}P$	30	0	7	15	− 8	0
${}^{4}_{3}P$	15	0	− 8	0	− 7	0
${}^{2}_{1}D$	60	16	− 9	− 5	− 21	− 9
${}^{2}_{3}D$	140	0	−49	45	21	− 25
${}^{2}_{3}F$	70	0	28	− 10	7	− 25
${}^{4}_{3}F$	5	0	− 1	0	4	0
${}^{2}_{3}G$	42	0	0	− 10	21	11
${}^{2}_{3}H$	2	0	0	0	− 1	1

Table 2[5]b. $|(d^4;\alpha SL\{|d^3\alpha' S'L', d)|^2$ as defined in 6[5]5. The signs of the square roots are indicated. N is the normalization constant as presented in 6[5]3.

d^4 \ d^3	N	${}^{2}_{3}P$	${}^{4}_{3}P$	${}^{2}_{1}D$	${}^{2}_{3}D$	${}^{2}_{3}F$	${}^{4}_{3}F$	${}^{2}_{3}G$	${}^{2}_{3}H$
${}^{1}_{0}S$	1	0	0	1	0	0	0	0	0
${}^{1}_{4}S$	1	0	0	0	1	0	0	0	0
${}^{3}_{2}P$	360	− 14	− 64	135	− 35	− 56	− 56	0	0
${}^{3}_{4}P$	90	25	− 14	0	10	− 25	16	0	0
${}^{1}_{2}D$	280	− 42	0	105	45	28	0	− 60	0
${}^{1}_{4}D$	140	42	0	0	20	63	0	15	0
${}^{3}_{4}D$	210	− 14	49	0	60	− 21	− 21	45	0
${}^{5}_{4}D$	10	0	3	0	0	0	7	0	0
${}^{1}_{4}F$	560	120	0	0	200	− 105	0	− 3	− 132
${}^{3}_{2}F$	840	16	− 56	315	15	− 14	224	90	110
${}^{3}_{4}F$	1680	− 200	− 448	0	120	− 175	− 112	− 405	220
${}^{1}_{2}G$	504	0	0	189	− 25	70	0	66	− 154
${}^{1}_{4}G$	1008	0	0	0	88	385	0	− 507	− 28
${}^{3}_{4}G$	1680	0	0	0	200	315	− 560	297	308
${}^{3}_{4}H$	60	0	0	0	0	5	20	− 9	26
${}^{1}_{4}I$	10	0	0	0	0	0	0	3	7

Table 2^5c. $|(d^5; \alpha SL\{|d^4\alpha'S'L', d)|^2$ as defined in $6^5$5. The signs of the square roots are indicated. N is the normalization constant as presented in $6^5$3.

d^5 \\ d^4	N	$_0^1S$	$_4^1S$	$_2^3P$	$_4^3P$	$_2^1D$	$_4^1D$	$_4^3D$
$_5^2S$	5	0	0	0	0	0	– 2	3
$_5^6S$	1	0	0	0	0	0	0	0
$_3^2P$	150	0	0	14	25	30	15	10
$_3^4P$	300	0	0	– 64	14	0	0	35
$_1^2D$	50	6	0	– 9	0	– 5	0	0
$_3^2D$	350	0	– 14	– 49	– 14	45	– 10	60
$_5^2D$	700	0	– 56	0	126	0	90	60
$_5^4D$	700	0	0	0	126	0	0	– 135
$_3^2F$	2800	0	0	448	– 200	– 160	180	120
$_5^2F$	2800	0	0	0	360	0	– 100	600
$_3^4F$	700	0	0	– 56	– 16	0	0	– 15
$_3^2G$	8400	0	0	0	0	– 800	– 100	600
$_5^2G$	18480	0	0	0	0	0	1452	968
$_5^4G$	420	0	0	0	0	0	0	25
$_3^2H$	1100	0	0	0	0	0	0	0
$_5^2I$	550	0	0	0	0	0	0	0

Some useful properties of the coefficients of fractional parentage are

$$(a^n\alpha SL\{|a^{n-1}\alpha'S'L', a) = (a^{n-1}\alpha'S'L', a|\}a^n\alpha SL) \tag{1}$$

$$\sum_{\alpha'L'S'} (a^n\alpha SL\{|a^{n-1}\alpha'S'L', a)(a^{n-1}\alpha'S'L', a|\}a^n\alpha''SL) = \delta(\alpha,\alpha'') \tag{2}$$

$$(a^n\alpha SL\{|a,a^{n-1}\alpha'S'L') = (-1)^{L+L'-l_a+S+S'-1/2}$$
$$\times (a^n\alpha SL\{|a^{n-1}\alpha'S'L', a) \tag{3}$$

The coefficients of fractional parentage for the configurations with two equivalent electron orbitals are

$$(a^2SL\{|a\, s_a l_a; a) = \frac{1}{2}[1 + (-1)^{L+S}] \tag{4}$$

${}^{5}_{4}D$	${}^{1}_{4}F$	${}^{3}_{2}F$	${}^{3}_{4}F$	${}^{1}_{2}G$	${}^{1}_{4}G$	${}^{3}_{4}G$	${}^{3}_{4}H$	${}^{1}_{4}I$
0	0	0	0	0	0	0	0	0
1	0	0	0	0	0	0	0	0
0	− 15	− 16	− 25	0	0	0	0	0
− 75	0	− 56	56	0	0	0	0	0
0	0	− 21	0	− 9	0	0	0	0
0	35	21	− 21	− 25	− 11	45	0	0
0	35	0	189	0	99	45	0	0
−175	0	0	− 84	0	0	180	0	0
0	105	112	− 175	−400	275	− 405	220	0
0	− 525	0	− 315	0	495	− 9	−396	0
−175	0	224	14	0	0	− 90	− 110	0
0	− 7	1680	945	880	845	891	924	− 728
0	−2541	0	4235	0	−1215	−5577	− 308	−2184
−105	0	0	− 70	0	0	− 66	154	0
0	35	− 220	55	220	− 5	− 99	286	182
0	0	0	0	0	− 45	99	231	− 175

Tables 1^5 and 2^5a, 2^5b, and 2^5c present the squares of the coefficients of fractional parentage, in the seniority scheme, for the p^3, d^3, d^4, and d^5 configurations, respectively[1c]. Signs of the coefficients in Racah's phase convention are also indicated. The coefficients of fractional parentage for the p^n, d^n, and f^n configurations are tabulated by Nielson and Koster [2].

8^5. Seniority scheme

As a preliminary to the discussion of the seniority scheme, let us consider the coupling of two states $|a^{n-2}; S'M_S' \; L'M')$ and $|a^2; S''M_S'' \; L''M'')$ to give a final state $|a^n; SM_SLM)$. By $3^3 1$ this is

$$|a^n; SM_SLM) = \sum_{M'M''} \sum_{M_S' M_S''} |a^{n-2}(S'L'), a^2(S''L''); M_S''M''M_S'M') \times (M_S''M''M_S'M'|SM_SLM)$$

where

$$(M_S''M''M_S'M'|SM_SLM) = (M_S'M_S''|SM_S)(M'M''|LM)$$

For $L'' = M'' = S'' = M_S'' = 0$,

$$(M'0|LM) = (M_S'0|SM_S) = 1$$

giving

$$|a^n; SM_SLM) = |a^n; SL) = |a^{n-2}(SL), a^2(00); SL) \quad (1)$$

That is, the wave function for a given Russell-Saunders term of the a^n configuration can be produced by the vector coupling of the $|a^{n-2}; SL)$ and the $|a^2; {}^1S)$ states. Racah's seniority number scheme is based on this coupling procedure. The lowest $\nu = n, (n-2), (n-4), (n-6), \ldots$ number of the equivalent orbitals, in the a^n configuration, where the given Russell-Saunders term first appears is called the *seniority number* and is indicated by a prefix under the multiplicity number of each term.

A seniority number is not needed for the specification of the Russell-Saunders terms of the p^3 configuration. Such a need first arises to classify the two ${}^2D$ terms of the d^3 configuration. Therefore, we consider this case in detail. A single d electron ($\nu = 1$) gives rise simply to the ${}^2D$. By coupling this $|d; {}^2D)$ state to the $|d^2; {}^1S)$ we obtain one of the ${}^2D$ states of the d^3 configuration. The seniority number of this term is one and is indicated as ${}^2_1D$. These states, written in the zero-order scheme, are

$$|d; {}^2D) = \{0^+\} \quad \text{for} \quad M = 0, M_S = \frac{1}{2}$$

$$|d^2; {}'S) = \frac{1}{\sqrt{5}}[\{2^+, \bar{2}^-\} - \{2^-, \bar{2}^+\} - \{1^+, \bar{1}^-\} + \{1^-, \bar{1}^+\} + (0^+, 0^-\}] \quad \text{for} \quad M = M_S = 0$$

and

$$|d^3, {}^2D) = \frac{1}{2}[\{0^+, 2^+, \bar{2}^-\} - \{0^+, 2^-, \bar{2}^+\} - \{0^+, 1^+, \bar{1}^-\} + \{0^+, 1^-, \bar{1}^+\}] \quad \text{for} \quad M = 0, \quad M_S = \frac{1}{2}$$

where $\{0^+, 0^+, 0^-\}$ is forbidden by the exclusion principle. The $|d^5; {}^2_1D), |d^7; {}^2_1D)$ and $|d^9; {}^2_1D)$ states can be obtained easily in the same way.

The single ${}^2D$ term is the only term in the d^3 configuration that can be assigned seniority $\nu = 1$. Similarly, all the other terms first appear in d^3 configuration, so they have a prefix 3 to indicate their seniority number: ${}^2_3P$, ${}^4_3P$, ${}^2_3D$, ${}^2_3F$, ${}^4_3F$, ${}^2_3G$, ${}^2_3H$. For the d^5 configuration there are three ${}^2D$ terms with seniority numbers 1, 3, and 5.

Racah originally defined the seniority number in terms of the proper values of the scalar product $Q = \sum_{i<j} q_{ij}$, where

$$(a^2LM|q_{ij}|a^2LM) = (2l_a + 1)\delta(L, 0) \tag{2}$$

Then for each term of the a^n configuration with nonvanishing q, there is a corresponding term of the same kind in the a^{n-2} configuration. Thus the seniority number is again the lowest value of $\nu = n$, $(n-2)$, $(n-4)$, $(n-6)$, ... where the given term first appears. Conjugate terms (Sec. 14^5) have the same seniority.

The seniority scheme as defined here suffices for the d^n configuration, but in the f^n configurations complete classification of the terms requires three labels in addition to S and L. Note that the definition of seniority depends on identifying the parent state $|a^{n-2}; S'M'_SL'M')$ and relating seniority numbers to terms of the parent-state configuration. An alternative definition based on group representations is given in Section 2^7.

9^5. Energy matrices for the a^n configurations

The present method is applied to find the energy matrix elements of one- and two-body interactions. For one-body operators we have, by 7^45,

$$\left(a^n; \alpha SL \left| \sum_i^n f_i \right| a^n; \alpha' S'L' \right) = n(a^n; \alpha SL|f|a^n; \alpha' S'L')$$

and by 6^55,

$$\begin{aligned}\left(a^n; \alpha SL \left| \sum_i^n f_i \right| a^n; \alpha' S'L' \right) &= n \sum_{\alpha_1 L_1 S_1} (a^n; \alpha SL\{|a^{n-1}\alpha_1 S_1 L_1, a) \\ &\times \left(S_1 L_1 \frac{1}{2} l_a; SM_S LM \,|f|\, S_1 L_1 \frac{1}{2} l_a; SM_S LM \right) \\ &\times (a^{n-1}\alpha_1 S_1 L_1, a|\}a^n; \alpha' S'L') \end{aligned} \tag{1}$$

For the kinetic energy and one-body Coulomb interaction the matrix elements are

$$\begin{aligned}\left(S_1 L_1 \frac{1}{2} l_a; SM_S LM \left| \frac{p}{2m} - \frac{Ze^2}{r} \right| S_1 L_1 \frac{1}{2} l_a; S'M'_S L'M' \right) \\ = \delta(L, L')\delta(M, M')\delta(S, S')\delta(M_S, M'_S) I(n_a l_a) \end{aligned} \tag{2}$$

where $I(n_a l_a)$ was defined in 8^45.

To find the matrix elements of the two-body symmetric operator, g_{ij}, we expand the wave function $|a^{n-1}a; (S'L')SL)$ in 6^55 in terms of the states $|[a^{n-2}(S'L')a]a;[S'L']SL)$. This is done in the same way as the function $|a^n, \alpha SL)$ is expanded in terms of $|a^{n-1}a;\ (S'L')SL)$, thus obtaining

$$|a^n; SL) = \sum_{\alpha_1 L_1 S_1} \sum_{\alpha_2 L_2 S_2} (a^n; \alpha SL\{|a^{n-1}\alpha_1 S_1 L_1, a)\,(a^{n-1}; \alpha_1 S_1 L_1\{|a^{n-2}\alpha_2 S_2 L_2, a) \times |[a^{n-2}(S_2L_2)a]a;\ [S_1L_1]SL) \qquad (3)$$

Racah's *coefficient of fractional parentage for two particles* [1c]

$$|a^n; \alpha SL) = \sum_{\alpha_2 L_2 S_2} \sum_{L_3 S_3} (a^n \alpha SL\{|a^{n-2}\alpha_2 S_3 L_3)|a^{n-2}(S_2L_2)a^2(S_3L_3);\ SL)$$

can be obtained from (3), recoupling the state as $a^{n-2}(aa)$.

Then,

$$\left(a^n; \alpha SL\left|\sum_{i<j}^{n} g_{ij}\right|a^n; \alpha' SL\right) = \frac{n(n-1)}{2} \sum_{\alpha_1 L_1 S_1} \sum_{\alpha_2 L_2 S_2} \sum_{\alpha'_1 L'_1 S'_1} (a^n; \alpha SL\{|a^{n-1}\alpha_1 S_1 L_1, a)$$
$$\times (a^{n-1}; \alpha_1 S_1 L_1\{|a^{n-2}\alpha_2 S_2 L_2,\ a)\,([a^{n-2}(\alpha_2 S_2 L_2)a_i]a_j;$$
$$[\alpha_1 S_1 L_1]\alpha SL|g_{ij}|[a^{n-2}(\alpha_2 S_2 L_2)a_i]a_j;\ [\alpha'_1 S'_1 L'_1]\alpha' SL)$$
$$\times (a^{n-2}\alpha_2 S_2 L_2,\ a|\}a^{n-1};\ \alpha'_1 S'_1 L'_1)(a^{n-1}\alpha'_1 S'_1 L'_1,\ a|\}a^n;\ \alpha SL) \qquad (4)$$

where $n(n-1)/2$ is the number of electron pairs that can be formed from n electrons. Finally, we recouple the state $|[a^{n-2}(\alpha_2 S_2 L_2)a_i]a_j;$ $[\alpha_1 S_1 L_1]SL)$ as $|a^{n-2}(\alpha_2 S_2 L_2)[a_i a_j];\ [S_3 L_3]SL)$, obtaining

$$\left(a^n; \alpha SL\left|\sum_{i<j}^{n} g_{ij}\right|a^n; \alpha' SL\right) = \frac{n(n-1)}{2} \sum_{\alpha_1 S_1 L_1} \sum_{\alpha_2 S_2 L_2} \sum_{\alpha'_1 S'_1 L'_1} (a^n, \alpha SL\{|a^{n-1}\alpha_1 S_1 L_1, a)$$
$$\times (a^{n-1}; \alpha_1 S_1 L_1\{|a^{n-2}\alpha_2 S_2 L_2,\ a)(a^{n-2}\alpha_2 S_2 L_2,\ a|\}a^{n-1};\ \alpha'_1 S'_1 L'_1)$$
$$\times (a^{n-1}\alpha'_1 S'_1 L'_1,\ a|\}a^n;\ \alpha SL)$$
$$\times \sum_{L_3 S_3}([a^{n-2}(\alpha_2 S_2 L_2)a]a;\ [\alpha_1 S_1 L_1]\alpha SL|a^{n-2}(\alpha_2 S_2 L_2)[a_i a_j][S_3 L_3];\ \alpha SL)$$
$$\times (a^{n-2}(\alpha_2 S_2 L_2)[a_i a_j][S_3 L_3]\alpha' SL|[a^{n-2}(\alpha_2 S_2 L_2)a]a;\ [\alpha'_1 S'_1 L'_1]\alpha' SL)$$
$$\times (a_i a_j;\ S_3 L_3|g_{ij}|a_i a_j;\ S_3 L_3) \qquad (5)$$

where by $9^3 7$

$$([a^{n-2}(\alpha_2 S_2 L_2)a]a;[\alpha_1 S_1 L_1]\alpha SL|a^{n-2}(\alpha_2 S_2 L_2)[a_i a_j][S_3 L_3];\ \alpha SL)$$
$$= (-1)^{L_2+L+S_2+S+1}\,[(2L_1+1)(2L_3+1)\,(2S_1+1)(2S_3+1)]^{1/2}$$
$$\times \begin{Bmatrix} S & 1/2 & S_1 \\ 1/2 & S_2 & S_3 \end{Bmatrix} \begin{Bmatrix} L & l_a & L_1 \\ l_a & L_2 & L_3 \end{Bmatrix}$$

The matrix elements $(a_i a_j,\ SL|g_{ij}|a_i a_j,\ SL)$ are found by the method used in the previous section for a specified operator g_{ij}. For the Coulomb interaction they are given by $2^5 5'$.

Equation (5) becomes very complicated when it is applied to configurations containing more than three equivalent-electron orbitals. In such cases, it is more convenient to express a given two-body operator in terms of one-body operators and then to use (1). We will demonstrate how this is done for the Coulomb interaction in Section 10.

For the configuration with three equivalent orbitals, $n = 3$, by 7[5]3, the coefficients of fractional parentage for two particles are related to the one particle coefficients as

$$(a^n\alpha SL\{|a^{n-2}\alpha_1 S_1 L_1, a^2 S_2 L_2) = (a^3\alpha SL\{|a, a^2 S_2 L_2)$$
$$= (-1)^{L+S-l_a-1/2}(a^3\alpha SL\{a^2 S_2 L_2, a)$$

where the $(a^3\alpha SL\{|a^2 S_2 L_2, a)$ are known. Then, for the p^3 configuration the Coulomb interaction matrix elements (5) reduce to

$$\left(a^3\alpha SL\Big| \sum_{i>j}^{3} g_{ij}\Big| a^3\alpha' SL \right) = 3 \sum_{L_2 S_2} (a^3\alpha SL\{|a^2 S_2 L_2, a)$$
$$\times (a^2 S_2 L_2|g|a^2 S_2 L_2)(a^2 S_2 L_2, a|\}a^3\alpha' SL) \qquad (5')$$

where, because it is always even for three-electron configurations, the phase factor $(2L - 2l_a + 2S - 1)$ is omitted. The matrix elements $(a^2 S_2 L_2|g|a^2 S_2 L_2)$ for $l_a = 1$ were computed in Section 10[4]. Then the Coulomb interaction matrix elements are

$$W(^4S) = 3 \times \left[F^0(pp) - \frac{1}{5}F^2(pp)\right] = 3F^0(pp) - \frac{3}{5}F^2(pp)$$
$$W(^2D) = \frac{3}{2}\left\{\left[F^0(pp) - \frac{1}{5}F^2(pp)\right] + \left[F^0(pp) + \frac{1}{25}F^2(pp)\right]\right\}$$
$$= 3F^0(pp) - \frac{6}{25}F^2(pp)$$
$$W(^2P) = \frac{3}{18}\left\{4\left[F^0(pp) + \frac{2}{5}F^2(pp)\right] + 9\left[F^0(pp) - \frac{1}{5}F^2(pp)\right]\right.$$
$$\left. + 5\left[F^0(pp) + \frac{1}{25}F^2(pp)\right]\right\} = 3F^0(pp)$$

Therefore, the term energies relative to the configuration average energy become

$$E(^4S) = -\frac{9}{25}F^2(pp)$$
$$E(^2D) = 0$$
$$E(^2P) = \frac{6}{25}F^2(pp)$$

For d^3 the matrix elements $(d^2 S_2 L_2|g|d^2 S_2 L_2)$ in (5) are known from the computations in Section 2[5]. They are, writing F^k for $F^k(dd)$,

$$\left(d^2\,{}^3F\Big| \sum_{i>j} \frac{e^2}{r_{ij}}\Big| d^2\,{}^3F \right) = W(^3F) = F^0 - \frac{8}{49}F^2 - \frac{9}{441}F^4$$
$$W(^3P) = F^0 + \frac{7}{49}F^2 - \frac{84}{441}F^4$$

$$W({}^1G) = F^0 + \frac{4}{49}F^2 + \frac{1}{441}F^4$$

$$W({}^1D) = F^0 - \frac{3}{49}F^2 + \frac{36}{441}F^4$$

$$W({}^1S) = F^0 + \frac{14}{49}F^2 + \frac{126}{441}F^4$$

For the Coulomb interaction matrix elements of the d^3 configuration:

$$W({}^4_3F) = \frac{3}{5}\left\{\left[F^0 + \frac{7}{49}F^2 - \frac{84}{441}F^4\right] + 4\left[F^0 - \frac{8}{49}F^2 - \frac{9}{441}F^4\right]\right\}$$

$$= 3F^0 - \frac{15}{49}F^2 - \frac{72}{441}F^4$$

$$W({}^4_3P) = \frac{3}{15}\left\{8\left[F^0 + \frac{7}{49}F^2 - \frac{84}{441}F^4\right] + 7\left[F^0 - \frac{8}{49}F^2 - \frac{9}{441}F^4\right]\right\}$$

$$= 3F^0 - \frac{147}{441}F^4$$

$$W({}^2_3H) = \frac{3}{2}\left\{\left[F^0 - \frac{8}{49}F^2 - \frac{9}{441}F^4\right] + \left[F^0 + \frac{4}{49}F^2 + \frac{1}{441}F^4\right]\right\}$$

$$= 3F^0 - \frac{6}{49}F^2 - \frac{12}{441}F^4$$

$$W({}^2_3G) = \frac{3}{42}\left\{10\left[F^0 - \frac{3}{49}F^2 + \frac{36}{441}F^4\right] + 21\left[F^0 - \frac{8}{49}F^2 - \frac{9}{441}F^4\right]\right.$$

$$\left. + 11\left[F^0 + \frac{4}{49}F^2 + \frac{1}{441}F^4\right]\right\} = 3F^0 - \frac{11}{49}F^2 + \frac{13}{441}F^4$$

$$W({}^2_3F) = \frac{3}{70}\left\{28\left[F^0 + \frac{7}{49}F^2 - \frac{84}{441}F^4\right] + 10\left[F^0 - \frac{3}{49}F^2 + \frac{36}{441}F^4\right]\right.$$

$$\left. + 7\left[F^0 - \frac{8}{49}F^2 - \frac{9}{441}F^4\right] + 25\left[F^0 + \frac{4}{25}F^2 + \frac{1}{441}F^4\right]\right\}$$

$$= 3F^0 + \frac{9}{49}F^2 - \frac{87}{441}F^4$$

$$W({}^2_1D) = \frac{3}{60}\left\{16\left[F^0 + \frac{14}{49}F^2 + \frac{126}{441}F^4\right] + 9\left[F^0 + \frac{7}{49}F^2 - \frac{84}{441}F^4\right]\right.$$

$$+ 5\left[F^0 - \frac{3}{49}F^2 + \frac{36}{441}F^4\right] + 21\left[F^0 - \frac{8}{49}F^2 - \frac{9}{441}F^4\right]$$

$$\left. + 9\left[F^0 + \frac{4}{49}F^2 + \frac{1}{441}F^4\right]\right\} = 3F^0 + \frac{7}{49}F^2 + \frac{63}{441}F^4$$

$$W({}^2_3D) = \frac{3}{140}\left\{49\left[F^0 + \frac{7}{49}F^2 - \frac{84}{441}F^4\right] + 45\left[F^0 - \frac{3}{49}F^2 + \frac{36}{441}F^4\right]\right.$$

$$\left. + 21\left[F^0 - \frac{8}{49}F^2 - \frac{9}{441}F^4\right] + 25\left[F^0 + \frac{4}{49}F^2 + \frac{1}{441}F^4\right]\right\}$$

$$= 3F^0 + \frac{3}{49}F^2 - \frac{57}{441}F^4$$

$$W({}^2_1D{:}{}^2_3D) = \frac{3}{\sqrt{60}\sqrt{140}}\left\{21\left[F^0 + \frac{7}{49}F^2 - \frac{84}{441}F^4\right]\right.$$
$$- 15\left[F^0 - \frac{3}{49}F^2 + \frac{36}{441}F^4\right] - 21\left[F^0 - \frac{8}{49}F^2 + \frac{9}{441}F^4\right]$$
$$\left. + 15\left[F^0 + \frac{4}{49}F^2 + \frac{1}{441}F^4\right]\right\} = 3\sqrt{21}\left[\frac{F^2}{49} - \frac{5}{441}F^4\right]$$
$$W({}^2_3P) = \frac{3}{30}\left\{7\left[F^0 + \frac{7}{49}F^2 - \frac{84}{441}F^4\right] + 15\left[F^0 - \frac{3}{49}F^2 + \frac{36}{441}F^4\right]\right.$$
$$\left. + 8\left[F^0 - \frac{8}{49}F^2 - \frac{9}{441}F^4\right]\right\} = 3F^0 - \frac{6}{49}F^2 - \frac{12}{441}F^4$$

The term energies relative to the configuration average energy are given in Appendix 3.

10^5. Alternative method for the a^n configurations

The matrix elements of Coulomb interaction for a pair of electrons given by $2^5 5$ may be written as

$$W(ab; SL) = \sum_k f_k(l_a l_b L)F^k(ab) + g_k(l_a l_b L)G^k(ab) \quad (1)$$

where

$$f_k(l_a l_b L) = (-1)^{l_a+l_b+L}\langle l_a\|C^{(k)}(1)\|l_a\rangle\langle l_b\|C^{(k)}(2)\|l_b\rangle \begin{Bmatrix} l_a & l_a & k \\ l_b & l_b & L \end{Bmatrix} \quad (2a)$$

and

$$g_k(l_a l_b L) = -\frac{1}{2}(1 + 4\mathbf{s}_1\cdot\mathbf{s}_2)|\langle l_a\|C^{(k)}\|l_b\rangle|^2 \begin{Bmatrix} l_a & l_b & k \\ l_a & l_b & L \end{Bmatrix} \quad (2b)$$

The generalization of (2a) to configurations with n-electron orbitals is simple, because each spherical tensor operator acts on a definite orbital. For (2b this is not so, and it is convenient to express $g_k(l_a l_b L)$ as a sum of scalar products of unit tensor operators each acting on definite orbitals. By $9^3 13$a,

$$\begin{Bmatrix} l_a & l_a & k \\ l_b & l_b & L \end{Bmatrix} = \sum_r (-1)^{L+k+r}(2r+1)\begin{Bmatrix} l_a & l_a & r \\ l_b & l_b & k \end{Bmatrix}\begin{Bmatrix} l_a & l_a & r \\ l_b & l_b & L \end{Bmatrix}$$

and, by $1^5 6$,

$$(ab; LM|\mathbf{u}^{(r)}(1)\cdot\mathbf{u}^{(r)}(2)|ab; LM)$$
$$= (-1)^{l_a+l_b+L}\begin{Bmatrix} l_a & l_a & r \\ l_b & l_b & L \end{Bmatrix}\langle l_a\|u^{(r)}(1)\|l_a\rangle\langle l_b\|u^{(r)}(2)\|l_b\rangle$$

where

$$\langle nl\|\mathbf{u}^{(r)}\|n'l'\rangle = \delta(nn')\delta(ll') \quad (3)$$

Then,

$$\begin{Bmatrix} l_a & l_a & r \\ l_b & l_b & L \end{Bmatrix} = \sum_r (-1)^{l_a+l_b+L}(ab;LM|\mathbf{u}^{(r)}(1)\cdot\mathbf{u}^{(r)}(2)|ab;LM) \tag{4}$$

By substituting (4) into (2a) and into (2b) and using the fact that $(l_a + l_b + k)$ is even, we have

$$f_k(l_a l_b L) = \langle l_a\|C^{(k)}\|l_a\rangle\langle l_b\|C^{(k)}\|l_b\rangle(ab;LM|\mathbf{u}^{(k)}(1)\cdot\mathbf{u}^{(k)}(2)|ab;LM) \tag{5a}$$

$$g_k(l_a l_b L) = -\frac{1}{2}(1 + 4\mathbf{s}_1\cdot\mathbf{s}_2)|\langle l_a\|C^{(k)}\|l_b\rangle|^2 \sum_r(-1)^r(2r+1)$$
$$\times (ab;LM|\mathbf{u}^{(r)}(1)\cdot\mathbf{u}^{(r)}(2)|ab;LM)\begin{Bmatrix} l_a & l_a & r \\ l_b & l_b & k \end{Bmatrix} \tag{5b}$$

and consequently

$$W(ab;\,SL) = \sum_k\langle l_a\|C^{(k)}\|l_a\rangle\langle l_b\|C^{(k)}\|l_b\rangle(ab;\,LM|\mathbf{u}^{(k)}(1)\cdot\mathbf{u}^{(k)}(2)|ab;\,LM)F^k(ab)$$
$$-\frac{1}{2}(1 + 4\mathbf{s}_1\cdot\mathbf{s}_2)\sum_k|\langle l_a\|C^{(k)}\|l_b\rangle|^2\sum_r(-1)^r(2r+1)$$
$$\times (ab;LM|\mathbf{u}^{(k)}(1)\cdot\mathbf{u}^{(k)}(2)|ab;\,LM)\begin{Bmatrix} l_a & l_a & r \\ l_b & l_b & k \end{Bmatrix}G^k(ab) \tag{1'}$$

An *operator of the Coulomb interaction* for an electron pair may be defined as

$$\mathbf{W}(l_a l_b) = \mathbf{f}_k + \mathbf{g}_k$$

where

$$\mathbf{f}_k = \langle l_a\|C^{(k)}\|l_a\rangle\langle l_b\|C^{(k)}\|l_b\rangle\mathbf{u}^{(k)}(1)\cdot\mathbf{u}^{(k)}(2) \tag{5a'}$$

$$\mathbf{g}_k = -\frac{1}{2}(1 + 4\mathbf{s}_1\cdot\mathbf{s}_2)|\langle l_a\|C^{(k)}\|l_b\rangle|^2$$
$$\times\sum_r(-1)^r(2r+1)\mathbf{u}^{(r)}(1)\cdot\mathbf{u}^{(r)}(2) \tag{5b'}$$

A *double tensor* of order (K, k) is defined [**21³b**, p. 295] as a quantity that behaves as an irreducible tensor of the order K with respect to one part of a given system and as an irreducible tensor of the order k with respect to the rest of the system. For example, **ls** is a double vector, and $(\mathbf{l}\cdot\mathbf{s})$ is the scalar part of it. The quantities $(\mathbf{s}_1\cdot\mathbf{s}_2)(\mathbf{u}_1^{(k)}\cdot\mathbf{u}_2^{(k)})$ are scalar products of double tensors $\mathbf{s}_i\mathbf{u}_i^{(k)}$.

When n'-electron orbitals of a given configuration are equivalent it does not make any sense to consider a tensor $\mathbf{u}_{(i)}^{(k)}$ or a double tensor $\mathbf{v}^{(Kk)}(i)$, which operates on the ith orbital, because the n'-electron orbitals are indistinguishable. As a result, one has to consider the sum

$$\mathbf{U}^{(k)} = \mathbf{u}^{(k)}(1) + \mathbf{u}^{(k)}(2) + \ldots + \mathbf{u}^{(k)}(n') = \sum_i^{n'}\mathbf{u}^{(k)}(i) \tag{6}$$

$$\mathbf{V}^{(Kk)} = \mathbf{v}^{(Kk)}(1) + \mathbf{v}^{(Kk)}(2) + \ldots + \mathbf{v}^{(Kk)}(n') = \sum_i^{n'}\mathbf{v}^{(Kk)}(i) \tag{7}$$

The reduced-matrix elements of the tensor operator $\mathbf{U}^{(k)}$ for the a^n configuration, by $9^5 1$ and $1^5 8$, are

$$\begin{aligned}(a^n; \alpha SL\| U^{(k)}\| a^n; \alpha' S' L') &= \delta(S,S')n \sum_{\alpha_1 L_1 S_1} (a^n; \alpha SL\{| a^{n-1}\, \alpha_1 S_1 L_1, a) \\ &\quad \times (a^{n-1}, a; (\alpha_1 S_1 L_1)\alpha SL\| u^{(k)}(n)\| a^{n-1}, a; (\alpha_1 S_1 L_1)\alpha' S' L') \\ &\quad \times (a^{n-1}\alpha_1 S_1 L_1, a|\} a^n; \alpha' S' L') \\ &= \delta(S,S')n[(2L+1)(2L'+1)]^{1/2} \sum_{\alpha_1 L_1 S_1} (-1)^{L+L_1+l_a+k} \\ &\quad \times (a^n, \alpha SL\{| a^{n-1}\alpha_1 S_1 L_1, a)\, (a^{n-1}\alpha_1 S_1 L_1, a|\} a^n, \alpha' S' L') \begin{Bmatrix} L & L' & k \\ l_a & l_a & L_1 \end{Bmatrix}\end{aligned} \tag{8}$$

One can define the double tensor $\mathbf{v}^{(1k)}(i)$ as

$$\mathbf{v}^{(1k)}(i) = \mathbf{s}_i \mathbf{u}^{(k)}(i) \tag{9}$$

The reduced-matrix elements of the sum of the double tensors, $\mathbf{V}^{(1k)}$, for the a^n configuration, are found by repeating the previous calculation, once for the spin and once for the angular part. They are

$$\begin{aligned}(a^n; \alpha SL\| V^{(1k)}\| a^n; \alpha' S' L') &= n \left(\frac{3}{2}\right)^{1/2} [(2L+1)(2S+1)]^{1/2} \\ &\times [(2L'+1)(2S'+1)]^{1/2} \sum_{\alpha_1 L_1 S_1} (-1)^{L+L_1+l_a+k+S+S_1+3/2} \\ &\times (a^n; \alpha SL\{| a^{n-1}\alpha_1 S_1 L_1, a)(a^{n-1}\alpha_1 S_1 L_1, a|\} a^n, \alpha' S' L') \\ &\times \begin{Bmatrix} S & S' & 1 \\ 1/2 & 1/2 & S_1 \end{Bmatrix} \begin{Bmatrix} L & L' & k \\ l_a & l_a & L_1 \end{Bmatrix}\end{aligned} \tag{10}$$

where, for the reduced-matrix elements of the spin, the relation

$$\langle s\| s^{(1)}\| s' \rangle = \delta(ss')[s(s+1)(2s+1)]^{1/2} = \delta(ss') \left(\frac{3}{2}\right)^{1/2} \tag{11}$$

is used.

Returning to the configurations with n equivalent-electron orbitals, the matrix elements of the Coulomb interaction are

$$\begin{aligned}\left(a^n; \alpha SL \middle| \sum_{i<j} \frac{e^2}{r_{ij}} \middle| a^n; \alpha' SL'\right) &= \sum_k \left(a^n \alpha SL \middle| \sum_{i<j} \mathbf{C}^{(k)}(i) \cdot \mathbf{C}^{(k)}(j) \middle| a^n \alpha' SL' \right) F^k(aa) \\ &= \sum_k |\langle l_a\| C^{(k)}\| l_a\rangle|^2 \left(a^n \alpha SL \middle| \sum_{i<j} \mathbf{u}^{(k)}(i) : \mathbf{u}^{(k)}(j) \middle| a^n \alpha' S L' \right) F^k(aa)\end{aligned}$$

By considering

$$(\mathbf{A} + \mathbf{B}) \cdot (\mathbf{A} + \mathbf{B}) = \mathbf{A} \cdot \mathbf{A} + 2\mathbf{A} \cdot \mathbf{B} + \mathbf{B} \cdot \mathbf{B}$$

the summation of the unit vectors is written as

$$2\sum_{i<j} \mathbf{u}^{(k)}(i)\cdot\mathbf{u}^{(k)}(j) = \sum_i \mathbf{u}^{(k)}(i)\cdot\sum_j \mathbf{u}^{(k)}(j) - \sum_i \mathbf{u}^{(k)}(i)\cdot\mathbf{u}^{(k)}(i)$$
$$= \mathbf{U}^{(k)}\cdot\mathbf{U}^{(k)} - \sum_i \mathbf{u}^{(k)}(i)\cdot\mathbf{u}^{(k)}(i) \qquad (12)$$

By the Wigner-Eckart theorem and the definition of scalar products,

$$(\alpha LM|\mathbf{U}^{(k)}\cdot\mathbf{U}^{(k)}|\alpha'L'M') = \sum_{\alpha''L''M''}\sum_q (-1)^q(\alpha LM|U_q^{(k)}|\alpha''L''M'')$$
$$\times(\alpha''L''M''|U_{-q}^{(k)}|\alpha'L'M')$$
$$= \sum_{\alpha''L''}(-1)^{L+L''-2M}(\alpha L\|U^{(k)}\|\alpha''L'')(\alpha''L''\|U^{(k)}\|\alpha'L')$$
$$\times\sum_{qM''}\begin{pmatrix} L & k & L'' \\ -M & q & M''\end{pmatrix}\begin{pmatrix} L'' & k & L' \\ -M'' & -q & M'\end{pmatrix}$$

where, by 9³11b,

$$\sum_{qM''}\begin{pmatrix} L & k & L'' \\ -M & q & M''\end{pmatrix}\begin{pmatrix} L'' & k & L' \\ -M'' & -q & M'\end{pmatrix} = \sum_{qM''}\begin{pmatrix} L & K & L'' \\ -M & q & M''\end{pmatrix}\begin{pmatrix} L' & k & L'' \\ -M' & q & M''\end{pmatrix}$$
$$= \frac{\delta(L,L')\delta(M,M')}{(2L+1)}$$

And, finally,

$$(a^n;\alpha SL|\mathbf{U}^{(k)}\cdot\mathbf{U}^{(k)}|a^n;\alpha'S'L')$$
$$= \frac{1}{2L+1}\sum_{\alpha''S''L''}(-1)^{L+L''}(a^n\alpha SL\|U^{(k)}\|a^n\alpha''S''L'')$$
$$\times(a^n\alpha''S''L''\|U^{(k)}\|a^n\alpha'S'L')\delta(LL')\delta(SS') \qquad (13)$$

where the reduced-matrix elements of the tensor operator $\mathbf{U}^{(k)}$ are given by (8).

The second term in (12) is a tensor of zero order. That is

$$\mathbf{T}^{(0)} = \sum_i \mathbf{u}^{(k)}(i)\cdot\mathbf{u}^{(k)}(i) = \sum_i \mathbf{t}^0(i) \qquad (14)$$

By the Wigner-Eckart theorem,

$$(a^n;\alpha SM_SLM|T^{(0)}|a^n;\alpha'S'M'_SL'M')$$
$$= \delta(SS')\delta(M_SM'_S)(-1)^{L-M}\begin{pmatrix} L & 0 & L' \\ -M & 0 & M'\end{pmatrix}(a^n;\alpha SL\|T^{(0)}\|a^n;\alpha'S'L')$$
$$= \frac{(a^n;\alpha SL\|T^{(0)}\|a^n;\alpha'S'L')}{[2L+1]^{1/2}}\delta(\alpha\alpha')\delta(LL')\delta(MM')\delta(SS')\delta(M_SM'_S)$$
$$(15)$$

where the last step is taken by 1⁵4. The reduced-matrix elements of the zero-order tensor operator $\mathbf{T}^{(0)}$ for the a^n configuration are found by setting $k = 0$ and multiplying both sides by $\delta(\alpha\alpha')\delta(LL')\delta(MM')$ in (8). Then

$$(a^n;\alpha SL\|T^{(0)}\|a^n;\alpha SL) = n\,(2L+1)\sum_{\alpha_1 S_1 L_1}(-1)^{L+L_1+l_a}(a^n;\alpha SL\{|a^{n-1}\alpha_1 S_1 L_1,a)$$
$$\times\,(a^{n-1}\alpha_1 S_1 L_1,\,a|\}a^n;\alpha SL)\begin{Bmatrix} L & L & 0\\ l_a & l_a & L_1\end{Bmatrix}\langle l_a\|t^{(0)}\|l_a\rangle \tag{16}$$

where the quantity $\langle l_a\|t^{(0)}\|l_a\rangle$ has to be evaluated. By (15), we have

$$\langle l_a m_a|t^{(0)}|l_a m_a\rangle = \frac{\langle l_a\|t^{(0)}\|l_a\rangle}{[2l_a+1]^{1/2}}$$

and, by (13),

$$\langle l_a m_a|\mathbf{u}^{(k)}(1)\cdot\mathbf{u}^{(k)}(1)|l_a m_a\rangle$$
$$= \frac{1}{2l_a+1}\sum_{l_a}(-1)^{l_a-l'_a}\langle l_a\|u^{(k)}\|l'_a\rangle\langle l'_a\|u^{(k)}\|l_a\rangle = \frac{1}{2l_a+1}$$

Then, by (14),

$$\langle l_a\|t^{(0)}\|l_a\rangle = \frac{1}{[2l_a+1]^{1/2}} \tag{17}$$

so that

$$(a^n;\alpha SL\|T^{(0)}\|a^n;\alpha' SL) = n\,\frac{2L+1}{[2l_a+1]^{1/2}}\sum_{\alpha_1 S_1 L_1}(-1)^{L+L_1+l_a}(a^n;\alpha SL\{|a^{n-1}\alpha_1 S_1 L_1,\,a)$$
$$\times\,(a^{n-1}\alpha_1 S_1 L_1,\,a|\}a^n;\alpha' SL)\begin{Bmatrix} L & L & 0\\ l_a & l_a & L_1\end{Bmatrix}$$

This further simplifies by

$$\begin{Bmatrix} L_1 & l_a & L\\ 0 & L & l_a\end{Bmatrix} = \frac{(-1)^{L+L_1+l_a}}{[(2L+1)(2l_a+1)]^{1/2}} \tag{18}$$

and, by using $7^5 2$,

$$(a^n;\alpha SM_S LM\|T^{(0)}\|a^n;\alpha' SM_S LM) = \frac{n}{2l_a+1}\,\delta(\alpha\alpha') \tag{16'}$$

Finally, we combine (13) and (16′) to obtain

$$(a^n;\alpha SL\left|\sum_{i<j}^{n}\frac{e^2}{r_{ij}}\right|a^n;\alpha' SL) = W(a^2,\alpha SL{:}\alpha' SL)$$
$$= \sum_k \frac{1}{2}|\langle l_a\|C^{(k)}\|l_a\rangle|^2 F^k(aa)$$
$$\times\left[\frac{1}{2L+1}\sum_{\alpha'' S'' L''}(-1)^{L+L''}(a^n;\alpha SL\|U^{(k)}\|a^n;\alpha'' S'' L'')\right.$$
$$\left.\times\,(a^n;\alpha'' S'' L'')\|U^{(k)}\|a^n;\alpha' SL) - \frac{n}{2l_a+1}\,\delta(\alpha\alpha')\right] \tag{19}$$

The reduced matrix elements of the tensor $\mathbf{U}^{(k)}$ for the a^n configuration, given by (8) are easy to calculate when $k = 0$. By 7⁵2 and (18), they reduce to

$$(a^n; \alpha SL\|U^{(0)}\|a^n; \alpha' S'L') = n\left(\frac{2L+1}{2l_a+1}\right)^{1/2} \delta(\alpha\alpha')\delta(LL')\delta(SS') \qquad (20)$$

Furthermore,

$$(a^n; \alpha SL\|U^{(k)}\|a^n; \alpha' S'L') = (-1)^{L+L'}(a^n; \alpha' S'L'\|U^{(k)}\|a^n; \alpha SL) \qquad (21)$$

Tables 3⁵, 4⁵, and 5⁵ give the reduced-matrix elements of $\mathbf{U}^{(2)}$ for the d^3, d^4, and d^5 configurations, respectively [lc].

The term energies, relative to the configuration average, for the a^n configuration are

$$E(a^n; \alpha SL{:}\ \alpha' SL) = W(a^n; \alpha SL{:}\ \alpha' SL) - \frac{n(n-1)}{2}\mathrm{Av}(a, a) \qquad (22)$$

where $\mathrm{Av}(a, a)$ is given by 9⁴23.

11⁵. Energy matrices for the $a^n b$ configurations

The matrix elements of the Coulomb interaction for the $a^n b$ configurations may be divided into two parts: the matrix elements for the a^n core (Secs. 9⁵ and 10⁵) and the matrix elements of the b electron with the core, which, with the help of the coefficients of fractional parentage, can be written as

$$W'(\alpha_1 S_1 L_1, \tfrac{1}{2}\, l_b; SL{:}\ \alpha_1' S_1' L_1', \tfrac{1}{2}\, l_b; SL) = n \sum_k \sum_{\alpha_2 S_2 L_2} (a^n \alpha_1 S_1 L_1\{|a^{n-1}\alpha_2 S_2 L_2, a)$$
$$\times (a^{n-1}\alpha_2 S_2 L_2, a|\}a^n \alpha_1' S_1' L_1')\Big([a^{n-1}(\alpha_2 S_2 L_2)a]b; [\alpha_1 S_1 L_1]SL$$
$$\Big| e^2 \frac{r_<^k}{r_>^{k+1}} \mathbf{C}^{(k)}(i)\cdot \mathbf{C}^{(k)}(i+1)\Big| [a^{n-1}(\alpha_2 S_2 L_2)a]b; [\alpha_1' S_1' L_1']SL\Big)$$

where the index i specifies the coordinates of the Nth core electron. By using 9³7, the states can be recoupled as $a^{n-1}(ab)$ to obtain

$$W'(\alpha_1 S_1 L_1, \tfrac{1}{2}\, l_b; SL{:}\ \alpha_1' S_1' L_1', \tfrac{1}{2}\, l_b; SL) = n \sum_k \sum_{\alpha_2 S_2 L_2} \sum_{L_{ab} L'_{ab}} \sum_{S_{ab} S'_{ab}} [2L_1+1]^{1/2}$$
$$\times [(2S_1+1)(2L_1'+1)(2S_1'+1)(2L_{ab}+1)(2S_{ab}+1)(2L'_{ab}+1)(2S'_{ab}+1)]^{1/2}$$
$$\times \begin{Bmatrix} S & 1/2 & S_1 \\ 1/2 & S_2 & S_{ab} \end{Bmatrix} \begin{Bmatrix} S & 1/2 & S_1' \\ 1/2 & S_2 & S'_{ab} \end{Bmatrix} \begin{Bmatrix} L & l_b & L_1 \\ l_a & L_2 & L_{ab} \end{Bmatrix} \begin{Bmatrix} L & l_b & L_1' \\ l_a & L_2 & L'_{ab} \end{Bmatrix}$$
$$\times (a^n \alpha_1 S_1 L_1\{|a^{n-1}\alpha_2 S_2 L_2, a)\,(a^{n-1}\alpha_2 S_2 L_2, a|\}a^n \alpha_1' S_1' L_1')\Big(a^{n-1}(\alpha_2 S_2 L_2)[ab]$$
$$[S_{ab} L_{ab}]; SL\Big| e^2 \frac{r_<^k}{r_>^{k+1}} \mathbf{C}^{(k)}(N)\cdot \mathbf{C}^{(k)}(N+1)\Big| a^{n-1}(\alpha_2 S_2 L_2)[ab][S'_{ab} L'_{ab}]; SL\Big) \qquad (1)$$

Table 3[5]. $(d^3\alpha SL\|35U^{(2)}\|d^3\alpha' S'L')$ as defined in (8).

$S=1/2, 3/2$	${}^2_3P$	${}^4_3P$	${}^2_1D$	${}^2_3D$	${}^2_3F$	${}^4_3F$	${}^2_3G$	${}^2_3H$
${}^2_3P$	$-2\sqrt{21}$	0	$-\frac{21}{2}\sqrt{10}$	$\frac{1}{2}\sqrt{210}$	$-4\sqrt{21}$	0	0	0
${}^4_3P$	0	$7\sqrt{21}$	0	0	0	$-14\sqrt{6}$	0	0
${}^2_1D$	$\frac{21}{2}\sqrt{10}$	0	$\frac{35}{2}$	$\frac{15}{2}\sqrt{21}$	$-7\sqrt{15}$	0	$-15\sqrt{7}$	0
${}^2_3D$	$-\frac{1}{2}\sqrt{210}$	0	$\frac{15}{2}\sqrt{21}$	$\frac{15}{2}$	$-9\sqrt{35}$	0	$-5\sqrt{3}$	0
${}^2_3F$	$-4\sqrt{21}$	0	$7\sqrt{15}$	$9\sqrt{35}$	$7\sqrt{6}$	0	$2\sqrt{210}$	$-\sqrt{2310}$
${}^4_3F$	0	$-14\sqrt{6}$	0	0	0	$-7\sqrt{6}$	0	0
${}^2_3G$	0	0	$-15\sqrt{7}$	$-5\sqrt{3}$	$-2\sqrt{210}$	0	$3\sqrt{22}$	$-\sqrt{462}$
${}^2_3H$	0	0	0	0	$-\sqrt{2310}$	0	$\sqrt{462}$	$\sqrt{3003}$

Table 4[5]. $(d^4\alpha SL\|35U^{(2)}\|d^4\alpha' S'L')$ as defined in (8).

$S=0$	${}^1_0S$	${}^1_4S$	${}^1_2D$	${}^1_4D$	${}^1_4F$	${}^1_2G$	${}^1_4G$	${}^1_4I$
${}^1_0S$	0	0	$7\sqrt{30}$	0	0	0	0	0
${}^1_4S$	0	0	$3\sqrt{70}$	$4\sqrt{35}$	0	0	0	0
${}^1_2D$	$7\sqrt{30}$	$3\sqrt{70}$	-5	$-30\sqrt{2}$	0	$4\sqrt{5}$	$8\sqrt{55}$	0
${}^1_4D$	0	$4\sqrt{35}$	$-30\sqrt{2}$	-15	$10\sqrt{14}$	$10\sqrt{10}$	$2\sqrt{110}$	0
${}^1_4F$	0	0	0	$-10\sqrt{14}$	$\frac{35}{2}\sqrt{6}$	$-7\sqrt{70}$	$-\frac{1}{2}\sqrt{770}$	0
${}^1_2G$	0	0	$4\sqrt{5}$	$10\sqrt{10}$	$7\sqrt{70}$	$5\sqrt{22}$	$5\sqrt{2}$	$-2\sqrt{455}$
${}^1_4G$	0	0	$8\sqrt{55}$	$2\sqrt{110}$	$\frac{1}{2}\sqrt{770}$	$5\sqrt{2}$	$-\frac{125}{22}\sqrt{22}$	$-\frac{8}{11}\sqrt{5005}$
${}^1_4I$	0	0	0	0	0	$-2\sqrt{455}$	$-\frac{8}{11}\sqrt{5005}$	$\frac{35}{11}\sqrt{143}$
$S=1, 2$	${}^3_2P$	${}^3_4P$	${}^3_4D$	${}^5_4D$	${}^3_2F$	${}^3_4F$	${}^3_4G$	${}^3_4H$
${}^3_2P$	$-\frac{7}{3}\sqrt{21}$	$\frac{14}{3}\sqrt{6}$	$-\frac{28}{3}\sqrt{15}$	0	$\frac{14}{3}\sqrt{6}$	$\frac{28}{3}\sqrt{6}$	0	0
${}^3_4P$	$\frac{14}{3}\sqrt{6}$	$\frac{19}{3}\sqrt{21}$	$-\frac{4}{3}\sqrt{210}$	0	$\frac{22}{3}\sqrt{21}$	$\frac{8}{3}\sqrt{21}$	0	0
${}^3_4D$	$\frac{28}{3}\sqrt{15}$	$\frac{4}{3}\sqrt{210}$	5	0	$-4\sqrt{35}$	$4\sqrt{35}$	$20\sqrt{3}$	0
${}^5_4D$	0	0	0	-35	0	0	0	0
${}^3_2F$	$\frac{14}{3}\sqrt{6}$	$\frac{22}{3}\sqrt{21}$	$4\sqrt{35}$	0	$\frac{7}{3}\sqrt{6}$	$\frac{49}{3}\sqrt{6}$	$-3\sqrt{210}$	$\frac{2}{3}\sqrt{2310}$
${}^3_4F$	$\frac{28}{3}\sqrt{6}$	$\frac{8}{3}\sqrt{21}$	$-4\sqrt{35}$	0	$\frac{49}{3}\sqrt{6}$	$\frac{77}{6}\sqrt{6}$	$-\frac{1}{2}\sqrt{210}$	$\frac{2}{3}\sqrt{2310}$
${}^3_4G$	0	0	$20\sqrt{3}$	0	$3\sqrt{210}$	$\frac{1}{2}\sqrt{210}$	$-\frac{3}{2}\sqrt{22}$	$-2\sqrt{462}$
${}^3_4H$	0	0	0	0	$\frac{2}{3}\sqrt{2310}$	$\frac{2}{3}\sqrt{2310}$	$2\sqrt{462}$	$\frac{1}{3}\sqrt{3003}$

Table 5[5]. $(d^5\alpha SL\|35U^{(2)}\|d^5\alpha' S'L')$ as defined in (8).

S=1/2,3/2	${}^2_5S$	${}^2_3P$	${}^4_3P$	${}^2_1D$	${}^2_3D$	${}^2_5D$	${}^4_5D$
${}^2_5S$	0	0	0	0	$4\sqrt{70}$	0	0
${}^2_3P$	0	0	0	$-7\sqrt{30}$	0	$5\sqrt{105}$	0
${}^4_3P$	0	0	0	0	0	0	$7\sqrt{15}$
${}^2_1D$	0	$7\sqrt{30}$	0	0	$15\sqrt{7}$	0	0
${}^2_3D$	$4\sqrt{70}$	0	0	$15\sqrt{7}$	0	$5\sqrt{2}$	0
${}^2_5D$	0	$-5\sqrt{105}$	0	0	$5\sqrt{2}$	0	0
${}^4_5D$	0	0	$-7\sqrt{15}$	0	0	0	0
${}^2_3F$	0	0	0	$14\sqrt{5}$	0	0	0
${}^2_5F$	0	$4\sqrt{105}$	0	0	$-10\sqrt{7}$	0	0
${}^4_3F$	0	0	0	0	0	0	$-8\sqrt{35}$
${}^2_3G$	0	0	0	$-10\sqrt{21}$	0	$20\sqrt{6}$	0
${}^2_5G$	0	0	0	0	$-6\sqrt{55}$	0	0
${}^4_5G$	0	0	0	0	0	0	0
${}^2_3H$	0	0	0	0	0	0	0
${}^2_5I$	0	0	0	0	0	0	0

Because the spherical tensor operators $\mathbf{C}^{(k)}(N)$ and $\mathbf{C}^{(k)}(N+1)$ in (1) act only on the nth and $(n+1)$th electron coordinates, the matrix elements of the Coulomb interaction depend only on the quantum labels of these electrons, giving

$$\left(a^{n-1}(\alpha_2 S_2 L_2)[ab][S_{ab}L_{ab}];SL\left|e^2\frac{r_<^k}{r_>^{k+1}}\mathbf{C}^{(k)}(n)\cdot\mathbf{C}^{(k)}(n+1)\right|a^{n-1}(\alpha_2 S_2 L_2)[ab][S'_{ab}L'_{ab}];SL\right)$$
$$= \left(ab;S_{ab}L_{ab}\left|e^2\frac{r_<^k}{r_>^{k+1}}\mathbf{C}_1^{(k)}\cdot\mathbf{C}_2^{(k)}\right|ab;S'_{ab}L'_{ab}\right)\delta(S_{ab}S'_{ab})\delta(L_{ab}L'_{ab})$$
$$= f_k(l_a l_b L_{ab})F^k(ab) + g_k(l_a l_b L_{ab})G^k(ab) \tag{2}$$

where

$$f_k(l_a l_b L_{ab}) = (-1)^{L_{ab}}(2l_a+1)(2l_b+1)\begin{pmatrix} l_a & k & l_a \\ 0 & 0 & 0\end{pmatrix}\begin{pmatrix} l_b & k & l_b \\ 0 & 0 & 0\end{pmatrix}\begin{Bmatrix} l_a & l_a & k \\ l_b & l_b & L_{ab}\end{Bmatrix}$$

$$g_k(l_a l_b L_{ab}) = (-1)^{S_{ab}}(2l_a+1)(2l_b+1)\begin{pmatrix} l_a & k & l_b \\ 0 & 0 & 0\end{pmatrix}^2\begin{Bmatrix} l_a & l_b & k \\ l_a & l_b & L_{ab}\end{Bmatrix}$$

^2_3F	^2_5F	^4_3F	^2_3G	^2_5G	^4_5G	^2_3H	^2_5I
0	0	0	0	0	0	0	0
0	$4\sqrt{105}$	0	0	0	0	0	0
0	0	0	0	0	0	0	0
$-14\sqrt{5}$	0	0	$-10\sqrt{21}$	0	0	0	0
0	$10\sqrt{7}$	0	0	$-6\sqrt{55}$	0	0	0
0	0	0	$20\sqrt{6}$	0	0	0	0
0	0	$8\sqrt{35}$	0	0	0	0	0
0	$-7\sqrt{30}$	0	0	0	0	0	0
$-7\sqrt{30}$	0	0	$4\sqrt{42}$	0	0	$-2\sqrt{462}$	0
0	0	0	0	0	$15\sqrt{14}$	0	0
0	$-4\sqrt{42}$	0	0	$9\sqrt{30}$	0	0	$4\sqrt{273}$
0	0	0	$9\sqrt{30}$	0	0	$6\sqrt{70}$	0
0	0	$-15\sqrt{14}$	0	0	0	0	0
0	$-2\sqrt{462}$	0	0	$-6\sqrt{70}$	0	0	$7\sqrt{13}$
0	0	0	$4\sqrt{273}$	0	0	$-7\sqrt{13}$	0

When (2) is substituted in (1),

$$W'\left(\alpha_1 S_1 L_1, \tfrac{1}{2}\, l_b;\ SL:\ \alpha_1' S_1' L_1', \tfrac{1}{2}\, l_b;\ SL\right) = n \sum_k \sum_{\alpha_2 L_2 S_2} \sum_{L_{ab}S_{ab}} (2l_a + 1)(2l_b + 1)$$

$$\times\ (2L_{ab} + 1)(2S_{ab} + 1)[(2L_1 + 1)(2S_1 + 1)(2L_1' + 1)(2S_1' + 1)]^{1/2}$$

$$\times \begin{Bmatrix} S & 1/2 & S_1 \\ 1/2 & S_2 & S_{ab} \end{Bmatrix} \begin{Bmatrix} S & 1/2 & S_1' \\ 1/2 & S_2 & S_{ab} \end{Bmatrix} \begin{Bmatrix} L & l_b & L_1 \\ l_a & L_2 & L_{ab} \end{Bmatrix} \begin{Bmatrix} L & l_b & L_1' \\ l_a & L_2 & L_{ab} \end{Bmatrix}$$

$$\times\ (a^n \alpha_1 S_1 L_1 \{\ a^{n-1} \alpha_2 S_2 L_2,\ a)(a^{n-1}\alpha_2 S_2 L_2,\ a\ \}a^n \alpha_1' S_1' L_1')$$

$$\times \left[(-1)^{L_{ab}} \begin{pmatrix} l_a & k & l_a \\ 0 & 0 & 0 \end{pmatrix} \begin{pmatrix} l_b & k & l_b \\ 0 & 0 & 0 \end{pmatrix} \begin{Bmatrix} l_a & l_a & k \\ l_b & l_b & L_{ab} \end{Bmatrix} F^k(ab) \right.$$

$$\left. +\ (-1)^{S_{ab}} \begin{pmatrix} l_a & k & l_b \\ 0 & 0 & 0 \end{pmatrix}^2 \begin{Bmatrix} l_a & l_b & k \\ l_a & l_b & L_{ab} \end{Bmatrix} G^k(ab) \right]$$

By employing the summation relations $9^3 12$ on the spin, $9^3 13b$ on the angular part of the coefficient for the $F^k(ab)$ integrals, $9^3 13a$ on the spin, and 9³21 on the angular part of the coefficient for the $G^k(ab)$ integrals, this reduces to

$$W'\left(\alpha_1S_1L_1, \frac{1}{2}\,l_b;\ SL{:}\ \alpha_1'S_1'L_1', \frac{1}{2}\,l_b;\ SL\right) = n(2l_a+1)(2l_b+1)[(2L_1+1)$$
$$\times (2L_1'+1)]^{1/2}\sum_k \sum_{\alpha_2S_2L_2} (a^n\alpha_1S_1L_1\{|a^{n-1}\alpha_2S_2L_2, a)(a^{n-1}\alpha_2S_2L_2, a|\}a^n\alpha_1'S_1'L_1')$$
$$\times\Bigg[(-1)^{L+L_1+L'_1+L_2+K}\times\delta(S_1,S_1')\begin{Bmatrix} l_b & k & l_b\\ L_1 & L & L_1'\end{Bmatrix}\begin{Bmatrix} L_1 & l_a & L_2\\ l_a & L_1' & k\end{Bmatrix}\begin{pmatrix} l_a & k & l_a\\ 0 & 0 & 0\end{pmatrix}$$
$$\times\begin{pmatrix} l_b & k & l_b\\ 0 & 0 & 0\end{pmatrix}F^k(ab) + (-1)^{S_1+S'_1}[(2S_1+1)(2S_1'+1)]^{1/2}\begin{Bmatrix} S_2 & 1/2 & S_1'\\ S & 1/2 & S_1\end{Bmatrix}$$
$$\times\begin{Bmatrix} L_2 & l_a & L_1'\\ l_a & k & l_b\\ L_1 & l_b & L\end{Bmatrix}\begin{pmatrix} l_a & k & l_b\\ 0 & 0 & 0\end{pmatrix}^2 G^k(ab)\Bigg] \qquad (3)$$

The term energies, relative to the configuration average, for the a^nb configuration, are

$$E(a^nb;\ \alpha SL{:}\ \alpha' SL) = E(a^n;\ \alpha_1S_1L_1{:}\ \alpha_1'S_1L_1)$$
$$+\ W'(\alpha_1S_1L_1, \frac{1}{2}\,l_b;\ SL{:}\ \alpha_1'S_1'L_1, \frac{1}{2}\,l_b;\ SL) - n\ \mathrm{Av}(ab) \qquad (3')$$

where $E(a^n;\ \alpha_1S_1L_1{:}\ \alpha'S_1L_1)$ is given by 10⁵22 and Av(ab) by 9⁴21.

12⁵. Van Vleck's theorem

When the added electron b is an s electron, that is, $l_b = 0$, considerable simplification is possible in 11⁵3. For the coefficient of the $F^k(ab)$ integrals, by the triangular conditions discussed in Section 3³, the only nonzero value comes from $k = 0$, and then it is easy to evaluate the $3j$ and $6j$ symbols, because each of them contains at least one zero. By employing the relations 1⁵4, 10⁵18, and 7⁵2, the direct part of the Coulomb interaction matrix elements reduces to $nF^0(l_a, s)$. For the exchange part the $3j$ symbol is zero unless $k = l_a$; furthermore, the $9j$ symbol, by 1⁵5, reduces to a $6j$ symbol, which can be evaluated easily by using the relation 10⁵18, so that we have

$$W'\left(\alpha_1S_1L_1, \frac{1}{2}\,0;\ SL{:}\ \alpha_1'S_1'L_1', \frac{1}{2}\,0;\ SL\right)$$
$$= nF^0(l_a, s)\delta(L_1L_1')\delta(S_1S_1') + n\sum_{\alpha_2S_2L_2}\delta(L_1L_1')(-1)^{S_1+S'_1}$$
$$\times (a^n\alpha_1S_1L_1\{|\ a^{n-1}\alpha_2S_2L_2, a)\,(a^{n-1}\alpha_2S_2L_2|\}a^n\alpha_1'S_1'L_1')$$
$$\times\sqrt{(2S_1+1)(2S_1'+1)}\begin{Bmatrix} 1/2 & S_2 & S_1'\\ 1/2 & S & S_1\end{Bmatrix}\frac{G^{l_a}(l_a, s)}{2l_a+1}$$
$$= \delta(L_1L_1')\delta(S_1S_1')\left[nF^0(l_a, s) \mp \frac{2S_1+1\pm(n-1)}{2(2l_a+1)}G^{l_a}(l_a, s)\right] \qquad (1)$$

Then the term energies relative to the configuration average become

$$\begin{aligned}
&E(a^n b,\, l_b = 0;\, \alpha SL{:}\, \alpha' SL) \\
&= W'\left(\alpha_1 S_1 L_1, \frac{1}{2} 0;\, SL{:}\, \alpha_1' S_1' L_1', \frac{1}{2} 0;\, SL\right) - \mathrm{Av}(a, ns) \\
&= W'\left(\alpha_1 S_1 L_1, \frac{1}{2} 0;\, SL{:}\, \alpha_1' S_1' L_1', \frac{1}{2} 0;\, SL\right) - nF^0(l_a, s) + \frac{nG^{l_a}(l_a, s)}{2(2l_a + 1)} \\
&= E(a^n;\, \alpha_1' S_1 L_1{:}\, \alpha_1 S_1 L_1) - S_1 \frac{G^{l_a}(l_a, s)}{2l_a + 1} \quad \text{for} \quad S = S_1 + \frac{1}{2} \\
&= E(a^n;\, \alpha_1' S_1 L_1{:}\, \alpha_1 S_1 L_1) + (S_1 + 1) \frac{G^{l_a}(l_a, s)}{2l_a + 1} \quad \text{for} \quad S = S_1 - \frac{1}{2} \qquad (2)
\end{aligned}$$

These results are known as Van Vlecks's [3] theorem. The method was originally suggested by Dirac [4] and later improved and applied by Van Vleck, with Serber's [5] modification for the equivalent-electron orbitals. It takes advantage of the fact that when an s electron is added to a given configuration the angular part of the total Hamiltonian remains unchanged. That is, the result depends solely on the spin. Then it is natural to consider the spin-product functions and a spin-dependent Hamiltonian to find the Coulomb interaction matrix elements arising from the added s electron. This technique may be expanded [6] to the $a^n bc$ type configurations if $l_c = 0$. The part that is added to the elements of the energy matrix of the $a^n b$ configuration to give the final term energies is

$$\begin{aligned}
E(a^n bc,\, l_c = 0;\, \alpha SL{:}\, \alpha' SL) &= E(a^n b;\, \alpha_2 S_2 L_2{:}\, \alpha' S_2 L_2) - S_1 \frac{G^{l_a}(l_a, s)}{2l_a + 1} \\
&\quad - \frac{G^{l_b}(l_b, s)}{2(2l_b + 1)} \quad \text{for} \quad S = S_1 + 1 \quad \text{and} \quad S_2 = S_1 + \frac{1}{2} \\
E(a^n bc,\, l_c = 0;\, \alpha SL{:}\, \alpha' SL) &= E(a^n b;\, \alpha_2 S_2 L_2{:}\, \alpha_2' S_2 L_2) + \frac{2S_1 + 3}{2S_1 + 1} \\
&\quad \times \left[S_1 \frac{G^{l_a}(l_a, s)}{2l_a + 1} + \frac{G^{l_b}(l_b, s)}{2(2l_b + 1)} \right] \\
&\quad \text{for} \quad S = S_1 \quad \text{and} \quad S_2 = S_1 + \frac{1}{2} \qquad (3) \\
E(a^n bc,\, l_c = 0;\, \alpha SL{:}\, \alpha' SL) &= E(a^n b;\, \alpha_2 S_2 L_2{:}\, \alpha_2' S_2 L_2) + \frac{2S_1 - 1}{2S_1 + 1} \\
&\quad \times \left[-(S_1 + 1) \frac{G^{l_a}(l_a, s)}{2l_a + 1} + \frac{G^{l_b}(l_b, s)}{2(2l_b + 1)} \right] \\
&\quad \text{for} \quad S = S_1 \quad \text{and} \quad S_2 = S_1 - \frac{1}{2}
\end{aligned}$$

$$E(a^n bc, l_c = 0; \alpha SL: \alpha' SL) = E(a^n b; \alpha_2 S_2 L_2: \alpha_2' S_2 L_2) + (S_1 + 1)\frac{G^{l_a}(l_a, s)}{2l_a + 1} - \frac{G^{l_b}(l_b, s)}{2(2l_b + 1)}$$

$$\text{for} \quad S = S_1 - 1 \quad \text{and} \quad S_2 = S_1 - \frac{1}{2} \tag{4}$$

$$W[a^n(\alpha_1 S_1 L_1)b, \alpha_2 S_2 L_2, \frac{1}{2} 0; SL: a^n(\alpha_1 S_1 L_1)b, \alpha_2' S_2' L_2', \frac{1}{2} 0; SL] = \frac{[2S_1(2S_1 + 2)]^{1/2}}{2S_1 + 1}\left[\frac{G^{l_a}(l_a, s)}{2l_a + 1} - \frac{G^{l_b}(l_b, s)}{2l_b + 1}\right]\delta(L_2 L_2''')$$

$$\text{for} \quad S_2 = S_2' \pm \frac{1}{2} \tag{5}$$

13^5. Alternative method for the $a^n b$ configurations

The matrix elements of the Coulomb interaction for the $a^n b$ configuration may also be written

$$W'(\alpha_1 S_1 L_1, \frac{1}{2} l_b; \alpha SL: \alpha_1' S_1' L_1', \frac{1}{2} l_b; \alpha' SL) = \sum_k (a^n b; (\alpha_1 S_1 L_1)\alpha SL|\mathbf{f}'_k F^k(ab) - \mathbf{g}'_k G^k(ab)|a^n b; (\alpha_1' S_1' L_1')\alpha' SL)$$

where the operators $\mathbf{f}_k'$ and $\mathbf{g}_k'$ are given in terms of the operator of the Coulomb interaction of an electron pair ($10^5$5a) and ($10^5$5b), as

$$\mathbf{f}_k' = \langle l_a||C^{(k)}||l_a\rangle\langle l_b||C^{(k)}||l_b\rangle \sum_{i=1}^{n} [\mathbf{u}^{(k)}(i) \cdot \mathbf{u}^{(k)}(j)] \tag{1a}$$

$$\mathbf{g}_k' = |\langle l_a||C^{(k)}||l_b\rangle|^2 \sum_r (-1)^r(2r + 1)\begin{Bmatrix} l_a & l_a & r \\ l_b & l_b & k \end{Bmatrix} \times \sum_{i=1}^{n} \frac{1 + 4\mathbf{s}_i \cdot \mathbf{s}_i}{2}[\mathbf{u}^{(r)}(i) \cdot \mathbf{u}^{(r)}(j)] \tag{1b}$$

The index i runs over the coordinates of the core electrons. The index j has a fixed value, $j = N + 1$, and specifies the coordinates of the added electron. By $10^5$6,

$$\sum_{i=1}^{n} [\mathbf{u}^{(k)}(i)\mathbf{u}^{(k)}(j)] = \mathbf{U}^{(k)} \cdot \mathbf{u}^{(K)} \tag{2}$$

and, by $10^5$7 and $10^5$9,

$$\sum_{i=1}^{n} 2\mathbf{s}_i\mathbf{s}_j[\mathbf{u}^{(r)}(i)\mathbf{u}^{(r)}(j)] = 2\mathbf{V}^{(1r)} \cdot \mathbf{v}^{(1r)} \tag{3}$$

so that, by using $1^5$6 and $10^5$3,

$$(a^n b; (\alpha_1 S_1 L_1)\alpha SL|\mathbf{f}'_k|a^n b; (\alpha'_1 S'_1 L'_1)\alpha' SL)$$
$$= (-1)^{L+L'_1+l_b}\langle l_a\|C^{(k)}\|l_a\rangle\langle l_b\|C^{(k)}\|l_b\rangle \begin{Bmatrix} L_1 & l_b & L \\ l_b & L'_1 & k \end{Bmatrix} (a^n\alpha_1 S_1 L_1\|U^{(k)}\|a^n\alpha'_1 S_1 L'_1) \tag{4}$$

where the reduced-matrix elements, $(a^n\alpha_1 S_1 L_1\|U^{(k)}\|a^n\alpha'_1 S_1 L'_1)$ are given in $10^5 8$.

By $1^5 6$ and $10^5 11$,

$$(a^n b; (\alpha_1 S_1 L_1)\alpha SL|\mathbf{g}'_k|a^n b; (\alpha'_1 S'_1 L'_1)\alpha' SL)$$
$$= |\langle l_a\|C^{(k)}\|l_b\rangle|^2 \sum_r (-1)^r(2r+1) \begin{Bmatrix} l_a & l_a & r \\ l_b & l_b & k \end{Bmatrix} \Bigg[(-1)^{L+L'_1+l_b}\frac{1}{2} \begin{Bmatrix} L_1 & l_b & L \\ l_b & L'_1 & r \end{Bmatrix}$$
$$\times (a^n\alpha_1 S_1 L_1\|U^{(r)}\|a^n\alpha'_1 S_1 L'_1) + 2(-1)^{L+S+L'_1+S'_1+l_b+1/2}$$
$$\times \left[\frac{3}{2}\right]^{1/2} \begin{Bmatrix} L_1 & l_b & L \\ l_b & L'_1 & r \end{Bmatrix} \begin{Bmatrix} S_1 & 1/2 & S \\ 1/2 & S'_1 & 1 \end{Bmatrix} (a^n\alpha_1 S_1 L_1\| V^{(1r)}\| a^n\alpha'_1 S'_1 L'_1) \Bigg] \tag{5}$$

where the reduced-matrix elements $(a^n\alpha_1 S_1 L_1\|V^{(1r)}\|a^n\alpha'_1 S'_1 L'_1)$ are given by $10^5 10$. The elements of the energy matrix of the Coulomb interaction for the $a^n b$ configuration are then

$$W'(\alpha_1 S_1 L_1, \tfrac{1}{2}\, l_b; \alpha SL\colon \alpha'_1 S'_1 L'_1, \tfrac{1}{2}\, l_b; \alpha' SL)$$
$$= \sum_k \Bigg[(-1)^{L+L'_1+l_b}\langle l_a\|C^{(k)}\| l_a\rangle\langle l_b\| C^{(k)}\| l_b\rangle \begin{Bmatrix} L_1 & l_b & L \\ l_b & L'_1 & k \end{Bmatrix}$$
$$\times (a^n\alpha_1 S_1 L_1\| U^{(k)}\|a^n\alpha'_1 S_1 L'_1) \Bigg] F^k(ab) - \sum_k \Bigg\{ |\langle l_a\|C^{(k)}\| l_b\rangle|^2 \sum_r (-1)^r(2r+1)$$
$$\times \begin{Bmatrix} l_a & l_a & r \\ l_b & l_b & k \end{Bmatrix} \Bigg[(-1)^{L+L'_1+l_b}\frac{1}{2} \begin{Bmatrix} L_1 & l_b & L \\ l_b & L'_1 & r \end{Bmatrix} (a^n\alpha_1 S_1 L_1\| U^{(r)}\| a^n\alpha' S_1 L'_1)$$
$$+ (-1)^{L+S+L'_1+S'_1+l_b+1/2} 2 \left[\frac{3}{2}\right]^{1/2} \begin{Bmatrix} S_1 & 1/2 & S \\ 1/2 & S'_1 & 1 \end{Bmatrix} \begin{Bmatrix} L_1 & l_b & L \\ l_b & L'_1 & r \end{Bmatrix}$$
$$\times (a^n\alpha_1 S_1 L_1\| V^{(1r)}\| a^n\alpha'_1 S'_1 L'_1) \Bigg] \Bigg\} G^k(ab) \tag{6}$$

The elements of the energy matrix with respect to the configuration average are given by $11^5 3'$.

To demonstrate the use of (6) and $11^5 3'$ the $d^2 p$ configuration is used as an example. First, the reduced-matrix elements of the tensor operator $\mathbf{U}^{(k)}$ and the double tensor operator $\mathbf{V}^{(1r)}$, for the d^2 core; must be determined. Because the Russell-Saunders terms of the d^2 configuration are ${}^1G$, ${}^3F$, ${}^1D$, ${}^3P$, ${}^1S$, the nonzero reduced-matrix elements of the zero-rank tensor operator, $\mathbf{U}^{(0)}$, by $10^5 20$, become

$$(d^2; {}^1G\| U^{(0)}\| d^2; {}^1G) = \frac{6}{\sqrt{5}}$$

$$(d^2; {}^3F\|U^{(0)}\|d^2; {}^3F) = \frac{2\sqrt{7}}{\sqrt{5}}$$
$$(d^2; {}^1D\|U^{(0)}\|d^2; {}^1D) = 2$$
$$(d^2; {}^3P\|U^{(0)}\|d^2; {}^3P) = \frac{2\sqrt{3}}{\sqrt{5}}$$
$$(d^2; {}^1S\|U^{(0)}\|d^2; {}^1S) = \frac{2}{\sqrt{5}}$$

The nonzero reduced matrix elements of $\mathbf{U}^{(1)}$, by $10^5 8$, are

$$(d^2; {}^1G\|U^{(1)}\|d^2; {}^1G) = -\,2 \times 9 \times \begin{Bmatrix} 4 & 4 & 1 \\ 2 & 2 & 2 \end{Bmatrix} = \sqrt{6}$$
$$(d^2; {}^3F\|U^{(1)}\|d^2; {}^3F) = \frac{\sqrt{14}}{\sqrt{5}}$$
$$(d^2; {}^1D\|U^{(1)}\|d^2; {}^1D) = 1$$
$$(d^2; {}^3P\|U^{(1)}\|d^2; {}^3P) = \frac{1}{\sqrt{5}}$$

Finally, for the $\mathbf{U}^{(2)}$ nonzero reduced-matrix elements are the following

$$(d^2; {}^1G\|U^{(2)}\|d^2; {}^1G) = 2 \times 9 \times \begin{Bmatrix} 4 & 4 & 2 \\ 2 & 2 & 2 \end{Bmatrix} = \frac{3}{7}\sqrt{22}$$
$$(d^2; {}^1G\|U^{(2)}\|d^2; {}^1D) = (d^2; {}^1D\|U^{(1)}\|d^2; {}^1G) = \frac{12}{7\sqrt{5}}$$
$$(d^2; {}^3F\|U^{(2)}\|d^2; {}^3F) = \frac{\sqrt{6}}{5}$$
$$(d^2; {}^3F\|U^{(2)}\|d^2; {}^3P) = (d^2; {}^3P\|U^{(2)}\|d^2; {}^3F) = \frac{2\sqrt{6}}{5}$$
$$(d^2; {}^1D\|U^{(2)}\|d^2; {}^1D) = -\frac{3}{7}$$
$$(d^2; {}^1D\|U^{(2)}\|d^2; {}^1S) = (d^2; {}^1S\|U^{(2)}\|d^2; {}^1D) = \frac{2}{\sqrt{5}}$$
$$(d^2; {}^3P\|U^{(2)}\|d^2; {}^3P) = -\,\frac{\sqrt{21}}{5}$$

For the reduced-matrix elements of the double tensor $\mathbf{V}^{(1r)}$, $10^5 10$ reduces to a simple relation when $r = 0$, giving

$$(a^n\alpha SL\|V^{(10)}\|a^n\alpha' S' L') = [S(S+1)(2S+1)]^{1/2}\left(\frac{2L+1}{2l_a+1}\right)^{1/2}\delta(\alpha\alpha')\delta(LL')\delta(SS') \quad (7)$$

so that the nonzero elements for the d^2 configuration, are:

$$(d^2\ {}^3F)\|V^{(10)}\|d^2\ {}^3F) = \frac{\sqrt{42}}{5}$$

$$(d^2\ {}^3P\|V^{(10)}\|d^2\ {}^3P) = \frac{3\sqrt{2}}{5}$$

For $r = 1$ the nonzero values, by $10^5 10$, are

$$(d^2;\ {}^1G\|V^{(11)}\|d^2;\ {}^3F) = -2\frac{\sqrt{3}}{\sqrt{2}}3\sqrt{21}\begin{Bmatrix} 0 & 1 & 1 \\ 1/2 & 1/2 & 1/2 \end{Bmatrix}\begin{Bmatrix} 4 & 3 & 1 \\ 2 & 2 & 2 \end{Bmatrix}$$
$$= -\frac{3}{\sqrt{10}}$$

$$(d^2;\ {}^3F\|V^{(11)}\|d^2;\ {}^3F) = \frac{\sqrt{21}}{\sqrt{5}}$$

$$(d^2;\ {}^3F\|V^{(11)}\|d^2;\ {}^1G) = (d^2;{}^1G\|V^{(11)}\|d^2;{}^3F) = -\frac{3}{\sqrt{10}}$$

$$(d^2;\ {}^3F\|V^{(11)}\|d^2;\ {}^1D) = (d^2;\ {}^1D\|V^{(11)}\|d^2;\ {}^3F) = \frac{\sqrt{6}}{\sqrt{5}}$$

$$(d^2;\ {}^1D\|V^{(11)}\|d^2;\ {}^3P) = (d^2;\ {}^3P\|V^{(11)}\|d^2;\ {}^1D) = -\frac{\sqrt{21}}{2\sqrt{5}}$$

$$(d^2;\ {}^3P\|V^{(11)}\|d^2;\ {}^3P) = \frac{\sqrt{3}}{\sqrt{10}}$$

$$(d^2;\ {}^3P\|V^{(11)}\|d^2;\ {}^1S) = (d^2;\ {}^1S\|V^{(11)}\|d^2;\ {}^3P) = \frac{\sqrt{3}}{\sqrt{5}}$$

Finally, for $r = 2$ nonzero values of the reduced-matrix elements are the following

$$(d^2;\ {}^1G\|V^{(12)}\|d^2;\ {}^3F) = (d^2;\ {}^3F\|V^{(12)}\|d^2;\ {}^1G) = -\frac{3\sqrt{3}}{\sqrt{14}}$$

$$(d^2;\ {}^3F\|V^{(12)}\|d^2;\ {}^3F) = \frac{3}{5}$$

$$(d^2;\ {}^3F\|V^{(12)}\|d^2;\ {}^1D) = (d^2;\ {}^1D\|V^{(12)}\|d^2;\ {}^3F) = \frac{4\sqrt{3}}{\sqrt{35}}$$

$$(d^2;\ {}^3F\|V^{(12)}\|d^2;\ {}^3P) = (d^2;\ {}^3P\|V^{(12)}\|d^2;\ {}^3F) = \frac{6}{5}$$

$$(d^2;\ {}^1D\|V^{(12)}\|d^2;\ {}^3P) = (d^2;\ {}^3P\|V^{(12)}\|d^2;\ {}^1D) = -\frac{3}{2\sqrt{5}}$$

$$(d^2;\ {}^3P\|V^{(12)}\|d^2;\ {}^3P) = -\frac{3}{5}\frac{\sqrt{7}}{\sqrt{2}}$$

The reduced-matrix elements of the spherical tensor operators are given by 2⁵2. For $l_a = 2$ and $l_b = 1$ they become

$$\langle l_a \| C^{(0)} \| l_a \rangle = \sqrt{5}$$
$$\langle l_a \| C^{(1)} \| l_a \rangle = \langle l_a \| C^{(3)} \| l_a \rangle = 0$$
$$\langle l_a \| C^{(2)} \| l_a \rangle = -\frac{\sqrt{10}}{\sqrt{7}}$$
$$\langle l_b \| C^{(0)} \| l_b \rangle = \sqrt{3}$$
$$\langle l_b \| C^{(1)} \| l_b \rangle = \langle l_b \| C^{(3)} \| l_b \rangle = 0$$
$$\langle l_b \| C^{(2)} \| l_b \rangle = -\frac{\sqrt{6}}{\sqrt{5}}$$
$$\langle l_a \| C^{(0)} \| l_b \rangle = \langle l_a \| C^{(2)} \| l_b \rangle = 0$$
$$\langle l_a \| C^{(1)} \| l_b \rangle = \sqrt{2}$$
$$\langle l_a \| C^{(3)} \| l_b \rangle = -\frac{3}{\sqrt{7}}$$

In (6) the upper limit for r, from the triangular conditions for the $6j$ symbols, is determined by $l_<$, the smaller of l_a or l_b. The upper limit for r is $2l_<$. In this example, r runs from 0 up to and including 2.

The $6j$ symbols in (6) are evaluated easily by 10⁵18 if they contain at least one zero argument. The $6j$ symbols are computed from the special relations given in Appendix 2. When $L_1 = L_1' = S_1 = S_1' = 1$ and $L = 0$; $S = 1/2, 3/2$, the needed $6j$ symbols with nonzero arguments are

$$\begin{Bmatrix} 1 & 1 & 1 \\ 1/2 & 1/2 & 1/2 \end{Bmatrix} = -\frac{1}{3} \qquad \begin{Bmatrix} 1 & 1 & 1 \\ 1/2 & 1/2 & 3/2 \end{Bmatrix} = -\frac{1}{6}$$
$$\begin{Bmatrix} 2 & 2 & 1 \\ 1 & 1 & 1 \end{Bmatrix} = -\frac{1}{2\sqrt{5}} \qquad \begin{Bmatrix} 2 & 2 & 2 \\ 1 & 1 & 1 \end{Bmatrix} = \frac{1}{10}\frac{\sqrt{7}}{\sqrt{3}}$$
$$\begin{Bmatrix} 2 & 2 & 1 \\ 1 & 1 & 3 \end{Bmatrix} = \frac{1}{3\sqrt{5}} \qquad \begin{Bmatrix} 2 & 2 & 2 \\ 1 & 1 & 3 \end{Bmatrix} = \frac{1}{5\sqrt{21}}$$

Then,

$$W'[d^2(^3P)p;\, ^1S\colon d^2(^3P)p;\, ^1S] = 2F^0(dp) - \frac{2}{5}F^2(dp) - \frac{1}{5}G^1(dp) + \frac{3}{35}G^3(dp)$$
$$W'[d^2(^3P)p;\, ^3S\colon d^2(^3P)p;\, ^3S] = 2F^0(dp) - \frac{2}{5}F^2(dp) + \frac{2}{5}G^1(dp) - \frac{6}{35}G^3(dp)$$
$$W'[d^2(^3P)p;\, ^4P\colon d^2(^3P)p;\, ^4P] = 2F^0(dp) + \frac{1}{5}F^2(dp) - \frac{2}{5}G^1(dp) - \frac{9}{35}G^3(dp)$$
$$W'[d^2(^3F)p;\, ^4F\colon d^2(^3F)p;\, ^4F] = 2F^0(dp) - \frac{3}{35}F^2(dp) - \frac{1}{15}G^1(dp) - \frac{48}{245}G^3(dp)$$

$$W'[d^2(^3F)p;\,^4G\colon d^2(^3F)p;\,^4G] = 2F^0(dp) + \frac{1}{35}F^2(dp) - \frac{3}{5}G^1(dp) - \frac{12}{245}G^3(dp)$$

$$W'[d^2(^1G)p;\,^2H\colon d^2(^1G)p;\,^2H] = 2F^0(dp) + \frac{4}{35}F^2(dp) - \frac{2}{5}G^1(dp) - \frac{3}{245}G^3(dp)$$

$$W'[d^2(^1S)p;\,^2P\colon d^2(^1S)p;\,^2P] = 2F^0(dp) - \frac{2}{15}G^1(dp) - \frac{3}{35}G^3(dp)$$

$$W'[d^2(^3P)p;\,^2P\colon d^2(^3P)p;\,^2P] = 2F^0(dp) + \frac{1}{5}F^2(dp) + \frac{1}{5}G^1(dp) + \frac{9}{70}G^3(dp)$$

$$W'[d^2(^1D)p;\,^2P\colon d^2(^1D)p;\,^2P] = 2F^0(dp) - \frac{3}{35}F^2(dp) + \frac{1}{15}G^1(dp) - \frac{69}{490}G^3(dp)$$

$$W'[d^2(^1S)p;\,^2P\colon d^2(^3P)p;\,^2P] = \sqrt{3}\left[-\frac{1}{5}G^1(dp) + \frac{3}{35}G^3(dp)\right]$$

$$W'[d^2(^1S)p;\,^2P\colon d^2(^1D)p;\,^2P] = \sqrt{7}\left[\frac{4}{35}F^2(dp) - \frac{1}{15}G^1(dp) - \frac{3}{245}G^3(dp)\right]$$

$$W'[d^2(^3P)p;\,^2P\colon d^2(^1D)p;\,^2P] = \sqrt{21}\,\frac{3}{98}G^3(dp)$$

$$W'[d^2(^3P)p;\,^2D\colon d^2(^3P)p;\,^2D] = 2F^0(dp) - \frac{1}{25}F^2(dp) + \frac{4}{25}G^1(dp) + \frac{3}{50}G^3(dp)$$

$$W'[d^2(^1D)p;\,^2D\colon d^2(^1D)p;\,^2D] = 2F^0(dp) + \frac{3}{35}F^2(dp) - \frac{2}{15}G^1(dp) - \frac{57}{490}G^3(dp)$$

$$W'[d^2(^3F)p;\,^2D\colon d^2(^3F)p;\,^2D] = 2F^0(dp) + \frac{12}{175}F^2(dp) - \frac{2}{75}G^1(dp) + \frac{219}{1225}G^3(dp)$$

$$W'[d^2(^3P)p;\,^2D\colon d^2(^1D)p;\,^2D] = \frac{\sqrt{21}}{\sqrt{5}}\left[-\frac{1}{5}G^1(dp) + \frac{27}{490}G^3(dp)\right]$$

$$W'[d^2(^3P)p;\,^2D\colon d^2(^3F)p;\,^2D] = \frac{\sqrt{14}}{5}\left[\frac{12}{35}F^2(dp) + \frac{1}{5}G^1(dp) + \frac{9}{245}G^3(dp)\right]$$

$$W'[d^2(^1D)p;\,^2D\colon d^2(^3F)p;\,^2D] = \frac{\sqrt{6}}{\sqrt{5}}\left[\frac{1}{15}G^1(dp) + \frac{33}{245}G^3(dp)\right]$$

$$W'[d^2(^3P)p;\,^4D\colon d^2(^3P)p;\,^4D] = 2F^0(dp) - \frac{1}{25}F^2(dp) - \frac{8}{25}G^1(dp) - \frac{3}{25}G^3(dp)$$

$$W'[d^2(^3F)p;\,^4D\colon d^2(^3F)p;\,^4D] = 2F^0(dp) + \frac{12}{175}F^2(dp) + \frac{4}{75}G^1(dp) - \frac{438}{1225}G^3(dp)$$

$$W'[d^2(^3P)p;\,^4D\colon d^2(^3F)p;\,^4D] = \frac{\sqrt{14}}{5}\left[\frac{12}{35}F^2(dp) - \frac{2}{5}G^1(dp) - \frac{18}{245}G^3(dp)\right]$$

$$W'[d^2(^1D)p;\,^2F:\,d^2(^1D)p;\,^2F] = 2F^0(dp) - \frac{6}{245}F^2(dp) - \frac{23}{105}G^1(dp) - \frac{69}{1715}G^3(dp)$$

$$W'[d^2(^3F)p;\,^2F:\,d^2(^3F)p;\,^2F] = 2F^0(dp) - \frac{3}{35}F^2(dp) + \frac{1}{30}G^1(dp) + \frac{24}{245}G^3(dp)$$

$$W'[d^2(^1G)p;\,^2F:\,d^2(^1G)p;\,^2F] = 2F^0(dp) + \frac{11}{49}F^2(dp) - \frac{1}{70}G^1(dp) - \frac{372}{1715}G^3(dp)$$

$$W'[d^2(^1D)p;\,^2F:\,d^2(^3F)p;\,^2F] = \frac{\sqrt{3}}{\sqrt{7}}\left[-\frac{2}{3}G^1(dp) + \frac{6}{49}G^3(dp)\right]$$

$$W'[d^2(^1D)p;\,^2F:\,d^2(^1G)p;\,^2F] = \frac{\sqrt{3}}{7}\left[\frac{24}{35}F^2(dp) - \frac{2}{5}G^1(dp) - \frac{18}{245}G^3(dp)\right]$$

$$W'[d^2(^3F)p;\,^2F:\,d^2(^1G)p;\,^2F] = \frac{1}{2\sqrt{7}}\left[\frac{3}{5}G^1(dp) + \frac{162}{245}G^3(dp)\right]$$

$$W'[d^2(^3F)p;\,^2G:\,d^2(^3F)p;\,^2G] = 2F^0(dp) + \frac{1}{35}F^2(dp) + \frac{3}{10}G^1(dp) + \frac{6}{245}G^3(dp)$$

$$W'[d^2(^1G)p;\,^2G:\,d^2(^1G)p;\,^2G] = 2F^0(dp) - \frac{11}{35}F^2(dp) + \frac{1}{10}G^1(dp) - \frac{18}{245}G^3(dp)$$

$$W'[d^2(^3F)p;\,^2G:\,d^2(^1G)p;\,^2G] = \sqrt{15}\left[-\frac{1}{10}G^1(dp) + \frac{3}{245}G^3(dp)\right]$$

The elements of the energy matrix, relative to the configuration average, for the a^nb configuration ($11^53'$) are the sum of the elements for the a^n core, relative to the average of the a^n configuration, and those for the a^n core and b-added electron interaction relative to the average of this interaction. In the d^2p configuration, the matrix of the d^2 core does not contain any off-diagonal elements. The diagonal elements relative to the configuration average [that is, $E(a^n;\,\alpha_1S_1L_1{:}\ \alpha_1'S_1L_1) = E(d^2;\,SL)$] were found in Section 2^5. The elements of the d^2 core and added p-electron interaction energy matrix, W''s are given above. To find the energy matrix elements relative to the average of the d^2 core and added p-electron interaction it is necessary to subtract

$$n\,\mathrm{Av}(ab) = 2\mathrm{Av}(dp) = 2F^0\,(dp) - \frac{2}{15}G^1\,(dp) - \frac{3}{35}G^3(dp)$$

from these elements, so that

$$E'[d^2(\alpha_1S_1L_1)p;\,\alpha SL{:}\ d^2(\alpha_1'S_1'L_1')p;\,\alpha' SL] = W'[d^2(\alpha_1S_1L_1)p;\,\alpha SL{:}\ d^2(\alpha_1'S_1'L_1')p;\,\alpha' SL] - 2\mathrm{Av}(dp)$$

when $E(d^2;\ \alpha_1S_1L_1{:}\ \ \alpha_1'S_1L_1)$ and $E'[d^2(\alpha_1S_1L_1)p;\ \ \alpha SL{:}\ d^2(\alpha_1'S_1'L_1')p;\,\alpha' SL]$ are added together the same elements for the d^2p

Table 6^5. $(d^3\alpha SL\|\sqrt{30}V^{(11)}\|d^3\alpha' S'L')$

	${}^2_3P$	${}^4_3P$	${}^2_1D$	${}^2_3D$	${}^2_3F$	${}^4_3F$	${}^2_3G$	${}^2_3H$
${}^2_3P$	2	$-2\sqrt{14}$	$-\frac{1}{2}\sqrt{42}$	$\frac{9}{2}\sqrt{2}$	0	0	0	0
${}^4_3P$	$2\sqrt{14}$	$\sqrt{10}$	$-4\sqrt{3}$	0	0	0	0	0
${}^2_1D$	$\frac{1}{2}\sqrt{42}$	$-4\sqrt{3}$	$\frac{3}{2}\sqrt{5}$	$-\frac{1}{2}\sqrt{105}$	$\sqrt{42}$	$-\sqrt{42}$	0	0
${}^2_3D$	$-\frac{9}{2}\sqrt{2}$	0	$-\frac{1}{2}\sqrt{105}$	$-\frac{1}{2}\sqrt{5}$	$\sqrt{2}$	$5\sqrt{2}$	0	0
${}^2_3F$	0	0	$-\sqrt{42}$	$-\sqrt{2}$	$-\frac{1}{2}\sqrt{14}$	$-\sqrt{14}$	$-\frac{3}{2}\sqrt{10}$	0
${}^4_3F$	0	0	$-\sqrt{42}$	$5\sqrt{2}$	$\sqrt{14}$	$2\sqrt{35}$	$-3\sqrt{10}$	0
${}^2_3G$	0	0	0	0	$\frac{3}{2}\sqrt{10}$	$-3\sqrt{10}$	$\frac{9}{10}\sqrt{30}$	$\frac{6}{5}\sqrt{55}$
${}^2_3H$	0	0	0	0	0	0	$-\frac{6}{5}\sqrt{55}$	$\frac{3}{5}\sqrt{55}$

Table 7^5. $(d^3\alpha SL\|70V^{(12)}\|d^3\alpha' S'L')$

	${}^2_3P$	${}^4_3P$	${}^2_1D$	${}^2_3D$	${}^2_3F$	${}^4_3F$	${}^2_3G$	${}^2_3H$
${}^2_3P$	$-19\sqrt{14}$	-28	0	$8\sqrt{35}$	$-8\sqrt{14}$	$-22\sqrt{14}$	0	0
${}^4_3P$	28	$14\sqrt{35}$	0	$-28\sqrt{10}$	56	$-28\sqrt{10}$	0	0
${}^2_1D$	0	0	$35\sqrt{6}$	0	0	0	0	0
${}^2_3D$	$-8\sqrt{35}$	$-28\sqrt{10}$	0	$-5\sqrt{6}$	$-4\sqrt{210}$	$4\sqrt{210}$	$-60\sqrt{2}$	0
${}^2_3F$	$-8\sqrt{14}$	-56	0	$4\sqrt{210}$	-77	-98	$3\sqrt{35}$	$-4\sqrt{385}$
${}^4_3F$	$22\sqrt{14}$	$-28\sqrt{10}$	0	$4\sqrt{210}$	98	$-14\sqrt{10}$	$-18\sqrt{35}$	$4\sqrt{385}$
${}^2_3G$	0	0	0	$-60\sqrt{2}$	$-3\sqrt{35}$	$-18\sqrt{35}$	$3\sqrt{33}$	$12\sqrt{77}$
${}^2_3H$	0	0	0	0	$-4\sqrt{385}$	$-4\sqrt{385}$	$-12\sqrt{77}$	$-\sqrt{2002}$

configuration, relative to the configuration average as were given in Section 5^5, result.

The reduced-matrix elements of the double tensors $\mathbf{V}^{(11)}$ and $\mathbf{V}^{(12)}$ for the d^3, d^4, and d^5 configurations are presented in Tables 6^5, 7^5, 8^5, 9^5, 10^5, and 11^5, respectively [1c]. In Tables 9^5 and 11^5 the reduced matrix elements of the double tensor $\mathbf{V}^{(12)}$ are listed only for a fixed value of the seniority number. The others can be found from the following relations given by Racah:

$$(a^n\alpha\nu SL\|V^{(Kk)}\|a^n\alpha'\nu S'L') = (a^{n-2}\alpha\nu SL\|V^{(Kk)}\|a^{n-2}\alpha'\nu S'L') \quad (8)$$

where ν is the seniority number and α stands for any other label, in addition to ν, L, and S, to specify a given term:

Table 8[5]. $(d^4\alpha SL\|\sqrt{30}V^{(11)}\|d^4\alpha' S'L')$

	${}^1_0S$	${}^1_4S$	${}^3_2P$	${}^3_4P$	${}^1_2D$	${}^1_4D$	${}^3_4D$	${}^5_4D$
${}^1_0S$	0	0	$3\sqrt{3}$	0	0	0	0	0
${}^1_4S$	0	0	$-\sqrt{7}$	$2\sqrt{2}$	0	0	0	0
${}^3_2P$	$3\sqrt{3}$	$-\sqrt{7}$	1	$-2\sqrt{14}$	$-\frac{1}{2}\sqrt{14}$	$2\sqrt{7}$	0	$-4\sqrt{5}$
${}^3_4P$	0	$2\sqrt{2}$	$-2\sqrt{14}$	2	2	$\frac{1}{2}\sqrt{2}$	$\frac{9}{2}\sqrt{2}$	$\frac{1}{2}\sqrt{70}$
${}^1_2D$	0	0	$-\frac{1}{2}\sqrt{14}$	2	0	0	$2\sqrt{10}$	0
${}^1_4D$	0	0	$2\sqrt{7}$	$\frac{1}{2}\sqrt{2}$	0	0	$-\sqrt{5}$	0
${}^3_4D$	0	0	0	$-\frac{9}{2}\sqrt{2}$	$-2\sqrt{10}$	$\sqrt{5}$	$-\frac{1}{2}\sqrt{5}$	$\frac{5}{2}\sqrt{7}$
${}^5_4D$	0	0	$-4\sqrt{5}$	$\frac{1}{2}\sqrt{70}$	0	0	$-\frac{5}{2}\sqrt{7}$	$\frac{15}{2}$
${}^1_4F$	0	0	0	0	0	0	$-2\sqrt{5}$	0
${}^3_2F$	0	0	0	0	2	$\sqrt{2}$	$-5\sqrt{2}$	$-\sqrt{70}$
${}^3_4F$	0	0	0	0	4	$-4\sqrt{2}$	$-\sqrt{2}$	$\sqrt{70}$
${}^1_2G$	0	0	0	0	0	0	0	0
${}^1_4G$	0	0	0	0	0	0	0	0
${}^3_4G$	0	0	0	0	0	0	0	0
${}^3_4H$	0	0	0	0	0	0	0	0
${}^1_4I$	0	0	0	0	0	0	0	0

$$(a^n\alpha\nu SL\|V^{(Kk)}\|a^n\alpha'\nu S'L') = \frac{2l_a+1-n}{2l_a+1-\nu}(a^\nu\alpha\nu SL\|V^{(Kk)}\|a^\nu\alpha'\nu S'L') \quad \text{when} \quad K+k \quad \text{is even} \qquad (9a)$$

$$(a^n\alpha\nu SL\|V^{(Kk)}\|a^n\alpha'\nu S'L') = (a^\nu\alpha\nu SL\|V^{(Kk)}\|a^\nu\alpha'\nu S'L') \quad \text{when} \quad K+k \quad \text{is odd} \qquad (9b)$$

$$(a^n\alpha\nu SL\|V^{(Kk)}\|a^n\alpha'(\nu-2)S'L') = 0 \quad \text{when} \quad K+k \quad \text{is odd} \qquad (10)$$

The energy matrices for the d^n and d^nl_b and configuration can be computed by using 10[5]19 and (6) with the tabulated values for the reduced-matrix elements of $\mathbf{V}^{(Kk)}$ and $\mathbf{U}^{(k)}$.

14[5]. Configurations containing almost-closed shells

Computation of the electrostatic energy matrix elements becomes laborious as the number of electrons in open shells increases. However, the Pauli exclusion principle reduces the complexity of the spectrum when an open shell is more than half filled. This was emphasized at the

^1_4F	^3_2F	^3_4F	^1_2G	^1_4G	^3_4G	^3_4H	^1_4I
0	0	0	0	0	0	0	0
0	0	0	0	0	0	0	0
0	0	0	0	0	0	0	0
0	0	0	0	0	0	0	0
0	2	4	0	0	0	0	0
0	$\sqrt{2}$	$-4\sqrt{2}$	0	0	0	0	0
$-2\sqrt{5}$	$5\sqrt{2}$	$\sqrt{2}$	0	0	0	0	0
0	$-\sqrt{70}$	$\sqrt{70}$	0	0	0	0	0
0	$\sqrt{35}$	$\frac{1}{2}\sqrt{35}$	0	0	$-\frac{9}{2}$	0	0
$-\sqrt{35}$	$\sqrt{14}$	$-\sqrt{14}$	$-\sqrt{3}$	$\sqrt{33}$	$3\sqrt{10}$	0	0
$-\frac{1}{2}\sqrt{35}$	$-\sqrt{14}$	$-\frac{1}{2}\sqrt{14}$	$5\sqrt{3}$	$-\frac{1}{2}\sqrt{33}$	$-\frac{3}{2}\sqrt{10}$	0	0
0	$-\sqrt{3}$	$5\sqrt{3}$	0	0	3	$-\sqrt{66}$	0
0	$\sqrt{33}$	$-\frac{1}{2}\sqrt{33}$	0	0	$\frac{3}{2}\sqrt{11}$	$2\sqrt{6}$	0
$-\frac{9}{2}$	$-3\sqrt{10}$	$\frac{3}{2}\sqrt{10}$	-3	$-\frac{3}{2}\sqrt{11}$	$\frac{9}{10}\sqrt{30}$	$\frac{6}{5}\sqrt{55}$	0
0	0	0	$-\sqrt{66}$	$2\sqrt{6}$	$-\frac{6}{5}\sqrt{55}$	$\frac{3}{5}\sqrt{55}$	$-\frac{3}{2}\sqrt{26}$
0	0	0	0	0	0	$-\frac{3}{2}\sqrt{26}$	0

end of Section 6[4], as was also the fact that the configurations a^n and a^{4l_a+2-n} give rise to the same Russell-Saunders terms. We now consider a configuration which contains, in addition to closed shells, a shell $\mathcal{S}$, which is complete except for ϵ *missing* electrons or *holes*, and in addition η other electrons in a shell $\mathcal{S}'$, where ϵ and η are smaller than half of the number of electrons that would fill these open shells. Such a configuration is called *configuration* $\mathcal{R}$, because it occurs mainly for elements near the right side of the periodic table. We *correlate* to $\mathcal{R}$ a *configuration* $\mathcal{L}$, which contains ϵ number of electrons in the shell $\mathcal{S}$ in addition to the same number of closed shells and η other electrons of the $\mathcal{R}$ configuration. When $\eta = 0$ the $\mathcal{R}$ configuration becomes the *conjugate* of the configuration $\mathcal{L}$. Terms of the $\mathcal{R}$ and $\mathcal{L}$ configurations with the same quantum labels are said to be *conjugated,* generally defined later in this section. Finding relations between the electrostatic energy matrices for the conjugated terms involving correlated configurations is now discussed.

Early treatment of the problem in the zero-order scheme was undertaken by Shortley [7] and Johnson [8]. Because the electrostatic

Table 9[5]. $(d^4 4SL \| 70V^{(12)} \| d^4 4S'L')$

	^1_4S	^3_4P	^1_4D	^3_4D	^5_4D	^1_4F	^3_4F	^1_4G	^3_4G	^3_4H	^1_4I
^1_4S	0	0	0	$4\sqrt{210}$	0	0	0	0	0	0	0
^3_4P	0	$6\sqrt{14}$	$-15\sqrt{35}$	$-3\sqrt{35}$	-105	$-12\sqrt{35}$	$12\sqrt{14}$	0	0	0	0
^1_4D	0	$-15\sqrt{35}$	0	$5\sqrt{6}$	0	0	0	0	$60\sqrt{2}$	0	0
^3_4D	$-4\sqrt{210}$	$3\sqrt{35}$	$-5\sqrt{6}$	$-\frac{15}{2}\sqrt{6}$	$\frac{15}{2}\sqrt{210}$	$-10\sqrt{21}$	$9\sqrt{210}$	$6\sqrt{165}$	$15\sqrt{2}$	0	0
^5_4D	0	-105	0	$-\frac{15}{2}\sqrt{210}$	$-\frac{35}{2}\sqrt{30}$	0	$35\sqrt{6}$	0	$15\sqrt{70}$	0	0
^1_4F	0	$12\sqrt{35}$	0	$-10\sqrt{21}$	0	0	$-21\sqrt{10}$	0	$12\sqrt{14}$	$-6\sqrt{154}$	0
^3_4F	0	$12\sqrt{14}$	0	$-9\sqrt{210}$	$35\sqrt{6}$	$21\sqrt{10}$	-42	0	$-12\sqrt{35}$	$6\sqrt{385}$	0
^1_4G	0	0	0	$-6\sqrt{165}$	0	0	0	0	$27\sqrt{10}$	$6\sqrt{210}$	0
^3_4G	0	0	$-60\sqrt{2}$	$15\sqrt{2}$	$-15\sqrt{70}$	$12\sqrt{14}$	$12\sqrt{35}$	$-27\sqrt{10}$	$-6\sqrt{33}$	$-6\sqrt{77}$	$-12\sqrt{91}$
^3_4H	0	0	0	0	0	$6\sqrt{154}$	$6\sqrt{385}$	$6\sqrt{210}$	$6\sqrt{77}$	$-3\sqrt{2002}$	$-7\sqrt{39}$
^1_4I	0	0	0	0	0	0	0	0	$12\sqrt{91}$	$-7\sqrt{39}$	0

energy matrix for configuration $\mathcal{R}$ is in general quite different from that for $\mathcal{L}$, it must be separately computed. The individual sets of the closed shell $\mathcal{S}$ are denoted by $a_1, a_2, a_3, \ldots, a_\epsilon, a_{\epsilon+1}, \ldots, a_m$, where $m = 4l_a + 2$. For a given state A, in the nlm quantization, the sets $a_1, a_2, \ldots, a_\epsilon$ are holes, or missing sets, and $a_{\epsilon+1}, a_{\epsilon+2}, \ldots, a_m$ are the sets occurring in A. The rest of the sets in A, outside of the number of closed shells, are denoted by $b_1, b_2, \ldots, b_\eta$. All possible zero-order states of the configuration are obtained by writing all possible combinations of missing sets that satisfy the exclusion principle.

Simple cases where the term energies may be computed by the diagonal sum rule are considered first, so the nondiagonal elements 7[4]11, which arise for the cases in which two states A and B differ by one or two sets, are not needed. The diagonal element 9[4]14 may be divided into two parts: (1) terms in which both of the individual sets are in complete shells or one individual set is contained in a complete shell and the other in an open shell; sums of these terms are common to all diagonal elements, (Sec. 9[4]), and (2) terms in which individual sets are both in open shells. The terms of type (2) are

$$\sum_{i>j=\epsilon+1}^{m} [J(a_i, a_j) - K(a_i, a_j)] + \sum_{i=\epsilon+1}^{m} \sum_{j=1}^{\eta} [J(a_i, b_j) - K(a_i, b_j)] + \sum_{i>j=1}^{\eta} [J(b_i, b_j) - K(b_i, b_j)] \quad (1)$$

The long summation over the range $\epsilon + 1$ to m is now reduced to one over the range 1 to ϵ by rewriting the first term as

$$\sum_{i>j=\epsilon+1}^{m} W(a_i, a_j) = \sum_{i>j=1}^{m} W(a_i, a_j) - \sum_{i=1}^{m} \sum_{j=1}^{\epsilon} W(a_i, a_j) + \sum_{i=1}^{\epsilon} \sum_{j=1}^{\epsilon} W(a_i, a_j) - \sum_{i>j=1}^{\epsilon} W(a_i, a_j)$$

where $W(a_i, a_j)$ stands for $[J(a_i, a_j) - K(a_i, a_j)]$. Because $J(a_i, a_i) = K(a_i, a_i)$ and $J(a_i, a_j) = J(a_j, a_i)$, $K(a_i, a_j) = K(a_j, a_i)$,

$$\sum_{i=1}^{\epsilon} \sum_{j=1}^{\epsilon} W(a_i, a_j) - \sum_{i>j=1}^{\epsilon} W(a_i, a_j) = \sum_{i>j}^{\epsilon} W(a_i, a_j)$$

We have

$$\sum_{i>j=\epsilon+1}^{m} W(a_i, a_j) = \sum_{i>j=1}^{m} W(a_i, a_j) - \sum_{i=1}^{m} \sum_{j=1}^{\epsilon} W(a_i, a_j) + \sum_{i>j=1}^{\epsilon} W(a_i, a_j) \quad (2)$$

Table 10^5. $(d^5\alpha SL\|\sqrt{30}V^{(11)}\|d^5\alpha' S'L')$

	$^{2}_{5}S$	$^{6}_{5}S$	$^{2}_{3}P$	$^{4}_{3}P$	$^{2}_{1}D$	$^{2}_{3}D$	$^{2}_{5}D$	$^{4}_{5}D$
$^{2}_{5}S$	0	0	-4	$\sqrt{14}$	0	0	0	0
$^{6}_{5}S$	0	0	0	$3\sqrt{10}$	0	0	0	0
$^{2}_{3}P$	4	0	0	0	$-\sqrt{14}$	0	1	$2\sqrt{2}$
$^{4}_{3}P$	$\sqrt{14}$	$3\sqrt{10}$	0	0	-8	0	$-2\sqrt{14}$	$-\sqrt{70}$
$^{2}_{1}D$	0	0	$\sqrt{14}$	-8	0	$-\sqrt{35}$	0	0
$^{2}_{3}D$	0	0	0	0	$-\sqrt{35}$	0	$\sqrt{10}$	$-4\sqrt{5}$
$^{2}_{5}D$	0	0	-1	$-2\sqrt{14}$	0	$\sqrt{10}$	0	0
$^{4}_{5}D$	0	0	$2\sqrt{2}$	$\sqrt{70}$	0	$4\sqrt{5}$	0	0
$^{2}_{3}F$	0	0	0	0	$-2\sqrt{14}$	0	-8	$4\sqrt{2}$
$^{2}_{5}F$	0	0	0	0	0	$2\sqrt{10}$	0	0
$^{4}_{3}F$	0	0	0	0	$-2\sqrt{14}$	0	-2	$4\sqrt{5}$
$^{2}_{3}G$	0	0	0	0	0	0	0	0
$^{2}_{5}G$	0	0	0	0	0	0	0	0
$^{4}_{5}G$	0	0	0	0	0	0	0	0
$^{2}_{3}H$	0	0	0	0	0	0	0	0
$^{2}_{5}I$	0	0	0	0	0	0	0	0

The first sum on the right is a sum over the closed shell $\mathcal{S}$, given by 9^422, and the second sum is ϵ times the average interaction energy Av(a, a) given in 9^423. These are common to all diagonal elements and do not depend on the state A. Only the third term on the right depends on the particular state A of the $\mathcal{R}$ configuration, with the range of summation from 1 to ϵ as desired.

The only other term, in (1), that needs to be simplified can be written as

$$\sum_{i=\epsilon+1}^{m}\sum_{j=1}^{\eta} W(a_i, b_j) = \sum_{i=1}^{m}\sum_{j=1}^{\eta} W(a_i, b_j) - \sum_{i=1}^{\epsilon}\sum_{j=1}^{\eta} W(a_i, b_j) \qquad (3)$$

of which the first sum on the right is given by η times the average interaction energy, Av(a, b). It is common to all the diagonal elements. The second term on the right is the only term that depends on a given state. Therefore, the part of the diagonal term, $(A|G|A)$, which varies within the configuration is

$$W'(a, b) = \sum_{i>j=1}^{\epsilon} W(a_i, a_j) - \sum_{i=1}^{\epsilon}\sum_{j=1}^{\eta} W(a_i, b_j) + \sum_{i>j=1}^{\eta} W(b_i, b_j) \qquad (4)$$

${}^{2}_{3}F$	${}^{2}_{5}F$	${}^{4}_{3}F$	${}^{2}_{3}G$	${}^{2}_{5}G$	${}^{4}_{5}G$	${}^{2}_{3}H$	${}^{2}_{5}I$
0	0	0	0	0	0	0	0
0	0	0	0	0	0	0	0
0	0	0	0	0	0	0	0
0	0	0	0	0	0	0	0
$2\sqrt{14}$	0	$-2\sqrt{14}$	0	0	0	0	0
0	$-2\sqrt{10}$	0	0	0	0	0	0
8	0	-2	0	0	0	0	0
$4\sqrt{2}$	0	$-4\sqrt{5}$	0	0	0	0	0
0	$-\frac{1}{2}\sqrt{70}$	0	0	$-\frac{1}{2}\sqrt{66}$	$5\sqrt{6}$	0	0
$-\frac{1}{2}\sqrt{70}$	0	$-\sqrt{70}$	$\frac{9}{2}\sqrt{2}$	0	0	0	0
0	$\sqrt{70}$	0	0	$-\sqrt{66}$	$\sqrt{15}$	0	0
0	$-\frac{3}{2}\sqrt{2}$	0	0	$-\frac{3}{2}\sqrt{22}$	$-3\sqrt{2}$	0	0
$\frac{1}{2}\sqrt{66}$	0	$-\sqrt{66}$	$-\frac{3}{2}\sqrt{22}$	0	0	$-4\sqrt{3}$	0
$5\sqrt{6}$	0	$2\sqrt{15}$	$3\sqrt{2}$	0	0	$-2\sqrt{33}$	0
0	0	0	0	$4\sqrt{3}$	$-2\sqrt{33}$	0	$-3\sqrt{13}$
0	0	0	0	0	0	$3\sqrt{13}$	0

Table 11^5. $(d^5 5SL \,|70V^{(12)}\|\, d^5 5S'L')$

	${}^{2}_{5}S$	${}^{6}_{5}S$	${}^{2}_{5}D$	${}^{4}_{5}D$	${}^{2}_{5}F$	${}^{2}_{5}G$	${}^{4}_{5}G$	${}^{2}_{5}I$
${}^{2}_{5}S$	0	0	$-4\sqrt{210}$	$-6\sqrt{105}$	0	0	0	0
${}^{6}_{5}S$	0	0	0	$70\sqrt{3}$	0	0	0	0
${}^{2}_{5}D$	$-4\sqrt{210}$	0	$15\sqrt{6}$	$60\sqrt{3}$	$-20\sqrt{21}$	-4	$-20\sqrt{15}$	0
${}^{4}_{5}D$	$6\sqrt{105}$	$-70\sqrt{3}$	$-60\sqrt{3}$	$10\sqrt{15}$	0	$8\sqrt{330}$	$-40\sqrt{3}$	0
${}^{2}_{5}F$	0	0	$20\sqrt{21}$	0	-105	$\sqrt{1155}$	$14\sqrt{105}$	0
${}^{2}_{5}G$	0	0	$-4\sqrt{165}$	$-8\sqrt{330}$	$-\sqrt{1155}$	$\frac{125}{11}\sqrt{33}$	$-10\sqrt{3}$	$\frac{8}{11}\sqrt{30030}$
${}^{4}_{5}G$	0	0	$20\sqrt{15}$	$-40\sqrt{3}$	$14\sqrt{105}$	$10\sqrt{3}$	$-10\sqrt{330}$	$-2\sqrt{2730}$
${}^{2}_{5}I$	0	0	0	0	0	$\frac{8}{11}\sqrt{30030}$	$2\sqrt{2730}$	$-\frac{35}{11}\sqrt{858}$

The first term is the hole–hole Coulomb interaction, the second is the electron–hole interaction, and the third is the electron–electron interaction. Finally, the Russell-Saunders term energies relative to the configuration average are

$$E(SL) = W'(ab, SL) - \frac{\epsilon(\epsilon - 1)}{2}\,\mathrm{Av}(a, a) + \epsilon\eta\,\mathrm{Av}(a, b) - \frac{\eta(\eta - 1)}{2}\,\mathrm{Av}(b, b) \tag{5}$$

where $W'(ab, SL)$'s are the term energies, found by properly summing the diagonal elements (4), relative to the total closed shell energy computed as if the $\mathcal{S}$ shell were complete.

Hence diagonal elements are computed by using the quantum numbers of the missing electrons, or holes, of shell $\mathcal{S}$ exactly as for a *simple* configuration except that the sign of the interaction of the holes with an electron in another open shell is reversed. In obtaining the term energies the M and M_S values of the quantum state A are given by the negatives of the m and m_s values for the holes plus the m_b and m_{sb} values for the electrons of the other open shell.

Application of (4) and (5) can be illustrated by taking the p^5d configuration as an example. In the notation $(m_a^{m_{sa}}\, m_b^{m_{sb}})$ the labels m_a and m_{sa} now refer to a hole instead of an electron. In (4), $\epsilon = 1$ and $\eta = 1$, so that the first and the last terms do not contribute. By the diagonal sum rule, the term energies, in terms of the elements of the electrostatic interaction matrix of the p hole with the d electron, in the zero-order scheme, are

$$\begin{aligned}
W'(p^5d; {}^3F) = {}^3F &: (\bar{1}^-, 2^+)\\
{}^1F + {}^3F &: (\bar{1}^-, 2^-) + (\bar{1}^+, 2^+)\\
{}^3D + {}^3F &: (\bar{1}^-, 1^+) + (0^-, 2^+)\\
{}^1D + {}^3D + {}^1F + {}^3F &: (\bar{1}^-, 1^-) + (0^-, 2^-) + (\bar{1}^+, 1^+) + (0^+, 2^+)\\
{}^3P + {}^3D + {}^3F &: (\bar{1}^-, 0^+) + (0^-, 1^+) + (1^-, 2^+)\\
{}^1P + {}^3P + {}^1D + {}^3D + {}^1F + {}^3F &: (\bar{1}^-, \bar{1}^-) + (0^-, 0^-) + (1^-, 1^-) + (\bar{1}^+, \bar{1}^+)\\
&\quad + (0^+, 0^+) + (1^+, 1^+)
\end{aligned}$$

By using relations $9^4 12$, $9^4 13$, $9^4 14$, and (4) and Table 6^4 we obtain

$$\begin{aligned}
W'(p^5d; {}^3F) = {}^3F &= -F^0(pd) - \frac{2}{35}F^2(pd)\\
{}^1F + {}^3F &= -2F^0(pd) - \frac{4}{35}F^2(pd) + \frac{18}{49}G^3(pd)\\
{}^3D + {}^3F &= -2F^0(pd) + \frac{1}{7}F^2(pd)
\end{aligned}$$

$$^1D + {}^3D + {}^1F + {}^3F = -4F^0(pd) + \frac{2}{7}F^2(pd) + \frac{18}{49}G^3(pd)$$

$$^3P + {}^3D + {}^3F = -3F^0(pd) - \frac{2}{35}F^2(pd)$$

$$^1P + {}^3P + {}^1D + {}^3D + {}^1F + {}^3F = -6F^0(pd) - \frac{4}{35}F^2(pd) + \frac{4}{3}G^1(pd) + \frac{18}{49}G^3(pd)$$

And, finally,

$$W'(p^5d;\, {}^3F) = {}^3F = -F^0(pd) - \frac{2}{35}F^2(pd)$$

$$^1F = -F^0(pd) - \frac{2}{35}F^2(pd) + \frac{18}{49}G^3(pd)$$

$$^3D = -F^0(pd) + \frac{1}{5}F^2(pd)$$

$$^1D = -F^0(pd) + \frac{1}{5}F^2(pd)$$

$$^3P = -F^0(pd) - \frac{1}{5}F^2(pd)$$

$$^1P = -F^0(pd) - \frac{1}{5}F^2(pd) + \frac{4}{3}G^1(pd)$$

By (5), the Russell-Saunders term energies relative to the configuration average,

$$\mathrm{Av}(pd) = F^0(pd) - \frac{1}{15}G^1(pd) - \frac{3}{70}G^3(pd)$$

are

$$E(p^5d;\, {}^3F) = {}^3F = -\frac{2}{35}F^2(pd) - \frac{1}{15}G^1(pd) - \frac{3}{70}G^3(pd)$$

$$^1F = -\frac{2}{35}F^2(pd) - \frac{1}{15}G^1(pd) + \frac{159}{490}G^3(pd)$$

$$^3D = \frac{1}{5}F^2(pd) - \frac{1}{15}G^1(pd) - \frac{3}{70}G^3(pd)$$

$$^1D = \frac{1}{5}F^2(pd) - \frac{1}{15}G^1(pd) - \frac{3}{70}G^3(pd)$$

$$^3P = -\frac{1}{5}F^2(pd) - \frac{1}{15}G^1(pd) - \frac{3}{70}G^3(pd)$$

$$^1P = -\frac{1}{5}F^2(pd) + \frac{19}{15}G^1(pd) - \frac{3}{70}G^3(pd)$$

In the same way the electrostatic energies of the p^5p' configuration, correlated to two nonequivalent-electron configuration, are

$$W'(p^5p';{}^3D) = {}^3D = -F^0(pp') - \frac{1}{25}F^2(pp')$$

$${}^1D + {}^3D = -2F^0(pp') - \frac{2}{25}F^2(pp') + \frac{12}{25}G^2(pp')$$

$${}^3P + {}^3D = -2F^0(pp') + \frac{4}{25}F^2(pp')$$

$${}^1P + {}^3P + {}^1D + {}^3D = -4F^0(pp') + \frac{8}{25}F^2(pp') + \frac{12}{25}G^2(pp')$$

$${}^3S + {}^3P + {}^3D = -3F^0(pp') - \frac{6}{25}F^2(pp')$$

$${}^1S + {}^3S + {}^1P + {}^3P + {}^1D + {}^3D = -6F^0(pp') - \frac{12}{25}F^2(pp') + 6G^0(pp') + \frac{12}{25}G^2(pp')$$

and

$$W'(p^5p';{}^3D) = {}^3D = -F^0(pp') - \frac{1}{25}F^2(pp')$$

$${}^1D = -F^0(pp') - \frac{1}{25}F^2(pp') + \frac{12}{25}G^2(pp')$$

$${}^3P = -F^0(pp') + \frac{1}{5}F^2(pp')$$

$${}^1P = -F^0(pp') + \frac{1}{5}F^2(pp')$$

$${}^3S = -F^0(pp') - \frac{2}{5}F^2(pp')$$

$${}^1S = -F^0(pp') - \frac{2}{5}F^2(pp') + 6G^0(pp')$$

The term energies relative to the configuration average, which is

$$\mathrm{Av}(pp') = F^0(pp') - \frac{1}{6}G^0(pp') - \frac{1}{15}G^2(pp')$$

are

$$E(p^5p';{}^3D) = {}^3D = W'(p^5p';{}^3D) - \mathrm{Av}(pp')$$
$$= -\frac{1}{25}F^2(pp') - \frac{1}{6}G^0(pp') - \frac{1}{15}G^2(pp')$$

$${}^1D = -\frac{1}{25}F^2(pp') - \frac{1}{6}G^0(pp') + \frac{31}{75}G^2(pp')$$

$$^3P = \frac{1}{5}F^2(pp') - \frac{1}{6}G^0(pp') - \frac{1}{15}G^2(pp')$$

$$^1P = \frac{1}{5}F^2(pp') - \frac{1}{6}G^0(pp') - \frac{1}{15}G^2(pp')$$

$$^3S = -\frac{2}{5}F^2(pp') - \frac{1}{6}G^0(pp') - \frac{1}{15}G^2(pp')$$

$$^1S = -\frac{2}{5}F^2(pp') + \frac{35}{6}G^0(pp') - \frac{1}{15}G^2(pp')$$

These general statements may be made with regard to one-hole and one-electron configurations:

Coulomb interaction energies of two configurations $a^{m-1}b$ and $b^{m'-1}a$, where $m = 4l_a + 2$ and $m' = 4l_b + 2$, correlated to the two-electron configuration ab, are the same because the states of the $b^{m'-1}a$ configuration, in the nlm quantization, may be obtained from those for the $a^{m-1}b$ configuration by replacing the state $\{-m_a^{-m_{sa}}, m_b^{m_{sb}}\}$ by $\{-m_b^{-m_{sb}}, m_a^{m_{sa}}\}$. From the relations $9^4 12$, $9^4 12'$, $9^4 13$, and $9^4 13'$ and the properties of $3j$ symbols (Sec. 3^3),

$$\begin{aligned} J(m_a, m_{sa}; m_b, m_{sb}) &= J(-m_a, -m_{sa}; -m_b, -m_{sb}) \\ K(m_a, m_{sa}; m_b, m_{sb}) &= K(-m_a, -m_{sa}; -m_b, -m_{sb}) \end{aligned} \tag{6}$$

indicating that, by $9^4 14$, the diagonal Coulomb energy matrix elements are unchanged when the signs of the magnetic quantum numbers, m and m_s, are reversed; that is,

$$(m_a^{m_{sa}}, m_b^{m_{sb}}) = (-m_a^{-m_{sa}}, -m_b^{-m_{sb}}) \tag{7}$$

The term energies, which are appropriate sums of the diagonal matrix elements, are the same for the one-hole and one-electron configurations $a_h b$ or $b_h a$, where the subscript h stands for hole.

Because $J(a, b)$ depends on the absolute values of the magnetic quantum numbers, m and m_s, the terms in the Coulomb interaction energies involving F^k integrals for the $a^{m-1}b$ configuration will be, by (4), the negatives of those for the ab configuration. Because $K(a, b)$ vanishes unless $m_{sa} = m_{sb}$, the G^k integrals appear only for the states having $M_S = 0$, singlets; none of the triplets has G^k integrals when their energy is expressed relative to the energy of the closed shells. Then for each singlet, at most only one G^k integral has a nonvanishing coefficient, namely, that with $k = L$. Because, by $9^4 13'$, for even $l_a + l_b$ the coefficients of the G^k integrals with odd k all vanish, and for odd $l_a + l_b$ the coefficients of G^k integrals with even k vanish, every other singlet has the same energy as the corresponding triplet.

When the term energies are expressed relative to the average of the one-hole and one-electron configuration, as in (5), the first two of these statements about the appearance of the G^k integrals do not hold. But the last statement, namely, the fact that every other Russell-Saunders singlet has the same Coulomb energy as the corresponding triplet, is still valid.

Two configurations are called *conjugated* if both have the same closed and open shells and if the number of electrons that one configuration contains in its corresponding open shells equals the number of holes in the other. Thus if one of the configurations has $q(a)$, $q(b)$, $q(c)$, ... and the other $4l_a + 2 - q(a)$, $4l_b + 2 - q(\mathrm{b})$, $4l_c + 2 - q(c)$, ... electrons in the $a, b, c, \ldots$ open shells, then these two configurations are conjugated. The relation between the conjugate term energies is found by setting $\eta = 0$ in (4):

$$W'(a, a) = \sum_{i>j=1}^{\epsilon} W_h(a_i, a_j)$$

because, by (7),

$$\begin{aligned} W(a_i, a_j) &= W(-a_i, -a_j) = W_h(a_i, a_j) \\ W'(a, a) &= \sum_{i>j=1}^{\epsilon} W_h(a_i, a_j) = \sum_{i>j=1}^{\epsilon} W(a_i, a_j) \end{aligned} \tag{4'}$$

using a subscript h for holes. Therefore, the part of the diagonal element that varies within the configuration is the same for the conjugated a^{ϵ} and $a^{4l_a+2-\epsilon}$ configurations, so that the conjugate terms have, up to a constant, the same energies. This statement is true, in general, for any such configurations. For instance, the $a^{\epsilon}b$ and $a^{4l_a+2-\epsilon}b$ configurations have the same term energies when $l_b = 0$. As a special case, the singlet–triplet separation for the configurations p^5s, d^9s, $f^{13}s$ is equal to those of the ps, ds, and fs configurations, respectively. These statements are also true when the term energies are expressed relative to their configuration averages. The conjugate configurations however have different configuration averages as given by 9^425.

Quite often the wave function of a shell with one electron missing, in the nlm quantization, is needed to couple it with the other functions, in a defined way, to obtain the total wave function of the system in the desired coupling scheme. In the foregoing it was stated, without proof, that it suffices to change the sign of the magnetic quantum numbers to obtain the zero order states for missing electrons, or holes. It is convenient to replace the spin part of the one electron wave function in (1^42b), that is, $\delta(m_{si}, S_j)$ by $\chi(+)$ and $\chi(-)$ for the spin-up and spin-down states, respectively. We correlate up spin, $m_s = 1/2$, with odd values of i and down spin, $m_s = -1/2$, with even values of i. To do so we need to make the distinction between even and odd i's. This is accomplished by writing $2q + 1$ for the odd and $2q + 2$ for the even values of i. It is conventional to order the one-electron states, in a specified shell, such that the largest positive value of the angular magnetic quantum number; that is, $m = l$, is associated with $i = 1$, and then m decreases with increasing values of i by reaching the largest negative value, $m = -l$, for the upper limit of i. This is done by replacing m with $l - q$; that is,

$$m = l - q$$

so that for the one-electron wave functions we have

$$\begin{aligned} \phi(i = 2q + 1; j) &= R(nl; r_j)Y(l, l - q; \theta_j, \phi_j)\chi(+) \\ \phi(i = 2q + 2; j) &= R(nl; r_j)Y(l, l - q; \theta_j, \phi_j)\chi(-) \end{aligned} \tag{8}$$

The shift operators (Sec. 1[3]) acting on the above one-electron wave functions give

$$\begin{aligned} &S_+\phi(i, j) = \phi(i - 1, j); \qquad S_-\phi(i, j) = 0 \qquad\qquad i = 2q + 2 \\ &S_+\phi(i, j) = 0; \qquad S_-\phi(i, j) = \phi(i + 1, j) \qquad\qquad i = 2q + 1 \\ &L_+\phi(i, j) = \sqrt{(l - m_i)(l + m_i + 1)}\,\phi(i - 2, j) \\ &L_-\phi(i, j) = \sqrt{(l + m_i)(l - m_i + 1)}\,\phi(i + 2, j) \end{aligned}$$

The wave function for a closed shell may be written, by 7[2]9 and 7[2]2, as

$$\Phi = \frac{1}{\sqrt{n!}} \sum_P (-1)^p P[\phi(1, 1)\phi(2, 2) \ldots \phi(n, n)]$$

In the same manner the wave function of a shell with one electron; that is, the ith one missing is

$$\Phi(i) = \frac{1}{\sqrt{(n - 1)!}} \sum_P (-1)^p P[\phi(1, 1)\phi(2, 2) \ldots \phi(i - 1, i - 1)\phi(i + 1, i) \ldots \phi(n, n - 1)]$$

when the shift operators are applied on the wave function $\phi(i)$:

$$\begin{aligned} &L_Z\Phi(i) = -m_i\Phi(i) \\ &S_z\Phi(i) = (-1)^i \frac{1}{2}\Phi(i) \\ &S_+\Phi(i) = 0; \qquad S_-\Phi(i) = \Phi(i - 1) \qquad\qquad i = 2q + 2 \\ &S_+\Phi(i) = \Phi(i + 1); \qquad S_-\Phi(i) = 0 \qquad\qquad i = 2q + 1 \\ &L_+\Phi(i) = -\sqrt{[l + (-m_i) + 1][l - (-m_i)]}\,\Phi(i + 2) \end{aligned}$$

Therefore, the wave function of a shell with one electron missing may be written, Möller [9], in the nlm quantization as

$$\Phi(i) = (-1)^{m_i}|l, -m_i, 1/2, (-1)^i\, 1/2\rangle \tag{9}$$

This wave function may be coupled to the wave function for a single electron to obtain the state of the one-hole and one-electron configuration in the LSM quantization, obtaining:

$$|SM_SLM) = \sum_{m_h, m} \sum_{m_{sh}, m_s} (-1)^{m_h} (1/2, 1/2, -m_{sh}, m_s | SM_s) \times (l_h, l, -m_h, m | LM) \{-m_h^{-m_{sh}}, m^{m_s}\} \tag{10}$$

where the subscript h stands for hole, and quantum labels without a subscript refer to the electron. The above relation may be rewritten, by $3^3 4$, in terms of $3j$ symbols as

$$|SM_SLM) = \sqrt{(2L+1)(2S+1)} \sum_{m_h m} \sum_{m_{sh} m_s} (-1)^{l_h - l + M + M_s + m_h} \times \begin{pmatrix} 1/2 & 1/2 & S \\ -m_{sh} & m_s & -M_s \end{pmatrix} \begin{pmatrix} l_h & l & L \\ -m_h & m & -M \end{pmatrix} \{-m_h^{-m_{sh}}, m^{m_s}\} \tag{10'}$$

This is the LSM-coupled wave function corresponding to the one given in $2^5 3$ for the two-electron configurations.

The term energies are found in the same way as for the two-electron configuration (Sec. 2^5). These are

$$\begin{aligned} W'(SL) &= \left(SM_SLM \left| \sum_{i>j} \frac{e^2}{r_{ij}} \right| S'M_S'L'M' \right) \\ &= \delta(L, L')\delta(S, S')\delta(M, M')\delta(M_S, M_S') \\ &\quad \times \left[-\sum_k f_k F^k(l_h, l) + \delta(S, 0) \sum_k g_k G^k(l_h, l) \right] \end{aligned} \tag{11}$$

where

$$\begin{aligned} f_k &= (-1)^L (2l_h + 1)(2l + 1) \begin{pmatrix} l_h & l_h & k \\ 0 & 0 & 0 \end{pmatrix} \begin{pmatrix} l & l & k \\ 0 & 0 & 0 \end{pmatrix} \begin{Bmatrix} l_h & l & L \\ l & l_h & k \end{Bmatrix} \\ g_k &= \frac{2}{2L+1} (2l_h + 1)(2l + 1) \begin{pmatrix} l_h & l & L \\ 0 & 0 & 0 \end{pmatrix}^2 \delta(L, k) \end{aligned} \tag{12}$$

These equations agree with the statements made previously for the one-hole and one-electron configurations. The energies of the Russell-Saunders terms relative to the configuration average are

$$E(SL) = W'(SL) + \mathrm{Av}(l_h, l) \tag{13}$$

where $\mathrm{Av}(l_h, l)$ may be obtained from $9^4 21$ replacing l_a, l_b by l_h, l, respectively.

The elements of the Coulomb energy matrix for the two-equivalent-hole and one-electron configuration are

$$\begin{aligned} &W'[l_h^2 l; (S_{hh}L_{hh})SL: (S'_{hh}L'_{hh})S'L'] \\ &= \left(l_h^2 l; (S_{hh}L_{hh})SL \left| \sum_{i,j}{}' \frac{e^2}{r_{ij}} \right| l_h^2 l; (S'_{12}L'_{12})S'L' \right) \\ &= \delta(L, L')\delta(S, S')\delta(M, M')\delta(M_S, M_S') 2 \left[-\sum_k f_k F^k(l_h, l) + \sum_k g_k G^k(l_h, l) \right] \end{aligned} \tag{14}$$

where

$$f_k = (-1)^{L_{hh}+L'_{hh}+l_h+L}\delta(S_{hh}, S'_{hh})(2l_h + 1)(2l + 1) \times [(2L_{hh} + 1)(2L'_{hh} + 1)]^{1/2} \begin{pmatrix} l_h & l_h & k \\ 0 & 0 & 0 \end{pmatrix} \begin{pmatrix} l & l & k \\ 0 & 0 & 0 \end{pmatrix} \times \begin{Bmatrix} L_{hh} & L'_{hh} & k \\ l & l & L \end{Bmatrix} \begin{Bmatrix} L_{hh} & L'_{hh} & k \\ l_h & l_h & l_h \end{Bmatrix} \tag{15}$$

and

$$g_k = (-1)^{S_{hh}+S'_{hh}}\delta(S, 1/2)(1/2)(2l_h + 1)(2l + 1) \times [(2L_{hh} + 1)(2S_{hh} + 1)(2L'_{hh} + 1)(2S'_{hh} + 1)]^{1/2} \times \begin{pmatrix} l_h & l & k \\ 0 & 0 & 0 \end{pmatrix}^2 \begin{pmatrix} l_h & l_h & L_{hh} \\ l & L & k \end{pmatrix} \begin{Bmatrix} l_h & l_h & L'_{hh} \\ l & L & k \end{Bmatrix} \tag{15$'$}$$

The prime on the summation in (14) indicates that the hole–hole Coulomb interaction is excluded. When the matrix elements are expressed relative to the configuration average, we have

$$\begin{aligned} E\,[(S_{hh}L_{hh})SL] &= E[l_h^2l; (S_{hh}L_{hh})SL: (S_{hh}L_{hh})SL] \\ &= E[l_h^2; S_{hh}L_{hh}] \\ &\quad + W'[l_h^2l; (S_{hh}L_{hh})SL: (S_{hh}L_{hh})SL] \\ &\quad + 2\,\mathrm{Av}(l_h, l) \end{aligned} \tag{16}$$

and

$$E[l_h^2 l; (S_{hh} L_{hh})SL: (S'_{hh}L'_{hh})SL] = W'[l_h^2l; (S_{hh}L_{hh})SL: (S'_{hh}L'_{hh})SL] \tag{16$'$}$$

where $E[l_h^2; S_{hh}L_{hh}]$ may be obtained from 2⁵7 and 2⁵5′ by replacing l_a, L, and S by l_h, L_{hh}, and S_{hh}, respectively.

The d^8p configuration is used as an example of the application of (14), (15), and (16). For the $(^3P)^2S$ term $l_h = 2$, $l = 1$; $L_{hh} = S_{hh} = L'_{hh} = S'_{hh} = 1$ and $L = 0$, $S = 1/2$, so that

$$\begin{aligned} &W'[d^8p; (^3P)^2S: (^3P)^2S] \\ &= W'[(^3P)^2S] \\ &= 2\left\{-5 \times 3 \times \sqrt{3 \times 3}\left[\begin{pmatrix} 2 & 2 & 0 \\ 0 & 0 & 0 \end{pmatrix} \begin{pmatrix} 1 & 1 & 0 \\ 0 & 0 & 0 \end{pmatrix} \begin{Bmatrix} 1 & 1 & 0 \\ 1 & 1 & 0 \end{Bmatrix} \begin{Bmatrix} 1 & 1 & 0 \\ 2 & 2 & 2 \end{Bmatrix}\right.\right. \\ &\quad \left. \times F^0(dp) + \begin{pmatrix} 2 & 2 & 2 \\ 0 & 0 & 0 \end{pmatrix} \begin{pmatrix} 1 & 1 & 2 \\ 0 & 0 & 0 \end{pmatrix} \begin{Bmatrix} 1 & 1 & 2 \\ 1 & 1 & 0 \end{Bmatrix} \begin{Bmatrix} 1 & 1 & 2 \\ 2 & 2 & 2 \end{Bmatrix} F^2(dp)\right] \\ &\quad + \frac{1}{2} \times 5 \times 3 \times \sqrt{3 \times 3 \times 3 \times 3}\left[\begin{pmatrix} 2 & 1 & 1 \\ 0 & 0 & 0 \end{pmatrix}^2 \begin{Bmatrix} 2 & 2 & 1 \\ 1 & 0 & 1 \end{Bmatrix} \begin{Bmatrix} 2 & 2 & 1 \\ 1 & 0 & 1 \end{Bmatrix}\right. \\ &\quad \left.\left. \times G^1(dp) + \begin{pmatrix} 2 & 1 & 3 \\ 0 & 0 & 0 \end{pmatrix}^2 \begin{Bmatrix} 2 & 2 & 1 \\ 1 & 0 & 3 \end{Bmatrix} \begin{Bmatrix} 2 & 2 & 1 \\ 1 & 0 & 3 \end{Bmatrix} G^3(dp)\right]\right\} \\ &= -2F^0(dp) + \frac{2}{5}F^2(dp) \end{aligned}$$

Similarly:

$$W[d^8p;\,(^3P)^4S\colon(^3P)^4S] = W'[(^3P)^4S] = -2F^0(dp) + \frac{2}{5}F^2(dp) = W'[(^3P)^2S]$$

$$W'[(^3P)^4P] = -2F^0(dp) - \frac{1}{5}F^2(dp)$$

$$W'[(^3F)^4F] = -2F^0(dp) + \frac{3}{35}F^2(dp)$$

$$W'[(^3F)^4G] = -2F^0(dp) - \frac{1}{35}F^2(dp)$$

$$W'[(^1G)^2H] = -2F^0(dp) - \frac{4}{35}F^2(dp) + \frac{9}{49}G^3(dp)$$

$$W'[(^1S)^2P] = -2F^0(dp) + \frac{2}{15}G^1(dp) + \frac{3}{35}G^3(dp)$$

$$W'[(^3P)^2P] = -2F^0(dp) - \frac{1}{5}F^2(dp) + \frac{9}{10}G^1(dp) + \frac{9}{35}G^3(dp)$$

$$W'[(^1D)^2P] = -2F^0(dp) + \frac{3}{35}F^2(dp) + \frac{7}{30}G^1(dp) + \frac{3}{245}G^3(dp)$$

$$W'[d^8p;\,(^1S)^2P\colon(^3P)^2P] = \sqrt{3}\left[\frac{1}{5}G^1(dp) - \frac{3}{35}G^3(dp)\right]$$

$$W'[d^8p;\,(^1S)^2P\colon(^1D)^2P] = \sqrt{7}\times\left[-\frac{4}{35}F^2(dp) + \frac{1}{15}G^1(dp) + \frac{3}{245}G^3(dp)\right]$$

$$W'[d^8p;\,(^3P)^2P\colon(^1D)^2P] = \sqrt{21}\left[\frac{1}{10}G^1(dp) - \frac{3}{245}G^3(dp)\right]$$

$$W'[(^3P)^2D] = -2F^0(dp) + \frac{1}{25}F^2(dp) + \frac{9}{50}G^1(dp) + \frac{54}{175}G^3(dp)$$

$$W'[(^1D)^2D] = -2F^0(dp) - \frac{3}{35}F^2(dp) + \frac{7}{30}G^1(dp) + \frac{18}{245}G^3(dp)$$

$$W'[(^3F)^2D] = 2F^0(dp) - \frac{12}{175}F^2(dp) + \frac{28}{25}G^1(dp) + \frac{27}{1225}G^3(dp)$$

$$W'[d^8p;\,(^3P)^2D\colon(^1D)^2D] = \frac{\sqrt{21}}{\sqrt{5}}\left[\frac{1}{10}G^1(dp) - \frac{18}{245}G^3(dp)\right]$$

$$W'[d^8p;\,(^3P)^2D\colon(^3F)^2D] = \frac{\sqrt{14}}{5}\left[-\frac{12}{35}F^2(dp) + \frac{3}{5}G^1(dp) + \frac{27}{245}G^3(dp)\right]$$

$$W'[d^8p;\,(^1D)^2D\colon(^3F)^2D] = \frac{\sqrt{6}}{\sqrt{5}}\left[\frac{7}{15}G^1(dp) - \frac{9}{245}G^3(dp)\right]$$

$$W'[(^3P)^4D] = -2F^0(dp) + \frac{1}{25}F^2(dp)$$

$$W'[(^3F)^4D] = -2F^0(dp) - \frac{12}{175}F^2(dp)$$

$$W'[d^8p;\,(^3P)^4D{:}\,(^3F)^4D] = -\frac{12}{175}\sqrt{14}\,F^2(dp)$$

$$W'[(^1D)^2F] = -2F^0(dp) + \frac{6}{245}F^2(dp) + \frac{2}{105}G^1(dp) + \frac{216}{1715}G^3(dp)$$

$$W'[(^3F)^2F] = -2F^0(dp) + \frac{3}{35}F^2(dp) + \frac{2}{5}G^1(dp) + \frac{81}{490}G^3(dp)$$

$$W'[(^1G)^2F] = -2F^0(dp) - \frac{11}{49}F^2(dp) + \frac{18}{35}G^1(dp) + \frac{9}{3430}G^3(dp)$$

$$W'[d^8p;\,(^1D)^2F{:}\,(^3F)^2F] = \frac{\sqrt{3}}{\sqrt{7}}\left[\frac{2}{15}G^1(dp) - \frac{54}{245}G^3(dp)\right]$$

$$W'[d^8p;\,(^1D)^2F{:}\,(^1G)^2F] = \frac{\sqrt{3}}{7}\left[-\frac{24}{35}F^2(dp) + \frac{2}{5}G^1(dp) + \frac{18}{245}G^3(dp)\right]$$

$$W'[d^8p;\,(^3F)^2F{:}\,(^1G)^2F] = \frac{1}{\sqrt{7}}\left[\frac{6}{5}G^1(dp) - \frac{27}{490}G^3(dp)\right]$$

$$W'[(^3F)^2G] = -2F^0(dp) - \frac{1}{35}F^2(dp) + \frac{45}{98}G^3(dp)$$

$$W'[(^1G)^2G] = -2F^0(dp) + \frac{11}{35}F^2(dp) + \frac{3}{98}G^3(dp)$$

$$W'[d^8p;\,(^3F)^2G{:}\,(^1G)^2G] = -\frac{3}{98}\sqrt{15}\,G^3(dp)$$

Because

$$2\,\mathrm{Av}(dp) = 2F^0(dp) - \frac{2}{15}G^1(dp) - \frac{3}{35}G^3(dp)$$

the elements of the Coulomb interaction energy matrix for the d^8p configuration may be expressed relative to the configuration average, by (16), and are given in Appendix 3.

The relation between the coefficients of fractional parentage for the terms of a more than half-filled shell a^{4l_a+2-n} and those for the terms of a^{n+1} is given by Racah [1c] as

$$\begin{aligned}(a^{4l_a+2-n};\,\alpha SL\{|\,a^{4l_a+1-n}\,\alpha' S'L',\,a) &= (-1)^{L+S+L'+S'-l_a-1/2}\left[\frac{(n+1)(2L'+1)(2S'+1)}{(4l_a+2-n)(2L+1)(2S+1)}\right]^{1/2} \\ &\quad\times (a^{n+1};\,\alpha' S'L'\,\{|\,a^n\alpha SL,\,a)\end{aligned} \qquad (17)$$

Therefore, with the use of the above equation, the values for the $(d^7;\,\alpha' S'L'\{|d^6\alpha SL,\,d)$ and the $(d^6;\,\alpha' S'L'\{|d^5\alpha SL,\,d)$ may be found from the tabulated $(d^4;\,\alpha SL\{|d^3\alpha' S'L',\,d)$ and $(d^5;\,\alpha SL\{|d^4\alpha' S'L',\,d)$ values, in Tables 2^5b and 2^5c, respectively.

Reduced-matrix elements of the double tensor operator $\mathbf{T}^{(kr)}$ for the conjugate configurations are related [1b] as follows:

$$(a^n; \alpha SL \| T^{(kr)} \| a^n; \alpha' S' L') \\ = (-1)^{k+r+1} (a^{4l_a+2-n}; \alpha SL \| T^{(kr)} \| a^{4l_a+2-n}; \alpha' S' L') \quad (18)$$

Hence reduced-matrix elements of the tensor operators $\mathbf{U}^{(1)}$, $\mathbf{U}^{(3)}$, . . . ; $\mathbf{V}^{(12)}$ $\mathbf{V}^{(14)}$, . . . for the a^{4l_a+2-n} and for the a^n configurations are the same. On the other hand, reduced-matrix elements of the operators $\mathbf{U}^{(2)}$, $\mathbf{U}^{(4)}$, . . . ; $\mathbf{V}^{(13)}$, $\mathbf{V}^{(15)}$ . . . for the a^{4l_a+2-n} configurations are the negative of those for the a^n configurations. Therefore, as a third alternative, the term energies for the a^{4l_a+2-n} and the $a^{4l_a+2-n}b$ configurations may be found by using the values of the reduced-matrix elements of the operators $\mathbf{U}^{(k)}$ and $\mathbf{V}^{(kr)}$ for the a^n configurations and fixing their signs properly, by (18), in the relations, given in Sections 10^5 and 11^5 for the term energies of the a^n and a^nb configurations, respectively.

Finally, for the scalar operator $\mathbf{T}^{(00)}$

$$(a^{4l_a+2-n}; \alpha SL \| T^{(00)} \| a^{4l_a+2-n}; \alpha' SL) \\ = \frac{4l_a + 2 - n}{n} (a^n; \alpha SL \| T^{(00)} \| a^n; \alpha' SL) \quad (19)$$

6. Group-theoretical methods

> It is now natural for us to try to derive the laws of nature and to test their validity by means of the laws of invariance, rather than to derive the laws of invariance from what we believe to be the laws of nature.
>
> E. P. Wigner, *Symmetries and Reflections* (Bloomington: Indiana University Press, 1967), p. 5.

The basic way in which the theory of groups enters quantum mechanics was indicated in Sections 6^2 and 7^2. Much of the content of Chapter 3 is an application of the theory of the rotation group, although there it was not explicitly developed in this way.

When TAS was being written nearly four decades ago, it was decided, purely on pedagogical grounds, not to use group-theoretic methods. This was done not because of any lack of appreciation of their significance, or power, or beauty. On this we find the following statement (p. 11):

> When a physicist is desirous of learning of new theoretical developments in this subject, one of the greatest barriers is that it generally involves new mathematical techniques with which he is apt to be unfamiliar. Relativity theory brought the necessity of learning tensor calculus and Riemannian geometry. Quantum mechanics forces him to a more careful study of boundary value problems and matrix algebra. Hence if we can minimize the amount of new mathematics he must learn in order to penetrate a new field we do him a real service. . . . Many things which are done here could be done more simply if the theory of groups were part of the ordinary mathematical equipment of physicists. But as it is not, it seems like putting unnecessary obstacles in the way to treat the subject by a method which requires this equipment. . . .
>
> This does not mean that we underestimate the value of group theory for atomic physics nor that we feel that physicists should omit the study of that branch of mathematics now that it has been shown to be an important tool in the new theory.

Historically the situation here is analogous to that which obtained in the late nineteenth century in the treatment particularly of electrodynamics. This was always done by Cartesian component methods, rather than by the methods of vector analysis. In Maxwell's great treatise (3rd ed., p. 9) we read:

The introduction of coordinate axes into geometry by Descartes was one of the greatest steps in mathematical progress, for it reduced the methods of geometry to calculations performed on numerical quantities. . . . But for many purposes of physical reasoning, as distinguished from calculations, it is desirable to avoid explicitly introducing the Cartesian coordinates, and to fix the mind at once on a point of space instead of its three coordinates, and on the magnitude and direction of a force instead of its three components. . . . As the methods of Descartes are still the most familiar to students of science, and as they are really the most useful for purposes of calculation, we shall express all our results in the Cartesian form. I am convinced, however, that the introduction of the ideas, as distinguished from the operations and methods of Quaternions, will be of great use to us in the study of all parts of our subject, and especially in electrodynamics, where we have to deal with a number of physical quantities, the relations of which to each other can be expressed far more simply by a few expressions of Hamilton's, than by the ordinary equations.

This resembles the reason given in TAS for not using the theory of groups. This style continued for decades. For example, the widely used Jeans's *Electricity and Magnetism*, published in 1920, made no use of vector analysis. There followed a period in which the books on electrodynamics used vector methods but felt obliged to introduce a chapter on vector analysis, because the authors knew that most physicists had not studied the subject. Finally, newer books on electrodynamics began to appear in which knowledge of vector analysis was presupposed.

We are in a similar middle period with respect to group theory in relation to atomic structure. TAS made no explicit use of group theory. This work does so, but we feel it necessary to expound the main ideas of the subject briefly in this chapter, because the theory of groups is still not widely studied by physicists. Perhaps future books on atomic structure will presuppose knowledge of the elements of group theory.

In this chapter we present an outline of the theory of groups [1, 2] followed, in Chapter 7, by the application of these methods to the more complex configurations as studied first by Racah and later principally by Judd and by Wybourne. The same methods also find wide application in the shell theory of the atomic nucleus.

1^6. Abstract group theory

Enough has been said already about groups to make it desirable now to introduce the subject in terms of abstract definitions that can later be illustrated and applied by giving specific examples.

A *group* G consists of a set of *elements* $a, b, c, \ldots, x$, for which a binary rule of composition called multiplication is defined, such that ab and ba are members of the set. The rule of composition is associative, so that

$$(ab)c = a(bc) \tag{1}$$

Repeated application of this rule shows that a product of more elements, as $abc \ldots h$, need not be written with parentheses, because the result is the same no matter how the partial products are formed. Thus

$$abc \ldots h = (ab)(c \ldots h) = (abc)(d \ldots h)$$

in all possible ways, as long as the order of the factors is preserved. The group is called *commutative* or *Abelian* if $ab = ba$ for every pair of elements.

The set also includes a unit element e (Einheit), which has the property that its compositions with any other element, x, leave that element unaltered:

$$xe = ex = x \tag{2}$$

To each element in the group there exists a reciprocal element, x^{-1}, with the property that

$$xx^{-1} = x^{-1}x = e \tag{3}$$

A *finite group* consists of a finite number, h, of elements, in which case h is called the *order* of the group. If the number of elements is infinite, it may be denumerably infinite, or the elements may require r continuous real parameters for their identification, in which case the group is called an r-parameter *continuous* group. Groups that need an infinite number of parameters for their specification are called *infinite-parameter* groups.

A group is completely specified by giving its *multiplication table,* that is, by telling which element corresponds to ab for each a and each b. For a finite group this can be shown by a double entry table in which the element ab is entered in the ath row and the bth column. For an r-parameter continuous group an element a is labeled by parameters $a_1 \ldots a_r$, and the product element ab is labeled by $(ab)_1(ab)_2 \ldots (ab)_r$. The multiplication table then requires the specification of r functions of the parameters of a and b giving each $(ab)_s$ as functions of the $2r$ parameters $a_1 \ldots a_r$ and $b_1 \ldots b_r$,

$$(ab)_s = \phi_s(\mathrm{a}_1 \ldots \mathrm{a}_r; \mathrm{b}_1 \ldots \mathrm{b}_r) \tag{4}$$

We write $\bar{a}_s$ for the parameters which specify the reciprocal a^{-1} of a, so that

$$e_s = \phi_s(a_1 \ldots a_r; \bar{a}_1 \ldots \bar{a}_r) \tag{5}$$

If, in addition, the functions ϕ_s in (4), and the $\bar{a}_t$ as functions of the a_s defined by (5) are *analytic* functions of their arguments, the r-parameter continuous group is called an r-parameter *Lie group.*

It is convenient to visualize a space called the *group manifold* in which the elements of G are associated with points. For a finite group it is any set of h points in one-to-one correspondence with the h elements. For an r-parameter continuous group the manifold consists of a region of r-dimensional space. This region may consist of several disconnected "pieces" and of regions of lower dimensionality than r, including isolated points. The group manifold is *closed* when it contains all its limit points.

A group is called *compact* when its parameter space is bounded and closed. For example, the one-parameter Lorentz group can be parametrized by the number v/c, which is bounded but not closed (because v cannot equal c). Therefore the Lorentz group is noncompact. Compact subsets of a manifold are closed.

The labeling of the elements of a group is arbitrary. For a finite group of order h it is natural to use the integers 1 to h to label elements, but these can be associated with the individual elements in any way. Instead of these integers, any other h distinguishable marks may be used.

Two groups, G and G', are said to be *isomorphic*, $G \approx G'$, if it is possible to set each element x of G in a one-to-one correspondence with an element x' of G', in such a way that $x'y'$ corresponds to xy for each pair (x, y) or (x', y'); that is, $(xy)' = x'y'$ for all x and y.

If one or several elements in G correspond to the *same* element in G', but products are preserved, then the correspondence or mapping from G to G' is called a *homomorphism*. In such a relation the identity e of G is associated with e', because $ae = a$ implies $(ae)' = a'e' = a'$ so e' is the identity in G'. Suppose that the set of elements $(a_1, a_2, \ldots, a_k)$ (one of which is e) in G is associated with e' in G', then the elements $(ba_1, ba_2, \ldots, ba_k)$ are all associated with b' in G' for $(ba_s)' = b'e' = b'$, so that equal numbers (k) of elements of G are associated with each element of G'.

In the group G we can associate with each element, x, another element, uxu^{-1}, where u is a fixed element of the group. Such an association of elements of a group is called an *automorphism* of G.

A specific geometrical or physical interpretation of the elements of an abstract group G, is called a *realization* of G. The groups of interest in applications are those which originate in terms of a realization, the structure of the group being determined by the geometric or physical properties of the realization.

Every abstract group G has a faithful realization, called the *regular realization,* in the group manifold itself obtained by associating with the element a the translations from x to x', defined by

$$x' = ax \qquad (6)$$

where x is each element in G.

Of particular importance are the realizations of G by means of linear transformations in an n-dimensional vector space. In such a space a vector $\mathbf{x}$ is given by its components $(x_1, x_2 \ldots, x_n)$ in a particular basis.

The group element a is then associated with a transformation from vector $\mathbf{x}$ to $a\mathbf{x}$ by the linear transformation

$$(ax)_\mu = \sum_{\nu=1}^{n} D_{\mu\nu}(a)x_\nu; \qquad \mu = 1, 2, \ldots, n \tag{7}$$

in which the element a is represented by the $n \times n$ matrix, $D(a)$ whose elements are $D_{\mu\nu}(a)$. Such a realization is called a *representation of dimension n.* The determination of the representations of a given group is one of the most important problems in the entire theory.

An important concept is that of a group G, which is formed as the *direct product* of two groups, G' and G'', written

$$G = G' \times G'' \tag{8}$$

Denoting the elements of G' by $a', b' \ldots$ and of G'' by $a'', b'', \ldots$, the elements of G are defined as all possible pairs (a', b'') with the law of combination

$$(a', b'')(a''', b^{\text{iv}}) = (a'a''', b''b^{\text{iv}}) \tag{9}$$

of any element of G' with any element of G''. If G' and G'' are finite of order h' and h'', respectively, then the order of G is $h = h'h''$. This is easily generalized to define the direct product $G_1 \times G_2 \times \ldots \times G_r$ of r groups. Let s_k and t_k be arbitrary elements of the group G_k. Then an element of the product group is labeled by the ordered set $(s_1 \ldots s_r)$ with the law of combination

$$(t_1 \ldots t_r)(s_1 \ldots s_r) = (t_1 s_1 \ldots t_r s_r)$$

The order of the direct product is $h_1 h_2 \ldots h_r$. Similarly, if G' and G'' are continuous groups, the group manifold of G is the product space of the group manifolds of G' and G''. The rule of composition for G is defined as

$$(a', a'')(b', b'') = (a'b', a''b'') \tag{10}$$

In particular the identity in G is (e', e'') and the element reciprocal to (a', a'') is

$$(a', a'')^{-1} = (a'^{-1}, a''^{-1}) \tag{11}$$

If H consists of a subset of elements of G, which itself forms a group under the *same* rule of composition as holds for G, then H is called a *subgroup* of G. Two trivial cases are called *improper subgroups:* that in which H consists only of e, the trivial finite group of order 1, and that in which H contains all the elements of G. All other subgroups are called *proper subgroups*. An element of G may belong to more than one subgroup, in particular e belongs to all of the subgroups.

Lagrange's theorem says that the order h' of a subgroup H of a finite group is a divisor of the order h of the group G. This is immediately evident for the two improper subgroups of order 1 and h, respectively. Consider the set of elements, aH, where a is any element of G and H is the set of elements of the subgroup H. Such a set for $a \neq e$ is not a subgroup of G: it is called a *left coset* of H in G. If a is in H, then the set aH is the same as the set H because H is a subgroup. If a is not in H, then the products ah_1, ah_2, . . . are all different, because $ah_i = ah_j$ implies $h_i = h_j$; and none of them is contained in H, because $ah_i = h_j$ implies $a = h_j h_i^{-1}$ which is in H, contrary to hypothesis. Continuing, we may choose an element b, which is not in H or aH; then members of the set bH are not in H or aH, by the argument just given. In this way all of the h elements of G are exhibited in the sets $H, aH, bH, cH, \ldots$ in which each element c is not in any of the sets that precede it. Each set contains h' elements and therefore h is an integral multiple of h'. The integer h/h' is called the *index* of H in G. In a similar way, the elements of G can be exhibited as a set of *right cosets*, He, Ha', Hb',. . . .

Symbolically we write

$$G = H + aH + bH + cH + \ldots \tag{12}$$

where addition of sets means the set of all elements contained in some summand such that there is no element common to all elements in the summand sets.

The *order n of an element a* (as contrasted with the order of a group) is the least positive value of n for which

$$a^n = e \tag{13}$$

If the a^n are different for all n the element is of infinite order. Every element of finite order defines an Abelian group of order n, called a *cyclic* group, whose elements are

$$e, a, a^2, \ldots, a^{n-1} \tag{14}$$

Therefore, by Lagrange's theorem, all of the elements of a finite group G are of finite order, each order being a divisor of h. It follows that in case h is *prime*, the only finite group of order h is the cyclic group of order h.

Elements in G are called *conjugate* to one another; that is, b is conjugate to a if an element u in G exists such that

$$b = uau^{-1} \tag{15}$$

from which it is easily seen that if a is conjugate to b, and if c is conjugate to b, then c is conjugate to a because

$$c = vbv^{-1} = (vu)a(vu)^{-1}$$

and vu is in G if v and u are in G.

All the elements of G that are conjugate to each other are said to belong to the same *class*. The elements of G can thus be separated into sets of *conjugate classes.*

If u is a particular element of G and H is a subgroup of G, then the set of elements uHu^{-1}, where H stands for each element of the subgroup, is also a subgroup of G, called a *conjugate subgroup* of H in G. Different conjugate subgroups of H in G are obtained by making different choices of u. If uHu^{-1} is the same as H for all u, then H is called an *invariant subgroup* in G. Otherwise expressed, H is an invariant subgroup in G if $uH = Hu$ for all u; that is, the right and left cosets of H formed with any u are equal.

If H is an invariant subgroup in G, then for any element h in H, all elements in the same class as h are in H; that is, H consists of complete classes in G.

The set Z of elements of the group G, which commute with every element of G, is called the *center* of G. Z is also an invariant subgroup of G.

A group is called *simple* if it has no proper invariant subgroups. A group is called *semisimple* if none of its invariant subgroups is Abelian.

A class is called *ambivalent* if it contains with every element its reciprocal as well [4^2b, p. 78; 4^2c, p. 146]. If all the classes of a group are ambivalent, and if the direct product of any two irreducible representations contains no representation more than once, the group is called *simply reducible* [4^2c, p. 152].

The product of two cosets uH and vH of an invariant subgroup H in G is again a coset, because

$$(uH)(vH) = u(Hv)H = (uv)H$$

By the product of two sets A and B is meant the set of elements produced by combining any element of A with any element of B.

The coset $eH = H$ has the property that

$$H(uH) = uH$$

so eH as a coset may be regarded as the identity element. Moreover, the coset $u^{-1}H$, when multiplied by uH, gives H and so the coset $u^{-1}H$ is the reciprocal of uH. Thus the cosets uH may be regarded as group elements with coset multiplication as the rule of composition, in which case the cosets form a group. This group is called the *factor group* of H in G, written G/H, and having the order h/h' (for finite groups), the index of H in G.

2^6. Group representations

The idea of group representation [3] was introduced in Section 6^2 and again in 1^67. Each representation consists of a set of h $n \times n$ matrices

$D(g_i)$ which obey the group multiplication law and therefore constitute a homomorphism of G. If every $D(g_i)$ is different, they constitute an isomorphism of G, and the representation is called *faithful.*

It is customary to designate different representations by Γ_α, Γ_β, ... with the corresponding matrices denoted by $D^{(\alpha)}(g_i)$, $D^{(\beta)}(g_i)$, ... Every group has a trivial representation of dimension $n = 1$, in which each element is represented by the unity. Conventionally this is Γ_1,

$$D^{(1)}(g_i) = I \tag{1}$$

Two representations Γ_α and Γ_β are called *equivalent* if they are of the same dimension n, and if there exists a matrix A such that for all i,

$$D^{(\beta)}(g_i) = AD^{(\alpha)}(g_i)A^{-1} \tag{2}$$

in which A need not represent a group element. When acting on an n vector $\mathbf{X}$, A may be regarded as giving the components of the same vector in a transformed basis, in which case $D^{(\beta)}(g_i)$ is the form of the operator for the element g_i appropriate to the new basis. In other words, $D^{(\beta)}(g_i)$ acting on $A\mathbf{X}$ gives the components of $D^{(\alpha)}(g_i)$ in the new basis. This is sometimes called a *similarity* transformation. Applied to quantum mechanical operators it was called a *canonical* transformation in 1²29.

If a matrix A can be found that transforms each $D(g_i)$ into the form

$$D(g_i) = \begin{bmatrix} D^{(1)}(g_i) & D_{12}(g_i) & D_{13}(g_i) & D_{14}(g_i) & \cdots \\ 0 & D^{(2)}(g_i) & D_{23}(g_i) & D_{24}(g_i) & \cdots \\ 0 & 0 & D^{(3)}(g_i) & D_{34}(g_i) & \cdots \\ \cdots & \cdots & \cdots & \cdots & \cdots \end{bmatrix} \tag{3}$$

the steps $D^{(\alpha)}(g_i)$ having dimensions n_α whose sum over α is n, the representation is called *reducible.* If no such A can be found, the representation is *irreducible.* If it is possible to find a transformation matrix A, such that it transforms each $D(g_i)$ into step-matrix form, that is, makes all off-diagonal matrices $D_{mn}(g_i) = 0$, then the representation is called *fully reducible.* A basis in which all of the $D(g_i)$ have the form (3) is said to be *adapted* to the reduction. The step form implies that if a vector $\mathbf{X}$ lies wholly in any one of the subspaces, then $D(g_i)\mathbf{X}$ does also. By carrying out the reduction as far as possible, each step consists of the matrices of irreducible representations, some of which may be repeated. Thus every reducible representation may be indicated symbolically by

$$\Gamma = \sum_\alpha q_\alpha \Gamma_\alpha \quad \text{and} \quad n = \sum_\alpha q_\alpha n_\alpha \tag{4}$$

the sum extending over all irreducible representations and the q_α being nonnegative integers that give the number of repetitions of Γ_α

occurring in each $D(g_i)$. The matrices $D(g_i)$ are taken to be nonsingular.

When the h elements $g_i (i = 1, 2, \ldots, h)$ of the group G are multipled from the left by any element g_j of the same group (because $g_j g_i = g_k$), the initial order of the elements is changed with no element remaining in its original position, unless the element chosen, g_j, happens to be the identity e. If we regard h elements in the initial order as a row matrix, each multiplication of all elements by another element of the same group may be represented by an $h \times h$ transformation matrix. When the element that multiplies the others is the identity e, the matrix becomes the $h \times h$ unit matrix I_h. For all the other elements these matrices have a single unit element in every row and column and zeros everywhere else; in particular, all diagonal elements are zero. The set of these $h \times h$ matrices is called the *regular representation.*

An important result, sometimes called the *Schur-Auerbach theorem* is that for compact groups an A in (2) can always be found that transforms each $D(g_i)$ to unitary form. Thus attention can be restricted to representations by unitary matrices, which will be assumed from now on.

Another result of great importance, known as *Schur's lemma* [4²b, p. 75], says that any matrix M that commutes with each of the $D(g_i)$ of an irreducible representation is a multiple of the unit matrix. It is usually used in converse form: If a matrix M can be found, which is not a multiple of I, and which commutes with each $D(g_i)$, the representation is reducible.

With respect to the matrices $D^{(\alpha)}(g_i)$ and $D^{(\beta)}(g_i)$ of two representations of dimensions n_α and n_β, respectively, let M be an $n_\beta \times n_\alpha$ matrix such that for all i,

$$MD^{(\alpha)}(g_i) = D^{(\beta)}(g_i)M \tag{5}$$

then [4²b, p. 77], for $n_\beta \neq n_\alpha$, $M = 0$ is the only solution, and for $n_\beta = n_\alpha$, $M = 0$ satisfies (5), but there are nonsingular solutions for M, in which case the representations Γ_β and Γ_α are equivalent.

Orthogonality relations [4²b, p. 79] hold between the matrix elements of irreducible inequivalent representations:

$$\sum_{i=1}^{h} D_{\mu\nu}^{(\alpha)}(g_i)^* D_{\rho\tau}^{(\beta)}(g_i) = \frac{h}{n}\,\delta_{\alpha\beta}\delta_{\mu\rho}\delta_{\nu\tau};\ n = n_\alpha = n_\beta \tag{6}$$

In (6) there appear products of individual matrix elements, not matrix multiplication.

Matrix elements $D_{\mu\nu}^{(\alpha)}(g_i)$ for fixed (α, μ, ν) and $i = 1 \ldots h$ may be regarded as components of a vector in an h-dimensional space. Then (6) shows that these vectors are mutually orthogonal. For each representation of dimension n_α there are n_α^2 such vectors. The vectors completely span the space [4²b, p. 115], and

$$\sum_\alpha n_\alpha^2 = h \tag{7}$$

in which the sum is over the inequivalent irreducible representations, Γ_α.

A. Direct product of representations

Corresponding to the concept of direct product of two groups (1⁶8), the *direct* or *Kronecker* product of two square matrices D^α of dimension n and D^β of dimension m is defined as a matrix of dimension nm whose structure is

$$D^{(\alpha)} \times D^{(\beta)} = \begin{bmatrix} D_{11}^{(\alpha)}D^{(\beta)} & \cdots & D_{1n}^{(\alpha)}D^{(\beta)} \\ \vdots & \cdots & \vdots \\ D_{n1}^{(\alpha)}D^{(\beta)} & \cdots & D_{nn}^{(\alpha)}D^{(\beta)} \end{bmatrix} \tag{8}$$

and

$$D_{\mu\nu}^{(\alpha)} \times D^{(\beta)} = \begin{bmatrix} D_{\mu\nu}^{(\alpha)}D_{11}^{(\beta)} & \cdots & D_{\mu\nu}^{(\alpha)}D_{1m}^{(\beta)} \\ \vdots & \cdots & \vdots \\ D_{\mu\nu}^{(\alpha)}D_{m1}^{(\beta)} & \cdots & D_{\mu\nu}^{(\alpha)}D_{mm}^{(\beta)} \end{bmatrix}$$

Direct products have the properties

$$\begin{aligned} D^{(\alpha)} \times (D^{(\beta)} \times D^{(\gamma)}) &= (D^{(\alpha)} \times D^{(\beta)}) \times D^{(\gamma)} \\ (D^{(\alpha)} \times D^{(\beta)})(D^{(\gamma)} \times D^{(\delta)}) &= (D^{(\alpha)}D^{(\gamma)}) \times (D^{(\beta)}D^{(\delta)}) \end{aligned} \tag{9}$$

for nonsingular matrices

$$(D^{(\alpha)} \times D^{(\beta)})^{-1} = (D^{(\alpha)})^{-1} \times (D^{(\beta)})^{-1} \tag{10}$$

and for unitary matrices

$$\begin{aligned} (D^{(\alpha)} \times D^{(\beta)}) \times (D^{(\alpha)} \times D^{(\beta)})^\dagger &= I \\ (D^{(\alpha)} \times D^{(\beta)})^\dagger &= (D^{(\alpha)})^\dagger \times (D^{(\beta)})^\dagger \end{aligned} \tag{11}$$

The set of matrices $D^{(\alpha)}(g_i) \times D^{(\beta)}(g_i)$ corresponding to the representations Γ_α and Γ_β of G constitute a representation of G that is denoted by $\Gamma_\alpha \times \Gamma_\beta$. In case $D^{(\alpha)}(g_i)$ is a representation of G_1 and $D^{(\beta)}(g_j)$ of G_2, where G_1 and G_2 are two different groups then the $D^{(\alpha)}(g_i) \times D^{(\beta)}(g_j)$ matrices give a representation of $G_1 \times G_2$ of dimension nm.

The idea of direct product relates to that of factor group. If $G = G_1 \times G_2$, then $G_1 = G/G_2$ and $G_2 = G/G_1$ so G_1 and G_2 are factor groups of G.

In order to infer that $G = G_1 \times G_2$, one has to know both the relations $G_1 = G/G_2$ and $G_2 = G/G_1$.

Starting with G_1 and G_2 such that $G_2 = G/G_1$, G is called the *extension* of G_1 by G_2. If G_2 is isomorphic to subgroup H of G, the extension is called *inessential,* and if G_2 is isomorphic to an invariant subgroup of G, the extension is called *trivial*. When the extension is inessential, G is called the *semidirect* product of G_1 and G_2 written

$$G = G_1 \wedge G_2, \quad \text{also written} \quad G_1 \boxed{S} G_2 \quad \text{or} \quad G_1 \times G_2 \qquad (12)$$

B. Integer and half-integer representations

The conditions under which a given irreducible representation of the group G can be brought to real form are important. For finite groups, as well as for the compact simple Lie groups, the representations are unitary; that is, in terms of the representation matrices

$$D^\dagger(g)D(g) = 1 \quad \text{or, equally,} \quad \tilde{D}^{-1}(g = D^*(g)$$

where $\tilde{D}(g)$ is the transpose of matrix $D(g)$ obtained by interchanging rows and columns and $\tilde{D}^{-1}(g)$ is its inverse. The additional condition that the representations are real,

$$D_{ij}(g) = D^*_{ij}(g)$$

implies that the *character,* that is, the set of matrix element consisting of the traces of the matrices $D(g)$ for a given representation (see Sec. 3^6), is also real. Conversely, the representations associated with real characters are either real or equivalent to their complex conjugate. On the other hand, if the character of a representation denoted here by $\chi(g)$ is complex, such a representation can not be equivalent to its complex conjugate. Therefore, the irreducible representations of a group G may be divided as follows:

1. The character $\chi(g)$ is real: (a) $D(g)$ is real or can be brought to real form; or (b) $D(g)$ is equivalent to its complex conjugate but cannot be brought to real form.
2. The character $\chi(g)$ is complex: $D(g)$ is not equivalent to its complex conjugate.

In cases 1(a) and 1(b),

$$D(g) \approx \tilde{D}^{-1}(g)$$

and

$$\chi(g) = \chi(g^{-1})$$

This implies

$$\tilde{D}^{-1}(g) = SD(g)S^{-1} \qquad (13)$$

where S is a nonsingular unitary matrix,

$$S = \tilde{D}SD$$

That is, the matrix S is invariant under all transformations of the type

$$X' = D(g)X$$

Further investigation shows that [4²c, pp. 140–1]

$$S^{-1}S = qI \qquad q = \pm 1 \tag{14}$$

where I is a unit matrix and that furthermore for case 2 S must be a null matrix. Therefore

$$\tilde{S} = qS \tag{15}$$

The cases $q = 1$, $q = -1$, and $q = 0$ imply that the representations associated with the matrix S by (13), are of the type 1(a), 1(b), and 2, respectively.

Similar criteria may be given [4²c, p. 142] in terms of the characters χ^{α} *as*

$$\sum_{G} \chi^{q}(g^2) = q^{\alpha} h \tag{16}$$

where the sum over G implies that all group elements g must be considered; the cases $q^{\alpha} = 1$, $q^{\alpha} = -1$, and $q^{\alpha} = 0$ indicate that the representation matrices $D^{(\alpha)}$ are of the type 1(a), 1(b), and 2, respectively.

The representations of type 1(a) are called the *integer representations,* and the representations of type 1(b) are called *half integer representations.*

Let $\eta(s)$ be the number of solutions of the equation

$$g^2 = s$$

Combining this with (16),

$$\sum_{G} \eta(s)\chi^{\alpha}(s) = q^{\alpha} h$$

which results [4²c, p. 143] in

$$\eta(s) = \sum_{\alpha} q^{\alpha}\chi^{\alpha}(s) \tag{17}$$

The number of solutions $\eta(s)$ of the equation $g^2 = s$ are then given by the above equation. If $\chi^{\alpha}(s)$ is complex, then $q^{\alpha} = 0$; if $\chi^{\alpha}(s)$ is real, then $q^{\alpha} = 1$ or $q^{\alpha} = -1$, depending on whether the representation matrix $D^{(\alpha)}$ is or is not equivalent to a real representation matrix.

C. Classical Lie groups

The Lie groups associated with the algebras that were classified originally by Cartan [4] came to be known as *classical Lie groups* and found wide application in physics.

1. The *full linear groups, $GL(n)$*, have as elements all nonsingular $n \times n$ matrices, D, with matrix elements in the field of complex numbers, hence $GL(n)$ contains $2n^2$ parameters. Sometimes this group is designated as $GL(n, C)$, in contrast with $GL(n, R)$ in which the matrix elements are real, so that there are n^2 parameters.

2. The *unimodular groups, $SL(n)$*, or, more explicitly, $SL(n, C)$ and $SL(n, R)$, are subgroups of $GL(n, C)$ and $GL(n, R)$, respectively, in which the determinant of each matrix equals $+1$. S derives from the name "special" linear group.

3. The *unitary groups, $U(n)$*, have as elements all $n \times n$ unitary matrices; that is, $D^\dagger D = I$.

4. The *special unitary groups, $SU(n)$*, are subgroups of the $U(n)$ in which the determinant of each unitary matrix equals $+1$.

5. The *orthogonal groups, $O(n)$*, or, more explicitly $O(n, C)$ and $O(n, R)$ are those for which $\tilde{D}D = 1$, in which $\tilde{D}$ is written for the transpose of D. In most applications $O(n)$ means $O(n, R)$ unless the group is specifically called *complex orthogonal* when it refers to $O(n, C)$.

6. The *unimodular orthogonal groups, $SO(n)$* are the subgroups of $O(n)$ for which the determinant of each matrix equals $+1$. $SO(n)$ is sometimes written $O^+(n)$.

Pseudo-unitary, pseudo-orthogonal, and symplectic groups are also used in physics:

7. The *pseudo-unitary groups, $U(p, q)$* are those whose matrices satisfy the restriction

$$D^\dagger \beta D = I \quad \text{in which} \quad \beta = \begin{bmatrix} I_p & 0 \\ 0 & -I_q \end{bmatrix} \quad q = \text{even}$$

where I_p is the $p \times p$ unit matrix and I_q is the $q \times q$ unit matrix and $p + q = n$.

8. The *unimodular pseudo-unitary groups, $SU(p,q)$* are the subgroups of $U(p,q)$ in which each matrix has a determinant equal to $+1$.

9. The *pseudo-orthogonal groups, $O(p, q)$*, are those whose matrices satisfy

$$\tilde{D}\beta D = \beta$$

with β as in 7 and $\tilde{D}$ the transpose of D.

10. The *unimodular pseudo-orthogonal groups, $SO(p, q)$* are the subgroups of 9 with the determinant of each matrix equal to $+1$.

11. The *symplectic groups, Sp*$(2n, C)$ or $SP(2n, R)$ are made up of $2n \times 2n$ matrices with complex or real elements, respectively, in which the matrices satisfy

$$D\dagger\gamma D = \gamma$$

in which

$$\gamma = \begin{bmatrix} 0 & I_n \\ -I_n & 0 \end{bmatrix}$$

Other specialized subgroups of $GL(n, C)$ are discussed in [5].

3⁶. Group characters

The trace of $D(g_i)$ is called the *characteristic* of g_i in the representation Γ of G, and is denoted by $\chi(g_i)$,

$$\chi(g_i) = \sum_{\mu=1}^{n} D_{\mu\mu}(g_i) \tag{1}$$

The set of characteristics of the h matrices $g_i \ldots g_h$ is called the *character* of the representation [6].

The characteristic of g_i is invariant to a similarity transformation $D(g_i) \to SD(g_i)S^{-1}$, as noted in 1²38 in another context. Therefore, all equivalent representations have the same character.

Likewise, all group elements in the same class have the same characteristic. If the group consists of r classes, C_i, in which the class C_i contains h_i elements, and χ_i is the characteristic of each of its elements, the representation Γ, and all representations that are equivalent to it, has a character that is specified by giving the r values of the class characteristics, χ_i, with $i = 1, \ldots, r$.

The character of a reducible representation is called *compound;* that of each of the irreducible representations is called *primitive.* Literature on group theory proves that the number of inequivalent irreducible representations is equal to the number of classes.

As in 2⁶4, a general representation is represented symbolically as

$$\Gamma = \sum_{\alpha=1}^{r} q_\alpha \Gamma_\alpha \tag{2}$$

in which each q_α is a nonnegative integer. It follows that the character of Γ is given by

$$\chi_i = \sum_{\alpha=1}^{r} q_\alpha \chi_i^{(\alpha)}; \qquad i = 1, 2 \ldots, r \tag{3}$$

with the same q's as in (2).

Thus G can be described by an r by r *character table* of the $\chi_i^{(\alpha)}$, which is usually presented by listing the irreducible representations α in rows and the class characteristics in columns.

From $2^6 6$ we have, for unitary inequivalent irreducible representations,

$$\sum_{i=1}^{h} \chi^{(\alpha)}(g_i)^* \chi^{(\beta)}(g_i) = \sum_i \left[\sum_\mu D^{(\alpha)}_{\mu\mu}(g_i)\right]^* \left[\sum_\sigma D^{(\beta)}_{\sigma\sigma}(g_i)\right]$$
$$= \frac{h}{n}\,\delta_{\alpha\beta}\delta_{\mu\sigma}; \qquad n = n_\alpha = n_\beta \tag{4}$$

The sum on the left can be written as a sum over the r class characteristics of a character, giving the orthogonality property

$$\sum_{i=1}^{r} h_i (\chi_i^{(\alpha)})^* \chi_i^{(\beta)} = h\delta_{\alpha\beta} \tag{5}$$

Thus in an r-dimensional space one set of r orthogonal axes can be associated with the classes, and another set with the irreducible representations, interpreting (5) as indicating that $\sqrt{h_i/h}\,\chi_i^{(\alpha)}$ gives the component along the ith class axis of the unit vector representing the αth representation. This shows that the character table also has an orthogonality property by columns

$$\sum_{\alpha=1}^{r} (\chi_i^{(\alpha)})^* \chi_j^{(\alpha)} = \frac{h}{h_i}\,\delta_{ij} \tag{6}$$

The numbering of the classes and of the representations is arbitrary. Usually Γ_1 is taken to represent the trivial one-dimensional representation in which each g_i is represented by unity, for which each class characteristic is therefore unity. The first row of the character table then consists of ones and (5) gives, for $\beta \neq 1$

$$\sum_{i=1}^{r} h_i \chi_i^{(\beta)} = 0 \tag{7}$$

for the other rows. Similarly, C_1 is usually chosen to be the class ($h_i = 1$) that contains only the identity, so that each $\chi_1^{(\alpha)}$ is equal to n_α, the dimension of each of the representations, which fully determines the first column of the character table.

Applying (6) to C_1 we find

$$\sum_{\alpha=1}^{r} n_\alpha^2 = h \tag{8}$$

For an Abelian group each element is in a class by itself; therefore, from (8), $n_\alpha = 1$ for each of them.

Multiplying (3) by h_i times the complex conjugate of (3) gives, because the q_α are real,

$$\sum_i \frac{h_i}{h} \chi_i^* \chi_i = \sum_{\alpha=j}^{r} q_\alpha^2 \tag{9}$$

This relation provides a criterion of reducibility for the representation whose character is given by the χ_i: if it is irreducible, all $q_\alpha = 0$ except one of them which is equal to unity. The test can be made even if the character table is not known.

Similarly, suppose (9) is formed for two different possibly reducible representations giving

$$\sum_i \frac{h_i}{h} (\chi_i^{(1)})^* \chi_i^{(2)} = \sum_\alpha q_\alpha^{(1)} q_\alpha^{(2)} \tag{9'}$$

If Γ_1 and Γ_2 have no irreducible representations in common, the right side, and hence the left side, of (9′) vanishes.

When the character table is known, the individual q's can be found: Multiply (3) by $(h_i/h)(\chi_i^{(\beta)})^*$ and sum on i; then, by (5),

$$q_\beta = \sum_{i=1}^{r} \frac{h_i}{h} (\chi_i^{(\beta)})^* \chi_i \tag{10}$$

By the product of two classes, C_iC_j, is meant the set of all products g_ig_j for which g_i belongs to C_i and g_j to C_j. If an element of class C_k occurs in this set, perhaps several times, every element of C_k must occur the same number of times. In consequence the set C_iC_j can be written as the sum of sets C_k with nonnegative integral coefficients,

$$C_iC_j = \sum_{k=1}^{r} h_{ij,k} C_k \tag{11}$$

where the $h_{ij,k}$ are called *class multiplication coefficients*. These are evaluated by direct calculation from the group multiplication table. The complete determination [1h, p. 61] of the character table from the group multiplication table requires that the elements be analyzed into classes. This determines their number, r, and the h_i, the number of elements in each of them, and the class multiplication coefficients, $h_{ij,k}$.

The ith-*class matrix* D_i is defined in any representation as

$$D_i = \sum_{g_i} D(g_i) \tag{12}$$

that is, the sum of the matrices representing the elements g_i in the class C_i. From the definition, it is easily seen that

$$D_i^{(\alpha)} D_j^{(\alpha)} = \sum_{k=1}^{r} h_{ij,k} D_k^{(\alpha)} \tag{13}$$

with the same coefficients as in (11). For each $D^{(\alpha)}(g)$ there is $D^{(\alpha)}(g)^{-1} D_i^{(\alpha)} D^{(\alpha)}(g) = D_i^{(\alpha)}$, because the sum on the left is that in (12) with the summands permuted. Therefore, by Schur's lemma, because $D_i^{(\alpha)}$ commutes with each $D^{(\alpha)}(g)$, it is a multiple of the unit matrix,

$$D_i^{(\alpha)} = \frac{h_i}{n_\alpha} \chi_i^{(\alpha)} I \tag{14}$$

Combining (13) and (14), a relation valid for each irreducible representation is obtained:

$$(h_i \chi_i^{(\alpha)})(h_j \chi_j^{(\alpha)}) = n \sum_{k=1}^{r} h_{ij,k} (h_k \chi_k^{(\alpha)}) \tag{15}$$

which is useful in checking computations of characters.

4⁶. Group algebra

A linear vector space that is closed (that is, in which the product of two vectors belong to the space) under some multiplication law is called an *algebra*. If the multiplication law is taken from the multiplication law of a group, then the algebra is said to be the *group algebra* of that group, G. Because group multiplication is associative, the group algebra is an *associative algebra*.

Any representation of the group automatically gives a representation of the algebra and vice versa. If one of these representations is reducible or irreducible, so is the other.

If a linear vector space B is contained in A and is closed under the law of multiplication of the algebra A, then B is said to be a *subalgebra* of the algebra A.

Let a and b be any elements of the algebra A and of its subalgebra B, respectively. B is called a *left ideal* if it contains ab as an element.

In the regular representation, a left ideal $\mathscr{L}_1$ is an invariant subspace. This is because, for any element a of A, $a\mathscr{L}_1 = \mathscr{L}_1$. On the other hand, because the regular representation is fully reducible, the space A must be a direct sum of left ideals:

$$A = \mathscr{L}_1 + \mathscr{L}_2 \tag{1}$$

where

$$a\mathscr{L}_1 = \mathscr{L}_1 \quad \text{and} \quad a\mathscr{L}_2 = \mathscr{L}_2 \tag{2}$$

Every element of the algebra A is uniquely expressible as the sum of an element in $\mathscr{L}_1$ and an element in $\mathscr{L}_2$. That is, $a = a_1 + a_2$. Only the element zero (the origin) is common to both $\mathscr{L}_1$ and $\mathscr{L}_2$. The matrices $D(a)$ of the regular representation are also reducible.

$$D(a) = D_1(a) + D_2(a) \tag{3}$$

where $D_i(a)$ is the matrix of the linear transformation, which is induced in $\mathscr{L}_i$ by left multiplication with a.

The unit element e of the group G is also an element of the group algebra A. On the other hand,

$$ea = ae = a \tag{4}$$

for all elements of A. When the group algebra is a direct sum of two left ideals, $\mathscr{L}_1$ and $\mathscr{L}_2$, the unit element e may be expressed as

$$e = e_1 + e_2 \tag{5}$$

where e_1 is in $\mathscr{L}_1$ and e_2 is in $\mathscr{L}_2$. Multiplying the above equation on the left by a, and using the fact that $a = a_1 + a_2$, produces

$$a = ae_1 + ae_2 \tag{6a}$$

$$a_1 = ae_1 \text{ and } a_2 = ae_2 \tag{6b}$$

Therefore, ae_1 is in $\mathscr{L}_1$ and ae_2 is in $\mathscr{L}_2$. Furthermore,

$$\text{if } \; ae_1 = 0, \text{ then } \; a_2 = a \qquad (a \text{ is in } \mathscr{L}_2) \tag{7a}$$

$$\text{if } \; ae_2 = 0, \text{ then } \; a_1 = a \qquad (a \text{ is in } \mathscr{L}_1) \tag{7b}$$

If the element a is equal to e_1 or e_2 when it is in $\mathscr{L}_1$ or $\mathscr{L}_2$, respectively, the above equations become

$$e_2e_1 = 0; \qquad e_2^2 = e_2 \tag{8a}$$

$$e_1e_2 = 0; \qquad e_1^2 = e_1 \tag{8b}$$

A matrix D is said to be an *idempotent matrix* if $D^2 = D$. Therefore, the above elements e_1 and e_2 are *idempotents*. Because ae_1 is in $\mathscr{L}_1$ and ae_2 is in $\mathscr{L}_2$ for all a of A, e_1 and e_2 are the *generators* of the left ideals $\mathscr{L}_1$ and $\mathscr{L}_2$, respectively. That is, the resolution of the unit element into its components in $\mathscr{L}_1$ and $\mathscr{L}_2$ (5) produces the generators of these left ideals. If e_1e_2 is equal to e_2e_1 (if e_1 and e_2 commute), then e_1e_2 and e_2e_1 must be zero.

The left ideals of an algebra may contain subalgebras of their own. If these subalgebras are also left ideals, they are called *left subideals*. If a left ideal contains no proper subideal it is said to be a *minimal*. Therefore, a given algebra A may be expressed as a direct sum of minimal left ideals as

$$A = \sum_i \mathscr{L}_i \tag{9}$$

where the left ideal $\mathscr{L}_i$ is generated by the idempotent e_i, which satisfies the general relation

$$e_i e_j = \delta_{ij} e_i \tag{10}$$

The generators e_i of the left ideals $\mathscr{L}_i$ are obtained by the resolution of the unit element e into its components in the linear vector spaces $\mathscr{L}_i$. That is,

$$e = \sum_i e_i \tag{11}$$

where e_i is in $\mathscr{L}_i$.

An idempotent is called *primitive* if it cannot be resolved any further into components that satisfy relation (10). To make the distinction, a primitive idempotent is denoted by the Greek letter ϵ.

The left ideals generated by primitive generators are all minimal. Conversely, any generating unit of a minimal left ideal is primitive.

5^6. Symmetric (permutation) groups S_n

The $n!$ permutations of n numbers or symbols $(i_1, i_2, i_3, \ldots, i_n)$ among themselves form a group of order $h = n!$. The group is called the *symmetric* or *permutation group,* designated by S_n. Study of this group finds application in the quantum mechanics of indistinguishable particles as introduced in Sections 6^2 and 7^2. The identity element e of the group corresponds to no permutation. That is,

$$g_0 = e = \begin{pmatrix} 1 & 2 & 3 & \ldots & n \\ 1 & 2 & 3 & \ldots & n \end{pmatrix} \tag{1}$$

Any other element, denoted by g_i, is given as

$$g_i = \begin{pmatrix} 1 & 2 & 3 & \ldots & n \\ i_1 & i_2 & i_3 & \ldots & i_n \end{pmatrix} \tag{2}$$

where i_k gives the integer occupying the place of k after the permutation. The elements of the group do not, in general, commute. For example,

$$g_1 g_2 = \begin{pmatrix} 1 & 2 & 3 & 4 \\ 3 & 2 & 4 & 1 \end{pmatrix} \begin{pmatrix} 1 & 2 & 3 & 4 \\ 4 & 1 & 2 & 3 \end{pmatrix} = \begin{pmatrix} 1 & 2 & 3 & 4 \\ 1 & 3 & 2 & 4 \end{pmatrix}$$

$$g_2 g_1 = \begin{pmatrix} 1 & 2 & 3 & 4 \\ 4 & 1 & 2 & 3 \end{pmatrix} \begin{pmatrix} 1 & 2 & 3 & 4 \\ 3 & 2 & 4 & 1 \end{pmatrix} = \begin{pmatrix} 1 & 2 & 3 & 4 \\ 2 & 1 & 3 & 4 \end{pmatrix}$$

showing that $g_i g_j \neq g_j g_i$. Multiplication is here defined from the right. In the multiplication, start with a number of the upper row of the last permutation and proceed to the left. For instance, for $g_1 g_2$ start with 1 in the upper row of g_2 in which 1 is permuted to 4 and then 4 is permuted to 1 in g_1. The final result indicates that in the product, g_1g_2 1 remains unpermuted.

The inverse, g_i^{-1}, of an element g_i is given by

$$g_i^{-1} = \begin{pmatrix} i_1 & i_2 & i_3 & \dots & i_n \\ 1 & 2 & 3 & \dots & n \end{pmatrix}$$

For example,

$$g_1^{-1} = \begin{pmatrix} 3 & 2 & 4 & 1 \\ 1 & 2 & 3 & 4 \end{pmatrix} = \begin{pmatrix} 1 & 2 & 3 & 4 \\ 4 & 2 & 1 & 3 \end{pmatrix}$$

and

$$g_2^{-1} = \begin{pmatrix} 4 & 1 & 2 & 3 \\ 1 & 2 & 3 & 4 \end{pmatrix} = \begin{pmatrix} 1 & 2 & 3 & 4 \\ 2 & 3 & 4 & 1 \end{pmatrix}$$

Every finite group can be represented as a subgroup of S_n. This statement is known as *Cayley's theorem,* which states that *every group G of order n is isomorphic with a subgroup of* S_n [3e, pp. 16–7; 4²c, pp. 15–20].

Any permutation may be resolved into *cycles,* that is, permutations that affect only a smaller fixed number of elements. For example,

$$\begin{pmatrix} 1 & 2 & 3 & 4 & 5 & 6 & 7 & 8 & 9 \\ 2 & 3 & 1 & 5 & 4 & 6 & 8 & 7 & 9 \end{pmatrix} = (123)(45)(6)(78)(9)$$

where (123) is an abbreviation for $\begin{pmatrix} 1 & 2 & 3 \\ 2 & 3 & 1 \end{pmatrix}$. A cycle that contains only two numbers is called a *transposition.* Every cycle may be written as a product of transpositions as

$$(1234) = (14)(13)(12)$$

Cycles with no symbols in common commute:

$$(123)(45) = (45)(123)$$

When a permutation is expressed in terms of cycles, unpermuted symbols are usually omitted:

$$(123)(45)(6) = (123)(45)$$

In these cases it is necessary to know the *degree,* which is the number of symbols *n,* of the permutation.

Permutations that leave no symbol unchanged, or unpermuted, are called *regular permutations.* These together with the identity *e* form a *regular* subgroup of S_n. There are *n* possible regular permutations composed of cycles of equal length. Therefore, the subgroup that they form is of order $h' = n$. For instance, in S_4 the identity element *e,*

$$g_2 = \begin{pmatrix} 1 & 2 & 3 & 4 \\ 2 & 3 & 4 & 1 \end{pmatrix} \to (1234),$$

$$g_3 = \begin{pmatrix} 1 & 2 & 3 & 4 \\ 4 & 1 & 2 & 3 \end{pmatrix} \to (1432), \qquad g_4 = \begin{pmatrix} 1 & 2 & 3 & 4 \\ 3 & 4 & 2 & 1 \end{pmatrix} \to (1324)$$

are regular permutations that form a cyclic group of order 4.

The *decrement* of a permutation is defined as the difference between the degree of the permutation and the number of independent cycles into which the given permutation is resolved. A permutation is called *odd* or *even* according to whether its decrement is odd or even. In S_5 the first of the following two permutations is even and the second is odd: (12)(34)(5), (123)(45). The even permutations of degree n form a group called the *alternating group*, A_n, of order $h = n!/2$. Obviously A_n is a subgroup of S_n.

A transformation $gg_1g^{-1} = g_2$ conserves the length of the cycles of g_1. Therefore, the permutations with similar cycle structure form a class. For instance, the group S_3 has the three classes:

$$\begin{aligned} C_1(1) &= e \\ C_2(3) &= (12), (13), (23) \\ C_3(2) &= (123), (132) \end{aligned} \tag{3}$$

whereas the group S_4 has five classes:

$$\begin{aligned} C_1(1) &= e \\ C_2(6) &= (12), (13), (14), (23), (24), (34) \\ C_3(3) &= (12)(34), (13)(24), (14)(23) \\ C_4(8) &= (123), (132), (124), (142), (134), (143), (234), (243) \\ C_5(6) &= (1234), (1243), (1324), (1423), (1432), (1342) \end{aligned} \tag{4}$$

where the order h_i of the ith class is indicated as an argument.

Let α_1 denote the number of one-cycles and α_2 the number of two-cycles, in the independent cycles into which a permutation of degree n is resolved. Then

$$\alpha_1 + 2\alpha_2 + 3\alpha_3 + \ldots + n\alpha_n = n \tag{5}$$

This is verified for the permutation of degree 9:

$$\begin{pmatrix} 1 & 2 & 3 & 4 & 5 & 6 & 7 & 8 & 9 \\ 2 & 3 & 4 & 1 & 6 & 7 & 5 & 9 & 8 \end{pmatrix} \to (1234)(567)(89)$$

where

$$\alpha_1 = 0, \alpha_2 = 1, \alpha_3 = 1, \alpha_4 = 1, \text{and} \quad \alpha_n = 0 \quad \text{for} \quad n \geqslant 5$$

Therefore,

$$(2 \times 1) + (3 \times 1) + (4 \times 1) = 9$$

A class of a given symmetric group may be identified by its *cycle structure* that is designated as $(1^{\alpha_1}\, 2^{\alpha_2}\, 3^{\alpha_3} \ldots n^{\alpha_n})$.

A. Partitions and cycle structure

If n integers λ_n are defined in terms of α_i's, as

$$\begin{aligned} \alpha_1 + \alpha_2 + \alpha_3 + \ldots + \alpha_n &= \lambda_1 \\ \alpha_2 + \alpha_3 + \ldots + \alpha_n &= \lambda_2 \\ \ldots \quad \ldots\ldots & \\ \alpha_n &= \lambda_n \end{aligned} \tag{6}$$

then

$$\lambda_1 + \lambda_2 + \lambda_3 + \ldots + \lambda_n = n \tag{7}$$

and

$$\lambda_1 > \lambda_2 > \lambda_3 > \ldots > \lambda_n \geqslant 0 \tag{8}$$

which corresponds to a partition of n into $[\lambda_1, \lambda_2, \lambda_3, \ldots, \lambda_n]$. Because of the unique correspondence between the partition and the cycle structure, a partition may be used to identify a class C_i or an irreducible representation Γ_α of the group S_n. The number of independent partitions is equal to the number of classes. Partitions are differentiated from cycle structure by writing them in square brackets, when there is risk of confusion.

When partitions are known, the cycle structure can be found as

$$\begin{aligned} \alpha_1 &= \lambda_1 - \lambda_2 \\ \alpha_2 &= \lambda_2 - \lambda_3 \\ &\ldots \\ \alpha_n &= \lambda_n \end{aligned} \tag{9}$$

Then the order h_i of the ith class (the number of elements in the class) is given by

$$h_i = \frac{n!}{\alpha_1!1^{\alpha_1}\alpha_2!2^{\alpha_2} \ldots \alpha_n!n^{\alpha_n}} \tag{10}$$

Two special cases are important in several applications. If $\lambda_1 = n$, ($\lambda_2 = \lambda_3 = \ldots = \lambda_n = 0$), no symbol is permuted and the partition $[n]$ corresponds to the identity element e. The representation identified by the partition $[n]$ is called the *identity representation*. On the other hand, if $\lambda_1 = \lambda_2 = \ldots = \lambda_n$, every symbol is permuted and the representation designated by the partition $[1^n]$ is called the *alternating representation*.

Using the group S_4 as an example, the possible partitions of $n = 4$ subject to (8), are

$$\begin{gathered} [4], [3, 1] = [31], [2, 2] = [2^2], \\ [2, 1, 1] = [2, 1^2], [1, 1, 1, 1] = [1^4] \end{gathered}$$

Applying (9) to each partition, the cycle structures are found as

$$[4] \to \lambda_1 = \alpha_1 = 4 \to (1^4)$$
$$[3, 1] \to \lambda_1 = 3, \lambda_2 = 1; \alpha_1 = 2, \alpha_2 = 1 \to (1^2, 2^1) = (1^2, 2)$$
$$[2, 2] \to \lambda_1 = 2, \lambda_2 = 2; \alpha_1 = 0, \alpha_2 = 2 \to (2^2)$$
$$[2, 1, 1] \to \lambda_1 = 2, \lambda_2 = \lambda_3 = 1; \alpha_1 = 1, \alpha_2 = 0, \alpha_3 = 1 \to (1^1, 3^1) = (1, 3)$$
$$[1, 1, 1, 1] \to \lambda_1 = \lambda_2 = \lambda_3 = \lambda_4 = 1; \alpha_1 = \alpha_2 = \alpha_3 = 0, \alpha_4 = 1 \to (4^1) = (4)$$

Using the values of α_i in (10) the orders for the classes of S_4 are obtained

$$h_1 = 1, h_2 = 6, h_3 = 3, h_4 = 8, h_5 = 6$$

and

$$\sum_{i=1} h_i = 1 + 6 + 3 + 8 + 6 = 24 = 4!$$

as it should be.

B. Representation matrices and character tables

The representations of symmetric groups are easily found. Using the group S_3 as an example and labeling its six elements as,

$$g_1 = e = \begin{pmatrix} 1 & 2 & 3 \\ 1 & 2 & 3 \end{pmatrix} = (1)(2)(3) \qquad g_4 = \begin{pmatrix} 1 & 2 & 3 \\ 1 & 3 & 2 \end{pmatrix} = (1)(23)$$
$$g_2 = \begin{pmatrix} 1 & 2 & 3 \\ 2 & 1 & 3 \end{pmatrix} = (12)(3) \qquad g_5 = \begin{pmatrix} 1 & 2 & 3 \\ 2 & 3 & 1 \end{pmatrix} = (123)$$
$$g_3 = \begin{pmatrix} 1 & 2 & 3 \\ 3 & 2 & 1 \end{pmatrix} = (13)(2) \qquad g_6 = \begin{pmatrix} 1 & 2 & 3 \\ 3 & 1 & 2 \end{pmatrix} = (132)$$

the representation matrices are

$$D(g_1) = D(e) = \begin{pmatrix} 1 & 0 & 0 \\ 0 & 1 & 0 \\ 0 & 0 & 1 \end{pmatrix} \qquad D(g_4) = \begin{pmatrix} 1 & 0 & 0 \\ 0 & 0 & 1 \\ 0 & 1 & 0 \end{pmatrix}$$
$$D(g_2) = \begin{pmatrix} 0 & 1 & 0 \\ 1 & 0 & 0 \\ 0 & 0 & 1 \end{pmatrix} \qquad D(g_5) = \begin{pmatrix} 0 & 0 & 1 \\ 1 & 0 & 0 \\ 0 & 1 & 0 \end{pmatrix}$$
$$D(g_3) = \begin{pmatrix} 0 & 0 & 1 \\ 0 & 1 & 0 \\ 1 & 0 & 0 \end{pmatrix} \qquad D(g_6) = \begin{pmatrix} 0 & 1 & 0 \\ 0 & 0 & 1 \\ 1 & 0 & 0 \end{pmatrix}$$

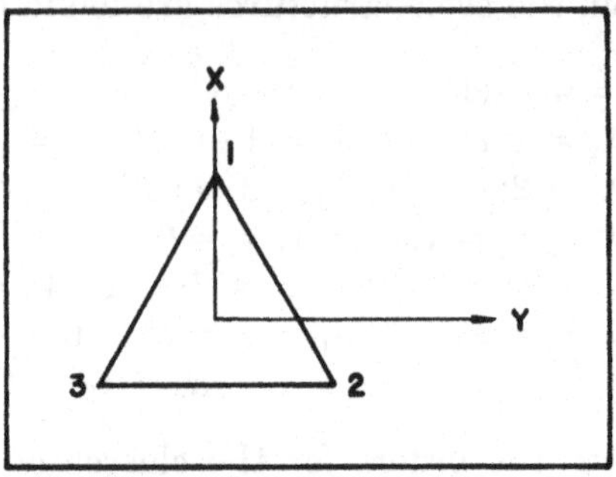

Figure 1[6]. An equilateral triangle that lies on the XY-plane with its center located at the origin of the Cartesian coordinate.

The compound characters are:

$$\chi(g_1) = \chi(e) = 3, \chi(g_2) = \chi(g_3) = \chi(g_4) = 1, \chi(g_5) = \chi(g_6) = 0$$

The group S_3 has three classes: C_1; which is composed of the identity element; C_2, which contains the three elements g_2, g_3, g_4; and C_3, which has the two elements g_5, g_6.

The partitions of $n = 3$ are: [3], [21], and [1^3]. The cycle structures corresponding to each of these partitions can be evaluated as

$$\begin{aligned}
[3] &\to \lambda_1 = \alpha_1 = 3 \to (1^3) \\
[2, 1] &\to \lambda_1 = 2, \lambda_2 = 1; \alpha_1 = 1, \alpha_2 = 1 \to (1, 2) \\
[1^3] &\to \lambda_1 = \lambda_2 = \lambda_3 = 1; \alpha_1 = \alpha_2 = 0, \alpha_3 = 1 \to (3)
\end{aligned}$$

Then the orders of the classes, found by (10), are

$$h_1 = 1, h_2 = 3, h_3 = 2$$

in agreement with the previous conclusion that the classes C_1, C_2, and C_3 have 1, 3, and 2 elements, respectively.

Because the sum of the squares of the dimensions of the irreducible representations is equal to the order of the group, and the number of irreducible representations for a finite group is equal to the number of classes, for the irreducible representations of the group S_3,

$$n_1^2 + n_2^2 + n_3^2 = 3! = 6$$

which gives $n_1 = 1$, $n_2 = 2$, $n_3 = 1$, on choosing the second irreducible representation to be two-dimensional. This representation is found by considering an equilateral triangle that lies on the XY plane, with the origin of the Cartesian coordinates located at its center (Fig. 1[6]) and its corners numbered 1, 2, 3, starting from the top and proceeding clockwise. If this corresponds to the standard ordering (that is, no rotation

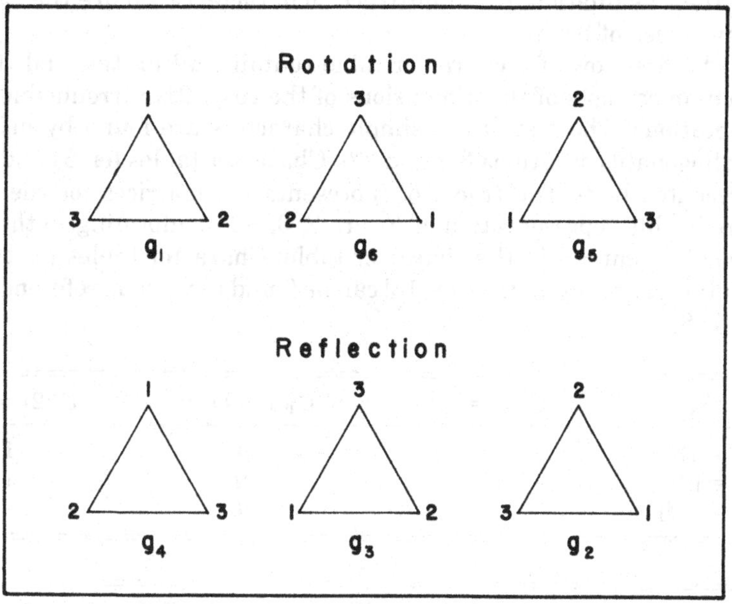

Figure 2⁶. Rotations of $\pm 2\pi/3$ around the Z-axis and reflections in the X-axis.

and no reflections), the group elements of S_3 may be interpreted either as a rotation of $\mp 2\pi/3$ around the Z-axis or as a reflection in the X-axis (Fig. 2⁶). Because the matrices of a rotation through β, around the Z-axis, and the reflection in the X-axis, respectively, are

$$\begin{pmatrix} \cos\beta & -\sin\beta \\ \sin\beta & \cos\beta \end{pmatrix} \quad \text{and} \quad \begin{pmatrix} 1 & 0 \\ 0 & -1 \end{pmatrix}$$

the irreducible representation matrices become

$$D^{(2)}(g_1) = D^{(2)}(e) = \begin{pmatrix} 1 & 0 \\ 0 & 1 \end{pmatrix} \qquad D^{(2)}(g_4) = \begin{pmatrix} 1 & 0 \\ 0 & 1 \end{pmatrix}$$

$$D^{(2)}(g_2) = \begin{pmatrix} -1/2 & 3^{1/2}/2 \\ 3^{1/2}/2 & 1/2 \end{pmatrix} \qquad D^{(2)}(g_5) = \begin{pmatrix} -1/2 & 3^{1/2}/2 \\ 3^{1/2}/2 & -1/2 \end{pmatrix}$$

$$D^{(2)}(g_3) = \begin{pmatrix} -1/2 & -3^{1/2}/2 \\ -3^{1/2}/2 & 1/2 \end{pmatrix} \qquad D^{(2)}(g_6) = \begin{pmatrix} -1/2 & -3^{1/2}/2 \\ 3^{1/2}/2 & -1/2 \end{pmatrix}$$

where the superscript 2 indicates the irreducible representation to which these matrices belong. The matrices for symmetric groups up to the fifth degree are listed by Hamermesh [4²c, pp. 224–30]. These matrices are not unique, because they depend on the choice of the basis in their evaluation, but their traces (characteristics) are the same for any choice of basis.

The first row of a character table contains all unities and the first column consists of the dimensions of the respective irreducible representations. The rest of the simple characters are found by using the orthogonality relations $3^{6}4$ and $3^{6}6$. Character tables for S_3 and S_4 are presented here. The traces of representation matrices for the second irreducible representation of S_3 are 2, 0, −1, conforming to the corresponding entries in the character table. Character tables for the permutation groups up to order 16! can be found in various references [4^2c, 6, 7, 8].

S_3	$C_1(1) = (1^3)$	$C_2(3) = (12)$	$C_3(2) = (3)$
$\Gamma_1 = [3]$	1	1	1
$\Gamma_2 = [21]$	2	0	−1
$\Gamma_3 = [1^3]$	1	−1	1

S_4	$C_1(1) = (1^4)$	$C_2(6) = (1^2 2)$	$C_3(3) = (2^2)$	$C_4(8) = (13)$	$C_5(6) = (4)$
$\Gamma_1 = [4]$	1	1	1	1	1
$\Gamma_2 = [31]$	3	1	−1	0	−1
$\Gamma_3 = [2^2]$	2	0	2	−1	0
$\Gamma_4 = [21^2]$	3	−1	−1	0	1
$\Gamma_5 = [1^4]$	1	−1	1	1	−1

C. Young diagrams

The partition of n into $[\lambda_1, \lambda_2, \lambda_3, \ldots, \lambda_n]$ may be represented uniquely by drawing λ_1 cells, where a cell is a square block for the first row and λ_2 cells for the second row, and so on. Condition (8) requires that no lower row can be longer than any of the upper ones. Diagrams obtained in this way are graphical representations of given partitions and are called *Young diagrams*. A particular diagram corresponding to a specified partition is called a *shape*. For example, the Young diagrams corresponding to the partitions [322], [321], and [31], respectively are

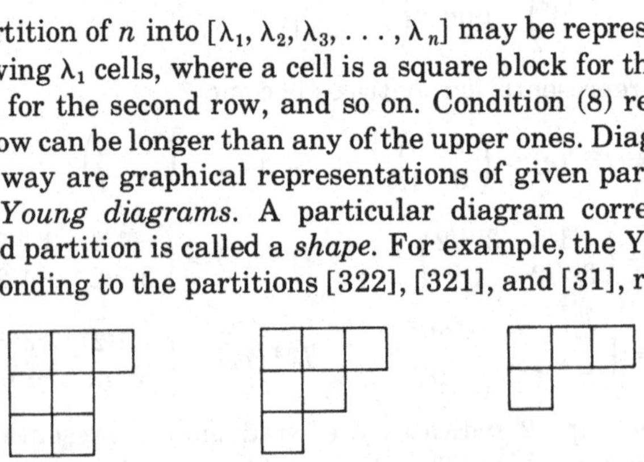

The shape formed by interchanging the rows and columns of a Young diagram, corresponding to a given partition $[\lambda_i]$ is said to be a *conju-*

gate, or *adjoint Young diagram,* and the corresponding partition is indicated as $[\widetilde{\lambda_i}]$. For instance,

$[21] =$ 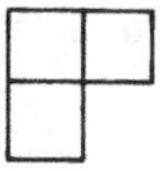 and $\widetilde{[21]} =$ 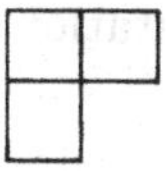

or

$[31] =$ 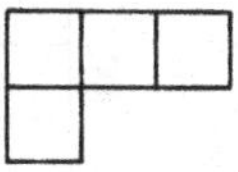 and $\widetilde{[31]} =$ 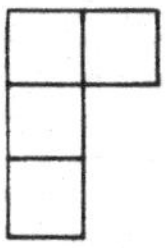

showing that $\widetilde{[21]} = [21]$ and $\widetilde{[31]} = [21^2]$. When the cells of a Young diagram are numbered by 1, 2, 3, . . . , n, it becomes a *Young tableau.* If the numbers in a Young tableau are arranged so that they increase from left to right and from top to bottom, the tableau is said to be a *standard Young tableau.* A number of standard Young tableaux correspond to a given partition. For example, the following three standard tableaux can be formed to correspond to the partition [31].

1	2	3
4		

1	3	4
2		

1	2	4
3		

When the cells of a Young diagram are numbered by the *hook lengths* the result is called a *hook graph* [9]. The hook length, l_{ij}, of a cell, identified as ith row and jth column cell of a given shape, is defined by

$$l_{ij} = m_i + m_j + 1 \tag{11}$$

where m_i is the number of cells to the right and m_j is the number of cells directly below the cell (i, j). For instance, the hook graph of the partition [432] is

6	5	3	1
4	3	1	
2	1		

Young diagrams, Young tableaux, and hook graphs are useful in many ways in handling the symmetric group and its application to physical problems. Their use is referred to as the *graphical method.*

An irreducible representation Γ_α of the symmetric group of nth degree S_n may be identified by one of the partitions of the number n, but there is a unique correspondence between a given partition and its Young diagram. Therefore, it is concluded that each shape corresponds to an irreducible representation of the group S_n. For $n = 4$, these are:

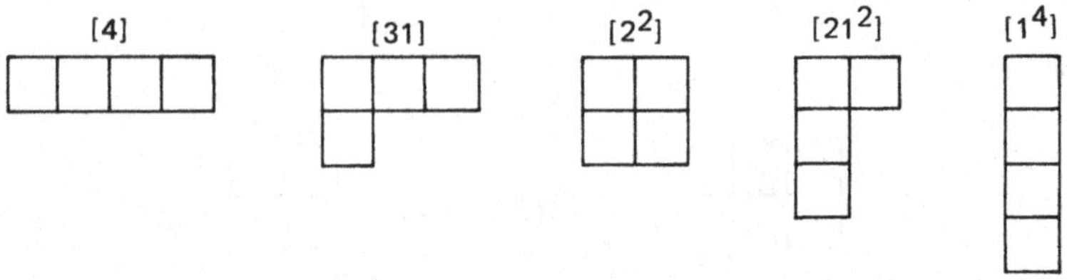

The total number of standard Young tableaux that can be generated from a given shape is equal to the dimension of the corresponding irreducible representation. For instance, for the group S_4, there is only one standard tableau for $[4]$ and $[1^4]$; there are three tableaux for $[31]$ and $[21^2]$ and two for $[2^2]$, thus indicating that the dimensions of the irreducible representations $[4]$, $[31]$, $[2^2]$, $[21^2]$, and $[1^4]$ are 1, 3, 2, 3, and 1, respectively.

D. Outer and inner products of two representations and their reductions

Two kinds of direct products are defined for the symmetric groups:

1. *outer product,* where the representations forming the direct product belong to different groups, denoted by $\otimes$
2. *inner product,* where the representations forming the direct product are irreducible representations of the same group, denoted by $\times$.

Sometimes the outer and inner products are called *direct* and *Kronecker* products, respectively.

Reduction is the expression of the direct product of two irreducible representations as a direct sum of product irreducible representations. Reduction of the outer product is easily performed by labeling the cells of different rows of the second shape in the product by different symbols. The first row is labeled a, the second row b, the third row c, and so on. The cells labeled by a and the first shape, in the direct product, are then used to form all possible regular shapes (no lower row can be longer than the upper ones) subject to the condition that no two identical symbols can appear in the same column. The process is repeated subject to the same conditions, with the resultant shapes and the cells labeled by b and continued by adding the cells labeled c, d, and so on, respectively, to the newly formed shapes. After all the labeled cells are exhausted the number of labels in each shape is counted, starting at the top and going by rows from right to left. The number of the first row label, a, should be larger than, or equal to, the number of the second row label, b, which in turn has to be larger than, or equal to, the third row label, c, and so on.

To demonstrate these steps, $\Gamma_1 = [31]$, and $\Gamma_2 = [21]$ are taken to find the reduction of the outer product $\Gamma_1 \otimes \Gamma_2$. First, the cells of Γ_2 are labeled

Γ_2

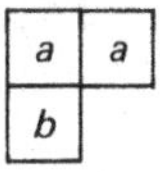

Then all admissible shapes with Γ_1 and the cells labeled by a are formed:

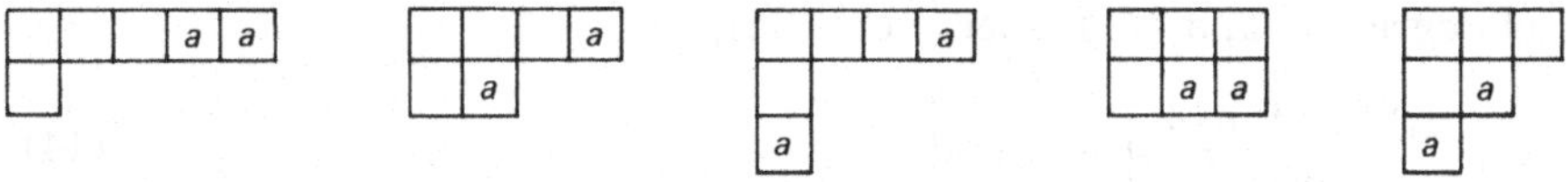

When cell b is added, the following permissible new shapes are obtained:

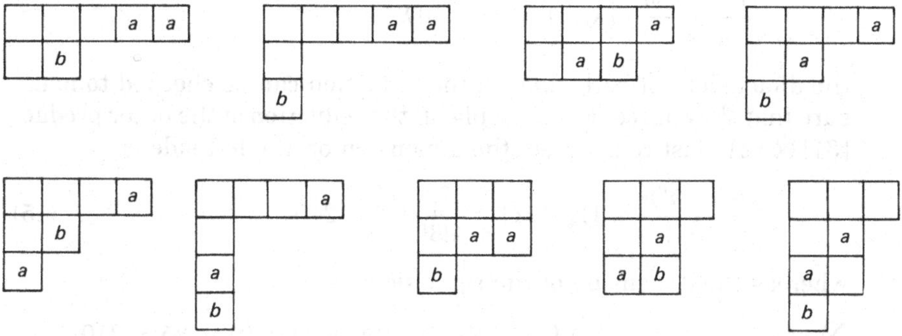

Because there are no more labeled cells to add, the final result in terms of partitions can be expressed as

$$[31] \otimes [21] = [52] + [51^2] + [43] + 2[421] + [41^3] + [3^21] + [32^2] + [321^2]$$

The dimension n_i of an irreducible representation Γ_i of the group S_n, or equally the degree $n[\lambda_i]$ of the corresponding partition $[\lambda_i]$, is [7a, p. 9]

$$n_i = n[\lambda_i] = \frac{n!}{M[\lambda_i]} \tag{12}$$

where $M[\lambda_i]$ is a number called the *product* of the hook lengths and is defined as

$$M[\lambda_i] = \frac{l_{11}!l_{12}! \ldots l_{1p}!}{\prod_{i<k} (l_{1i} - l_{1k})} \tag{13}$$

For example, the product of the hook lengths for the partition [432] is

$$M[432] = \frac{6!4!2!}{4 \cdot 2 \cdot 2} = 2160$$

and the degree of the same partition becomes

$$n[432] = \frac{9!}{2160} = 168$$

Because the degree of an outer product of two partitions, $[\lambda_1]$ and $[\lambda_2]$, of degree n' and n'', respectively [3b],

$$\frac{(n' + n'')!}{n'!n''!} n[\lambda_1]n[\lambda_2] \tag{14}$$

should be equal to the sum of degrees of the resultant partitions, labeled by ν, that is

$$\frac{(n' + n'')!}{n'!n''!} n[\lambda_1]n[\lambda_2] = \sum_\nu n[\lambda_i]_\nu$$

the dimensions of both sides in the reduction can be checked to make sure that they agree. For example, in the reduction of the outer product [31] ⊗ [21] just considered, the dimension on the left side is

$$\frac{(n' + n'')!}{n'!n''!} n[31]n[21] = \frac{7!}{4!3!} 3 \cdot 2 = 210 \tag{15}$$

whereas the dimension of the right side is:

$$\sum_\nu n[\lambda_i]_\nu = 14 + 15 + 14 + (2 \times 35) + 20 + 21 + 21 + 35 = 210$$

A generalized recipe does not exist [10] for the reduction of the inner product, so some special cases are presented here. More are given by Murnaghan [3a]:

$$[n-1, 1] \times [n-1, 1] = [n] + [n-1, 1] + [n-2, 2] + [n-2, 1^2] \tag{16a}$$

$$[n-1, 1] \times [n-2, 2] = [n-1, 1] + [n-2, 2] + [n-2, 1^2] + [n-3, 2, 1] + [n-3, 3] \tag{16b}$$

$$[n-1, 1] \times [n-2, 1^2] = [n-1, 1] + [n-2, 2] + [n-2, 1^2] + [n-3, 2, 1] + [n-3, 1^3] \tag{16c}$$

$$[n-2,2] \times [n-2,2] = [n] + [n-1,1] + 2[n-2,2] + [n-2,1^2] + [n-3,3] + 2[n-3,2,1] + [n-3,1^3] + [n-4,4] + [n-4,3,1] + [n-4,2^2] \quad (16d)$$

$$[n-1,1] \times [n-3,3] = [n-2,2] + [n-3,3] + [n-3,2,1] + [n-4,4] + [n-4,3,1] \quad (16e)$$

$$[n-1,1] \times [n-3,2,1] = [n-2,2] + [n-2,1^2] + [n-3,3] + 2[n-3,2,1] + [n-3,1^3] + [n-4,3,1] + [n-4,2^2] + [n-4,2,1^2] \quad (16f)$$

$$[n-1,1] \times [n-3,1^3] = [n-2,1^2] + [n-3,2,1] + [n-3,1^3] + [n-4,2,1^2] + [n-4,1^4] \quad (16g)$$

$$[n-2,2] \times [n-2,1^2] = [n-1,1] + [n-2,2] + 2[n-2,1^2] + [n-3,3] + 2[n-3,2,1] + [n-3,1^3] + [n-4,3,1] + [n-4,2,1^2] \quad (16h)$$

$$[n-2,1^2] \times [n-2,1^2] = [n] + [n-1,1] + 2[n-2,2] + [n-2,1^2] + [n-3,3] + 2[n-3,2,1] + [n-3,1^3] + [n-4,2^2] + [n-4,2,1^2] + [n-4,1^4] \quad (16i)$$

In addition to the above specific cases, the following relations are helpful [4²c, pp. 255–7]:

$$[\lambda] \times [1^n] = [\tilde{\lambda}] \quad (17)$$

$$([\lambda] \times [\mu]) \times [\nu] = [\lambda] \times ([\mu] \times [\nu]) = ([\lambda] \times [\nu]) \times [\mu] \quad (18)$$

$$([\lambda] \tilde{\times} [\mu]) = [\lambda] \times [\tilde{\mu}] = [\tilde{\lambda}] \times [\mu] \quad (19)$$

and

$$[\lambda] \times [\mu] = [\lambda] \times [\mu] \quad (20)$$

If $[\lambda] \times [\mu]$ contains $[\nu]$, then $[\lambda] \times [\nu]$ contains $[\mu]$, and $[\mu] \times [\nu]$ contains $[\lambda]$. The coefficients of the specified partitions which are contained in the reduction of the above mentioned inner products are always equal to the coefficient of the identity representation $[n]$ in $[\lambda] \times [\mu] \times [\nu]$. Thus the inner product $[\lambda] \times [\mu]$ contains the identity representation $[n]$ only once if $[\lambda] = [\mu]$. On the other hand, $[\lambda] \times [\mu]$ contains the alternating representation $[1^n]$ once if $[\lambda] = [\tilde{\mu}]$ [4²c, pp. 147–8].

A graphical method exists [10c] for the reduction of the inner products of type $[\lambda] \times [n-1, 1]$.

6⁶. Clebsch-Gordan coefficients; simply reducible groups

Let $D^{(\alpha)}$ and $D^{(\beta)}$ be two irreducible representation matrices. The reduction of their inner product

$$D^{(\alpha)} \times D^{(\beta)} = \sum_{\gamma} C_\gamma D^{(\gamma)} \quad (1)$$

where $D^{(\gamma)}$ is another irreducible representation matrix, is called the *Clebsch-Gordan series.* The integer expansion coefficient C_γ is the number of times that $D^{(\gamma)}$ is contained in $D^{(\alpha)} \times D^{(\beta)}$ and is given in terms of the characters of irreducible unitary representations Γ_α, Γ_β, and Γ_γ as

$$C_\gamma = \frac{1}{h} \sum_{i,j,k} \chi_i^{(\alpha)} \chi_j^{(\beta)} (\chi_k^{(\gamma)})^* \tag{2}$$

where each of the indices i, j, k run over the classes of the group.

Another convenient notation for the coefficients in the Clebsch-Gordan series is

$$C_\gamma = (\alpha\beta\gamma) \tag{3}$$

In physical applications the problem arises of finding the basis functions in the product space of two or more linear vector spaces in which the bases are known. Suppose that the n_α and n_β basis functions ψ_i^α and ϕ_j^β, where $i = 1, 2, \ldots, n_\alpha$ and $j = 1, 2, \ldots, n_\beta$, for the irreducible representations Γ_α and Γ_β, respectively, are known. The set of n_γ basis functions Ψ_k^γ, can be found as a sum of the product functions $\psi_i^\alpha \phi_j^\beta$, which form the basis for the representation Γ_γ, which exists only if $(\alpha\beta\gamma) \neq 0$. For $(\alpha\beta\gamma) > 1$ there are $(\alpha\beta\gamma)$ linear combinations of the product functions. Another index τ, which runs from 1 to $(\alpha\beta\gamma)$, is used to describe this multiplicity. The functions $\Psi_k^{\gamma\tau}$ are the sum of the product functions with proper coefficients:

$$\Psi_k^{\gamma\tau} = \sum_{i,j} \psi_i^\alpha \phi_j^\beta (\alpha i, \beta j | \gamma\tau k) \tag{4}$$

The coefficients $(\alpha i, \beta j | \gamma\tau k)$ are called *Clebsch-Gordan coefficients.*

Conversely, the product functions may be expressed as a sum of the basis functions $\psi_k^{\gamma\tau}$ as

$$\psi_i^\alpha \phi_j^\beta = \sum_{\gamma,\tau,k} (\alpha i, \beta j | \gamma\tau k) \Psi_k^{\gamma\tau} \tag{5}$$

Because the total number of functions $\Psi_k^{\gamma\tau}$ must be equal to the total number of product functions $\psi_i^\alpha \phi_j^\beta$,

$$\sum_\gamma (\alpha\beta\gamma)\, n_\gamma = n_\alpha n_\beta \tag{6}$$

which shows that the Clebsch-Gordan coefficients are the elements of an $(n_\alpha n_\beta) \times (n_\alpha n_\beta)$ matrix. From (4) and (5),

$$\sum_{\alpha,\beta} \sum_{i,j} (\gamma'\tau' k' | \alpha i, \beta j)(\alpha i, \beta j | \gamma\tau k) = \delta_{\gamma\gamma'} \delta_{\tau\tau'} \delta_{kk'} \tag{7a}$$

and

$$\sum_{\gamma,\tau} \sum_{k} (\alpha i', \beta j' | \gamma\tau k)(\gamma\tau k | \alpha i, \beta j) = \delta_{ii'} \delta_{jj'} \tag{7b}$$

For unitary representations

$$(\gamma\tau k|\alpha i, \beta j) = (\alpha i, \beta j|\gamma\tau k)^* \tag{8}$$

The Clebsch-Gordan coefficients have a higher degree of symmetry for the simply reducible groups. Some important groups, such as the symmetric groups S_3 and S_4, the two-dimensional special unitary group $SU(2)$, and the three-dimensional rotation group $SO(3)$, are simply reducible. Some of their properties are:

1. The characters of all irreducible representations are real.
2. All representations are integer or half-integer (Sec. 2).
3. The inner direct product of two-integer, or of two half-integer representations, contains only integer representations; the inner product of an integer and a half-integer representation contains only half-integer representations [4^2c, pp. 152–3].
4. When the inner product $D^{(\alpha)} \times D^{(\alpha)}$ is decomposed into symmetrized and antisymmetrized parts as

$$D^{(\alpha)} \times D^{(\alpha)} = [D^{(\alpha)} \times D^{(\alpha)}] + \{D^{(\alpha)} \times D^{(\alpha)}\} \tag{9}$$

if $D^{(\alpha)}$ is an integer representation, the irreducible representations in the symmetric part, $[D^{(\alpha)} \times D^{(\alpha)}]$, are called *even representations,* and the irreducible representations contained in the antisymmetric part, $\{D^{(\alpha)} \times D^{(\alpha)}\}$, are said to be *odd representations.* If $D^{(\alpha)}$, in (9), is a half-integer representation, the above definitions are reversed.

5. No representation can be both even and odd [4^2c, p. 155].

The coefficient $(\alpha\beta\gamma)$ in (3) now has the value zero or one and may be expressed as

$$\sum_{\alpha\beta\gamma} (\alpha\beta\gamma)^2 = \sum_{\alpha\beta\gamma} q^\alpha q^\beta q^\gamma(\alpha\beta\gamma) = h \sum_i \left(\frac{1}{h_i}\right)^2 \tag{10}$$

where q^α, q^β, q^γ are defined in $2^6$15 and $2^6$16. Now the index τ is not needed, and the Clebsch-Gordan coefficients simplify to the form

$$(\alpha i, \beta j|\gamma k)$$

For the three-dimensional rotation group $SO(3)$ the Clebsch-Gordan coefficients are the vector coupling coefficients of Chapter 3. For the symmetric groups S_n the recursion relations and the symmetry properties of these coefficients, as well as the numerical values for the specific examples: $[2, 1] \times [2, 1]$ and $[3, 1^2] \times [3, 1^2]$, are given in Hamermesh [4^2c, pp. 260–75].

The terminology by which (1) is called the Clebsch-Gordan series is apparently attributable to Weyl [11a] and also Van der Waerden [11b], particularly for the rotation group. A careful search of the publications of A. Clebsch and Paul Gordan has failed to reveal anything directly related to what Weyl calls the Clebsch-Gordan series. The two references that seem most directly relevant are Clebsch [11c] and Gordan [11d] in which they are studying the invariants of $GL(3, C)$, but not in connection with group representations. Many physicists have adopted Weyl's terminology and the extension to Clebsch-Gordan coefficients (often misspelling Gordan as Gordon), and giving TAS as a reference for tables of the coefficients (first evaluated by Wigner), even though TAS does not refer to the series or the coefficients by this name. Interesting personal insights into the mathematical contributions of Clebsch and Gordan are given by F. Klein [11e].

7^6. Young operators

The concepts of "symmetrizer" and "antisymmetrizer" were introduced (Sec. 7^2) in discussing the symmetry properties of product wave functions. A technique is now considered, based on group algebra and the concept of Young tableaux, for treating the same subject.

Horizontal permutations, which interchange the symbols in the same row of a Young tableau, are denoted by p. The vertical permutations in the same column are denoted by q. The quantities

$$P = \sum_p p \tag{1}$$

and

$$Q = \sum_q (-1)^q q \tag{2}$$

are called the *symmetrizer* and *antisymmetrizer*, respectively. The above summations include all possible horizontal and vertical permutations.

The *Young operator* is defined by

$$Y = QP \tag{3}$$

where the order of Q and P in the multiplication is important. All three quantities, P, Q, and Y, are essentially idempotents. They generate left ideals that provide the irreducible representations of the group S_n. The irreducible representations obtained for different shapes are inequivalent, but for different tableaux of the same shape they are all equivalent. On the other hand, the irreducible representations obtained for standard Young tableaux belong to the regular representation of S_n.

For example, there are two standard tableaux for $n = 2$:

$$\begin{array}{|c|c|}\hline 1 & 2 \\ \hline \end{array} \quad \text{and} \quad \begin{array}{|c|}\hline 1 \\ \hline 2 \\ \hline \end{array}$$

Therefore

$$P = e + (12) \tag{4a}$$

and

$$Q = e - (12) \tag{4b}$$

are the corresponding symmetrizer and antisymmetrizer. The resolution of the unit element into idempotents is

$$e = \frac{e + (12)}{2} + \frac{e - (12)}{2} \tag{4c}$$

The first generator gives the identity representation and the second generator produces the alternating representation. For $n = 3$, the four standard tableaux with corresponding P, Q, and Y operators are

$$\begin{array}{|c|c|c|}\hline 1&2&3\\\hline\end{array}: \; P = e + (12) + (23) + (13) + (123) + (132) \tag{5a}$$

$$\begin{array}{|c|c|}\hline 1&2\\\hline 3\\\cline{1-1}\end{array} \; : P = e + (12)$$

$$Q = e - (13)$$

$$Y = QP = e + (12) - (13) - (123) \tag{5b}$$

$$\begin{array}{|c|c|}\hline 1&3\\\hline 2\\\cline{1-1}\end{array} \; : P' = e + (13)$$

$$Q' = e - (12)$$

$$Y' = Q'P' = e - (12) + (13) - (132) \tag{5c}$$

$$\begin{array}{|c|}\hline 1\\\hline 2\\\hline 3\\\hline\end{array} \; : Q = e - (12) - (23) - (13) + (123) + (132) \tag{5d}$$

The unit element is expressed in terms of these idempotents as

$$e = \frac{1}{6}P + \frac{1}{6}Q + \frac{1}{3}Y + \frac{1}{3}Y' \tag{5e}$$

Although these generators could be used to obtain the matrices of all irreducible representations of the groups S_2 and S_3 [3e, pp. 86–90], the construction of product wave functions with a specified symmetry is considered.

In order to produce functions that possess a specified symmetry with respect to the coordinate labels, it is sufficient to multiply the product function

$$\psi = \phi(a; 1)\phi(b; 2) \ldots \phi(n; N) \tag{6}$$

by the operators (generators) P, Q, and Y of an irreducible representation of the group S_N. This process also gives the basis functions of the irreducible representation to which the idempotents P, Q, and Y belong.

As an example, there are only two operators for $N = 2$: one, the symmetrizer; another, the antisymmetrizer. They generate the following completely symmetric and antisymmetric functions:

$$\psi_S = \mathcal{P}\psi = \frac{1}{\sqrt{2}}[e + (12)]\phi(a; 1)\phi(b; 2)$$
$$= \frac{1}{\sqrt{2}}[\phi(a; 1)\phi(b; 2) + \phi(a; 2)\phi(b; 1)]$$
$$\psi_A = \mathcal{Q}\psi = \frac{1}{\sqrt{2}}[e - (12)]\phi(a; 1)\phi(b; 2)$$
$$= \frac{1}{\sqrt{2}}\begin{bmatrix} \phi(a; 1) & \phi(a; 2) \\ \phi(b; 1) & \phi(b; 2) \end{bmatrix}$$

where the normalized operators $\mathcal{P}$ and $\mathcal{Q}$ are defined, in terms of P and Q, as

$$\mathcal{P} = \frac{P}{\sqrt{n_p^\alpha!}} \tag{7}$$
$$\mathcal{Q} = \frac{Q}{\sqrt{n_q^\alpha!}} \tag{8}$$

n_p^α and n_q^α are the number of horizontal and vertical cells, respectively, which are permuted to obtain P and Q, of the standard Young tableau that corresponds to the αth irreducible representation of S_N. The normalized Young operator in terms of $\mathcal{P}$ and $\mathcal{Q}$ becomes

$$\mathcal{Y} = \mathcal{Q}\mathcal{P} \tag{9}$$

For $N = 3$:

$$\psi_S = \frac{1}{\sqrt{6}}[\phi(a; 1)\phi(b; 2)\phi(c; 3)$$
$$+ \phi(a; 2)\phi(b; 1)\phi(c; 3) + \phi(a; 1)\phi(b; 3)\phi(c; 2)$$
$$+ \phi(a; 3)\phi(b; 2)\phi(c; 1) + \phi(a; 2)\phi(b; 3)\phi(c; 1)$$
$$+ \phi(a; 3)\phi(b; 1)\phi(c; 2)]$$

$$\psi_A = \frac{1}{\sqrt{6}}\begin{bmatrix} \phi(a; 1) & \phi(a; 2) & \phi(a; 3) \\ \phi(b; 1) & \phi(b; 2) & \phi(b; 3) \\ \phi(c; 1) & \phi(c; 2) & \phi(c; 3) \end{bmatrix}$$

$$\psi_1 = \mathcal{Y}\psi = \frac{1}{2}[\phi(a; 1)\phi(b; 2)\phi(c; 3) + \phi(a; 2)\phi(b; 1)\phi(c; 3)$$
$$- \phi(a; 3)\phi(b; 2)\phi(c; 1) - \phi(a; 2)\phi(b; 3)\phi(c; 1)]$$

$$\psi_2 = y'\psi = \frac{1}{2}[\phi(a;1)\phi(b;2)\phi(c;3) - \phi(a;2)\phi(b;1)\phi(c;3) + \phi(a;3)\phi(b;2)\phi(c;1) - \phi(a;3)\phi(b;1)\phi(c;2)]$$

where ψ_S and ψ_A are completely symmetric and antisymmetric with respect to all coordinate labels 1, 2, 3. On the other hand, ψ_1 is symmetric with respect to 1 and 2 and antisymmetric in 1 and 3. ψ_2 is symmetric with respect to the coordinate labels 1 and 3 and antisymmetric in 1 and 2.

In obtaining these results it is assumed that each of the single-particle wave functions are different, which is not usually the case. (For instance, because only two spin states are available, there are only two different spin functions.) Assuming that in (6) $a = b = \ldots = n$,

$$\psi = \phi(a;1)\phi(a;2)\ldots\phi(a;N) \tag{10}$$

which is completely symmetric. Therefore, the only representation that can occur is the one that corresponds to the partition $[N]$. In an N-particle system where there are m particles of one kind labeled by a, and $(N - m)$ particles of another kind labeled by b, the product wave function becomes

$$\psi = \phi(a;1)\ldots\phi(a;m)\phi(b;m+1)\ldots\phi(b;N) \tag{11}$$

Such a function can be antisymmetrized with respect to only two coordinate labels. Therefore, the standard Young tableau, which corresponds to the irreducible regular representation of the group S_N, can have only two rows. This is the case for spin functions.

A product wave function, which is the basis of an irreducible representation, is identified by the partition $[\lambda_i]$, of the group S_n. This function has either a mixed symmetry (that is, it is symmetric with respect to some coordinate labels and antisymmetric in others) or total symmetry (that is, it is symmetric with respect to all coordinate labels). Then the outer product $[\lambda_i] \otimes [\tilde{\lambda}_i]$ gives a wave function that is completely antisymmetric. For example, in the $N = 3$ case:

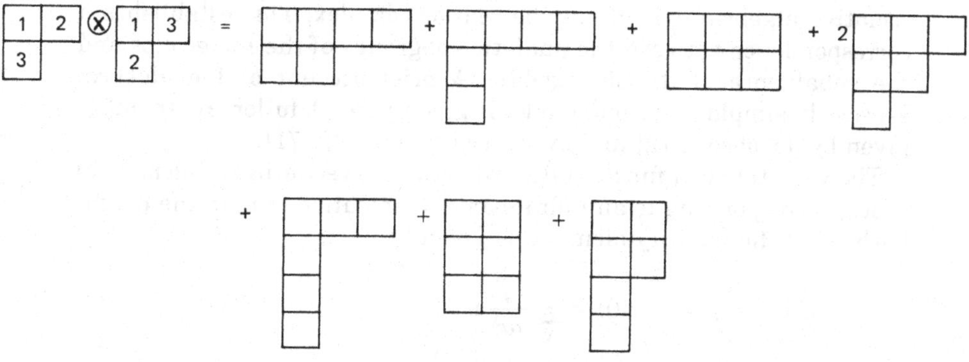

or equally,

$$[2, 1] \otimes \widetilde{[2, 1]} = [4, 2] + [4, 1^2] + [3^2] + 2[3, 2, 1] + [3, 1^3] + [2^3] + [2^2, 1^2]$$

is completely antisymmetric in all coordinate labels 1, 2, and 3.

8⁶. The Lie algebras of simple and semisimple Lie groups

Lie groups are extraordinarily important in atomic and molecular physics as well as in nuclear and elementary particle physics. Some examples of Lie groups (Sec. 2⁶) are: $GL(n, C)$, $SL(n, C)$, $SL(n, R)$, $U(n)$, $SU(n)$, $SO(n)$, and $Sp(2n)$.

A group is said to be continuous if the following three conditions are satisfied [3e, pp. 132–43]:

1. Any group element may be designated by a certain number of continuous parameters. The space spanned by them is called the *parameter space*. Assuming that the number of parameters, r, is finite and that the parameters are real they define an r-dimensional real space denoted by R^r. It is possible to associate a vector **a**, in R^r with each element g_a of the group G. The parameters are assumed to be *essential;* that is, no smaller number of parameters will suffice to label the elements of G, and it is not possible to express any one parameter in terms of the others.

2. If the vectors **a** and **b** in R^r are images of the group elements g_a and g_b, respectively, there is another vector **ab** in R^r, which corresponds to $g_a g_b$. The vector **ab** is a continuous function of a and b; that is,

$$\mathbf{ab} = \phi(\mathbf{a}, \mathbf{b}) \tag{1}$$

3. The vector $\mathbf{a}^{-1}$, in R^r, which corresponds to an inverse element g_a^{-1} of g_a is a continuous function of **a**; that is,

$$\mathbf{a}^{-1} = f(\mathbf{a}) \tag{2}$$

With every Lie group it is possible to associate a Lie algebra (nonassociative algebra), which may be real or complex. This establishes a correspondence between the analytic subgroups of the Lie groups and the subalgebra of its Lie algebra. A brief account of Lie algebra, especially simple and semisimple, is given here. A fuller treatment is given by Jacobson [12] and by Bacry [3e, pp. 142–71].

The variation of a function $f(\mathbf{a})$, where **a** is a vector in r-dimensional space, corresponding to an infinitesimal translation of **a**, in the neighborhood of the identity element, is given by

$$df(\mathbf{a}) = f(\mathbf{a}) - f(0) = \sum_{\lambda} \frac{\partial f}{\partial a^{\lambda}} a^{\lambda} \tag{3}$$

where a^λ is written for the λth component, to show that it is contravariant.

By introducing an *infinitesimal operator* and a *generator* defined by, respectively,

$$X_\lambda = \frac{\partial}{\partial a^\lambda}, \quad J_\lambda = -iX_\lambda \tag{4}$$

the translation of **a** may be expressed by a *translation operator*

$$T(\mathbf{a}) = 1 + \sum_\lambda a^\lambda X_\lambda = 1 + i \sum_\lambda a^\lambda J_\lambda \tag{5}$$

and

$$f(\mathbf{a}) = T(\mathbf{a})f(0) \tag{6}$$

Then, for the successive translations $\mathbf{U} = \mathbf{b}^{-1}\mathbf{a}^{-1}\mathbf{ba}$

$$T(\mathbf{U}) = T(\mathbf{b}^{-1})T(\mathbf{a}^{-1})T(\mathbf{b})T(\mathbf{a}) \tag{7a}$$

and

$$T(\mathbf{U}) = 1 + \sum_{\lambda,\mu} \{b^\lambda X_\lambda a^\mu X_\mu - a^\mu X_\mu b^\lambda X_\lambda\} \tag{7b}$$

A *Lie bracket* is defined as

$$[X_\lambda, X_\mu] = X_\lambda X_\mu - X_\mu X_\lambda \tag{8}$$

Then (7b) becomes

$$T(\mathbf{U}) = 1 + \sum_{\lambda,\mu} b^\lambda a^\mu [X_\lambda, X_\mu] \tag{9}$$

Because

$$U^\sigma = \sum_{\lambda,\mu} f_{\lambda\mu}{}^\sigma b^\lambda a^\mu \tag{10}$$

$$T(\mathbf{U}) = 1 + \sum_\sigma U^\sigma X_\sigma \tag{11}$$

it follows that

$$[X_\lambda, X_\mu] = \sum_\sigma f_{\lambda\mu}{}^\sigma X_\sigma \tag{12}$$

where the operators X_α are infinitesimal operators, and the f's are called the *structure constants*. From the definition of the vector **U** and the relation (10) we infer that if the group is Abelian (g_a, g_b commute) the structure constants are all zero. The structure constants also obey the following relations:

$$\sum_\gamma \{f_{\alpha\beta}{}^\gamma f_{\gamma\nu}{}^\mu + f_{\beta\nu}{}^\gamma f_{\gamma\alpha}{}^\mu + f_{\nu\alpha}{}^\gamma f_{\gamma\beta}{}^\mu\} = 0 \tag{13}$$

and

$$f_{\alpha\beta}{}^\gamma = -f_{\beta\alpha}{}^\gamma \tag{14}$$

The infinitesimal operators X_α form a basis of an r-dimensional vectorial space $\mathcal{A}$ on which their commutation relations (12) give the structure of an algebra. Because

$$[X_\alpha, X_\beta] = -[X_\beta, X_\alpha] \tag{15a}$$

and

$$[[X_\alpha, X_\beta], X_\gamma] + [[X_\beta, X_\gamma], X_\alpha] + [[X_\gamma, X_\alpha], X_\beta] = 0 \tag{15b}$$

the algebra is nonassociative. Relation (15b) is called the *Jacobi identity*. If a Lie group is real, so also is the corresponding algebra. The algebras corresponding to simple and semisimple Lie groups, respectively, are called simple or semisimple Lie algebras. An algebra of dimension larger than unity and containing no ideals is called simple. The necessary and sufficient condition for a Lie algebra to be semisimple, as given by Cartan [4, 13], is that its *metric tensor,* defined as

$$g_{\alpha\beta} = g_{\beta\alpha} = \sum_{\nu,\gamma} f_{\alpha\gamma}{}^{\nu} f_{\beta\nu}{}^{\gamma} \tag{16}$$

must be regular; that is, det $g \neq 0$.

The indices of the structure constants $f_{\alpha\beta}{}^{\gamma}$ may be raised or lowered by using the metric tensor. For example,

$$f_{\alpha\beta\gamma} = f_{\alpha\beta}{}^{\nu} g_{\nu\gamma} \tag{17}$$

where the structure constants $f_{\alpha\beta\gamma}$ are now completely antisymmetric.

The metric tensor may be considered as the scalar product of the basis vectors X_α and X_β:

$$(X_\alpha, X_\beta) = g_{\alpha\beta} \tag{18}$$

The elements of the Lie algebra are vectors of the form

$$\mathbf{A} = \sum_{\alpha} a^\alpha X_\alpha \tag{19}$$

If, now, **A** and **B** are two elements of the algebra,

$$(\mathbf{A}, \mathbf{B}) = \sum_{\alpha,\beta} (a^\alpha X_\alpha, b^\beta X_\beta) = \sum_{\alpha,\beta} a^\alpha b^\beta g_{\alpha\beta} \tag{20}$$

An operator Q_A that transforms an element **B** of the algebra into the element [**A, B**], that is

$$Q_A \mathbf{B} = [\mathbf{A}, \mathbf{B}] \tag{21}$$

is called the *adjoint operator*. The scalar product of two elements **A** and **B** in terms of the adjoint operators, is

$$(\mathbf{A}, \mathbf{B}) = \mathrm{Tr}(Q_A Q_B) \tag{22}$$

From the Jacobi identity

$$[Q_A, Q_B] = Q_{[A,B]} \tag{23}$$

where $Q_{[A,B]}$ is the operator that transforms an element **C** of the algebra into the element [[**A**, **B**], **C**]. The adjoint operators define a representation of the Lie algebra which is called the *adjoint representation.*

The necessary and sufficient condition for a semisimple group to be compact is that its metric tensor must be negative definite.

A semisimple Lie algebra may be expressed as a direct sum of simple Lie algebras. Conversely, a semisimple Lie algebra may be decomposed into its subalgebras, which are simple.

All reducible representations of a compact semisimple Lie algebra are fully reducible (Weyl's theorem). Noncompact algebras have reducible representations that are not fully reducible.

Cartan's classification [4, 13] of the simple and semisimple Lie algebras based on the adjoint representation was later perfected by Dynkin [14].

Cartan sought a solution for the proper-value equation:

$$Q_A\mathbf{X} = [\mathbf{A}, \mathbf{X}] = a\mathbf{X}; \qquad \mathbf{X} = \sum_\alpha x^\alpha X_\alpha \tag{24}$$

where a and $\mathbf{X}$ are the proper value and the proper vector of Q_A, respectively. In general, the proper values of a vector may be complex, necessitating a complex Lie algebra. Here real algebras are considered first. The real algebra is extended to obtain the complex one. The process is called the *complexification* of the real algebra. This resembles considering an n-dimensional complex space as a $2n$-dimensional real space.

Cartan's theorem states that if **A** is an element of a semisimple Lie algebra and is so chosen that the secular equation has the maximum number of distinct roots, that is, the algebra possesses the maximum number of proper values, then only the proper-value zero is degenerate. There are l linearly independent proper vectors, $H_1, H_2, \ldots, H_l$ corresponding to the zero proper value, where l is the degree of degeneracy. The algebra generated by $H_1, H_2 \ldots, H_l$ is called a *nilpotent* or *maximal (Cartan) subalgebra;* l is said to be the *rank,* or *order,* of the algebra. From the above definition of H_i,

$$[\mathbf{A}, H_i] = 0; \qquad i = 1, 2, \ldots, l$$

Because $\mathbf{A} = \sum_i a^i H_i$

$$[H_i, H_j] = 0 \tag{25}$$

That is, the subalgebra generated by H_i is Abelian. The structure constants are all zero:

$$f_{ij}{}^\sigma = 0 \tag{26}$$

where the ranges of the indices are: $i, j = 1, 2, \ldots, l; \sigma = 1, 2, \ldots, h$.

Suppose that E_α, $\alpha = l+1, l+2, \ldots, h$, is a proper vector of H_1 with the proper-value α_1, that is,

$$[H_1, E_\alpha] = \alpha_1 E_\alpha \tag{27}$$

The proper value of H_i is found by applying the Jacobi identity to H_1, H_i and E_α and by using relations (25) and (27):

$$\begin{aligned}[H_1,[H_i, E_\alpha]] &= [H_i,[H_1, E_\alpha]] + [[H_1, H_i], E_\alpha] \\ &= \alpha_1[H_i, E_\alpha]\end{aligned} \tag{28}$$

which indicates that $[H_i, E_\alpha]$ is also a proper-vector of H_1 with the proper value α_1. Because α_1 is nondegenerate and the vector $[H_i, E_\alpha]$ is collinear with the vector E_α, (27) can be generalized by writing

$$[H_i, E_\alpha] = \alpha_i E_\alpha \tag{29}$$

and consequently

$$f_{i\alpha}{}^{\sigma} = \alpha_i \delta_\alpha{}^{\sigma} \tag{30}$$

Therefore, a proper-vector E_α of H_i is characterized by the l proper values $\alpha_1, \alpha_2, \ldots, \alpha_l$. The ensemble of these l numbers is called a *root* and may be considered as the components of a *root vector* $\boldsymbol{\alpha}$. The roots of H_i are all zero.

When the Jacobi identity is applied to H_i, E_α, and E_β, where $\alpha, \beta = l+1, l+2, \ldots h$, it gives

$$\begin{aligned}[H_i,[E_\alpha, E_\beta]] &= [E_\alpha,[H_i, E_\beta]] + [[H_i, E_\alpha], E_\beta] \\ &= (\alpha_i + \beta_i)[E_\alpha, E_\beta]\end{aligned} \tag{31}$$

in which the following two possibilities exist: First, if $\alpha + \beta$ is a root and $\beta = -\alpha$, then

$$[E_\alpha, E_\beta] = f_{\alpha,-\alpha}{}^{i} H_i \tag{32}$$

If $\alpha + \beta$ is a root and $\beta \neq -\alpha$, then

$$[E_\alpha, \mathrm{E}_\beta] = f_{\alpha\beta}{}^{\alpha+\beta}\, E_{\alpha+\beta} = N_{\alpha\beta} E_{\alpha+\beta} \tag{33}$$

That is, $[E_\alpha, E_\beta]$ is proportional to a vector $E_{\alpha+\beta}$ and $N_{\alpha\beta}$ is the proportionality constant.

Second, if $\alpha + \beta$ is not a root, then

$$[E_\alpha, E_\beta] = 0 \tag{34}$$

The structure constants in the space defined by H_i and E_α are given by the relations (26), (30), (32), and (33). The last two may be rewritten as

$$(\alpha + \beta) f_{\alpha\beta}{}^{i} = 0 \tag{35}$$
$$(\gamma - \alpha - \beta) f_{\alpha\beta}{}^{\gamma} = 0 \tag{36}$$

From these relations it is inferred that if α is a root, so is $-\alpha$.

The nonzero components of the metric tensor are:

$$g_{ij} = \sum_{\alpha} \alpha_i \alpha_j \tag{37}$$

and

$$g_{\alpha,-\alpha} = \sum_{i} 2\alpha_i f_{\alpha,-\alpha}{}^{i} + \sum_{\beta,\alpha+\beta} f_{\alpha,\beta}{}^{\alpha+\beta} f_{-\alpha,\alpha+\beta}{}^{\beta} \tag{38}$$

It is convenient to normalize the vectors E_α so that

$$g_{\alpha,-\alpha} = 1 \tag{39}$$

The metric tensor g may be divided into four parts according to the range of the indices i, j, α, and β, as follows:

$$g = \left[\begin{array}{c|c} g_{ij} & g_{i\alpha} \\ \hline g_{\beta j} & g_{\beta\alpha} \end{array}\right] \tag{40}$$

where $g_{i\alpha} = g_{\beta j} = 0$ and $g_{\alpha\beta} = \delta_{-\alpha,\beta}$. Therefore the condition, $\det g \neq 0$, which is necessary and sufficient for the Lie algebra to be semisimple, reduces to $\det g_{ij} \neq 0$. This, with the help of other relations pertaining to the structure constants, results in

$$f_{\alpha,-\alpha}{}^{i} = f_{\alpha,-\alpha,j}\, g^{ij} = f_{j,\alpha,-\alpha}\, g^{ij} = g_{\alpha,-\alpha}\alpha^{i} = \alpha^{i} \tag{41}$$

Thus the relation (32) becomes

$$[E_\alpha, E_{-\alpha}] = \alpha^i H_i \tag{32'}$$

The ensemble of H_i and E_α that obeys the relations (25), (29), (32′), (33), and (34) is called the *standard, Cartan,* or *canonical basis.*

9⁶. The classification of simple Lie algebras; root figures and Dynkin diagrams, and the generators of the classical Lie groups

Simple and semisimple Lie groups may be classified by their *root figures* [15]. The following two theorems about the roots are useful:

1. If H_i are the elements of the subalgebra, chosen so that they render the matrices of g_{ij} positive definite, the roots are all positive, and the root space is real.

2. If $\boldsymbol{\alpha}$ and $\boldsymbol{\beta}$ are any two root vectors, the ratio $2(\boldsymbol{\alpha}, \boldsymbol{\beta})/(\boldsymbol{\alpha}, \boldsymbol{\alpha})$ is a positive integer and $\boldsymbol{\beta} - [2(\boldsymbol{\alpha}, \boldsymbol{\beta})/(\boldsymbol{\alpha}, \boldsymbol{\alpha})]\boldsymbol{\alpha}$ is also a root vector.

Some simple relations exist for the angles between roots and the ratios of their lengths. Consider two vectors $\boldsymbol{\alpha}$ and $\boldsymbol{\beta}$, denoting the angle between them by $\theta_{\alpha\beta}$. Because the root vectors are symmetric with respect to the origin it suffices to consider only the region

$$0 < \theta_{\alpha\beta} \leqslant \frac{\pi}{2}$$

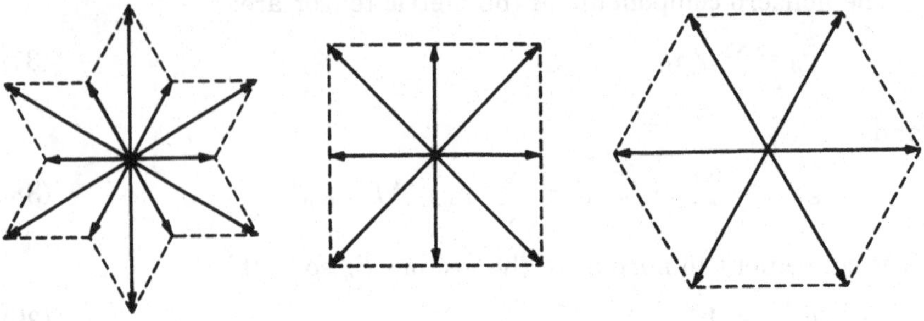

Figure 3[6]. The root figures in the two-dimensional space.

From the second of the above theorems

$$2(\boldsymbol{\alpha}, \boldsymbol{\beta}) = n(\boldsymbol{\alpha}, \boldsymbol{\alpha}) = m(\boldsymbol{\beta}, \boldsymbol{\beta})$$

Assuming that $n > m$,

$$\cos^2 \theta_{\alpha\beta} = \frac{(\boldsymbol{\alpha}, \boldsymbol{\beta})^2}{(\boldsymbol{\alpha}, \boldsymbol{\alpha})(\boldsymbol{\beta}, \boldsymbol{\beta})} = \frac{nm}{4} \tag{1}$$

The following specific values are easily obtained

m	n	$\theta_{\alpha\beta}$	$(\boldsymbol{\beta}, \boldsymbol{\beta})/(\boldsymbol{\alpha}, \boldsymbol{\alpha})$	$[(\boldsymbol{\beta}, \boldsymbol{\beta})/(\boldsymbol{\alpha}, \boldsymbol{\alpha})]^{1/2}$
0	0	$\pi/2$	arbitrary	arbitrary
1	1	$\pi/3$	1	1
1	2	$\pi/4$	2	$\sqrt{2}$
1	3	$\pi/6$	3	$\sqrt{3}$

where $[(\boldsymbol{\beta}, \boldsymbol{\beta})/(\boldsymbol{\alpha}, \boldsymbol{\alpha})]^{1/2}$ is the ratio of lengths of the roots. The cases $m = 0, n \neq 0$; $n = m = 2$ cannot occur because the first case means $(\boldsymbol{\alpha}, \boldsymbol{\alpha}) = 0$, which is not possible; $\boldsymbol{\alpha} \neq 0$, and the second case implies $\boldsymbol{\alpha} = \boldsymbol{\beta}$, which cannot be because $\boldsymbol{\alpha}$ is nondegenerate. The case $m = 1, n = 4$ also gives unacceptable results.

The root figures in the two-dimensional space, $l = 2$, corresponding to the angles $\theta_{\alpha\beta} = \pi/3, \pi/4$, and $\pi/6$ are shown in Figure 3[6]. In case $\theta_{\alpha\beta} = \pi/3$ there are six roots in two-dimensional space. Therefore, the Lie group associated with this figure must have $h = 6 + 2 = 8$ elements. For the cases $\theta_{\alpha\beta} = \pi/4$ and $\theta_{\alpha\beta} = \pi/6$ the corresponding Lie groups have 10 and 14 elements, respectively. These three groups are $SU(3)$, $SO(5)$, and G_2.

Consider the root vectors $\boldsymbol{\alpha}$ and $\boldsymbol{\alpha}'$ in an l-dimensional space, each having l components denoted by $\alpha_1, \alpha_2, \ldots, \alpha_l$ and $\alpha'_1, \alpha'_2, \ldots, \alpha'_l$ ordered in some specific manner. A root $\boldsymbol{\alpha}$ is said to be *positive* if its first nonzero component is positive. The root $\boldsymbol{\alpha}$ is called bigger than the root $\boldsymbol{\alpha}'$ if the first nonzero component of the vector $\boldsymbol{\beta} = \boldsymbol{\alpha} - \boldsymbol{\alpha}'$ is positive. If a positive root can not be decomposed in a sum of two positive roots, it is *simple.*

There are l linearly independent simple roots of an algebra of rank l. The diagrams associated with the simple roots are called *Dynkin diagrams.* Any set of linearly independent vectors $\boldsymbol{\alpha}_1, \boldsymbol{\alpha}_2, \ldots, \boldsymbol{\alpha}_l$ in a real, Euclidean, space is called an *allowable configuration* if

$$4\cos^2\theta_{ij} = \frac{4(\boldsymbol{\alpha}_i, \boldsymbol{\alpha}_j)}{(\boldsymbol{\alpha}_i, \boldsymbol{\alpha}_i)(\boldsymbol{\alpha}_j, \boldsymbol{\alpha}_j)} = 0, 1, 2, 3 \tag{2}$$

and if

$$\cos\theta_{ij} \leqslant 0 \tag{3}$$

for every i and j where $i \neq j$. Thus for an allowable configuration

$$\cos\theta_{ij} = 0, -\frac{1}{2}, -\frac{\sqrt{2}}{2}, -\frac{\sqrt{3}}{2}$$

and correspondingly

$$\theta_{ij} = \pi/2, 2\pi/3, 3\pi/4, 5\pi/6$$

It is customary to replace $\boldsymbol{\alpha}_i$ by a unit vector $\hat{\mathbf{e}}_i$, which is a positive multiple of $\boldsymbol{\alpha}_i$. The conditions for allowable configurations, now in terms of unit vectors $\hat{\mathbf{e}}_i$, become

$$(\hat{\mathbf{e}}_i, \hat{\mathbf{e}}_i) = 1 \tag{4}$$
$$4(\hat{\mathbf{e}}_i, \hat{\mathbf{e}}_j)^2 = 0, 1, 2, 3 \tag{5}$$
$$(\hat{\mathbf{e}}_i, \hat{\mathbf{e}}_j) \leqslant 0 \tag{6}$$

where $i \neq j$ and $i = 1, 2, \ldots, l$. The Dynkin diagram of an allowable configuration is a collection of points e_i and lines connecting these according to the following rules: The points e_i and e_j are not connected if $(\hat{\mathbf{e}}_i, \hat{\mathbf{e}}_j) = 0$ and they are connected by $4(\hat{\mathbf{e}}_i, \hat{\mathbf{e}}_j)^2 = 1, 2, 3$ lines otherwise. Usually the points are represented by small circles, connected by lines, the number of which specifies the angle θ_{ij} between the roots, that is, no line for $\theta_{ij} = \pi/2$, one line for $\theta_{ij} = 2\pi/3$, two lines for $\theta_{ij} = 3\pi/4$, and three lines for $\theta_{ij} = 5\pi/6$. For example, the Dynkin diagram  represents three roots: starting at the left the angles between the first and second roots, the second and third roots, and the first and third roots are $2\pi/3$, $3\pi/4$, and $\pi/2$, respectively. Jacobson [12, pp. 128–35] and Hausner and Schwartz [2i, pp. 120–2] give detailed treatments of the Dynkin diagrams.

Because not all of the simple roots have the same length it is necessary to introduce a way to distinguish the small circles in a Dynkin diagram when they correspond to the roots of different length. This is done by blackening the small circles associated with the longer

Table 1[6]. Simple Lie algebras

Cartan's notation	Dimension	Dynkin diagram	Compact group	Roots
$A_l (l > 1)$	$l(l+2)$		$SU(l+1)$	$\hat{\mathbf{e}}_i - \hat{\mathbf{e}}_j$ $(i, j = 1, 2, \ldots, l+1; i \neq j)$
$B_l (l > 2)$	$l(2l+1)$		$SO(2l+1)$	$\pm\hat{\mathbf{e}}_i, \pm\hat{\mathbf{e}}_i \pm \hat{\mathbf{e}}_j$ $(i, j = 1, 2, \ldots, l; i \neq j)$
$C_l (l > 3)$	$l(2l+1)$		$Sp(2l)$	$\pm 2\hat{\mathbf{e}}_i, \pm\hat{\mathbf{e}}_i \pm \hat{\mathbf{e}}_j$ $(i, j = 1, 2, \ldots, l; i \neq j)$
$D_l (l > 4)$	$l(2l-1)$	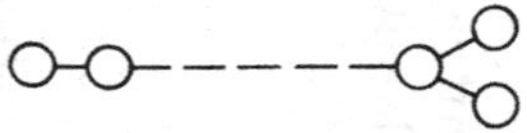	$SO(2l)$	$\pm\hat{\mathbf{e}}_i \pm \hat{\mathbf{e}}_j$ $(i, j = 1, 2, \ldots, l; i \neq j)$
E_6	78	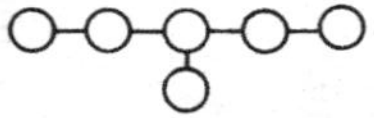		$\hat{\mathbf{e}}(A_5), \pm\sqrt{2}\hat{\mathbf{e}}_7, (1/2)\left[\sum_{i=1}^{6} \pm \hat{\mathbf{e}}_i\right]' \pm \frac{\hat{\mathbf{e}}_7}{\sqrt{2}}$
E_7	133	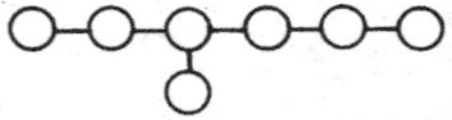		$\hat{\mathbf{e}}(A_7), (1/2)\left[\sum_{i=1}^{8} \pm \hat{\mathbf{e}}_i\right]'$
E_8	248	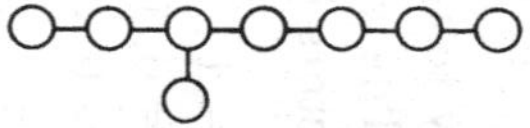		$\hat{\mathbf{e}}(D_8), (1/2)\left[\sum_{i=1}^{8} \pm \hat{\mathbf{e}}_i\right]''$
F_4	52			$\hat{\mathbf{e}}(B_4), (1/2)(\pm\hat{\mathbf{e}}_1 \pm \hat{\mathbf{e}}_2 \pm \hat{\mathbf{e}}_3 \pm \hat{\mathbf{e}}_4)$
G_2	14			$\hat{\mathbf{e}}_i - \hat{\mathbf{e}}_j$ $(i, j = 1, 2, 3; i \neq j)$ $\pm 2\hat{\mathbf{e}}_i \mp \hat{\mathbf{e}}_j \mp \hat{\mathbf{e}}_k$ $(i, j, k = 1, 2, 3; i \neq j \neq k)$

Table 2^6. Maximal subalgebras of the simple Lie algebras

Algebra	Maximal subalgebra
B_n	$D_k + B_{n-k}$
C_n	$C_k + C_{n-k}$
D_n	$D_k + D_{n-k}$
E_6	$A_5 + A_1$; $A_2 + A_2 + A_2$
E_7	$D_6 + A_1$; $A_5 + A_2$; $A_3 + A_3 + A_1$; A_7
E_8	$A_1 + E_7$; $A_2 + E_6$; $A_3 + D_5$; $A_4 + A_4$; $A_1 + A_2 + A_5$; $A_1 + A_7$; D_8; A_8
F_4	$A_1 + C_3$; $A_2 + A_2$; $A_3 + A_1$; B_4
G_2	$A_1 + A_1$; A_2

roots. For instance, in a diagram , the circle at the right signifies the longer root.

The classification of the root figures, as well as the Dynkin diagrams, produces a classification of the simple and semisimple Lie algebras. A basic theorem about the compact Lie algebras states that [12, p. 147] if $\mathscr{L}$ is a complex Lie algebra there exists a compact real Lie algebra $\mathscr{L}_R$; furthermore, the complexification of the compact real algebra $\mathscr{L}_R$ produces the complex one, $\mathscr{L}$.

The complexification of a real Lie algebra, with the real basis X_1, X_2, X_3 may be accomplished by taking the linear combinations of the quantities X_1, X_2, X_3 with complex coefficients and defining the commutation relation for the basis functions as

$$[X_1 + iX_2, X_3] = [X_1, X_3] + i[X_2, X_3] \tag{7}$$

The simple Lie algebras, as classified by Cartan, are presented in Table 1^6. Their Dynkin diagrams, ranks, dimensions, as well as the unit root vectors, are also included. Here $\hat{\mathbf{e}}(A_l)$ stands for all the roots of algebra A_l. One prime on a bracket, which contains a summation, indicates that in the sum plus and minus signs should be equal in number. Two primes show that the number of plus signs must be even. The ensemble of $\hat{\mathbf{e}}_1, \hat{\mathbf{e}}_2, \ldots, \hat{\mathbf{e}}_k$ forms an orthonormal basis. The five algebras E_6, E_7, E_8, F_4, and G_2 are called *exceptional.*

The simple Lie algebras may be expressed as a direct sum of their maximal subalgebras. These are listed in Table 2^6 as derived by Dynkin [14b].

The generators and the properties of the groups which are the simplest realizations of the A_l, B_l, C_l, D_l Lie algebras (Table 1^6) are now discussed.

The full linear group in $(l + 1)$ dimensions $GL(l + 1)$, whose elements depend on $2(2l + 1)^2$ real parameters, may be generated by the infinitesimal operators ($8^6$4),

$$X^i_{\ k} = x^i \frac{\partial}{\partial x^k}; \ i, k = 1, 2, \dots, l + 1 \tag{8a}$$

in which x^i, as defined in $8^6$24, is written for the ith component to show that it is contravariant. The Lie brackets ($8^6$8) for the group become

$$[X^i_{\ k}, X^m_{\ n}] = \delta^m_k X^i_{\ n} - \delta^i_n X^m_{\ k} \tag{8b}$$

Therefore, the structure constants are either one or zero, depending on the value of indices i, k, m, n. The operator $\sum_i X^i_{\ i}$ commutes with every operator of the group, and as a result the subgroup that it generates is Abelian. (Because here the metric tensor $g_{ij} = \delta_{ij}$, $X_{ij} = g_{ik}X^k_{\ j} = X^i_{\ j}$.)

The full linear group is not semisimple. To have a semisimple group the discussion must be restricted to the unimodular group in $(l + 1)$ dimensions, which is also called the special linear group $SL(l + 1)$. For this group the number of parameters is $2(l + 2)$, whereas its real subgroup $SL(l + 1, R)$ depends on half this many parameters. The operators $X^i_{\ i}$ are now no longer infinitesimal operators of the Abelian subgroup but should be replaced by

$$Y^i_{\ i} = X^i_{\ i} - \frac{1}{(l + 1)} \sum_j X^j_{\ j} \tag{9}$$

which still obey the commutation relation (8). If the identifications

$$\begin{aligned} H_i &= Y^i_{\ i} \\ E_{ik} &= X^i_{\ k} \end{aligned} \tag{10}$$

are made, the set of operators H_i and E_{ik} forms the canonical basis for the algebra A_l. Because

$$\sum_i^{l+1} Y^i_{\ i} = \sum_i^{l+1} H_i = 0 \tag{11}$$

among $(l + 1)$ operators H_i, only l of them are linearly independent.

The scalar product $(\mathbf{x}, \mathbf{x}')$ of two vectors $\mathbf{x}$ and $\mathbf{x}'$ in the $(2l + 1)$-dimensional space, with components $(x^{-l}, \dots, x^{-1}, x^0, x^1 \dots, x^l)$ and $(x^l, \dots x^1, x^0, x^{-1}, \dots, x^{-l})$, respectively, is invariant under orthogonal transformations. Consequently, the orthogonal group in $(2l + 1)$ dimensions, which is a $l(2l + 1)$ parameter group, leaves the quadratic form

$$\sum_{k=-l}^{l} x^k x^{-k} = (x^0)^2 + 2 \sum_{k=1}^{l} x^k x^{-k} \tag{12}$$

invariant. The infinitesimal operators

$$X^i{}_k = -X^k{}_i = x^i \frac{\partial}{\partial x^{-k}} - x^k \frac{\partial}{\partial x^{-i}}; \quad i, k = 0, \pm 1, \ldots, \pm l \tag{13}$$

with the commutation relations

$$[X^i{}_k, X^m{}_n] = \delta[(k+m), 0]\, X^i{}_n - \delta[(k+n), 0] X^{im} - \delta[(i+m), 0] X_{kn} - \delta[(i+n), 0] X^m{}_k \tag{14}$$

generate the orthogonal group in $(2l + 1)$ dimensions. If

$$\left.\begin{aligned} H_i &= X^i{}_{-i};\ i > 0 \\ E_k &= X^0{}_{\pm k};\ k > 0 \\ E_{ik} &= X^{\pm i}{}_{\pm k};\ i, k > 0 \end{aligned}\right\} \tag{15}$$

the set of operators H_i, E_k, and E_{ik} forms the canonical basis for the algebra B_l.

The symplectic group in $2l$ dimensions leaves the antisymmetric bilinear form

$$\sum_{k=1}^{l} (x^k y^{-k} - x^{-k} y^k) \tag{16}$$

invariant. The infinitesimal operators that generate the group $Sp(2l)$ are

$$X^i{}_k = X^k{}_i = \epsilon^i x^i \frac{\partial}{\partial x^{-k}} + \epsilon^k x^k \frac{\partial}{\partial x^{-i}}; \qquad i, k = \pm 1, \pm 2, \ldots, \pm l \tag{17}$$

and the Lie brackets become

$$[X^i{}_k, X^m{}_n] = \epsilon^m \delta[(k+m), 0] X^i{}_n + \epsilon^n \delta[(k+n), 0] X^{im} + \epsilon^m \delta[(i+m), 0] X_{kn} + \epsilon^n \delta[(i+n), 0] X^m{}_k \tag{18}$$

where

$$\begin{aligned} \epsilon^k &= +1 \quad \text{for} \quad k > 0 \\ \epsilon^k &= -1 \quad \text{for} \quad k < 0 \end{aligned}$$

If

$$\begin{aligned} H_i &= X^i{}_{-i};\ i > 0 \\ E_{ik} &= X^{\pm i}{}_{\pm k};\ i, k > 0 \end{aligned} \tag{19}$$

the set of operators H_i and E_{ik} forms the canonical basis for the algebra C_l.

The orthogonal group in $2l$ dimensions leaves the quadratic form

$$\sum_{k=1}^{l} x^k x^{-k} \tag{20}$$

invariant. The infinitesimal operators defined in (13), with the commutation relations (14), generate the group if the condition

$$i, k \neq 0$$

is imposed. Then the set of operators H_i and E_{ik} given in (15) forms the canonical basis for the algebra D_l.

The unitary group in $(l + 1)$ dimensions, $U(l + 1)$, depends on $(l + 1)^2$ real parameters and leaves the quadratic form

$$\sum_{k=1}^{l+1} x^k x^{-k} \tag{21}$$

invariant. Its subgroup $SU(l + 1)$ is a $[(l + 1)^2 - 1]$ parameter group. The operators defined in (10) can also be taken as the generators of the group $SU(l + 1)$.

10^6. Classification of the irreducible representations of the simple and semisimple Lie groups

The h-dimensional representation space of a Lie group, or algebra, has l commuting operators H_i and $(h - l)$ noncommuting operators E_α. Because each of the operators H_i commutes with the others, they can be diagonalized simultaneously. In the representation space there exists at least one simultaneous proper-vector $|m\rangle$ of the operators H_i with the proper value m_i. That is,

$$H_i|m\rangle = m_i|m\rangle \tag{1}$$

where, instead of $|m\rangle$, any basis functions, denoted by ϕ or ψ in previous sections, could be used. The proper values m_i define a vector $\mathbf{m}$,

$$\mathbf{m} = (m_1, m_2, \ldots, m_l) \tag{2}$$

which is called the *weight* of the proper vector $|m\rangle$. Thus the root vector $\boldsymbol{\alpha}$, defined in $8^6 29$, is a weight in the adjoint representations. Furthermore, because

$$[H_i, E_\alpha]|m\rangle = \alpha_i E_\alpha|m\rangle$$

we have

$$H_i E_\alpha|m\rangle = (m_i + \alpha_i)E_\alpha|m\rangle \tag{3}$$

indicating that the weight of the vector $E_\alpha|m\rangle$ is $(\mathbf{m} + \boldsymbol{\alpha})$.

If $\mathbf{m}$ is one of the weights of an irreducible representation and $\boldsymbol{\alpha}$ is a root vector, then

$$2\frac{(\mathbf{m}, \boldsymbol{\alpha})}{(\boldsymbol{\alpha}, \boldsymbol{\alpha})} \tag{4}$$

is an integer, and

$$\mathbf{m} - 2\frac{(\mathbf{m}, \boldsymbol{\alpha})}{(\boldsymbol{\alpha}, \boldsymbol{\alpha})} \tag{5}$$

is another weight of the same representation [15b, p. 47]. Thus a diagram constructed for the weights of an irreducible representation has the same symmetry as the root figure of the group to which the representation belongs. However, in constrast to the roots, the weights can be degenerate. Then the number of proper vectors associated with the same weight is said to be the *multiplicity* of the degenerate weight. Proper vectors with different weights are linearly independent [15b, p. 46].

The following definitions and properties of weights are useful:

1. **m** is *positive* if its first nonvanishing component is positive.
2. **m** is *higher* then **m**′ if the first nonvanishing component of the vector (**m** − **m**′) is positive.
3. **m** is *equivalent* to all the other roots defined by (5), for all α.
4. **m** is *dominant* if it is not lower than all of its equivalents.
5. **m** is *simple* if it is nondegenerate.
6. $\mathbf{m}^0$ stands for the highest weight.

The following three theorems and four lemmas are given without proof [15b, pp. 48–51].

Theorem 1. The highest weight $\mathbf{m}^0$ of an irreducible representation Γ_i is simple.

Theorem 2. The necessary and sufficient condition for two irreducible representations Γ_i and Γ_j to be equivalent is that they should have the same highest weight $\mathbf{m}^0$.

Theorem 3. If $\mathbf{m}^0$ is the highest weight and $\boldsymbol{\alpha}^0$ is a simple root, then the quantity

$$k_\alpha = 2\frac{(\mathbf{m}^0, \boldsymbol{\alpha}^0)}{(\boldsymbol{\alpha}^0, \boldsymbol{\alpha}^0)} \tag{6}$$

is a positive integer.

Lemma 1. The weights **m** and $\mathbf{m} - [2(\mathbf{m}, \boldsymbol{\alpha})/(\boldsymbol{\alpha}, \boldsymbol{\alpha})]\boldsymbol{\alpha}$ have the same multiplicity.

Lemma 2. Every semisimple Lie group of rank l has l fundamental weights $\mathbf{m}^{(1)}, \mathbf{m}^{(2)}, \ldots \mathbf{m}^{(l)}$. Therefore, the dominant weight is given by

$$\mathbf{m} = \sum_i k_i \mathbf{m}^{(i)} \tag{7}$$

Lemma 3. The l fundamental representations $\Gamma_1, \Gamma_2, \ldots, \Gamma_l$ of a semisimple Lie group of rank l are defined as the representations which have the fundamental weights $\mathbf{m}^{(1)}, \mathbf{m}^{(2)}, \ldots, \mathbf{m}^{(l)}$ as their highest weights.

Lemma 4. For any given dominant weight there exists an irreducible representation having this weight as its highest weight.

Because the Dynkin diagrams, defined in terms of simple roots, may be used to classify the semisimple Lie groups; because the irreducible representations of a semisimple Lie group can be labeled by their highest weights (Theorem 2); and because (6) implies that we may associate a positive integer k_α with each of the l highest weights, a Dynkin diagram in which every small circle, corresponding to a simple root α^0, is marked by the positive integer k_α uniquely classifies the irreducible representations of a simple or semisimple Lie group.

Alternatively, the l positive integers k_α may be ordered in the same manner as defined in 5[6]8, resulting in the unique partition

$$[k_1, k_2, \ldots, k_l] \tag{8}$$

where $k_1 \geqslant k_2 \ldots \geqslant k_l$, each k_i corresponding to a simple root. This partition also may be used to classify the irreducible representations of a simple or a semisimple Lie group.

Using the roots of the simple Lie algebras, as listed in Table 1[6], the conditions imposed by (5) may be deduced for some of the algebras.

The algebra A_l has $l(l + 1)$ roots $(\hat{\mathbf{e}}_i - \hat{\mathbf{e}}_j)$ formed by the $(l + 1)$ unit vectors $\hat{\mathbf{e}}_i$ [15b, p. 42]. The roots are situated on a plane which is perpendicular to the vector $(\hat{\mathbf{e}}_1 + \hat{\mathbf{e}}_2 + \ldots + \hat{\mathbf{e}}_{l+1})$. This condition reduces the dimensions of the root diagrams to $(l - 1)$ [3e, p. 172]. Then there are $(l + 1)$ components of the weight $\mathbf{m}$ in an l-dimensional weight space. In order to have all components fall in the weight space, only l of the $(l + 1)$ components should be linearly independent; that is,

$$\sum_{i=1}^{l+1} m_i = 0 \tag{9}$$

This condition can be obtained from the relation 9[6]11 which, in turn, was a result of 9[6]9, necessitated by the unimodular condition. Thus the condition 9[6]9 ensures that all $(l + 1)$ components of the weight $\mathbf{m}$ fall in the l-dimensional weight space. Because

$$(\boldsymbol{\alpha}, \boldsymbol{\alpha}) = |\hat{\mathbf{e}}_i - \hat{\mathbf{e}}_j|^2 = 2$$

a weight

$$\mathbf{m} = m_1\hat{\mathbf{e}}_1 + m_2\hat{\mathbf{e}}_2 + \ldots + m_{l+1}\hat{\mathbf{e}}_{l+1}$$

must satisfy, in addition to the condition (9), the condition that

$$\mathbf{m} \cdot (\hat{\mathbf{e}}_i - \hat{\mathbf{e}}_j) = m_i - m_j \tag{10}$$

be an integer. Therefore, the components of the weight $\mathbf{m}$ are fractions with the denominator $(l + 1)$, which differ by integers.

In the algebra B_l two conditions result: The quantities $(m_i - m_j)$ and

$2(\mathbf{m}, \hat{\mathbf{e}}_i)$ are integers. Therefore, the components of any weight are either all integers or all half-integers.

For the algebra C_l the quantities $(m_i - m_j)$ and $(\mathbf{m}, 2\hat{\mathbf{e}}_i)/2$ have to be integers, which implies that all components of the weight $\mathbf{m}$ are integers.

In the algebra D_l both $(m_i - m_j)$ and $(m_i + m_j)$ must be integers. Therefore, the components of $\mathbf{m}$ can be either integer or half-integer.

11^6. Homomorphism and isomorphism of Lie algebras and Lie groups; Casimir operators; invariant integration; infinitesimal matrices

We discuss here the concepts of homomorphism, isomorphism, and automorphism defined in Section 1^6 as applied to Lie algebras. If A and B are two ensembles and if for each element x in A, there corresponds an element x' in B, an *application* or mapping ϕ of A in B is defined such that

$$x \xrightarrow{\phi} x' \text{ or } x' = \phi(x)$$

Any application ϕ of a Lie algebra A in another Lie algebra B that preserves the commutation relation

$$\phi([X_\alpha, X_\beta]) = [\phi(X_\alpha), \phi(X_\beta)] \tag{1}$$

and

$$\phi(mX_\alpha + nX_\beta) = m\phi(X_\alpha) + n\phi(X_\beta) \tag{2}$$

where X_α and X_β are the basis of algebra A, and m, n are constants, is a *homomorphism*. If there is a one-to-one correspondence between X_α, X_β and $\phi(X_\alpha)$, $\phi(X_\beta)$, respectively, then ϕ is an *isomorphism*. That is, two Lie algebras are isomorphic when the number of parameters and the law of combination of these parameters are the same. If the Lie algebras A and B are identical, the isomorphism becomes an *automorphism*. If B is an algebra associated with the linear operators of a vectorial space, then the words *representation* and *faithful representation* are respectively substituted for homomorphism and isomorphism.

The following are theorems about homomorphism and isomorphism:

Theorem 1. The kernel $N = \phi^{-1}(0)$ of a homomorphism $A \xrightarrow{\phi} B$ is an ideal of A.

Theorem 2. There exists an isomorphism between the algebra of A/N and B.

The *center* C of an algebra A is defined as the set of operators $\mathbf{c}$ which commute with all the basis elements $\mathbf{x}$ of the algebra

$$[\mathbf{c}, \mathbf{x}] = 0 \tag{3}$$

The center of A is also an ideal of A.

The following isomorphisms between the compact Lie groups are useful:

$$\begin{aligned} SO(2) &\approx U(1) \\ SU(2) &\approx Sp(2) \\ SO(3) &\approx \frac{SU(2)}{Z_2} \\ SO(4) &\approx SU(2) \times \frac{SU(2)}{Z_2} \\ SO(5) &\approx \frac{Sp(4)}{Z_2} \\ SO(6) &\approx \frac{SU(4)}{Z_2} \end{aligned} \tag{4}$$

where Z_2 is the cyclic group of two dimensions and is also the center of the groups $SO(2l + 1)$ and $Sp(2l)$. The cyclic group Z_n is one of the realizations of the abstract cyclic group whose elements are defined in 1[6]14.

The quadratic *Casimir operator* [16], which is very useful in discussing Lie algebras (such as to characterize the representation and the state space), is defined as

$$C^2 = \sum_{\alpha\beta} g^{\alpha\beta} X_\alpha X_\beta \tag{5}$$

where $\Sigma g^{\alpha\beta} g_{\beta\gamma} = \delta^\alpha_\gamma$. The Casimir operator commutes with all the operators of a Lie algebra [4[2]c, pp. 317–9]. That is,

$$[C^2, X_\gamma] = 0 \tag{6}$$

The minimum number of such operators is equal to the rank of the algebra. However, for the remaining operators a general relation like (5) cannot be given. It is necessary to specify the Lie algebra to evaluate them.

The following theorem is attributable to Racah [15b, pp. 53–6]: For every semisimple Lie group there exists a set of l functions $F_i(X_\alpha)$, which commute with every operator of the group; that is,

$$[F, X_\alpha] = 0 \tag{7}$$

and whose proper values characterize the irreducible representations.

Although the Casimir operators commute with every operator of the group, there are cases where their proper values are not sufficient to label the irreducible representations. Therefore, in general, the functions $F_i(X_\alpha)$ in (7), are called the "generalized Casimir operators," and are not the same as the ordinary Casimir operators. However, for most cases they are identical.

For compact groups ($g^{\alpha\beta} = -\delta^{\alpha\beta}$), the quadratic Casimir operator becomes

$$C^2 = \sum_{\alpha} X_{\alpha}^2 \tag{8}$$

Before the theory of continuous groups that has been presented so far can be applied, it is necessary to generalize some of the relations given for the finite groups, such as the orthogonality of group characters, to the continuous groups in general and to the Lie groups in particular. Because the product of two elements is another element $g_b g_a = g_c$, the weights attached to an element g_a and to the product of two elements $g_b\, g_a$ are equal. It is possible to associate a *differential volume measure* $d\tau_a$ with the set of group elements in the neighborhood of g_a. Then the differential volume measure $d\tau_{ba}$ associated with the set of elements in the neighborhood of $g_b^{-1} g_a$ are equal to $d\tau_a$ if these two sets have the same weight:

$$d\tau_{ba} = d\tau_a \tag{9}$$

The new element $g_b^{-1} g_a$ is the left translation of g_a by g_b and therefore this relation establishes a *left-invariant* measure. When $f(g_a)$ and $f(g_b^{-1} g_a)$ are two functions defined in the group manifold, for finite groups

$$\sum_G f(g_a) = \sum_G f(g_b^{-1} g_a) \tag{10}$$

This is because each summation contains all the same functions of the group elements in different order. The generalization of the above relation to the Lie groups is accomplished by replacing sums with integrals and integrating over the differential volume $d\tau_a$. That is,

$$\int_G f(g_a)\, d\tau_a = \int_G f(g_b^{-1} g_a)\, d\tau_a \tag{11}$$

In (10) the range of summation could be limited to a subset, assuming that a given subset $\mathfrak{M}$ is left transferred to another subset $\mathfrak{M}'$ under the left multiplication by g_b^{-1}. Then

$$\sum_{\mathfrak{M}} f(g_a) = \sum_{\mathfrak{M}'} f(g_b^{-1} g_a) \tag{10'}$$

and correspondingly for the Lie groups

$$\int_{\mathfrak{M}} f(g_a)\, d\tau_a = \int_{\mathfrak{M}'} f(g_b^{-1} g_a)\, d\tau_a \tag{11'}$$

In the parameter space where the group elements g_a and g_b are designated by a set of continuous parameters a^i and b^i (or equally by vectors **a** and **b**) the volumes occupied by these sets are da and db, respectively. The left translation of the set a^i by the set b^i results in another set of parameters c^i given by

$$c^i = \phi^i(a, b) \tag{12}$$

Thus, after the translation, the volume occupied by the set is dc. The density functions $\rho(a)$ and $\rho(c)$ in the parameter space are defined such that

$$d\tau_{\mathfrak{M}} = \rho(a)\, da = \rho(c)\, dc = d\tau_{\mathfrak{M}'} \tag{13}$$

That is, the measure of the group elements in two subsets $\mathfrak{M}$ and $\mathfrak{M}'$ of the group manifold is preserved. The value of the density function in the neighborhood of the origin of the parameter space, corresponding to the identity element e; that is, $\rho(0)$ may be fixed arbitrarily as unity:

$$\rho(0) = 1 \tag{14}$$

A left translation of a set in the neighborhood of identity by the set b^i gives the same set, b^i:

$$b^i = \phi^i(0, b) \tag{15}$$

The volume element associated with the ith component of the set b is obtained as

$$db^i = \sum_j \left[\frac{\partial \phi^i(a, b)}{\partial a^j}\right]_{a=0} da^j = \phi^i{}_j\,(0, b)'\, da^j \tag{16}$$

Then

$$db = \begin{bmatrix} \phi^1{}_1\,(0, b)' & \phi^2{}_1\,(0, b)' & \ldots & \phi^n{}_1\,(0, \mathrm{b})' \\ \phi^1{}_2\,(0, b)' & \phi^2{}_2\,(0, b)' & \ldots & \phi^n{}_2\,(0, \mathrm{b})' \\ \ldots & \ldots & \ldots & \ldots \\ \phi^1{}_n\,(0, b)' & \phi^2{}_n\,(0, b)' & \ldots & \phi^n{}_n\,(0, \mathrm{b})' \end{bmatrix} da = J(b)\, da \tag{17}$$

where $J(b)$ is the *Jacobian*.
Then the condition

$$\rho(b)\, db = \rho(0)\, da$$

reduces to

$$\rho(b) = \frac{\rho(0)}{J(b)} = \frac{1}{J(b)} \tag{18}$$

For an r parameter Lie group of linear transformations the *infinitesimal matrix* σ_k is given as

$$[\sigma_k]_{ij} = -i\left[\frac{\partial}{\partial a^k} g_{ij}\right]_{a^k=0} \tag{19}$$

where g_{ij} stands for the matrix elements of the group element g. In the neighborhood of the identity each element g_{ij} may be expanded in series in powers of the parameters a^k. Because the parameters are small, only the terms that are linear in a^k need be retained, giving the result

$$g = e + i\sum_k a^k\sigma_k = e^{ia^k\sigma_k} \tag{20}$$

The infinitesimal matrices satisfy the commutation relation

$$[\sigma_k, \sigma_l] = -i\sum_m f_{kl}{}^m\sigma_m \tag{21}$$

Also the infinitesimal matrices A_k of a representation can be defined as

$$A_k = -i\left[\frac{\partial D}{\partial a^k}\right]_{a^k=0} \tag{22}$$

where $D = D(a^1, a^2, \ldots, a^r)$ is a representation matrix. Again the infinitesimal matrices A_k obey the commutation relation

$$[A_k, A_l] = -i\sum_m f_{kl}{}^m A_m \tag{23}$$

Any left translation of a set a^i by b^i to the new set c^i may be accomplished in two steps by using these relations: first, going to the origin with an inverse translation, which involves a^{-1}, and then translating to the set c^i. In general, the left invariant measure is not the same as the right invariant measure, but for the compact Lie groups they are the same [4²c, pp. 316–7].

Many of the relations given for finite groups may now be generalized for the compact Lie groups by changing sums to integrals and integrating over the infinitesimal volume measure $d\tau$, as applied later to specific Lie groups. Some of the general statements made for finite groups may also be generalized to compact Lie groups. In particular, (1) every representation is equivalent to a unitary representation; (2) every representation is fully reducible to a sum or irreducible representations of finite dimensions; (3) the regular representation contains all irreducible representations of the group.

12[6]. Two- and three-dimensional rotation groups and their representations

Two- and three-dimensional rotations in connection with the symmetries of the nonrelativistic Hamiltonian were discussed in Chapter 3, specifically Section 1. They are discussed here to demonstrate the use of group theoretical formalism developed in the previous sections.

A rotation in the XY plane may be expressed by the matrix

$$R(\theta) = \begin{bmatrix} \cos\theta & -\sin\theta \\ \sin\theta & \cos\theta \end{bmatrix} \tag{1}$$

where θ is the amount of rotation. A group formed by such matrices is isomorphic to $SO(2)$.

Because

$$R'(\theta) = BR(\theta)B^{-1} = \begin{bmatrix} e^{i\theta} & 0 \\ 0 & e^{-i\theta} \end{bmatrix} \tag{2}$$

is a rotation matrix, where

$$B = \frac{1}{\sqrt{2}}\begin{bmatrix} 1 & i \\ 1 & -i \end{bmatrix} \tag{3}$$

the group $SO(2)$ is fully reducible.

The transformation (2) may be considered as a change of basis from the orthonormal base

$$\hat{\mathbf{e}}_1,\ \hat{\mathbf{e}}_2, \text{ where } \quad (\hat{\mathbf{e}}_i, \hat{\mathbf{e}}_j) = \delta_{ij} \tag{4}$$

to the *canonical base*

$$\hat{\mathbf{e}}_+ = \frac{1}{\sqrt{2}}(\hat{\mathbf{e}}_1 + i\hat{\mathbf{e}}_2), \qquad \hat{\mathbf{e}}_- = \frac{1}{\sqrt{2}}(\hat{\mathbf{e}}_1 - i\hat{\mathbf{e}}_2) \tag{5a}$$

where

$$(\hat{\mathbf{e}}_+, \hat{\mathbf{e}}_+) = (\hat{\mathbf{e}}_-, \hat{\mathbf{e}}_-) = 0; \quad (\hat{\mathbf{e}}_+, \hat{\mathbf{e}}_-) = 1 \tag{5b}$$

The group $SO(2)$ is Abelian; therefore, its irreducible representation matrices are all one-dimensional:

$$R(\theta) \to D^{(m)} = e^{im\theta} \tag{6}$$

For a single-valued representation m can take negative as well as positive integer values. Only the representations Γ_1 and Γ_{-1} associated with the matrices $D^{(1)}$ and $D^{(-1)}$ are faithful. The representation Γ_0 is trivial. The representations Γ_m and $\Gamma_{m'}$ are distinct for $m' \neq m$.

The plane rotations may be labeled by a single continuous parameter θ. Therefore, the infinitesimal operator defined in 8^{6}4 becomes

$$X_\theta = \frac{\partial}{\partial\theta} = x\frac{\partial}{\partial y} - y\frac{\partial}{\partial x} \tag{7}$$

The only generator of the group is

$$J_\theta = -i\frac{\partial}{\partial\theta} \tag{8}$$

which results in the proper value equation

$$J_\theta\, e^{im\theta} = me^{im\theta} \tag{9}$$

Because the representation matrices are one-dimensional, their values are also equal to the characters of the group; that is,

$$\chi^{(m)}(\theta) = e^{im\theta} \tag{10}$$

Thus the characters $\chi^{(m)}$ and $\chi^{(-m)}$ are complex conjugates. Two dimensional rotations may be associated with transformations of the type

$$\mathbf{x}' = \mathbf{x} + \mathbf{b}$$

In the parameter space this becomes

$$\mathbf{c} = \mathbf{a} + \mathbf{b}$$

Then the density function

$$\rho(b) = \left(\frac{1}{\partial c/\partial a}\right)_{a=0} = 1$$

is uniform. Thus the orthogonality property of finite group characters, ($3^6$4) takes the form

$$\int_0^{2\pi} (\chi^{(m)})^* \chi^{(m')}\, d\theta = 2\pi\delta_{mm'} \tag{11}$$

where, because $\rho(\theta) = 1$, the integration over the group reduces to integration over θ.

The Laplace operator is invariant under two-dimensional rotation. In the complex plane, where $z = x + iy = re^{i\theta}$, the Laplace equation takes the form

$$\frac{\partial^2\phi}{\partial z \partial z^*} = 0 = \frac{\partial^2\phi}{\partial r^2} + \frac{1}{r}\frac{\partial\phi}{\partial r} + \frac{1}{r^2}\frac{\partial\phi}{\partial\theta^2} \tag{12}$$

A solution is a polynomial of degree m,

$$P_m = r^m(A\cos m\theta + B\sin m\theta) \tag{13}$$

The ensemble of these polynomials constitute a representation space of $SO(2)$ for $D^{(m)}$.

The irreducible and real representations of $SO(2)$, besides the trivial representation for $m = 0$, may be obtained from the direct sum of one dimensional irreducible and complex representations. Therefore, two-dimensional real representations have the form $D^{(m)} + D^{(-m)}$ $(m \neq 0)$.

$$D_R^{(m)} = \begin{bmatrix} \cos m\theta & -\sin m\theta \\ \sin m\theta & \cos m\theta \end{bmatrix} \tag{14}$$

The infinitesimal matrix σ_θ ($11^6$19) associated with the group element given in (1) is one of the Pauli matrices ($2^3$6′):

$$\sigma_\theta = \begin{bmatrix} 0 & i \\ -i & 0 \end{bmatrix} \tag{15}$$

The three-dimensional rotational group $SO(3)$ is a three-parameter group. The parameters usually are chosen as the axis of rotation $\hat{\mathbf{n}}$, a unit vector, which may be specified by an azimuthal and polar angle and the rotation ω about this axis, the sense being fixed by the right-hand rule. If a vector $\boldsymbol{\omega}$ is defined as

$$\boldsymbol{\omega} = \omega\hat{\mathbf{n}}$$

the parameter space, corresponding to all rotations about a point, becomes a sphere of radius ω where ω varies between zero and π. The parameter space is finite, so this group is compact. Two points on the surface of a sphere of radius π, at opposite ends of a diameter, correspond to the same rotation through π.

The vector $\mathbf{x}'$, obtained by rotating the vector $\mathbf{x}$ about a unit vector $\hat{\mathbf{n}}$ by an angle ω, is given by

$$\begin{aligned}\mathbf{x}' &= \hat{\mathbf{n}}(\hat{\mathbf{n}}\cdot\mathbf{x}) - \hat{\mathbf{n}}\,(\hat{\mathbf{n}}\times\mathbf{x})\cos\omega + (\hat{\mathbf{n}}\times\mathbf{x})\sin\omega \\ &= \mathbf{x}\cos\omega + \hat{\mathbf{n}}(\hat{\mathbf{n}}\cdot\mathbf{x})\,(1-\cos\omega) + (\hat{\mathbf{n}}\times\mathbf{x})\sin\omega\end{aligned} \tag{16}$$

Therefore the elements of the transformation matrix $R(\hat{\mathbf{n}}, \omega)$ defined as

$$\mathbf{x}' = R(\hat{\mathbf{n}}, \omega)\mathbf{x} \tag{17}$$

are

$$R^{ij} = \delta^{ij}\cos\omega + n^i n^j(1-\cos\omega) + n^k\epsilon_{ijk}\sin\omega \tag{18}$$

in which

$$\begin{aligned}\epsilon_{ijk} &= 0 \quad \text{if any two of the indices are equal} \\ \epsilon_{ijk} &= 1 \quad \text{for indices 123 or an even permutation} \\ \epsilon_{ijk} &= -1 \quad \text{for an odd permutation of 123}\end{aligned}$$

The matrix $R(\hat{\mathbf{n}}, \omega)$ is orthogonal, $\widetilde{R} = R^{-1}$, and its determinant is $+1$. The group formed by the matrices $R(\hat{\mathbf{n}}, \omega)$ is isomorphic to the group $SO(3)$. For rotation around the X-axis, $\hat{\mathbf{n}}^1 = \hat{\mathbf{e}}_1$, and $\hat{n}^2 = \hat{n}^3 = 0$, $R(\hat{\mathbf{n}}, \omega)$ reduces to

$$R(\hat{\mathbf{e}}_1, \omega) = \begin{bmatrix} 1 & 0 & 0 \\ 0 & \cos\omega & -\sin\omega \\ 0 & \sin\omega & \cos\omega \end{bmatrix} \tag{19}$$

A matrix B

$$B = \begin{bmatrix} 1 & 0 & 0 \\ 0 & 1/\sqrt{2} & i/\sqrt{2} \\ 0 & 1/\sqrt{2} & -i/\sqrt{2} \end{bmatrix} \tag{20}$$

diagonalizes $R(\hat{\mathbf{e}}_1, \omega)$:

$$R'(\hat{\mathbf{e}}_1, \omega) = BR(\hat{\mathbf{e}}_1, \omega)B^{-1} = \begin{bmatrix} 1 & 0 & 0 \\ 0 & e^{i\omega} & 0 \\ 0 & 0 & e^{-i\omega} \end{bmatrix}$$

Rotations through the same angle about any axis are equivalent and belong to the same class. Therefore, $SO(3)$ has an uncountably infinite number of classes. Because the group elements $R(\hat{\mathbf{n}}, \omega)$ do not commute, $SO(3)$ is not Abelian.

Denoting the three parameters in the neighborhood of the identity by $\delta\omega^1$, $\delta\omega^2$, $\delta\omega^3$, the transformation matrix for an infinitesimal rotation is

$$S = \begin{bmatrix} 1 & -\delta\omega^3 & \delta\omega^1 \\ \delta\omega^3 & 1 & -\delta\omega^2 \\ -\delta\omega^1 & \delta\omega^2 & 1 \end{bmatrix} \tag{21}$$

where $\delta\omega^1 = \hat{n}^1\delta\omega$, $\delta\omega^2 = \hat{n}^2\delta\omega$, $\delta\omega^3 = \hat{n}^3\delta\omega$

The infinitesimal matrices σ_1, σ_2, σ_3, as defined in $11^6$19, for $R(\hat{\mathbf{e}}_1, \omega)$, $R(\hat{\mathbf{e}}_2, \omega)$, $R(\hat{\mathbf{e}}_3, \omega)$, respectively, are

$$\sigma_1 = -i\left.\frac{\partial R(\hat{\mathbf{e}}_1, \omega)}{\partial\omega}\right|_{\omega=0} = \begin{bmatrix} 0 & 0 & 0 \\ 0 & 0 & i \\ 0 & -i & 0 \end{bmatrix} \tag{22a}$$

$$\sigma_2 = -i\left.\frac{\partial R(\hat{\mathbf{e}}_2, \omega)}{\partial\omega}\right|_{\omega=0} = \begin{bmatrix} 0 & 0 & -i \\ 0 & 0 & 0 \\ i & 0 & 0 \end{bmatrix} \tag{22b}$$

$$\sigma_3 = -i\left.\frac{\partial R(\hat{\mathbf{e}}_3, \omega)}{\partial\omega}\right|_{\omega=0} = \begin{bmatrix} 0 & i & 0 \\ -i & 0 & 0 \\ 0 & 0 & 0 \end{bmatrix} \tag{22c}$$

In terms of the infinitesimal matrices the transformation matrix S is

$$S = 1 + i(\sigma_1\hat{n}^1 + \sigma_2\hat{n}^2 + \sigma_3\hat{n}^3)\delta\omega = \exp i(\sigma_1\hat{n}^1 + \sigma_2\hat{n}^2 + \sigma_3\hat{n}^3)\delta\omega \tag{23}$$

Consequently, for a vector $\mathbf{x}$ undergoing an infinitesimal rotation

$$\mathbf{x}' = \{I + i(\sigma_1\hat{n}^1 + \sigma_2\hat{n}^2 + \sigma_3\hat{n}^3)\delta\omega\}\mathbf{x} \tag{24}$$

The density function $\rho(\theta)$ for the group $SO(3)$ [4^2c, p. 329; 3e, pp. 221–2] is

$$\rho(\omega) = 2\,\frac{1 - \cos\omega}{\omega^2} \tag{25}$$

Integration of a function $f(\hat{\mathbf{n}}, \omega)$ over the group becomes

$$\int f(\hat{\mathbf{n}}, \omega)\rho(\omega)\omega^2\, d\omega\, d\Omega \tag{26a}$$

where $d\Omega$ is the element of solid angle at the direction $\hat{\mathbf{n}}$. For the characters $\chi(\omega)$ of a representation, this integral is reduced to

$$8\pi\int_0^\pi \chi(\omega)(1 - \cos\omega)\, d\omega \tag{26b}$$

Consequently, the orthogonality relation for the characters of the ith and jth irreducible representations of $SO(3)$ may be written as

$$\frac{1}{\pi}\int_0^\pi (\chi^i)^* \chi^j (1 - \cos\omega)\, d\omega = \delta_{ij} \tag{27}$$

The total "volume" V of the group is given by

$$V = \int \rho(\omega)\omega^2\, d\omega\, d\Omega = 8\pi \int_0^\pi (1 - \cos\omega)\, d\omega = 8\pi^2 \tag{28}$$

The three generators ($8^6 4$) of the group $SO(3)$ are Hermitian,

$$J_1 = -i\frac{\partial}{\partial\omega^1};\ J_2 = -i\frac{\partial}{\partial\omega^2};\ J_3 = -i\frac{\partial}{\partial\omega^3} \tag{29}$$

Because the group is compact any of its operators may be chosen to be unitary. A new basis for the Lie algebra can be formed from the generators

$$J^+ = J_1 + iJ_2, \quad J^- = J_1 - iJ_2 \tag{30}$$

The quadratic Casimir operator for the group is

$$C^2 = J_1^2 + J_2^2 + J_3^2 \tag{31}$$

and the commutation relations are easily obtained:

$$[J_3, J^+] = J^+; [J_3, J^-] = -J^-; [J^+, J^-] = 2J_3 \tag{32}$$

Also

$$\begin{aligned} J^-J^+ &= C^2 - J_3(J_3 + 1) \\ J^+J^- &= C^2 - J_3(J_3 - 1) \end{aligned} \tag{33}$$

J^+ and J^- are the shift operators ($1^3 16$), and $C^2 = \mathbf{J}^2$ ($1^3 15$). The proper vectors $|j, m\rangle$ satisfy

$$\begin{aligned} C^2|j, m\rangle &= |j, m\rangle\, j(j + 1) \\ J_3|j, m\rangle &= |j, m\rangle\, m \end{aligned} \tag{34}$$

in which the range of the label m is ($1^3 19$)

$$-j \leqslant m \leqslant +j$$

For a two-dimensional rotation about the $\hat{n}^3 = \hat{\mathbf{e}}_3$ axis, $\omega = \theta$ and the representation matrix was found to be

$$D^m(\theta) = e^{im\theta}$$

where m is the proper value of $J_\theta = J_3$. This may be generalized to find the representation matrix for the group $SO(3)$, as

$$D^{(j)}(\hat{\mathbf{e}}_3, \theta) = \begin{bmatrix} e^{ij\theta} & & & & \\ & e^{i(j-1)\theta} & & & \\ & & \cdot & & \\ & & & \cdot & \\ & & & & \cdot \\ & & & & e^{-ij\theta} \end{bmatrix} \tag{35}$$

Thus the dimension of the jth irreducible representation Γ_j of $SO(3)$ is $2j + 1$.

The character of the jth class is found to be

$$\chi_j(\theta) = \sum_{m=-j}^{m=j} e^{im\theta} = \frac{\sin(j + 1/2)\theta}{\sin \theta/2} \tag{36}$$

The Laplace equation provides the irreducible and real representation of $SO(2)$ and of $SO(3)$. The Laplacian operator, in spherical polar coordinates, is invariant under rotation. Any solution ψ of the Laplace equation $\nabla^2\psi = 0$ is also invariant under the operations of $SO(3)$, thus providing the basis functions for the group. The general solution is

$$\psi(r, \theta, \phi) = \sum_{j=0}^{\infty} \sum_{m=-j}^{j} [A_{jm} r^j + B_{jm} r^{-(j+1)}] Y(j, m; \theta, \phi) \tag{37}$$

where j is allowed to take only integral values, and A_{jm} and B_{jm} are fixed by the boundary conditions, $Y(j, m; \theta, \phi)$ is a spherical harmonic ($4^3 11$), tabulated in Table 6^3

For a given j, the spherical harmonics give the basis for a $(2j + 1)$-dimensional representation. The rows are labeled by the index m. Because most of the results in Chapter 3 were presented in terms of the Euler angles, these angles are used as group parameters in the rest of this section. Because the representation matrix of a rotation around X-axis, $R(\alpha, 0, 0)$, may be given as $D_m(R) = e^{im\alpha}$,

$$R(\alpha, 0, 0) Y(j, m) = \sum_m e^{im\alpha} Y(j, m) \tag{38}$$

Then for a general rotation:

$$\begin{aligned} R(\alpha, \beta, \gamma) Y(j, m; \theta, \phi) &= Y(j, m; \theta', \phi') \\ &= \sum_{m'} Y(j, m'; \theta, \phi) D^{(j)}_{m'm}(\alpha, \beta, \gamma) \end{aligned} \tag{39}$$

where $D^{(j)}_{m'm}$ is the matrix of the rotation $R(\alpha, \beta, \gamma)$, in the representation Γ_j, based on the spherical harmonics of order j. The representation matrix $D^{(j)}_{m'm}$ was introduced in $2^3 9$, and many of its properties were investigated in Sections 2^3, 3^3, and 6^3.

$SO(3)$ is simply reducible; consequently the properties and definitions given in Section 6 for simply reducible groups are valid. By rewriting $6^6$4,

$$\Psi_{\gamma k} = \sum_{i,j} \psi_{\alpha i}\psi_{\beta j} \langle \alpha i, \beta j | \gamma k)$$

and substituting the spherical harmonics for the basis $\psi_{\alpha i}$ and $\psi_{\beta j}$,

$$\Psi_{jm}(\theta_1, \phi_1; \theta_2, \phi_2) = \sum_{m_1}\sum_{m_2} Y(j_1, m_1; \theta_1, \phi_1) Y(j_2, m_2; \theta_2, \phi_2)\langle j_1 m_1, j_2 m_2 | jm) \tag{40}$$

The new basis Ψ_{jm} belongs to the irreducible representation Γ_j.

When an operator of a group element acts on the new basis

$$\begin{aligned} R(\alpha, \beta, \gamma)\Psi_{jm} &= \sum_{m_3} \Psi_{jm_3} D^{(j)}_{m_3 m}(\alpha, \beta, \gamma) \\ &= \sum_{m_1}\sum_{m_2}\sum_{m_3} Y(j_1, m_1) Y(j_2, m_2) \langle j_1 m_1, j_2 m_2 | j m_3) D^{(j)}_{m_3 m} \end{aligned} \tag{41a}$$

or

$$\begin{aligned} R(\alpha, \beta, \gamma)\Psi_{jm} &= \sum_{m_3}\sum_{m_4} R(\alpha, \beta, \gamma)[Y(j_1, m_3; \theta_1, \phi_1) Y(j_2, m_4; \theta_2, \phi_2)]\langle j_1 m_3, j_2 m_4 | j \\ &= \sum_{m_3}\sum_{m_4}\sum_{m_1}\sum_{m_2} [Y(j_1, m_1) Y(j_2, m_2) D^{(j_1)}_{m_1 m_3} D^{(j_2)}_{m_2 m_4}]\langle j_1 m_3, j_2 m_4 | jm) \end{aligned} \tag{41b}$$

Because $Y(j_1, m_1)Y(j_2, m_2)$ are linearly independent functions, it is inferred that

$$\sum_{m_3}\sum_{m_4} D^{(j_1)}_{m_1 m_3} D^{(j_2)}_{m_2 m_4} \langle j_1 m_3, j_2 m_4 | jm) = \sum_{m_3} \langle j_1 m_1, j_2 m_2 | j m_3) D^{(j)}_{m_3 m} \tag{42}$$

If both sides of the above equation are multiplied by $(jm | j_1 m_3, j_2 m_4\rangle$ and summed over the indices j and m, with the help of $6^6$7b,

$$D^{(j_1)}_{m_1 m_3} D^{(j_2)}_{m_2 m_4} = \sum_{j,m,m_3} \langle j_1 m_1, j_2 m_2 | j m_3) D^{(j)}_{m_3 m} (jm | j_1 m_3, j_2 m_4\rangle \tag{43}$$

This result was presented in $3^3$20. From the developments in Section 3^3, the connections between the vector coupling coefficients, $3j$ symbols, and Clebsch-Gordan coefficients for the group $SO(3)$ are obvious.

13^6. Spinor representations and the unimodular unitary group $SU(2)$.

In the previous section the index j was assumed to be an integer. These representations are single-valued. In this section the case of $j = 1/2$ is investigated, because the acceptable values for j can be half-integrals

(1³18). The representations $D_{1/2}$ of the group $SO(3)$ is called the *two-valued representation* or the *spin representation.*

The Pauli matrices, 2³6′, are Hermitian and satisfy the following relations:

$$\begin{aligned}
&\sigma_i\sigma_j = \delta_{ij}I + \epsilon_{ijk}\sigma_k \\
&\boldsymbol{\sigma} \times \boldsymbol{\sigma} = 2i\boldsymbol{\sigma} \\
&\sigma_j\sigma_k\sigma_l = \delta_{kl}\sigma_j - \delta_{jl}\sigma_k + \delta_{jk}\sigma_l + i\epsilon_{jkl} \\
&\mathrm{Tr}(\sigma_i) = 0 \\
&\mathrm{Tr}(\sigma_i\sigma_j) = 2\delta_{ij} \\
&\mathrm{Tr}(\sigma_j\sigma_k\sigma_l) = 2i\epsilon_{jkl}
\end{aligned} \tag{1}$$

The scalar product of the vector **x** and the matrix $\boldsymbol{\sigma}$, denoted by X,

$$X = x^i\sigma_i = \begin{bmatrix} x^3 & x^1 - ix^2 \\ x^1 + ix^2 & -x^3 \end{bmatrix} \tag{2}$$

is called a spinor representation of the vector **x** in the two-dimensional *spinor space* defined by the matrices σ_i. The operator X is Hermitian and its trace is zero; that is,

$$X^\dagger = X \quad \text{and} \quad \mathrm{Tr}X = 0 \tag{3}$$

If U is a 2×2 unitary matrix, the trace of X is unchanged under the unitary transformation

$$X' = UXU^\dagger \tag{4}$$

Therefore, $\mathrm{Tr}(UXU^\dagger) = 0$ and X' is Hermitian.

These indicate that X' may be written in terms of a vector $\mathbf{x}'$ and the matrix $\boldsymbol{\sigma}$ as

$$X' = x'^i\sigma_i \tag{5}$$

where $\mathbf{x}'$ is obtained by a rotation from **x**; that is,

$$\mathbf{x}' = R(\hat{\mathbf{n}}, \omega)\mathbf{x}; \quad x'^i = R_{ij}x^j \tag{6}$$

Then

$$\begin{aligned}
x'^i\sigma_i &= UXU^\dagger = R_{ij}x^j\sigma_i \\
Ux^j\sigma_jU^\dagger &= R_{ij}x^j\sigma_i \quad \text{for all } \mathbf{x}
\end{aligned}$$

thus giving

$$U\sigma_jU^\dagger = \sigma_iR_{ij} \tag{7}$$

Multiplying both sides by σ_i and taking the trace, using the fifth relation in (1), gives:

$$R_{ij}(U) = \frac{1}{2}\mathrm{Tr}(\sigma_iU\sigma_jU^\dagger) \tag{8}$$

This is the spinor representation of the matrix of a three-dimensional rotation. $+U$ or $-U$ corresponds to the same R:

$$\pm U \rightarrow R$$

There is thus a two-to-one correspondence between U and R; consequently U and R are homomorphic but not isomorphic.

The most general form of a unitary matrix U is

$$U = aI + \boldsymbol{\sigma} \cdot \mathbf{b}$$

where a and $\mathbf{b}$ are complex. Because

$$UU^\dagger = U^\dagger U = I$$

we have

$$|a|^2 + \boldsymbol{\sigma} \cdot (a\mathbf{b}^* + a^*\mathbf{b}) \pm i(\mathbf{b} \times \mathbf{b}^*) \cdot \boldsymbol{\sigma} = I$$

which implies that

$$a\mathbf{b}^* + a^*\mathbf{b} = 0, \quad \mathbf{b} \times \mathbf{b}^* = 0$$

$$a = e^{i\phi} \cos\frac{\omega}{2}, \quad b = -ie^{i\phi} \sin\frac{\omega}{2}$$

Therefore,

$$\begin{aligned} U &= e^{i\phi}[I \cos(\omega/2) - i\boldsymbol{\sigma} \cdot \hat{\mathbf{n}} \sin(\omega/2)] \\ &= e^{i\phi} \begin{bmatrix} \cos(\omega/2) - i\hat{n}^3 \sin(\omega/2) & -i(\hat{n}^1 - i\hat{n}^2)\sin(\omega/2) \\ -i(\hat{n}^1 + i\hat{n}^2)\sin(\omega/2) & \cos(\omega/2) + i\hat{n}^3 \sin(\omega/2) \end{bmatrix} \end{aligned} \tag{9}$$

where the angles ϕ and ω, as well as the unit vector $\hat{\mathbf{n}}$, are shown in Figure 4[6].

These 2×2 unitary matrices U form the four-parameter group $U(2)$, the four parameters being ϕ and the three parameters of the three dimensional group. The determinant of the unitary matrix U is $e^{i\phi}$, consequently the matrix

$$U^+ = \frac{U}{e^{i\phi}} \tag{10}$$

is unimodular, determinant $= \pm 1$, and unitary. The plus sign as a right superscript indicates that U is unimodular.

The group formed by all 2×2 unimodular unitary matrices U^+ is the three-parameter group $SU(2)$. The quotient group $SU(2)/Z_2$ is isomorphic to $SO(3)$. Because Z_2 is its center (1[6]), the group $SU(2)$ is called the *covering group* of $SO(3)$. The idea of a covering group, obtained by forming G'/Z' where Z' is the center of G', is used to reduce the homomorphism to the isomorphism of two groups G and G'.

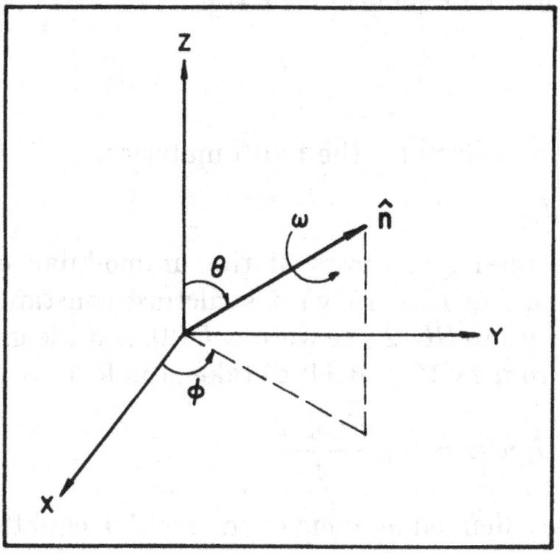

Figure 4[6]. Rotation around a given direction specified by the unit vector $\hat{\mathbf{n}}$.

The matrix U may be expressed in terms of the rotation matrix $R(\hat{\mathbf{n}}, \omega)$ by

$$U = \frac{1}{2} e^{i\phi} \frac{(I + \sigma_j R^{jk} \sigma_k)}{\pm[\mathrm{Tr} R + 1]^{1/2}} \tag{11}$$

For the unimodular unitary matrix U^+ this reduces to

$$U^+ = \frac{I + \sigma_j R^{jk} \sigma_k}{\pm 2[\mathrm{Tr} R + 1]^{1/2}} \tag{11'}$$

For infinitesimal ϕ and ω the first equation in (9) becomes

$$U = (1 + i\phi)(I + i\frac{\omega}{2}\boldsymbol{\sigma} \cdot \hat{\mathbf{n}}) \tag{12}$$

Then the infinitesimal matrices (11[6]19)

$$\left.\frac{\partial U}{\partial \omega^k}\right|_{\omega^k=0} = \frac{1}{2} i\sigma_k \quad \text{and} \quad \left.\frac{\partial U}{\partial \phi}\right|_{\phi=0} = iI \tag{13}$$

For the unimodular matrices these reduce to

$$U^+ = I + i\frac{\omega}{2}\boldsymbol{\sigma} \cdot \hat{\mathbf{n}} \tag{12'}$$

and

$$\left.\frac{\partial U^+}{\partial \omega^k}\right|_{\omega^k=0} = \frac{1}{2} i\sigma_k \tag{13'}$$

Consequently the infinitesimal generator (5^64) is

$$J_k = \frac{1}{2}\sigma_k \tag{14}$$

From the commutation relations for the Pauli matrices,

$$[J_k, J_l] = i\epsilon_{klm}J_m \tag{15}$$

Therefore, the infinitesimal generators of the unimodular unitary group $SU(2)$ generate a Lie algebra with structure constants ϵ_{klm}, showing again that the group $SU(2)$, as well as $U(2)$, is a Lie group.

The matrix $R(\hat{\mathbf{n}}, \omega)$ from 11^619 and 11^620 takes the form

$$R(\hat{\mathbf{n}}, \omega) = \exp iJ_k\hat{n}^k\omega = \exp\frac{i\sigma_k\hat{n}^k\omega}{2} \tag{16}$$

The proper values, λ, are defined as roots of the secular equation

$$\det(U - \lambda) = 0 \tag{17}$$

Diagonalization of the matrix U amounts to making the rotation axis $\hat{\mathbf{n}}$ coincide with the Z-axis. The matrix V, which diagonalizes the matrix U, is given by

$$V = \begin{bmatrix} \cos(\theta/2) & e^{i\phi}\sin(\theta/2) \\ -e^{-i\phi}\sin(\theta/2) & \cos(\theta/2) \end{bmatrix} \tag{18}$$

where the angles θ and ϕ are shown in Figure 4^6. The matrix V is unitary and unimodular, and the proper vectors, that is, the rows of V, are

$$|V_1\rangle = \begin{bmatrix} \cos(\theta/2) \\ e^{i\phi}\sin(\theta/2) \end{bmatrix} \quad \text{and} \quad |V_2\rangle = \begin{bmatrix} -e^{-i\phi}\sin(\theta/2) \\ \cos(\theta/2) \end{bmatrix} \tag{19}$$

The properties of V ensure that the proper vectors are othogonal and orthonormal.

After diagonalization, the diagonal matrices U' and U'^+ correspond to rotations about the Z-axis by an angle ω (see also 5^37); that is,

$$U' = e^{i\phi}\begin{bmatrix} \exp(+i\omega/2) & 0 \\ 0 & \exp(-i\omega/2) \end{bmatrix} = e^{i\phi}\exp(+i\omega/2)\sigma_3 \tag{20}$$

$$U'^+ = \begin{bmatrix} \exp(+i\omega/2) & 0 \\ 0 & \exp(-i\omega/2) \end{bmatrix} = \exp(+i\omega/2)\sigma_3 \tag{21}$$

For the unimodular unitary matrix U'^+ the generator now is

$$J_3 = \frac{1}{2}\sigma_3 = \begin{bmatrix} 1/2 & 0 \\ 0 & -1/2 \end{bmatrix} \tag{22}$$

Therefore, the proper values associated with the operator J_3 are 1/2 and $-1/2$, that is, projection of the spin angular momentum on the Z axis, showing the connection between $SU(2)$ and the spin of an electron.

The components of the vector $\hat{\mathbf{n}}$ can be evaluated. The proper values of the scalar product of two vectors $\hat{\mathbf{n}}$ and $\boldsymbol{\sigma}$ are ± 1; that is,

$$\hat{\mathbf{n}} \cdot \boldsymbol{\sigma} \,|V_1\rangle = \pm |V_1\rangle \tag{23}$$

By multiplying both sides from the left by $\langle V_1|\hat{\mathbf{n}}$ and using $\hat{\mathbf{n}} \cdot \hat{\mathbf{n}} = 1$,

$$\langle V_1|\sigma_i|V_1\rangle = \hat{n}^i \tag{24}$$

The two components ψ_1 and ψ_2 of the basis function ψ in the spinor space may be written as

$$\psi \equiv \begin{bmatrix} \psi_1 \\ \psi_2 \end{bmatrix} = \psi_1|1\rangle + \psi_2|2\rangle \tag{25}$$

where

$$|1\rangle = \begin{bmatrix} 1 \\ 0 \end{bmatrix} \quad \text{and} \quad |2\rangle = \begin{bmatrix} 0 \\ 1 \end{bmatrix} \tag{26}$$

The transformation matrix V is then given by

$$V = \langle V_i|k\rangle \tag{27}$$

where $|k\rangle$ is either $|1\rangle$ or $|2\rangle$.

The matrix of two-valued representation in terms of the Euler angles is given by Lomont [1k, p. 150]. (See also $6^3$6a.)

$$D^{(1/2)}(\alpha, \beta, \gamma) = \begin{bmatrix} \exp (i/2)(\alpha + \gamma) \cos (\beta/2) & \exp (i/2)(\alpha - \gamma) \sin (\beta/2) \\ -\exp -(i/2)(\alpha - \gamma) \sin (\beta/2) & \exp -(i/2)(\alpha + \gamma) \cos (\beta/2) \end{bmatrix} \tag{28}$$

The reduction of the inner direct product of two representations follows the same laws whether the label j is an integer or a half-integer. However, spherical harmonics $Y(j, m)$ as basis functions can not be defined in terms of the polar coordinates for half-integral j.

7. Applications of group theory

> We wish finally to make a few remarks concerning the place of the theory of groups in the study of the quantum mechanics of atomic spectra. The reader will have heard that this mathematical discipline is of great importance for the subject. We manage to get along without it. When Dirac visited Princeton in 1928 he gave a seminar report on his paper showing the connection of exchange energy with the spin variables of the electrons. In the discussion following the report, Weyl protested that Dirac had said that he would derive the results without the use of group theory, but, as Weyl said, all of Dirac's arguments were really applications of group theory. Dirac replied, "I said I would obtain the results without previous knowledge of group theory."
>
> E. U. Condon and G. H. Shortley, *The Theory of Atomic Spectra*, (New York: Cambridge University Press, 1935), pp. 10–11.

We should view the above incident in quite a positive way. It shows the intimate relation between the formalism of quantum mechanics and the theory of group representations in linear spaces. It has become clearer in recent years that not only quantum kinematics but also quantum dynamics can be expressed in terms of representations of symmetry and *generalized* symmetry groups and the corresponding harmonic analysis. One can, in some general sense, even equate quantum theory with group representations. It is thus not surprising to find quantum mechanical techniques in the construction of group representations and vice versa. From this point of view, the systematic study of group representations can only enrich quantum theory because of the existence of a large and established body of mathematical knowledge. New mathematical concepts and results find more and more applications in many areas of physics. Physicists have come a long way in recent years from simple compact symmetry groups to noncompact groups, infinite parameter groups, topological groups, and so on and their representations.

This chapter demonstrates the power and elegance of group theory and group representations. Even though it is possible to handle most of the problems of atomic spectra without group theory, its methods make the problems simpler, and, more importantly, inherent regularities stand out more clearly. Moreover, there are cases, such as the labeling of the terms of complex spectra, in which the use of group theory is almost a necessity.

At present, the use of dynamical groups in attacking some of the most complex problems of atomic structure seems very promising. Therefore, we have devoted the last section to this subject. However, because it is a very active field, up-to-date information must be found in recent literature. Group theory may also encourage new ways of looking at atomic structure, such as, for example, the search for global quantum numbers as suggested by the dynamical group structure of the periodic systems.

1^7. The generators constructed from creation and destruction operators

The existence of a symmetry of a system implies that the generators J_k of the symmetry group commute with the Hamiltonian; that is,

$$[H, J_k] = 0 \tag{1}$$

The subset of mutually commuting generators is therefore the conserved physical observables. The symmetries possessed by the Hamiltonian of a free N-electron atom were first mentioned in Section 6^2. In Section 12^6 the importance of rotational symmetries and the connection between the theory of angular momentum and the group $SO(3)$ were presented.

The problems of Chapter 4, which motivated the study of group theory as a helpful tool, are different in nature from those mentioned above. There is a need to find meaningful labels with which to classify the terms of the configurations $(nl)^q$ for $l > 2$ and $q > 3$. Also, evaluation of the matrix elements for the configurations $(nl)^q$ with $l > 3$ and $q > 3$ is prohibitively laborious, even for the simplest interactions, requiring that some simplifications be introduced. The three-dimensional rotation group $SO(3)$ does not provide the answer. For these reasons Racah [1^5d] investigated more general symmetries than the geometrical ones to evaluate the term energies for the f^q configurations. This approach was later elaborated, most notably by Judd [1, 22^3], Wybourne [7^6a], and Moshinsky [2].

The irreducible tensor operators (Sec. 7^3) proved to be useful in Chapter 5. In Section 7^2 the creation and destruction operators were used to build the determinantal product states as well as one- and two-body operators. Here the information is based on second quantization, used in atomic spectroscopy by Judd [1] and others, which is more elegant than Racah's conventional technique [15^6b, 3].

The fermion creation and destruction operators, $\eta^\dagger_a$ and η_a, anticommute ($7^2$22). For fixed n_a and l_a, the other quantum numbers can take $2(2l_a + 1)$ values. For a given shell there are, therefore, $4(2l_a + 1) =$

$8l_a + 4$ creation and destruction operators. Because the commutators $[\eta^\dagger_i, \eta^\dagger_j]$, $[\eta_i, \eta_j]$ and $[\eta^\dagger_i, \eta_j]$, where $1 \leq i, j \leq 4l_a + 2$, are neither zero nor equal to another creation or destruction operator, the set of $(8l_a + 4)$ individual operators, $\eta^\dagger_a$, η_a, is not closed with respect to commutation, as required by the relations 8[6]25, 29, 32′, 33, and 34. Thus the set of individual operators $\eta^\dagger_a$ and η_a does not form the basis for a simple Lie algebra. However, a set of operators that is closed with respect to commutation can be set up by adding $(4l_a + 2)(8l_a + 3)$ distinct nonzero commutators to the set of $(8l_a + 4)$ individual operators. The choice of operators determines the corresponding simple Lie algebra, if there is one, for which they form a basis.

A. The group $SO(8l_a + 5)$

The following set is closed with respect to commutation; that is, the commutator of any pair is equal to a linear combination (in this case comprising a single one) of the operators in the set:

$$H_i = (1/2)[\eta^\dagger_i, \eta_i] \qquad 1 \leq i \leq 4l_a + 2 \tag{2}$$

$$\left.\begin{aligned} E^1_j &= \eta^\dagger_j \\ E^2_j &= \eta_j \end{aligned}\right\} \qquad 1 \leq j \leq 4l_a + 2 \tag{3}$$

$$\left.\begin{aligned} E^3_{j,k} &= (1/2)[\eta^\dagger_j, \eta^\dagger_k] = \eta^\dagger_j \eta^\dagger_k \\ E^4_{j,k} &= (1/2)[\eta_j, \eta_k] = \eta_j \eta_k \\ E^5_{j,k} &= (1/2)[\eta^\dagger_j, \eta_k] = \eta^\dagger_j \eta_k \end{aligned}\right\} \qquad \begin{matrix} 1 \leq j, k \leq 4l_a + 2 \\ j \neq k \end{matrix} \tag{4}$$

In order that the operators H_i form part of the basis for a simple Lie algebra, they must commute among themselves and the commutators $[H_i, E_\alpha]$ must be equal to the operators E_α multiplied by a scaler that is one of the components of the root vector $\boldsymbol{\alpha}$ [(8[6]25), (8[6]29)]. Then, because the set is closed under commutation, the relations 8[6]32′, 33, and 34 are satisfied.

Using 7[2]22 these commutation relations are easily found:

$$\begin{aligned} [H_i, H_j] &= 0 \\ [H_i, E^1_j] &= \delta_{ij} E^1_j \\ [H_i, E^2_j] &= -\delta_{ij} E^2_j \\ [H_i, E^3_{j,k}] &= (\delta_{ij} + \delta_{ik}) E^3_{j,k} \\ [H_i, E^4_{j,k}] &= -(\delta_{ij} + \delta_{ik}) E^4_{j,k} \\ [H_i, E^5_{j,k}] &= (\delta_{ij} - \delta_{ik}) E^5_{j,k} \end{aligned} \tag{5}$$

Therefore, the roots are $\pm\hat{\mathbf{e}}_i$ and $\pm\hat{\mathbf{e}}_i \pm \hat{\mathbf{e}}_j$, where all combinations of signs are permissible. Comparison with the roots of simple Lie algebras as tabulated in Table 1[6], shows that the set of operators defined in (2), (3), and (4) forms the canonical basis for the algebra B_n, where $n = 4l_a + 2$. Thus the associated simple and compact Lie group is $SO(2n + 1) = SO(8l_a + 5)$.

The states produced by the creation and destruction operators $n\dagger_a$ and η_a acting on any state of a given shell (n_a, l_a) belong to the same shell. These form the basis for an irreducible representation of the group $SO(8l_a + 5)$, which may be identified by the highest weight (Sec. 10[6]). To obtain the weight of the determinantal product state $\{a, b, \ldots, n\}$ is acted on by the operator H_i. By using 7[2]31 and (2)

$$H_i\{a, b, \ldots, n\} = (\eta\dagger_i\eta_i - 1/2)\eta\dagger_n \ldots \eta\dagger_b\eta\dagger_a|0\rangle \tag{6}$$

in which there are two possibilities according to whether a specified value of i is or is not contained in the set $(a, b, \ldots, n)$ with the corresponding results:

$$H_i\{a, b, \ldots, n\} = \pm(1/2)\{a, b, \ldots, n\}$$

where the sign is + or − according to whether i is contained in the set $(a, b, \ldots, n)$ or not. Therefore, the components of the weight of the determinantal product state are $\pm 1/2$ for a fixed value of i. Because i takes values from 1 to $4l_a + 2$, all possible determinantal product states of a given shell possess a weight, which, with a set of $(4l_a + 2)$ components, $\pm 1/2$, is given as

$$\mathbf{m} = (\pm 1/2, \pm 1/2, \ldots, \pm 1/2) \tag{7}$$

Each determinantal product state has a different set of values, for $a, b, \ldots, n$, so no two weights in (7) belong to the same state. The highest weight corresponds to the sets of $(2l_a + 1)$ components with positive signs:

$$\mathbf{m}^0 = (1/2, 1/2, \ldots, 1/2) \tag{7'}$$

For $l_a = 0$ (that is, for an s-shell) the group $SO(2n + 1) = SO(5)$, whose root diagram is given in Figure 3[6]. Because the space is two-dimensional, $n = 2$, the weight $\mathbf{m}$ has only two components. There are four weights belonging to the irreducible representation, of the group $SO(5)$, labeled by the highest weight $\mathbf{m}^0 = (1/2, 1/2)$. These weights are given in terms of their components as: $(-1/2, -1/2)$, $(1/2, -1/2)$, $(-1/2, 1/2)$, $(1/2, 1/2)$. From (3) and (5) the unit vectors $\hat{\mathbf{e}}_i$ and $-\hat{\mathbf{e}}_i$ are associated with the creation and destruction operators, $\eta\dagger_i$ and η_i, respectively. Equations (4) and (5) show that the vectors $(\hat{\mathbf{e}}_i + \hat{\mathbf{e}}_j)$, $-(\hat{\mathbf{e}}_i + \hat{\mathbf{e}}_j)$, and $(\hat{\mathbf{e}}_i - \hat{\mathbf{e}}_j)$ correspond to the operators $\eta\dagger_i\eta\dagger_j$, $\eta_i\eta_j$, and $\eta\dagger_i\eta_j$. In the (1/2, 1/2) irreducible representation of the group $SO(5)$, indices i and j depend only on the quantum label m_s; that is, $i, j = m_s$. The creation and destruction operators can be identified by the m_s values according to

$$\begin{aligned} \eta\dagger_{\pm 1/2}|0\rangle &= |m_s = \pm 1/2\rangle \\ \eta_{\pm 1/2}|0\rangle &= |0\rangle;\ \eta_{\pm 1/2}|m_s = \pm 1/2\rangle = |0\rangle \end{aligned} \tag{8}$$

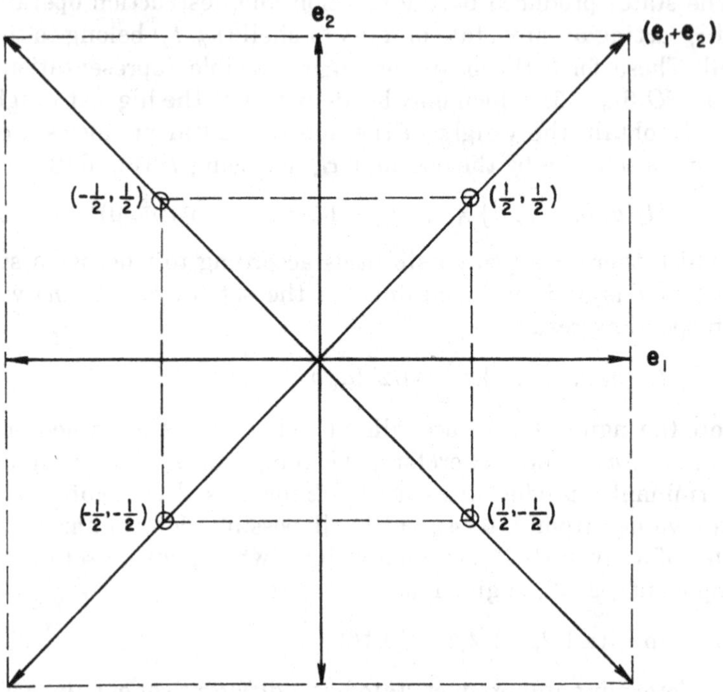

Figure 1[7]. The weights of the (1/2, 1/2) irreducible representation of the group $SO(5)$ superimposed on the root diagram of Figure 3[6].

The correspondence between the weights and the proper vectors becomes:

$$
\begin{aligned}
(-1/2, -1/2) &\rightarrow |q_s = 0;\ m_s = 0\rangle \\
(1/2, -1/2) &\rightarrow |q_s = 1;\ m_s = 1/2\rangle \\
(-1/2, 1/2) &\rightarrow |q_s = 1;\ m_s = -1/2\rangle \\
(1/2, 1/2) &\rightarrow |q_s = 2;\ m_s = 0\rangle
\end{aligned}
\tag{9}
$$

in which q_s is the number of s electrons. Figure 1[7] shows the weights of the (1/2, 1/2) irreducible representation of the group $SO(5)$ superimposed on the root diagram of Figure 3[6].

B. The subgroups of the group $SO(8l_a + 5)$; branching rule

Sets smaller than the one defined in (2), (3), and (4) may also form the basis for simple Lie algebras. Such sets will be a subset of the larger set. Therefore, the algebras, or groups that they define, will be subalgebras, or subgroups, of the ones defined by the larger set. A study of the operators of (2), (3), and (4) shows that the set composed of the operators H_i, $E^3_{j,k}$, $E^4_{j,k}$ and $E^5_{j,k}$ is closed with respect to commutation. From (5) the root vectors are: $\pm\mathbf{e}_i \pm \mathbf{e}_j$, which correspond, in Table 1⁶, to the simple Lie algebra D_n, where $n = 4l_a + 2$. The associated simple and compact Lie group is $SO(2n) = SO(8l_a + 4)$; showing that the group $SO(8l_a + 4)$ is a subgroup of $SO(8l_a + 5)$. This is indicated as

$$SO(8l_a + 5) \supset SO(8l_a + 4)$$

and read as "the group $SO(8l_a + 5)$ contains the group $SO(8l_a + 4)$ as a subgroup."

The group $SO(8l_a + 4)$ has two irreducible representations, labeled by the highest weights (1/2, 1/2, . . . , 1/2) and (1/2, 1/2, . . . , 1/2, −1/2), respectively, which correspond to even and odd numbers of electrons in the given shell. This is a natural result of the fact that the operators $E^3_{i,j}$, $E^4_{i,j}$, and $E^5_{i,j}$, which may be taken as "shift operators" in considering their effects on weights, 10⁶3, connect only the shells differing in either two electrons or none. Thus operators of the group $SO(8l_a + 4)$ either create or destroy two electrons, or leave the number unchanged. In any case the parity of q, the number of electrons, is unchanged. The complete set of states, which forms the basis for the irreducible representations of the group, may be identified by the evenness or oddness of q in the specified shell. Because every filled shell has an even number of electrons, and because, for filled shells, the highest weight has all components positive, the irreducible representation (1/2, 1/2, . . . , 1/2) is associated with the even number of electrons. For the representation corresponding to the odd number of electrons, the weight vectors $\mathbf{m}$ have an odd number of negative components. The highest weight, $\mathbf{m}^0$, is the one with a minimum number of negative values, that is, with only one.

In general, under the reduction of a group G into its subgroup G', the αth irreducible representation, Γ_α of G may be expressed as a direct sum of the irreducible representations Γ'_β, Γ'_α, . . . of G'. This is called the *branching rule* [4²c, pp. 208–12; 22³, pp. 136–42], indicated as

$$\begin{array}{l} G \rightarrow G' \\ \Gamma_\alpha \rightarrow \Gamma'_\beta + \Gamma'_\gamma + \dots \end{array} \tag{10}$$

The most direct method of finding the branching rules, for a given case, is to compute the characters of the irreducible representations, using the fact that

$$\chi^\alpha = \chi'^\beta + \chi'^\gamma + \dots \tag{11}$$

For continuous groups this is often too laborious to be practical. On the other hand, if the reductions of the inner products of the irreducible representations (Clebsch-Gordan series) of both group G and subgroup G' are known, the branching rules can be found more easily [22³, pp.

142–5]. Most of the branching rules, for the groups which are widely used in atomic spectroscopy, have been computed and tabulated by Butler [7[6]a, Tables C1–16].

From the foregoing discussion it follows that in the reduction

$$SO(8l_a + 5) \to SO(8l_a + 4)$$

the (1/2, 1/2, . . . , 1/2) irreducible representation of the group $SO(8l_a + 5)$ may be expressed as a direct sum of the (1/2, 1/2, . . . , 1/2) and the (1/2, 1/2, . . . , 1/2, −1/2) irreducible representations of the group $SO(8l_a + 4)$. That is,

$$(1/2, 1/2, \ldots, 1/2) \to (1/2, 1/2, \ldots, 1/2) + (1/2, 1/2, \ldots, 1/2, -1/2) \qquad (12)$$

It is possible to find still smaller sets, composed of the operators of (2)–(4), which form the basis for a simple Lie algebra. The set of operators H_i and E^5_{jk} is closed with respect to commutation; moreover, from (5), the roots are $(\mathbf{e}_i - \mathbf{e}_j)$, which correspond (Table 1[6]) to the roots of the algebra A_n, with $n = 4l_a + 1$. The simple and compact Lie group, associated with the algebra A_n, is then $SU(n + 1) = SU(4l_a + 2)$, indicating that

$$SO(8l_a + 4) \supset SU(4l_a + 2)$$

The components of the weights for the algebra A_n are (Sec. 10[6]) fractions with the denominator $(n + 1)$. This awkwardness can be avoided by removing the condition that the algebra be semisimple. The set of $(4l_a + 2)^2$ operators

$$\left.\begin{aligned} H_i &= \eta^\dagger{}_i\eta_i \\ E_{ij} &= \eta^\dagger{}_i\eta_j \end{aligned}\right\} \qquad \begin{aligned} &i, j = 1, 2, \ldots, 4l_a{+}2 \\ &i \neq j \end{aligned} \qquad (13)$$

generates the real full linear group in $(4l_a + 2)$ dimensions $GL(4l_a + 2, R)$. The full linear group is not semisimple (Sec. 9[6]), but because it is locally isomorphic to the direct product of a semisimple group with an Abelian group [15[6]b, p. 59], it shares many properties with the semisimple groups. The components of the weights, for the full linear group, are always integers.

C. Tensors as basis of the representations

A tensor T^f of rank, or degree, f, in the k-dimensional space whose symmetry is defined by the partition $[f_1, f_2, \ldots, f_k]$, where $f = \Sigma f_i$, may be taken as a basis of the representation of the full linear group, whose highest weight has the components f_i [3[6]e, pp. 105–12]. The Young diagram associated with the partition $[f_i, f_2, \ldots, f_k]$ consists of f cells in k rows, the ith row containing f_i cells. If the dimension of the space is larger than the number of components of f, some of the components are zero. In particular, if nonzero components are equal to unity, there are f number of components with unit value, and the remaining $(k - f)$ components are zero, resulting in the partition $[1, 1, \ldots, 1, 0, 0, \ldots, 0]$, corresponding to a completely antisymmetric Young diagram. Therefore the basis for the representation characterized by this partition is formed by totally antisymmetric tensors of rank f. This property of the full linear group holds also for its unitary subgroup.

Some restrictions on the set defined by (13) can be imposed so that the transformations are unitary to ensure that the states remain orthogonal. This can be done by introducing a parameter ϵ_{ij}^{kt}, resulting in

$$\left.\begin{aligned} H_i &= \eta^\dagger{}_i \eta_i \\ E_{ij} &= \sum_{k,t} \epsilon_{ij}^{kt} \eta^\dagger{}_k \eta_t \end{aligned}\right\} \qquad \begin{aligned} &i, j, k, t = 1, 2, \ldots, 4l_a + 2 \\ &i \neq j; \qquad k \neq t \end{aligned} \tag{14}$$

When the parameters ϵ_{ij}^{kt} are chosen so that the transformations are unitary, this set generates the unitary group $U(4l_a + 2)$.

The proper values of H_i for the determinantal product states $\{a, b, \ldots, n\}$, for a fixed value of the index i, are either unity or zero, depending on whether i is contained in the set $(a, b, \ldots, n)$. For a given configuration $(n_a l_a)^q$, as the index i runs through the $(4l_a + 2)$ possible single-electron states, H_i produces q proper values equal to one and $(4l_a + 2 - q)$ proper values equal to zero. There are $\binom{4l_a + 2}{q}$ ways of arranging these symbols in sequence (3⁴5), corresponding to the number of states of the configuration.

An irreducible representation of the unitary group $U(4l_a + 2)$ may be labeled by its highest weight $\mathbf{m}^0$, and the set of the components of $\mathbf{m}^0$ can also be interpreted as partitions defining certain symmetry. Therefore, it is customary to write them in brackets rather than in parentheses, as

$$\mathbf{m}^0 = [m_1, m_2, \ldots, m_{4l+2}]$$

The determinantal product states $\{a, b, \ldots, n\}$ form the basis for the irreducible representations of the group $U(4l_a + 2)$. For a configuration $(n_a l_a)^q$ the highest weight $\mathbf{m}^0$ is

$$\mathbf{m}^0 = [1, 1, \ldots, 1, 0, 0, \ldots, 0]$$

where there are q ones and $(4l_a + 2 - q)$ zeros. It is also customary to omit the components with zero values, reducing the bracket to

$$\mathbf{m}^0 = [1, 1, \ldots, 1]$$

when all of its components are zeros; the highest weight is shown simply as $\mathbf{m}^0 = 0$.

The p shell may serve as an example. Here $4l_a + 2 = 6$, so the weights have six components. Because one of the irreducible representations of the group $SO(8l_a + 4)$ corresponds to configurations with an even number of electrons and the other to those with an odd number, the branching rules are

$$SO(8l_a + 4) \to U(4l_a + 2)$$
$$(1/2, 1/2, 1/2, 1/2, 1/2, 1/2,)$$
$$\to [0] + [1, 1] + [1, 1, 1, 1] + [1, 1, 1, 1, 1, 1] \qquad (15)$$

and

$$(1/2, 1/2, 1/2, 1/2, 1/2, -1/2) \to [1] + [1, 1, 1] + [1, 1, 1, 1, 1]$$

Relations (12) and (15) give the decomposition of the p shell into the configurations $(np)^q$.

To find the subgroups of the group $U(4l_a + 2)$, a subset of the operators in (14) must be considered. This can be done either by selecting the quantum numbers i and j, each of which runs through all one-electron states, or by taking suitable linear combinations of the operators $\eta\dagger_i\eta_j$, which have well-defined properties with respect to the group $SO(3)$. The second approach is more convenient because it is related to ideas presented in Section 7[3] and in Chapter 5.

D. Creation and destruction operators as components of a double tensor

The $(2l_a + 1)$ components of the creation operator $\eta\dagger_i$, for a fixed value of m_{sa}, behave like the components of a tensor of order l_a in the orbital space. The two components of $\eta\dagger_i$, for m_a fixed, form a tensor of order s_a in the spin space. Therefore, from the definition of a double tensor following 10[5]5b′ the $(4l_a + 2)$ components of the creation operator $\eta\dagger_i$ may be taken as the components of a double tensor denoted as $\boldsymbol{\eta}\dagger$. When a new operator is defined in terms of the destruction operator η_j as

$$\bar{\eta}_k = (-1)^n\eta_j \qquad (16)$$

where $k \equiv (m_a, m_{sa})$; $n = (l + s) - (m_a + m_{sa})$; $j \equiv (-m_a, -m_{sa})$, the components of the operator $\bar{\eta}_k$ also form a double tensor of the order l_a and s_a with respect to the angular and spin parts of the system, denoted by $\boldsymbol{\eta}$ [1b, pp. 194–5].

The tensors $\boldsymbol{\eta}\dagger$ and $\boldsymbol{\eta}$ may be coupled (Sec. 3[3]) to give

$$[\boldsymbol{\eta}\dagger\boldsymbol{\eta}]^{(Kk)}_{m_sm} = T^{(Kk)}_{m_sm} = (-1)^{-m-m_s}[(2K + 1)(2k + 1)]^{1/2} \times \sum_{ij} \begin{pmatrix} 1/2 & 1/2 & K \\ m_{sa} & m'_{sa} & -m_s \end{pmatrix} \begin{pmatrix} l_a & l_a & k \\ m_a & m'_a & -m \end{pmatrix} \eta\dagger_i\bar{\eta}_j \qquad (17)$$

where

$$i \equiv (m_a, m_{sa}); \quad j \equiv (m'_a, m'_{sa})$$
$$-K \leqslant m_s \leqslant K; \quad -k \leqslant m \leqslant k$$
$$0 \leqslant K \leqslant 1; \quad 0 \leqslant k \leqslant 2l_a$$

The set of operators in (14) may be replaced now by the set of coupled tensors of (17); and the commutation relations

$$[\eta^\dagger_i \eta_j, \eta^\dagger_k \eta_l] = \delta_{jk}\eta^\dagger_i \eta_l - \delta_{il}\eta^\dagger_k \eta_j \tag{18}$$

become

$$\begin{aligned}
[T^{(Kk)}_{m_s m}, T^{(K'k')}_{m'_s m'}] &= (-1)^{-K-k}[(2K+1)(2K'+1)(2k+1)(2k'+1)]^{1/2} \\
&\times \sum_{K''k''} \sum_{m''_s m''} \{(-1)^{K'+k'-m''_s-m''-1} + (-1)^{K''+k''-K-k-m''_s-m''}\} \\
&\times \sqrt{(2K''+1)(2k''+1)} \begin{pmatrix} K & K' & K'' \\ m_s & m'_s & -m''_s \end{pmatrix} \begin{pmatrix} k & k' & k'' \\ m & m' & -m'' \end{pmatrix} \\
&\times \begin{Bmatrix} K & K' & K'' \\ 1/2 & 1/2 & 1/2 \end{Bmatrix} \begin{Bmatrix} k & k' & k'' \\ l_a & l_a & l_a \end{Bmatrix} T^{(K''k'')}_{m''_s m''}
\end{aligned} \tag{18'}$$

E. Decomposition for the $(n_a l_a)^q$ Configurations

Because the tensor $\mathbf{T}^{(10)}$ has order one with respect to spin and zero with respect to the orbital part of the system, and is, in addition, a sum of single-electron operators, it must be proportional to the total-spin operator $\mathbf{S}$. On the other hand, the spin operators may be taken as the generators of the three dimensional rotation group. Therefore, the subset $\mathbf{T}^{(10)}$ corresponds to the group $SO_s(3)$, where a subscript s indicates the kind of generators of the group. The group $SO_s(3)$ can be replaced by $SU(2)$ (Sec. 13^6). The tensors $\mathbf{T}^{(0k)}$ are scalars with respect to the spin and have $(2k+1)$ components in the orbital space. The foregoing arguments for $U(4l_a+2)$ show that the $(2l_a+1)^2$ components of the operators $\mathbf{T}^{(0k)}$ $(0 \leqslant k \leqslant 2l_a)$ generate $U(2l_a+1)$. If the operators of the different subsets commute, the set that is the sum of the subsets generates a group that is homomorphic to the direct products ($1^6$8) of the groups generated by the subsets [3^6e, pp. 158–9]. To see whether the subsets $\mathbf{T}^{(0k)}$ and $\mathbf{T}^{(10)}$ commute, the values $K = 0$, $K' = 1$, and $k' = 0$ are substituted in (18′). Then because of the triangular condition for the nonvanishing $6j$ symbols, $K'' = 1$ and $k'' = k$. For these values the phase factor in (18′) becomes

$$[(-1)^{K'+k'-m''_s-m''-1} + (-1)^{K''+k''-K-k-m''_s-m''}] = (-1)^{-m''_s-m''}[1 + (-1)^1] = 0$$

showing that the subsets $\mathbf{T}^{(0k)}$ and $\mathbf{T}^{(10)}$ commute. Together they form the generators of a group that is homomorphic to the direct product

$SO_s(3) \times U(2l_a + 1)$, or $SU(2) \times U(2l_a + 1)$, thus indicating that

$$U(4l_a + 2) \supset SO_s(3) \times U(2l_a + 1)$$

or

$$U(4l_a + 2) \supset SU(2) \times U(2l_a + 1)$$

Because the unimodular unitary group is a subgroup of the unitary group in the same dimension, that is,

$$U(2l_a + 1) \supset SU(2l_a + 1)$$

we can alternatively write

$$U(4l_a + 2) \supset SU(2) \times SU(2l_a + 1)$$

For $K = K' = 0$ and for odd values of k and k', the commutator in (18′) is a double tensor with odd k'', so the operators $\mathbf{T}^{(0k)}$ with odd k forms a subset. The group generated by this subset is the $(2l_a + 1)$-dimensional unimodular orthogonal group $SO(2l_a + 1)$ [15[6]b, p. 71]. Finally, the operators $\mathbf{T}^{(01)}$ are proportional to the angular momentum operator $\mathbf{L}$ and generate the three-dimensional rotation group $SO(3)$. By including the last two results the entire decomposition, from shell to term, for the $(n_a l_a)^q$ type configurations, may be summarized as

$$\begin{aligned} SO(8l_a + 5) \supset SO(8l_a + 4) \supset U(4l_a + 2) \supset SU(2) \times U(2l_a + 1) \\ \supset SU(2) \times SO(2l_a + 1) \supset SU(2) \times SO(3) \end{aligned} \tag{19}$$

This decomposition is not unique, because it is also possible to take the symplectic $Sp(2l_a + 2)$ as a subgroup of the unitary group $U(4l_a + 2)$ [4[2]c, pp. 403–12] and write

$$\begin{aligned} SO(8l_a + 5) \supset SO(8l_a + 4) \supset U(4l_a + 2) \supset Sp(4l_a + 2) \\ \supset SU(2) \times SO(2l_a + 1) \supset SU(2) \times SO(3) \end{aligned} \tag{19'}$$

{Systematic studies of the possible symmetry chains that are admitted for a particular group are given by M. Lorente and B. Gruber [*J. Math. Phys.* **13**, 1639 (1972)], and by B. Gruber and M. T. Samuel [*Group Theory and Its Applications,* Vol. 3, E. M. Loebl, ed., (New York: Academic Press, 1975), pp. 95–141].}

F. Decomposition for the mixed configurations

Feneuille studied [4] configurations of the type $l_a^{q_a}\, l_b^{q_b}$, where $q_a + q_b = q$. Wybourne [7[6]a] used the technique of *plethysm* [5] to treat the mixed configurations of the type $l_a^{q_a}\, l_b^{q_b}\, l_c^{q_c}$, where $q_a + q_b + q_c = q$. The decomposition of the equivalent electron configurations, l^q, (19′), may be generalized, up to a certain point, to such mixed configurations as

$$U[4(l_a + l_b + l_c + 3/2)] \supset Sp[4(l_a + l_b + l_c + 3/2)] \supset SU(2) \times SO[2(l_a + l_b + l_c + 3/2)] \quad (20)$$

The chain is started with the unitary group, because this is where the classification of the configurations begins. Instead of (19′), the decomposition given in (19) can also be taken in this generalization.

The subgroups of the unimodular group $SO[2(l_a + l_b + l_c + 3/2)]$ may be obtained by first choosing the coupling order of the configurations $l_a^{q_a}$, $l_b^{q_b}$, and $l_c^{q_c}$. By assuming that the configurations $l_b^{q_b}$ and $l_c^{q_c}$ are coupled first,

$$SO[2(l_a + l_b + l_c + 3/2)] \supset SO(2l_a + 1) \times SO[2(l_b + l_c + 1)] \quad (21a)$$

For the other choices of coupling order the results are

$$SO[2(l_a + l_b + l_c + 3/2)] \supset SO[2(l_a + l_b + 1)] \times SO(2l_c + 1) \quad (21b)$$

and

$$SO[2(l_a + l_b + l_c + 3/2)] \supset SO[2(l_a + l_c + 1)] \times SO(2l_b + 1) \quad (21c)$$

The right sides of these inclusions are similar in structure to that associated with the recoupling procedures of three angular momenta $\mathbf{J}_1$, $\mathbf{J}_2$, and $\mathbf{J}_3$ as defined in 9³7 and 9³7′. Here the transformation coefficents are, of course, not $6j$ symbols. In (21) the subgroups of the unimodular orthogonal groups $SO[2(l_b + l_c + 1)]$, $SO[2(l_a + l_b + 1)]$, and $SO[2(l_a + l_c + 1)]$, respectively, are

$$SO[2(l_b + l_c + 1)] \supset SO(2l_b + 1) \times SO(2l_c + 1) \supset SO_{l_b}(3) \times SO_{l_c}(3) \supset SO(3) \quad (22a)$$

$$SO[2(l_a + l_b + 1)] \supset SO(2l_a + 1) \times SO(2l_b + 1) \supset SO_{l_a}(3) \times SO_{l_b}(3) \supset SO(3) \quad (22b)$$

$$SO[2(l_a + l_c + 1)] \supset SO(2l_a + 1) \times SO(2l_c + 1) \supset SO_{l_a}(3) \times SO_{l_c}(3) \supset SO(3) \quad (22c)$$

When these inclusions are substituted in (21) and the inclusions of (21) are used in (20), the complete decompositions of the mixed configurations $l_a^{q_a}\, l_b^{q_b}\, l_c^{q_c}$ are obtained.

If the branching rules under the decompositions given in (19), (19′), (20), (21), and (22) are known, the components of the highest weights of the irreducible representations of the groups in each chain may be used as labels in classifying the terms.

Parity is a quantum label, and care should be taken in choosing the electron orbitals so that the individual configurations making up $(l_a + l_b + l_c)^q$ are of the same parity. This requires that the parities of l_a, l_b, l_c be the same, which constitutes a limitation of the method. This classification scheme does not keep track of the number of electrons q_a, q_b, and q_c in different shells; instead the total number of electrons $q = q_a + q_b + q_c$, should remain the same. To remove these limitations two unitary groups, $U(4l_a + 2)$ and $U[4(l_b + l_c + 3/2)]$, may be considered, where l_b and l_c have the same parity, opposite to l_a, such that their direct product is homomorphic to the unitary group $U[4(l_a + l_b + l_c + 3/2)]$. That is,

$$U[4(l_a + l_b + l_c + 3/2)] \rightarrow U(4l_a + 2) \times U[4(l_b + l_c + 1)]$$

The two configurations $l_a^{q_a}$ and $(l_b + l_c)^{q'}$ are classified separately where $q' = q - q_a$, and the results are coupled by use of angular momentum theory [7[6]a, pp. 81–3].

G. Equivalent f-electron configurations

For equivalent f-electron configurations the last two steps of the decompositions in (19) and (19′) are different. Here $l_a = 3$, so the $(2l_a + 1)$-dimensional unimodular orthogonal group becomes $SO(7)$. The generators of the group $SO(7)$ are $\mathbf{T}^{(01)}$, $\mathbf{T}^{(03)}$, and $\mathbf{T}^{(05)}$. The commutators

$$[T^{(05)}_{0m}, T^{(05)}_{0m'}]$$

do not produce any of the components of the tensor $\mathbf{T}^{(03)}$. This is because

$$\begin{Bmatrix} 5 & 5 & 3 \\ 3 & 3 & 3 \end{Bmatrix} = 0$$

Therefore, all the coefficients of the components of $\mathbf{T}^{(03)}$ vanish in the right side of (18′). Moreover, neither

$$[T^{(01)}_{0m}, T^{(01)}_{0m'}] \quad \text{nor} \quad [T^{(01)}_{0m}, T^{(05)}_{0m'}]$$

produce any of the components of the tensor $\mathbf{T}^{(03)}$. The subset they form may be taken as a basis of a simple Lie algebra. The group associated with the algebra is a subgroup of $SO(7)$. The roots of the algebra, for which the set of operators $\mathbf{T}^{(01)}$ and $\mathbf{T}^{(05)}$ forms the basis, can be found from (18′) by identifying the operators as

$$\begin{aligned} H_i &= T^{(0k)}_{00}; \quad k = 1, 5 \\ E_\alpha &= T^{(05)}_{0m}; \quad |m| > 1 \\ E'_\alpha &= A\, T^{(01)}_{01} + B\, T^{(05)}_{01} \end{aligned}$$

where A and B are positive or negative scalars chosen so that 8[6]29 is satisfied. The algebra and the corresponding simple Lie group are called G_2. For equivalent f electrons the chain in (19) and (19′) then becomes,

$$SO(29) \supset SO(28) \supset U(14) \supset SU(2) \times U(7) \supset SU(2) \times SO(7) \supset SU(2) \times G_2 \supset SU(2) \times SO(3) \quad (23)$$

$$SO(29) \supset SO(28) \supset U(14) \supset Sp(14) \supset SU(2) \times SO(7) \supset SU(2) \times G_2 \supset SU(2) \times SO(3) \quad (23')$$

The irreducible representations of G_2 provide the extra labels for the classification of the terms of the f^q configurations.

2^7. Seniority scheme; quasi-spin; classification of the terms for the $(n_a l_a)^q$ configurations

The symplectic group in $(4l_a + 2)$ dimensions $Sp(4l_a + 2)$ may be generated by the set of operators $\mathbf{T}^{(Kk)}$ for which $(K + k)$ is odd. An irreducible representation $[1, 1, \ldots, 1]$, where there are q ones, of the unitary group $U(4l_a + 2)$, breaks up into the direct sum of the irreducible representations of $Sp(4l_a + 2)$ given by the branching rules. An even number of the components of the highest weights of the irreducible representations, for the group $Sp(4l_a + 2)$, are zero and the rest are ones; that is,

$$\mathbf{m}^0 = (1, 1, \ldots, 1, 0, \ldots, 0)$$

If there are ν ones on the right side it can be written alternatively as

$$\mathbf{m}^0 = (\nu)$$

Therefore, the branching rules become

$$U(4l_a + 2) \to Sp(4l_a + 2)$$

$$[1, 1, \ldots, 1] \to \sum_{\nu} (\nu); \quad \nu = q, q-2, q-4, \ldots \quad (1)$$

The smallest value of ν is either one or zero, depending on whether q is odd or even. The number ν is called *seniority* and was discussed in Section 8^5.

The seniority is related to certain properties of the three-dimensional rotation group $SO(3)$, by defining the components of a vector $\mathbf{Q}$, called *quasi-spin,* as

$$Q_+ = \left(\frac{2l_a + 1}{2}\right)^{1/2} [\boldsymbol{\eta}^\dagger \boldsymbol{\eta}^\dagger]^{(00)}$$

$$Q_- = -\left(\frac{2l_a + 1}{2}\right)^{1/2} [\boldsymbol{\eta}^\dagger \boldsymbol{\eta}]^{(00)} \quad (2)$$

$$Q_Z = -\left(\frac{2l_a + 1}{8}\right)^{1/2} \{[\boldsymbol{\eta}\boldsymbol{\eta}]^{(00)} + [\boldsymbol{\eta}^\dagger \boldsymbol{\eta}^\dagger]^{(00)}\}$$

Then

$$[Q_+, Q_-] = 2Q_Z$$
$$[Q_Z, Q_\pm] = \pm Q_\pm \quad (3)$$

Thus the components of the quasi-spin obey the commutation rules of an angular momentum vector.

By using 1'18 the Z-component of quasi-spin may be written

$$Q_Z = (1/4) \sum_i (\eta^\dagger{}_i \eta_i - \eta_i \eta^\dagger{}_i)$$
$$= -(1/2)(2l_a + 1) + (1/2) \sum_i \eta^\dagger{}_i \eta_i \quad (4)$$

Then the proper value M_Q of Q_Z is

$$M_Q = -(1/2)(2l_a + 1 - q) \quad (5)$$

The operators Q_+, Q_-, and Q_Z commute with the operators $\mathbf{T}^{(Kk)}$ when $(K + k)$ is odd. Because the set of $\mathbf{T}^{(Kk)}$ with odd $(K + k)$ generates the group $Sp(4l_a + 2)$, the operators Q_+, Q_-, and Q_Z acting on the basis of a representation (ν) of $Sp(4l_a + 2)$ do not produce any different basis. That is, the seniority number ν is conserved. In view of 1'15 and (1) the smallest value of q for which the configuration $(n_a l_a)^q$ contains the basis (ν) is $q = \nu$. Therefore, the minimum value of M_Q becomes $-(1/2)(2l + 1 - \nu)$. On the other hand, because $\mathbf{Q}$ behaves like $\mathbf{L}$,

$$-Q \leqslant M_Q \leqslant Q$$

which gives

$$Q = (1/2)(2l_a + 1 - \nu) \quad (6)$$

Thus the labels Q and M_Q carry the same information as the number of electrons q and the seniority number ν, respectively.

Components of the vector $\mathbf{Q}$ are linear combinations of the generators of $SO(8l_a + 4)$ and may be taken also as generators of the three-dimensional rotation group $SO_Q(3)$. Because the generators of $SO_Q(3)$ and $Sp(4l_a + 2)$ commute, the set of operators $\mathbf{T}^{(Kk)}$ with odd $(K + k)$ and Q_+, Q_-, Q_Z form the basis of a Lie algebra. The corresponding group is homomorphic to the direct product $SO_Q(3) \times Sp(4l_a + 2)$ and is a subgroup of $SO(8l_a + 4)$. That is,

$$SO(8l_a + 4) \supset SO_Q(3) \times Sp(4l_a + 2)$$

For p electrons the two irreducible representations (1'12) of the group $SO(8l_a + 4) = SO(12)$ may be expressed as a direct sum of the irreducible representations of $SO_Q(3)$ and $Sp(4l_a + 2) = Sp(6)$ as

$$SO(12) \rightarrow SO_Q(3) \times Sp(6)$$
$$(1/2, 1/2, 1/2, 1/2, 1/2, 1/2) \rightarrow {}^4(0) + {}^2(2) \quad (7)$$
$$(1/2, 1/2, 1/2, 1/2, 1/2, -1/2) \rightarrow {}^3(1) + {}^1(3)$$

The irreducible representations Γ_Q of $SO_Q(3)$ are indicated by the value of $(2Q + 1)$ as a prefix to the irreducible representations (ν) of the group $Sp(6)$. Because of the homomorphism between $SO(3)$ and $SU(2)$, $SU_Q(2)$ can be substituted for $SO_Q(3)$ in these equations.

A. Decomposition of the $(n_a l_a)^q$ type configurations into the terms; classification of the p shell

The branching rules were given ($1^7$12, 15) for the reduction

$$SO(8l_a + 5) \rightarrow SO(8l_a + 4) \rightarrow U(4l_a + 2)$$

This chain corresponds to decomposition of the shells into the configurations. Branching rules for the rest of the chain can be obtained by identifying the operator $T^{(Kk)}_{00}$ with H_i and by finding the components of the highest weights that label the irreducible representations of the groups. But, because these rules are available in the literature, only the decomposition of the $(np)^q$ configurations into terms is discussed here. The results for the $(nd)^q$ and $(nf)^q$ configurations are also presented.

For any p shell the reduction $U(4l_a + 2) \rightarrow SU(2) \times U(2l_a + 1)$ becomes $U(6) \rightarrow SU(2) \times U(3)$. The commuting operators of the unitary group $U(3)$ are of the type $T^{(0k)}_{00}$ which, in terms of the creation and destruction operators,become

$$T^{(0k)}_{00} = \sum_m (-1)^m \sqrt{2k+1} \begin{pmatrix} 1 & 1 & k \\ m & -m & 0 \end{pmatrix} \times (\eta^\dagger_{1/2,m}\, \eta_{1/2,m} + \eta^\dagger_{-1/2,m}\, \eta_{-1/2,m}) \tag{8}$$

Instead of identifying the operators H_i with $T^{(0k)}_{00}$, which are the linear combinations of the quantities $(\eta^\dagger_{1/2,m}\, \eta_{1/2,m} + \eta^\dagger_{-1/2,m}\, \eta_{-1/2,m})$, it is more convenient to take H_i equal to these quantities themselves:

$$H_i = \eta^\dagger_{1/2,m}\, \eta_{1/2,m} + \eta^\dagger_{-1/2,m}\, \eta_{-1/2,m} \tag{9}$$

When the proper values of H_i act on the states (basis) of the configuration $(np)^q$ they produce the components of the weights. The usual notation $\{n_a, l_a, m_a, m_{sa};\ n_b, l_b, m_b, m_{sb};\ \ldots\}$ for the determinantal product state of the configuration $(abc \ldots n)$ can be simplified considerably, because $a = b = c = \ldots = n$, and it is sufficient to specify the quantum labels $n_a l_a$ once and to omit them in the notation for the states of $(n_a l_a)^q$. For example, one of the allowed states of the configuration p^4 is then

$$\phi = \{1, 1/2; 1, -1/2; 0, 1/2; -1, -1/2\} \tag{10}$$

The operators H_1, H_2, H_3 correspond in (9) to $m = 1$, 0, and -1, respectively. Thus when the proper values of H_i act on the state ϕ they are

$$H_1\phi = 2\phi,\ H_2\phi = \phi,\ H_3\phi = \phi \tag{11}$$

giving

$$\mathbf{m}^0 = [2, 1, 1]$$

which labels the state ϕ, as well as the representation for which the totality of states ϕ forms the basis. Because, for a fixed value of the quantum numbers S and M_S, there corresponds a linear combination of the determinantal product states that is the basis for the irreducible representations of the unitary group $U(3)$, the components of the highest weights associated with these states label the irreducible representations. For instance, the total quantum numbers $S = M_S = 1/2$ occur in the configurations with an odd number of electrons, p, p^3, and p^5. The components of the highest weights for these configurations, respectively, are $[1, 0, 0] = [1]$, $[2, 1, 0] = [2, 1]$, and $[2, 2, 1]$. Because of the Pauli exclusion principle the maximum value for the components of the highest weights, which is equal to the maximum number of times that a specified value of m can occur, is limited to two.

The angular momentum operator $\mathbf{L}$ may be expressed in terms of the creation and destruction operators by replacing L for F and l for f in $7^2$28:

$$\mathbf{L} = \sum_{i,j} \eta^\dagger{}_i \langle i|\mathbf{l}|j\rangle \eta_j \tag{12}$$

The Z-component of $\mathbf{L}$ is then

$$L_Z = \sum_{i,j} \eta^\dagger{}_i \langle i|l_z|j\rangle \eta_j \tag{13}$$

For the $(np)^q$ configurations, $l_z = 1, 0, -1$, and the indices i, j take 6 values $\pm 1/2, \pm 1, 0$, resulting in

$$\begin{aligned} L_Z = {} & (\eta^\dagger{}_{1/2,1}\eta_{1/2,1} + \eta^\dagger{}_{-1/2,1}\eta_{-1/2,1}) \\ & - (\eta^\dagger{}_{1/2,-1}\eta_{1/2,-1} + \eta^\dagger{}_{-1/2,-1}\eta_{-1/2,-1}) \end{aligned} \tag{14}$$

In the reduction $U(3) \to SO(3)$ the branching rules can be found by considering all possible weights of an irreducible representation labeled by $[m_1, m_2, m_3]$ and by calculating M from the equation $M = m_1 - m_3$. Total angular momentum L corresponds (Sec. 1^7) to the irreducible representations of the group $SO(3)$, characterized by the quantum number j in Section 12^6. For example, for the $[2, 1, 1]$ irreducible representation of the group $U(3)$, all possible different weights are $[2, 1, 1]$, $[1, 2, 1]$, and $[1, 1, 2]$. Thus $M = 1, 0, -1$, and

$$U(3) \to SO(3):\quad [2, 1, 1] \to D^{(1)}$$

Table 1⁷. Classification of the p shell

p^q	$U(6)$	$SU(2) \times U(3)$	v	$SU(2) \times SO(3)$
p	[1]	$^2[1]$	1	^2P
p^2	[1, 1]	$^1[2]$	0	^1S
			2	^1D
		$^3[1, 1]$	2	^3P
p^3	[1, 1, 1]	$^2[2, 1]$	1	^2P
			3	^2D
		$^4[1, 1, 1]$	3	^4S
p^4	[1, 1, 1, 1]	$^1[2, 2]$	0	^1S
			2	^1D
		$^3[2, 1, 1]$	2	^3P
p^5	[1, 1, 1, 1, 1]	$^2[2, 2, 1]$	1	^2P
p^6	[1, 1, 1, 1, 1, 1]	$^1[2, 2, 2]$	0	^1S

Similarly,

$$[2, 2] \rightarrow D^{(2)} + D^{(0)}$$

because all possible different weights of the [2, 2] representation are [2, 2, 0], [2, 0, 2], [0, 2, 2], [2, 1, 1], [1, 2, 1], and [1, 1, 2].

It is customary to indicate the irreducible representations Γ_L of $SO(3)$ by L alone. The value of $(2S + 1)$, the multiplicity of a term, is used to identify the irreducible representations Γ_S of the group $SU(2)$. It is put as a prefix to the label that classifies the representation of the group that is associated with the radial part of the system. Table 1⁷ gives the complete classification of the p shell.

B. Branching rules under the reduction $U(2l_a + 1) \rightarrow SO(2l_a + 1)$

In the classification of the terms for the d^q and f^q configurations the problem arose of finding the branching rules under the reduction $U(2l_a + 1) \rightarrow SO(2l_a + 1)$. The weights of the irreducible representations of the group $SO(2l_a + 1)$ are denoted by $\mathbf{W}$ instead of $\mathbf{m}$. Because the rank of the group is l_a, there are l_a components of the weight $\mathbf{W}$, that is

$$\mathbf{W} = (w_1, w_2, \ldots, w_{l_a})$$

The general equations for the dimensions N_W of the irreducible representation matrices $D^{(W)}$ for the groups $SO(5)$ and $SO(7)$ are given in [22³, p. 131]. Dimensions of the irreducible representations of the groups S_n, $U(3)$ through $U(18)$, $O(4)$, $O(6)$, $O(8)$, $SO(2l + 1)$ for l = 2-7, and $Sp(2l)$ for l = 2-9 are tabulated by Wybourne [7[6]a, Tables (A-1) through (A-34)].

Judd [22[3], p. 139] gives some rules, attributable to Littlewood [6[6]] using Young tableaux (7[6]) for the branching under the reduction $U(2l + 1) \rightarrow SO(2l + 1)$. They are applicable when there are fewer than l components, of the highest weights, for the irreducible representations of $U(2l_a + 1)$, with a nonzero value. It is necessary first to draw a shape which has two columns for the partition $[m_1, m_2, \ldots, m_{2l+1}]$, which labels a specified irreducible representation of the group $U(2l + 1)$, and then to apply the rules:

1. Leave the shape unchanged.
2. Omit the two cells at the bottom of the columns, if this is possible.
3. Omit the next two cells at the bottom of the columns.
4. Continue until a shape composed of a single column or no shape at all remains.

For example, the branching rules for the [2, 2, 1] irreducible representation of $U(7)$ under the reduction $U(7) \rightarrow SO(7)$ may be found as

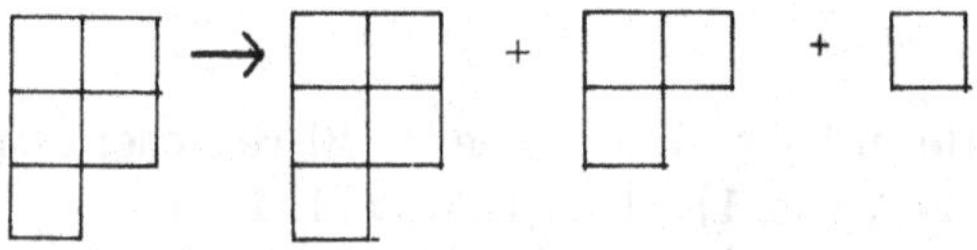

$$[2, 2, 1] \rightarrow (2, 2, 1) + (2, 1, 0) + (1, 0, 0)$$

The following equivalence relations [22[3], p. 140; 7[6]a, p. 141] may be used to reduce the number of nonzero components of the weights that label the representations of the group $U(2l_a + 1)$, having more than l nonzero components:

$$\begin{aligned} [m_1, m_2, \ldots, m_{2l+1}] &\equiv [(m_1 - m_{2l+1}), \\ &\qquad (m_2 - m_{2l+1}), \ldots, 0] \\ [m_1, m_2, \ldots, m_{2l+1}] &\equiv [(m_1 - m_{2l+1}), \\ &\qquad (m_1 - m_{2l}), \ldots, (m_1 - m_2), 0] \end{aligned} \tag{15}$$

These rules then become applicable for finding the branching under the reduction $U(2l_a + 1) \rightarrow SO(2l_a + 1)$. The branching rules for l = 1, 2, 3, 4, 5, 6, respectively, are given by Wybourne [7[6]a, Tables C1, 3, 5, 7, 8, and 9].

C. Classifications of the d and f shells

The final step in the classification of the states of d^q and f^q involves the branching rules under the reduction

$$SO(2l_a + 1) \rightarrow SO(3)$$

Table 2^7. Branching rules for the reduction $U(10) \to SU(2) \times U(5) \to SU(2) \times U(5)$

d^q	$U(10)$	$SU(2) \times U(5)$	$SU(2) \times SO(5)$
d	$[1]$	$^2[1]$	$^2(1, 0)$
d^2	$[1^2]$	$^1[2] + {}^3[1, 1]$	$^1(0, 0) + {}^1(2, 0) + {}^3(1, 1)$
d^3	$[1^3]$	$^2[2, 1] + {}^4[1, 1]$	$^2(1, 0) + {}^2(2, 1) + {}^4(1, 1)$
d^4	$[1^4]$	$^1[2, 2] + {}^3[2, 1, 1] + {}^5[1]$	$^1(0, 0) + {}^1(2, 0) + {}^1(2, 2) + {}^3(1, 1) + {}^3(2, 1) + {}^5(1, 0)$
d^5	$[1^5]$	$^2[2, 2, 1] + {}^4[2, 1, 1, 1] + {}^6[0]$	$^2(1, 0) + {}^2(2, 1) + {}^2(2, 2) + {}^4(1, 1) + {}^4(2, 0) + {}^6(0, 0)$
d^6	$[1^6]$	$^1[2, 2] + {}^3[2, 1, 1] + {}^5[1]$	$^1(0, 0) + {}^1(2, 0) + {}^1(2, 2) + {}^3(1, 1) + {}^3(2, 1) + {}^5(1, 0)$
d^7	$[1^7]$	$^2[2, 1] + {}^4[1, 1]$	$^2(1, 0) + {}^2(2, 1) + {}^4(1, 1)$
d^8	$[1^8]$	$^1[2] + {}^3[1, 1]$	$^1(0, 0) + {}^1(2, 0) + {}^3(1, 1)$
d^9	$[1^9]$	$^2[1]$	$^2(1, 0)$
d^{10}	$[1^{10}]$	$^1[0]$	$^1(0, 0)$

which it is possible to find by using the expressions for the characters of the irreducible representations for the groups $SO(2l_a + 1)$ and $SO(3)$ [22^3, pp. 137–8]. They are computed and tabulated by Wybourne [7^6a, Tables C-16 to C-24] for $l = 2 - 10$.

Table 2^7 gives the branching rules for the d^q configuration under the reduction

$$U(10) \to SU(2) \times U(5) \to SU(2) \times SO(5)$$

where the usual notation $[1, 1, \ldots, 1]$ for the irreducible representations of the group $U(4l + 2)$ is changed to $[1^q]$.

Examination of the irreducible representations of the direct product groups $SU(2) \times U(5)$ and $SU(2) \times SO(5)$ reveals a symmetry such that the branching rules for the irreducible representations $[1^q]$ and $[1^{q'}]$ of the group $U(4l_a + 2)$ are the same where $q' = (2l_a + 1) - q$. This is accomplished by using the equivalent representations (15), for the irreducible representations of the groups $SU(2)$ and $U(5)$. For instance, the following equivalences for the representations of the unitary group $U(5)$ are used:

$$[1, 1] \equiv [1, 1, 1]; \quad [1] \equiv [1, 1, 1, 1]; \quad [0] \equiv [1, 1, 1, 1, 1]$$
$$[2, 2] \equiv [2, 2, 2]; \quad [2, 1, 1] \equiv [2, 2, 1, 1,]; \quad [1] \equiv [2, 1, 1, 1, 1];$$
$$[2, 1] \equiv [2, 2, 2, 1]; \quad [1, 1] \equiv [2, 2, 1, 1, 1]; \quad [1] \equiv [2, 2, 2, 2, 1];$$
$$[0] = [2, 2, 2, 2, 2]$$

It suffices to consider only the part for which $q \leqslant (2l_a + 1)/2$, for any configuration of the type $(n_a l_a)^q$. Implications of this symmetry were discussed in Section 12^4 by using conventional methods.

Table 3^7. Classification of Russell-Saunders terms of the d^q configurations

d^q	$SU(2) \times SO(5)$	ν	$SU(2) \times SO(3)$
d	$^2(1, 0)$	1	^2D
d^2	$^1(0, 0)$	0	^1S
	$^1(2, 0)$	2	$^1D, {}^1G$
	$^3(1, 1)$	2	$^3P, {}^3F$
d^3	$^2(1, 0)$	1	^2D
	$^2(2, 1)$	3	$^2P, {}^2D, {}^2F, {}^2G, {}^2H$
	$^4(1, 1)$	3	$^4P, {}^4F$
d^4	$^1(0, 0)$	0	^1S
	$^1(2, 0)$	2	$^1D, {}^1G$
	$^1(2, 2)$	4	$^1S, {}^1D, {}^1F, {}^1G, {}^1I$
	$^3(1, 1)$	2	$^3P, {}^3F$
	$^3(2, 1)$	4	$^3P, {}^3D, {}^3F, {}^3G, {}^3H$
	$^5(1, 0)$	4	^5D
d^5	$^2(1, 0)$	1	^2D
	$^2(2, 1)$	3	$^2P, {}^2D, {}^2F, {}^2G, {}^2H$
	$^2(2, 2)$	5	$^2S, {}^2D, {}^2F, {}^2G, {}^2I$
	$^4(1, 1)$	3	$^4P, {}^4F$
	$^4(2, 0)$	5	$^4D, {}^4G$
	$^6(0, 0)$	5	^6S

Table 3^7 completes the classification of the terms of the configurations d^q where only the ones for which $q \leqslant 5$ are included.

In the decomposition of the f^q configurations the branching rules for the reductions

$$SU(2) \times SO(7) \rightarrow SU(2) \times G_2 \rightarrow SU(2) \times SO(3)$$

are involved. Because the exceptional group G_2 is of rank 2, the weights of its irreducible representations have only two components denoted as

$$\mathbf{U} = (u_1, u_2)$$

The dimensions of the irreducible representations of the group G_2 and the branching rules for the reduction $SO(7) \rightarrow G_2$, $G_2 \rightarrow SO(3)$ are tabulated by Wybourne [7^6a, Tables E-1 to E-3].

Table 4^7 presents the branching rules for the f^q configurations, under the reduction $SU(2) \times U(7) \rightarrow SU(2) \times SO(7) \rightarrow SU(2) \times G_2$, where the $[m_1, m_2, \ldots, m_q]$ representation of the group $U(7)$ is written as $[m_1^i, m_2^j, \ldots]$ if $m_1 = m_2 = \ldots = m_i \neq 0$, and $m_{i+1} = m_{i+2} = \ldots = m_{i+j} \neq 0$.

Table 4^7. Branching rules for the reduction $SU(2) \times U(7) \to SU(2) \times SO(7) \to SU(2) \times G_2$

f^q	$SU(2) \times U(7)$	$SU(2) \times SO(7)$	$SU(2) \times G_2$
f^1	${}^2[1]$	${}^2(1, 0, 0)$	${}^2(1, 0)$
f^2	${}^1[2]$	${}^1(0, 0, 0) + {}^1(2, 0, 0)$	${}^1(0, 0) + {}^1(2, 0)$
	${}^3[1^2]$	${}^3(1, 1, 0)$	${}^3(1, 0) + {}^3(1, 1)$
f^3	${}^2[2, 1]$	${}^2(1, 0, 0) + {}^2(2, 1, 0)$	${}^2(1, 0) + {}^2(1, 1) + {}^2(2, 0) + {}^2(2, 1)$
	${}^4[1^3]$	${}^4(1, 1, 1)$	${}^4(0, 0) + {}^4(1, 0) + {}^4(2, 0)$
f^4	${}^1[2^2]$	${}^1(0, 0, 0) + {}^1(2, 0, 0) + {}^1(2, 2, 0)$	${}^1(0, 0) + {}^1(2, 0) + {}^1(2, 0) + {}^1(2, 1) + {}^1(2, 2)$
	${}^3[2, 1^2]$	${}^3(1, 1, 0) + {}^3(2, 1, 1)$	${}^3(1, 0) + {}^3(1, 1) + {}^3(1, 0) + {}^3(1, 1) + {}^3(2, 0) + {}^3(2, 1) + {}^3(3, 0)$
	${}^5[1^4]$	${}^5(1, 1, 1)$	${}^5(0, 0) + {}^5(1, 0) + {}^5(2, 0)$
f^5	${}^2[2^2, 1]$	${}^2(1, 0, 0) + {}^2(2, 1, 0) + {}^2(2, 2, 1)$	${}^2(1, 0) + {}^2(1, 1) + {}^2(2, 0) + {}^2(2, 1) + {}^2(1, 0)$ $+ {}^2(1, 1) + {}^2(2, 0) + {}^2(2, 1) + {}^2(3, 0) + {}^2(3, 1)$
	${}^4[2, 1^3]$	${}^4(1, 1, 1) + {}^4(2, 1, 1)$	${}^4(0, 0) + {}^4(1, 0) + {}^4(2, 0) + {}^4(1, 0) + {}^4(1, 1) + {}^4(2, 0) + {}^4(2, 1) + {}^4(3, 0)$
	${}^6[1^5]$	${}^6(1, 1, 0)$	${}^6(1, 0) + {}^6(1, 1)$
f^6	${}^1[2^3]$	${}^1(0, 0, 0) + {}^1(2, 0, 0) + {}^1(2, 2, 0) + {}^1(2, 2, 2)$	${}^1(0, 0) + {}^1(2, 0) + {}^1(2, 0) + {}^1(2, 1) + {}^1(2, 2)$ $+ {}^1(0, 0) + {}^1(1, 0) + {}^1(2, 0) + {}^1(3, 0) + {}^1(4,0)$
	${}^3[2^2, 1^2]$	${}^3(1, 1, 0) + {}^3(2, 1, 1) + {}^3(2, 2, 1)$	${}^3(1, 0) + {}^3(1, 1) + {}^3(1, 0) + {}^3(1, 1) + {}^3(2, 0) + {}^3(2, 1) + {}^3(3, 0)$ $+ {}^3(1, 0) + {}^3(1, 1) + {}^3(2, 0) + {}^3(2, 1) + {}^3(3, 0) + {}^3(3, 1)$
	${}^5[2, 1^4]$	${}^5(1, 1, 1) + {}^5(2, 1, 0)$	${}^5(0, 0) + {}^5(1, 0) + {}^5(2, 0) + {}^5(1, 1) + {}^5(2, 0) + {}^5(2, 1)$
	${}^7[1^6]$	${}^7(1, 0, 0)$	${}^7(1, 0)$
f^7	${}^2[2^3, 1]$	${}^2(1, 0, 0) + {}^2(2, 1, 0) + {}^2(2, 2, 1) + {}^2(2, 2, 2)$	${}^2(1, 0) + {}^2(1, 1) + {}^2(2, 0) + {}^2(2, 1) + {}^2(1, 0) + {}^2(1, 1) + {}^2(2, 0) + {}^2(2, 1)$ $+ {}^2(3, 0) + {}^2(3, 1) + {}^2(0, 0) + {}^2(1, 0) + {}^2(2, 0) + {}^2(3, 0) + {}^2(4, 0)$
	${}^4[2^2, 1^3]$	${}^4(1, 1, 1) + {}^4(2, 1, 1) + {}^4(2, 2, 0)$	${}^4(0, 0) + {}^4(1, 0) + {}^4(2, 0) + {}^4(1, 0) + {}^4(1, 1) + {}^4(2, 0)$ $+ {}^4(2, 1) + {}^4(3, 0) + {}^4(2, 0) + {}^4(2, 1) + {}^4(2, 2)$
	${}^6[2, 1^5]$	${}^6(1, 1, 0) + {}^6(2, 0, 0)$	${}^6(1, 0) + {}^6(1, 1) + {}^6(2, 0)$
	${}^8[1^7]$	${}^8(0, 0, 0)$	${}^8(0, 0)$

Table 5[7]. Some of the branching rules for the reduction $G_2 \rightarrow SO(3)$

(u_1, u_2)	L
(0, 0)	S
(1, 0)	F
(1, 1)	P, H
(2, 0)	D, G, I
(2, 1)	D, F, G, H, K, L
(3, 0)	P, F, G, H, I, K, M
(2, 2)	S, D, G, H, I, L, N
(3, 1)	$P, D, F, F, G, H, H, I, I, K, K, L, M, N, O$
(4, 0)	$S, D, F, G, G, H, I, I, K, L, L, M, N, Q$

The branching rules for the reduction $SU(2) \times G_2 \rightarrow SU(2) \times SO(3)$ produce too many terms for a complete classification [2[5], pp. 2–3; 6, p. 15]. Table 5[7] gives the branching rules for the reduction $G_2 \rightarrow SO(3)$. These values, properly combined with those in Table 4[7], yield the classification of the terms for the f^q configurations. Because, in some cases, more than one term occurs with the same G_2 label, this group-theoretical classification is not sufficient to identify every term without ambiguity, and an arbitrary label τ is introduced. The labels W, U, τ, L, and S classify any term of the f^q configurations unambiguously, and a state is given as $|f^q; WU, LMSM_S\rangle$.

D. Symmetries of the spin basis functions

The general subject of symmetries of the basis functions and how to generate them with the help of the Young operators was introduced in Section 7[6]. A simple recipe now is given for the symmetries of the basis for the irreducible spin representations Γ_S for q electrons.

1. Generate all possible total spins by combining q spins.
2. The symmetries corresponding to the negative and positive total spin S (that is, $\pm S$) are the same, so the negative values can be ignored.
3. For total spin S count the number q^+ of individual positive spins, $+1/2$, and the number of q^- of individual negative spins, $-1/2$. The partition $[q^+, q^-]$, and the corresponding Young diagram, specifies the symmetry of the basis for the irreducible representations Γ_S.

For example, for $q = 5$, the total spin values are: $S = \pm 5/2, \pm 3/2, \pm 1/2$. The number of positive and negative individual spins which results in positive values of these total spins are

$S = 5/2$: $q^+ = 5, q^- = 0$
$S = 3/2$: $q^+ = 4, q^- = 1$
$S = 1/2$: $q^+ = 3, q^- = 2$

Table 6[7]. Symmetries of the spin-basis functions

S	$2\|S\|+1$	q = 1	2	3	4	5	6	7
0	1		$[1^2]$		$[2^2]$		$[3^2]$	
±1/2	2	[1]		[2, 1]		[3, 2]		[4, 3]
±1	3		[2]		[3, 1]		[4, 2]	
±3/2	4			[3]		[4, 1]		[5, 2]
±2	5				[4]		[5, 1]	
±5/2	6					[5]		[6, 1]
±3	7						[6]	
±7/2	8							[7]

Therefore, the partitions that specify the symmetries of the basis functions for the $\Gamma_{5/2}$, $\Gamma_{3/2}$, and $\Gamma_{1/2}$ become [5], [4, 1], and [3, 2], respectively. Table 6[7] presents the partitions associated with the symmetries of the spin basis functions for q = 1–7 electrons.

For configurations $(n_a l_a)^q$ the symmetries possessed by the basis functions of the irreducible representations of the group $U(2l_a + 1)$ are given by the partitions that also correspond to the components of the highest weights of these irreducible representations. The total basis functions have to be antisymmetric, which requires that the outer product of two irreducible representations, corresponding to the spin and orbital angular momentum parts of the system, should contain a partition $[1^q]$. This is accomplished by taking the partitions, which are associated with the irreducible representations of the spin and orbital angular momentum symmetry groups, as conjugate to each other. Tables 1[7] and 2[7] show that this requirement is satisfied for all the entries. Starting with a completely antisymmetric representation $[1^q]$ of the unitary group $U(4l_a + 2)$ takes into account the fact that the total basis functions must be antisymmetric with respect to all electron indices.

3[7]. Wigner-Eckart theorem; classification of the operators; Coulomb interaction

A. Wigner-Eckart Theorem

The discussion of the Wigner-Eckart theorem in Section 7[3] is in terms of $SO(3)$. Here it is recognized that the theorem can be generalized to any other compact semisimple Lie group [7].

Evaluation of the general behavior of the matrix elements

$$\langle a|V|b\rangle$$

where V is the perturbation potential, is a basic and important problem. The basis functions $\langle a|$ and $|b\rangle$ are supposed to correspond to the irreducible representations of certain groups. V needs to be expressed in terms of the operators $\mathbf{O}$, which have well-defined properties with respect to these groups. If $\langle a|$, $\mathbf{O}$, and $|b\rangle$ transform according to the irreducible representations Γ^*_1, Γ_2, and Γ_3 of the group G, the above matrix elements may be expressed as

$$\langle \tau_1\Gamma_1\alpha_1|\mathbf{O}(\tau_2\Gamma_2\alpha_2)|\tau_3\Gamma_3\alpha_3\rangle$$

where α_1, α_2, and α_3 specify the components of the irreducible representations Γ_1, Γ_2, and Γ_3. Therefore, they may also specify the irreducible representations of the subgroup G' of the group G. The labels τ_1, τ_2, and τ_3 are necessary to distinguish different states or operators that possess the same Γ and α labels.

The part $\mathbf{O}(\tau_2\Gamma_2\alpha_2)|\tau_3\Gamma_3\alpha_3\rangle$ transforms as the inner product $\Gamma_2 \times \Gamma_3$, and can be expressed as a direct sum of (unnormalized) basis functions $|\tau_4\beta\Gamma_4\alpha_4\rangle$ of an irreducible representation Γ_4 of the group G. That is,

$$\mathbf{O}(\tau_2\Gamma_2\alpha_2)|\tau_3\Gamma_3\alpha_3\rangle = \sum_{\beta,\alpha_4} |\tau_4\beta\Gamma_4\alpha_4\rangle(\beta\Gamma_4\alpha_4|\Gamma_2\alpha_2, \Gamma_3\alpha_3) \quad (1)$$

where $(\beta\Gamma_4\alpha_4|\Gamma_2\alpha_2, \Gamma_3\alpha_3)$ are the *Clebsch-Gordan coefficients*. Then the matrix elements are

$$\langle \tau_1\Gamma_1\alpha_1|\mathbf{O}(\tau_2\Gamma_2\alpha_2)|\tau_3\Gamma_3\alpha_3\rangle = \sum_{\beta,\Gamma_4,\alpha_4} \langle \tau_1\Gamma_1\alpha_1|\tau_4\beta\Gamma_4\alpha_4\rangle(\beta\Gamma_4\alpha_4|\Gamma_2\alpha_2, \Gamma_3\alpha_3) \quad (2)$$

It can be shown that < $\langle \tau_1\Gamma_1\alpha_1|\tau_4\beta\Gamma_4\alpha_4\rangle$ vanishes unless $\Gamma_1 = \Gamma_4$; and then it is independent of α. The proof is not trivial: it depends on Schur's lemma for a finite group G or, in the case of a compact Lie group G, an integration over the group manifold. Therefore the final result, which is called the Wigner-Eckart theorem, can be written

$$\langle \tau_1\Gamma_1\alpha_1|\mathbf{O}(\tau_2\Gamma_2\alpha_2)|\tau_3\Gamma_3\alpha_3\rangle = \sum_{\beta}\langle \tau_1\Gamma_1\|O_\beta(\tau_2\Gamma_2)\|\tau_3\Gamma_3\rangle(\beta\Gamma_1\alpha_1|\Gamma_2\alpha_2, \Gamma_3\alpha_3) \quad (3)$$

The quantity $\langle \tau_1\Gamma_1\|O_\beta(\tau_2\Gamma_2)\|\tau_3\Gamma_3\rangle$ is called the *reduced*-matrix element. The number of terms $C(\Gamma_1\Gamma_2\Gamma_3)$ in the sum over β is equal to the number of times Γ_1 occurs in the inner direct product $\Gamma_2 \times \Gamma_3$. Therefore, when $C(\Gamma_1\Gamma_2\Gamma_3)$ is zero the matrix elements vanish for all choices of τ and α. Determination of $C(\Gamma_1\Gamma_2\Gamma_3)$ is made by the methods of Section 5^6D.

In (3), if the labels α_1, α_2, and α_3 correspond to the irreducible representations of a subgroup G' of the group G, two additional labels are needed for the complete classification of the states and operators: one to distinguish a state or an operator when the corresponding irreducible representation of G' occurs more than once, and another to specify the components of α_i. Denoting these labels by γ_i and ρ_i, respectively, (3) may be rewritten

$$\langle\tau_1\Gamma_1\gamma_1\alpha_1\rho_1|\mathbf{O}(\tau_2\Gamma_2\gamma_2\alpha_2\rho_2)|\tau_3\Gamma_3\gamma_3\alpha_3\rho_3\rangle = \sum_\beta \langle\tau_1\Gamma_1\gamma_1\|O_\beta(\tau_2\Gamma_2\gamma_2)\|\tau_3\Gamma_3\gamma_3\rangle(\beta\Gamma_1\gamma_1\alpha_1\rho_1|\Gamma_2\gamma_2\alpha_2\rho_2, \Gamma_3\gamma_3\alpha_3\rho_3) \quad (4a)$$

where the Clebsch-Gordan coefficient $(\beta\Gamma_1\gamma_1\alpha_1\rho_1|\Gamma_2\gamma_2\alpha_2\rho_2, \Gamma_3\gamma_3\alpha_3\rho_3)$ can be written as [1b, p. 211]

$$(\beta\Gamma_1\gamma_1\alpha_1\rho_1|\Gamma_2\gamma_2\alpha_2\rho_2, \Gamma_3\gamma_3\alpha_3\rho_3) = \sum_\omega (\omega\alpha_1\rho_1|\alpha_2\rho_2, \alpha_3\rho_3)(\beta\Gamma_1\gamma_1\alpha_1|\Gamma_2\gamma_2\alpha_2 + \Gamma_3\gamma_3\alpha_3)_\omega \quad (4b)$$

The coefficient on the right of (4b) is called the *isoscalar factor*. This result may also be obtained by starting to factor out the ρ_i dependence first. That is [7[6]a, p. 93],

$$\langle\tau_1\Gamma_1\gamma_1\alpha_1\rho_1|\mathbf{O}(\tau_2\Gamma_2\gamma_2\alpha_2\rho_2)|\tau_3\Gamma_3\gamma_3\alpha_3\rho_3\rangle = \sum_\beta A_\beta(\beta\alpha_1\rho_1|\alpha_2\rho_2 + \alpha_3\rho_3) \quad (5)$$

where A_β is the reduced-matrix element

$$A_\beta = \langle\tau_1\Gamma_1\gamma_1\alpha_1\|O_\beta(\tau_2\Gamma_2\gamma_2\alpha_2)\|\tau_3\Gamma_3\gamma_3\alpha_3\rangle$$

Then applying the Wigner-Eckart theorem once more to A_β,

$$\langle\tau_1\Gamma_1\gamma_1\alpha_1\rho_1|\mathbf{O}(\tau_2\Gamma_2\gamma_2\alpha_2\rho_2)|\tau_3\Gamma_3\gamma_3\alpha_3\rho_3\rangle = \sum_{\beta,\theta} A_{\beta\theta}(\theta\Gamma_1\gamma_1\alpha_1|\Gamma_2\gamma_2\alpha_2 + \Gamma_3\gamma_3\alpha_3)(\beta\alpha_1\rho_1|\alpha_2\rho_2 + \alpha_3\rho_3) \quad (6a)$$

where

$$A_{\beta\theta} = (\tau_1\Gamma_1\gamma_1\|O_{\beta\theta}(\tau_2\Gamma_2\gamma_2)\|\tau_3\Gamma_3\gamma_3) \quad (6b)$$

B. Classification of the operators

Group-theoretical labels can be attached to a set of operators $\mathbf{O}_\alpha$ starting with the characteristic equations for linear transformations of the operators $\mathbf{O}_\alpha$ and the basis functions ϕ_α:

$$\mathbf{O}'_\alpha = U\mathbf{O}_\alpha U^{-1}; \quad \phi'_\alpha = U\phi_\alpha \quad (7)$$

For an infinitesimal transformation (8[6]5), where

$$U = 1 + \sum_\lambda a^\lambda X_\lambda$$

(7) may be written

$$(\mathbf{O}'_\alpha - \mathbf{O}_\alpha) = \sum_\lambda a^\lambda[X_\lambda, \mathbf{O}_\alpha]$$

$$(\phi'_\alpha - \phi_\alpha) = \sum_\lambda a^\lambda X_\lambda\phi_\alpha \quad (8)$$

Therefore, the transformation properties (which provide the group-theoretical labels) of the operators can be studied with the help of the transformation properties of the basis functions by replacing the product $X_\lambda \phi_\alpha$ with the commutator $[X_\lambda, \mathbf{O}_\alpha]$. The properties of many sets of basis functions are well known. One-to-one correspondence is established between the operators $\mathbf{O}_\alpha$ and the basis functions ϕ_α by verifying that in the linear expansion of the commutator $[X_\lambda, \mathbf{O}_\alpha]$ and the product $X_\lambda \phi_\alpha$ the coefficients are identical. That is,

$$A_{\lambda\beta} = B_{\lambda\beta}$$

in the equations

$$\begin{aligned} [X_\lambda, \mathbf{O}_\alpha] &= \sum_\beta A_{\lambda\beta} \mathbf{O}_\beta \\ X_\lambda \phi_\alpha &= \sum_\beta B_{\lambda\beta} \phi_\beta \end{aligned} \tag{9}$$

Because an operator $\mathbf{F}^{(1)}$ ($7^2 28$) is a sum of the quantities $\eta^\dagger{}_a \eta_b$ multiplied by the matrix elements of the single-particle operator **f**, it can be classified by the study of the operators $\eta^\dagger{}_a \eta_b$. In view of (9) and $1^7 16$ this can be accomplished by considering [1b, p. 208]

$$\begin{aligned} [(\eta^\dagger{}_h \eta_i + \eta^\dagger{}_i \bar{\eta}_h), \eta^\dagger{}_j \bar{\eta}_k] &= (-1)^{l+s}\{(-1)^{m_j+m_{sj}}[\delta(m_i, -m_j)\delta(m_{si}, -m_{sj})\eta^\dagger{}_h \eta_k \\ &\quad + \delta(m_h, -m_j)\delta(m_{sh}, -m_{sj})\eta^\dagger{}_i \eta_k] + (-1)^{m_k+m_{sk}}[\delta(m_h, -m_k) \\ &\quad \times \delta(m_{sh}, -m_{sk})\eta^\dagger{}_j \eta_i + \delta(m_i, -m_k)\delta(m_{si}, -m_{sk})\eta^\dagger{}_i \bar{\eta}_h]\} \end{aligned}$$

and

$$(\eta^\dagger{}_h \bar{\eta}_i + \eta^\dagger{}_i \bar{\eta}_h)\eta^\dagger{}_j \eta^\dagger{}_k |0\rangle = [(\eta^\dagger{}_h \bar{\eta}_i + \eta^\dagger{}_i \bar{\eta}_h), \eta^\dagger{}_j \bar{\eta}_k]|0\rangle$$

The one-to-one correspondence between the operators $\eta^\dagger{}_h \bar{\eta}_i$ and the basis functions $\eta^\dagger{}_h \eta^\dagger{}_i |0\rangle$ is obvious. This is indicated as

$$\eta^\dagger{}_h \bar{\eta}_i \leftrightarrow \eta^\dagger{}_h \eta^\dagger{}_i |0\rangle \tag{10}$$

This correspondence can be achieved only for the groups generated by the linear combinations of the operators $(\eta^\dagger{}_h \bar{\eta}_i + \eta^\dagger{}_i \bar{\eta}_h)$ or equally by the double tensor operators $\mathbf{T}^{(Kk)}$ with odd $(K + k)$. From Section 1^7 these groups are: $Sp(4l + 2)$, $SO(2l + 1)$, G_2, and $SO(3)$. When both sides of (10) are coupled, we obtain the correspondence

$$T^{(SL)}_{M_S M} \leftrightarrow [\boldsymbol{\eta}^\dagger \boldsymbol{\eta}^\dagger]^{(SL)}_{M_S M} |0\rangle$$

or, equally,

$$T^{(SL)}_{M_S M} \leftrightarrow |a^2; LM, SM_S\rangle \tag{11}$$

That is, the classification of the operators $\mathbf{T}^{(SL)}$ is accomplished by classifying the terms of the $(n_a l_a)^2$ configuration according to the chain of reductions

$U(6) \rightarrow Sp(6) \rightarrow SU(2) \times SO(3)$ for $l_a = 1$
$U(10) \rightarrow Sp(10) \rightarrow SU(2) \times SO(5) \rightarrow SU(2) \times SO(3)$ for $l_a = 2$
$U(14) \rightarrow Sp(14) \rightarrow SU(2) \times SO(7) \rightarrow SU(2) \times G_2 \rightarrow SU(2) \times SO(3)$
for $l_a = 3$

Classifications of the terms of the $(n_a l_a)^q$ configurations for $l_a = 1, 2, 3$ are tabulated in Tables 1^7, 3^7, 4^7, and 5^7. Classification of the operators $\mathbf{T}^{(SL)}$ corresponding to the terms of the $(n_a l_a)^2$ configuration may be found in these tables. The classification associated with the forbidden terms is found by starting with the completely symmetric representation [2], of the unitary group $U(4l_a + 2)$, which is not acceptable in the proper decomposition of the $(n_a l_a)^2$ configuration. These are given for the *p, d,* and *f* shells in Table 7^7 where the zeros are ignored in the representations of the group $Sp(4l_a + 2)$, and also for the groups $U(4l_a + 2)$ and $U(2l_a + 1)$.

C. Quasi-spin classification and seniority number selection rules

In the description of the operators $\mathbf{T}^{(Kk)}$ the representations of the symplectic group $Sp(4l_a + 2)$ may be used instead of those of the unimodular orthogonal group $SO(2l_a + 1)$, by assigning a quasi-spin order Q' to the operators in question. For an operator that leaves the number of electrons unchanged, the component $M_{Q'}$, associated with the order Q', is zero. The quasi-spin operator $\mathbf{Q}$ commutes with $\mathbf{T}^{(Kk)}$ when $(K + k)$ is odd; as a result, the order Q' is zero for these operators. The reduction used to find the quasi-spin classification for the operators $\mathbf{T}^{(Kk)}$ for which $(K + k)$ is even is

$$SO(8l_a + 4) \rightarrow SO_Q(3) \times Sp(4l_a + 2)$$

or equally

$$SO(8l_a + 4) \rightarrow SU_Q(2) \times Sp(4l_a + 2)$$

are used. If the representation, of the group $SO(8l_a + 4)$ that may be assigned to the operators $\mathbf{T}^{(Kk)}$ is known, the quasi-spin labels can be found with the help of the branching rules for this reduction. Therefore the representation of $SO(8l_a + 4)$ that classifies the operators $\mathbf{T}^{(Kk)}$ should be found first.

The commutators of a set of $(4l_a + 2)$ creation and destruction operators $\eta^\dagger_a$ and η_b together with the generators $(1/2)[\eta^\dagger_a, \eta_b]$ of the group $SO(8l_a + 4)$ form the same set of creation and destruction operators, indicating that the set transforms according to a representation of the group $SO(8l_a + 4)$. The determination of this representation

Table 7[7]. Classification of the operators $\mathbf{T}^{(SL)}$ for the p, d, and f shells

l_a	$U(4l_a+2)$	$SU(2)\times U(2l_a+1)$	$Sp(4l_a+2)$	$SU(2)\times SO(2l_a+1)$	$SU(2)\times G_2$	$SU(2)\times SO(3)$
1	$[1^2]$	$^1[2]$	(0)			^1S
			(1^2)			^1D
		$^3[1^2]$	(1^2)			^3P
	$[2]$.	$^3[2]$	(2)			^3S
			(2)			^3D
		$^1[1^2]$	(2)			^1P
2	$[1^2]$	$^1[2]$	(0)	$^1(0, 0)$		^1S
			(1^2)	$^1(2, 0)$		^1D, ^1G
		$^3[1^2]$	(1^2)	$^3(1, 1)$		^3P, ^3F
	$[2]$	$^3[2]$	(2)	$^3(0, 0)$		^3S
			(2)	$^3(2, 0)$		^3D, ^3G
		$^1[1^2]$	(2)	$^1(1, 1)$		^1P, ^1F
3	$[1^2]$	$^1[2]$	(0)	$^1(0, 0, 0)$	$^1(0, 0)$	^1S
			(1^2)	$^1(2, 0, 0)$	$^1(2, 0)$	^1D, ^1G, ^1I
		$^3[1^2]$	(1^2)	$^3(1, 1, 0)$	$^3(1, 0)$	^3F
					$^3(1, 1)$	^3P, ^3H
	$[2]$	$^3[2]$	(2)	$^3(0, 0, 0)$	$^3(0, 0)$	^3S
			(2)	$^3(2, 0, 0)$	$^3(2, 0)$	^3D, ^3G, ^3I
		$^1[1^2]$	(2)	$^1(1, 1, 0)$	$^1(1, 0)$	^1F
					$^1(1, 1)$	^1P, ^1H

Table 8⁷. Branching rules for the reduction $SO(8l_a + 4) \to SU_Q(2) \times Sp(4l_a + 2)$

$SO(8l_a + 4)$	$SU_Q(2) \times Sp(4l_a + 2)$
(0)	${}^1(0)$
(1^2)	${}^1(2)$
	${}^3(0) + {}^3(1^2)$
(1^4)	${}^1(0) + {}^1(1^2) + {}^1(2^2)$
	${}^3(1^2) + {}^3(2) + {}^3(21^2)$
	${}^5(0) + {}^5(1^2) + {}^5(1^4)$

may be achieved by considering the components of the weights **m** given by

$$[H_i, \eta^\dagger_a] = \delta_{i,a}\eta^\dagger_a$$
$$[H_i, \eta_b] = -\delta_{i,b}\eta_b$$

where

$$H_i = (1/2)[\eta^\dagger_i, \eta_i]$$

from which it is concluded that

$$\mathbf{m} = (0, 0, \ldots, \pm 1, 0, \ldots, 0)$$

The highest weight

$$\mathbf{W} = \mathbf{m}^{(0)} = (1, 0, 0, \ldots, 0)$$

corresponds to the irreducible representation (W) of the group $SO(8l_a + 4)$. Thus an operator that contains N number of creation and destruction operators must be associated with some component of a representation (W) of $SO(8l_a + 4)$, which occurs in the decomposition of $(1, 0, \ldots, 0)^N$. However, some representations (W) can be ignored if the numbers $C(\Gamma_1\Gamma_2\Gamma_3) = C(1/2, 1/2, \ldots, \pm 1/2)\Gamma_W(1/2, 1/2, \ldots, \pm 1/2)$, which are the numbers of terms in the sum over β in (3), are zero. These are the representations $(1^{N'})$ where $N' < N$ is an odd number.

Branching rules for the reduction

$$SO(8l_a + 4) \to SU_Q(2) \times Sp(4l_a + 2)$$

are presented in Table 8⁷ for (0), (1^2), and (1^4) representations of the group $SO(8l_a + 4)$, which are the only ones needed in the classification of one and two-body operators, as given by Judd [1b, p. 212].

The operators $\mathbf{T}^{(Kk)}$ with odd $(K + k)$ are labeled with the (2) representation of $Sp(4l_a + 2)$ in Table 7⁷. For the same representation in Table 8⁷, $(2Q' + 1) = 1 \to Q' = 0$, as expected. The (1^2) representation of the symplectic group $Sp(4l_a + 2)$ is associated with the operators $\mathbf{T}^{(Kk)}$ for which $(K + k)$ is even. In this case $Q' = 1$. The Wigner-Eckart theorem of (3) can be applied to the quasi-spin space. For instance, the matrix elements, diagonal in Q, of an operator labeled by $Q' = 1$ (that is, $\mathbf{O}(Q'M_{Q'}) = \mathbf{O}(10)$; $M_{Q'}$ is always zero), become

$$\langle QM_Q|\mathbf{O}(10)|QM_Q\rangle = \sum_{\beta} \langle Q\|O_\beta(10)\|Q\rangle(QM_A|10, QM_Q)$$

Because $C(Q1Q)$ is either unity or zero for a given Q, the summation over β can be lifted, resulting in

$$\langle QM_Q|\mathbf{O}(10)|QM_Q\rangle = (QM_Q|10, QM_Q)\langle Q\|O_\beta(10)\|Q\rangle \tag{12}$$

This indicates that the matrix elements, diagonal in Q, of the operator $\mathbf{O}(10)$ are proportional to $(QM_Q|10, QM_Q)$. In turn, the Clebsch-Gordan coefficients $(QM_Q|10, QM_Q)$ are proportional to M_Q, which is given by 2⁷5. The Clebsch-Gordan coefficients for the three-dimensional rotation group $SO_Q(3)$ can be expressed in terms of the $3j$-symbols as given in 3³4.

The form of the Wigner-Eckart theorem, in which the representations Γ_Q, $\Gamma_{Q'}$, and $\Gamma_{Q''}$ belong to the group $SO_Q(3)$, can be used as in 7³12. The matrix elements of an operator of quasi-spin order Q' then becomes

$$\langle QM_Q|T_0^{Q'}|Q''M_{Q''}\rangle = (-1)^{Q-M_Q}\begin{pmatrix} Q & Q' & Q'' \\ -M_Q & O & M_{Q''}\end{pmatrix} \times \langle Q\|T^{Q'}\|Q''\rangle\delta(M_Q, M\ ) \tag{13}$$

From the triangular conditions for the 3j symbols

$$(Q + Q'') \geq Q' \geq |Q - Q''|$$

Therefore, for $Q' = 0$, we have $Q = Q''$, indicating that the matrix elements of an operator of quasi-spin order zero are diagonal in Q or, equivalently, diagonal with respect to the seniority number ν. In view of 1⁵4, the matrix elements are independent of M_Q and, as a result of 2⁷5, of the number of electrons q of the configuration $(n_a l_a)^q$. For $Q' = 1$ the above triangular condition requires that $|Q - Q''| = 0, 1$. From 2⁷6 the seniority number selection rules become

$$\Delta\nu = 0, \pm 2$$

For the half-filled shells $q = 2l_a + 1$ and $M_Q = 0$. Then the only nonvanishing values of the $3j$ symbols come from the even $(Q + Q'' + 1)$, which requires that $Q \neq Q''$. Thus for half-filled shells the matrix elements of an operator of quasi-spin order Q' equal to 1 are all zero unless

$$\Delta\nu = \pm 2$$

Applying the Wigner-Eckart theorem to the quasi-spin space as in (13) and using the relations $2^7$5-6, we obtain the following results, which relate the matrix elements of the operators $\mathbf{T}^{(Kk)}$, where $(K + k)$ is even, for the configurations $(n_a l_a)^q$ and $(n_a l_a)^\nu$:

$$\frac{\langle l_a^q;\, \nu\theta \| T^{(Kk)} \| l_a^q;\, \nu\theta' \rangle}{\langle l_a^\nu;\, \nu\theta \| \mathrm{T}^{(Kk)} \| l_a^\nu;\, \nu\theta' \rangle} = \frac{2l_a + 1 - q}{2l_a + 1 - \nu} \tag{14a}$$

and

$$\frac{\langle l_a^q;\, \nu\theta \| T^{(Kk)} \| l_a^q;\, (\nu - 2)\theta' \rangle}{\langle l_a^\nu;\, \nu\theta \| T^{(Kk)} \| l_a^\nu;\, (\nu - 2)\theta' \rangle} = \frac{[(q + 2 - \nu)(4l_a + 4 - q - \nu)]^{1/2}}{2\sqrt{2l_a + 2 - \nu}} \tag{14b}$$

where θ stands for a set of quantum numbers.

D. Spin-orbit interaction

As for the quasi-spin, it is possible to find the general selection rules for a given operator without prior calculation of the matrix elements by applying the Wigner-Eckart theorem to the specified symmetry space, and by studying the values of the numbers $C(\Gamma_1\Gamma_2\Gamma_3)$. Many relations can be established [7^6a, pp. 90–5] between the matrix elements of the same operator for different configurations or between the matrix elements of differently classified operators for the same configurations. Spin-orbit interaction provides one of the best examples for one-body operators ($7^2$23). The selection rules are investigated here.

The single-particle operator f is

$$f = \xi(r)\, \mathbf{s}\cdot\mathbf{l} \tag{15}$$

where $\xi(r)$ depends on the central potential and is a function of radius alone. The creation and destruction operators in $7^2$28 are coupled then by using the Clebsch-Gordan coefficients produced in the evaluation of $\langle i|f|j\rangle$ for the $(n_a l_a)^q$ configuration, resulting in

$$\mathbf{F} = \zeta\left[\frac{l_a(l_a + 1)(2l_a + 1)}{2}\right]^{1/2} \mathbf{T}^{(11)0} \tag{16}$$

where

$$\zeta = \langle n_a l_a | f | n_a l_a \rangle \tag{17}$$

is a radial integral. The operators $\mathbf{T}^{(11)0}$ are of order one, both with regard to the spin and the angular momentum, and the final zero indicates that the spin and orbital orders are coupled to order zero [1b, p. 209].

Repeated application of the Wigner-Eckart theorem to the three-dimensional rotation space shows that the matrix elements of the

Table 9[7]. Numbers $C(W(1, 1)W')$

W	W' (0, 0)	(1, 0)	(1, 1)	(2, 0)	(2, 1)	(2, 2)
(0, 0)	0	0	1	0	0	0
(1, 0)	0	1	1	0	1	0
(1, 1)	1	1	1	1	1	1
(2, 0)	0	0	1	1	1	0
(2, 1)	0	1	1	1	2	1
(2, 2)	0	0	1	0	1	1

operators $\mathbf{T}^{(11)0}$ for the configurations $(n_a l_a)^q$ vanish if one of the three numbers $C(\Gamma_S\Gamma_1\Gamma_{S'})$, $C(\Gamma_L\Gamma_1\Gamma_{L'})$, and $C(\Gamma_J\Gamma_0\Gamma_{J'})$ vanishes. Then, because

$$C(\Gamma_S\Gamma_1\Gamma_{S'}) = 0 \quad \text{unless} \quad \Delta S = 0, \pm 1$$
$$C(\Gamma_L\Gamma_1\Gamma_{L'}) = 0 \quad \text{unless} \quad \Delta L = 0, \pm 1$$
$$C(\Gamma_J\Gamma_0\Gamma_{J'}) = 0 \quad \text{unless} \quad \Delta J = 0$$

the selection rules ΔS, $\Delta L = 0, \pm 1$ and $\Delta J = 0$ are obtained.

For p shells the symplectic and the three-dimensional rotation symmetries are the only ones involved, and the above selection rules combined with the ones for quasi-spin are all there are for the p_q configurations. For the d and f shells there are more symmetries to be taken into account which might result in additional selection rules. The operators $\mathbf{T}^{(11)0}$, for which $S = 1$, $L = 1$, are associated with the ^3P terms of the $(n_a l_a)^2$ configuration. That is, they are labeled by the $\Gamma_2 = (1, 1)$ representation of the group $SO(5)$ for d shells and the $\Gamma_2 = (1, 1)$, $\Gamma_2 = (1, 1, 0)$ representations of G_2, $SO(7)$, respectively, for f shells. The new selection rules associated with these symmetries can be found by examining the values that the numbers $C(\Gamma_1\Gamma_2\Gamma_3)$ take for each case. These numbers can be found from the reductions of inner direct products of the representations for the groups $SO(5)$, G_2, and $SO(7)$ which are tabulated by Wybourne [7[6]a, Tables D-1, E-4, and D-4, respectively].

For the d^q configurations the numbers $C(W(1, 1)W')$, which are the number of times the (1, 1) representation occurs in $W \times W'$; where (1, 1), W, and W' are the irreducible representations of the group $SO(5)$, are presented in Table 9[7]. A study of the cases for which $C(W(1, 1)W')$ is zero reveals that there are no new selection rules imposed by the $SO(5)$ symmetry.

For the f^q configurations the numbers $C(U(1, 1)U')$[the number of

Table 10^7. Numbers $C(U(1, 1)U')$

U	U' (0, 0)	(1, 0)	(1, 1)	(2, 0)	(2, 1)	(3, 0)	(2, 2)	(3, 1)	(4, 0)
(0, 0)	0	0	1	0	0	0	0	0	0
(1, 0)	0	1	0	1	1	0	0	0	0
(1, 1)	1	0	1	1	0	1	1	0	0
(2, 0)	0	1	1	1	1	1	0	1	0
(2, 1)	0	1	0	1	2	1	0	1	1
(3, 0)	0	0	1	1	1	1	1	1	1
(2, 2)	0	0	1	0	0	1	1	1	0
(3, 1)	0	0	0	1	1	1	1	2	1
(4, 0)	0	0	0	0	1	1	0	1	1

times the (1, 1) representation occurs in $U \times U'$; where (1, 1), U, and U' are the irreducible representations of the group G_2] are given in Table 10^7. The numbers $W(W(1, 1, 0)W')$ for the f^q configuration [the number of times the (1, 1, 0) representation occurs in $W \times W'$; where (1, 1, 0), W, and W' are the irreducible representations of the group $SO(7)$] are presented in Table 11^7. The contents of Tables 10^7 and 11^7 were given first by McLellan [8]. An investigation of the zero values for $C(W(1, 1, 0)W')$ (Table 11^7) shows that they produce no new selection rules. The zeros of Table 10^7, however, produce some new ones. For example, because $C((1, 1)(1, 1)(2, 1)) = 0$,

$$(f^q; \nu W(2, 1)SL\|T^{(11)0}\|f^q; \nu' W'(1, 1)S'L') = 0 \tag{18}$$

for all q, W, W', S, S', L, and L'.

E. Coulomb interaction

The Coulomb interaction is a two-body interaction as seen in Sections 9^5–11^5. Application of the group-theoretical method to obtain the general expressions for the Coulomb interaction matrix elements is attributable to Racah [1^5d]. All reduced-matrix elements needed in the computation of the term energies for the p^q, d^q, and f^q configurations are tabulated by Nielson and Koster [2^5]. Thus, following Judd [1b], it is now possible to classify the Coulomb interaction operators and to find their matrix elements differently from Racah's original analysis by referring to the tables of Nielson and Koster and to those of Wybourne and Butler [7^6a].

Here the method for the f_q configurations is demonstrated. The two-

Table 11^7. Numbers $C(W(1, 1, 0)W')$

W	W' (0, 0, 0)	(1, 0, 0)	(1, 1, 0)	(2, 0, 0)	(1, 1, 1)	(2, 1, 0)	(2, 1, 1)	(2, 2, 0)	(2, 2, 1)	(2, 2, 2)
(0, 0, 0)	0	0	1	0	0	0	0	0	0	0
(1, 0, 0)	0	1	0	0	1	1	0	0	0	0
(1, 1, 0)	1	0	1	1	1	0	1	1	0	0
(2, 0, 0)	0	0	1	1	0	0	1	0	0	0
(1, 1, 1)	0	1	1	0	1	1	1	0	1	0
(2, 1, 0)	0	1	0	0	1	2	1	0	1	0
(2, 1, 1)	0	0	1	1	1	1	2	1	1	1
(2, 2, 0)	0	0	1	0	0	0	1	1	1	0
(2, 2, 1)	0	0	0	0	1	1	1	1	2	1
(2, 2, 2)	0	0	0	0	0	0	1	0	1	1

body Coulomb interaction operator [7[4]2] is

$$g_{12} = \frac{e^2}{r_{12}} = e^2 \sum_{k,q} \left(\frac{4\pi}{2k+1} \right) \left(\frac{r_<^k}{r_>^{k+1}} \right) Y^*(k, q; \theta, \phi) Y(k, q; \theta', \phi')$$

in which the spherical harmonics $Y(k, q)$ correspond to the operators $T_{0q}^{(0k)}$. The matrix elements of the Coulomb interaction for the $(n_a l_a)^q$ configurations are zero unless k is even. The range of k for the f^q configurations is $k = 0, 2, 4, 6$. For $k = 0$, the operator $T_{00}^{(00)}$ is a scalar and the matrix elements of the Coulomb interaction are independent of L and S; that is, they are constant for all terms of a given $(n_a l_a)^q$ configuration. Their dependence on the number of electrons q is given by $q(q-1)/2$, which is the number of pairs. The three remaining operators $\mathbf{T}^{(02)}$, $\mathbf{T}^{(04)}$, and $\mathbf{T}^{(06)}$, respectively, are associated with the ^1D, ^1G, and ^1I terms of the f^2 configuration. In Table 7[7] these operators, or equally the corresponding spherical harmonics, are labeled by the (2, 0) representation of the group G_2, and by the (2, 0, 0) representation of $SO(7)$. The operators g_{ij}, which are proportional to the product of two spherical harmonics, have the symmetry of the inner direct product of the irreducible representations

$$(2, 0, 0) \times (2, 0, 0) \rightarrow (0, 0, 0) + (1, 1, 0) + (2, 0, 0) + (2, 2, 0) + (3, 1, 0) + (4, 0, 0) \qquad (19a)$$

$$(2, 0) \times (2,0) \rightarrow (0, 0) + (1, 0) + (1, 1) + 2(2, 0) + 2(2, 1) + (2, 2) + (3, 0) + (3, 1) + (4, 0) \qquad (19b)$$

where the reductions of the inner direct products are taken from Tables D-4 and E-4 of Wybourne [7[6]a], respectively. Because the central-field Hamiltonian has three-dimensional rotational symmetry, the Coulomb interaction is a scalar with respect to the group $SO(3)$. Consequently, the acceptable representations, contained in the reduction of the above inner direct products, are those which have the $\Gamma_{L=0}$ representation of $SO(3)$ when regarded as representations of the three-dimensional rotation group. These can be found by considering the branching rules for each representation on the right of (19) under the chain reductions $SO(7) \rightarrow G_2 \rightarrow SO(3)$.

Table E-4 of Wybourne [7[6]a] reveals that only the (0, 0)(2, 2), and (4, 0) representations of G_2 contain the $\Gamma_{L=0}$ representation of $SO(3)$ in their decomposition. From Table E-2 of Wybourne [7[6]a] it is found that these representations of G_2 in turn are contained in the decomposition of the (0, 0, 0), (2, 2, 0), and (4, 0, 0) representations, respectively, of $SO(7)$ in (19a). Disregarding a constant part, the Coulomb interaction for f-shell electrons can be represented as a sum of three operators, e_1, e_2, and e_3, labeled by $(0, 0, 0)(0, 0)\Gamma_{L=0}$, $(4, 0, 0)(4, 0)\Gamma_{L=0}$, and $(2, 2, 0)(2, 2)\Gamma_{L=0}$, respectively.

The simplest of the equivalent f-electron configurations which exhibits the two-particle effect is f^2. The terms are classified as

ν	W	U	Term
2	(1, 1, 0)	(1, 0)	^3F
		(1, 1)	^3P, ^3H
2	(2, 0, 0)	(2, 0)	^1D, ^1G, ^1I
0	(0, 0, 0)	(0, 0)	^1S

The matrix elements of $e_1[(0, 0, 0)(0, 0)00]$ for the ^1D, ^1G, and ^1I terms, which are classified as (2, 0, 0)(2, 0), becomes (4a)

$$\begin{aligned}&\langle f^2\nu_1(2, 0, 0)(2, 0)LM_L|e_1(\nu_2(0, 0, 0)(0, 0)00)|f^2;\ \nu_3(2, 0, 0)(2, 0)LM_L\rangle\\&\quad = \sum_\beta A_\beta(\beta(2, 0, 0)(2, 0)LM_L|(0, 0, 0)(0, 0)00,\ (2, 0, 0)(2, 0)LM_L)\end{aligned}$$

where

$$A_\beta = \langle f^2;\ \nu_1(2, 0, 0)(2, 0)\|e_1(\nu_2(0, 0, 0)(0, 0))\|f^2;\ \nu_3(2, 0, 0)(0, 0)\rangle\!\rangle$$

Because $C((2, 0)(0, 0)(2, 0)) = 1$, the summation over β can be lifted. By using (4b) the above Clebsch-Gordan coefficient may be factorized as

$$\begin{aligned}&((2, 0, 0)(2, 0)LM_L|(0, 0, 0)(0, 0)00,\ (2, 0, 0)(2,0)LM_L)\\&\quad = ((2, 0)LM_L|(0, 0)00,\ (2, 0)LM_L)\\&\quad\times((2, 0, 0)(2, 0)|(0, 0, 0)(0, 0) + (2, 0, 0)(2, 0))\\&\quad = (LM_L|00,\ LM_L)((2, 0)L|(0, 0)0 + (2, 0)\mathrm{L})\\&\quad\times((2, 0, 0)(2, 0)|(0, 0, 0)(0, 0) + (2, 0, 0)(2, 0))\end{aligned}$$

where $(LM_L|\ 00,\ LM_L) = \pm 1$. Thus the only L dependence of the matrix elements of operator e_1 is contained in the isoscalar factor $((2, 0)L|(0, 0)0 + (2, 0)L)$. In Section 10[5] the reduced-matrix elements of the tensor operator $\mathbf{U}^{(k)}$ are tabulated as defined by 10[5]8. Because the operators $\mathbf{U}^{(k)}$ and $\mathbf{T}^{(0k)}$ are related as

$$\mathbf{U}^{(k)} = \left(\frac{2}{2k + 1}\right)^{1/2}\mathbf{T}^{(0k)}$$

they have the same classification. The values of the isoscalar factors are easily found (5) by relating them to the reduced-matrix elements of $\mathbf{U}^{(k)}$.

$$\begin{aligned}(f^q;\ WUL\|\mathbf{U}^{(L)}\|f^q;\ W'U'L') &= (f^q;\ WUL\|U(WUL)\|f^q;\ W'U'L')\\&= \sum_\beta (WU\|U_\beta(WU)\|W'U')(\beta UL|UL + U'L')\end{aligned}$$

That is, for $C(UU'U) = 1$, the isoscalar factors, $(UL|UL + U'L')$, which may be substituted by $(UL|U'L' + UL)$, are proportional to the reduced-matrix element $(f^q;\ WUL\|\mathbf{U}^{(L)}\|f^q;\ W'U'L')$. Then the values of $((2, 0)L|(0, 0)0 + (2, 0)L)$ can be obtained, up to a proportionality factor, by finding the values of $(f^q;\ (2, 0, 0)(2, 0)^1L\|\mathbf{U}^{(L)}\|f^q;\ (0, 0, 0)(0, 0)^1S)$.

The matrix elements of the operator $e_1((0, 0, 0)(0, 0)\Gamma_{L=0})$ are independent of the W, U, and L labels, as can be checked from the tabulated values [2[5]]. Because they depend only on the quasi-spin and spin classifications, they are proportional to the matrix elements of $(\mathbf{Q}^2 - \mathbf{S}^2)$, within an additional constant α_1, or equally to the proper values $Q(Q + 1) - S(S + 1)$. For the ^1D, ^1G, and ^1I terms, $\nu = 2$, $S = 0$,

$$\langle f^2; \nu(2, 0, 0)(2, 0)^1LM_L|e_1|f^2; \nu(2, 0, 0)(2, 0)^1LM_L\rangle = [(35/4) - \alpha_1]\, E_1$$

where E_1 is a total proportionality constant.

The matrix elements of $e_2((4, 0, 0)(4, 0)00)$ for the same singlets are

$$\begin{aligned}\langle f^2; \nu_1(2, 0, 0)(2, 0)^1LM_L|e_2(\nu_2(4, 0, 0)(4, 0)00)|f^2; \nu_3(2, 0, 0)(2, 0)^1LM_L\rangle \\ = A((2, 0, 0)(2, 0)^1LM_L|(4, 0, 0)(4, 0)00, (2, 0, 0)(2, 0)^1LM_L) \\ A = \langle f^2; \nu_1(2, 0, 0)(2, 0)\|e_2(\nu_2\,(4, 0, 0)(4, 0))\|f^2; \nu_3(2, 0, 0)(2, 0)\rangle\end{aligned}$$

and

$$\begin{aligned}&((2, 0, 0)(2, 0)^1LM_L|(4, 0, 0)(4, 0)00, (2, 0, 0)(2, 0)^1LM_L) \\ &= ((2, 0)^1L|(4, 0)0 + (2, 0)^1L)((2, 0, 0)(2, 0)|(4, 0, 0)(4, 0) + (2, 0, 0)(2, 0))\end{aligned}$$

Because $C((2, 0, 0)(2, 0, 0)(2, 2, 2)) = 0$,

$$\begin{aligned}&(f^6; (2, 0, 0)(2, 0)^1L\|U^{(L)}\|f^6; (2, 2, 2)(4, 0)^1S) \\ &= (f^6; (2, 2, 0)(2, 0)\|U^{(L)}\|f^6; (2, 2, 2)(4, 0)) \times ((2, 0)^1L|(2, 0)^1L + (4, 0)0)\end{aligned}$$

is proportional to the isoscalar factor $((2, 0)^1L|(2, 0)^1L + (4, 0)0)$. Tables of Nielson and Koster [2[5]] provide the values

$$\begin{aligned}&(f^6; (2, 2, 0)(2, 0)^1L\|U^{(L)}\|f^6; (2, 2, 2)(4, 0)^1S) \\ &= \left(\frac{11 \cdot 13}{2 \cdot 3^3 \cdot 7^2}\right)^{1/2} \quad \text{for} \quad L = 2 \\ &= -\left(\frac{2 \cdot 5^2 \cdot 13}{3^3 \cdot 7^2 \cdot 11}\right)^{1/2} \quad \text{for} \quad L = 4 \\ &= \left(\frac{5^2}{2 \cdot 3^3 \cdot 11 \cdot 13}\right)^{1/2} \quad \text{for} \quad L = 6\end{aligned}$$

Therefore,

$$\begin{aligned}&\langle f^2; \nu_1(2, 0, 0)(2, 0)^1LM_L|e_2|f^2; \nu_3(2, 0, 0)(2, 0)^1LM_L\rangle \\ &= 143\alpha_2E_2 \quad \text{for} \quad L = 2 \\ &= -130\alpha_2E_2 \quad \text{for} \quad L = 4 \\ &= 35\alpha_2E_2 \quad \text{for} \quad L = 6\end{aligned}$$

where α_2 and E_2 are proportionality constants.

The matrix elements of $e_3((2, 2, 0)(2, 2)00)$ are

$$\begin{aligned}&\langle f^2; \nu_1(2, 0, 0)(2, 0)^1LM_L|e_3(\nu_2(2, 2, 0)(2, 2)00)|f^2; \nu_3(2, 0, 0)(2, 0)^1LM_L\rangle \\ &= B((2, 0, 0)(2, 0)^1LM_L|(2, 2, 0)(2, 2)00, (2, 0, 0)(2, 0)^1LM_L) \\ B &= \langle f^2; \nu_1(2, 0, 0)(2, 0)\|e_3(\nu_2(2, 2, 0)(2, 2)\|f^2; \nu_3(2, 0, 0)(2, 0)\rangle\end{aligned}$$

and

$$((2, 0, 0)(2, 0)^1LM_L|(2, 2, 0)(2, 2)00, (2, 0, 0)(2, 0)^1LM_L)$$
$$= ((2, 0)^1L|(2, 2)0 + (2, 0)^1L)((2, 0, 0)(2, 0)|(2, 2, 0)(2, 2) + (2, 0, 0)(2, 0))$$

Because

$$(f^6; (2, 0, 0)(2, 0)^1L\|U^{(L)}\|f^6; (2, 2, 0)(2, 2)^1S)$$
$$= (f^6; (2, 2, 0)(2, 0)\|U^{(L)}\|f^6; (2, 2, 0)(2, 2)) \times ((2, 0)^1L|(2, 0)^1L + (2, 2)0)$$

the isoscalar factors $((2, 0)^1L|(2, 2)0 + (2, 0)^1L)$ are proportional to the reduced matrix elements $(f^6; (2, 0, 0)(2, 0)^1L\|U^{(L)}\|f^6; (2, 2, 0)(2, 2)^1S)$. From the tables of Neilson and Koster [2[5]].

$$(f^6; (2, 0, 0)(2, 0)^1L\|U^{(L)}\|f^6; (2, 2, 0)(2, 2)^1S) = -\left(\frac{11}{5\cdot 7}\right)^{1/2} \quad \text{for} \quad L = 2$$
$$= -\frac{4}{(5\cdot 7\cdot 11)^{1/2}} \quad \text{for} \quad L = 4$$
$$= \left(\frac{7}{5\cdot 11}\right)^{1/2} \quad \text{for} \quad L = 6$$

Therefore

$$\langle f^2; \nu_1(2, 0, 0)(2, 0)^1LM_L|e_3|f^2; \nu_3(2, 0, 0)(2, 0)^1LM_L\rangle = -11\alpha_3E_3 \quad \text{for} \quad L = 2$$
$$= -4\alpha_3E_3 \quad \text{for} \quad L = 4$$
$$= 7\alpha_3E_3 \quad \text{for} \quad L = 6$$

where α_3 and E_3 are proportionality constants.

The matrix element of the operator e_1 for the ^1S term is

$$\langle f^2; 0(0, 0, 0)(0, 0)^1S|e_1|f^2; 0(0, 0, 0)(0, 0)^1S\rangle = \left(\frac{63}{4} - \alpha_1\right)E_1$$

Because $C((0, 0)(4, 0)(0, 0)) = C((0, 0)(2, 2)(0, 0)) = 0$, there are no contributions from the operators e_2 and e_3.

For the triplet terms the matrix element of e_1 becomes

$$\langle f^2; 2WU^3L|e_1|f^2; 2WU^3L\rangle = \left(\frac{27}{4} - \alpha_1\right)E_1$$

The vanishing of $C((1, 0)(4, 0)(1, 0))$, $C((1, 0)(2, 2)(1, 0))$, and $C((1, 1)(4, 0)(1, 1))$ prevents any contributions from the operators e_2 and e_3 for the ^3F term and from the operator e_2 for the ^3P and ^3H terms. Then the nonzero elements of e_3 for the ^3P and ^3H terms are written

$$\langle f^2; 2(1, 1, 0)(1, 1)^3LM_L|e_3|f^2; 2(1, 1, 0)(1, 1)^3LM_L\rangle = \beta_1E_3 \quad \text{for} \quad L = 1$$
$$\beta_2E_3 \quad \text{for} \quad L = 5$$

If α_3 is set equal to unity to agree with Racah's original derivation, the term energies of the f^2 configuration become

$$^1S = \left(\frac{63}{4} - \alpha_1\right) E_1$$
$$^3P = \left(\frac{27}{4} - \alpha_1\right) E_1 + \beta_1 E_3$$
$$^1D = \left(\frac{35}{4} - \alpha_1\right) E_1 + 143\alpha_2 E_2 - 11E_3$$
$$^3F = \left(\frac{27}{4} - \alpha_1\right) E_1$$
$$^1G = \left(\frac{35}{4} - \alpha_1\right) E_1 - 130\alpha_2 E_2 - 4E_3$$
$$^3H = \left(\frac{27}{4} - \alpha_1\right) E_1 + \beta_2 E_3$$
$$^1I = \left(\frac{35}{4} - \alpha_1\right) E_1 + 35\alpha_2 E_2 + 7E_3$$

The term energies of the $(n_a l_a)^2$ configurations that were obtained by conventional methods are given in terms of F^k integrals by $2^5 5'$. Thus there are seven parameters to evaluate and seven equations. Because the matrix elements of the operator $\mathbf{T}^{(00)}$ are not included in view of $10^5$19,20, the F^0 integrals should be deleted in the term energies. Then

$$\alpha_1 = \frac{225}{28}, \quad E_1 = \frac{14}{405} F^2(ff) + \frac{7}{297} F^4(ff) + \frac{350}{11583} F^6(ff)$$
$$\alpha_2 = 2, \quad E_2 = \frac{1}{2025} F^2(ff) - \frac{1}{3267} F^4(ff) + \frac{175}{1656369} F^6(ff)$$
$$E_3 = \frac{1}{135} F^2(ff) + \frac{2}{1089} F^4(ff) - \frac{175}{42471} F^6(ff)$$
$$\beta_1 = 33, \quad \beta_2 = -9$$

Therefore, the term energies for the f^2 configuration become

$$^1S = \frac{54}{7} E_1$$
$$^3P = -\frac{9}{7} E_1 + 33E_3$$
$$^1D = \frac{5}{7} E_1 + 286E_2 - 11E_3$$
$$^3F = -\frac{9}{7} E_1$$
$$^1G = \frac{5}{7} E_1 - 260E_2 - 4E_3$$

$$^3H = -\frac{9}{7}E_1 - 9E_3$$

$$^1I = \frac{5}{7}E_1 + 70E_2 + 7E_3$$

The term energies for any f^q configuration can be found by starting with those for the f^2 configuration and by using the following equation repeatedly:

$$\langle f^q;\, \nu_1 W_1 U_1 L_1 | e_i(\nu_1' W_1' U_1' L_1') | f^q;\, \nu_1'' W_1'' U_1'' L_1'' \rangle$$
$$= \frac{q}{q-2} \sum_{\theta_2, \theta_2''} (\theta_1\{|\theta_2) \langle f^{q-1};\, \theta_2 | e_i | f^{q-1};\, \theta_2'' \rangle (\theta_2''|\}\theta_1'') \qquad (20)$$

where $(\theta_1\{|\theta_2)$ and $(\theta_2'|\}\theta_1')$ are written for the coefficients of fractional parentage of Section 7[5], and θ_i stands for $\nu_i W_i U_i L_i$. In view of the Wigner-Eckart theorem,

$$\langle f^q;\, \nu_1 W_1 U_1 L_1 | e_i(\nu_1' W_1' U_1' L_1') | f^q;\, \nu_1'' W_1'' U_1'' L_1'' \rangle$$
$$= \sum_\beta A_\beta (\beta \nu_1 W_1 U_1 L_1 | \nu_1' W_1' U_1' L_1',\, \nu_1'' W_1'' U_1'' L_1'')$$

where

$$A_\beta = \langle f^q;\, \nu_1 W_1 U_1 \| e_i(\nu_1' W_1' U_1')_\beta \| f^q;\, \nu_1'' W_1'' U_1'' \rangle$$

The Clebsch-Gordan coefficient $(\beta \nu_1 W_1 U_1 L_1 | \nu_1' W_1' U_1' L_1',\, \nu_1'' W_1'' U_1'' L_1'')$ can be expressed, using (6a), in terms of the known isoscalar factors $(\nu_1 W_1 U_1 | \nu_1' W_1' U_1' + \nu_1'' W_1'' U_1'')$ and $(U_1 L_1 | U_1' L_1' + U_1'' L_1'')$. Then (20) determines the reduced-matrix elements A_β. This simplifies the process considerably.

The quasi-spin classification of the operators e_2 and e_3 also simplifies the computation of the coefficients of E_2 and E_3. The operators $\mathbf{T}^{(Kk)}$, with even ranks or, equally, the corresponding spherical harmonics, are labeled by the (1^2) representation of $Sp(14)$ for f-shell electrons. From Table D-13 of Wybourne [7[6]a]

$$(1^2) \times (1^2) \rightarrow (0) + (1^2) + (1^4) + (2) + (2, 1^2) + (2^2)$$

Since the operator e_2 is classified as $(4, 0, 0)(4, 0)\Gamma_{L=0}$, its symplectic symmetry is found by considering the decomposition of each representation at the right of the above equation and determining which ones contain the (2, 0, 0) representation of $SO(7)$. Tables C-17 and C-30 of Wybourne [7[6]a] reveal that only the (2^2) representation of $Sp(14)$ contains the (2, 0, 0) representation of $SO(7)$ in its decomposition under the reduction $Sp(14) \rightarrow SO(7)$. The operator e_2 therefore has (2^2) symplectic symmetry that corresponds to the quasi-spin rank $Q' = 0$. The matrix elements of e_2 are diagonal in seniority number, as discussed earlier.

The operator e_3 is labeled by $(2, 0, 0)(2, 2)\Gamma_{L=0}$, and the (2, 2, 0) representation of $SO(7)$ is contained in decomposition of the representations (1^4) and (2^2) of the group $Sp(14)$ under the reduction $Sp(14) \rightarrow$ SO(7). The corresponding quasi-spin ranks for these representations of $Sp(14)$ is $Q' = 2$ and $Q' = 0$, respectively. In order to have a well-defined quasi-spin classification, the operator e_3 has to be rewritten in two parts. This can be achieved by defining an operator $-\Omega$ of quasi-spin rank zero, such that

$$e_3 = (e_3 + \Omega) - \Omega$$

where the operator $(e_3 + \Omega)$ now has the quasi-spin rank $Q' = 2$. The explicit constructions of the operators $-\Omega$ and $(e_3 + \Omega)$ are given in Judd [22[3], p. 208]. The calculation of the coefficients of e_3 becomes very much simplified if the matrix elements of each part of the operator e_3 are computed separately.

Term energies for the f^q configurations are tabulated in Nielson and Koster [2[5], pp. 53–63]. The classification of the single-particle operators for the mixed configurations of the type $(l_a + l_b)^q$ and the treatment of the Coulomb interaction in $(d + s)^q$ configurations are given in Wybourne [7[6]a, pp. 117–21].

4[7]. Coefficients of fractional parentage for f^q configurations

The concept of the coefficients of fractional parentage is introduced in Section 7[5] and applied there to the p^q and d^q configurations. The approach of Section 7[5] produces so many entries that a complete listing is not practical for the f^q configurations. (Tabulation of these coefficients occupies 46 pages in Nielson and Koster [2[5]].) Therefore it is necessary to factorize them. This can be accomplished by using 3[7]4b or 3[7]5, following Racah [1[5]d]. His results are

$$\begin{aligned}(f^q v_1 W_1 U_1 \tau_1 S_1 L_1 \{| f^{q-1} v_2 W_2 U_2 \tau_2 S_2 L_2, f) \\ = (U_1 \tau_1 L_1 | U_2 \tau_2 L_2 + f)(W_1 U_1 | W_2 U_2 + f)(f^q; v_1 S_1 \{| f^{q-1}; v_2 S_2 + f) \qquad (1)\end{aligned}$$

where the labels $v_1 W_1 U_1 \tau_1 S_1 L_1$ and $v_2 W_2 U_2 \tau_2 S_2 L_2$ belong to f^q and f^{q-1} configurations, respectively. An arbitrary label τ is introduced to distinguish the terms with identical $vWUSL$ labels. Here f stands for the relevant quantum labels, $1(1, 0, 0)(1, 0)^2F$, of a single f electron. The notation for the coefficient of fractional parentage here is slightly different from that in 7[5].

The isoscalar factors $(U_1 \tau_1 L_1 | U_2 \tau_2 L_2 + f)$ and $(W_1 U_1 | W_2 U_2 + f)$ were originally computed by, but not all included in, Racah [1[5]d]. Wybourne [9] obtained some of the missing factors, namely, $((2, 2, 1)U | (2, 1, 1)U' + f)$, $((3, 1)L | (2, 1)L' + f)$, and $((3, 1)L | (3, 0)L' + f)$, which are needed in the computation of the spin-orbit interaction matrix elements. These factors are proportional to the reduced-matrix elements of the operators $\mathbf{T}^{(0k)}$, as explained in 3[7]. Their values are given in Tables 12[7],

Table 12[7]. $|(WU|W'U' + f)|^2$ for $W' = (000), (100), (110), (200), (111), (2\dot{1}0), (211)$. The signs of square roots are indicated. Each entry should be divided by the square of the normalization constant, namely, by N^2.

		W′	(0,0,0)	(1,0,0)	(1,1,0)		(2,0,0)	(1,1,1)			(2,1,0)			(2,1,1)				
W	U	N^2 \ U′	(0,0)	(1,0)	(1,0)	(1,1)	(2,0)	(0,0)	(1,0)	(2,0)	(1,1)	(2,0)	(2,1)	(1,0)	(1,1)	(2,0)	(2,1)	(3,0)
(0,0,0)	(0,0)	1	0	1	0	0	0											
(1,0,0)	(1,0)	3	3	0	1	2	3											
(1,1,0)	(1,0)	35	0	35	0	0	0	3	14	18	14	21	0					
	(1,1)	70	0	70	0	0	0	0	-7	-63	0	6	64					
(2,0,0)	(2,0)	105		105	0	0	0	0	0	0	14	27	64					
(1,1,1)	(0,0)	1			1	0	0	0	-1	0	0	0	0	1	0	0	0	0
	(1,0)	168			112	-56	0	-24	-63	81	0	0	0	-7	56	105	0	0
	(2,0)	5832			1296	-4536	0	0	729	-5103	0	0	0	1	-56	135	2560	3080
(2,1,0)	(1,1)	42			42	0	42	0	0	0	0	0	0	-7	0	15	20	0
	(2,0)	1701			1323	378	1701	0	0	0	0	0	0	98	448	270	-500	385
	(2,1)	672			0	672	672	0	0	0	0	0	0	0	-7	-60	220	385
(2,1,1)	(1,0)	5040						3888	-1134	18	-3024	2016	0	-350	2800	-1890	0	0
	(1,1)	630						0	567	-63	0	576	-54	175	0	-135	320	0
	(2,0)	2520						0	2205	315	840	720	-960	-245	-280	-867	-512	616
	(2,1)	5040						0	0	5040	945	-1125	2970	0	560	-432	-2816	1232
	(3,0)	315						0	0	315	0	45	270	0	0	27	64	-224
(2,2,0)	(2,0)	105									56	-48	1					
	(2,1)	56									-7	27	22					
	(2,2)	1									0	0	1					

Table 13[7]. $|(UL|U'L' + f)|^2$ for $U' = (0, 0), (1, 0), (1, 1), (2, 0)$. The signs of square roots are indicated. Each entry should be divided by the square of the normalization constant, namely, by N^2.

	U'		(0,0)	(1,0)	(1,1)		(2,0)		
U	L	N^2	S	F	P	H	D	G	I
(0,0)	S	1	0	-1	0	0	0	0	0
(1,0)	F	378	378	378	81	297	-70	-126	-182
(1,1)	P	21	0	21	0	0	10	-11	0
	H	2079	0	2079	0	0	220	585	-1274
(2,0)	D	49	0	49	-27	-22	-16	33	0
	G	3234	0	3234	1089	-2145	1210	-750	1274
	I	11	0	11	0	11	0	3	8
(2,1)	D	49			-22	27	33	16	0
	F	1386			-1089	297	-605	-144	637
	G	882			585	297	13	624	245
	H	297			0	297	143	-144	-10
	K	33			0	33	0	16	-17
	L	1			0	1	0	0	1
(3,0)	P	21					11	10	0
	F	4158					1573	-2340	245
	G	198					121	12	-65
	H	693					286	162	245
	I	11					0	8	-3
	K	33					0	17	16
	M	1					0	0	1

13[7], and 14[7], taken from Racah's original work [1[5]d]; the detailed explanation of how they are evaluated is in Judd [22[3], pp. 174–82]. Complete tables of the composite coefficients are given by Nielson and Koster [2[5]].

Because the reductions of the inner direct products $\Gamma_{S_2} \times \Gamma_{1/2}$ and $(1^{\nu_2}) \times (1)$, where $\nu_2 < 2l_a + 1$, are given by

$$\Gamma_{S_2} \times \Gamma_{1/2} \to \Gamma_{S_2+1/2} + \Gamma_{S_2-1/2}$$
$$(1^{\nu_2}) \times (1) \to (1^{\nu_2-1}) + (1^{\nu_2+1}) + (21^{\nu_2-1})$$

for a fixed value of ν_1 and S_1, there are at most four coefficients:

$$(f^q; \nu_1 S_1; \{|f^{q-1}; \nu_2 S_2 + f)$$

This makes possible an explicit algebraic construction for the last factor in (1) [1[5]d, pp. 1358–60; 22[3], pp. 182–4] with the following results:

$$
\begin{aligned}
&(f^q;\, \nu_1 S_1 \{| f^{q-1};\, \nu_2 S_2 + f)^2 \\
&= \frac{(16 - q - \nu)(2 + \nu + 2S)S}{2q(8 - \nu)(2S + 1)} \quad \text{for} \quad \nu_2 = \nu_1 - 1,\, S_2 = S_1 - \frac{1}{2} \\
&= \frac{(16 - q - \nu)(\nu - 2S)(S + 1)}{2q(8 - \nu)(2S + 1)} \quad \text{for} \quad \nu_2 = \nu_1 - 1,\, S_2 = S_1 + \frac{1}{2} \\
&= \frac{(q - \nu)(18 - \nu + 2S)S}{2q(8 - \nu)(2S + 1)} \quad \text{for} \quad \nu_2 = \nu_1 + 1,\, S_2 = S_1 - \frac{1}{2} \\
&= \frac{(q - \nu)(16 - \nu - 2S)(S + 1)}{2q(8 - \nu)(2S + 1)} \quad \text{for} \quad \nu_2 = \nu_1 + 1,\, S_2 = S_1 + \frac{1}{2}
\end{aligned}
\tag{2}
$$

The phases taken for the coefficients themselves (rather than their squares) are $(-1)^{S_2}$ if ν_1 is odd and $(-1)^{S_2 + (1/2)(\nu_2 - \nu_1)}$ if ν_1 is even. This is the convention established by Racah [1[5]d].

5[7]. Quasi-particle scheme

Disregarding a few ambiguities for the f shell, the continuous Lie groups introduced in the previous sections are enough to classify the Russell-Saunders terms of the l^n configurations for $l < 3$. (Because q is used here to denote the components of the operators, n is chosen to specify the number of electrons.) However, for the cases where $l > 3$, the branching rules for the reduction $SO(2l + 1) \rightarrow SO(3)$ contain a large number of repeated terms. As a result, Racah's scheme does not suffice for defining states for these configurations. Moreover, there are certain regularities or simplifications for which the previous approaches do not provide an explanation. A few of these are:

1. The matrix elements of certain operators vanish for no apparent reason.
2. Complete matrices are proportional to one another, in spite of the fact that the Wigner-Eckart theorem predicts no simple proportionality.

The quasi-particle approach that was introduced to atomic physics by Armstrong and Judd [10], following an idea first initiated by Shudeman [11] and elaborated by many others [12], seems to explain many of the above mentioned regularities or simplifications. Furthermore, in this new scheme it becomes possible to define without ambiguity all the states of the l^n configurations up to, and including, $l = 8$. This is done by paying a price. That is, in the new method the total spin, S, and the total number of electrons, $N = n$, are not good quantum numbers. The fact that the total number of electrons is not conserved gives the name "quasi-particles" to the scheme.

Table 14[7]. $|(UL|(21)L' + f)|^2$. The signs of square roots are indicated. Each entry should be divided by the square of the normalization constant, namely, by N^2.

U	L	N^2	D	F	G	H	K	L
(1,1)	P	1344	220	-539	-585	0	0	0
	H	4928	-270	147	-297	1078	1470	-1666
(2,0)	D	31360	8910	8085	351	-14014	0	0
	G	4312	330	147	1287	1078	-1470	0
	I	18304	0	-1911	1485	220	4590	10098
(2,1)	D	5390	375	1960	-1144	1911	0	0
	F	154	-40	49	-65	0	0	0
	G	630630	-74360	207025	226941	-51744	70560	0
	H	15730	-2535	0	1056	-3179	7260	1700
	K	2860	0	0	-192	968	-85	1615
	L	572	0	0	0	-40	-285	247
(3,0)	P	2688	1156	-245	1287	0	0	0
	F	112	-39	0	24	49	0	0
	G	7920	1375	2450	-858	1617	1620	0
	H	640640	-42250	-207025	3971	261954	490	124950
	I	9152	0	1274	2750	1320	2125	-1683
	K	1040	0	0	-204	-136	605	95
	M	64	0	0	0	0	-15	-49
(2,2)	S	1	0	1	0	0	0	0
	D	640	-130	195	297	18	0	0
	G	51480	18590	4225	-429	-23826	-4410	0
	H	6160	980	-2450	1078	-132	245	-1275
	I	18304	0	1105	9163	-2244	-4802	990
	L	364	0	0	0	152	-147	-65
	N	16	0	0	0	0	-5	11

A. Fourfold factorization of the l shell

If all the electrons with spin up ($m_s = 1/2$) are regarded as being in a different space than the electrons with spin down ($m_s = -1/2$), the 2^{4l+2} states of the l shell can be divided in two equidimensional spaces. Because the 2^{2l+1} states of either spin-up or spin-down space form the basis for the (1/2, 1/2, . . . , 1/2) representation of group $SO(4l + 3)$

(1⁷7′), this corresponds to the reduction

$$SO(8l+5) \to SO_+(4l+3) \times SO_-(4l+3) \tag{1}$$

where plus or minus signs assigned as subscripts to SO specify the spin orientation. The actual atomic states are thus obtained by coupling the spin up and down spaces.

Russell-Saunders terms of the l^n configurations, with maximum multiplicity (that is, all spins either up or down), show a remarkable repetition. For example, in the d^n configuration P and F terms appear, with all $m_s = 1/2$, for $n = 2, 3$; D term for $n = 1, 4$ and S term for $n = 0, 5$. This implies that the $(1/2, 1/2, \ldots, 1/2)$ representation of $SO(4l + 3)$ may be resolved into two equal parts. In view of $1^7 12$, the states of l^n with $n =$ even, and all spins up, belong to the $(1/2, 1/2, \ldots, 1/2, -1/2)$ representation of $SO_+(4l + 2)$; and those with $n =$ odd, and all spins up, belong to $(1/2, 1/2, \ldots, 1/2, 1/2)$. Therefore, both the spin-up and spin-down spaces factor into two equal parts. Mathematically this is accomplished by introducing the following normalized generators [10]:

$$\begin{aligned}
\lambda^\dagger{}_m &= \frac{\eta^\dagger{}_{1/2,m} + (-1)^{l-m}\eta_{1/2,-m}}{\sqrt{2}} \\
\mu^\dagger{}_m &= \frac{\eta^\dagger{}_{1/2,m} - (-1)^{l-m}\eta_{1/2,-m}}{\sqrt{2}} \\
\nu^\dagger{}_m &= \frac{\eta^\dagger{}_{-1/2,m} + (-1)^{l-m}\eta_{-1/2,-m}}{\sqrt{2}} \\
\xi^\dagger{}_m &= \frac{\eta^\dagger{}_{-1/2,m} - (-1)^{l-m}\eta_{-1/2,-m}}{\sqrt{2}}
\end{aligned} \tag{2}$$

where the index m stands for m_l. The reasoning which leads to the above combination is this [13]: In the theory of quasi-particles [14] a basic step involves linearly combining creation operators $\eta^\dagger{}_a$ with destruction operators $\eta_{a'}$, where a' is time reversed with respect to a. In the present case, because the time reversal changes m_s to $-m_s$, the spin projection numbers should not be allowed to time-reverse; in other words, they have to be fixed. Then the study of combinations of the kind

$$\theta^\dagger{}_m = A\eta^\dagger{}_{1/2,m} + B\eta_{1/2,-m}$$

shows that the operators θ with $A = B$ anticommute with the ones $A = -B$. Moreover, a phase factor $(-1)^{l-m}$, in front of the destruction operator, is needed to ensure that $(2l + 1)$ components of $\theta^\dagger{}_m$ form a tensor $\boldsymbol{\theta}^\dagger$. The commutation relations of the generators $\theta^\dagger{}_m(\theta^\dagger{}_m \equiv \lambda^\dagger{}_m, \mu^\dagger{}_m, \nu^\dagger{}_m$, or $\xi^\dagger{}_m)$ with the angular momentum vector $\boldsymbol{l}$ satisfy $7^3 13$, provided that $k = l$. Therefore $(2l + 1)$ operators $\theta^\dagger{}_m$ form the components of the tensors $\boldsymbol{\theta}\dagger$ with rank l. Since the components of different tensors $\boldsymbol{\lambda}\dagger$, $\boldsymbol{\mu}\dagger$, $\boldsymbol{\nu}\dagger$, and $\boldsymbol{\xi}\dagger$ anticommute, each of them defines a distinct $(2l + 1)$-dimensional space.

Following the same steps a new operator $\tilde{\theta}_m$ can be defined which also forms the components of a tensor $\boldsymbol{\theta}$ with rank l. This is easily done by defining $\tilde{\theta}_m$ as

$$\tilde{\theta}_m = (-1)^{l-m}\theta_{-m} \tag{3}$$

where θ_m stands for λ_m, μ_m, ν_m, or ξ_m. Then

$$\left.\begin{aligned}\boldsymbol{\theta} &= \boldsymbol{\theta}\dagger\,(\theta \equiv \lambda, \nu)\\ \boldsymbol{\theta} &= -\boldsymbol{\theta}\dagger\,(\theta \equiv \mu, \xi)\end{aligned}\right\} \tag{4}$$

indicating that the new tensors are proportional to the old ones. The coupled products $(1/2)(\boldsymbol{\theta}\dagger\boldsymbol{\theta})^{(k)}$ for odd k satisfy the same commutation relations among themselves as $\mathbf{T}^{(0k)}$. The operators $\mathbf{T}^{(0k)}$ with odd k are the generators of the group $SO(2l + 1)$. Then so are the products $(1/2)(\boldsymbol{\theta}\dagger\boldsymbol{\theta})^{(k)}$ with k = odd. Because the members from different sets commute, the operators in (2) are the generators of the direct product

$$SO_\lambda(2l + 1) \times SO_\mu(2l + 1) \times SO_\nu(2l + 1) \times SO_\xi(2l + 1)$$

Moreover, the following chains result:

$$\begin{aligned}&SO(8l + 4) \supset SO_\lambda(2l + 1) \times SO_\mu(2l + 1)\\ &\times SO_\nu(2l + 1) \times SO_\xi(2l + 1)\\ &SO_+(4l + 2) \supset SO_\lambda(2l + 1) \times SO_\mu(2l + 1)\\ &SO_-(4l + 2) \supset SO_\nu(2l + 1) \times SO_\xi(2l + 1)\end{aligned} \tag{5}$$

in which the last two relations express the factorization of the spin-up and spin-down spaces.

B. Generators; the vacuum state of quasi-particles

As a natural consequence one expects the basic operators of (2) to be the creation operators for quasi-particles. This is almost the case except for the fact that, because m_s is fixed, $\theta\dagger_m$ and $\theta\dagger_{m'}$ do not anticommute when $m = -m'$. Therefore, the anticommutation relations satisfied by the $\theta\dagger_m$ and θ_m are not quite the same as those appropriate to fermions. There are two ways to avoid this difficulty: either to consider only those $\theta\dagger_m$ for which $m > 0$ or to define a new operator $\Theta\dagger_q$ as

$$\Theta\dagger_q = (1/2)^{1/2}[\theta\dagger_q, \theta\dagger_0] \qquad (q > 0) \tag{6}$$

where now $\Theta = \Lambda, M, N, \Xi$ and $\theta = \lambda, \mu, \nu, \xi$, respectively. So for $q, q' > 0$ in the spin-up space we have

$$\begin{aligned} [\Theta\dagger_q, \Theta\dagger_{q'}]_+ &= 0 \\ [\Theta_q, \Theta_{q'}]_+ &= 0 \\ [\Theta\dagger_q, \Theta_{q'}]_+ &= \delta(q, q') \end{aligned} \tag{7}$$

indicating that the operators $\Theta\dagger_q$ are the creation operators for a fermion quasi particle with l components. When these operators act on any state of the configuration l^n they preserve the parity, unlike the operators $\theta\dagger_m$. Moreover, we can rewrite (6) in terms of the linear combinations of the generators $(\theta\dagger\theta)_q^{(k)}$ of group $SO_\theta(2l+1)$, subscript θ indicating the basis

$$\Theta\dagger_q = \pm\sqrt{2}\sum_k (l0, lq|kq)(\theta\dagger\theta)_q^{(k)}, \quad \text{for} \quad k = \text{odd} \tag{6'}$$

in which the positive sign should be taken for $\theta = \lambda, \nu$ and the negative sign for $\theta = \mu, \xi$, meaning that the operators $\Theta\dagger_q$ cannot connect different irreducible representations of $SO_\theta(2l+1)$.

The vacuum state of quasi-particles for $\theta = \lambda$ or μ, that is, in the spin-up space, can be chosen to be either of the following:

$$\begin{aligned} |O_+\rangle &= \prod_{m<O} \eta\dagger_{1/2,m}|0\rangle \\ |O'_+\rangle &= \prod_{m\leq O} \eta\dagger_{1/2,m}|0\rangle \end{aligned} \tag{8}$$

Then

$$\Theta_q|O_+\rangle = \Theta_q|O'_+\rangle = 0 \qquad (q > 0, \mathbf{\Theta} = \mathbf{\Lambda}, \mathbf{M})$$

and the creation operators $\Lambda\dagger_q$, $M\dagger_q$ generate quasi-particle states when they act on either $|O_+\rangle$ or $|O'_+\rangle$. Therefore the l fermion operators $\Theta\dagger_q$ (for $q > 0$), combined into 2^l distinct products $\Theta\dagger_{q_1}\Theta\dagger_{q_2}\ldots\Theta\dagger_{q_k}$, generate the 2^l states that belong to a single irreducible representation of $SO_\theta(2l+1)$. If $|O_+\rangle$ is taken as the vacuum state, the parity of n is equal to the parity of l; in the case of $|O'_+\rangle$ vacuum states n and l have opposite parities.

The operator H_i, defined in Section 8[6], can be expressed in terms of the commutator of the creation and destruction operators $\theta\dagger_q$ and θ_q as

$$H_{l-q+1} = (1/2)[\theta\dagger_q, \theta_q] \qquad (q = l, l-1, \ldots, 1) \tag{9}$$

When the operators H_i act either on a state

$$\Theta\dagger_{q_1}\Theta\dagger_{q_2}\ldots\Theta\dagger_{q_k}|O_+\rangle$$

or on

$$\Theta\dagger_{q_1}\Theta\dagger_{q_2}\ldots\Theta\dagger_{q_k}|O'_+\rangle$$

their proper values may be evaluated by bringing them to the right with repeated use of the commutation relation

$$[H_{l-q+1}, \Theta\dagger_{q_k}] = \Theta\dagger_{q_f}\delta(q, q_k) \tag{10}$$

to act on the vacuum and produce a proper value of $-1/2$. The final result is either $-1/2$ if none of the q_k values are equal to q, or $\pm 1/2$ if one of the q_k values is equal to q. Thus the proper values ($\pm 1/2$, $\pm 1/2$, . . . , $\pm 1/2$) are produced by varying the values of index q between l and 1. Such a presentation (Section 10[6]) is labeled by the highest weight, which is (1/2, 1/2, . . . , 1/2). Actually, in (9) the factor 1/2 is introduced to have the correct length of the roots.

Similar definitions of vacuum states $|O_-\rangle$ and $|O'_-\rangle$ may be given for spin-down space. If we wish to consider the cases with either spin orientation, then a vacuum state of the type

$$\Pi\eta\dagger_{1/2,m}\Pi\eta\dagger_{-1/2,m'}|0\rangle$$

must be introduced. Because, first, either m or m', and second, neither nor both may be permitted to take the value of zero, there are now four choices of vacuum states compared with two for the all spin-up or spin-down cases.

The total angular momentum $\mathbf{L}$ is a sum of $\mathbf{l}_\theta$ as

$$\mathbf{L} = \mathbf{l}_\lambda + \mathbf{l}_\mu + \mathbf{l}_\nu + \mathbf{l}_\xi \tag{11}$$

in which l_θ is defined by

$$l_\theta = \frac{1}{2}\frac{[l(l+1)(2l+2)]^{1/2}}{[3]^{1/2}}(\boldsymbol{\theta}\dagger\boldsymbol{\theta})^{(1)} \tag{11'}$$

Thus the Z component becomes

$$(\mathbf{l}_\theta)_z = \sum_{q>0} qH_{l-q+1} \tag{12}$$

C. Branching rules; states

The branching rules of the reductions

$$SO_\theta(2l+1) \rightarrow SO_\theta(3)$$

for half-integral representations of $SO(2l+1)$ are determined by the requirement

$$(1, 0, \ldots, 0) \rightarrow (l)$$

Geometrically this corresponds to the fact that when the $(2l + 1)$ weights

$$(\pm 1, 0, 0, \ldots, 0), (0, \pm 1, 0, \ldots, 0), \ldots, (0, 0, 0, \ldots, 0)$$

are projected on the line representing the one-dimensional space of $SO_\theta(3)$, they should give $(2l + 1)$ equally spaced points [13, p. 50]. This fixes the direction cosines of such a line. Then, in turn, the branching rules are found by projecting the weights of half-integer representation on this line.

Two examples demonstrate this: First, for $l = 2$, the requirement that the projection of five weights

$$(\pm 1, 0), (0, \pm 1), (0, 0)$$

on the one-dimensional space, a line, of $SO_\theta(3)$ should give five equally spaced points fixes the magnitudes of the directional cosines of the line as

$$\frac{2}{\sqrt{5}} \quad \text{and} \quad \frac{1}{\sqrt{5}}$$

Then, the projections of four weights (1/2, 1/2), (1/2, −1/2), (−1/2, 1/2), and (−1/2, −1/2), of $SO_\theta(5)$, on this line (in units of $1/\sqrt{5}$) are

$$\pm\frac{3}{2}, \pm\frac{1}{2}$$

which correspond to the 3/2, or equally to the $D^{3/2}$ representation of $SO_\theta(3)$. By proceeding in the same way for $l = 3$ the directional cosines of the line are found as

$$\frac{3}{\sqrt{14}}, \frac{2}{\sqrt{14}}, \frac{1}{\sqrt{14}}$$

so that the projection of eight weights (±1/2, ±1/2, ±1/2), in units of $1/\sqrt{14}$, are

$$\pm\frac{3}{2}, \pm\frac{2}{2}, \pm\frac{1}{2}$$

Then, the possible combinations of signs produce (in units of $1/\sqrt{14}$)

$$\pm 3, \pm 2, \pm 1, 0, 0$$

which correspond to the (3) and (0) representations of $SO_\theta(3)$. When this process is carried out for higher values of l the branching rules tabulated in Table 15[7], for the reduction $SO_\theta(2l + 1) \rightarrow SO_\theta(3)$, are obtained [10a].

The representation (1/2, 1/2) of $SO_\theta(5)$ reduces to 3/2 of $SO_\theta(3)$. This means that the terms of d^n with all $m_s = 1/2$, and the number of particles being either odd or even, may be obtained by coupling two 3/2 angular momenta; this accounts for the *S*, *P*, *D*, and *F* terms with maximum multiplicity.

Table 15^7. Branching rules for the reduction $SO_\theta(2l+1) \to SO_\theta(3)$

l	Half integral representation of $SO_\theta(2l+1)$	Representations of $SO_\theta(3)$: (l_θ)
1, p	$\left(\frac{1}{2}\right)$	$\left(\frac{1}{2}\right)$
2, d	$\left(\frac{1}{2}, \frac{1}{2}\right)$	$\left(\frac{3}{2}\right)$
3, f	$\left(\frac{1}{2}, \frac{1}{2}, \frac{1}{2}\right)$	(0), (3)
4, g	$\left(\frac{1}{2}, \frac{1}{2}, \frac{1}{2}, \frac{1}{2}\right)$	(2), (5)
5, h	$\left(\frac{1}{2}, \frac{1}{2}, \frac{1}{2}, \frac{1}{2}, \frac{1}{2}\right)$	$\left(\frac{5}{2}\right), \left(\frac{9}{2}\right), \left(\frac{15}{2}\right)$
6, i	$\left(\frac{1}{2}, \frac{1}{2}, \frac{1}{2}, \frac{1}{2}, \frac{1}{2}, \frac{1}{2}\right)$	$\left(\frac{3}{2}\right), \left(\frac{9}{2}\right), \left(\frac{11}{2}\right), \left(\frac{15}{2}\right), \left(\frac{21}{2}\right)$
7, k	$\left(\frac{1}{2}, \frac{1}{2}, \frac{1}{2}, \frac{1}{2}, \frac{1}{2}, \frac{1}{2}, \frac{1}{2}\right)$	(2), (4), (5), (7), (8), (9), (11), (14)
8, l	$\left(\frac{1}{2}, \frac{1}{2}, \frac{1}{2}, \frac{1}{2}, \frac{1}{2}, \frac{1}{2}, \frac{1}{2}, \frac{1}{2}\right)$	(0), (3), (4), (5), (6), (7), (8), (9), (10), (11), (12), (13), (15), (18)

Examination of Table 15^7 shows that in the reduction of the half-integral representations of $SO_\theta(2l+1)$ no l_θ occurs more than once for $l < 9$. Therefore the states of the l^n configuration are uniquely defined by

$$|(l_\lambda l_\mu)_p L_A (l_\nu l_\xi)_{p'} L_B; LM_L\rangle \tag{13}$$

in which the indices p and p' specify the parities of the number of particles in the spin-up and spin-down spaces, respectively. The angular momenta L_A and L_B, in two different spin spaces, are coupled to L. For the states of maximum multiplicity, in less than half-filled shells, the states (13) simplify to

$$|(l_\lambda l_\mu)_p LM_L\rangle = |p(l_\lambda l_\mu) LM_L\rangle \tag{14}$$

It is important to note that in the new scheme the states of the type $|(l_\lambda m_{l_\lambda})(l_\mu m_{l_\mu})\rangle\rangle$ cannot be factored into the product of two distinct parts $|l_\lambda m_{l_\lambda}\rangle$ and $|l_\mu m_{l_\mu}\rangle$. Because the angular momenta l_λ, l_μ, l_ν, and l_ξ appearing in the states (13) may be coupled without the restrictions that are enforced when coupling fermions or bosons, calculations of matrix elements are carried out without the need of coefficients of fractional parentage. Finally, the quasi-particle technique is not limited to configurations of the type l^n. Configurations $(l_1 + l_2)^n$ can be studied by replacing the group $SO_\theta(2l+1)$ with $SO_\Theta(2l_1+2l_2+2)$ and proceeding as explained in this section [12b, 12c].

D. Computation of the matrix elements

Although the representation of states, as given in (13), are very much simplified, the handling of the operators becomes more complicated. Therefore, for simplicity it is desirable to carry out the calculations in the spin-up space. This amounts to the evaluation of matrix elements within the state (14). First, the one- and two-body operators given in 7²28 and 7²29 must be expressed in terms of operators that act on the λ and μ spaces. This is done by substitutions obtained from (2), (3), and (4):

$$\begin{aligned} \eta^\dagger{}_a &= (1/2)^{1/2}(\lambda^\dagger{}_m + \mu^\dagger{}_m) \\ \eta_a &= (1/2)^{1/2}(\lambda_m + \mu_m) \end{aligned} \tag{15}$$

If the Coulomb interaction of electrons in the spin-up space is taken as an example, in view of 2⁵1, the two-body operator of 7²29 takes the form:

$$\sum_{i<j}\frac{e^2}{r_{ij}} = \frac{e^2}{2}\sum_{a,b,c,d,k}\eta^\dagger{}_a\eta^\dagger{}_b\langle a_1 b_2|\frac{r_<^k}{r_>^{k+1}}\mathbf{C}_1^{(k)}\cdot\mathbf{C}_2^{(k)}|c_1 d_2\rangle\eta_d\eta_c$$

Substituting the values for $\eta^\dagger$ and η as given by (15), and using the following relation [1a, p. 19]:

$$e^2\langle a_1 b_2|\frac{r_<^k}{r_>^{k+1}}\mathbf{C}_1^{(k)}\cdot\mathbf{C}_2^{(2)}|c_1 d_2\rangle = \sum_q(-1)^{m_{l_a}+m_{l_b}+q}\begin{pmatrix} l & k & l \\ -m_{l_a} & q & m_{l_c}\end{pmatrix} \times\begin{pmatrix} l & k & l \\ -m_{l_b} & -q & m_{l_d}\end{pmatrix}\langle l\|C^{(k)}\|l\rangle|^2F^k(l,l) \tag{16}$$

where $l = l_a = l_b = l_c = l_d$ and all $m_s = 1/2$,

$$\sum_{i<j}\frac{e^2}{r_{ij}} = \frac{1}{8}\sum_{a,b,c,d,k,q}(-1)^{a+b+q}(\lambda^\dagger{}_a+\mu^\dagger{}_a)(\lambda^\dagger{}_b+\mu^\dagger{}_b)(\lambda_d+\mu_d)(\lambda_c+\mu_c)|\langle l\|\mathbf{C}^{(k)}\|l\rangle|^2 \times\begin{pmatrix} l & k & l \\ -a & q & c\end{pmatrix}\begin{pmatrix} l & k & l \\ -b & -q & d\end{pmatrix}F^k(l,l)$$

in which a, b, c, d now stand for m_{l_a}, m_{l_b}, m_{l_c} and m_{l_d}. The sixteen terms of the above equation can be coupled to give [10b]

$$\sum_{i<j}\frac{e^2}{r_{ij}} = \frac{1}{8}\sum_k\left\{\left[1-(-1)^l\frac{4}{2l+1}(\boldsymbol{\lambda}^\dagger\cdot\boldsymbol{\mu}^\dagger)\right]\left[(2l+1)\delta(k,0)-1\right]\cdot +4\sum_{K=\mathrm{odd}}\begin{Bmatrix} l & l & K \\ l & l & k\end{Bmatrix}[(\boldsymbol{\lambda}^\dagger\boldsymbol{\lambda})^{(K)}\cdot(\boldsymbol{\mu}^\dagger\boldsymbol{\mu})^{(K)}]\right\}|\langle l\|C^{(k)}\|l\rangle|^2F^k(l,l) \tag{17}$$

which is valid for terms of maximum multiplicity. The different terms of (17) may be evaluated, within the state (14), by using the relations in Section 1[5]. By employing 1[5]6,

$$(p(l_\lambda l_\mu)L|(\boldsymbol{\lambda}\dagger \cdot \boldsymbol{\mu}\dagger)|p'(l'_\lambda l'_\mu)L') = (-1)^{l'_\lambda+l_\mu+L}\delta(L,L')\begin{Bmatrix} l_\lambda & l_\mu & L \\ l'_\mu & l'_\lambda & l \end{Bmatrix} \times \sum_{p''}\langle pl_\lambda\|\lambda\dagger\|p''l'_\lambda\rangle\langle p''l_\mu\|\mu\dagger\|p'l'_\mu\rangle \quad (17a)$$

Application of 1[5]9 gives

$$\langle pl_\lambda\|(\boldsymbol{\lambda}\dagger\boldsymbol{\lambda})^{(K)}\|p'l'_\lambda\rangle = (-1)^{l_\lambda+l'_\lambda+K}\sqrt{2K+1} \times \sum_{p'',l''_\lambda}\begin{Bmatrix} l & l & K \\ l_\lambda & l'_\lambda & l''_\lambda \end{Bmatrix}\langle pl_\lambda\|\lambda\dagger\|p''l''_\lambda\rangle\langle p''l''_\lambda\|\lambda\|p'l'_\lambda\rangle \quad (17b)$$

in which all terms with K = even are zero except for $K = 0$, then

$$(\boldsymbol{\lambda}\dagger\boldsymbol{\lambda})^{(0)} = 1/2(2l+1)^{1/2}$$

Before evaluating the reduced-matrix elements $\langle pl_\theta\|\theta\dagger\|p'l'_\theta\rangle$, $\langle pl_\theta\|\theta\|p'l'_\theta\rangle$, it is necessary to make a number of phase choices. The coefficients of fractional parentage and the coefficients in the expansion of the states (14) must be real. These conditions imply [10b]

$$\langle pl_\theta\|\theta\dagger\|p'l'_\theta\rangle = (-1)^{l+l_\theta-l'_\theta}\langle p'l'_\theta\|\theta\|pl_\theta\rangle \quad (18)$$

where $p \neq p'$, because the operators $\boldsymbol{\theta}\dagger$ and $\boldsymbol{\theta}$ connect states differing by one electron. Supposing that the λ and μ spaces can be interchanged according to the rules in Chapter 3, the matrix elements $[p(l_\lambda l_\mu L_A|(\boldsymbol{\mu}\dagger \cdot \boldsymbol{\lambda}\dagger)|p'(l'_\lambda l_\mu)L_A]$ may be evaluated either by writing $(\boldsymbol{\mu}\dagger \cdot \boldsymbol{\lambda}\dagger) = -(\boldsymbol{\lambda}\dagger \cdot \boldsymbol{\mu}\dagger)$ and using tensorial techniques, or by interchanging μ and λ in the bra and ket and employing the tensorial relations of Chapter 5. Both ways should produce the same result:

$$\langle pl_\theta\|\theta\dagger\|p'l'_\theta\rangle = (-1)^l\langle p'l_\theta\|\theta\|pl'_\theta\rangle \quad (19)$$

Finally, the phase choices made for $\boldsymbol{\mu}\dagger$ affect the number of possibilities open for $\boldsymbol{\lambda}\dagger$. These relations are [10b]

$$\begin{aligned} \langle p_u l_\tau\|\lambda\dagger\|p_g l_\omega\rangle &= (-1)^l\langle p_u l_\tau\|\mu\dagger\|p_g l_\omega\rangle \\ \langle p_g l_\tau\|\lambda\dagger\|p_u l_\omega\rangle &= (-1)^{l+1}\langle p_g l_\tau\|\mu\dagger\|p_u l_\omega\rangle \end{aligned} \quad (20)$$

where the indices τ and ω refer to the spaces λ and μ, and p_g, p_u stand for even and odd parities, respectively. The limited arbitrariness that remains in the choice of phases can be fixed by forcing the results to agree with those obtained by conventional methods.

Armstrong and Judd [10b] evaluated the reduced-matrix elements $\langle pl_\theta\|\theta\dagger\|p'l'_\theta\rangle$ for $l = 2, 3, 4, 5$, as given in Table 16[7]. The matrix elements for $\langle pl_\theta\|\theta\|p'l'_\theta\rangle$ can be found by (18). The other combinations are given in (19) and (20). The entries in Table 16[7] are the only ones needed in

Table 16^7. Reduced-matrix elements $\langle pl_\theta \| \theta^\dagger \| p'l'_\theta \rangle$

l	pl_θ	$p'l'_\theta$	$\langle pl_\theta \| \theta^\dagger \| p'l'_\theta \rangle$	l	pl_θ	$p'l'_\theta$	$\langle pl_\theta \| \theta^\dagger \| p'l'_\theta \rangle$
d, 2	$p_u3/2$	$p_g3/2$	$(10)^{1/2}$	h, 5	$p_g5/2$	$p_u5/2$	$(63/22)^{1/2}$
f, 3	p_g0	p_u0	0		$p_g5/2$	$p_u9/2$	$(45/2)^{1/2}$
	p_g0	p_u3	$-(7/2)^{1/2}$		$p_g5/2$	$p_u15/2$	$-(84/11)^{1/2}$
	p_g3	p_u0	$(7/2)^{1/2}$		$p_g9/2$	$p_u5/2$	$(45/2)^{1/2}$
	p_g3	p_u3	$-(21)^{1/2}$		$p_g9/2$	$p_u9/2$	$(165/26)^{1/2}$
g, 4	p_u2	p_g2	$(35/4)^{1/2}$		$p_g9/2$	$p_u15/2$	$(340/13)^{1/2}$
	p_u2	p_g5	$-(55/4)^{1/2}$		$p_g15/2$	$p_u5/2$	$(84/11)^{1/2}$
	p_u5	p_g2	$(55/4)^{1/2}$		$p_g15/2$	$p_u9/2$	$-(340/13)^{1/2}$
	p_u5	p_g5	$(143/4)^{1/2}$		$p_g15/2$	$p_u15/2$	$(7752/143)^{1/2}$

the computation of matrix elements of the Coulomb interaction. The simplification brought by the present technique becomes obvious when the number of fractional parentage coefficients required to carry out such a calculation is considered.

In order to relate the states (14) to the conventional states $|l^n; \gamma, S = M_s = n/2, L, M_L\rangle$, one needs the coupling coefficients $\langle p(l_\lambda l_\mu)L|l^n; S = n/2, L\rangle$. These can be found, in principle, by choosing a suitable operator that is not diagonal in the space of (14) but diagonal in the number of electrons, n, and in the total angular momenta, L. Then the diagonalization of the matrix, whose elements are evaluated within the states (14), yields the proper functions that correspond to the definite values of n. In the case that one value of γ is enough to specify a state, which is so for the configurations p^n, d^n, f^n, the element of the matrix (often chosen to be unitary to preserve orthonormality of the states) that diagonalizes the matrix of the operator are the coefficients that are sought.

The number operator

$$N = \sum_a \eta^\dagger{}_a \eta_a$$

is such an operator whose proper values are the number of electrons in the proper states. In the new scheme it is given as

$$N = \frac{1}{2} \sum_a (\lambda^\dagger{}_a + \mu^\dagger{}_a)(\lambda_a + \mu_a)$$
$$= \frac{1}{2}(2l + 1) + (-1)^{l+1}(\boldsymbol{\lambda}^\dagger \cdot \boldsymbol{\mu}^\dagger)$$

in which the matrix elements of $(\boldsymbol{\lambda}^\dagger \cdot \boldsymbol{\mu}^\dagger)$ may be found from (17a) and Table 16^7. The coefficients $\langle p(l_\lambda l_\mu)L|f^n; S = n/2, L\rangle$ are given in Armstrong and Judd [10b].

6^7. The four-dimensional orthogonal group and the hydrogen atom

It long has been known [15] that the symmetry group of the nonrelativistic Coulomb central field is that of $SO(4)$ in a projective momentum space, that is, in a four-dimensional space projected on a three-dimensional one.

Therefore, the properties of the four-dimensional rotational group are discussed first. Its generators are given by [16]

$$\begin{aligned} J_{\alpha\beta} &= -i(x^\alpha{}_\beta - x^\beta{}_\alpha);\ \alpha,\beta = 1, 2, 3, 4 \\ &= -i\left(x^\alpha \frac{\partial}{\partial x^\beta} - x^\beta \frac{\partial}{\partial x^\alpha}\right) \end{aligned} \tag{1}$$

which obey the relation

$$J_{\alpha\beta} = -J_{\beta\alpha} \tag{2}$$

Thus the number of independent generators is reduced from twelve to six. The commutation relations for these generators are

$$[J_{\alpha\beta}, J_{\gamma\nu}] = i\delta_{\alpha\gamma}J_{\beta\nu} + i\delta_{\beta\nu}J_{\alpha\gamma} + i\delta_{\alpha\nu}J_{\gamma\beta} + i\delta_{\beta\gamma}J_{\nu\alpha} \tag{3}$$

It is more convenient to define the two new operators as

$$\begin{aligned} \mathbf{L} &= \hat{\mathbf{i}}J_{23} + \hat{\mathbf{j}}J_{31} + \hat{\mathbf{k}}J_{12} \\ \mathbf{A} &= \hat{\mathbf{i}}J_{14} + \hat{\mathbf{j}}J_{24} + \hat{\mathbf{k}}J_{34} \end{aligned} \tag{4}$$

Then, the commutation relations (3) become

$$\begin{aligned} [L_i, L_j] &= i\epsilon_{ijk}L_k \\ [L_i, A_j] &= i\epsilon_{ijk}A_k \\ [A_i, A_j] &= i\epsilon_{ijk}L_k \end{aligned} \tag{3$'$}$$

Because the operators **A** and **L** do not commute, their linear combinations are considered:

$$\mathbf{J}_a = (1/2)(\mathbf{L} + \mathbf{A}),\ \mathbf{J}_b = (1/2)(\mathbf{L} - \mathbf{A}) \tag{5}$$

Now,

$$[\mathbf{J}_a, \mathbf{J}_b] = 0, \quad \mathbf{J}_a \times \mathbf{J}_a = i\mathbf{J}_a, \quad \mathbf{J}_b \times \mathbf{J}_b = i\mathbf{J}_b \tag{6}$$

indicating that the operators $\mathbf{J}_a$ and $\mathbf{J}_b$ generate an algebra. Because these relations are the commutation relations for two commuting three-dimensional angular momentum operators, the group so generated is isomorphic to $SO(3) \times SO(3)$.

Therefore, simultaneous proper functions of the following four operators can be constructed:

$$\mathbf{J}_a^2,\ J_{a3},\ \mathbf{J}_b^2,\ J_{b3}$$

These proper functions, which will be used as a basis for representations, may be labeled as $|j_a m_a, j_b m_b\rangle$. Then,

$$\begin{aligned}
\mathbf{J}_a^2|j_a m_a, j_b m_b\rangle &= j_a(j_a+1)|j_a m_a, j_b m_b\rangle \\
J_{a3}|j_a m_a, j_b m_b\rangle &= m_a|j_a m_a, j_b m_b\rangle \\
\mathbf{J}_b^2|j_a m_a, j_b m_b\rangle &= j_b(j_b+1)|j_a m_a, j_b m_b\rangle \\
J_{b3}|j_a m_a, j_b m_b\rangle &= m_b|j_a m_a, j_b m_b\rangle
\end{aligned} \tag{7}$$

It is now convenient to introduce Schwinger's boson spin creation and destruction operators [4[3]]:

$$\eta\dagger_a = (\eta\dagger_{a+}, \eta\dagger_{a-}), \qquad \eta_a = (\eta_{a+}, \eta_{a-}) \tag{8}$$

which obey the commutation relations (7[2]18). Because each operator has two components, it is convenient to change the notation:

$$a\dagger_i = \eta\dagger_a,\ a_i = \eta_a; \qquad b\dagger_i = \eta\dagger_b,\ b_i = \eta_b; \qquad i = 1, 2 \tag{9}$$

The state defined in 7[2]19 then becomes

$$|j_a m_a, j_b m_b\rangle = \frac{(a\dagger_1)^{j_a+m_a}(a\dagger_2)^{j_a-m_a}(b\dagger_1)^{j_b+m_b}(b\dagger_2)^{j_b-m_b}}{[(j_a+m_a)!(j_a-m_a)!(j_b+m_b)!(j_b-m_b)!]^{1/2}}|0\rangle \tag{10}$$

The operators $\mathbf{J}_a$ and $\mathbf{J}_b$ take the form

$$\mathbf{J}_a = \frac{1}{2}\,a\dagger\boldsymbol{\sigma}a,\ \mathbf{J}_b = \frac{1}{2}\,b\dagger\boldsymbol{\sigma}b \tag{11}$$

where $\boldsymbol{\sigma}$ are the Pauli spin matrices.

The lowering and raising operators, in complete analogy with the three-dimensional case, are defined as

$$\begin{aligned}
J_a^+ &= J_{a1} + iJ_{a2} = a\dagger_1 a_2;\ J_b^+ = J_{b1} + iJ_{b2} = b\dagger_1 b_2 \\
J_a^- &= J_{a1} - iJ_{a2} = a\dagger_2 a_1;\ J_b^- = J_{b1} - iJ_{b2} = b\dagger_2 b_1
\end{aligned} \tag{12}$$

Because $\mathbf{J}_a$ and $\mathbf{J}_b$ commute, their proper states can be coupled in the usual manner (Section 3[3]) to obtain the proper state of $\mathbf{L}^2 = (\mathbf{J}_a + \mathbf{J}_b)^2$ with definite total angular momentum l and its projection m:

$$|j_a, j_b; l, m\rangle = \sum_{m_a, m_b} C(j_a m_a, j_b m_b; lm)|j_a m_a, j_b m_b\rangle \tag{13}$$

These new states are usually labeled by p and q rather than by j_a and j_b, where

$$p = j_a + j_b,\ q = j_a - j_b \tag{14}$$

Although $\mathbf{J}_a^2$ and $\mathbf{J}_b^2$ commute with all the generators of $O(4)$, it is customary to call

$$C_1^2 = \mathbf{A}^2 + \mathbf{L}^2 = 2(\mathbf{J}_a^2 + \mathbf{J}_b^2) \tag{15}$$

the *Casimir operator* of $O(4)$. The scalar product of $\mathbf{A}$ and $\mathbf{L}$:

$$C_2^2 = \mathbf{A} \cdot \mathbf{L} = \mathbf{J}_a^2 - \mathbf{J}_b^2 \tag{16}$$

is called the *second Casimir operator*. The proper values of C_1^2 and C_2^2 are:

$$\begin{aligned} C_1^2|p, q; l, m\rangle &= [p(p+2) + q^2]|p, q; l, m\rangle \\ C_2^2|p, q; l, m\rangle &= q(p+1)|p, q; l, m\rangle \end{aligned} \tag{17}$$

The dimension of an irreducible representation, of $O(4)$, labeled by j_a, j_b is $(2j_a + 1)(2j_b + 1)$. The irreducible representations may also be identified by p, q. The dimension of such a representation in terms of p and q is found by replacing $j_a = (p + q)/2, j_b = (p - q)/2$ in $(2j_a + 1)(2j_b + 1)$, to be $(p + 1)^2 - q^2$. For every positive value of q there is a negative value which satisfies the dimension relation. The corresponding representations are called *conjugate representations*. When $q = 0$ they become *self-conjugate*.

The condition for representations to be self-conjugate, $q = 0$, is found from (16) and (17) to be

$$\mathbf{A} \cdot \mathbf{L} = 0 \tag{18a}$$

which also implies that

$$j_a = j_b \equiv j \tag{18b}$$

Thus the dimension of a self-conjugate representation is $(2j + 1)^2$.

Pauli [17] showed that the nonrelativistic central Coulomb field problem could be characterized by two three-dimensional vectors: the angular momentum

$$\mathbf{L} = \frac{\mathbf{r} \times \mathbf{p}}{\hbar}$$

and the Runge-Lenz vector [18]

$$\mathbf{R} = \frac{[\mathbf{L} \times \mathbf{p} - \mathbf{p} \times \mathbf{L}]}{2Ze^2\mu\hbar} + \hat{\mathbf{r}} \tag{19}$$

both of which commute with the Hamiltonian

$$H = \frac{p^2}{2\mu} - \frac{Ze^2}{r}$$

A new vector $\mathbf{A}$ is defined in terms of the Runge-Lenz vector as

$$\mathbf{A} = \frac{\mathbf{R}}{[-2H/Z^2e^2\mu]^{1/2}} \tag{20}$$

so the commutation relations between the angular momentum vector $\mathbf{L}$ and the new vector $\mathbf{A}$ are exactly those of (3′). The Hamiltonian of the system with this new scaled Runge-Lenz vector is

$$H = -\frac{Z^2 e^4 \mu}{2\hbar^2(\mathbf{A}^2 + \mathbf{L}^2 + 1)} \tag{21}$$

Because $\mathbf{A} \cdot \mathbf{L} = 0$, the angular momentum and the scaled Runge-Lenz vector generate self-conjugate representations of $O(4)$, provided that the scale factor

$$p_4 = \sqrt{-2\mu W} = \frac{Z\mu e^2}{\hbar n} \tag{22}$$

is positive. The principal quantum number n then becomes identical, as a label, with $2j + 1$. Consequently, there is one-to-one correspondence between n^2 proper states of hydrogen in the level n and the number of self-conjugate irreducible representations of $O(4)$ of dimension $(2j + 1)^2$.

Fock [15a] established the correspondence between hydrogenic momentum space wave functions, $\phi(\mathbf{p}; n, l, m)$, and the realization of irreducible representations of $O(4)$ on a space of functions $f(x_1, x_2, x_3, x_4)$. He showed that under the substitutions

$$\begin{aligned}
x_1 &= \frac{2p_4 p_x}{p_4^2 + p^2} = \sin\alpha \sin\nu \cos\varphi \\
x_2 &= \frac{2p_4 p_y}{p_4^2 + p^2} = \sin\alpha \sin\nu \sin\varphi \\
x_3 &= \frac{2p_4 p_z}{p_4^2 + p^2} = \sin\alpha \cos\nu \\
x_4 &= \frac{p_4^2 - p^2}{p_4^2 + p^2} = \cos\alpha \\
\Psi(\hat{\mathbf{x}}) &= \frac{\pi}{\sqrt{8}} p_4^{-5/2}(p_4^2 + p^2)^2 \psi(\mathbf{p}) \\
x_1^2 &+ x_2^2 + x_3^2 + x_4^2 = 1
\end{aligned} \tag{23}$$

the momentum space Schrödinger equation (2⁴16) is converted into a symmetric integral equation

$$p_4 \psi(\hat{\mathbf{x}}) = \frac{Z\mu e^2}{2\pi^2 \hbar} \int \frac{\psi(\hat{\mathbf{x}}')}{|\mathbf{x} - \mathbf{x}'|^2}\, d\Omega \tag{24}$$

which is invariant under $O(4)$. Here $d\Omega$ is the surface area element on the unit hypersphere in 4-space:

$$d\Omega = \sin^2\alpha \sin\nu \, d\alpha \, d\nu \, d\varphi \tag{25}$$

Moreover, $\psi(\hat{\mathbf{x}})$ is transferred into an $O(4)$ spherical harmonic $Y(n, l, m; \hat{\mathbf{x}})$.

The four-dimensional spherical harmonics have been calculated by Biedenharn [16]. A few of them are

$$\begin{aligned} Y(1, 0, 0; \hat{\mathbf{x}}) &= 1 \\ Y(2, 0, 0; \hat{\mathbf{x}}) &= 2 \cos \alpha \\ Y(2, 1, 0; \hat{\mathbf{x}}) &= i2 \sin \alpha \cos \nu \\ Y(3, 0, 0; \hat{\mathbf{x}}) &= 4 \cos^2 \alpha - 1 \\ Y(3, 1, 0; \hat{\mathbf{x}}) &= i\sqrt{6} \sin 2\alpha \cos \nu \\ Y(3, 2, 0; \hat{\mathbf{x}}) &= -\sqrt{2} \sin^2 \alpha \,(3 \cos^2 \nu - 1) \end{aligned} \tag{26}$$

where

$$\int Y^*(n', l', m'; \hat{\mathbf{x}}) Y(n, l, m; \hat{\mathbf{x}})\, d\Omega = 2\pi^2 \delta_{nn'} \delta_{ll'} \delta_{mm'} \tag{27}$$

and

$$Y^*(n, l, m; \hat{\mathbf{x}}) = (-1)^{l-m} Y(n, l, -m; \hat{\mathbf{x}}) \tag{28}$$

Because p_4 is a scale factor involved in the correspondence between hydrogenic momentum space functions and the irreducible representations of $O(4)$, an extra label in the hydrogenic wave functions for the identification of the scale factor is introduced. If $\psi(n, l, m; \mathbf{r}, p_\alpha)$ is a position function, then the corresponding momentum wave function is (Sec. 5[2])

$$\phi(n, l, m; \mathbf{p}, p_\alpha) = (2\pi)^{-3/2} \int e^{-i\mathbf{p}\cdot\mathbf{r}} \psi(n, l, m; \mathbf{r}, p_\alpha)\, d^3 r \tag{29}$$

In terms of the four-dimensional spherical harmonics this relation becomes

$$\phi(n, l, m; \mathbf{p}, p_\alpha) = (-1)^{n+1} \frac{\sqrt{8}}{\pi} p_\alpha^{5/2} (p_\alpha^2 + p^2)^{-2} Y(n, l, m; \hat{\mathbf{x}}) \tag{30}$$

Two momentum-space hydrogen wave functions with $p_{\alpha'} = p_\alpha = p_4$ satisfy the relation

$$(-1)^{n+n'} \int \phi^*(n', l', m'; \mathbf{p}, p_4) \left[\frac{p^2 + p_4^2}{2p_4^2}\right] \phi(n, l, m; \mathbf{p}, p_4)\, d^3 p = \delta_{nn'} \delta_{ll'} \delta_{mm'} \tag{31}$$

in which

$$d^3 p = \left[\frac{p^2 + p_4^2}{2p_4}\right]^3 d\Omega$$

and

$$p = p_4 \tan \alpha/2$$

For the case $n = n'$, $l = l'$, and $m = m'$ the relation (31) states the *virial theorem*. It also shows that the set of wave functions

$$\bar{\phi}(n, l, m; \mathbf{p}, p_4) = (-1)^{n+1}\left(\frac{p^2 + p_4^2}{2p_4^2}\right)^{1/2} \phi(n, l, m; \mathbf{p}, p_4) \qquad (32)$$

form an orthonormal basis set. From (31) one can obtain (in atomic units)

$$(-1)^{n+n'} \int \psi^*(n', l', m'; \mathbf{r}, p_4)\left[-\frac{\nabla^2}{2} + \frac{p_4^2}{2}\right] \psi(n, l, m; \mathbf{r}, p_4)\, d^3r$$

$$= p_4^2 \delta_{nn'} \delta_{ll'} \delta_{mm'} \qquad (33)$$

Therefore,

$$(-1)^{n+n'} \int \psi^*(n', l', m'; \mathbf{r}, p_4) \frac{1}{r} \psi(n, l, m; \mathbf{r}, p_4)\, d^3r$$

$$= \frac{p_4}{n} \delta_{nn'} \delta_{ll'} \delta_{mm'} \qquad (34)$$

indicating that the set of wave functions

$$\bar{\psi}(n, l, m; \mathbf{r}, p_4) = (-1)^{n+1}\left(\frac{n}{p_4 r}\right)^{1/2} \psi(n, l, m; \mathbf{r}, p_4) \qquad (35)$$

also forms an orthonormal basis.

The integrals involving hydrogenic functions of the same shell can be simplified if the operators occurring in them can be expressed in terms of the generators of $O(4)$ and commutators of the form

$$[f(\mathbf{A}^2 + \mathbf{L}^2), g(\mathbf{x}, \mathbf{p})] \qquad (36)$$

This commutator has vanishing matrix elements within a given shell provided that the functions $f(\mathbf{A}^2 + \mathbf{L}^2)$ and $g(\mathbf{x}, \mathbf{p})$ are not singular. Swamy, Kulkarni, and Biedenharn [19] used this technique in the evaluation of

$$\int_0^\infty P(n, l'; r) \frac{1}{r^q} P(n, l; r)\, dr = \left\langle n, l' \left| \frac{1}{r^q} \right| n, l \right\rangle \qquad (37)$$

Their results for $l' = l$ are presented in Table 2[4]. For $l' = l + 1$ they obtained (in atomic units)

$$\left\langle l+1 \left| \frac{1}{r} \right| l \right\rangle = \frac{1}{n^2}\left(\frac{n - l - 1}{n + l + 1}\right)^{1/2}$$

$$\left\langle l+1 \left| \frac{1}{r^2} \right| l \right\rangle = 0$$

$$\left\langle l+1 \left| \frac{1}{r^3} \right| l \right\rangle = \frac{2|1 - [(l+1)/n]^2|^{1/2}}{n^3(l+1)(2l+1)(2l+3)} \tag{38}$$

$$\left\langle l+1 \left| \frac{1}{r^4} \right| l \right\rangle = \frac{4|1 - [(l+1)/n]^2|^{1/2}}{n^3 l(l+1)(l+2)(2l+1)(2l+3)}$$

$$\left\langle l+1 \left| \frac{1}{r^5} \right| l \right\rangle = \frac{6[5n^2 - l(l+2)]|1 - [(l+1)/n]^2|^{1/2}}{n^5 l(l+1)(l+2)(2l-1)(2l+1)(2l+3)(2l+5)}$$

They also derived the Pasternack-Sternheimer result [20]. That is,

$$\int_0^\infty P(n, l+l'; r) \frac{1}{r^q} P(n, l; r)\, dr = 0$$

if

$$2 \leqslant q \leqslant l' + 1 \tag{39}$$

This result is useful in the evaluation of the Coulomb integral

$$\int \psi(n, l', m'; \mathbf{r}) \frac{1}{|\mathbf{r} - \mathbf{r}'|} \psi(n, l, m; \mathbf{r})\, d^3 r$$

When the $1/|\mathbf{r} - \mathbf{r}'|$ term is expanded in the multipole series, the coefficients of the angular integral in the qth term is obtained as

$$r'^q \int_0^\infty P(n, l'; r) \frac{1}{r^{q+1}} P(n, l; r)\, dr$$

$$- \int_0^{r'} P(n, l'; r) \left[\frac{r'^q}{r^{q+1}} - \frac{r^q}{r'^{q+1}} \right] P(n, l; r)\, dr$$

Because the first integral vanishes for $2 \leqslant q + 1 \leqslant |l' - l| + 1$, the Coulomb potential mixes functions differing in l to a severely limited extent. For example, when $n = 2, 3$ and $l' \neq l$ all the integrals are zero.

7⁷. Dynamical groups

The concept of dynamical groups was initiated by A. O. Barut [21], and it has been further developed and used, also under the name of *noninvariance groups* and *spectrum generating groups,* by Y. Dothan, M. Gell-Mann, and Y. Ne'eman [22], and by E. C. G. Sudarshan, N. Mukunda, and L. O'Raifeartaigh [23]. Review articles are also listed [24].

Up to now group theory and group representations have been used to characterize the symmetry or invariance properties of a quantum mechanical Hamiltonian. The basic idea was as follows:

Let $H(q_0)$ be the Hamiltonian of an isolated system with the set of *dynamical variables* q_i (discrete and/or continuous), that is, many body space coordinates, and consider the transformation

$$q \to g^{-1}q \tag{1}$$

in which g's are the elements of a group G of transformations – again; discrete or continuous – (in other words, reflections or rotations) under which the Hamiltonian is invariant. That is,

$$H(gq) = H(q) \tag{2}$$

Then,

$$D(g)\psi(q) = \psi(g^{-1}q) = \psi'(q) \tag{3}$$

is also a state of the system together with the state $\psi(q)$, both corresponding to the same energy W. The operation g is a group operation acting on the space of q. The operation $D(g)$ is, on the other hand, a representation of the group element g acting always on the element $\psi(q)$ of the Hilbert space of physical states of the system. Because H is also an operator in the Hilbert space,

$$D(g)H\psi = WD(g)\psi$$
$$D(g)HD^{-1}(g)D(g)\psi = WD(g)\psi$$

Thus

$$D(g)HD^{-1}(g) = H(gq) = H(q) \tag{4}$$

That is, $D(g)$ commutes with H, or H is an invariant operator of the representation $D(g)$ of G. G is called a *symmetry group* in this specific and narrow sense.

More generally, symmetry groups are associated with equivalent descriptions of a system from a passive point of view: One observer may describe the states by the set of wave functions $\{\psi(q)\}$ and another by the set $\{\psi(g^{-1}q)\}$, for example, the observer with a rotated frame of reference. These two descriptions are equivalent if

$$(\psi(q), \phi(q)) = (\psi(g^{-1}q), \phi(g^{-1}q)), \qquad \text{for all } g$$

Wigner has shown [4^2b, App. Chap. 20, pp. 233–6] that for all normalized wave functions ψ and ϕ wave functions representing equivalent descriptions may be related to each other by a canonical transformation:

$$D(g)\psi(q) = \psi(g^{-1}q)$$

which is identical to (3).

Knowledge of the symmetry groups gives only partial information about the quantum system, namely, only the degeneracy properties of the states of a given energy W (that is, the $(2j + 1)$ degenerate levels under the rotation symmetry group). To find which energy levels occur with what multiplicity, the specific form of the Hamiltonian must be used. Sometimes the symmetry group of a Hamiltonian is larger than the apparent geometrical symmetry of the physical situation. This is called *accidental degeneracy* or *dynamical symmetry,* a symmetry larger than the pure geometrical symmetry. The best-known examples are the Kepler problem (motion in a central inverse square field), which is treated later from a more general point of view, and the quantum mechanical three-dimensional oscillator.

Because even the dynamical symmetry is not enough to solve the complete problem by group theory, use is made of more general Lie groups or Lie algebras, which contain the symmetry groups as subgroups to provide the solution of the complete dynamical problems under certain conditions. These groups are then called *dynamical groups*. Thus the following hierarchy is established:

geometrical symmetry groups
↓
dynamical symmetry groups
↓
dynamical groups

As an illustrative example, for the one-dimensional oscillator with the Hamiltonian (in appropriate units, that is, $m = \hbar = \omega = 1$),

$$H = \frac{1}{2}(p^2 + q^2) \equiv 2L_3 \tag{5a}$$

Consider the two operators

$$L_2 = -\frac{1}{4}(qp + pq) = -\frac{i}{4}\left(q\frac{d}{dq} + \frac{d}{dq}q\right), \quad L_1 = \frac{1}{2}\left(H - 2q^2\right) \tag{5b}$$

The three operators L_1, L_2, L_3 have the commutation relations

$$\begin{aligned} [L_1, L_2] &= -iL_3 \\ [L_2, L_3] &= iL_1 \\ [L_3, L_1] &= iL_2 \end{aligned} \tag{6}$$

The negative sign in the first equation distinguishes these relations from the ordinary angular momentum commutation relations ($1^3$10). They are the commutation relations of the Lie algebra of the group $O(2, 1)$, that is, the two-dimensional *Lorentz group.* The operator

$$\mathbf{C}^2 = L_3^2 - L_2^2 - L_1^2 \tag{7}$$

commutes with L_i. Therefore, it is the quadratic Casimir operator of the algebra (6), analogous to the $\mathbf{L}^2$ of the angular momentum algebra, and characterizes its irreducible representations. From (5) it is then found that

$$\mathbf{C}^2 \equiv C(C+1) = -\frac{3}{16} \quad \text{and} \quad C = -\frac{1}{4} \tag{8}$$

Hence *all* the states of (5a) belong to an irreducible representation of $o(2, 1)$. Here small letters denote Lie algebras, and capital letters denote the corresponding groups [$o(2, 1)$ and $O(2, 1)$].

It is essential for this entire section to know the representations of noncompact algebras (and groups), in particular those of $o(2, 1)$.

A. Irreducible representations of $o(2, 1)$

It is physically instructive to use the boson spin creation and destruction operators given by 6⁷8. As in 6⁷9, the subscripts + and − are replaced by 1 and 2. Thus

$$a = \begin{pmatrix} a_1 \\ a_2 \end{pmatrix}, \quad a^\dagger = \widetilde{a^\dagger_1 \, a^\dagger_2} \tag{9}$$

The commutation relations (6) have a fundamental two-dimensional representation

$$L_1 = \frac{i}{2} a^\dagger \sigma_1 a, \quad L_2 = \frac{i}{2} a^\dagger \sigma_2 a, \quad L_3 = \frac{1}{2} a^\dagger \sigma_3 a \tag{10}$$

where σ_i are the Pauli matrices. The states of this two-dimensional representation are

$$a^\dagger_1|0\rangle \quad \text{and} \quad a^\dagger_2|0\rangle \tag{11}$$

The general irreducible representations are constructed by the tensor products of the states in (11). This is similar to the procedure followed in the theory of angular momentum. Considering the general states (6⁷10)

$$|\phi, m\rangle = \mathrm{N}^\phi_m (a^\dagger_1)^{\phi+m} (a^\dagger_2)^{\phi-m} |0\rangle \tag{12}$$

where N^ϕ_m is a normalization factor, the action of L_3 (10) and of $\mathbf{C}^2$(7) on this state is

$$\begin{aligned} L_3|\phi, m\rangle &= m|\phi, m\rangle \\ \mathbf{C}^2|\phi, m\rangle &= \phi(\phi+1)|\phi, m\rangle \end{aligned} \tag{13}$$

Thus ϕ plays the role of l in the angular momentum case. The lowering and raising operators, similar to 1³16,

$$L^\pm = \frac{1}{\sqrt{2}}\left(-L_1 \pm iL_2\right) \tag{14}$$

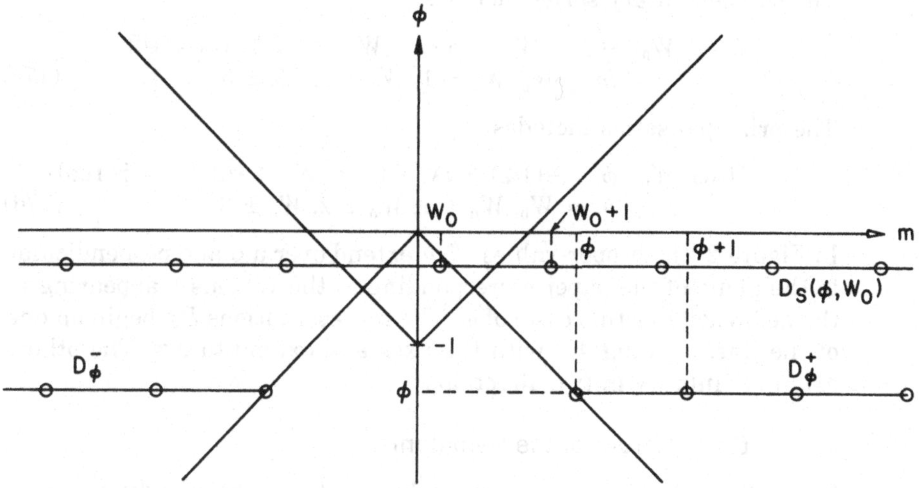

Figure 2[7]. Unitary irreducible representations of $o(2, 1)$.

have the properties

$$
\begin{aligned}
L^{+}\left|\phi, m\right\rangle &= \frac{i}{\sqrt{2}}\left(\phi - m\right)\frac{N_m^{\phi}}{N_{m+1}^{\phi}}\left|\phi, m+1\right\rangle \\
L^{-}\left|\phi, m\right\rangle &= \frac{i}{\sqrt{2}}\left(\phi + m\right)\frac{N_m^{\phi}}{N_{m-1}^{\phi}}\left|\phi, m-1\right\rangle
\end{aligned}
\tag{15}
$$

[For angular momentum the factors i in (15) are replaced by 1.]

The theory of $o(3)$ and $o(2, 1)$ differs only in the application of the hermiticity condition for L_i. For unitary representations of the group, the generators L_i must be Hermitian [This assumes that the parameters of the group (the analogs of the angles of rotation for $o(3)$) are real.]. Otherwise, as far as irreducible representations are concerned, every representation of $o(3)$ is also a representation of $o(2, 1)$ and vice versa. However, *unitary* representations of $o(3)$ are all *finite* dimensional, whereas the unitary representations of $o(2, 1)$ are all *infinite* dimensional.

The unitary conditions

$$L^{\dagger}_3 = L_3, \; (L^{+})^{\dagger} = L^{-} \tag{16}$$

in addition to (12)–(15) give the following classes of *unitary irreducible* representations of $o(2, 1)$ [25], shown in Figure 2[7]. The discrete series includes

$$D_{\phi}^{+}: \phi \text{ real negative, } m = -\phi, -\phi + 1, -\phi + 2, \ldots \tag{17a}$$
$$D_{\phi}^{-}: \phi \text{ real negative, } m = \phi, \phi - 1, \phi - 2, \ldots \tag{17b}$$

The supplementary series includes

$$D_s(\phi, W_0): -1 + |W_0| < \phi < -|W_0|, \ -1/2 < W_0 < 1/2$$
$$m = W_0, W_0 \pm 1, W_0 \pm 2, W_0 \pm 3, \ldots \tag{17c}$$

The principal series includes

$$D_p(\phi, W_0): \phi = -(1/2) + i\lambda, \ -1/2 < W_0 \leqslant 1/2 \qquad (\lambda \text{ is real})$$
$$m = W_0, W_0 \pm 1, W_0 \pm 2, W_0 \pm 3, \ldots \tag{17d}$$

In Figure 2[7] the proper-values of m extend over a plane perpendicular to the plane of the paper corresponding to the various λ appearing in the definition (for this case) of ϕ. The representations $D_\phi^\pm$ begin on one of the lines making 45° with the axes and extend to $\pm\infty$. The others extend to infinity in two directions.

B. Spectrum of the Hamiltonian

From (8) and (15) the explicit value of ϕ for the one-dimensional harmonic oscillator may be obtained:

$$\phi(\phi + 1) = -\frac{3}{16} \quad \text{or} \quad \phi = -\frac{1}{4} \tag{18}$$

Because the energy is positive, there is a discrete D_ϕ^+ representation. Therefore, the proper values m of L_3 are

$$m = 1/4, (1/4) + 1, \ldots, (1/4) + n, \ldots$$

From (5a),

$$W = (1/2) + 2n \tag{19}$$

The states are specified by the quantum number n and belong to the irreducible representations of $o(2, 1)$.

Algebras similar to (5a) and (5b) can be developed for other classes of Hamiltonians. Instead of discussing these further, a different line of development is presented that has rather far-reaching possibilities toward relativistic generalizations and toward creating external interactions of the system.

C. Wave equations in terms of $O(2, 1)$-generators

Consider the following operators [26]:

$$\Gamma_0 = \frac{1}{2}(rp^2 + r)$$
$$\Gamma_4 = \frac{1}{2}(rp^2 - r) \tag{20}$$
$$T = \mathbf{r} \cdot \mathbf{p} - i$$

in which $\mathbf{r}$ and $\mathbf{p}$ are the usual relative canonical variables of the system at rest as a whole. These operators are Hermitian in the new scalar product

$$(\tilde{\psi}, \tilde{\psi}) \equiv \int \tilde{\psi}^* \frac{1}{r} \tilde{\psi}\, d^3x \tag{21}$$

where $\tilde{\psi}$, as will be explained later, stands for the physical nonrelativistic proper functions. From (20), the commutation relations

$$\begin{aligned} [\Gamma_0, \Gamma_4] &= iT \\ [\Gamma_4, T] &= -i\Gamma_0 \\ [T, \Gamma_0] &= i\Gamma_4 \end{aligned} \tag{22}$$

are computed, which are again precisely of the form (6) and hence an $o(2, 1)$ algebra is obtained. The quadratic Casimir operator can also be evaluated, yielding

$$C^2 \equiv \Gamma_0^2 - \Gamma_4^2 - T^2 = (\mathbf{r} \times \mathbf{p})^2 = \mathbf{L}^2 \tag{23}$$

Thus the Casimir operators of the angular momentum algebra $\mathbf{L} = \mathbf{r} \times \mathbf{p}$ and that of the $o(2, 1)$ algebra coincide. Using the values given in (17), it is found that

$$\text{spectrum of } \Gamma_0\text{: } l + 1, l + 2, \ldots \tag{24}$$

So far this is independent of a specific physical problem.

Now consider the Hamiltonian for a quantum mechanical nonrelativistic Kepler problem:

$$H = \frac{1}{2m} p^2 - \frac{\alpha}{r} \quad \text{in which} \quad \hbar = 1, c = 1 \tag{25}$$

In view of the relation (22) and (25), the Lagrangian operator may be written as

$$\Theta = r(H - W) = \left(\frac{1}{2m} - W\right) \Gamma_0 + \left(\frac{1}{2m} + W\right) \Gamma_4 - \alpha \tag{26}$$

Then the calculation of the spectrum of the Hamiltonian can be reduced to the solution of the equation

$$\Theta\tilde{\psi} \equiv \left[\left(\frac{1}{2m} - W\right) \Gamma_0 + \left(\frac{1}{2m} + W\right) \Gamma_4 - \alpha\right] \tilde{\psi} = 0 \tag{27}$$

This is a wave equation but is written entirely in terms of the generators Γ_0 and Γ_4 of an $o(2, 1)$ algebra. In the angular momentum case it would correspond to the algebraic equation of the form

$$(aL_3 + bL_1 + c)\phi = 0$$

Equation (27) can be brought to a simpler form

$$\left[\left(\frac{-2W}{m}\right)^{1/2} \Gamma_0 - \alpha\right] \psi = 0 \tag{28}$$

by tilting $\tilde{\psi}$ as

$$\tilde{\psi} = e^{i\varphi T}\psi \tag{29}$$

in which

$$\tan \varphi = \frac{W + (1/2m)}{W - (1/2m)} \tag{30}$$

If $W < 0$, the coefficient of Γ_0 is positive. Denoting the discrete proper values of Γ_0 by n, we obtain

$$W_n = -\frac{1}{2}\frac{m\alpha^2}{n^2} \tag{31}$$

in units for which $\hbar = 1$, $c = 1$; or $W_n = -(1/2)mc^2(\alpha^2/n^2)$, the Balmer formula [3'1].

For $W > 0$, Γ_0 cannot be diagonalized. It is necessary to go back to (29) and use a different tilt

$$\tilde{\psi} = e^{i\varphi' T}\psi' \tag{32}$$

and choose

$$\tanh \varphi' = \frac{W - (1/2m)}{W + (1/2m)} \tag{33}$$

to obtain the reduced equation

$$\left[\left(\frac{2W}{m}\right)^{1/2}\Gamma_4 - \alpha\right]\psi' = 0 \tag{34}$$

The generator Γ_4 of O(2, 1) now has a *real* continuous spectrum λ, resulting in

$$W_\lambda = \frac{1}{2}\frac{m\alpha^2}{\lambda^2}, \quad (\hbar = 1, c = 1) \tag{35}$$

Thus there is a purely algebraic characterization of the complete dynamical problem. The states ψ are referred to as the *group states.* The normalized states $\tilde{\psi}$; that is,

$$\tilde{\psi} = \frac{1}{N}e^{i\varphi T}\psi \tag{36}$$

are not quite the Schrödinger states, because according to (21) they are normalized by

$$\langle\tilde{\psi}|(\Gamma_0 - \Gamma_4)|\tilde{\psi}\rangle = 1 \tag{37}$$

Because of (24), for each l the range of n is

$$n = l + 1, l + 2, \ldots \tag{38}$$

But, purely algebraically, which l values occur is not known yet.

D. The full dynamical group

In addition to the operators given in (20) now consider the following twelve operators:

$$\begin{aligned} \mathbf{L} &= \mathbf{r} \times \mathbf{p} \\ \mathbf{A} &= (1/2)\mathbf{r}p^2 - \mathbf{p}(\mathbf{r} \cdot \mathbf{p}) - (1/2)\mathbf{r} \\ \mathbf{M} &= (1/2)\mathbf{r}p^2 - \mathbf{p}(\mathbf{r} \cdot \mathbf{p}) + (1/2)\mathbf{r} \\ \mathbf{\Gamma} &= r\mathbf{p} \end{aligned} \tag{39}$$

Here **A** is essentially the *Runge-Lenz vector* (Sec. 6[7]) known from the accidental degeneracy of the Kepler problem. These generators, together with Γ_0, Γ_4, and T in (20), form the Lie algebra of the group $O(4, 2)$. This is a noncompact form of the six-dimensional rotation group $O(6)$ with fifteen generators (i.e., 15-parameter group). The operators **L** and **A** generate an $O(4)$ group responsible for the well-known $O(4)$ symmetry of the Kepler problem. The Casimir operators of $o(4, 2)$ in the representation, defined by (20) and (39), are

$$\begin{aligned} C_2^2 &= \mathbf{L}^2 + \mathbf{A}^2 - \mathbf{M}^2 - \mathbf{\Gamma}^2 + \Gamma_0^2 - \Gamma_4^2 - T^2 = -3 \\ C_3^2 &= 0 \\ C_4^2 &= 0 \end{aligned} \tag{40}$$

where C_3^2 and C_4^2 are invariant operators of third and fourth order in the generators. Consequently, there is a single irreducible representation of $o(4, 2)$, characterized by (40), to which *all* levels of the system belong, no matter what energy.

Figure 3[7] shows the weight diagram corresponding to the representation characterized by (40) when Γ_0, L^2, and L_3 are diagonalized with proper values n, l, and m, respectively.

The group states $|nlm\rangle$ are in one-to-one correspondence with the physical states through (36). For each n, clearly, $l = 0, 1, 2, \ldots, n - 1$, or the n^2 degeneracy of the levels.

Because the $o(4, 2)$ representation contains all the quantum numbers and all the levels of the system, it is called the dynamical group of the system. The representations of the dynamical symmetry group $O(4)$ occur, in $o(4, 2)$ representation, with dimensions n^2; $n = 1, 2, 3. \ldots$ In each $o(4)$ representation, in turn, the representations of the geometrical symmetry group $O(3)$ occur, satisfying the condition $l = 0, 1, 2, \ldots, n - 1$.

E. Problems with broken symmetry

The algebraic solution of the Kepler problem presented in the previous subsections is not restricted to the $O(4)$-symmetric nonrelativistic Kepler problem. The group states $|nlm\rangle$, represented in Figure 3[7], give only the quantum numbers and their range. The actual values of the energy or the mass of the states is determined by the wave equation

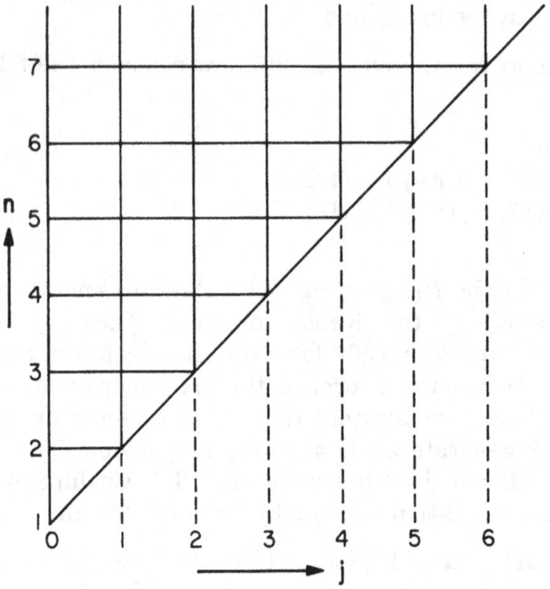

Figure 3[7]. The weight diagram for the representation, specified by (40), of the group $O(4, 2)$.

(27). Thus equations could be postulated with different coefficients than those in (27), for example, which would result in different energy formulas but the same quantum number n, l, m. A physical realization of this situation is provided by inclusion of the spin-orbit terms into the Kepler problem. In this case one no longer has $O(4)$ symmetry or n^2 degeneracy of the level but rather a splitting of the levels with the same n but different j values. (Here $\mathbf{L}$ and l are replaced by $\mathbf{J}$ and j, respectively.)

F. Fully covariant equations

In the $o(4, 2)$ Lie algebra, (20) and (39), the generators $\mathbf{J}$ and $\mathbf{M}$ have the commutation relations of the homogeneous Lorentz group, where $\mathbf{J}$ is the angular momentum and $\mathbf{M}$ is the generator of pure Lorentz transformations (boosts). It is also possible to identify the generators of the Galilean boosts, namely, $\mathbf{M} - \mathbf{A} = \mathbf{r}$.

With respect to the Lorentz algebra ($\mathbf{J}$, $\mathbf{M}$), the operators $\Gamma_\mu = (\Gamma_0, \mathbf{\Gamma})$ form a four-vector, as can be seen by their commutation relations. The operator Γ_4 is a Lorentz scalar. Equation (27) is the zero component of a four-vector equation of the form

$$(\alpha\Gamma^\mu P_\mu + \beta\Gamma_4 + \gamma)\psi(P) = 0 \tag{41}$$

Indeed, in the rest frame $P_\mu = (P_0, 0)$, where P_μ is the total-energy momentum four-vector of the system as a whole (p_μ is written for electron total-energy-momentum four-vector),

$$(\alpha M\Gamma_0 + \beta\Gamma_4 + \gamma)\psi(0) = 0 \tag{42}$$

which is identical to (27).

The situation here is completely analogous to the Dirac theory of the electron in which there is a vector operator γ_μ and a wave equation in the momentum space:

$$(\gamma^\mu p_\mu - K)\psi(p) = 0 \tag{43}$$

In the rest frame (43) becomes

$$(\gamma^0 m - K)\psi(0) = 0 \tag{44}$$

The difference between (41) and (43) is that the γ^μ's are 4×4 matrices, whereas the Γ^μ's are infinite dimensional matrices, both of $o(4, 2)$.

In (41) the relative coordinates have been eliminated, and the whole atom can be treated as a single particle with internal degrees of freedom. These internal degrees of freedom come from the spectrum Γ_0. The Dirac wave function has four components, and the atom wave function has infinitely many components denoted by ψ_{njm}. Therefore, equations of the type (41) are called *infinite-component wave equations*. The degrees of freedom in the Dirac equation are two particles (positron and electron), both having spin 1/2, with spin components $m = \pm 1/2$.

The complete solution of (41) for the hydrogen atom is obtained by Barut and his students [27]. The binding energy they found does not only contain relativistic corrections but the recoil term as well, in agreement with the relativistic perturbation theory to order α^2[11^4, p. 195].

The infinite-component wave function for the atom has been further applied to the explicit evaluation of relativistic form factors [28] and of relativistic and nonrelativistic Coulomb-scattering amplitudes [29].

The solution of relativistic equations (Dirac and Klein-Gordon) using dynamical group methods is treated in Barut and Bornzin [26], where other references may also be found.

G. Physical interpretation

The dynamical group G, like $O(4, 2)$, is not the symmetry group of the stationary Hamiltonian H_0, because some of the generators of G do not commute with H_0. In fact, the dynamical group G gives rise to transitions from one energy value to another. These transitions are caused physically by the external agents. The presence of the external interactions reveal a larger "symmetry" of the physical system. When external forces are present, the Hamiltonian is no longer H_0 but $H = H_0 + H_1(t)$. In such a general situation, the constants of the motion L_i (e.g., generators of G) satisfy the more general equation

$$\frac{dL_i}{dt} = [H, L_i] + \frac{\partial L_i}{\partial t} = 0 \tag{45}$$

instead of simply commuting with the Hamiltonian: the system plus interactions has a larger geometrical symmetry than an isolated system. These considerations need further elaborations, but a certain analogy with the point of view of general relativity toward a geometrization of dynamics is recognizable.

The application of dynamical groups to atomic and molecular physics is reviewed by Wulfman [30].

Dynamical group methods have also been used to study the symmetry properties of hydrogenic radial wave functions [31, 32]. The basic procedure introduced by Armstrong [31] is to study the $O(2, 1) \times O(3)$ subgroup of the $O(4, 2)$ dynamical group of the hydrogen atom. One of the results of this study is an explanation of the Pasternack-Sternheimer selection rule 6^7 39. It turns out that the coupling coefficients of the noncompact group $O(2, 1)$ are very similar to those of the three dimensional rotation group 3^3 4. Therefore, the matrix elements of r^{-k}, for $k \leqslant l' + l - 2$ and for varying n, can be expressed as [33]

$$\left\langle nl' \left| \frac{1}{r^k} \right| nl \right\rangle = (-1)^{l'-n} \frac{A(l', l, k)}{n^{k+1}} \begin{pmatrix} l' & k-2 & l \\ -n & 0 & n \end{pmatrix} \tag{46}$$

where A, as indicated, is a function of l', l, and k but not of n. Then the selection rule 6^7 39 corresponds to the violation of a triangular condition on the triad $(l', k - 2, l)$. For the special case $l' = l + 1$

$$\left\langle n, l+1 \left| \frac{1}{r^k} \right| n, l \right\rangle = \left\langle l+1 \left| \frac{1}{r^k} \right| l \right\rangle$$
$$= (-1)^{1+l-n} \frac{A(l+1, l, n)}{n^{k+1}} \begin{pmatrix} l+1 & k-2 & l \\ -n & 0 & n \end{pmatrix}$$

Thus the specific values, corresponding to different k, given by 6^7 38 can be obtained from this general relation.

8. Self-consistent fields: Hartree-Fock theory; Thomas-Fermi model

> . . . for an atom of N electrons, the solution is a function of $3N$ variables, and even if it were possible to evaluate such a solution to any degree of numerical accuracy required, no satisfactory way of presenting the results, either in tabular or graphical form, is known. It has been said that the tabulation of one variable requires a page, of two variables a volume, and of three variables a library; but the full specification of a single wave function of neutral Fe is a function of seventy-eight variables. It would be rather crude to restrict to ten the number of values of each variable at which to tabulate this function, but even so, full tabulation of it would require 10^{78} entries, and even if this number could be reduced somewhat from considerations of symmetry, there would still not be enough atoms in the whole solar system to provide the material for printing such a table.
>
> D. R. Hartree, *Reports on Progress in Physics,* Vol. 11 (London: The Physical Society, 1947), p. 113.

Ways for obtaining the central fields, $V(r)$, and the radial functions, $P(n, l; r)$, that are used for specific atomic structure calculations in the central field approximation (Sec. 3^4) are considered in this chapter. In order that we be too greatly overawed by Hartree's estimate of the complexity of the problem, we note that the normal configuration for the 26 electrons of Fe is $1s^2 2s^2 2p^6 3s^2 3p^6 3d^6 4s^2$. Instead of a wave function of $3 \times 26 = 78$ variables, the central-field approximation uses seven radial functions for the shells actually present in this configuration. The enormous reduction from 78 to seven variables comes about because of using the spherical harmonic factors $Y(l, m; \theta, \phi)$ for the dependence of the wave function on the $2 \times 26 = 54$ angles and also because of assuming that the radial functions are alike for all electrons in the same shell.

The simplification introduced by the central-field approximation is actually much greater than such consideration of a single example implies, because we are interested in calculating, describing, and comparing the wave functions of all of some hundred atoms, in all stages of ionization, a total of some 5000 species, each of which has a high order of infinity of excited states. When we approach this gigantic task by the central-field approximation, we are helped not only by the fact that the states are described by a comparatively few radial functions but also that the $P(n, l; r)$ depend smoothly on the parameters Z and N and on the nl's of the configurations involved. This makes possible a good deal of interpolation for the dependence of the $P(n, l; r)$ on these parameters.

Despite these favorable factors, it must not be forgotten that the central-field approximation is an *approximation*. The central-field approximation as presented here does not include the effects of interaction between alternative configurations and provides only an approximate treatment of electron exchange effect. Because it is based on the Hamiltonian $2^2 9$, it also neglects all relativistic effects.

The basic problem is the determination of the effective potential energy $V(r)$ for every configuration of every (Z, N). The $V(r)$ depends on the radial functions $P(n, l; r)$, which in turn depend on the $V(r)$, so both have to be determined together in an iterative way. In the early days (1928–38) work of this kind was done using analytical expressions for the $P(n, l; r)$ containing parameters, but it quickly became clear that real progress could only be made by numerical integration of the differential equations for the radial functions. At first this was laboriously done by mechanical desk calculators, and few physicists had the stamina to make such calculations. Calculations at first were made by the self-consistent field method of Hartree, which was later improved by Fock (Sec. 2^8) by the inclusion of exchange effects but at the price of a great complication in the computations. Slater showed in 1951 how exchange effects can be taken into account, approximately without much extra complication over the Hartree method. So the resulting Hartree-Fock-Slater method has been much favored for finding the $V(r)$ and $P(n, l; r)$.

The 1950s brought an increasing amount of numerical self-consistent field calculations using improved numerical techniques and an alternative numerical procedure based on expansion of the radial wave functions in terms of known analytic functions. Both of these developments were greatly facilitated by the increasing availability of automatic digital electronic computers.

Systematic computations for the normal configurations of all or nearly all neutral atoms are now available. These include the numerical Hartree-Fock-Slater (HFS) solutions published by F. Herman and S. Skillman, the Hartree-Fock (HF) solutions by Charlotte Froese Fischer, and the center-of-gravity Hartree-Fock (CGHF) solutions of Froese Fischer and Joseph B. Mann. In addition, systematic computations for the normal (and some excited) configurations of neutral atoms and positive (and a few negative) ions using the analytic methods for all elements with $Z \leqslant 54$ have been published by Enrico Clementi. Recently, an extensive work by Fraga, Karwowski, and Saxena, numerical Hartree-Fock solutions for the normal configurations of neutral atoms and positive (as well as negative) ions, appeared first as a report and then as a book.

Such computations account satisfactorily for many specific details of the level structure of atoms. Based as they are on various simplifying assumptions, the level structure computed in this way does not agree exactly with observation. Even where close agreements with observed values are found, it is still necessary to show that higher approximations do not spoil the agreement.

1^8. The Thomas-Fermi field

The earliest quantitative procedure for obtaining an approximation for an effective central field, $V(r)$, was developed independently by both Thomas [1] and Fermi [2], in 1927. The method was extended to positive ions by Baker with corrections by Guth and Peierls [3]. The method has a large literature which has been thoroughly surveyed by Gombás [4].

The electrons are treated semiclassically as a degenerate gas obeying Fermi-Dirac statistics [5]. Although attracted to the nucleus, they have a distribution of finite density, both because of the exclusion principle, and because of their mutual Coulomb repulsion. The Coulomb interaction is treated as in classical electrostatics, regarding the distribution of N discrete charges as a continuous density that is related to the potential by Poisson's equation.

The resulting density is a smooth function of N that takes no account of the shell structure of the states. It is therefore not accurate for the distribution of charge and potential at the outermost part of the atom. Even so, it gives a good general account of the main trends of the central potential $V(r)$ from which valuable insights into the general behavior of the radial wave functions $P(n, l; r)$ with changing N are obtained.

For electrons having a maximum momentum, p, and contained in volume dv, the corresponding volume in phase space is $(4/3)\pi p^3 dv$. There can be at most two electrons, with opposite spins, in each phase space volume h^3. Therefore the number density, n, of electrons whose maximum kinetic energy is T is

$$n = \frac{(2mT)^{3/2}}{3\pi^2\hbar^3} \tag{1}$$

Let $\phi(r)$ be the electrostatic potential at r, because of the combined fields of nucleus, Ze/r, and the space charge, $-en(r)$, of the electrons, so the potential energy of an electron at r is $-e\phi(r)$. The total energy, W, of the highest occupied state, often called the *Fermi level,* is independent of r. As the electrons are presumed to be bound, $W < 0$. Hence

$$T = e\phi(r) + W \tag{2}$$

gives the maximum kinetic energy of the electrons at distance r. This is a decreasing function of r, which becomes equal to zero at the value of r for which $W = -e\phi(r_0)$. For $r > r_0$ the density of electrons is set equal to zero.

We now relate ϕ to n through the Poisson equation $\nabla^2\phi = 4\pi ne$,

$$\nabla^2(e\phi + W) = \frac{4e^2}{3\pi\hbar^3}[2m(e\phi + W)]^{3/2} = 0 \tag{3}$$

where the lower line applies for $r > r_0$.

We introduce a new dependent function

$$\Phi(x) = \frac{r}{Ze^2}(e\phi + W) \quad \text{with} \quad \Phi(0) = 1 \tag{4}$$

and a scale change

$$r = xb \quad \text{with} \quad b = \frac{1}{4}\left(\frac{9\pi^2}{2}\right)^{1/3}\left(\frac{a}{Z}\right)^{1/3} = 0.8853\left(\frac{a}{Z}\right)^{1/3} \tag{5}$$

where $a = \hbar^2/me^2$. This has the effect that Φ satisfies the same differential equation, known as the *Thomas-Fermi equation,* for all Z (in atomic units)

$$\begin{aligned} \Phi > 0 \qquad & \Phi''(x) = \frac{\Phi^{3/2}}{x^{1/2}} \\ \Phi < 0 \qquad & = 0 \end{aligned} \tag{6}$$

Solutions are found by various approximation and numerical integration procedures [6]. The solutions that satisfy $\Phi(0) = 1$, form a one-parameter family, characterized by the slope $\Phi'(0)$ at the origin. This slope is negative for all solutions of physical interest. The general features are shown in Figure 1[8]. The critical solution, $\Phi_0(x)$, is the one for which

$$\Phi'_0(0) = -1.588070972 \tag{7}$$

This solution approaches zero asymptotically, $\Phi_0(x) \to 0$, as $x \to \infty$. Table 1[8] gives values of $\Phi_0(x)$ as compiled by Gombás [4a, p. 45] from various computations by Fermi, Bush and Caldwell, and Miranda.

From (6) we see that $\Phi''(x) \to \infty$ as $x \to 0$, so $\Phi(x)$ cannot be expressed in a Taylor series. Baker [3a, error corrected in 4a, p. 43] has given a semiconvergent series for $\Phi(x)$,

$$\Phi(x) = 1 - \alpha x + \frac{4}{3}x^{3/2} - \frac{2}{5}\alpha x^{5/2} + \frac{1}{3}x^3 + \frac{3}{70}\alpha^2 x^{7/2} - \frac{2}{15}\alpha x^4 + \frac{4}{63}\left(\frac{7}{6} + \frac{1}{16}\alpha^3\right)x^{9/2} + \ldots$$

in which $\alpha = -\Phi'(0)$. For $\Phi_0(x)$ Latter [7] has obtained an empirical formula,

$$\Phi_0(x) = [1 + 0.02747x^{1/2} + 1.243x - 0.1486x^{3/2} + 0.2302x^2 + 0.007298x^{5/2} + 0.006944x^3]^{-1}$$

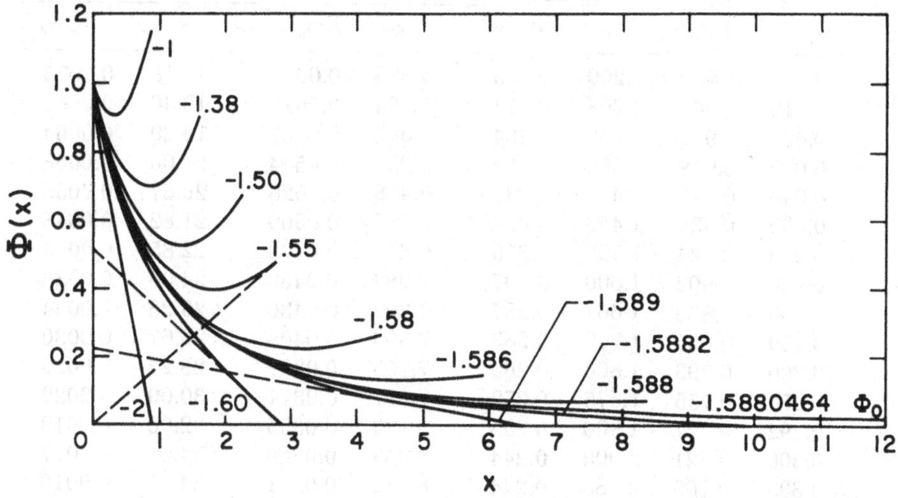

Figure 1[8]. Solutions for Thomas-Fermi $\Phi(X)$, for various values of initial slope $\Phi(0)$ as labeled (from Gombás [4]).

for which the maximum error is 0.3 percent. For $\Phi'(0)$ more negative than (7), $\Phi(x) = 0$ at a finite value of $x = x_0$. Then, in accordance with (6), the solution continues in a straight line of the same slope:

$$x > x_0 \qquad \Phi(x) = \Phi'(x_0)(x - x_0) \tag{8}$$

The solution $\Phi_0(x)$ has had the most widespread application. Strictly speaking, the solutions that have greater negative initial slope than (7) should be used for positive ions, but in fact the solution $\Phi_0(x)$ has been used in this way in many instances, which is permissible because the positive ion solutions do not differ much from $\Phi_0(x)$ for low stages of ionization.

The total number of electrons N corresponding to a particular solution is

$$N = 4\pi \int_0^{r_0} n(r) r^2\, dr = Z \int_0^{x_0} x\Phi''(x)\, dx = Z\,[1 + x_0\Phi'_0(x_0)] \tag{9}$$

For $\Phi = \Phi_0$, the limiting value of $x_0\Phi'_0(x_0)$ is zero, so this solution corresponds to the neutral atom. Using different approximation procedures, Fermi, and also Sommerfeld [8], have given some values of x_0 for various values of N/Z:

N/Z	0.99	0.98	0.97	0.95	0.92	0.90	
x_0 (Fermi)	35.5	25.4	21.1	16.2	12.9	11.2	(10)
x_0 (Sommerfeld)	34.23	24.95	20.54	15.81	12.18	10.64	

Table 1[8]. Thomas-Fermi $\Phi_0(\chi)$ as defined in (4) and (7) [4a].

x	$\Phi_0(x)$	x	$\Phi_0(x)$	x	$\Phi_0(x)$	x	$\Phi_0(x)$
0	1.000	1.200	0.375	5.625	0.0656	17.78	0.0075
0.010	0.985	1.208	0.374	5.834	0.0619	18.46	0.0069
0.020	0.972	1.250	0.364	6.042	0.0587	19.20	0.0064
0.030	0.959	1.300	0.353	6.250	0.0554	20.00	0.0058
0.040	0.947	1.400	0.333	6.458	0.0526	20.87	0.0053
0.050	0.935	1.458	0.322	6.667	0.0500	21.82	0.0048
0.060	0.924	1.500	0.315	6.875	0.0473	22.85	0.0043
0.080	0.902	1.600	0.297	7.083	0.0450	24.00	0.0038
0.100	0.882	1.667	0.287	7.292	0.0430	25.26	0.0034
0.150	0.835	1.700	0.283	7.500	0.0408	26.67	0.0030
0.200	0.793	1.800	0.268	7.708	0.0389	28.24	0.0026
0.250	0.755	1.875	0.259	7.917	0.0371	30.00	0.0022
0.292	0.727	1.900	0.255	8.125	0.0355	32.00	0.0019
0.300	0.721	2.000	0.244	8.333	0.0340	34.00	0.0017
0.333	0.700	2.083	0.234	8.542	0.0321	34.29	0.0016
0.350	0.691	2.200	0.221	8.750	0.0310	36.00	0.0015
0.375	0.675	2.292	0.212	8.958	0.0298	36.92	0.0011
0.400	0.660	2.400	0.202	9.167	0.0287	38.00	0.0013
0.417	0.651	2.500	0.193	9.375	0.0275	40.00	0.0011
0.458	0.627	2.600	0.185	9.583	0.0265	45.00	0.00079
0.500	0.607	2.708	0.176	9.792	0.0255	50.00	0.00061
0.542	0.582	2.800	0.170	10.000	0.0244	55.00	0.00049
0.584	0.569	2.918	0.162	10.22	0.0235	60.00	0.00039
0.600	0.562	3.000	0.157	10.44	0.0225	65.00	0.00031
0.625	0.552	3.125	0.150	10.67	0.0216	70.00	0.00026
0.667	0.535	3.200	0.145	10.92	0.0206	75.00	0.00022
0.700	0.521	3.333	0.138	11.16	0.0198	80.00	0.00018
0.709	0.518	3.400	0.134	11.43	0.0189	85.00	0.00015
0.750	0.502	3.500	0.130	11.72	0.0180	90.00	0.00012
0.792	0.488	3.542	0.127	12.01	0.0171	95.00	0.00011
0.800	0.485	3.600	0.125	12.31	0.0163	100.00	0.00010
0.833	0.475	3.750	0.118	12.63	0.0155	200.00	$0.0^{4}15$
0.875	0.461	3.800	0.116	12.97	0.0147	300.00	$0.0^{5}46$
0.900	0.453	3.960	0.110	13.33	0.0139	400.00	$0.0^{5}20$
0.917	0.449	4.000	0.108	13.72	0.0131	500.00	$0.0^{5}10$
0.958	0.436	4.167	0.102	14.12	0.0123	600.00	$0.0^{6}61$
1.000	0.425	4.375	0.0956	14.55	0.0116	700.00	$0.0^{6}39$
1.042	0.414	4.583	0.0895	15.01	0.0109	800.00	$0.0^{6}26$
1.083	0.406	4.792	0.0837	15.48	0.0102	900.00	$0.0^{6}18$
1.100	0.398	5.000	0.0788	16.00	0.0094	1000.00	$0.0^{6}13$
1.125	0.393	5.209	0.0739	16.56	0.0088		
1.167	0.382	5.418	0.0695	17.14	0.0081		

The electrostatic potential $\phi(r)$ is

$$\phi(r) = \frac{Ze}{r} - e\int_0^{r_0} \frac{n(r')}{|\mathbf{r} - \mathbf{r}'|}\, d\tau'$$

Then, because

$$\int_0^{r_0} n(r')\, d\tau' = N$$

we have

$$\phi(r) = \frac{(Z - N)e}{r} \quad \text{for} \quad r > r_0 \tag{11}$$

and therefore

$$W = -\frac{(Z - N)e^2}{r_0}$$

so the Fermi level for a neutral atom is equal to zero.

It is not correct to use $Vr) = -(Ze^2/r)\Phi_0(r/b)$ for the motion of individual electrons. This does not have the correct asymptotic behavior, going too rapidly to zero at large r, because the use of the Poisson equation in effect assumes that the charge is subdividable into infinitesimal portions instead of occurring in discrete charges of amount $-e$. Various modifications have been used to allow for this fact. In most of the work of Fermi and his students the form used was

$$V(r) = \frac{Ce^2}{r} - \frac{(N - 1)e^2}{r}\Phi_0(r/b) \tag{12}$$

or

$$Z(r) = C + (N - 1)\,\Phi_0(r/b); \quad C = Z - N + 1 \tag{12'}$$

for the effective nuclear charge. A different modification, called the *Latter cutoff*, is recommended by Latter [7] in his extensive study of wave functions based on the Thomas-Fermi potential:

$$\begin{aligned} V(r) &= -\frac{Ze^2}{r}\Phi_0(r/b) \qquad 0 < r < r_0 \\ &= -\frac{Ce^2}{r} \qquad\qquad\quad r_0 < r \end{aligned} \tag{13}$$

in which r_0 is the value of r at which these two expressions are equal. In this case $Z(\mathrm{r})$ is the larger of

$$Z\Phi_0(r/b) \quad \text{or} \quad C \tag{13'}$$

Figure 2^8 shows the trend of the $Z(r)$ curves for various values of Z on a log-log scale. For neutral atoms the effective field is in all cases Coulombic for $r > 3a$.

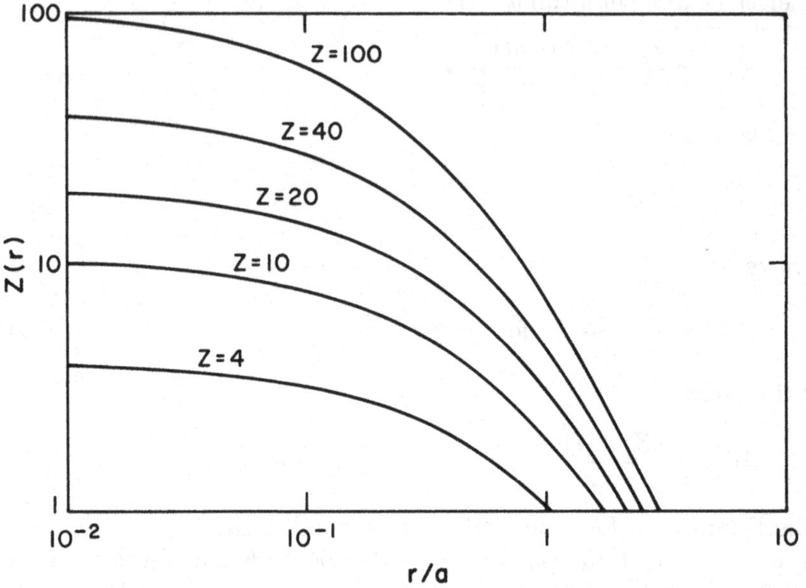

Figure 2[8]. Effective Thomas-Fermi $Z(r)$ with Latter cutoff.

The foregoing discussion neglects exchange effects. These were first considered by Dirac and later by Jensen [9]. The density in (1) is replaced by

$$n = (2m)^{3/2}\frac{[(T + \tau^2)^{1/2} + \tau]^3}{3\pi^2\hbar^3} \quad \text{where} \quad \tau^2 = \frac{e^2}{2\pi^2 a} \tag{14}$$

which reduces to (1) on neglecting the small finite value of τ. In this case the (Z, N) ion has a finite radius, at which the density has a finite limiting value

$$n_0 = \frac{125}{192\pi^5}a^{-3} = 2.127 \times 10^{-3}a^{-3} \tag{15}$$

For $r > r_0$ the density drops discontinuously to zero. The corresponding alteration of (11) is

$$W = -\frac{(Z - N)e^2}{r_0} - \frac{15}{16}\tau^2 \tag{16}$$

To find the potential distribution and the value of r_0 it is necessary to solve a modified form of (3) in which the density expression (14) is used in place of (1). Making the same scale change as in (5) and defining

$$\beta = \tau\left(\frac{b}{Ze^2}\right)^{1/2} = 0.2118Z^{-2/3} \tag{17}$$

and writing

$$\Psi(x) = r\frac{[e\phi(r) + W + \tau^2]}{Ze^2} \quad \text{with} \quad \Phi(0) = 1 \tag{18}$$

the equation for Ψ is

$$\Psi''(x) = x\left\{\left[\frac{\Psi(x)}{x}\right]^{1/2} + \beta\right\}^3 \tag{19}$$

which is known as the *Thomas-Fermi-Dirac equation*. The boundary condition at $r_0 = x_0 b$ is

$$\frac{\Psi(x_0)}{x_0} = \frac{\beta^2}{16} \quad \text{and} \quad \frac{Z - N}{Z} = x_0\left[\frac{\beta^2}{16} - \Psi'(x_0)\right] \tag{20}$$

The numerical solution for this model is considerably more complicated than for that in which exchange is neglected. Because of the appearance in (19) of the Z-dependent β a different solution has to be integrated for each atom. The literature contains a rather complete set of numerical results [10].

The effective potential energy functions for hydrogen are shown in Figure 1^4. When the Thomas-Fermi field (13) is used in place of $-e^2/r$, these curves are distorted downward in the general range $0 < r < 2a$ by amounts that rapidly increase with Z. This has the effect that the $1s$ level has large negative values of the order $-Z^2$ Rydbergs.

All of the $V_l(r)$ curves for $l > 0$ tend to ∞ as $r \to 0$, but the qualitative behavior of the $V_p(r)$ curves is quite different from that of the $V_d(r)$ and $V_f(r)$ curves, because of the smaller magnitude of the centrifugal barrier for the p states. The contrasting behavior is illustrated in Figures 3^8 and 4^8, which show, respectively, the $V_p(r)$ and $V_d(r)$ curves for several values of Z. In the former, the centrifugal barrier is so small that even for a moderately small value such as $Z = 10$ there is practically no maximum to the curve near $r = 2a$ the $V_p(r)$ curves dropping to a deep minimum, rising to positive values only for $r < 0.1a$.

The Thomas-Fermi or Thomas-Fermi-Dirac potential, corrected to provide a more realistic asymptotic behavior, as given by (13), may be used as an effective central potential $V(r)$ in $1^4 3$. This procedure defines sets of one-electron energy levels $\epsilon(nl)$ and single-electron radial wave functions whose dependence on Z is parametric. Solutions of $1^4 3$ for $\epsilon(nl$ and $P(nl; r)$ for various values of Z have been obtained by Latter [7].

Very close to $r = 0$, the radial wave functions $P(ns; r)$ are linear in r and the $P(np; r)$ are quadratic, but their values are so small that the weak centrifugal barrier in this range does not have much effect on the wave function. The oscillatory part of the wave function in both cases develops in the general range $0.1a < r < 2a$, followed by an exponential tail in the outer range where $W < V(r)$.

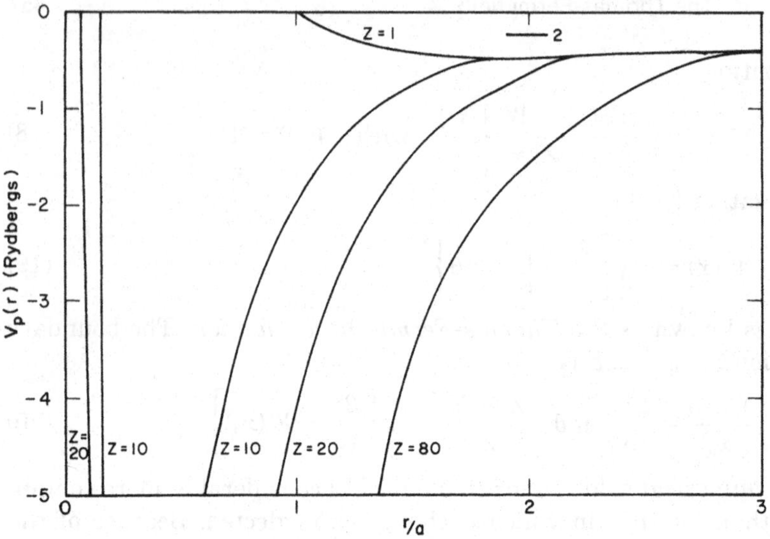

Figure 3[8]. Effective $V_p(r)$ with Thomas-Fermi $Z(r)$ for p states of neutral atoms. Hydrogenic value of energy for $n = 2$ (0.25 Rydbergs) is also shown.

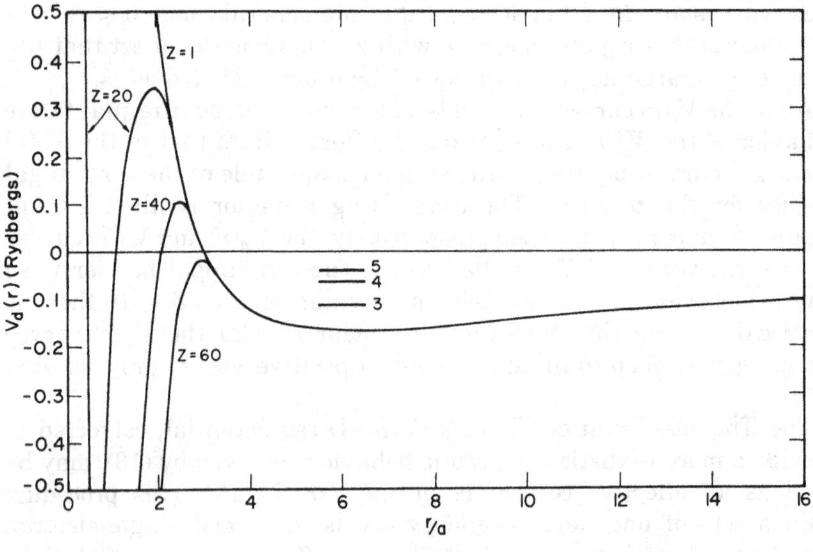

Figure 4[8]. Effective $V_d(r)$ with Thomas-Fermi $Z(r)$ for d states of neutral atoms. Hydrogenic values of energy for $n = 3, 4, 5$ are also shown.

For the d states, and even more so for the f states, the centrifugal barrier is so great that there is a marked development of two minima in the $V_d(r)$ and $V_f(r)$ curves. As Figure 4^8 shows, one minimum of $V_d(r)$ is at $r = 6a$, the same value at which it occurs for $Z = 1$. At smaller radii a marked potential barrier develops, reaching to positive values of $V_d(r)$ for $Z < 60$. This is followed by a deep and narrow potential well in the range $r < 2a$, with $V_d(r)$ going to ∞ for very small values of r. Similarly, in the case of $V_f(r)$, the outer part has a minimum at $r = 12a$, the same as for hydrogen. Another minimum exists for small values of r. For a given Z these are separated by a higher barrier than in the case of $V_d(r)$. Alternatively, higher values of Z are required to pull down the barrier separating the two minima. The consequences of these features of the $V_d(r)$ and $V_f(r)$ curves were studied by Mayer [10g].

The dependence on Z of the $\epsilon(nl)$ are shown in Figure 5^8, based on the calculations of Latter [7] using the field (13). A log-log plot is used because of the great range of values involved. The curves for particular values of (nl) start at the nth hydrogenic value for $Z = 1$. The single $1s$ level starts dropping at once with increasing Z.

The $2s$ and $2p$ states also become more tightly bound with increasing Z as both the nuclear charge and the number of screening electrons increase. But the curves show quite clearly that the $\epsilon(2s)$ drops much more rapidly than $\epsilon(2p)$, corresponding to the fact that s states penetrate close to the origin, whereas p states are held away from the origin by the centrifugal barrier.

Similar trends are shown for the $3s$, $3p$, and $3d$ states and for the states for higher n. A striking feature is the square shoulder on the ϵ $(3d)$ curve, whereby this level remains almost constant at the hydrogenic value out to $Z = 26$, when it drops abruptly to much lower values. Another interesting feature is the step-wise behavior of the $\epsilon(4d)$ and $\epsilon(5d)$ curves: $\epsilon(4d)$ remains essentially constant at the hydrogenic values for $n = 4$ out to the value of Z at which $\epsilon(3d)$ drops abruptly. Then $\epsilon(4d)$ drops abruptly to near the $n = 3$ hydrogenic value at which it levels off until at a higher Z it drops abruptly. Two such steps are shown by the $\epsilon(5d)$ curve, and a similar step-wise behavior is shown for the $4f$ and $5f$ states. This is connected with the double-minimum character of the $V_d(r)$ and $V_f(r)$ curves

If these curves represented accurately the relative tightness of binding of the shells, they would indicate the order in which shells are filled in the periodic table (Sec. 11^1). The results shown in Figure 5^8 are correct up to $Z = 20$. Even for $Z = 10$ the $\epsilon(4s)$ level is lower than $\epsilon(3d)$, which agrees with the observed fact that, after argon, the $4s$ shell fills before the $3d$ shell starts to fill. However, Figure 5^8 is in error in that the curves here would indicate that the $4p$ shell should also fill before the $3d$, contrary to fact. Qualitatively, the curves do

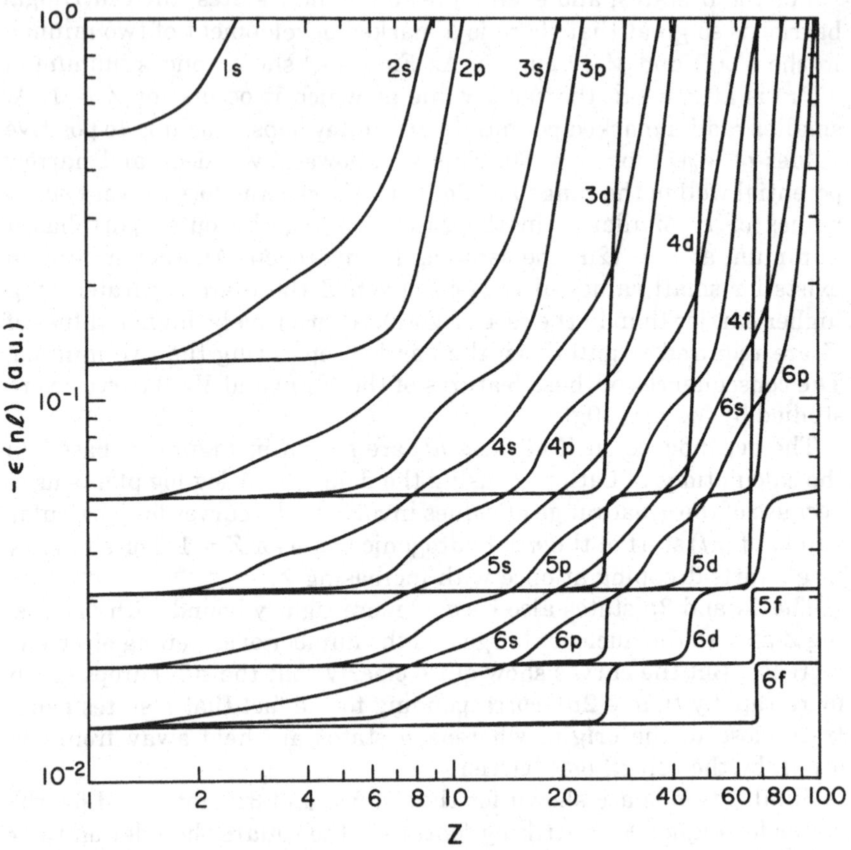

Figure 5[8]. $-\epsilon(nl)$ as functions of Z using the Latter cutoff form of Thomas-Fermi $Z(r)$.

show that in the range $20 < Z < 30$, the $3d$, $4s$, and $4p$ levels are fairly nearly equal in energy.

After $Z = 36$ with its filled $4p$ shell, the Thomas-Fermi results correctly indicate that the $5s$ shell fills next but incorrectly indicate that the $5p$ shell would fill before the $4d$ shell.

The $\epsilon(4f)$ curve indicates that these levels should start to be tightly bound at approximately $Z = 65$, whereas the $4f$ shell actually starts to fill at $Z = 58$.

The same error is made again with regard to the relative order of $5d$ and $6p$ shell filling. All of these defects appear to have a common origin in that the Z-dependence at the $3d$, $4d$, and $5d$ binding energies are not actually as square shouldered as the Thomas-Fermi field indicates. The order of filling of shells hinges on energy differences of the order 0.1 Rydberg. Evidently the relatively crude considerations on which this field is based are not accurate enough to give correct results on this point.

The Thomas-Fermi model has two major defects. First, the electron density (therefore, the kinetic energy) becomes infinite at the nucleus. Second, at large distances the electron density vanishes as $1/r^6$ instead of the exponential vanishing of the wave-mechanical electron density.

Many attempts have been made to improve the Thomas-Fermi model while keeping it simple enough to be useful. These improvements can be classified in two major groups as potential and kinetic energy corrections. For example, the Fermi and Amaldi [5c] and the Dirac [9a] corrections are of the potential energy type. The correlation correction, introduced by Gombás and Lewis [11] also falls in the same category. These corrections are well founded and quite satisfactory.

Kinetic energy corrections arise from the fact that the electron gas can not be regarded as free, especially near the nucleus. Various corrections of this type have been considered, but none seems to be satisfactory. Weizsäcker [12] obtained constant density near the nucleus by modifying the electron-wave functions, which are plane waves $e^{i\mathbf{k}\cdot\mathbf{r}}$ in the Thomas-Fermi model, using $(1 + \boldsymbol{\alpha} \cdot \mathbf{r})e^{i\mathbf{k}\cdot\mathbf{r}}$. This modification also gives an electron density that decreases exponentially at large distances from the nucleus. In spite of this correct behavior at the boundaries, the energies so obtained do not agree well with the wave-mechanical values. Yonei and Tomishima [13] have shown empirically that if the Weizsäcker correction is reduced by a factor of 1/5 excellent agreement with wave-mechanical energies is obtained. Kirzhnits [14a], Golden [14b], and Baraff and Borowitz [14c] introduced quantum corrections to the Thomas-Fermi model by using a perturbation expansion. Kirzhnits has shown that the factor by which the Weizsäcker correction should be reduced is 1/9. The work of Baraff and Borowitz is based on an expansion of the many-electron Green's function in powers of $\hbar$, of which the lowest-order term gives the Thomas-Fermi model. The new higher-order term arises from exchange and inhomogeneity effects. The numerical calculations of Schey and Schwartz [15] show that these corrections give little improvement. Plaskett and March [16] included the centrifugal barrier in the Thomas-Fermi potential by applying the WBKJ approximation. Greatly improved energies are obtained by replacing the centrifugal term $l(l + 1)$ by $(l + 1/2)^2$. This makes the radial density $4\pi r^2 n(r)$ vanish at some inner cutoff radius. The normally high Thomas-Fermi model density is reduced, which leads to improved energies. Ashby and Holzman [17] corrected the form of the electron density near the nucleus by using the idea that in this region the total density is nearly equal to the K-shell electron density that can be obtained easily. Their semistatistical model consists of retaining the Thomas-Fermi model for radii r greater than some small radius r_0 but using a wave-mechanical density for radii $r < r_0$. At r_0 the two densities are joined smoothly. They express the Thomas-Fermi equations as integral equations, which are solved by a numerical iteration technique. They also incorporate relativistic effects in their semistatistical model.

The statistical theories do not show the maximum densities at K-, L-, M-, . . . shell positions, because this is averaged over. The grouping of electrons into shells is a quantum mechanical effect. The incorporation of the shell structure into the statistical theory has been accomplished by Gombás and associates [18] by treating the electrons of different shells separately from the beginning. The Pauli exclusion principle is taken into account by using a statistical repulsive potential [19]. Another statistical atom model, based on a local Coulomb approximation, which shows the inner-shell structure, has been developed by Ashby and Holzman [20].

2^8. Hartree and Hartree-Fock fields

The earliest formulation of the self-consistent field method in quantum mechanics is attributable to Hartree [21]. Later Fock [22] extended the method to take exchange interaction into account. Slater [23] showed how both methods can be based on the variation principle (Sec. 8^2). Detailed accounts have been given by Hartree [21d], Slater [24a, Chap. 17] and Bethe and Jackiw [24b, Chap. 4].

As in $7^4 1$, 2 the nonrelativistic Hamiltonian is written

$$H = \sum_i f(i) + \sum_{i<j} g(i,j)$$
$$f(i) = \frac{p_i^2}{2m} - \frac{Ze^2}{r_i}; \qquad g(i,j) = \frac{e^2}{r_{ij}} \tag{1}$$

A trial form for the N-electron wave function Φ is adopted, which may contain unknown functions or unknown parameters. With this form the mean energy W is calculated:

$$W = \frac{\int \Phi^* H \Phi \, d\tau}{\int \Phi^* \Phi \, d\tau} \tag{2}$$

where the integrals are understood to include summation over the N-spin coordinates. In this way an expression for W is obtained in terms of the unknown functions or parameters in Φ. These are then determined by the requirement that $\delta W = 0$ when these unknowns in Φ are varied. With these unknowns so determined, the corresponding value of W is an approximation to a proper energy, and Φ is an approximation to a corresponding proper state of H.

If the unknowns in Φ are subject to k constraints of the form $C_\alpha =$

constant ($\alpha = 1, 2, \ldots, k$), these are taken into account by the method [25] of Lagrange multipliers. We form

$$W' = W - \sum_{\alpha} C_\alpha \epsilon_\alpha \tag{3}$$

and find the unknowns in Φ and the values of ϵ_α by the requirement that $\delta W' = 0$.

The accuracy of Φ and W as proper state and proper value of H depends on the choice of the trial form for Φ. The method usually uses the independent-particle approximation (Sec. 7^2). Hartree used for Φ the simple product of N one-electron functions $7^2 6$

$$\Phi = \phi(a; 1)\phi(b;2) \ldots \phi(n; N) \tag{4}$$

Here each ϕ is normalized, but the one-electron functions need not be orthogonal. The N normalization conditions introduce N Lagrange multipliers, $\epsilon_a, \epsilon_b, \ldots, \epsilon_n$.

The Hartree-Fock method takes exchange effects into account by using the determinant form, $7^2 9$, for Φ, and requiring the one-electron functions ϕ to be orthonormal, so now there is a Lagrange multiplier, ϵ_{ab}, associated with each pair of one-electron states, that is, one-electron states having the same spatial dependence but different spins. In this case W is expressed as in $10^4 1$ in terms of the I, J, and K integrals:

$$W = \sum_{a} I(a) + \sum_{pairs} [J(a, b) - K(a, b)] \tag{5}$$

in which each pair is counted once with $a \neq b$: and

$$I(a) = \int \phi^*(a; 1)f(1)\phi(a; 1)\, d\tau_1$$
$$J(a, b) = \iint |\phi(a; 1)|^2 g(1, 2)|\phi(b, 2)|^2\, d\tau_1\, d\tau_2$$
$$K(a, b) = \iint \phi^*(a; 1)\phi(b; 1)g(1, 2)\phi^*(b; 2)\phi(a; 2)\, d\tau_1\, d\tau_2$$

When the simple product form (4) is used, W is given by (5) with omission of the $K(a, b)$ exchange integrals.

Because $\int\phi^*(a; 1)\phi(b; 1)\, d\tau_1 = 0$ is the same condition as $\int\phi^*(b; 1)\phi(a; 1)\, d\tau_1 = 0$, it follows that $\epsilon_{ab} = \epsilon^*_{ba}$ may be assumed for the Lagrange multipliers.

Assuming that each ϕ can be labeled by an exact m_s, that is, that the spin dependence of each $\phi(a; 1)$ is through a factor $\delta(m_{sa}; s_1)$; the spin sums give a unit factor on the I and J integrals in (5) and a factor $\delta(m_{sa}, m_{sb})$ on the K integrals, as in Section 9^4. The discussion in Section 9^4 assumes that each wave function $\phi(a; 1)$ is represented by central-field wave functions (Eq. $1^4 2$b), but this is not assumed here.

On varying the ϕ's, subject to the constraint that all ϕ's are orthogonal, each $\delta\phi$ and $\delta\phi^*$ is treated as independent, resulting in the following N Hartree-Fock (HF) equations, one for each $a, b, \ldots, n$,

$$f(1)\phi(a; 1) + \sum_b \int g(1, 2)|\phi(b; 2)|^2 \, d\tau_2 \, \phi(a; 1)$$
$$- \sum_b \delta(m_{sa}, m_{sb}) \int g(1, 2)\phi^*(b; 2)\phi(a; 2) \, d\tau_2 \phi(b; 1)$$
$$= \sum_b \epsilon_{ab}\delta(m_{sa}, m_{sb})\phi(b; 1) \qquad (6)$$

Actually, Equation 6 represents a set of coupled integro-differential equations which must be solved for the orbitals $\phi(a; 1)$ and the matrix of energy parameters ϵ_{ab}. In general the solution of this set of equations is not unique. This is not surprising because Φ is taken as a single determinant with paired spin states. Row or column transformations among orbitals with the same spin will leave Φ unaltered but will change the form of individual orbitals. Thus, if a given set of ϕ's satisfy (6) an alternative set of orbitals ψ will also satisfy the same equation where

$$\phi(a; \quad) = \sum_\alpha \psi(\alpha; \quad)C_{\alpha a}$$

with each $C_{\alpha a}$ containing a factor $\delta(m_{s\alpha}, m_{sa})$. Alternative solutions of (6) are thus related to each other by unitary spin-preserving transformations of the individual orbitals. When the orbitals are so transformed it may be shown directly from (6) [24b, p. 48] that the energy matrix ϵ_{ab} transforms according to the relation

$$\epsilon'_{\alpha\beta} = \sum_{a,b} C\dagger_{\alpha a}\epsilon_{ab}C_{b\beta}$$

Because the ϵ_{ab} form a Hermitian matrix, the $C_{\alpha a}$ may be chosen in such a way that $\epsilon'_{\alpha\beta}$ is diagonal. Therefore, there exists a solution of (6) for which all off diagonal elements of $\epsilon'_{\alpha\beta}$ are zero. The orbitals for this solution remain orthogonal to each other [24a, p. 7].

In (6) the $b = a$ terms in the sums over b on the first and second lines are equal, with opposite sign, and so drop out. The remaining terms in the first line correspond to the electron in state a being in a field of local potential energy with interaction with the other $(N - 1)$ electrons,

$$V_a(1) = \int g(1, 2) \sum_b{}' \left|\phi(b; 2)\right|^2 d\tau_2 \qquad (7)$$

The terms on the second line, which arise from the exchange integrals $K(a, b)$, can be expressed as a nonlocal potential energy

$$U_a(1, 2) = g(1, 2) \sum_b{}' \delta(m_{sa}, m_{sb})\phi^*(b; 2)\phi(b; 1) \qquad (8)$$

where the prime on the summation signifies omission of the $b = a$ term. In this way the N Hartree-Fock equations take the form

$$f(1)\phi(a;1) + V_a(1)\phi(a;1) - \int U_a(1,2)\phi(a;2)\,d\tau_2 = \epsilon'_{aa}\phi(a;1) \qquad (9)$$

The Hartree equations are obtained from (9) by omitting the nonlocal potential terms. As $V_a(1)$ may be different for the various a states, the ϕ's are then no longer orthogonal. The nonlocal potential energy terms in (9) greatly complicate the task of finding solutions for the HF equations.

Up to this point the only assumption made about the form of the $\phi(a; 1)$ is that the spin dependence is through the factor $\delta(m_{sa}; s_1)$. In the central-field approximation the further assumption is made that the orbitals have the form $1^4$2b, in which the radial functions, $P(a; 1)$ depend on n_a, l_a but not on (m_a, m_{sa}).

The validity of the central-field assumptions for the $\phi(a; 1)$ depends on the accuracy with which the V and U in (9) are invariant under rotations. Such spherical symmetry is possessed by configurations consisting of closed shells only, as shown by Delbrück [26], but the fields depart from spherical symmetry to some extent in other cases. For configurations having one or more open shells the use of a central field is an approximation whose validity depends on smallness of the departure of the fields from spherical symmetry.

The central-field approximation (Chaps. 4 and 5) underlies the major part of the theory of atomic energy levels. Making this assumption the results of Chapter 4 et seq. may be used to find and label the terms associated with a single configuration, and the results $8^4$5, $9^4$12, and $9^4$13 may be used to express the I, J, and K integrals in terms of single and double radial integrals.

The HF equations for the radial functions of a configuration consisting only of closed shells can be worked out explicitly by using the results of Section 8^4 and 9^4. The form for W' is, with $Q_a = 2(2l_a + 1)$,

$$\begin{aligned} W' = \sum_a \Big[Q_a I(a) + \frac{1}{2} Q_a(Q_a - 1)F^0(a,a) + {\sum_b}' Q_a Q_b F^0(a,b) \Big] \\ - \sum_{\substack{a \\ k \neq 0}} \langle l_a \| C^{(k)} \| l_a \rangle^2 F^k(a,a) - 2\sum_{a,b,k} \langle l_a \| C^{(k)} \| l_b \rangle^2 G^k(a,b) \\ - \sum_{a,b} \epsilon_{ab}\delta(l_a, l_b) \int_0^\infty P(a;r)P(b;r)\,dr \end{aligned} \qquad (10)$$

in which the $\langle l_a \| C^{(k)} \| l_b \rangle^2$ are related to the $c^k(a, b)$ of Table 6^4 and $6^3$21 by

$$\langle l_a \| C^{(k)} \| l_b \rangle^2 = \frac{1}{2}\sqrt{Q_a Q_b}\, c^k(l_a, 0; l_b, 0) \qquad (11)$$

and $F^k(a, b)$ and $G^k(a, b)$ are given by $9^4 12''$ and $9^4 13''$. The variation of W' involves the following expressions for variations of integrals in which $P(a; r)$ occurs (limits on all integrals are from zero to infinity):

$$\delta \int P^2(a; r)\, dr = 2 \int \delta P(a; r) P(a; r)\, dr$$
$$\delta \int P(a; r)P(b; r)\, dr = \int \delta P(a; r)P(b; r)\, dr$$
$$\delta I(a) = 2 \int \delta P(a; r) \times \left[-\frac{\hbar^2}{2m}\frac{d^2}{dr^2} - \frac{Ze^2}{r} + \frac{\hbar^2}{2m}\frac{l_a(l_a+1)}{r^2} \right] P(a; r)\, dr$$
$$\delta F^k(a, a) = 4 \iint \delta P(a; r) P(a; r) \frac{e^2 r_<^k}{r_>^{k+1}} P^2(a; s)\, dr\, ds$$
$$\delta F^k(a, b) = 2 \iint \delta P(a; r) P(a; r) \frac{e^2 r_<^k}{r_>^{k+1}} P^2(b; s)\, dr\, ds$$
$$\delta G^k(a, b) = 2 \iint \delta P(a; r) P(b; r) \frac{e^2 r_<^k}{r_>^{k+1}} P(a; s) P(b; s)\, dr\, ds \qquad (12)$$

Using these to find $\delta W'$ and equating the coefficient of each $\delta P(a; r)$ to zero one finds a set of HF equations, for the radial functions, one for each closed shell, which can be written

$$\left[-\frac{\hbar^2}{2m}\left(\frac{d^2}{dr^2} - \frac{l_a(l_a+1)}{r^2} \right) - \frac{Z_a(r)e^2}{r} \right] P(a; r) - \sum_b \int U(a, b; r, s) P(a; s)\, ds$$
$$= \frac{\epsilon_{aa}}{Q_a} P(a; r) + \sum_b \frac{\epsilon_{ab}}{Q_a} \delta(l_a, l_b) P(b; r) \qquad (13)$$

Here $Z_a(r)$ is given by

$$Z_a(r) = Z - \sum_b (Q_b - \delta_{ab}) r \int \frac{1}{r_>} P^2(b; s)\, ds \qquad (14)$$

and the kernel for the nonlocal potential energy for $b \neq a$ is

$$U(a, b; r, s) = \frac{2}{Q_a} \sum_k \langle l_a \| C^{(k)} \| l_b \rangle^2 \frac{e^2 r_<^k}{r_>^{k+1}} P(b; r) P(b; s) \qquad (15a)$$

and for $b = a$

$$U(a, a; r, s) = \frac{2}{Q_a} \sum_{k=2}^{2l_a} \langle l_a \| C^{(k)} \| l_a \rangle^2 \frac{e^2 r_<^k}{r_>^{k+1}} P(a; r) P(a; s) \qquad (15b)$$

The nondiagonal ϵ_{ab} in (13) may be dropped by a spin-preserving unitary transformation, which diagonalizes the ϵ matrix, as discussed previously, giving

$$\left[-\frac{\hbar^2}{2m}\left(\frac{d^2}{dr^2}-\frac{l_a(l_a+1)}{r^2}\right)-\frac{Z_a(r)}{r}e^2\right]P(a;r)-\sum_b\int U(a,b;r,s)P(a;s)\,ds$$
$$=\epsilon(a)P(a;r) \qquad (16)$$

in which $\epsilon(a) = \epsilon'_{aa}/Q_a$, with ϵ'_{aa} the diagonal elements after the transformation. That is, in the HF formalism for closed-shell configurations the nondiagonal elements, ϵ_{ab}, can be completely disregarded. This does not hold in dealing with open-shell configurations. The nondiagonal elements ϵ_{ab}, of the energy matrix, do play a role in the HF theory in such cases as discussed in the next section.

Omission of the nonlocal potential term in (16) reduces the radial HF equation to the Hartree equation:

$$\left\{-\frac{\hbar^2}{2m}\left[\frac{d^2}{dr^2}-\frac{l_a(l_a+1)}{r^2}\right]-\frac{Z_a(r)}{r}e^2\right\}P(a;r)=\epsilon(a)P(a;r) \qquad (17)$$

which is also valid for open-shell configurations, provided that in (14) Q_b is changed to q_b for the open shells.

A somewhat different HF formalism (a matrix representation of Hartree-Fock equations), developed by Roothaan [27] for closed-shell molecular systems, is also applicable to atoms with closed shells.

3[8]. Hartree-Fock equations for open-shell configurations

For configurations involving one or more open shells, it is not in general possible to represent the terms by single determinants of one-electron functions. Instead they are linear combinations of determinants (Secs. 11[4], 12[4]).

The *restricted* or *conventional* HF method uses one-electron functions, which are products of a radial function by a spherical harmonic by a spin factor, in which the radial function is independent of m_a and m_{sa}. In the literature various alternative HF methods have been developed. If some of the one-electron functions are not assumed to have the usual spherical harmonic and spin dependence, the method is called *unrestricted* HF. When the spherical harmonic angular dependence is retained but the radial functions are allowed to depend on m_s, the method is called *spin-polarized* HF [28], or sometimes it is also called unrestricted. The method in which each (m_s, m) state is allowed to have a different radial function (that is, a truly unrestricted HF method) is called the *extended method* by Jucys [29]. When the trial wave functions are properly symmetrized before being subjected to the variational calculations the scheme is called *extended* HF [30].

In the conventional HF method for configurations involving one or more open shells in addition to closed shells [21d], the integrals occurring in the expression $2^8 3$ for W can be classified in three groups:

1. those involving wave functions of closed shells only
2. the I integrals for open shells and the J and K integrals involving one closed-shell wave function with one open-shell wave function
3. the J and K integrals in which both wave functions belong to open shells

Using the indices a and c to refer to closed shells and b to refer to open shells, and writing q_b for the actual number of electrons in the bth open shell the results of Section 9^4 show that the contributions of group 1 to W are

$$\sum_a Q_a I(a) + \sum_a \frac{1}{2} Q_a(Q_a - 1)F^0(a, a) - \sum_{a,k\neq 0} \langle l_a \| C^{(k)} \| l_a \rangle^2 F^k(a, a) + {\sum_{a,c}}' Q_a Q_c F^0(a, c) - 2 \sum_{a,c,k} \langle l_a \| C^{(k)} \| l_c \rangle^2 G^k(a, c) \qquad (1)$$

and the contributions of group 2 to W are

$$\sum_b q_b \left[I(b) + \sum_a Q_a F^0(a, b) - \frac{2}{Q_b} \sum_{a,k} \langle l_a \| C^{(k)} \| l_b \rangle^2 G^k(a, b) \right] \qquad (2)$$

The contributions of group 3 cannot be written in closed form. For $q_b = 2$ or 3 they are given by $2^5 5$ and the sum of $4^5 3$ a, b, c, respectively. For more complicated cases the method of Section 11^5–14^5 must be used.

Three kinds of Lagrange multipliers are associated with the three kinds of contributions of groups 1, 2, and 3. The multipliers related to the closed shells (that is, resulting from the contributions of group 1) can be diagonalized by unitary transformations. But the off-diagonal multipliers that couple the closed and open shells or two open shells cannot be eliminated. This is because a unitary transformation of orbitals in this case does not leave full determinantal wave function unchanged. This can be shown by a simple example. Consider the ground state of Li, $1s^2 2s\ ^2S$. The total wave function may be represented by a single determinant

$$\Phi(1s^2, 2s) = \frac{1}{\sqrt{3}} \begin{vmatrix} \phi(1s^+, 1) & \phi(1s^+, 2) & \phi(1s^+, 3) \\ \phi(1s^-, 1) & \phi(1s^-, 2) & \phi(1s^-, 3) \\ \phi(2s^-, 1) & \phi(2s^-, 2) & \phi(2s^-, 3) \end{vmatrix}$$

where + and −, as superscripts, indicate spin states of different orbitals. The total energy of the ground state will be

$$W(1s^2 2s;\ ^2S) = 2I(1s) + I(2s) + F^0(1s, 1s) + 2F^0(1s, 2s) - G^0(1s, 2s)$$

when the functions $P(1s; r)$ and $P(2s; r)$ in

$$W' = W - \epsilon_{11} \int P^2(1s; r)\, dr - \epsilon_{12} \int P(1s; r)P(2s; r)\, dr - \epsilon_{22} \int P^2(2s; r)\, dr$$

are varied, one at a time, and $\delta W'$ is set to zero; the radial HF equations

$$\left[-\frac{\hbar^2}{2m}\frac{d^2}{dr_1^2} - Z\frac{e^2}{r_1} + \int \frac{e^2}{r_>} P^2(1s; r_2)\, dr_2 + \int \frac{e^2}{r_>} P^2(2s; r_2)\, dr_2 \right] \times P(1s; r_1) - \frac{1}{2}\left[\int \frac{e^2}{r_>} P(1s; r_2)P(2s; r_2)\, dr_2 \right] P(2s; r_1) = \frac{1}{2}\epsilon_{11}P(1s; r_1) + \frac{1}{4}\epsilon_{12}P(2s; r_1) \tag{3a}$$

and

$$\left[-\frac{\hbar^2}{2m}\frac{d^2}{dr_1^2} - Z\frac{e^2}{r_1} + 2\int \frac{e^2}{r_>} P^2(1s; r_2)\, dr_2 \right] P(2s; r_1) - \left[\int \frac{e^2}{r_>} P(1s; r_2)P(2s; r_2)\, dr_2 \right] P(1s; r_1) = \epsilon_{22}P(2s; r_1) + \frac{1}{2}\epsilon_{12}P(1s; r_1) \tag{3b}$$

result. The equations are similar to $2^8 13$ for closed shells. Here, however, the nondiagonal parameters cannot be eliminated, because the $1s$ orbital in (3) represents the radial behavior of both $1s$ orbitals with different spins, whereas the $2s$ orbital refers to a specific spin state. Thus any unitary transformation, for eliminating the nondiagonal energy parameters, of the $1s$ and $2s$ orbitals will not preserve the initial determinant.

In case there is but one open shell the contributions (group 3), take the simple form

$$\frac{1}{2} q_b(q_b - 1)F^0(b, b) - \sum_{k>0} C_{lk}F^k(b, b) \tag{4}$$

where the coefficients C_{lk} for the $(np)^q$, $(nd)^q$, and $(nf)^q$ configurations are given in Appendix 3.

Hartree [21d, pp. 111–3] has given a method for treating configurations with closed shells plus a single open shell. To find $P(b; r)$ we must calculate the variation of $2^8 3$ for an arbitrary $\delta P(b; r)$. This involves only groups 2 and 3, in which $P(b; r)$ occurs. By comparing (2) with the average energy of a pair, $9^4 21$, the terms in (2) are q_b/Q_b times the total interaction of a closed shell b with the closed shell a.

Following the methods used in Section 2^8 the equation for the radial function $P(b; r)$ of the open shell, analogous to $2^8 13$, becomes

$$\left\{-\frac{\hbar^2}{2m}\left[\frac{d^2}{dr^2}-\frac{l_b(l_b+1)}{r^2}\right]-Z_b(r)\frac{e^2}{r}\right\}P(b;r)-\sum_a \int \tilde{U}(b,a;r,s)P(b;s)\,ds$$
$$=\frac{\epsilon_{bb}}{q_b}P(b;r)+\sum_a \frac{\epsilon_{ab}}{q_b}\delta(l_a,l_b)P(a;r) \qquad (5)$$

in which $Z_b(r)$ is essentially the same as $Z_a(r)$ defined in $2^8 14$, but now the summation is over index a and for $a = b$ Q_b is replaced by q_b. $\tilde{U}$ is a modified form of $2^8 15a$,

$$\tilde{U}(b,a;r,s) = U(b,a;r,s) + \frac{2}{q_b}\sum_{k>0} C_{lk}\, e^2 \frac{r_<^k}{r_>^{k+1}} P(b;s)P(b;r) \qquad (6)$$

Because the C_{lk} appearing in (6) may be different for different terms of the configuration, $P(b; r)$ may be different for different terms of the same configuration.

The contributions to the radial equation for a particular closed shell, $P(c; r)$, which depend on the radial function $P(b; r)$ of the open shell, are

$$q_b\left[Q_c F^0(b;c) - \frac{2}{Q_b}\sum_k \langle l_b \| C^{(k)} \| l_c\rangle^2 G^k(b,c)\right] \qquad (7)$$

which is equal to q_b/Q_b times what it is when the b shell is closed. Thus the closed-shell radial equations for the case of one open shell are of the form $2^8 13$, index a changed to c, but now in $2^8 14$ Q_b should be replaced by q_b for $b \neq c$, and the nonlocal contribution for $b \neq c$, equation $2^8 15a$, is multiplied by q_b/Q_b. Because the $P(b; r)$ are different for different terms, this produces a slight (SL) term dependence of the radial functions for the closed shell.

Roothaan [31a] developed a method of calculation in matrix notation which takes the nondiagonal Lagrange multipliers into account exactly for configurations with one open shell. This was generalized to configurations having two open shells by Huzinaga [31b]. Roothaan and Bagus [31c] extended the method to several open shells, provided that no two open shells have the same l.

Instead of using the individual term energies in $2^8 3$, Slater has recommended the use of a variant in which the configuration average energy W_{Av} (defined in $9^4 25$) is varied to obtain equations for the radial functions. This procedure is considerably simpler computationally than one that allows the radial functions to vary from term to term within a configuration. It is called the center-of-gravity Hartree-Fock (CGHF) method [32]. Values for the total configuration average energies, the one-electron energy parameters, $\epsilon(a)$, the F^k and G^k integrals, the radial expectation values, the screening parameters, and the radial-wave functions were calculated by this method for all the normal configurations of neutral atoms by Froese [33a], and later by Mann [33b].

The total energy of a configuration with closed shells is given by (1). Then we can compute the energy required to remove one electron from the outermost shell, *supposing* that the orbitals of the configuration with a missing electron are the same as before. To do so we need the total energy of such a configuration, which is

$$W(Q_c, Q_a - 1) = \sum_c (Q_c - \delta_{ac}) I(c) + \sum_c \frac{1}{2}(Q_c - \delta_{ac})(Q_c - 1 - \delta_{ac}) F^0(c, c)$$
$$- \sum_{c,k\neq 0} \frac{(Q_c - 2\delta_{ac})}{Q_c} \langle l_c \| C^{(k)} \| l_c \rangle^2 F^k(c, c) + \sum_{a,c}{}' (Q_a - 1) Q_c F^0(a, c)$$
$$- 2 \sum_{c\neq a,k} \frac{(Q_a - 1)}{Q_a} \langle l_a \| C^{(k)} \| l_c \rangle G^k(a, c)$$

where subscript a indicates the outermost shell and c stands for other closed shells. The energy required to remove an electron (ionization potential: *IP*); that is, the negative of the binding energy is given by the difference

$$IP(a) = W(Q_c, Q_a) - W(Q_c, Q_a - 1)$$

where $W(Q_c, Q_a)$ is used for the total energy of a configuration with all closed shells. Keeping in mind that the radial-wave functions used in the evaluation of I, F^k, and G^k integrals in the expressions for $W(Q_c, Q_a)$ and $W(Q_c, Q_a - 1)$ are the *same*, the ionization potential of an outermost-shell electron can also be written as

$$IP(a) = I(a) + (Q_a - 1) F^0(a, a) - \frac{2}{Q_a} \sum_{k>0} \langle l_a \| C^{(k)} \| l_a \rangle^2 F^k(a, a)$$
$$+ \sum_{c\neq a} Q_c F^0(a, c) - \frac{2}{Q_a} \sum_{c\neq a,k} \langle l_a \| C^{(k)} \| l_c \rangle^2 G^k(a, c)$$

Now multiply both sides of $2^8 16$ by $P(a; r)$ and then integrate with respect to r:

$$\int_0^\infty P(a; r) \left\{ -\frac{\hbar^2}{2m} \left[\frac{d^2}{dr^2} - \frac{l_a(l_a + 1)}{r^2} \right] - \frac{Z}{r} e^2 \right\} P(a; r) dr$$
$$-\sum_c (Q_c - \delta_{ac}) \int_0^\infty \frac{1}{r_>} P^2(a; r) P^2(b; s)\, dr\, ds - \sum_c \int_0^\infty \frac{2}{Q_a}$$
$$\times \sum_k \langle l_a \| C^{(k)} \| l_c \rangle^2 e^2 P(a; r) P(c; r) \frac{r_<^k}{r_>^{k+1}} P(c; s) P(a; s)\, dr\, ds$$
$$= \epsilon(a) \int_0^\infty P^2(a; r)\, dr$$

This expression can be simplified, using the definition of I, F^k, and G^k integrals and the fact that the radial-wave functions are orthonormal, as

$$I(a) - \sum_c (Q_c - \delta_{ac})F^0(a, c) - \frac{2}{Q_a}\sum_{k>0} \langle l_a \| C^{(k)} \| l_a \rangle^2 F^k(a, a)$$
$$- \frac{2}{Q_a}\sum_{c\,:\,a,k} \langle l_a \| C^{(k)} \| l_c \rangle^2 G^k(a, c) = \epsilon(a)$$

When this relation is compared with the expression found for $IP(a)$, one finds that

$$IP(a) = \epsilon(a)$$

The same result can be demonstrated for a configuration that contains one electron outside a closed shell by simply using the expression found for the total energy of the normal configuration of Li. If the $1s^2$ core is taken to be frozen, as in the previous example, and both sides of (3b) are multiplied by $P(2s; r_1)$ and integrated with respect to r_1, we obtain

$$I(2s) + 2F^0(1s, 2s) - G^0(1s, 2s) = \epsilon_{22}$$

Because the left side of this equation is formally equal to the ionization energy (the difference in total energies of $1s^2 2s$ and $1s^2$ configurations) of an electron in $2s$ shell, again

$$IP(2s) = \epsilon_{22}$$

The interpretation of one-electron parameters $\epsilon(a) \equiv \epsilon_{aa}$ as the ionization potential, as in the above examples, is known as Koopmans's theorem [34a].

Actually, as pointed out by Charlotte Froese Fischer in a private communication, Koopmans proved more than the above. In general, as shown in the previous section, solutions to the Hartree-Fock equations ($2^8 6$) are not unique when spin-preserving unitary transformations are possible. Koopmans also showed [34a] that when such transformations are possible the solutions with the nondiagonal terms of ϵ_{ab} equal to zero should be selected. His reasoning was that of all possible solutions the solution that minimizes the core energy, that is, the energy obtained when one electron is removed from the system, is the best one, because only in this case does the core energy come closest to the total energy of the ion. These are the solutions with the nondiagonal terms equal to zero [34b].

For cases involving open shells alternative terms arise from a given configuration. The true ionization potential is defined as the difference in term values between the lowest term of a given open-shell configuration and the lowest term of that configuration containing one less electron. Koopmans's theorem does not take account of such term splittings and therefore provides only a rough estimate of the ionization potential when such splittings occur. Strictly speaking, it should only be applied to the centers of gravity of the terms of a given configuration. In spite of these limitations it provides a valuable estimate of ionization potentials.

The expression of the term energies in terms of F^k and G^k integrals, as in Chapters 4 and 5, assumes that the radial integrals remain the same for all terms of a configuration. This approximation is inherent in Koopmans's theorem. In one case the excitation and in the other the removal of an outer electron is supposed not to alter the remaining radial functions.

Ionization potentials may also be estimated by direct calculation of the difference in total energies obtained by separate Hartree-Fock calculations of an atom or ion and of the ion with the same charge but with one less electron. Tables 2^8a and 2^8b give the values of the ionization potential obtained by this procedure and the one-electron energy parameter, $\epsilon(a)$, in atomic units, for the neutral atoms and the singly ionized ions with $Z = 3$–20 as computed by Clementi [35] and by Fraga, Karwowski, and Saxena [46]. These tables establish the following:

1. $-\epsilon(a)$ is always larger than the theoretical ionization potential. It often is closer to the true ionization potential than $W' - W$.
2. The interpretation of $-\epsilon(a)$ as a theoretical ionization potential is almost exact when the final configuration is composed of closed shells, as expected from Koopmans's theorem.
3. Agreement between $-\epsilon(a)$ and the theoretical ionization potential is poorest when the final configuration contains a half-filled l_a shell.
4. For other intermediate cases the interpretation of $-\epsilon(a)$ as a theoretical ionization potential is a rough approximation.

4^8. Analytic and numerical solutions of the Hartree-Fock equations

The Hartree-Fock equations may be solved by two alternative methods classified as numerical and analytic. In the numerical case the radial equations $2^8$16 are treated as a set of coupled integro-differential equations in which the radial-wave functions $P(a;\, r)$ are variables. Because the radial wave functions must obey the boundary conditions $3^4$2 and $3^4$3, solutions of these equations exist only for particular values of the parameters $\epsilon(a)$. A solution can thus be obtained numerically by "integrating to self-consistency," that is, by repeatedly solving the equations numerically and adjusting the parameters $\epsilon(a)$ until a solution satisfying the boundary conditions is obtained.

In the analytic case the radial functions are expressed in analytic form. The standard approach is to express each radial-wave function in

Table 2[8]a. HF values (a.u.) of $IP(a; S, L) = W'(b; S', L') - W(a; S, L)$ and $-\epsilon(a; S, L)$ for the first- and second-row elements. S', L' and S, L refer to the lowest normal configuration term of the singly ionized ion and the neutral atom, respectively, and a stands for the outermost electron of the initial configuration. The first entry is taken from Clementi [35] and the second from Fraga, Karwowski, and Saxena [46].

Element	$IP = W' - W$	$-\epsilon(a)$	Observed	Element	$IP = W' - W$	$-\epsilon(a)$	Observed
Li	0.19631	0.19632	0.19814	Mg	0.24300	0.25296	0.28099
	0.19630	0.19632			0.2426	0.25305	
Be	0.29568	0.30927	0.34260	Al	0.20222	0.21003	0.21997
	0.29563	0.30927			0.2021	0.20996	
B	0.29149	0.30987	0.30495	Si	0.28144	0.29711	0.29957
	0.29147	0.30986			0.2811	0.29712	
C	0.39637	0.43316	0.41381	P	0.36910	0.39182	0.38537
	0.39643	0.43336			0.3690	0.39171	
N	0.51293	0.56771	0.53412	S	0.33184	0.43719	0.38072
	0.51295	0.56760			0.3318	0.43737	
O	0.43679	0.63177	0.50045	Cl	0.43340	0.50638	0.47655
	0.43721	0.63192			0.4335	0.50640	
F	0.57760	0.73000	0.64028	Ar	0.54276	0.59071	0.57915
	0.57764	0.73003			0.5431	0.59103	
Ne	0.72933	0.85020	0.79248	K	0.14736	0.14742	0.15952
	0.7293	0.85043			0.1471	0.14748	
Na	0.18181	0.18199	0.18886	Ca	0.18815	0.19550	0.22465
	0.1820	0.18210			0.1881	0.19550	

Table 2[8]b. HF values (a.u.) of $IP'(a; S', L') = W''(b; S'', L'') - W'(a; S', L')$ and $-\epsilon'(a; S', L')$ for the first- and second-row elements. S'', L'' and S', L' refer to the lowest normal configuration term of the doubly and singly ionized ion, respectively, and a stands for the outermost electron of the initial configuration. The first entry is taken from Clementi [35] and the second from Fraga, Karwowski, and Saxena [46].

Element	$IP' = W'' - W'$	$-\epsilon'(a)$	Observed	Element	$IP' = W'' - W'$	$-\epsilon'(a)$	Observed
Li	2.73641	2.79234	2.77972	Mg	0.54081	0.54121	0.55254
	2.73643	2.79237			0.5410	0.54144	
Be	0.66607	0.66614	0.66925	Al	0.64384	0.65261	0.69193
	0.66609	0.66615			0.6439	0.65230	
B	0.86166	0.87386	0.92442	Si	0.57760	0.58614	0.60070
	0.86160	0.87383			0.5773	0.58588	
C	0.88373	0.90475	0.89607	P	0.70491	0.72039	0.72662
	0.88373	0.90476			0.7047	0.71998	
N	1.07219	1.10991	1.08783	S	0.84060	0.86231	0.85773
	1.07223	1.11008			0.8403	0.86179	
O	1.27239	1.32659	1.29057	Cl	0.82235	0.93926	0.87513
	1.27205	1.32692			0.8221	0.93879	
F	1.22271	1.44339	1.28515	Ar	0.97015	1.04631	1.01537
	1.22279	1.44414			0.9702	1.04534	
Ne	1.44571	1.60612	1.50537	K	1.12581	1.17004	1.16221
	1.4457	1.60662			1.1259	1.17055	
Na	1.67955	1.79629	1.73776	Ca	0.41629	0.41631	0.43628
	1.6796	1.79721			0.4155	0.41631	

terms of normalized radial basis functions called Slater-type orbitals (STOs) [27, 31a]:

$$P_b(r) = \frac{1}{\sqrt{2n_b!}} (2\zeta_b)^{n_b+1/2} r^{n_b} e^{-\zeta_b r} \tag{1}$$

where ζ_b is a parameter, by the relation

$$P(a;\, r) = \sum_b C_{ab} P_b(r) \tag{2}$$

The same set of STOs is used for each value of l. Other types of basis functions may also be used [36].

Using the analytic form of the radial wave functions the expression for the total energy W, $2^8 5$, may be expressed in analytic form since the integrals I(a), $J(a, b)$, and $K(a, b)$ can be evaluated. The derivation of the Hartree-Fock equations is then accomplished by treating each of the C_{ab}'s in (2) as an unknown parameter and varying these parameters while requiring the individual orbitals $\phi(a)$ to be orthonormal.

The end result is a set of equations for the parameters C_{ab} (actually, a matrix representation of HF equations) which, like the numerical Hartree-Fock equations $2^8 16$, have solutions only for particular values of the parameters $\epsilon(a)$.

These equations can be solved numerically by standard matrix diagonalization procedures. But as in the case of the numerical method an iterative self-consistency procedure is required, because the coefficients of the C_{ab}'s depend on these parameters. Solutions of the Hartree-Fock equations must be obtained by numerical techniques using either the numerical or analytic methods. The advantage of using analytic methods is that various physical and chemical properties can be evaluated with great ease. On the other hand, because less than complete basis sets (a finite range for index b) must be used, the analytic form assumed for the radial basis functions represents a certain restriction, which imposes a limitation on the method. Therefore, the choice of good incomplete radial basis sets for a given problem is very important. It is usually accomplished by taking advantage of the results of previous calculations. Details of how calculations are performed using the analytic method are described by Roothaan and Bagus [31c]. There exist many analytic HF solutions based on either Nesbet's [28e] or Roothaan's [31a] analysis [35, 37]. Tables 3^8a and 3^8b give the parameters ζ_b and the expansion coefficients C_{ab} (proper vectors) of the functions (1) and (2), respectively, as taken from Clementi [35], for the terms of the normal configurations of the first-row elements.

In the analytic method it is relatively easy to optimize the basis sets for the light atoms or ions. But the basis functions for heavy atoms must be more extensive, and therefore, the method becomes unmanageable. Here numerical methods are considerably faster.

Table 3[8]a. Parameters ζ_b of the radial-basis functions for the terms of the normal configuration of the first-row elements

			Li	Be	B	C			N			O			F	Ne
l_a	n	b	^2S	^1S	^2P	^3P	^1D	^1S	^4S	^2D	^2P	^3P	^1D	^1S	^2P	^1S
0	1	1	2.4803	3.4703	4.4661	5.4125	5.4300	5.3842	6.4595	6.4730	6.5035	7.6160	7.6105	7.5333	8.5126	9.5735
		2	4.7071	6.3681	7.8500	9.2863	9.1500	9.0600	10.8389	10.9700	11.1318	13.3243	13.2681	12.7015	14.4130	15.4496
	2	3	0.3500	0.7516	0.8320	1.0311	1.2255	1.2100	1.4699	1.2745	1.2721	1.7582	1.7582	1.7098	1.8599	1.9550
		4	0.6600	0.9084	1.1565	1.5020	1.6142	1.5929	1.91613	1.8034	1.8038	2.5627	2.5630	2.4512	2.7056	2.8462
		5	1.0000	1.4236	1.91197	2.58975	2.6990	2.5964	3.1560	3.1159	3.0120	4.2832	4.2754	4.0761	4.9019	4.7746
		6	1.7350	2.7616	3.5213	4.2595	4.2131	4.2500	5.0338	5.0338	5.2338	5.9445	5.9445	5.9445	6.4440	7.7131
1	2	1			0.8783	0.9554	0.9372	1.1060	1.1937	1.0906	1.0580	1.1536	1.0626	1.0555	1.2655	1.4700
		2			1.3543	1.4209	1.4147	0.5074	1.7124	1.6446	1.6279	1.7960	1.7405	1.6985	2.0301	2.3717
		3			2.2296	2.5873	2.5545	2.3590	3.0112	3.0200	3.0243	3.4379	3.4198	3.3517	3.9106	4.4545
		4			5.3665	6.3438	6.3021	6.2000	7.1018	7.1650	7.2029	7.9070	7.8890	7.6690	8.6363	9.4550

Table 3[8]b. Proper vectors C_{ab} for the terms of the normal configuration of the first-row elements [as defined in (2)])

		Li	Be	B	C			N			O			F	Ne
a	b	^2S	^1S	^2P	^3P	^1D	^1S	^4S	^2D	^2P	^3P	^1D	^1S	^2P	^1S
1s	1	0.89778	0.91792	0.92109	0.92695	0.91965	0.91962	0.92787	0.93112	0.93232	0.93835	0.93800	0.93397	0.94205	0.93093
	2	0.11031	0.08742	0.07847	0.07665	0.07951	0.08620	0.06535	0.06148	0.05675	0.03825	0.03909	0.05014	0.04499	0.04610
	3	−0.00006	0.00147	−0.00036	0.00073	−0.00090	0.00068	−0.00067	−0.00033	−0.00020	−0.00097	−0.00097	−0.00071	−0.00010	−0.00085
	4	0.00109	−0.00267	0.00085	−0.00167	0.00220	−0.00137	0.00188	0.00122	0.00091	0.00439	0.00440	0.00313	0.00153	0.00321
	5	−0.00144	0.00222	0.00002	0.00539	−0.00043	0.00466	0.00017	0.00048	0.00095	−0.00829	−0.00820	−0.00316	−0.00076	−0.00180
	6	0.00937	0.00597	0.01146	0.00210	0.01067	−0.00019	0.01665	0.01732	0.02105	0.04171	0.04100	0.02883	0.02389	0.03537
2s	1	−0.14426	0.17065	−0.19401	−0.20786	−0.20727	−0.20999	−0.21744	−0.21838	−0.21658	−0.21979	−0.22033	−0.22210	−0.22924	−0.23171
	2	−0.01584	0.01469	−0.01243	−0.01175	−0.01235	−0.01333	−0.00843	−0.00809	−0.00841	−0.00573	−0.00588	−0.00738	−0.00534	−0.00442
	3	0.00202	−0.11551	0.01800	0.06494	0.23681	0.19885	0.27744	0.08811	0.08958	0.42123	0.41542	0.33802	0.27178	0.18201
	4	0.97901	−0.67835	0.71580	0.74109	0.61238	0.61835	0.54808	0.70830	0.68905	0.54368	0.54890	0.58983	0.65367	0.66106
	5	0.09077	−0.30265	0.38903	0.34626	0.31624	0.33657	0.33901	0.36978	0.36685	0.23061	0.23101	0.25681	0.33031	0.32372
	6	−0.08964	0.09232	−0.11030	−0.13208	−0.14709	−0.13361	−0.14212	−0.14451	−0.12646	−0.17856	−0.17837	−0.16495	−0.23130	−0.14244
2p	1			0.54005	0.24756	0.31917	0.68350	0.29731	0.25796	0.27164	0.16371	0.13324	0.15046	0.17003	0.22430
	2			0.38245	0.57774	0.50063	0.05446	0.48388	0.52425	0.51301	0.57600	0.60066	0.56887	0.55982	0.51826
	3			0.13208	0.23563	0.25045	0.35041	0.28079	0.28935	0.29307	0.33392	0.34397	0.36196	0.34875	0.33902
	4			0.00957	0.01090	0.01097	0.01237	0.01352	0.01260	0.01208	0.01495	0.01500	0.01715	0.01691	0.01765

Hartree [21b] gave the first detailed account of the numerical procedures for equations without exchange, $2^8$17. Later Hartree and Black [38] and also Hartree and Hartree [39] modified these for equations with exchange, $2^8$16. For many years these methods, sometimes with minor variations, were used for atomic structure calculations. An excellent survey of these computations is given by Hartree [21c]. With the availability of computers some new techniques (that is, some new modifications to Hartree's methods), also became feasible. These are described by Ridley [40] for Hartree equations, and by Piper [41], Froese [42], and Worsley [43] for HF equations. Commonly used computer programs [33, 44, 45, 46] are based on these modified Hartree methods.

Although the Hartree-Fock equations for any term of a given configuration may be written, the task of finding physically meaningful solutions to the equations using either the numerical or analytic method is not straightforward. The basic difficulty is that the variation method applied to the mean energy W in Section 2^8 and used to derive the Hartree-Fock equations assures us only that a solution represents an approximation to a proper-value of H and not necessarily that solution desired. Thus solutions of a given set of Hartree-Fock equations exist in general, which correspond to states other than for which they were derived. Because the method of solution involves iteration to self-consistency, it is possible for a numerical procedure to oscillate between alternative solutions and fail to converge.

The problems of instability of numerical solutions are especially severe for the case of excited states and has been discussed by Griffin, Cowan, and Andrew [47] and more recently by Froese Fisher [48].

For the ground configurations the problems of instability present a lesser problem. Convergence to the proper-state is obtained in ground state calculations by imposing additional constraints on the system. These constraints are: (1) each radial wave function, $P(n,l;\ r)$, is required to have $n - l - 1$ nodes, and (2) the energy parameters $-\epsilon(a)$ are constrained to be approximately equal to the binding energies of inner-shell electrons in accord with Koopmans's theorem. Whereas application of these criteria is usually adequate to assure convergence to the desired solution, they must be applied judiciously because they are violable. For example, extra nodes may appear in radial wave functions for open-shell systems because of the exchange forces, and the energy parameters $-\epsilon(a)$ may in some cases deviate considerably from known values of the binding energies.

Finally, because Hartree-Fock solutions are always obtained by numerical methods, so they are numerical approximations. Because calculations are normally carried out to only six- or seven-decimal figure accuracy, the accuracy of results can be no better than this and may be less because of rounding errors in the numerical processes used.

The quality, that is, the goodness, of solutions can be determined according to how well they satisfy the following two conditions:

1. the virial theorem, $V/KE = 2$, which indicates the accuracy of the method in including the non-diagonal energy parameters in the computation,
2. the cusp condition,

$$(l_a + 1)\left[\frac{f'(l_a; r)}{f(l_a; r)}\right]_{r\to 0} = -Z; \quad f(a; r) = \frac{P(a; r)}{r^{l+1}}, f' = \frac{df}{dr} \tag{3}$$

which is a measure of correct behavior of the radial function near the nucleus.

At this point some numerical comparisons of results obtained for ground state configurations by various Hartree-Fock calculations will be instructive. Tables 4^8a and 4^8b give the parameters ζ_b of (1) used in the basis sets for the normal configuration terms of the second-row elements in two different analytic HF solutions [35, 37p]. These tables reveal the previously mentioned fact that HF functions for the different terms of a configuration are different although the differences are usually slight. Note that the basis sets are different for each calculation. In spite of this the total energies and one electron energies obtained for the two calculations (Tables 5^8 and 6^8) agree to approximately six-figure accuracy. This illustrates the fact that there is no unique choice for the radial basis set. The different sets used should be viewed as alternative analytic representations of the radial orbitals of (1).

Table 5^8 compares the total energies obtained for the lowest terms of the ground configurations for elements $Z = 3$–36 as taken from the calculations of Clementi [35], Watson [37d], and Watson and Freeman [37e], all done using the analytic method with the total energies obtained by Froese [44a] using the numerical method.

Because the Hartree-Fock equations have been derived using the Rayleigh-Ritz variation method (8^2) subject to constraints as described in this chapter, the total energy obtained by the Hartree-Fock procedure represents a rigorous upper bound to the true total energy of the nonrelativistic Hamiltonian $2^8$1 for the normal states of atoms. By this criterion the numerical Hartree-Fock solutions of Froese provide a better or equivalent upper bound to the total energy in each case. This is what would be expected, because the analytic method imposes the additional restriction of finite basis sets for the radial orbitals. All calculations predict the same total energies to better than five significant figures and may thus be considered as alternative numerical approximations to the normal configurations of these elements. Nothing can be said about the relative accuracy of the wave functions and of physical parameters derived from them obtained by the alternative Hartree-Fock procedures.

Table 6^8 compares the one-electron energy parameters for the nor-

Table 4^8a. The parameters ζ_b of the radial-basis functions for the terms of the normal configuration of the second-row elements, as taken from Clementi [35].

			Na	Mg	Al		Si			P			S		Cl	Ar
l_a	n	b	^2S	^1S	^2P	^3P	^1D	^1S	^4S	^2D	^2P	^3P	^1D	^1S	^2P	^1S
0	1	1	11.0000	12.0000	13.0000	14.0000	14.0000	14.0000	15.0000	15.0000	15.0000	16.0000	16.0000	16.0000	17.0000	18.0000
	3	2	12.3685	13.5552	14.7112	15.9821	15.6200	15.3200	16.8783	17.0197	16.7110	18.1208	18.0248	18.0248	18.4649	21.2848
		3	8.0254	9.2489	9.9000	10.5969	11.2701	11.0000	11.4903	11.9378	11.8521	12.8168	12.8818	12.8818	13.8529	15.5021
		4	5.7059	6.5517	7.1088	7.7715	8.0722	8.3113	8.4471	8.8304	8.4741	9.6270	9.5644	9.5644	10.0968	11.2367
		5	3.6310	4.2008	4.7467	5.2394	5.2927	5.2592	5.7587	5.8237	5.7587	6.4124	6.4031	6.4031	6.9392	7.5066
		6	2.1537	2.4702	2.8768	2.9000	3.2246	2.8992	3.6000	3.5000	3.6123	3.7742	3.7743	4.2766	4.4364	4.7029
		7	1.1081	1.4331	1.9029	2.0892	2.0836	2.1369	2.4121	2.4137	2.3472	2.7095	2.8284	2.5949	2.9057	3.2138
		8	0.7083	0.8783	1.1140	1.2945	1.2945	1.2986	1.4900	1.4867	1.4683	1.6709	1.6943	1.6294	1.8190	1.9931
1	2	1	5.5000	6.0000	6.5000	7.0000	7.0000	7.0000	7.5000	7.5000	7.5000	8.0000	8.0000	8.0000	8.5000	9.0000
	4	2	8.3937	7.9884	10.1887	15.8461	15.8800	14.8826	13.6161	13.7400	12.4904	14.1259	14.0267	14.0722	15.0124	15.0000
		3	5.4206	5.3197	7.8527	10.5146	10.1729	14.3826	11.5855	12.0899	10.5823	11.5718	12.7661	11.7197	12.3257	11.9644
		4	3.5646	3.7168	5.9826	6.8346	7.1260	7.4766	7.6093	7.7224	7.4297	7.9171	8.0659	7.7864	8.3724	8.7924
		5	2.2833	2.5354	4.0417	4.7281	4.9147	4.9697	5.2743	5.3682	5.1871	5.6039	5.7427	5.5376	6.1092	6.3011
		6	—	—	1.9559	2.3010	2.2635	2.2557	2.6270	2.6127	2.5512	2.8906	2.8891	2.8274	3.1931	3.4327
		7	—	—	1.1918	1.3013	1.2248	1.1365	1.4963	1.4515	1.3964	1.6275	1.5878	1.5339	1.7863	1.9409
		8	—	—	0.8630	0.7734	0.6000	—	0.7583	0.7865	0.6928	0.8665	0.6961	0.6771	0.9293	1.0309

Table 4^8b. The parameters ζ_b of the radial-basis functions for the terms of the normal configuration of the second-row elements, as taken from Malli [37p].

			Na	Mg	Al	Si			P			S			Cl	Ar
l_a	n	b	^2S	^1S	^2P	^3P	^1D	^1S	^4S	^2D	^2P	^3P	^1D	^1S	^2P	^1S
0	1	1	11.000	12.000	13.000	14.000	14.000	14.000	15.000	15.000	15.000	16.000	16.000	16.000	17.000	18.000
	3	2	12.600	23.000	14.634	16.080	16.080	16.080	17.240	17.240	17.240	18.400	18.400	18.400	19.560	20.720
		3	9.400	13.760	9.900	11.920	11.920	11.920	12.760	12.760	12.760	13.600	13.600	13.600	14.440	15.280
		4	5.700	10.240	6.914	7.596	7.551	7.665	8.389	8.322	8.415	9.071	9.068	8.863	9.943	10.006
		5	3.400	6.261	4.602	4.922	4.875	4.947	5.491	5.407	5.506	6.020	6.026	5.932	6.610	6.849
		6	1.15875	3.898	2.877	2.049	2.075	2.028	2.325	2.325	2.326	2.618	2.608	2.619	2.840	3.200
		7	0.72201	1.455	1.742	1.287	1.306	1.277	1.465	1.464	1.466	1.649	1.643	1.650	1.795	2.008
		8	—	0.886	1.086	—	—	—	—	—	—	—	—	—	0.900	—
1	2	1	5.500	6.000	6.500	7.000	7.000	7.000	7.500	7.500	7.500	8.000	8.000	8.000	8.500	9.000
	4	2	9.000	9.710	9.382	15.000	15.500	15.024	17.013	16.017	16.972	21.116	17.979	19.250	20.000	15.554
		3	5.4495	5.993	8.387	8.344	8.500	8.342	9.074	8.885	9.054	9.5458	9.4128	9.453	10.2193	9.879
		4	3.503	3.945	5.857	5.612	5.675	5.612	6.086	6.047	6.095	6.424	6.433	6.429	7.06286	7.830
		5	2.267	2.441	3.880	3.488	3.488	3.444	3.512	3.445	3.435	3.5445	3.4597	3.512	3.992	5.969
		6	—	—	1.801	2.120	2.143	2.116	2.248	2.276	2.203	2.170	2.131	2.126	2.4904	3.389
		7	—	—	0.987	1.260	1.253	1.210	1.402	1.392	1.337	1.3455	1.3399	1.288	1.5667	1.976
		8	—	—	—	0.737	0.737	0.731	0.800	0.861	0.825	—	—	—	1.1204	1.155

Table 5[8]. Comparison of the total energies (a.u.) for the lowest-lying term (except for Cu) of the normal configuration of the elements with $Z = 3 - 36$, obtained by the analytic [35, 37d, e, p] and numerical [44a, 46] methods. The first entry of the analytic method is taken from Clementi [35], second from Watson [37d] or Watson and Freeman [37e], and the third from Malli [37p]. The first entry of the numerical method is taken from Froese Fischer [44a] and the second from Fraga, Karwowski, and Saxena 46].

	Method				
Z	Analytic			Numerical	
3, Li (^2S)	−7.4327	—	—	−7.4328	−7.4327
4, Be (^1S)	−14.5730	—	—	−14.5731	−14.5730
5, B (^2P)	−24.5290	—	—	−24.5291	−24.5291
6, C (^3P)	−37.6886	—	—	−37.6888	−37.6887
7, N (^4S)	−54.4009	—	—	−54.4011	−54.4010
8, O (^3P)	−74.8094	—	—	−74.8096	−74.8095
9, F (^2P)	−99.4093	—	—	−99.4096	−99.4094
10, Ne (^1S)	−128.5470	—	—	−128.5474	−128.5472
11, Na (^2S)	−161.8589	—	−161.8599	−161.8599	−161.8591
12, Mg (^1S)	−199.6146	—	−199.6146	−199.6154	−199.6145
13, Al (^2P)	−241.8766	−241.8692	−241.8766	−241.8775	−241.8768
14, Si (^3P)	−288.8543	−288.8536	−288.8543	−288.8550	−288.8544
15, P (^4S)	−340.7187	−340.7177	−340.7187	−340.7192	−340.7188
16, S (^3P)	−397.5047	−397.5031	−397.5047	−397.5061	−397.5050
17, Cl (^2P)	−459.4819	−459.4797	−459.4820	−459.4829	−459.4822
18, Ar (^1S)	−526.8173	−526.8140	−526.8174	−526.8189	−526.8178
19, K (^2S)	−599.1645	—	—	−599.1655	−599.1648
20, Ca (^1S)	−676.7580	—	—	−676.7595	−676.7580
21, Sc (^2D)	−759.7355	—	—	−759.7376	−759.7359
22, Ti (^3F)	−848.4052	−848.4011	—	−848.4090	−848.4059
23, V (^4F)	−942.8829	−942.8803	—	−942.8856	−942.8846
24, Cr (^5D)	−1043.3061	−1043.3036	—	−1043.313	−1043.310
25. Mn (^6S)	−1149.8651	−1149.8620	—	−1149.868	−1149.866
26, Fe (^5D)	−1262.4425	−1262.4367	—	−1262.444	−1262.444
27, Co (^4F)	−1381.4136	−1381.4045	—	−1381.415	−1381.415
28, Ni (^3F)	−1506.8690	−1506.8569	—	−1506.885	−1506.872
29, Cu (^2D)	−1638.9491	−1638.9310	—	−1638.954	−1638.951
30, Zn (^1S)	−1777.8470	−1777.8234	—	−1777.847	−1777.849
31, Ga (^2P)	−1923.2602	—	—	−1923.265	−1923.271
32, Ge (^3P)	−2075.3590	—	—	−2075.372	−2075.360
33, As (^4S)	−2234.2382	—	—	−2234.247	−2234.236
34, Se (^3P)	−2399.8669	—	—	−2399.875	−2399.866
35, Br (^2P)	−2572.4406	—	—	−2572.445	−2572.443
36, Kr (^1S)	−2752.0543	—	—	−2752.058	−2752.054

Table 6[8]. Negative of the one-electron energy parameters (a.u.), $-\epsilon(a)$, for the lowest term of the normal configuration of the second-row elements. The first entry is taken from Clementi [35], the second from Watson and Freeman [37e], the third from Malli [37p], and the fourth from Froese [44a].

Z / $-\epsilon(a)$	11, Na (^2S)	12, Mg (^1S)	13, Al (^2P)	14, Si (^3P)	15, P (^4S)	16, S (^3P)	17, Cl (^2P)	18, Ar (^1S)
$-\epsilon(1s)$	40.47849	49.03165	58.50128	68.81244	79.96982	92.00461	104.88465	118.61064
	—	—	58.4880	68.7954	79.9553	91.9923	104.8766	118.606
	40.47840	49.03164	58.50095	68.81240	79.96972	92.00462	104.88447	118.61025
	40.47875	49.03201	58.50130	68.81265	79.96995	92.00475	104.88465	118.61080
$-\epsilon(2s)$	2.79702	3.76768	4.91094	6.15659	7.51119	9.00446	10.60775	12.32244
	—	—	4.9069	6.1501	7.5062	8.9996	10.6040	12.319
	2.79697	3.76768	4.91060	6.15651	7.51109	9.00442	10.60754	12.32209
	2.79706	3.76774	4.91068	6.15653	7.51108	9.00431	10.60748	12.32216
$-\epsilon(2p)$	1.51813	2.28219	3.21858	4.25611	5.40105	6.68268	8.07250	9.57176
	—	—	3.2145	4.2500	5.3963	6.6780	8.0688	9.568
	1.51808	2.28217	3.21823	4.25604	5.40095	6.68262	8.07231	9.57142
	1.51816	2.28222	3.21830	4.25603	5.40093	6.68251	8.07220	9.57144
$-\epsilon(3s)$	0.18211	0.25304	0.39348	0.53991	0.69645	0.87963	1.07310	1.27752
	—	—	0.3940	0.5389	0.6955	0.8785	1.0717	1.276
	0.18209	0.25305	0.39334	0.53985	0.69639	0.87956	1.07288	1.27731
	0.18210	0.25305	0.39342	0.53985	0.69641	0.87953	1.07291	1.27736
$-\epsilon(3p)$	—	—	0.21017	0.29709	0.39170	0.43739	0.50652	0.59116
	—	—	0.1990	0.2965	0.3911	0.4363	0.5051	0.589
	—	—	0.20992	0.29712	0.39169	0.43739	0.50637	0.59096
	—	—	0.20995	0.29711	0.39171	0.43736	0.50640	0.59102

mal states of second-row elements obtained from the same four calculations. The table illustrates the fact that agreement between alternative calculations is relatively good for deep shells but becomes poorer for less tightly bound electrons. Again this is expected; the inner electrons make the major contribution to the total energy, which is what is being minimized in the Hartree-Fock procedure.

5⁸. Approximations to the Hartree-Fock method

Even though the HF procedure is the most accurate method, in the single-configuration approximation (no correlation), of computing the total energies of free atoms or ions, it is often desirable to employ less accurate methods to reduce numerical complexities. Wave functions and energy parameters obtained in this way may also be used as starting values for solving the HF equations.

Slater [49] developed a method by which the nonlocal potential can be approximated by a local potential

$$V_{XS}(r) = -k_x\left[\frac{3}{\pi}\rho(r)\right]^{1/3}, \quad k_x = \frac{3}{2}, \quad \text{(in a.u.)} \tag{1}$$

where ρ is the local electron density in the atom at distance r,

$$\rho(r) = \frac{1}{4\pi r^2}\sum_{a=1}^{N}|P(a;\, r)|^2 \tag{2}$$

This potential results from equating the exchange energy to that of an electron in a zero-temperature uniform free electron gas which is assumed to be represented locally by (2). Essentially the same approximation is used in treating exchange in the Thomas-Fermi-Dirac equation (1^8). The set of Hartree-Fock-Slater (HFS) equations, obtained by substituting the potential (1) in the HF equations for the exchange potential, are as simple to solve as the Hartree equations.

Hartree noted [21d, p. 61] that the HFS wave functions would agree better with the HF ones if the coefficient $k_x = 3/2$ in (1) were made smaller. Later Gáspár [50] and Kohn and Sham [51] showed that when one-electron radial HFS equations are derived by applying the variational principle to minimize the total energy, obtained by using the free-electron exchange approximation, the coefficient k_x in (1) becomes unity. This question was studied in detail by Cowan [52].

The central field potential $V(r)$ can be expressed in terms of the local electron density $\rho(r)$ as

$$V(r) = -\frac{Z}{r} + \frac{1}{r}\int_0^r \sum_a q_a|P(a;\, s)|^2\, ds + \int_r^\infty \frac{1}{s}\sum_a q_a\, |P(a;\, s)|^2\, ds$$
$$- k_x\left[\frac{3}{4\pi^2 r^2}\sum_a q_a|P(a;\, r)|^2\right]^{1/3} \tag{3}$$

which is approximation to the potentials used in the Hartree-Fock radial equations 2^816. The nonlocal terms corresponding to 2^815b have been replaced by the effective exchange potential given by (1), and the terms corresponding to 2^815a have been dropped. Because of the incorrect asymptotic behavior of the exchange part, this potential approaches zero exponentially at large distances from the nucleus, that is, faster than $1/r$. Herman and Skillman [53] adopted the same method used by Latter [7] in treating the Thomas Fermi model to obtain a realistic potential at large radial distances; that is,

$$\begin{aligned} V_{XL}(r) &= V_{XS}(r) \quad \text{for} \quad 0 < r < r_0 \\ &= -\frac{C}{r} \quad \text{for} \quad r_0 < r \end{aligned} \tag{4}$$

as in Equation 1^813. Thus the potential (3) is cut off at its proper asymptotic value, and a Coulomb tail is substituted.

The HFS radial-wave functions $P(a;\, r)$ are solutions to the one-electron HFS equations

$$[-(1/2)\nabla^2 + V(r)]\, P(a;\, r) = \epsilon(a) P(a;\, r)$$

where the central-field potential $V(r)$ is given either by (3) or by (4). The approximations made reduce the Hartree-Fock equations, 2^816, to a system in which each radial orbital $P(a;\, r)$ is subject to the *same* local potential $V(r)$. The set of equations must, however, still be solved self-consistently for the proper values of the binding energies, because $V(r)$ depends on the radial functions $P(a;\, r)$. The orbitals $P(a;\, r)$ and $P(b;\, r)$ with $l_a = l_b$ will be automatically orthogonal, because they satisfy the same radial equation. The energy $w(a)$ of an electron a and the total energy of the system, respectively, are

$$w(a) = I(a) + \sum_b (q_b - \delta_{ab})\, [J(a, b) - K(a, b)]$$

$$W = \sum_a q_a I(a) + \sum_{\text{pairs}} [J(a, b) - K(a, b)]$$

The energy $w(a)$ can be interpreted as the negative of the binding energy of electron a provided that, when electron a is removed, the spin orbitals for the other electrons remain unchanged. Here the one-electron energy parameters $\epsilon(a)$, of the HFS scheme, cannot be taken as ionization potentials, because $w(a) \neq \epsilon(a)$. Therefore, a correction to the energy parameter $\epsilon(a)$ given by

$$\begin{aligned} w(a) &= \epsilon(a) + \delta\epsilon(a) \\ \delta\epsilon(a) &= -\left\langle a \left| \frac{Z}{r} + V(r) \right| a \right\rangle + \sum_b (q_b - \delta_{ab})\, [J(a, b) - K(a, b)] \end{aligned} \tag{5}$$

is needed. Then the total energy becomes

$$W = W' + (1/2) \sum_a q_a \delta\epsilon(a) \tag{6a}$$

Table 7[8]. Ionization potentials, $IP(a) = -w(a)$, and total energy (a.u.) for the normal configuration of neutral Ar, obtained by various methods [56a]: HF, HFS, HFS′ (HFS stands for the method in which the potential $V(r)$ of (3) with $k_x = 3/2$ is used; in HFS′ $k_x = 1$).

	HF	HFS	HFS′	Observed*
$IP(1s)$	118.606	117.830	118.678	117.85
$IP(2s)$	12.321	11.901	12.345	12.15
$IP(2p)$	9.571	9.125	9.607	9.15
$IP(3s)$	1.277	1.146	1.292	1.074
$IP(3p)$	0.591	0.473	0.603	0.583
$-W$	526.818	526.679	526.795	529.31

*Measured from the center of gravity of the configuration (found by using the observed values [57]).

in which

$$W' = \sum_a q_a \epsilon(a) - (1/2) \sum_a q_a \left\langle a \left| \frac{Z}{r} + V(r) \right| a \right\rangle \tag{6b}$$

The following example illustrates the magnitude of this correction term (in a.u.)

	Al	K	Cu
W'_{HFS} =	−240.271	−596.571	−1634.922
W_{HFS} =	−241.789	−599.027	−1638.705
$W_{\mathrm{HF}} = W'_{\mathrm{HF}}$ =	−241.876	−599.164	−1638.961

where W'_{HFS} and W_{HFS} are computed by Snow, Canfield and Waber [54], and by Lindgren [55], respectively, by using Herman-Skillman [53] values. The HF value for the total energy, $W_{\mathrm{HF}} = W'_{\mathrm{HF}}$, is taken from Clementi's calculations [37k].

The effect of the different suggested values of the coefficient k_x, in (3), on the ionization potentials $-w(a)$ was investigated by Cowan, Larson, Liberman, Mann, and Waber [56a] and by Rosén and Lindgren [56b]. The first group concluded that for neutral Ar $k_x=1$ produces better agreement with the HF results, whereas Rosén and Lindgren found that the optimum value of k_x depends on Z, approaching unity (Gáspár-Kohn-Sham value) for large Z. Because of the statistical nature of the Slater exchange potential, for light elements ($Z < 10$) the HFS approximation is less justified. Comparison of the ionization potentials, $-w(a)$, for neutral Ar, as obtained by using two different values of k_x in the exchange potential, is given in Table 7[8].

Various elaborations of the statistical exchange approximation have been introduced. Lindgren [55] used three adjustable parameters A, n,

m and represented the exchange potential $V_X(r)$ as:

$$-rV_X(r) = \frac{3}{2}A\left[\frac{3}{4\pi^2}\right]^{1/3} r^{n/3}\,[4\pi r^2\rho(r)]^{m/3} \tag{7}$$

The parameters are determined by varying them until the minimum value for the total energy is reached. Such calculations for neutral Al, K, and Cu give the respective sets $A = 0.84$, $n = 1.20$, $m = 1$; $A = 0.72$, $n = 1.01$, $m = 1$; $A = 0.85$, $n = 1.24$, $m = 1$, indicating that only the parameters A and n are needed.

The virial theorem is not satisfied when the exchange potential (1) is used [58]. Berrondo and Goscinski [59] multiplied V_{XS} by a parameter α and varied the parameter to satisfy the virial theorem. Potentials obtained by this scheme, which is usually called the $X\alpha$ method, produce results that are closer to HF. Wood [60] computed the ionization potentials and total energies for the iron group elements by finding the optimum values of α. His results agree well with the CGHF computations of Mann [33b].

Cowan [52] introduced an exchange potential V_{XH}, by modifying (3), such that the potential vanishes for any single electron, and, since the exchange arises only for parallel spin, for two-electron shells ns^2 also. Then the potential V_{XH} depends on orbital label and has the form

$$V_{XH}(a; r) = \frac{1}{r}\int_0^r 4\pi s^2\rho(a; s)\,ds + \int_r^\infty 4\pi s\rho(a; s)\,ds + k_1\left(\frac{\rho_{XH}}{\rho}\right)^{k_0}\left(\frac{3\rho}{\pi}\right)^{1/3} \times \frac{\rho_{XH}}{\{\rho_{XH} + [k_2/(n-l)]\}} f(r) \tag{8}$$

in which

$$\rho_{XH} = \rho(r) - \rho(a; r) \quad \text{for a single electron}$$
$$= \rho(r) - 2\rho(a; r) \quad \text{for the } ns^2 \text{ shell}$$
$$\rho(a; r) = \frac{1}{4\pi r^2}\left|P(a; r)\right|^2$$

and

$$f(r) = 1 \qquad \text{for} \quad r \geqslant r_0$$
$$= 1 + k_3\left(1 - \frac{r}{r_0}\right) \qquad \text{for} \quad r < r_0$$

Here r_0 is the location of the kth node of the radial wave function; k is the number of orbitals with $l = l_a$ and $n < n_a$. In most cases the factors $\rho_{XH}/[\rho_{XH} + k_2/(n-l)]$ and $f(r)$ in (8) are very close to unity. The constants k_0, k_1, k_2, k_3 are determined empirically to satisfy the conditions required. Cowan chose as optimum values $k_0 = 1$, $k_1 = 0.7$, $k_2 = 0.5$, $k_3 = 0.7$. Thus the central-field potential (3) to be used in the

Table 8[8]. Ionization energies, $IP(a) = -w(a)$, and total energy (a.u.) for the normal and the $3p^5 3d$-excited configurations of neutral Ar derived by various methods [52]: HF, HFS_L, HX (HFS_L is the method in which $V_{XL}(r)$ of (4) with $k_x = 1$, is used, HX is the method where $V_{XH}(a; r)$ of (8) is employed).

Conf.:	$3p^6$			$3p^5 3d$		
Method:	HF	HFS_L	HX	HF	HFS_L	HX
$IP(1s)$	118.606	118.775	118.415	119.004	119.059	118.808
$IP(2s)$	12.321	12.433	12.241	12.706	12.713	12.623
$IP(2p)$	9.571	9.696	9.492	9.953	9.974	9.872
$IP(3s)$	1.277	1.341	1.270	1.581	1.578	1.569
$IP(3p)$	0.591	0.639	0.585	0.915	0.911	0.906
$IP(3d)$				0.058	0.058	0.058
$-W$	526.818	526.768	526.839	526.332	526.311	526.353

radial HFS equations is different for each orbital (shell) and given by

$$V(a; r) = -\frac{Z}{r} + \frac{1}{r}\int_0^r 4\pi s^2 \rho(s)\, ds + \int_r^\infty 4\pi s \rho(s)\, ds - V_{XH}(a, r) \qquad (9)$$

A comparison of the ionization potentials obtained for neutral Ar in the ground and $3p^5 3d$ configurations by this method and by others is given in Table 8[8].

Herman, Van Dyke, and Ortenburger [61] developed an improved expression for the free-electron exchange by adding an inhomogeneity correction term, similar to that of Weizsäcker [12] for the statistical model, given by

$$\beta\left[2\left(\frac{\nabla\rho}{\rho}\right)^2 - 3\frac{\nabla^2\rho}{\rho}\right]\left(\frac{3}{\pi\rho}\right)^{1/3}$$

Consequently, the exchange potential is

$$V_{X\alpha\beta}(r) = \alpha V_{XS}(r) - \beta\left[2\left(\frac{\nabla\rho}{\rho}\right)^2 - 3\frac{\nabla^2\rho}{\rho}\right]\left(\frac{3}{\pi\rho}\right)^{1/3} \qquad (10)$$

in which the optimum value of α is found to be 2/3 for all atoms, and the optimum value of β differs slightly from one atom to another. That is, in this scheme (the $X\alpha\beta$ method) β is the only parameter to be optimized. The optimized values of β for neutral Ar, Kr, and Xe are found to be 0.004, 0.005, and, 0.005 respectively. Because of the presence of the $\nabla^2\rho/\rho^{4/3}$ term in (10) the exchange potential behaves like r^{-1} at $r = 0$. Therefore, special care must be taken in the evaluation of wave functions with $l = 0$ at the nucleus. The total energies obtained

Table 9[8]. One-electron energy parameters and total energies (a.u.) for the ^2P and ^4P terms of the $1s^22s^22p^43p$-excited configuration of Ne^+. The first entry is obtained by an approximate HF method [66b] and the second entry by an exact HF method [67].

	$(^3P)^2P$	$(^1D)^2P$	$(^1S)^2P$	$(^3P)^4P$
$\epsilon(2p)$	−4.16203	−4.16203	−4.16203	−4.16203
	−4.43914	−4.32652	−4.14612	−4.40064
$\epsilon(3p)$	−0.69866	−0.69866	−0.69866	−0.69866
	−0.70154	−0.70954	−0.71874	−0.75101
W	−126.7144	−126.6007	−126.4201	−126.7422
	−126.7163	−126.6001	−126.4200	−126.7474

by the $X\alpha\beta$ method agree with HF energies better than the HFS, with $k_x = 1$ and $k_x = 3/2$, and the $X\alpha$ methods [61].

Kmetko [62] investigated the behavior of the optimum values of α, in the $X\alpha$ method, for elements with $Z = 1 - 100$, and concluded that:

1. Addition of electrons with $l = 0$ and $l = 1$ causes the value of α to decrease, electrons with $l = 0$ having the stronger effect.
2. Addition of electrons with $l = 2$ and $l = 3$ causes the value of α to increase.
3. α has its largest value for $Z = 2$ (He), $\alpha = 1.004$, and approaches 2/3 as Z becomes larger.
4. Omission of the added Coulomb tail (Latter correction) invariably results in lower α values.

In the self-consistent energy band calculations for transition metals, the occupation numbers, which produce the energy bands in agreement with the observed Fermi surfaces, come out to be nonintegers [63]. Following this idea Slater, Mann, Wilson, and Wood [32a] introduced the use of fractional occupation numbers, n_a, in atomic systems. They specifically considered configurations of type $3d^{n+x}\ 4s^{2-x}$ in which x is treated as a variable. They also defined the modified one-electron energy $\epsilon'(a)$ as

$$\epsilon'(a) = \frac{\partial W_{\mathrm{Av}}}{\partial n_a} \tag{11}$$

where n_a is $n + x$ or $2 - x$. For the configurations under investigation the condition for minimum energy, with respect to the variation of x, turns out to be $\epsilon'(n + a) = \epsilon'(2 - x)$. The subject of the fractional occupation numbers has been further studied by Slater and Wood [64] and by Slater [65].

Finally, Jucys and his collaborators [66] suggested an approximation to the HF method that is independent of the coupling scheme, that is, no L and S dependence as is also the case in the CGHF method. The approximation amounts to dropping the terms, in the total energy, which are attributable to the nonspherical part of

1. the direct interaction between electrons of the same open shell (that is, all $F^k(b, b)$ integrals with $k > 0$, where b denotes an open shell)
2. the direct interaction between electrons of different open shells (that is, all $F^k(b, c)$ integrals with $k > 0$, where b and c are open shells)
3. the exchange interaction between electrons of different open shells (that is, all $G^k(b, c)$ integrals, where b and c denote open shells)

Then the quantity W' to be varied reduces to

$$\begin{aligned} W' &= \sum_a q_a I(a) + \sum_a \frac{1}{2} q_a(q_a - 1)F^0(a, a) + \sum_{a,b \neq a} q_a q_b F^0(a, b) \\ &- \frac{q_a}{Q_a} \sum_{a,k\neq 0} \langle l_a \| C^{(k)} \| l_a \rangle^2 F^k(a, a) - \frac{q_a}{Q_a} \sum_{a,b\neq a} \sum_k 2\frac{q_b}{Q_b} \langle l_a \| C^{(k)} \| l_b \rangle^2 G^k(a, b) \\ &- \sum_{n_a, n_b, l} \epsilon_{n_a l, n_b l} \int_0^\infty P(n_a l; r) P(n_b l; r)\, dr \end{aligned} \qquad (12)$$

which is equal to the HF expression ($2^8 10$) if a, b shells are all closed; that is, $q_a = Q_a$. It is also exact for the case where all a shells are closed and $q_b = 1$. The deviation increases with increasing number of open-shell electrons.

In Table 9^8 the one-electron and total energies obtained by this approximation [66b] are compared with the ones of the HF method [67] for the ${}^2P$ and ${}^4P$ terms of the $1s^2 2s^2 2p^4 3p$-excited configuration of Ne^+.

9. Configuration structure

> The theoretical treatment of many-electron configurations is generally based on the assumption that the electron may be regarded to a first approximation as moving independently in a *central field* which arises as a net result of the positive charge of the nucleus and the average distribution of the negative charges of the electrons. The state of an atom is then characterized by assigning a definite n and l to each electron. Although the validity of this *electron-configuration* concept is limited, it gives a useful starting point for the description."
>
> Bengt Edlén, *Encyclopedia of Physics,* Vol. 27 (Berlin: Springer-Verlag, 1964), p. 99.

The aim of any theory is to predict observed quantities correctly and consistently. Although the accuracy of a prediction can give a fair idea about the validity of the theory and the approximations involved, one must be aware that the agreement may have come about through the cancellation of errors or merely by chance. For example, if a crude method gives a result that compares more closely with observation than a more sophisticated method, some explanation is needed. Otherwise the theory comes into question.

In this chapter, results obtained with several approximations based on the central-field theory will be examined by comparing them with observation. Because the magnetic effects, such as spin-orbit, spin-spin, spin-other orbit, have not yet been introduced, the discussion will be limited to term energies that represent averages over term splittings arising from these effects.

In all theoretical predictions, the approximations involved in obtaining them, described in the previous chapter, must be considered. For the sake of uniformity and simplicity, all equations and data, theoretical as well as observed, will be given in atomic units in this chapter.

1⁹. Observed regularities

Let us begin with a study of observed regularities. The large body of spectral data available on free atoms and ions when classified in an orderly fashion provides a great deal of information on the underlying atomic structure. The regularities that emerge from the data may be considered in various ways; it is thus possible to study the data as both nuclear charge and the total number of electrons is increased (along periods) or as a function of nuclear charge for a constant number of electrons (along isoelectronic sequences). Alternatively, information may be obtained by comparing spectra of neighboring elements in the periodic table. Examples of this type of analysis are given here. A more complete study of observed regularities may be found in the review article of Edlén [1].

The concept of term values, defined in $8^1 1$, plays a central role in experimental spectroscopy. What is measured experimentally are wavelengths that are inversely proportional to energy differences, so term values must be referred to some reference point. The common practice as in Moore's tables (8^4) is to take the ground (normal) state energy as the zero point; that is, $T = [W(\text{normal state}) - W(t)]/hc$. Terms defined in this way are called *relative term values*. Alternatively *absolute term values* may be defined by taking the ionization limit as the reference point; that is, $T = -W(t)/hc$, which has the advantage that all term values, including the normal state, may then be interpreted as binding energies.

For atoms and ions containing more than two open-shell electrons there are usually several ionization limits because of the genealogical character of the LS-coupled terms (Sec. 5^4). For example, the excited terms of the oxygen atom are characterized by the four quantum numbers L', S', L, S, where L' and S' belong to the ${}^4S$, ${}^2D$, ${}^2P$ parent terms. When an oxygen atom loses one electron, the ion can be in any one of these three terms. Because the energies of these terms are all different, so are the ionization limits for the terms of oxygen characterized by them. This is shown in Figure 1^9. The spectrum is the sum of three core plus one-electron spectra, each of which converges to one of the three terms of the $2p^3$ configuration. The series (fixed l and increasing n) characterized by the lowest parent term is called the first series, and its limit the first ionization limit. All terms and ionization limits considered below are of this kind.

The behavior of terms along periods indicates the order in which electron shells are being filled. In Figures 2^9a and 2^9b the binding energies of the ground and excited terms for the neutral elements with $Z = 2$–11 and $Z = 10$–19, respectively, are plotted. Where there is more than one term for a specified configuration, the binding energies of the highest and lowest terms are indicated. The energy scale is determined by the term energies of hydrogen, identified by the value of the principal quantum number n. The Z scale is arbitrary. Figures 2^9a and 2^9b show the filling of the $2s$ and $2p$ and of the $3s$ and $3p$ subshells with increasing Z by the rapid increase in binding energy of terms corresponding to electrons in these subshells. Physically this rapid increase in binding energy can be interpreted as caused by a change in

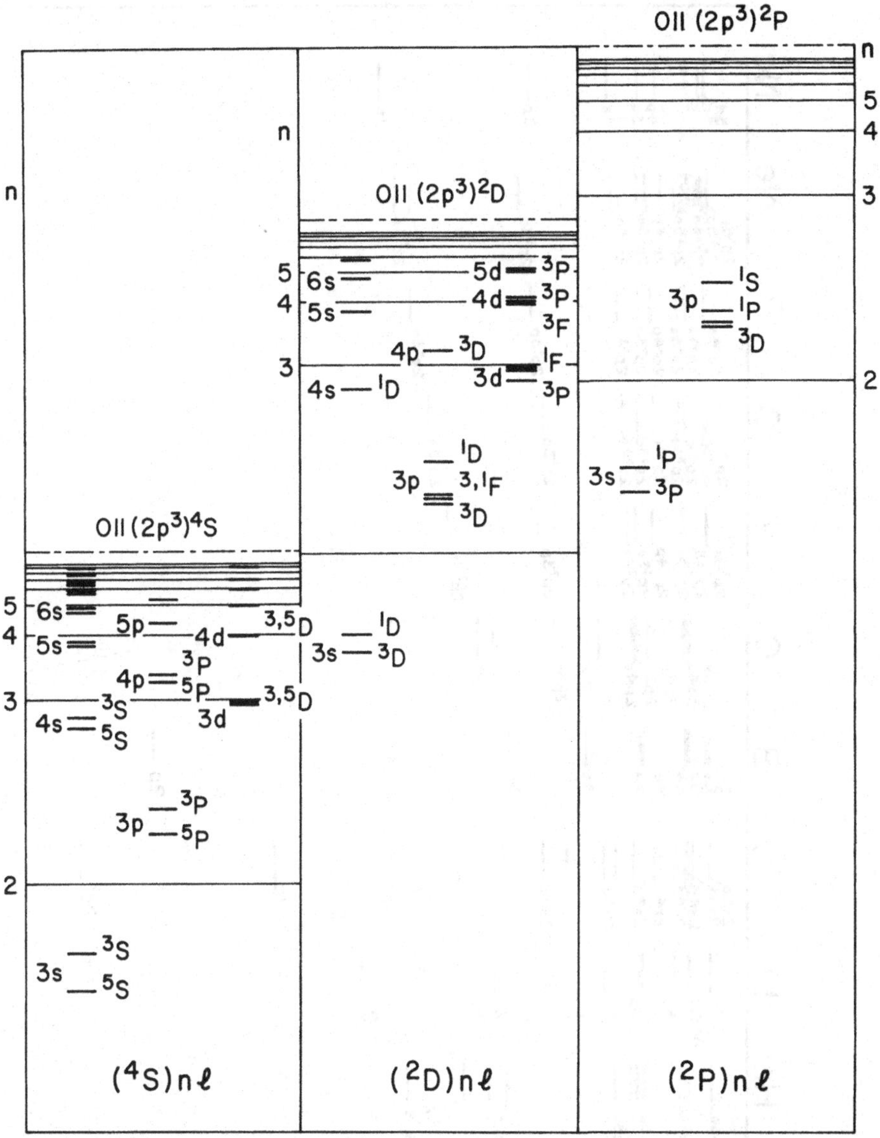

Figure 1[9]. The observed term energies of neutral oxygen, illustrating the concept of parentage and the different ionization limit associated with them. The term energies of the hydrogen atom, specified by the quantum number n, are taken as the energy scale.

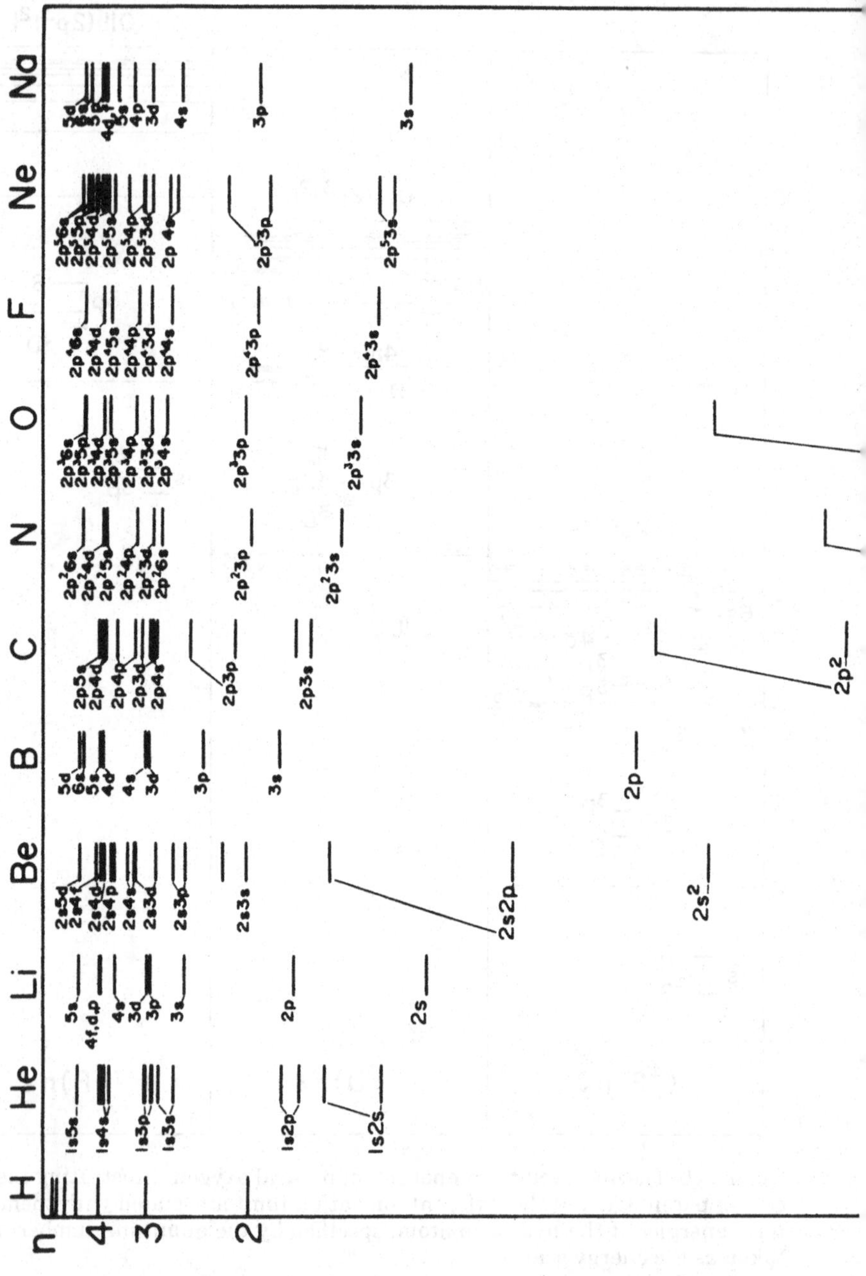

Figure 2[9]a. The observed energies of the lowest and highest terms of the normal and excited configurations for the elements with Z = 2–11. When the highest term is not observed, or lies above the first ionization limit, only the

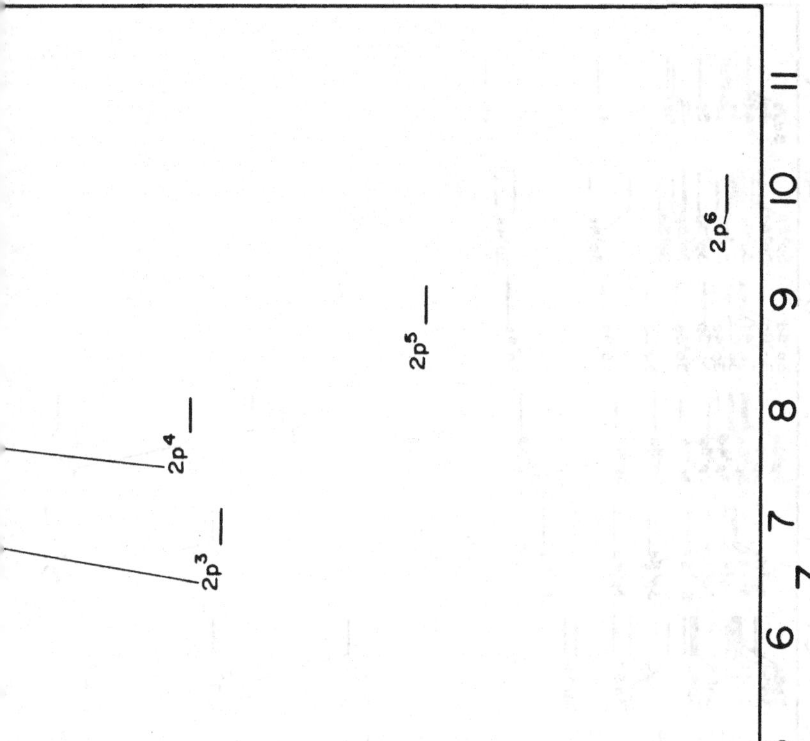

lowest term is shown. The Z scale is linear and chosen arbitrarily. The term energies of the hydrogen atom, specified by the quantum number n, are taken as the energy scale.

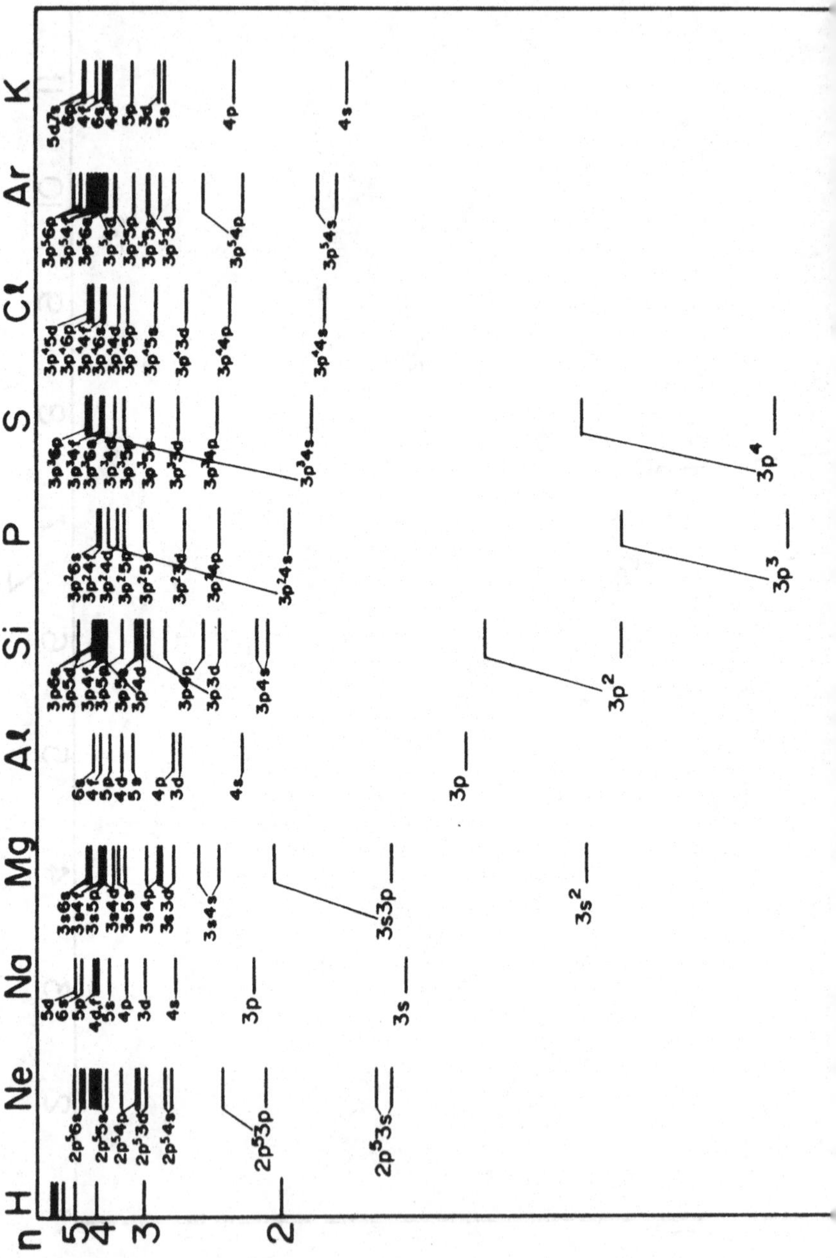

Figure 2⁹b. The observed energies of the lowest and highest terms of the normal and excited configurations for the elements with Z = 10–19. When the highest term either is not observed or lies above the first ionization limit, only

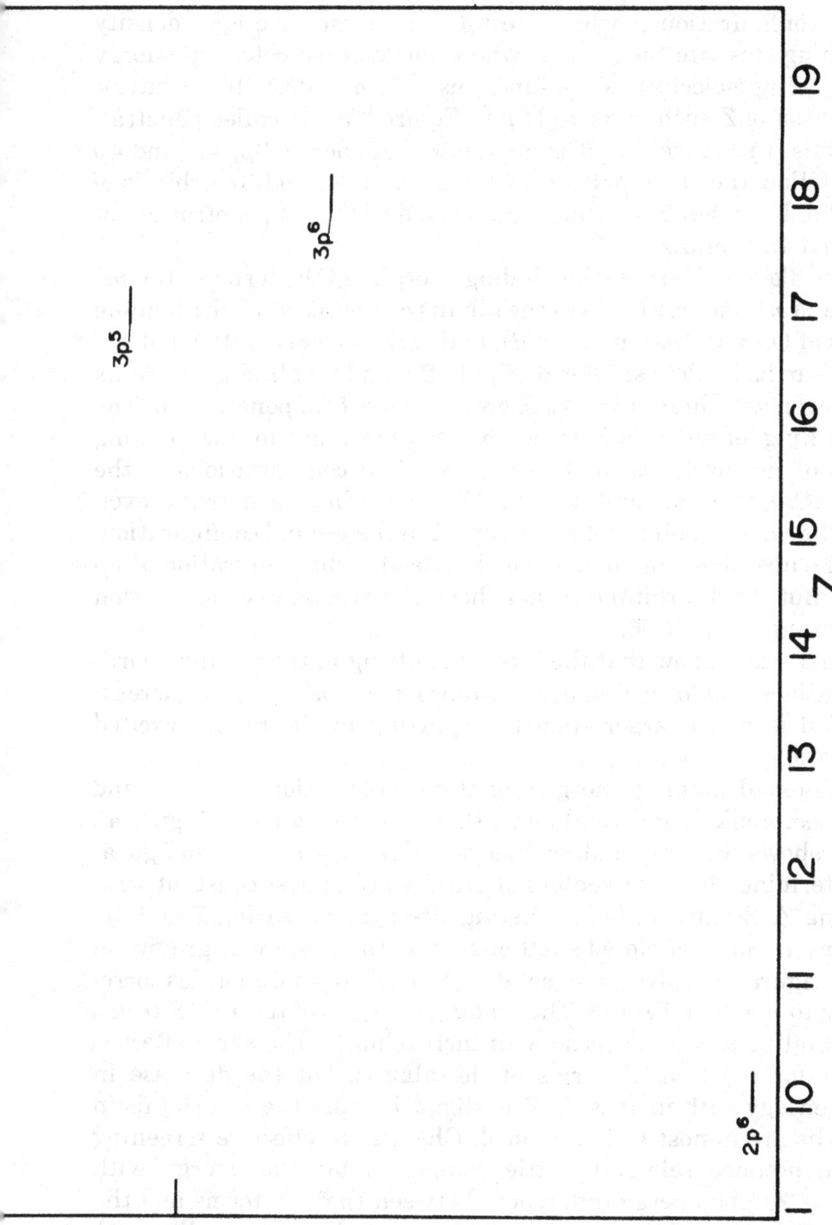

the lowest term is given. The Z scale is linear and chosen arbitrarily. The term energies of the hydrogen atom, specified by the quantum number n, are taken as the energy scale.

the radial distribution of the wave functions of electrons in the outer or valence configurations, which results in their average density approaching closer to the nucleus where the average potential energy is larger. Single-electron wave functions that manifest this behavior with increasing Z such as $2s$ and $2p$ in Figure 2[9]a, are called penetrating orbitals. In Figure 2[9]b, it is apparent that the $3s$, $3p$, $4s$, and $4p$ orbitals follow the same pattern as the $2s$, $2p$, $3s$, and $3p$ orbitals of Figure 2[9]a. Note, however, that there is no evidence of penetration for the $3d$ and $4d$ orbitals.

Figures 3[9]a and 3[9]b show the binding energies of the terms of the *ns*, *np*, *nd*, and *nf* configurations of the alkalis compared with the binding energies of $(nsn's)$, $(nsn'p)$, $(nsn'd)$, and $(nsn'f)$ configurations of the alkaline earths in plots similar to Figure 2[9]a and 2[9]b. In Figure 3[9]a *ns* orbitals even with high n values show some effects of penetration. The most striking effect, however, is the large change in the binding energies of *nd* electrons in the lowest excited configurations of the pairs (K, Ca), (Rb, Sr), and (Cs, Ba). The *nd* orbital penetrates even though the *nd* subshell is not yet occupied in the ground configuration. Figure 3[9]b also shows on an expanded scale of Z the penetration of *np* orbitals. But the $4f$ orbital does not show any appreciable penetration over the range of $Z \leqslant 56$.

Figure 3[9]a also show that the ${}^1D-{}^3D$ splitting in the alkaline earth configurations $(nsn'd)$ is roughly constant for $n = n' + 1$ with increasing Z and is much larger than the splitting in the higher excited configurations $n \leqslant n'$.

The observed binding energies of the configurations ns, ns^2, and $ns(n + 1)s$ for alkali and alkaline earth atoms are plotted in Figure 4[9]. The plot shows that the binding energies of the $ns(n + 1)s$ configurations determined by their centers of gravity are almost constant with increasing Z. Because of the competing effects of increasing Z and the increasing number of closed-shell electrons, the centers of gravity of these configurations always lie between the hydrogenic energies corresponding to $n = 2$ and $n = 3$. The binding energies of the $ns^2\ {}^1S$ terms in the alkaline earths decrease with increasing Z. The same effect is apparent for the $(ns)\ {}^2S$ terms of the alkalis, but the decrease in binding energy with increasing Z is slight, because the *ns* orbitals in the alkalis are almost fully screened. Changes in effective screening therefore produce relatively little change in binding energy with increasing Z. The energy differences between $(ns^2)\ {}^1S$ terms and the centers of gravity of the $ns(n + 1)\ {}^{1,3}S$ terms become smaller with increasing n and Z. This indicates that the difference in the electron interactions (3[8]1–3[8]3) between the closed core and these configurations becomes smaller with increasing n and Z.

The absolute term values of a series (that is, a set of terms with constant l values and increasing quantum number n) are related to the Rydberg constant R, effective charge C, and quantum defect Δ_{nl} (Sec. 9^1) by

$$T_{nl} = -R \frac{C^2}{(n - \Delta_{nl})^2} \tag{1}$$

(1) may be viewed in alternative ways. Because states of high excitation (large n) correspond physically to a single electron well removed in space from a core of more tightly bound electrons, Δ_{nl} measures the departure of the orbitals of such electrons from hydrogenic orbitals for which the absolute term values are given by the Rydberg formula ($9^1 5$), which is merely (1) with $\Delta_{nl} = 0$.

The magnitude of Δ_{nl} thus measures the deviation of an orbital describing a single electron in such a highly excited state from a hydrogenic orbital of the same n and l values because of the interactions with the more tightly bound electrons. The magnitude of Δ_{nl} is found experimentally to remain nearly constant with increasing n, a fact that allows ionization potentials to be determined from the term values using (1). The procedure may be improved by assuming a linear dependence of the quantum defect on T as in the Ritz formula [2]

$$\Delta_{nl} = \alpha_l + \beta_l T_{nl} \tag{2}$$

or by adding higher terms in a power series expansion in T_{nl}

$$\Delta_{nl} = \alpha_l + \beta_l T_{nl} + \gamma_l T_{nl}^2 + \cdots \tag{3}$$

The procedures for determining ionization potentials from measured term values using these relations are given by Edlén [1, p. 124].

(1) may also be used to determine quantum defects for the lowest members of a given group of series and their properties studied as a function of l or Z. Figure 5^9 taken from Edlén [1, p. 88] shows a plot of quantum defects versus Z for the lowest terms corresponding to $s, p, d,$ and f series of the alkalis. The figure shows that the quantum defects for $4f$ are always near zero for $Z \leqslant 56$, whereas those for $3d$ remain near zero for Li, Na, and K but increase by approximately unit steps for Rb and Cs. The quantum defects for ns, np, and nd also increase by approximately unit steps for increasing Z. The increase in unit steps reflects the well-known fact that binding energies of the lowest excited states of the alkalis are approximately constant.

Examination of the behavior of quantum defects for different series and for different values of n in a given case produces information about the behavior of the orbitals. For example, the binding energies for the terms of doubly ionized lanthanum (La III) with the observed transitions (Grotrian diagram) are given in Figure 6^9, taken from Odabasi [3]. The quantum defects corresponding to the computed value of the ionization potential are plotted against these observed term energies

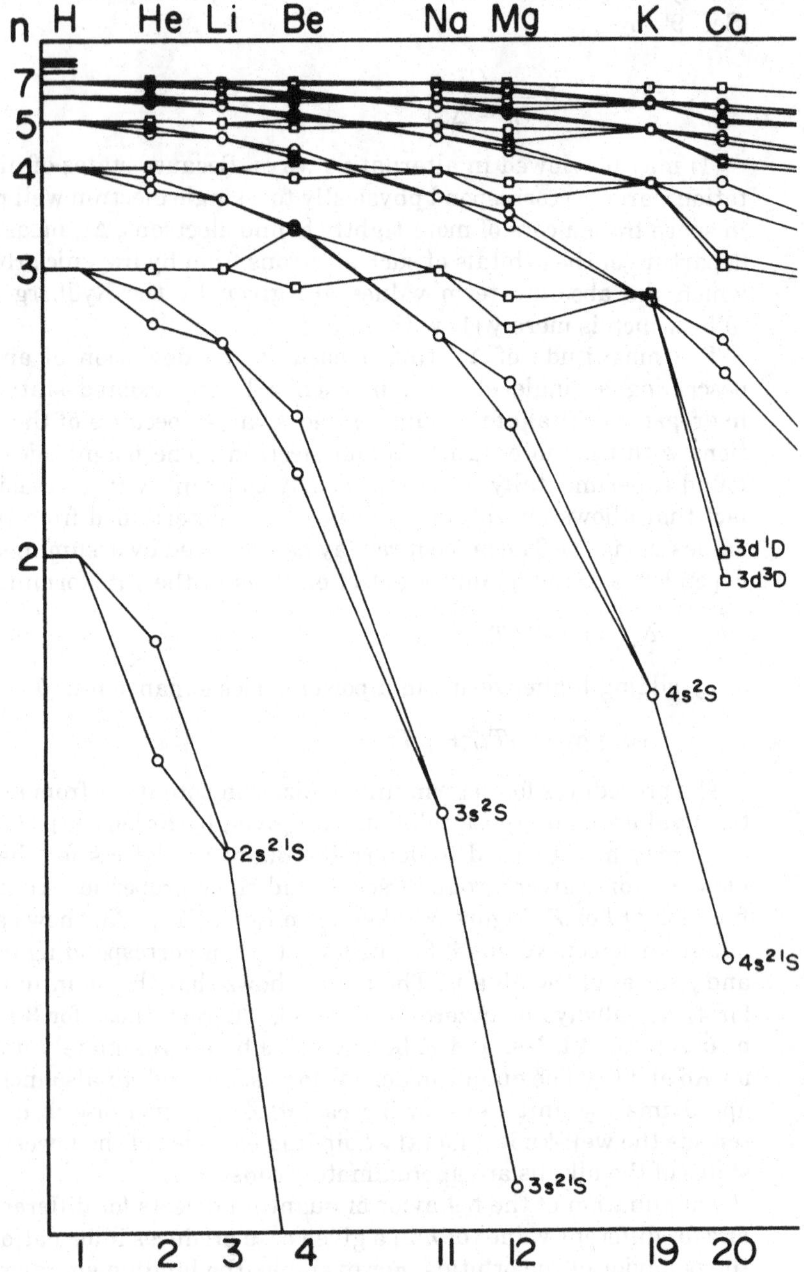

Figure 3⁹a. The observed energies of the terms of the $n's$ and $n'd$ configurations for alkalis and of the $nsn's$ and $nsn'd$ configurations for alkaline earths. The Z scale is not linear and is chosen arbitrarily. The term energies of the

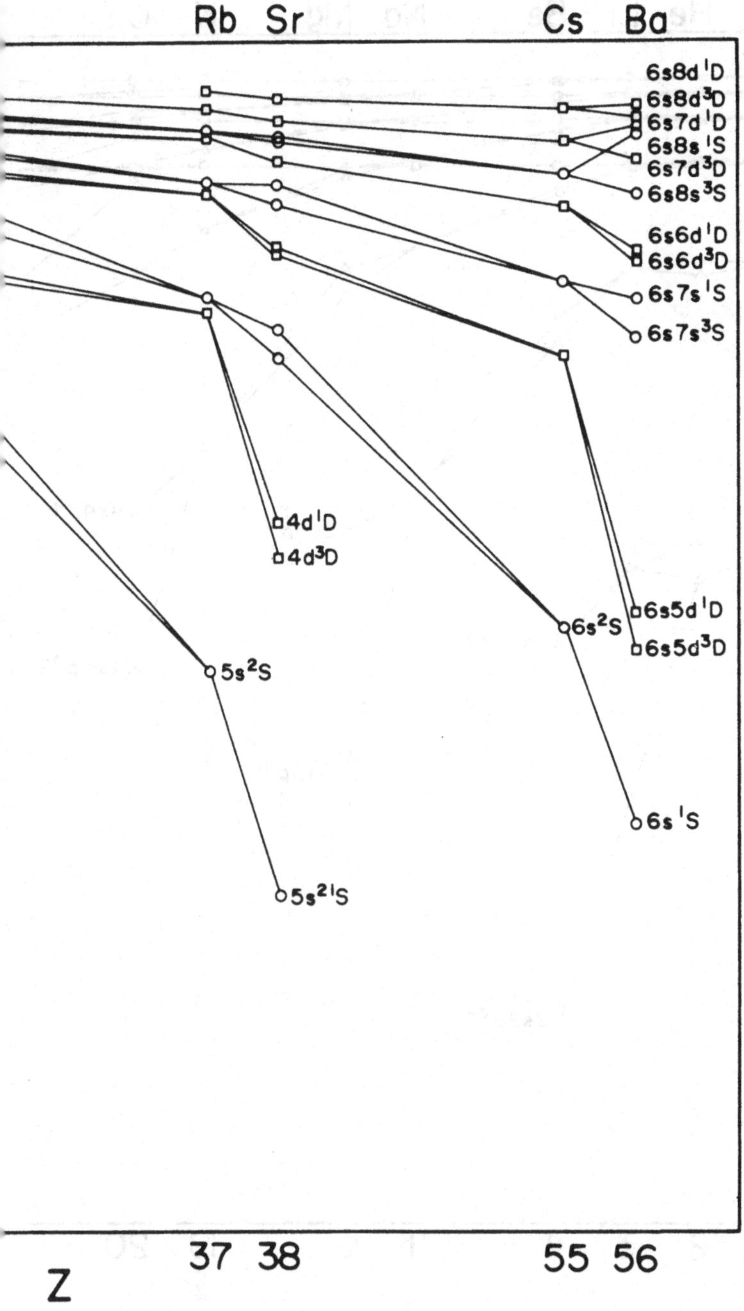

hydrogen atom, specified by the quantum number n, are taken as the energy scale.

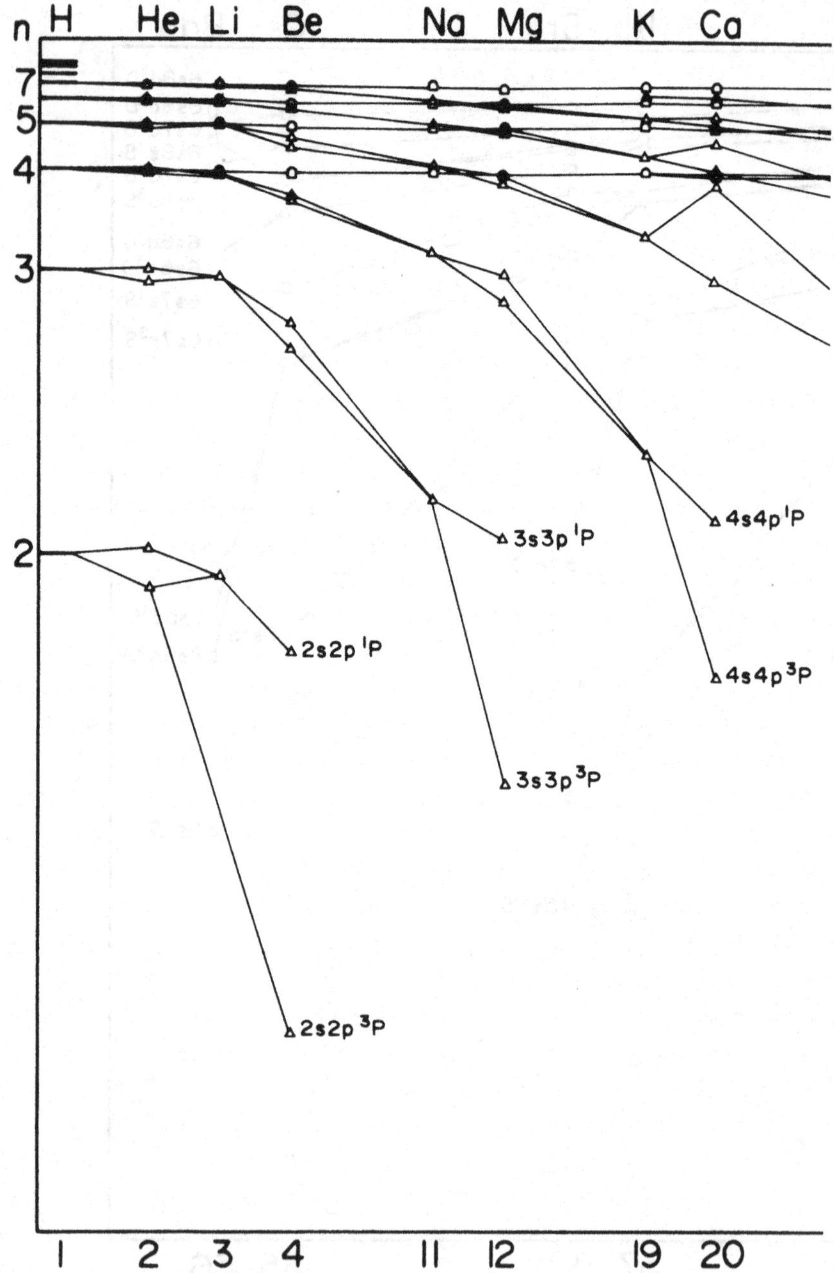

Figure 3[9]b. The observed energies of the terms of the $n'p$ and $n'f$ configurations for alkalis and of the $nsn'p$ and $nsn'f$ configurations for alkaline earths. The Z scale is not linear and is chosen arbitrarily. The term energies of the

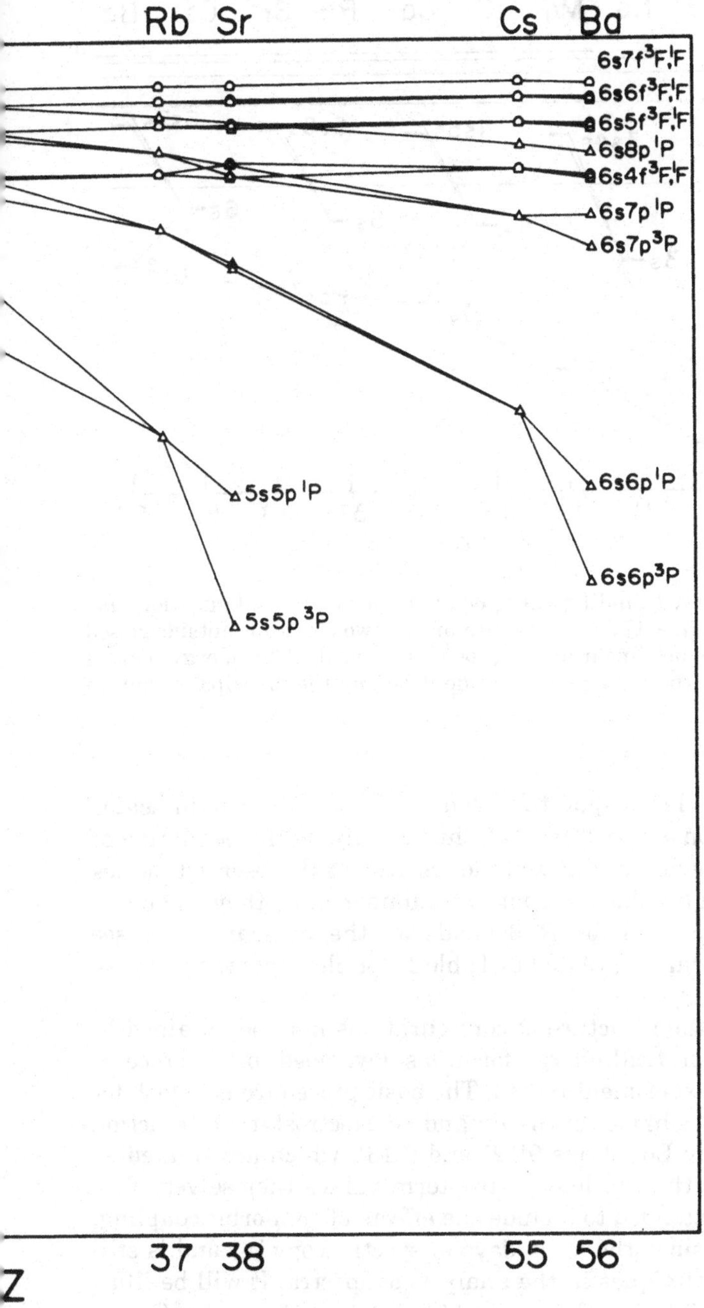

hydrogen atom, specified by the quantum number n, are taken as the energy scale.

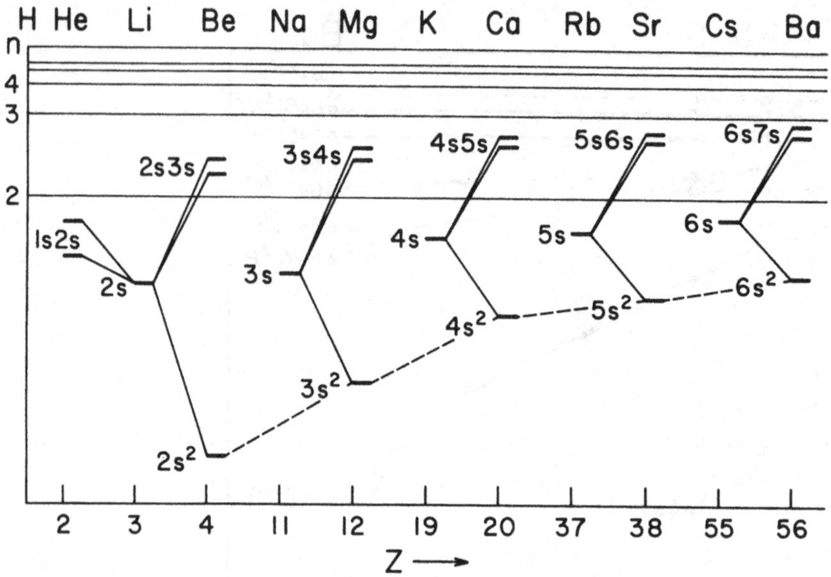

Figure 4[9]. The observed binding energies of the terms of configurations *ns*, ns^2, and *nsn*'*s* ($n' = n + 1$) for atoms with one or two electrons outside closed shells. The Z scale is not linear and is chosen arbitrarily. The energy scale is the term energies of the hydrogen atom identified by the principal quantum numbers *n*.

in Figure 7[9]. The large quantum defect of the 4*f* term indicates unusual penetration of the 4*f* orbital, thus signifying the beginning of the rare-earth elements (lanthanide series). All observed series become Ritz for high values of quantum number *n*, as they should.

In (1) the value used for R depends on the nuclear mass (see Appendix 1). These are tabulated in Table 1[9] for elements with $Z = 1$–20.

Information on the structure of configurations may be obtained by comparing the theoretical energy formulas developed in the proceeding chapters with experimental data. The basic procedure is to look for regularities in the behavior of the F^k and G^k electrostatic interaction integrals defined by Equations 9[4]12″ and 9[4]13″, which are treated as parameters rather than to look at the term values themselves. This procedure, when extended to include the effects of spin-orbit coupling, has been pursued since the early days of spectroscopy [4] and is still one of the basic techniques in the analysis of spectra. It will be illustrated here by considering the energy differences between configurations of the type s^2p^q, sp^{q+1}, and p^{q+2} ($q = 0$–4) both as functions of q and of increasing Z (along isoelectronic sequences).

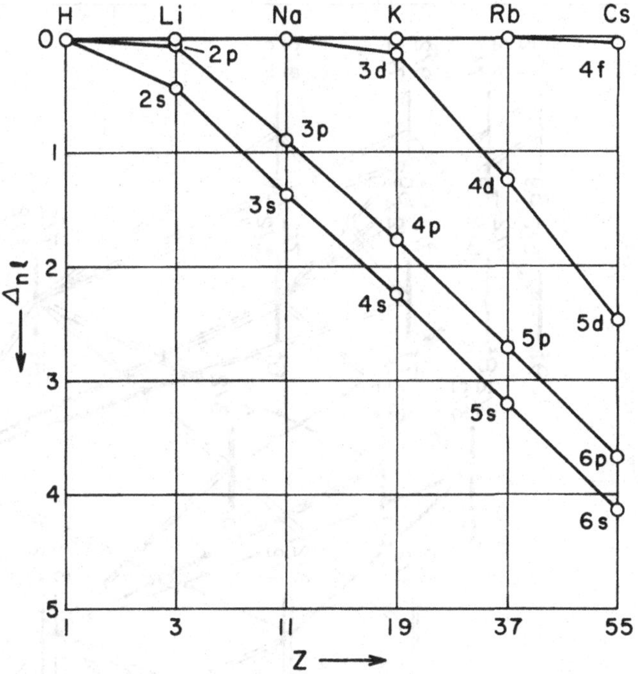

Figure 5⁹. Observed quantum defects of the *ns, np, nd,* and *nf* orbitals in the alkalis. The Z scale is not linear and is chosen arbitrarily.

The configuration average energies (9⁴21, 23) of the above configurations are

$$W_{\mathrm{Av}}(s^2p^q) = W(\mathrm{c.c.}) + F^0(s, s) + \frac{q(q-1)}{2}\left[F^0(p, p) - \frac{2}{25}F^2(p, p)\right] + 2q\left[F^0(s, p) - \frac{1}{6}G^1(s, p)\right] \tag{4a}$$

$$W_{\mathrm{Av}}(sp^{q+1}) = W(\mathrm{c.c.}) + \frac{q(q+1)}{2}\left[F^0(p, p) - \frac{2}{25}F^2(p, p)\right] + (q+1)\left[F^0(s, p) - \frac{1}{6}G^1(s, p)\right] \tag{4b}$$

$$W_{\mathrm{Av}}(p^{q+2}) = W(\mathrm{c.c.}) + \frac{(q+1)(q+2)}{2}\left[F^0(p, p) - \frac{2}{25}F^2(p, p)\right] \tag{4c}$$

in which W(c.c) stands for the energy of the closed core, that is, all shells interior to the ns^rnp^q configurations under consideration. If it is assumed that W(c.c) as well as the electrostatic integrals remain the

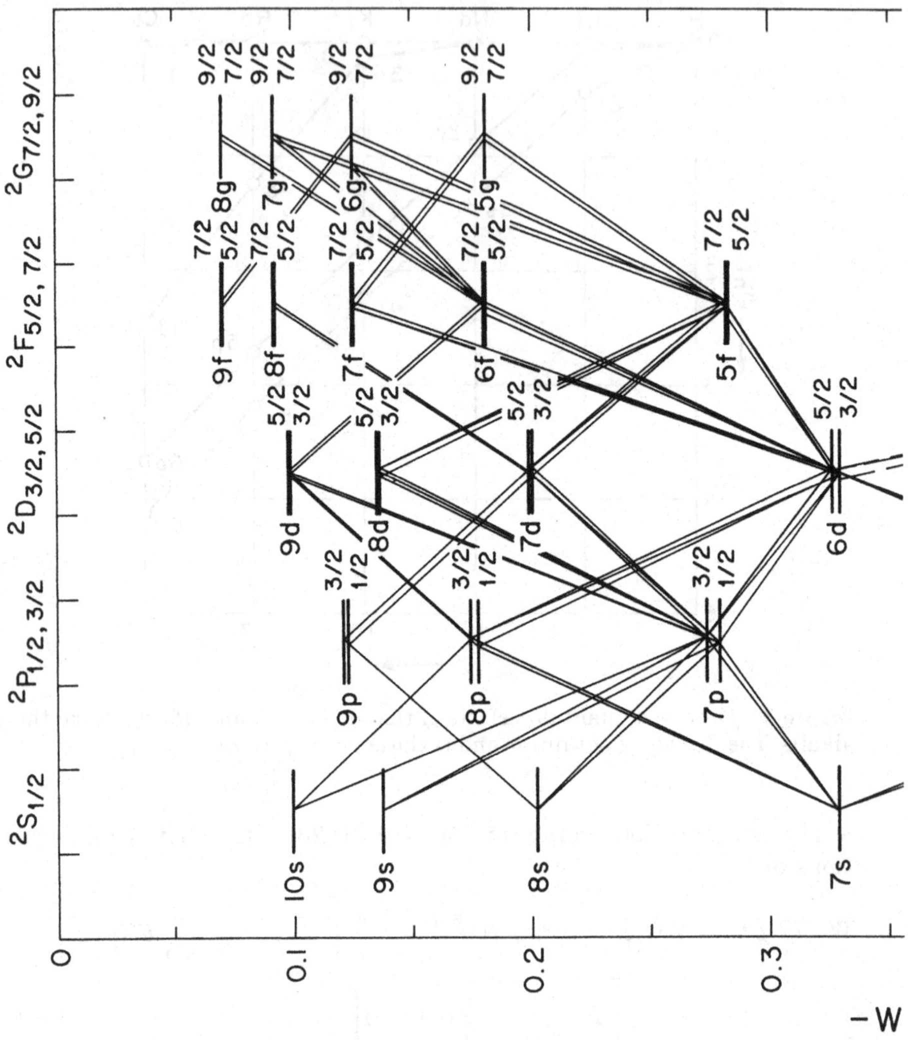

Figure 6[9]. The observed energies and transitions for the levels of La III.

same for each of these configurations, the following relations are obtained:

$$W_{Av}(sp^{q+1}) - W(s^2p^q) = q\left[F^0(p,p) - \frac{2}{25}F^2(p,p) - F^0(s,p) + \frac{1}{6}G^1(s,p)\right]$$
$$+\left[F^0(s,p) - \frac{1}{6}G^1(s,p) - F^0(s,s)\right] \qquad \text{(5a)}$$

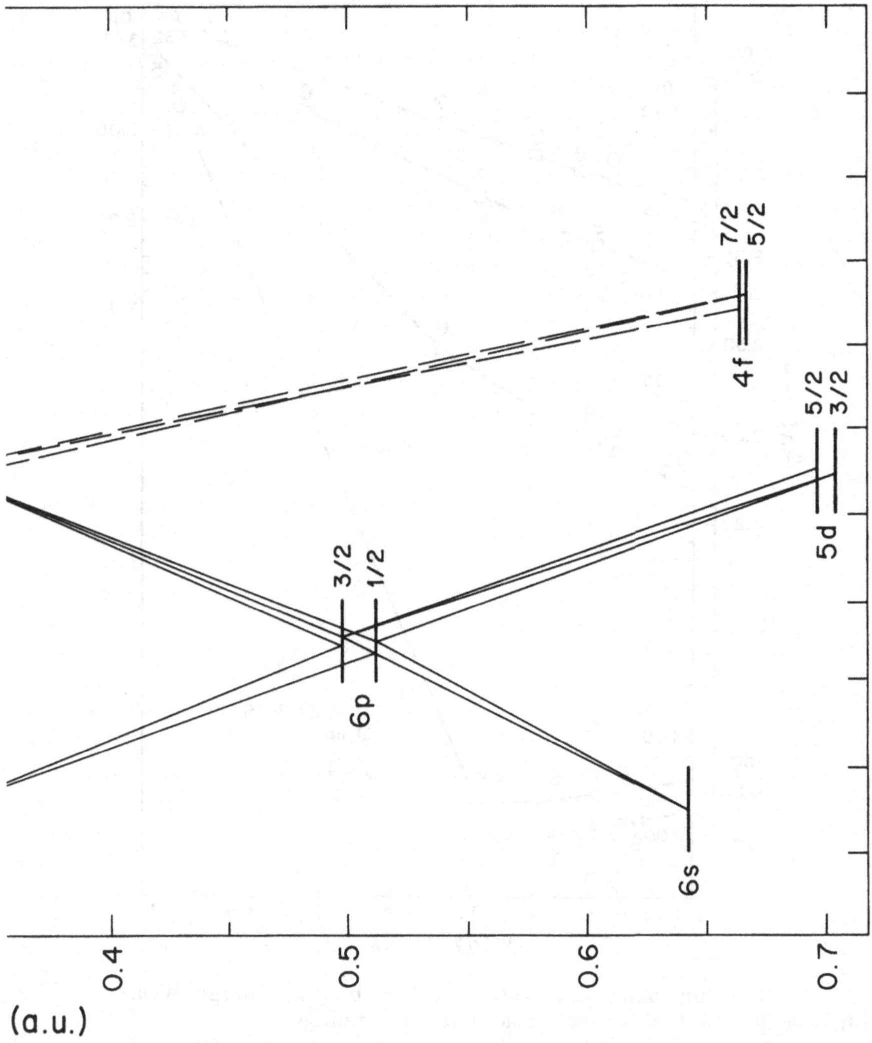

$$W_{\text{Av}}(p^{q+2}) - W_{\text{Av}}(sp^{q+1}) = (q+1) \times \left[F^0(p,p) - \frac{2}{25}F^2(p,p) - F^0(s,p) + \frac{1}{6}G^1(s,p) \right] \quad \text{(5b)}$$

These relations would predict a linear dependence on q of the energy differences, provided the electrostatic integrals are the same for each

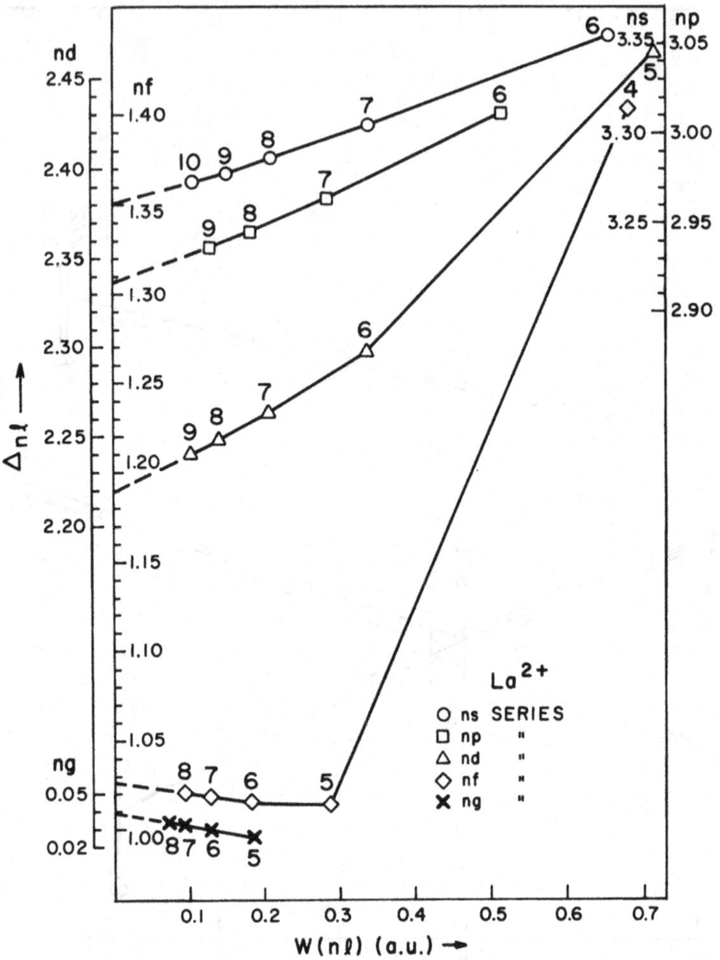

Figure 7[9]. Quantum defect Δ_{nl} plotted against binding energy $W(nl)$. The ordinate scale is shifted for each series but is not changed.

value of q as well as the same in each configuration. Early work by Bacher and Goudsmit [4] concentrated on linear relations of this type and was extended by Meshkov and Ufford [5a] and Trees [5b].

In a detailed investigation of the configuration average intervals given in (5), Edlén found that the configuration average intervals depend quadratically rather than linearly on q. This in turn means that the electrostatic integrals must vary linearly with q. The observed values of the energy differences $[W_{AV}(2s2p^{q+1}) - W_{AV}(2s^2 2p^q)]$ as taken from Edlén (1) are given in Table 2[9]. Also shown

Table 1^9. Value of the Rydberg constant (cm^{-1}) for elements with Z = 1–20

Z	Element	R	Z	Element	R
1	^{1}H	109677.58	11	^{23}Na	109734.69
2	^{4}He	109722.27	12	^{24}Mg	109734.80
3	^{7}Li	109728.73	13	^{27}Al	109735.08
4	^{9}Be	109730.63	14	^{28}Si	109735.16
5	^{11}B	109731.84	15	^{31}P	109735.37
6	^{12}C	109732.29	16	^{32}S	109735.43
7	^{14}N	109733.01	17	^{35}Cl	109735.59
8	^{16}O	109733.55	18	^{40}Ar	109735.80
9	^{19}F	109734.14	19	^{39}K	109735.77
10	^{20}Ne	109734.30	20	^{40}Ca	109735.80

are energy differences obtained from a calculation of the configuration average energies using the HFS method [6a] for the boron and carbon isoelectronic sequences. A study of Table 2^9 shows that the agreement between the HFS values and the observed ones is better for the carbon than for the boron isoelectronic sequence. The observed values of Table 2^9 are plotted against q in Figure 8^9. The q dependence for each value of Z is roughly quadratic.

In deriving (5) it was assumed that both the core energy W(c.c.) and all of the electrostatic interaction integrals remain constant for the configurations s^2p^q, sp^{q+1}, and p^{q+2} for fixed Z. The validity of this assumption can be tested by direct computation of these energy parameters using the HFS method. Tables 3^9, 4^9a, and 4^9b present the HFS core energies and electrostatic interaction integrals, respectively, as taken from Condon and Odabasi [6a] and Odabasi [6b], for the $2s^22p^2$, $2s2p^3$ and $3s^23p^2$, $3s3p^3$ configurations. The entries of Table 3^9 reveal that the assumption of constancy of core energies is better, understandably, for the $(1s^22s^22p^6)$ core than the $(1s^2)$ core. However, the variation in all cases considered is small enough to take them as constants. The study of Tables 4^9a and 4^9b indicates that most of the time the variation of the electrostatic integrals, from one of the configurations to the other, is smaller than 1 percent. The largest variations usually occur at the beginning of the isoelectronic sequences, that is, for small values of Z when N is kept constant.

The linear q dependence of the parameters $F^2(p, p)$ and $G^1(s, p)$ for the configurations can be investigated with the help of data deduced from observation. The expressions for the term energies of the config-

Table 2[9]. Observed configuration average energy intervals (a.u.) between configurations $2s2p^{q+1}$ and $2s^22p^q$: $W_{\mathrm{Av}}(2s2p^{q+1}) - W_{\mathrm{Av}}(2s^22p^q)$. In the boron and carbon isoelectronic sequences the HFS values [6a] are given as second entries. The conversion from cm^{-1} to a.u. is performed by employing R_∞.

Z	$2s2p - 2s^2$	$2s2p^2 - 2s^22p$	$2s2p^3 - 2s^22p^2$	$2s2p^4 - 2s^22p^3$	$2s2p^5 - 2s^22p^4$	$2s2p^6 - 2s^22p^5$
4	0.12348					
5	0.21096	0.21050				
		0.19544				
6	0.29571	0.32213	0.31666			
		0.29544	0.32374			
7	0.37909	0.43012	0.45472	—		
		0.39403	0.45050			
8	0.46201	0.53574	0.58636	0.60736	—	
		0.49156	0.57590			
9	0.54494	0.64335	0.71853	0.76683	0.79098	—
		0.58894	0.70104			
10	—	—	0.85021	0.92539	0.97187	0.98781
		0.68554	0.82588			
11	0.71170	0.85659	0.98189	1.08304	1.15731	1.20287
		0.78190	0.94994			
12	—	0.96366	1.11448	1.24206	1.34275	1.41611
		0.87360	1.07413			
13	—	—	—	1.40426	1.52911	1.62980
		0.97617	1.19778			

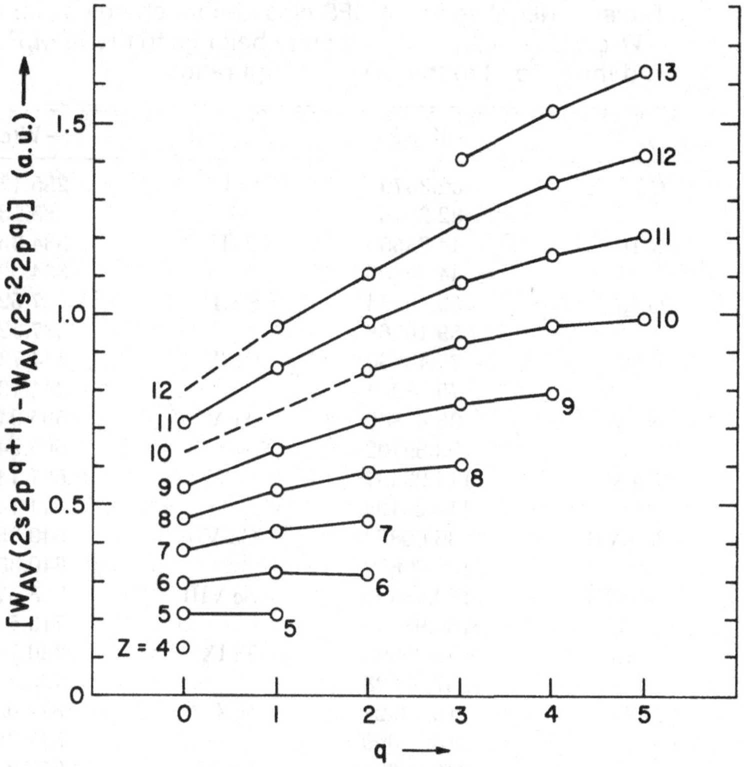

Figure 8⁹. The observed configuration average energy intervals between configurations $2s2p^{q+1}$ and $2s^22p^q$ plotted against q values.

urations $nsnp$, $nsnp^2$, $nsnp^3$ (cases 2, 10, 21 of Appendix 3) result in the following relations for the $G^1(s, p)$ parameters

$$
\begin{aligned}
nsnp,\ nsnp^5&: \quad G^1(s,p) = \frac{3}{2}\,({}^1P - {}^3P)\\
nsnp^2,\ nsnp^4&: \quad G^1(s,p) = \quad ({}^2P - {}^4P)\\
nsnp^3&: \text{(a)}\ G^1(s,p) = \frac{3}{4}\,({}^3S - {}^5S)\\
&\quad\ \text{(b)}\ G^1(s,p) = \frac{3}{2}\,({}^1D - {}^3D)\\
&\quad\ \text{(c)}\ G^1(s,p) = \frac{3}{2}\,({}^1P - {}^3P)
\end{aligned}
\tag{6}
$$

which are given in Table 5⁹, for $n = 2$, as derived from the observed term intervals by Edlén [1, p. 72], with the HFS values [6a] for

Table 3[9]. Negative of the HFS closed-core energies (a.u.), $-W$(c.c.), as in (4). The first entry belongs to the ns^2np^2 and the second to the $nsnp^3$ configuration.

$n = 2$	$-W$(c.c.)	$n = 3$	$-W$(c.c.)
C I	32.35792	Si I	285.13573
	32.35762		285.13488
N II	44.73350	P II	334.33033
	44.73257		334.32998
O III	59.10854	S III	387.52945
	59.10768		387.52955
F IV	75.48360	Cl IV	444.73092
	75.48254		444.73126
Ne V	93.85860	Ar V	505.94126
	93.85702		505.94188
Na VI	114.23367	K VI	571.14808
	114.23194		571.14879
Mg VII	136.60879	Ca VII	640.35900
	136.60669		640.35987
Al VIII	160.98391	Sc VIII	713.57178
	160.98147		713.57262
Si IX	187.35908	Ti IX	790.78592
	187.35628		790.78667
P X	215.73427	V X	872.00153
	215.73099		872.00227
S XI	246.10971	Cr XI	957.21787
	246.10580		957.21850

comparison. The observed values of Table 5[9], are plotted against q in Figure 9[9]. Although the q dependence is not quite linear for low values of C, it becomes so with increasing degree of ionization. In plotting the $G^1(s, p)$ values for the $2s2p^3$ configuration, only the values obtained by relation (a) of (6) are considered, because the observed ^3S and ^5S terms are perturbed less than the others. The theoretical values of $G^1(2s, 2p)$ are not plotted, but they fall on almost straight lines for fixed Z.

The parameters $F^2(p, p)$ for the s^2p^q configurations can be derived from the observed data by using equations that take the spin-orbit interaction into account as well [1, p. 115]. Simpler relations can be derived from the formulas given in cases 3, 9 of Appendix 3, if one considers the weighted average of the observed energy levels, that is, the term energies. In obtaining these parameters it is necessary to consider the least perturbed terms. For example, among the terms of the np^3 configurations the ^2D and ^4S are more reliable, less perturbed,

Table 4⁹a. HFS values (a.u.) of some electrostatic interaction integrals for the $2s^22p^2$ and the $2s2p^3$ configurations. The first entry belongs to $2s^22p^2$ and the second to $2s2p^3$.

	$F^0(2s, 2p)$	$F^0(2p, 2p)$	$F^2(2p, 2p)$	$G^1(2s, 2p)$
C I	0.56724	0.54254	0.23855	0.34325
	0.57205	0.54943	0.24231	0.34665
N II	0.74438	0.74911	0.33957	0.45233
	0.74286	0.74710	0.33856	0.45150
O III	0.91472	0.94401	0.43439	0.55207
	0.91035	0.93816	0.43128	0.54980
F IV	1.08224	1.13420	0.52672	0.64776
	1.07626	1.12623	0.52245	0.64485
Ne V	1.24817	1.32186	0.61771	0.74118
	1.24115	1.31251	0.61269	0.73794
Na VI	1.41315	1.50800	0.70791	0.83321
	1.40541	1.49771	0.70238	0.82979
Mg VII	1.57750	1.69314	0.79760	0.92433
	1.56922	1.68216	0.79168	0.92080
Al VIII	1.74143	1.87762	0.88694	1.01481
	1.73273	1.86610	0.88075	1.01121
Si IX	1.90503	2.06160	0.97602	1.10482
	1.89600	2.04966	0.96958	1.10118
P X	2.06838	2.24518	1.06490	1.19447
	2.05910	2.23293	1.05828	1.19082
S XI	2.23153	2.42846	1.15362	1.28386
	2.22206	2.41596	1.14687	1.28020

than the ^2P [1, pp. 202–3]. In Table 6⁹ the values of $F^2(np, np)$, for $n = 2, 3$, as derived from the observed energy levels by Edlén [1, p. 175, 203], as well as the theoretical values computed by the HFS [6] and by the HF [46⁸] methods for the ns^2np^2, ns^2np^3, ns^2np^4 configurations, are presented. The agreement between theory and experiment is better for the cases with $n = 2$. The HFS values are 40 percent larger for the worst and 10 percent larger for the best case. The HF values of $F^2(np, np)$, as taken from Fraga, Karwowski, and Saxena [46⁸], for the normal configurations of atoms and ions in Table 6⁹, are in general somewhat smaller than the HFS values. Even though the use of a more sophisticated method, such as the HF, to obtain the theoretical values does not improve the agreement with observation considerably, the important thing is that it brings, in general, an improvement, however small, as it should. Remembering the approximations involved in the HF and in the HFS method as given in Chapter 8, it is not surprising to see that

Table 4[9]b. HFS values (a.u.) of some electrostatic interaction integrals for the $3s^23p^2$ and the $3s3p^3$ configurations. The first entry belongs to $3s^23p^2$ and the second to $3s3p^3$.

	$F^0(3s, 3p)$	$F^0(3p, 3p)$	$F^2(3p, 3p)$	$G^1(3s, 3p)$
Si I	0.37775	0.33700	0.16408	0.23830
	0.37830	0.34918	0.17131	0.24712
P II	0.46998	0.43745	0.21892	0.30832
	0.47539	0.44381	0.22257	0.31263
S III	0.55345	0.52651	0.26617	0.36811
	0.55684	0.53042	0.26835	0.37062
Cl IV	0.63269	0.61031	0.30988	0.42294
	0.63494	0.61289	0.31127	0.42452
Ar V	0.70942	0.69110	0.35154	0.47484
	0.71093	0.69283	0.35244	0.47585
K VI	0.78449	0.76996	0.39191	0.52481
	0.78549	0.77111	0.39247	0.52545
Ca VII	0.85839	0.84749	0.43139	0.57342
	0.85902	0.84820	0.43172	0.57378
Sc VIII	0.93144	0.92405	0.47025	0.62102
	0.93178	0.92444	0.47039	0.62118
Ti IX	1.00386	0.99990	0.50864	0.66785
	1.00396	1.00001	0.50864	0.66786
V X	1.07576	1.07517	0.54666	0.71407
	1.07568	1.07508	0.54656	0.71397
Cr XI	1.14726	1.14998	0.58440	0.75981
	1.14702	1.14972	0.58420	0.75961

the agreement between theoretical and observed values is not very good. The observed $F^2(np, np)$ values for the configurations $ns^2np^q(n = 2, 3)$ are plotted against q in Figure 10[9], where the linear q dependence of $F^2(np, np)$ is obvious.

2[9]. The order of electron shells for atoms and ions

If the configuration average energies for atoms and ions were given by hydrogenic relations (that is, if the quantum defects were zero), the electron shells would be filled in the order of the quantum numbers– first n and then l. This prediction was made by Bohr [91[1]] in the early days of quantum mechanics (Section 11[1]) even before the Pauli exculsion principle [92[1]]. As shown in the previous section, the terms of a given configuration in general do not possess hydrogenic behavior. Therefore, the shells are filled in an order other than n and l. It is very important to note that for ions the electron shells are filled in an order

Table 5[9]. Electrostatic interaction integrals $G^1(2s, 2p)$ (a.u.) for the $2s2p^q$ configurations. For the cases $2s2p$, $2s2p^2$ the first entry is derived from observed energy levels [1] and the second entry is the HFS value [6a]. For the $2s2p^3$ configuration the first three entries are observed values derived by using the relations (a), (b), (c), respectively, in (6), and the last entry is the HFS value. For the $2s2p^4$ and $2s2p^5$ configurations no theoretical value is available. The conversion from cm^{-1} to a.u. is performed by using R_∞.

	Be I	B II	C III	N IV	O V	F VI	Ne VII
$2s2p$	0.14065	0.24673	0.34111	0.43242	0.52325	0.61305	—
	0.19556	0.31113	0.41217	0.50778	0.60073	0.69220	0.78274
	B I	C II	N III	O IV	F V	Ne VI	Na VII
$2s2p^2$	0.19929	0.30796	0.40406	0.49687	0.58722	—	0.76984
	0.27437	0.38270	0.48191	0.57704	0.66997	0.76158	0.85232
	C I	N II	O III	F IV	Ne V	Na VI	Mg VII
$2s2p^3$	0.24604	0.36994	0.46732	0.56024	0.65261	0.74447	0.83703
	—	0.35460	0.45769	0.55359	0.64736	0.73840	0.83015
	—	0.39318	0.46491	0.56098	0.65349	0.74447	0.83474
	0.34665	0.45150	0.54980	0.64485	0.73794	0.82979	0.92080
	N I	O II	F III	Ne IV	Na V	Mg VI	Al VII
$2s2p^4$	—	0.42229	0.52226	—	0.71385	0.80603	0.89614
	O I	F II	Ne III	Na IV	Mg V	Al VI	Si VII
$2s2p^5$	—	0.50980	0.58000	0.67968	0.77454	0.86867	0.95951

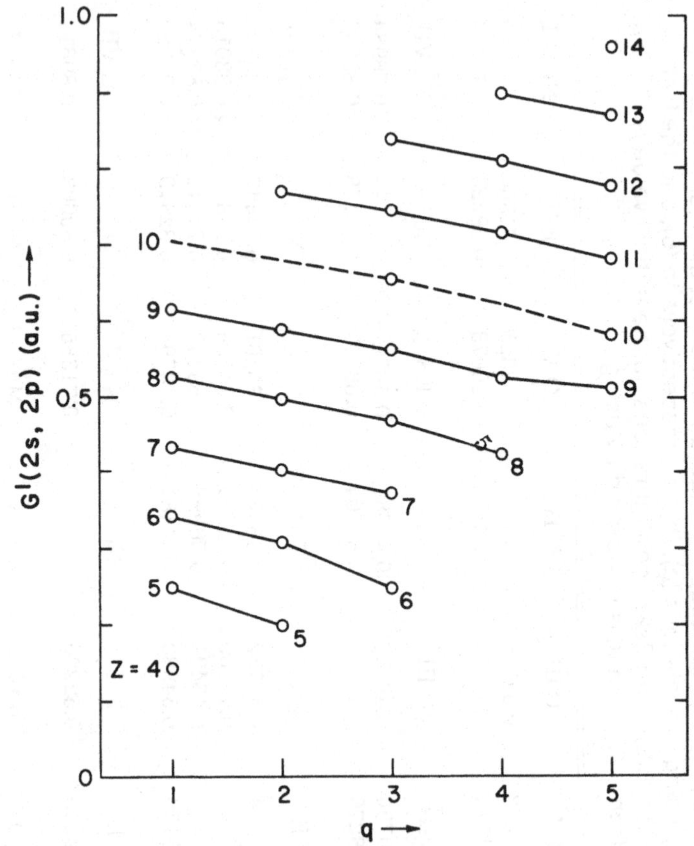

Figure 9[9]. The parameters $G^1(2s, 2p)$ for the configurations $2s2p^q$, as derived from the observed term intervals [1], plotted against q values.

different from that for atoms, the hydrogenic order being the limit for highly ionized cases. The following rule for atoms, purposely given in two parts, was discovered empirically by Madelung [7]:

1. When consecutive atoms are considered, the electron shells are filled in the order of the sum of two quantum numbers n and l, that is, $(n + l)$.
2. Shells with equal $(n + l)$ numbers are filled in the order of quantum number n.

This rule predicts that the $4s$ shell will be filled before the $3d$ shell and that the $4f$ and $5f$ shells will become occupied for the first time at $Z = 57$ and $Z = 89$. Madelung's rule should be interpreted as applying to the configuration average energy.

The rule is approximate in the sense that it completely ignores not

Table 6[9]. Electrostatic interaction integrals $F^2(np, np)$ (a.u.) for the configurations ns^2np^2, ns^2np^3, ns^2np^4, where $n = 2, 3$. The first entry is experimental, as derived from observed energy levels by Edlén [1], and the second and third are theoretical calculated by the HFS [6] and the HF [46[8]] methods. The conversion from cm^{-1} to a.u. is carried out by using R_∞.

	C I	N II	O III	F IV	Ne V	Na VI	Mg VII
$2s^22p^2$	0.19278	0.28867	0.37999	0.46982	0.55790	0.64485	0.73070
	0.23855	0.33957	0.43439	0.52672	0.61771	0.70791	0.79760
	0.24330	0.33986	0.43261	0.52372	0.61394	0.70360	0.79288
	N I	O II	F III	Ne IV	Na V	Mg VI	Al VII
$2s^22p^3$	0.24331	0.33933	0.43137	0.52124	0.61180	—	—
	0.29241	0.39173	0.48610	0.57831	0.66929	0.75955	0.84930
	0.29542	0.39082	0.48325	0.57428	0.66448	0.75417	0.84350
	O I	F II	Ne III	Na IV	Mg V	Al VI	Si VII
$2s^22p^4$	0.29972	0.39295	0.48426	0.57425	0.66327	0.75261	0.84089
	0.34414	0.44240	0.53654	0.62368	0.71970	0.81001	0.89982
	0.33606	0.43165	0.52434	0.61560	0.70602	0.79589	0.88537
	Si I	P II	S III	Cl IV	Ar V	K VI	Ca VII
$3s^23p^2$	0.11665	0.16240	0.20376	0.24291	0.28077	0.31794	0.35457
	0.16408	0.21892	0.26617	0.30988	0.35154	0.39191	0.43130
	0.16593	0.21709	0.26262	0.30533	0.34634	0.38624	0.42534
	P I	S II	Cl III	Ar IV	K V	Ca VI	Sc VII
$3s^23p^3$	0.14398	0.18829	0.22907	0.26803	0.30596	—	—
	0.19649	0.24808	0.29400	0.33704	0.37833	0.41848	0.45783
	0.19675	0.24549	0.29001	0.33219	0.37292	0.41266	0.45170
	S I	Cl II	Ar III	K IV	Ca V	Sc VI	Ti VII
$3s^23p^4$	0.17140	0.21403	0.25425	0.29294	0.33072	0.36787	0.40458
	0.22638	0.27590	0.32092	0.36348	0.40452	0.44452	0.48379
	0.22018	0.26826	0.31258	0.35473	0.39549	0.43530	0.47442

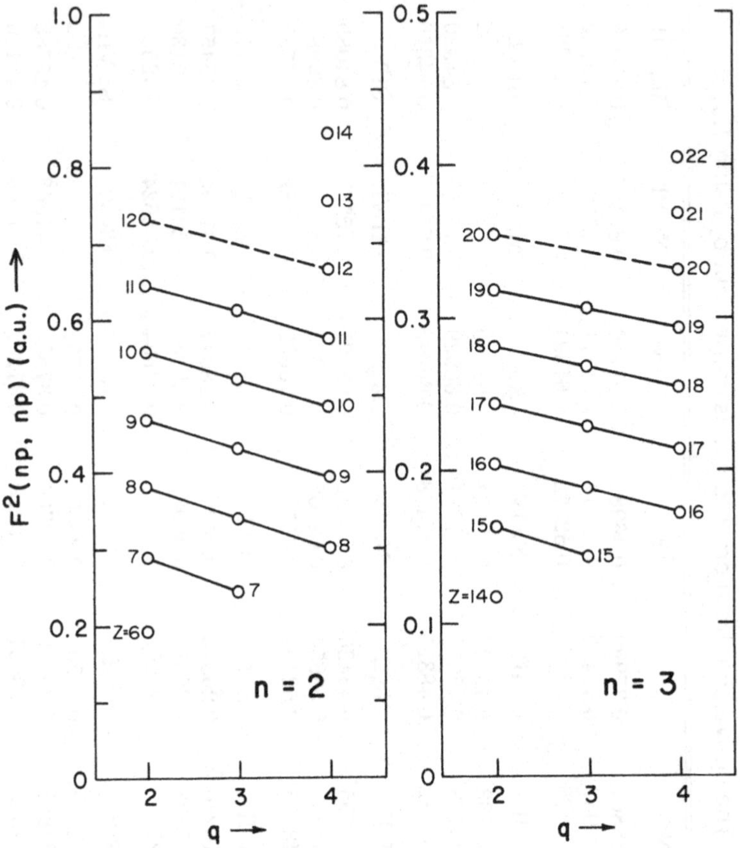

Figure 10[9]. The parameters $F^2(np, np)$ for the configurations $ns^2np^q(n = 2, 3)$, as derived from observed energy levels [1], plotted against q values.

only the multiplet structure of terms but also that different configurations containing two or more open shells may lie close in energy. The normal state of a system, which is defined as the lowest-term value and thus is not necessarily predicted by the rule. It also ignores the magnetic effects and the effects of configuration interaction. These effects play an important role for large values of Z and in cases where two or more open shells might be expected to lie close in energy. When these considerations become important, as in the rare-earth and actinide series of elements, the question of order of electrons in shells becomes somewhat meaningless. Here these effects will be ignored, and the discussion will be limited to comparisons with calculations, which do not include them, performed by the nonrelativistic CGHF method.

Despite the above qualifications Madelung's rule provides a remarkably consistent picture of the uniform filling of electron shells and provides the basis for existing periodic tables such as Table 3^1. The rule also provides some insight concerning where deviations might be expected to occur, namely, when the ordering predicted differs for that predicted by the hydrogenic ordering of increasing n and then l. Deviations of this type occur for $4s$ and $3d$, for $5s$ and $4d$, and for $6s$, $4f$, and $5d$ shells. The rule thus reflects the fact that the binding energies of electrons in these shells are of comparable magnitude. Consequently, alternative configurations composed of electrons in these shells lie close in energy.

CGHF calculations may be used to provide theoretical predictions of such deviations. A complete set of such calculations for all atoms for $Z = 2 - 86$ has been made by Froese Fischer [33^8a], assuming that the order of filling of shells is that predicted by Madelung's rule. The configurations represented by these calculations are not necessarily the normal configurations for any particular atom. Similar CGHF calculations may be made for alternative configurations and the energies of these calculations compared with the previous calculations to predict deviations from the normal ordering.

Selected samples of calculations of this type are shown in Table 7^9, which is based on data provided by Froese Fischer (private communication).

The table shows that most of the cases considered involving a single electron outside closed shells agree with Madelung's rule. Whereas the total configuration average energies actually determine the energy ordering, the one-electron energies follow the same order and thus predict the relative binding in different shells. This is not true for the case of La, however. The CGHF results for this case predict that the normal configuration is $6s5d^2$ even though the $\epsilon(4f)$ parameter in $6s^24f$ is larger in magnitude than $\epsilon(5d)$ in either the $6s^25d$ or $6s5d^2$ configurations. Note that the total configuration average energies are almost equal for all three of these configurations, which accounts for the deviation from Madelung's rule. The CGHF results also disagree with observation in this case, which identifies $6s^25d$ as the normal configuration, with the center of gravity of the $6s5d^2$ configuration lying immediately above it.

There are also some other deviations. Table 8^9 presents the CGHF total average energies for alternative configurations, for elements where deviations might be expected. The CGHF results are mostly in agreement with the observed ones. But there are also a few disagreements, at least seemingly so; the theory favors the $3d^44s^2$, $4d^35s^2$, $5d^26s$, $4f^26s^2$, $4f^86s^2$, $4f^96s^2$ configurations as opposed to the observed $3d^54s$, $4d^45s$, $5d6s^2$, $4f5d6s^2$, $4f^75d6s^2$, $4f^85d6s^2$ normal configurations for the elements Cr, Nb, La, Ce, Gd, Tb, respectively. Some of

Table 7[9]. CGHF values of the one-electron energy parameters, $\epsilon(nl)$, where nl specifies the outermost electron and the total configuration average energies (both in a.u.) for different configurations of the elements K, Sc, Rb, Y, Cs, La, Lu, and Ac.

Z, Element				
	$3p^6 4p$	$3p^6 3d$	$3p^6 4s$	
19, K	$-\epsilon(4p) = 0.09558$	$-\epsilon(3d) = 0.05815$	$-\epsilon(4s) = 0.14748$	
	$-W_{Av} = 599.11314$	$-W_{Av} = 599.07571$	$-W_{Av} = 599.16479$	
	$4s^2 4p$	$4s^2 3d$		
21, Sc	$-\epsilon(4p) = 0.17561$	$-\epsilon(3d) = 0.34371$		
	$-W_{Av} = 759.63168$	$-W_{Av} = 759.73572$		
	$4p^6 5p$	$4p^6 4d$	$4p^6 5s$	
37, Rb	$-\epsilon(5p) = 0.09016$	$-\epsilon(4d) = 0.06018$	$-\epsilon(5s) = 0.13786$	
	$-W_{Av} = 2938.3101$	$-W_{Av} = 2938.2800$	$-W_{Av} = 2938.3574$	
	$5s^2 5p$	$5s^2 4d$		
39, Y	$-\epsilon(5p) = 0.15880$	$-\epsilon(4d) = 0.24985$		
	$-W_{Av} = 3331.6264$	$-W_{Av} = 3331.6842$		
	$5p^6 6p$	$5p^6 5d$	$5p^6 6s$	
55, Cs	$-\epsilon(6p) = 0.08408$	$-\epsilon(5d) = 0.06782$	$-\epsilon(6s) = 0.12367$	
	$-W_{Av} = 7553.894$	$-W_{Av} = 7553.877$	$-W_{Av} = 7553.934$	
	$6s^2 6p$	$6s^2 5d$	$6s5d^2$	$6s^2 4f$
57, La	$-\epsilon(6p) = 0.14619$	$-\epsilon(5d) = 0.26885$	$-\epsilon(5d) = 0.19931$	$-\epsilon(4f) = 0.37863$
	$-W_{Av} = 8220.9731$	$-W_{Av} = 8221.0667$	$-W_{Av} = 8221.0730$	$-W_{Av} = 8221.0638$
	$6s^2 6p$	$6s^2 5d$		
71, Lu	$-\epsilon(6p) = 0.15858$	$-\epsilon(5d) = 0.24335$		
	$-W_{Av} = 13851.754$	$-W_{Av} = 13851.808$		
	$7s^2 7p$	$7s^2 6d$		$7s^2 5f$
89, Ac	$-\epsilon(7p) = 0.13900$	$-\epsilon(6d) = 0.25149$		$-\epsilon(5f) = 0.29167$
	$-W_{Av} = 23722.104$	$-W_{Av} = 23722.192$		$-W_{Av} = 23722.170$

Table 8^9. CGHF values (a.u.) of the total average energies for alternative normal configurations of selected elements (asterisks indicate the configuration containing the lowest observed level)

24, Cr		29, Cu		41, Nb	
$3d^4 4s^2$	$3d^5 4s$*	$3d^9 4s^2$	$3d^{10} 4s$*	$4d^3 5s^2$	$4d^4 5s$*
−1043.1762	−1043.1418	−1638.9501	−1638.9637	−3753.4918	−3753.4914
42, MO		44, Ru		45, Rh	
$4d^4 5s^2$	$4d^5 5s$*	$4d^6 5s^2$	$4d^7 5s$*	$4d^7 5s^2$	$4d^8 5s$*
−3975.3422	−3975.3686	−4441.3715	−4441.4559	−4685.7217	−4685.8369
46, Rd			47, Ag		
$4d^8 5s^2$	$4d^9 5s$	$4d^{10}$*	$4d^9 5s^2$	$4d^{10} 5s$*	
−4937.7830	−4937.8896	−4937.9210	−5197.5179	−5197.6985	
57, La			58, Ce		
$4f6s^2$	$5d6s^2$*	$5d^2 6s$	$4f5d6s^2$*	$4f^2 6s^2$	
−8221.0638	−8221.0667	−8221.0730	−8566.8484	−8566.877	
64, Gd		65,Tb			
$4f^8 6s^2$	$4f^7 5d6s^2$*	$4f^9 6s^2$	$4f^8 5d6s^2$*		
−10820.275	−10820.127	−11226.301	−11226.141		

these result because the observed normal configurations are assigned according to the lowest-lying level rather than the average energy. For example, the CGHF total configuration average energies listed in Table 8^9 indicate that for neutral chromium the $3d^4 4s^2$ configuration average lies lower than the $3d^5 4s$ configuration average. The energies of the lowest terms of these configurations, relative to their configuration averages, are given (Appendix 3, cases 25, 41) as

$$E[(3d^4 4s^2)\,^5D] = -\frac{105}{441}F^2(3d, 3d) - \frac{105}{441}F^4(3d, 3d)$$

$$E[(3d^5 4s)\,^7S] = -\frac{175}{441}F^2(3d, 3d) - \frac{175}{441}F^4(3d, 3d) - \frac{1}{2}G^2(3d, 4s)$$

in which the CGHF values (in a.u.) of the electrostatic interaction integrals, as computed by Froese Fischer [8], are $F^2(3d, 3d) = 0.3422$, $F^4(3d, 3d) = 0.2122$ for the $3d^4 4s^2$ configuration, and $F^2(3d, 3d) = 0.2999$, $F^4(3d, 3d) = 0.1844$, $G^2(3d, 4s) = 0.0386$ for the $3d^5 4s$ configuration. Then $E[(3d^4 3s^2)\,^5D] = -0.1320$, $E[(3d^5 4s)\,^7S] = -0.2115$; consequently,

$$W[(3d^4 4s^2)\,^5D] = -1043.3082$$
$$W[(3d^5 4s)\,^7S] = -1043.3533$$

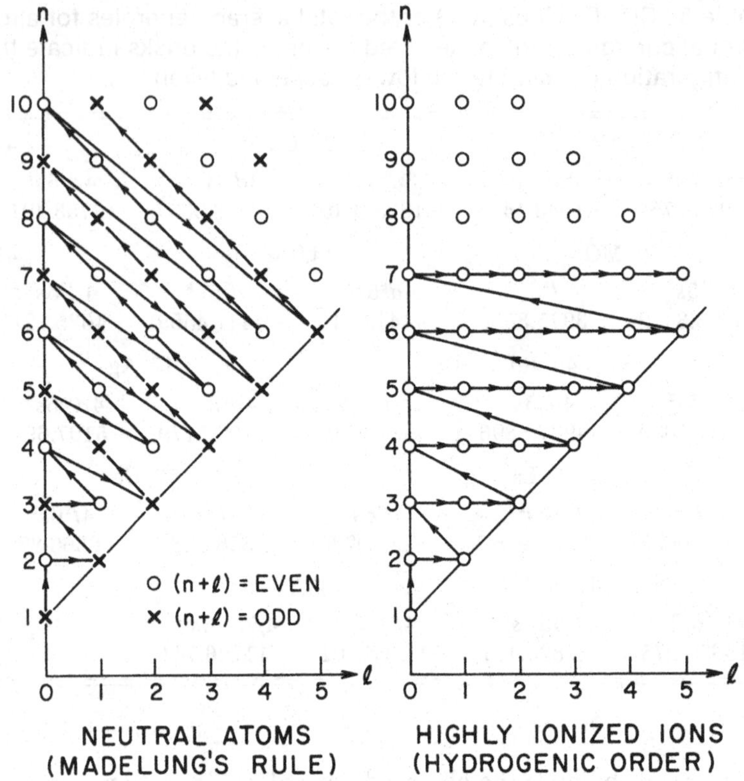

Figure 11^9. The order in which electron shells are filled.

showing that the lowest term of the $3d^54s$ configuration lies lower than the lowest term of $3d^44s^2$ configuration.

The order of electron shells for atoms (Madelung's rule), as well as for highly ionized ions (hydrogenic order) is shown schematically in Figure 11^9. Even though the order for atoms is quite different from the hydrogenic case, the symmetries involved in both cases are very remarkable. This has been studied by Neubert [9a], who also proposed a new periodic system that exhibits a double-shell structure horizontally and vertically, by Löwdin [9b], and by Novaro and Berrondo [9c].

The dynamical group-theoretical treatment of the periodic system of elements is attributable to Barut [10]. Starting from the fact that a special representation of the dynamical group $SO(4, 2)$ provides the quantum numbers n, l, m for the hydrogen atom [8^7], and taking each atom or ion in its normal configuration as a "state" he demonstrated that hydrogenic order (for highly ionized ions) can be explained by the chain

$$SO(4, 2) \rightarrow SO(4, 1) \rightarrow SO(4) \rightarrow SO(3) \tag{1}$$

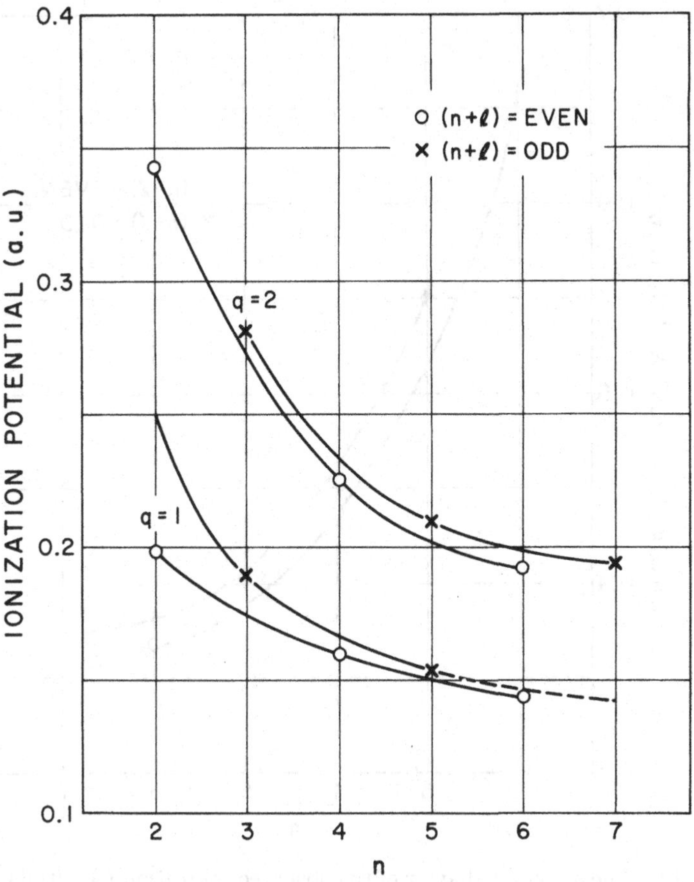

Figure 12[9]. The smooth behavior of the observed ionization potentials, for the $(ns)^q$ configurations, with changing n and fixed l quantum numbers, when they are considered in two groups according to whether $(n + l)$ is odd or even.

whereas Madelung's rule is given by the reduction

$$SO(4, 2) \rightarrow SO(3, 2) \rightarrow SO(3) \otimes SO(2) \tag{2}$$

Because two representations of $SO(3, 2)$ occur [11] in the particular representation of $SO(4, 2)$ whose multiplicity pattern coincides with the structure of the periodic table, the neutral elements are sorted in two groups according to whether $(n + l)$ is even or odd (Fig. 11[9]). When considered separately, redefining the quantum numbers as

$$(n, l, m) \rightarrow \left(n' = \frac{n + l + 1}{2}, l, m \right); \quad \text{for odd} \quad (n + l)$$

$$(n, l, m) \rightarrow \left(n' = \frac{n + l}{2} + 1, l, m \right); \quad \text{for even} \quad (n + l)$$

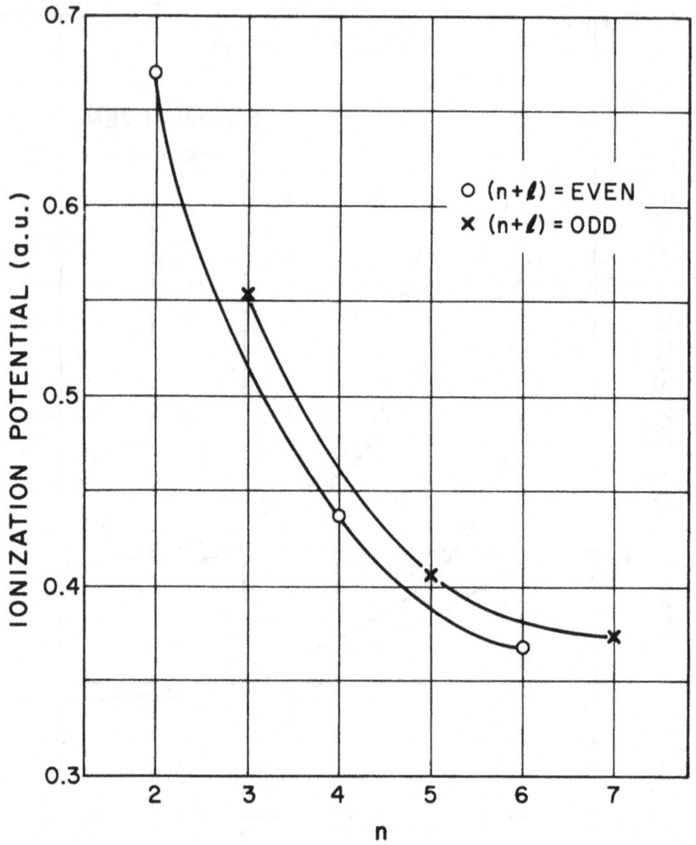

Figure 13⁹. The smooth behavior of the observed ionization potentials, for the *ns* configuration of singly ionized ions, with changing n and fixed l quantum numbers, when they are considered in two groups according to whether $(n + l)$ is odd or even.

each group is filled in hydrogenic order. The order of filling is such that after one shell belonging to one group is filled the next shell of the other group begins to fill.

Consequently, we expect that the properties of the neutral or lightly ionized elements can be considered in two groups according to whether $(n + l)$ is even or odd, and that a certain property, in a given group, shows smooth behavior with changing n and fixed l numbers. Indeed this seems to be the case [12]. For example, the observed ionization potentials and the radial expectation values $\langle ns^q|r^{-2}|ns^q\rangle$, as taken from Mann's calculations [33⁸b], for the alkali atoms and alkaline earth atoms and singly ionized ions fall on separate smooth curves according to whether $(n + l)$ is even or odd, as shown in Figures 12⁹, 13⁹, and 14⁹.

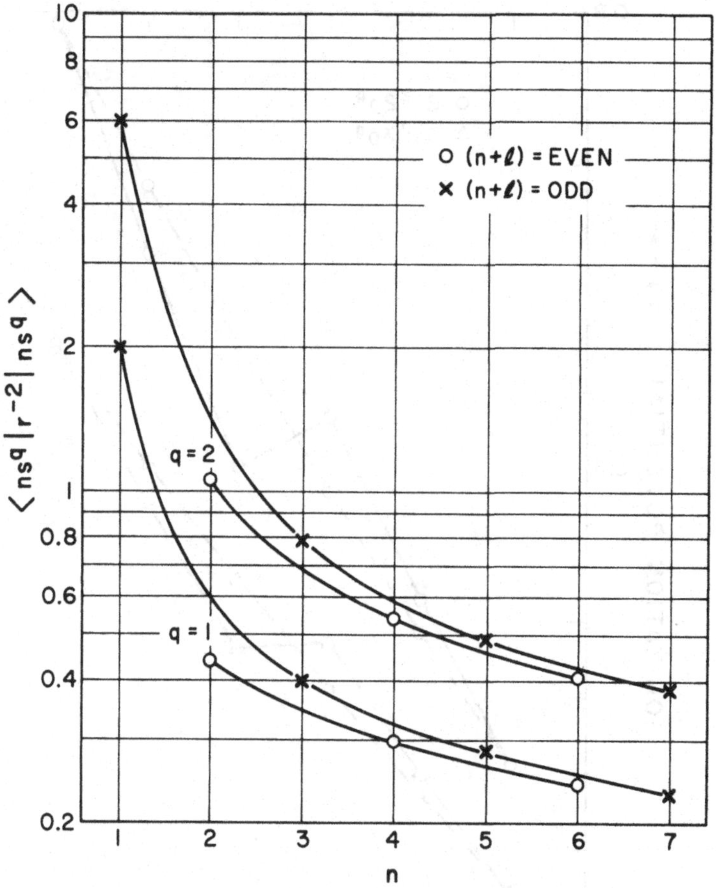

Figure 14[9]. The smooth behavior of the radial expectation values $\langle ns^q|r^{-2}|ns^q\rangle$ (in units of the radius of first Bohr orbit), as taken from Mann [33[8]b], with changing n and fixed l quantum numbers, when they are considered in two groups according to whether $(n + l)$ is odd or even.

3[9]. Observed and HF ionization potentials

It is instructive to see how well the HF approximation results for ionization potentials, which are the negative of binding energies, agree with observation. Here we concentrate on comparing the HF results for atoms and ions with $Z \leqslant 30$, because the total term energies for ions in this region are available from Clementi's computation [35[8]] and also because, in spite of the fact that recently more extensive HF calculations [46[8]] became available, relativistic effects makes it meaningless to compare the conventional HF binding energies, which do not contain any relativistic corrections, with the observed ones for atoms and ions with $Z \geqslant 30$.

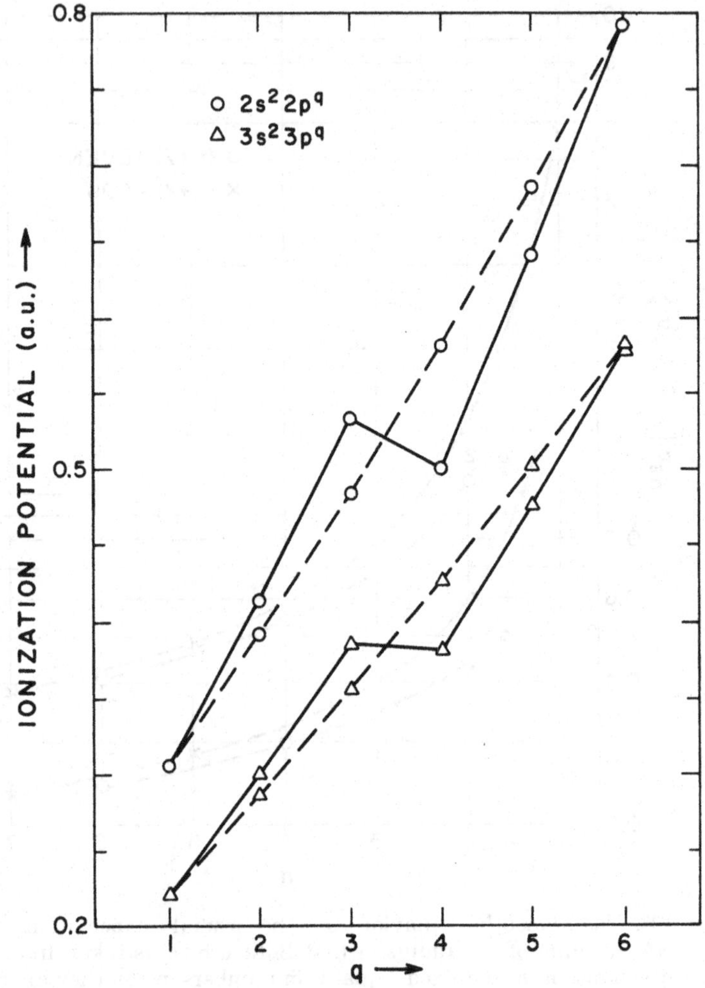

Figure 15[9]. The variation of observed ionization potentials with q for the ns^2np^q configurations (given in Table 9[9]). Values corresponding to the points joined by solid lines are measured from the lowest level of the atom to the lowest level of the ion. Values corresponding to the points joined by broken lines are from the configuration average to the configuration average.

The observed ionization potentials (absolute term values) exhibit regularities along periods as well as along isoelectronic sequences. The regularities in the variation along periods can be demonstrated by taking the ns^2np^q normal configurations as examples. When the observed values of ionization potentials, from Moore's tables [8[4], 9[4]f], are plotted against q, the curve connecting the points consists of two smooth branches separated by a discontinuity at $q = 3$ (solid lines in

Table 9⁹. The observed values of ionization potentials for the ns^2np^q configurations. The first entry is measured from the lowest level of atom to the lowest level of ion, and the second entry is from the configuration average to the configuration average (in a.u.).

	q					
n	1	2	3	4	4	6
2	0.3050	0.4138	0.5341	0.5005	0.6403	0.7925
	0.3050	0.3919	0.4843	0.5823	0.6855	0.7937
3	0.2200	0.2996	0.3854	0.3807	0.4765	0.5792
	0.2200	0.2857	0.3547	0.4268	0.5024	0.5813

Figure 15⁹). But when the differences of the average energies of normal configurations are considered, all points fall on a smooth curve as shown by broken lines in Figure 15⁹. These values are also tabulated in Table 9⁹. Theoretically this is expected since the average interaction of a p-shell electron with remaining electrons is given by 9⁴21 and 9⁴23, in the approximation of Koopman's theorem (Sec. 3⁸)

$$w(p, q) = W_0(p) - 2F^0(s, p) + \frac{1}{3} G^1(s, p) - (q - 1)F^0(p, p) + 2\frac{q-1}{25} F^2(p, p) \qquad (1)$$

in which $W_0(p)$ is the energy that contains the interaction of a p-shell electron with the nucleus and the core electrons. Then the average binding energy, $w(p,q)$, is a smoothly varying function of q, because all F^k and G^k integrals are (Sec. 1⁹). Now the ionization potentials can be expressed in terms of $w(p, q)$ as (App. 3)

$$\begin{aligned}
W(s^2p;\,^2P) - W(s^2;\,^1S) &= -w(p, 1) \\
W(s^2p^2;\,^3P) - W(s^2p;\,^2P) &= -w(p, 2) - \frac{3}{25} F^2(p, p) \\
W(s^2p^3;\,^4S) - W(s^2p^2;\,^3P) &= -w(p, 3) - \frac{6}{25} F^2(p, p) \\
W(s^2p^4;\,^3P) - W(s^2p^3;\,^4S) &= -w(p, 4) + \frac{6}{25} F^2(p, p) \\
W(s^2p^5;\,^2P) - W(s^2p^4;\,^3P) &= -w(p, 5) + \frac{3}{25} F^2(p, p) \\
W(s^2p^6;\,^1S) - W(s^2p^5;\,^2P) &= -w(p, 6)
\end{aligned} \qquad (2)$$

which clearly indicates the discontinuity at $q = 3$.

The regularities in the behavior of ionization potentials along iso-

Table 10[9]. Ionization potentials for the normal configurations of the elements in $N = 2, 3, 4$ isoelectronic sequences. The first entry is the observed and the second the HF value (in a.u.).

Z, Element	N = 2		3		4	
	IP	IP/C^2	IP	IP/C^2	IP	IP/C^2
2, He	0.90357	0.90357	—	—	—	—
	0.86168	0.86168	—	—	—	—
3, Li	2.77972	0.69493	0.198142	0.198142	—	—
	2.73641	0.68410	0.196313	0.196313	—	—
4, Be	5.65559	0.62840	0.669247	0.167312	0.342600	0.342600
	5.61126	0.62347	0.666073	0.166518	0.295685	0.295685
5, B	9.53186	0.59574	1.393924	0.154880	0.924423	0.231106
	9.48619	0.59289	1.389712	0.154412	0.861654	0.215413
6, C	14.40893	0.57636	2.370107	0.148132	1.759843	0.195538
	14.36115	0.57445	2.364823	0.147801	1.682507	0.186945
7, N	20.28825	0.56356	3.597395	0.143896	2.847098	0.177944
	20.23613	0.56211	3.590650	0.143626	2.755527	0.172220
8, O	27.17003	0.55449	5.075793	0.140994	4.185708	0.167428
	27.11112	0.55329	5.066876	0.140746	4.079705	0.163188
9, F	35.05557	0.54774	6.805473	0.138887	5.775709	0.160436
	34.98611	0.54666	6.793345	0.138640	5.654580	0.157072
10, Ne	43.94586	0.54254	8.786717	0.137292	7.617243	0.155454
	43.86110	0.54149	8.76998	0.137031	7.47992	0.152651
11, Na	53.84248	0.53842	11.020405	0.136054	9.708636	0.151697
	53.73609	0.53736	10.99672	0.135762	9.55557	0.149306

electronic sequences may be understood by the following argument. The ionization potential of a hydrogenic ion of charge Z is equal to Z^2R. To the extent that an electron moving in a central field may be described by an effective screening charge $Z-S$, an approximation that becomes valid in the limit of large Z, its ionization potential, will be given by $(Z - S)^2R$.

The behavior of the observed ionization potentials along isoelectronic sequences are investigated in detail by Edlén [1] and the regularities in the behavior are used to obtain accurate ionization potentials for atoms and ions with 3–10 electrons [13].

The observed ionization potentials that are compared with Hartree-Fock values [35[8]] below have been obtained from a tabulation by C. E. Moore [9[4]]. In the conventional nonrelativistic HF calculations the magnetic interactions are not included. Therefore, the HF ionization potentials are the energy differences between the lowest terms before

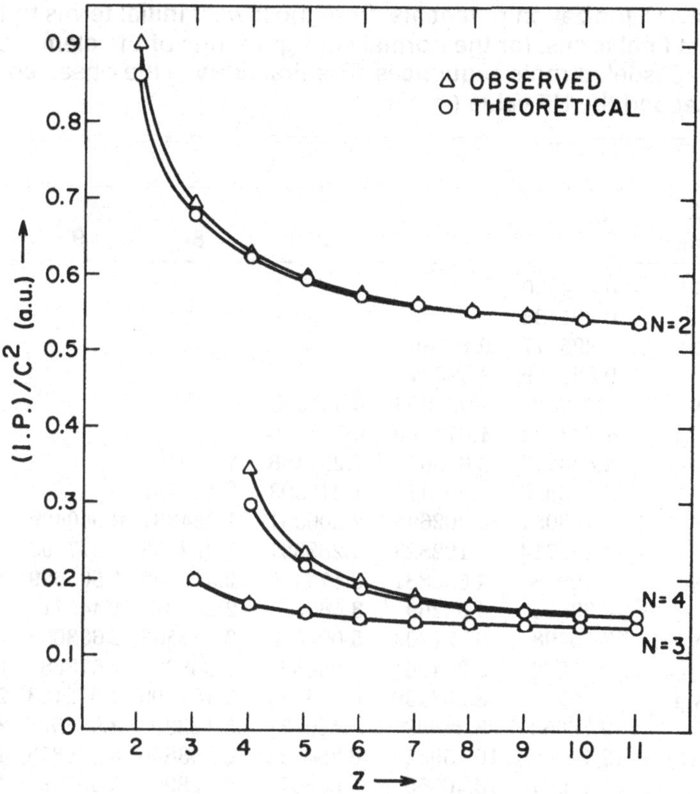

Figure 16⁹. Normalized ionization potentials, $(IP)/C^2$ for the normal configurations of the elements in $N = 2, 3, 4$ isoelectronic sequences (Table 10⁹).

and after ionization. On the other hand, the observed ionization potentials are measured from the lowest level to the lowest level. Consequently, when comparing these two values, the observed ones should be corrected accordingly. That is, the energy differences ΔT_i and ΔT_f of the lowest levels with the weighed averages of all levels belonging to the lowest (normal) terms, before and after ionization, should be subtracted and added, respectively, to the observed ionization potentials. For the $N = 2, 3, 4, 11, 12$ isoelectronic sequences no such correction is needed. In the other isoelectronic sequences this correction to the observed ionization potentials, regardless of how small, has been carried out. All observed values are converted to atomic units by using R_∞.

The observed and the HF values of ionization potentials in the He, Li, Be isoelectronic sequences are listed in Table 10⁹. Agreement between theory and observation is better than 5 percent for the He, 1

Table 11^9. Ionization potentials, from the lowest initial terms to the lowest final terms, for the normal configurations of the elements in $N = 5$–10 isoelectronic sequences. The first entry is the observed and the second the HF value (in a.u.).

Z, Element	N: 5	6	7	8	9	10
5, B	0.304900					
	0.291494					
6, C	0.895877	0.413866				
	0.883729	0.396371				
7, N	1.743198	1.087952	0.534523			
	1.733473	1.072192	0.512930			
8, O	2.843729	2.019070	1.291486	0.500099		
	2.836990	2.005472	1.272393	0.436791		
9, F	4.196081	3.202698	2.306381	1.284387	0.640429	
	4.192714	3.192823	2.289331	1.222708	0.577603	
10, Ne	5.79988	4.638881	3.572420	2.330489	1.505629	0.793667
	5.79983	4.63269	3.56061	2.26781	1.44571	0.729330
11, Na	7.65498	6.327203	5.091543	3.632566	2.633052	1.739838
	7.65788	5.32426	5.08468	3.56833	2.57368	1.67955
12, Mg	9.76194	8.267436	6.863244	5.187509	4.015183	2.948641
	9.76663	8.26699	6.86067	5.12239	3.95752	2.89025
13, Al	12.12049	10.45958	8.886432	6.993800	5.649813	4.414846
	12.12587	10.46055	8.88807	6.92890	5.59529	4.35735
14, Si	14.73127	12.90403	11.16189	9.046793	7.541002	6.136466
	14.73549	12.90479	11.16645	8.98721	7.48582	6.07876
15, P	17.59489	15.60136	13.68979	11.35225	9.681048	8.111746
	17.59539	15.59952	13.69564	11.29587	9.62837	8.05325

percent for the Li and 15 percent for the Be isoelectronic sequence. The general rule is that the theoretical values are always smaller than the observed ones and that agreement gets better, at least percentage-wise, with increasing degree of ionization.

The normalized ionization potentials, defined as the ionization potentials divided by $C^2 = (Z - N + 1)^2$, are also included in Table 10^9 and plotted in Figure 16^9. This figure shows that these values approach, first rapidly and then slowly, their hydrogenic limit (that is, 0.5 a.u. for $n = 1$, 0.125 a.u. for $n = 2$) smoothly, as expected.

The observed and the HF values of ionization potentials in the B, C, N, O, F, Ne isoelectronic sequences are presented in Table 11^9. For the B, C, N sequences the HF values are at worst 5 percent smaller than

Table 12^9. Ionization potentials for the normal configurations of the elements in $N = 11$, 12 isoelectronic sequences. The first entry is observed and the second is HF value (in a.u.).

Z, Element	N = 11 IP	N = 11 IP/C^2	N = 12 IP	N = 12 IP/C^2
11, Na	0.188857	0.188857		
	0.18181	0.18181		
12, Mg	0.552536	0.138134	0.280994	0.280994
	0.54081	0.13520	0.24300	0.24300
13, Al	1.045431	0.116159	0.691927	0.172982
	1.03023	0.11447	0.64384	0.16096
14, Si	1.658930	0.103683	1.230845	0.136761
	1.64046	0.10253	1.17412	0.13046
15, P	2.389615	0.095585	1.887746	0.117984
	2.36763	0.09471	1.82568	0.11410
16, S	3.235836	0.089884	2.670923	0.106837
	3.20981	0.08916	2.59471	0.10379
17, Cl	4.196616	0.085645	3.565787	0.099050
	4.16583	0.08502	3.47919	0.09664
18, Ar	5.272043	0.082376	4.568773	0.093240
	5.23497	0.08180	4.47792	0.09139
19, K	6.461201	0.079768	5.691317	0.088927
	6.41681	0.07922	5.59006	0.08734
20, Ca	7.764221	0.077642	6.928817	0.085541
	7.71094	0.07711	6.81507	0.08414
21, Sc	9.181377	0.075879	8.280681	0.082807
	9.11717	0.07535	8.15256	0.08153
22, Ti	10.712579	0.074393	9.747365	0.080557
	10.63532	0.07386	9.60234	0.07936

the observed ones. This degree of disagreement, for the worst case, increases to 7 percent in the O, 11 percent in the F, and 9 percent in the Ne isoelectronic sequences.

The observed and the HF values of ionization potentials in the Na and Mg isoelectronic sequences are reported in Table 12^9. The theoretical values for neutral sodium and magnesium are 4 percent and 15 percent smaller than the observed ones. Even though the difference between the HF and the observed values increases along the isoelectronic sequences, because of the increasing magnitude of the ionization potentials with increasing degree of ionization, the agreement gets better percentage-wise. For example, the theoretical values for singly

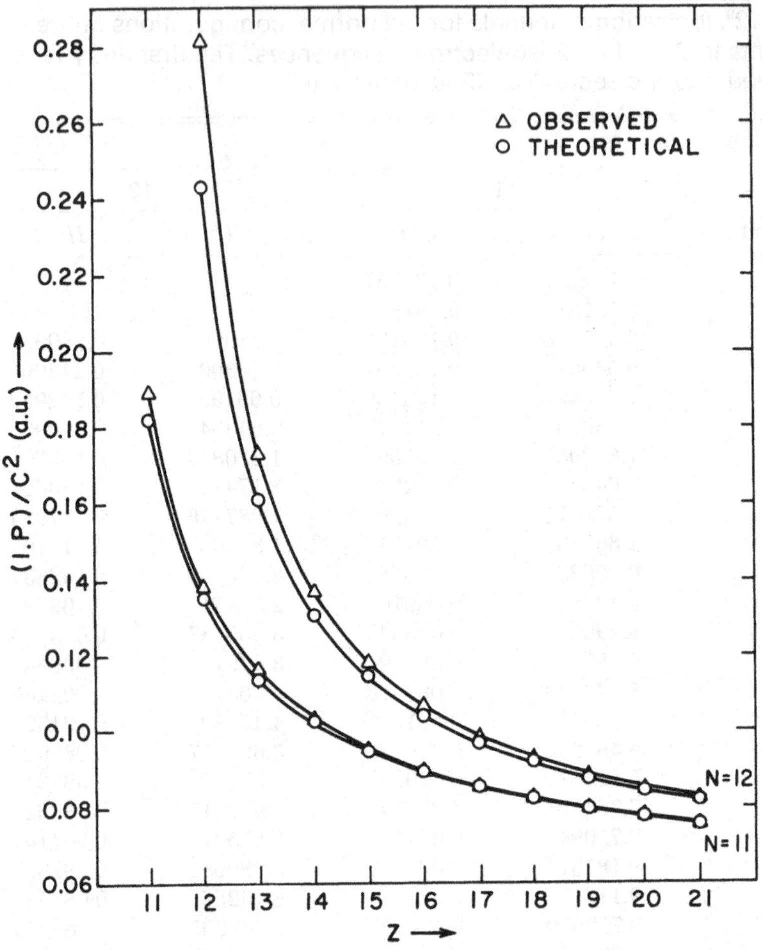

Figure 17[9]. Normalized ionization potentials, $(IP)/C^2$, for the normal configurations of the elements in $N = 11$, 12 isoelectronic sequencies (Table 12[9]).

ionized Mg and Al are 2 percent and 7.5 percent smaller than the observed ones.

The HF and the observed values of the normalized ionization potentials in the Na and Mg sequences are also included in Table 12[9]. The same data are plotted in Figure 17[9], which indicates that, although their behavior is like that of the He, Li, Be sequences, the convergence to the hydrogenic limit, 0.0555 a.u., requires many more degrees of ionization. In the chosen scale no deviation from smooth behavior is detectable along the isoelectronic sequence.

The observed and theoretical ionization potentials in the Al, Si, P, S,

Table 13⁹. Ionization potentials, from the lowest initial terms to the lowest final terms, for the normal configurations of the elements in $N = 13–18$ isoelectronic sequences. The first entry is the observed and the second the HF value (in a.u.).

Z, Element	N = 13	14	15	16	17	18
13, Al	0.219632					
	0.20222					
14, Si	0.599827	0.299759				
	0.57760	0.28144				
15, P	1.107312	0.725175	0.386812			
	1.08193	0.70491	0.36910			
16, S	1.735544	1.280202	0.860061	0.379826		
	1.70794	1.25552	0.84060	0.33184		
17, Cl	2.487783	1.965202	1.459889	0.873574	0.476773	
	2.45202	1.92716	1.43773	0.82235	0.43340	
18, Ar	3.337824	2.757444	2.204276	1.494727	1.015681	0.581329
	3.31205	2.71669	2.15529	1.43939	0.97015	0.54276
19, K	4.310798	3.675535	3.046829	2.234809	1.680804	1.165501
	4.28681	3.62210	2.99049	2.17760	1.63274	1.12581
20, Ca	5.398030	4.693311	4.010517	3.096640	2.466467	1.875759
	5.37547	4.64225	3.94147	3.03402	2.41628	1.83394
21, Sc	6.598296	5.832221	5.089042	4.075135	3.369281	2.707245
	6.57732	5.77631	5.00716	4.00677	3.31800	2.66283
22, Ti	7.911946	7.098477	6.215259	5.166136	4.387390	3.648143
	7.89195	7.02369	6.18671	5.09457	4.33604	3.60994
23, V	9.343396	8.469879	7.593755	6.370475	5.519949	4.711473
	9.31909	8.38393	7.47957	6.29659	5.46927	4.67340
24, Cr	10.914548	9.947621	9.021004	7.674293	6.790701	5.889253
	10.85839	9.8567	8.8854	7.6123	6.7168	5.8520

Cl, Ar isoelectronic sequences are given in Table 13⁹. (The observed ionization potentials given in the Ar isoelectronic sequence are mostly taken from the recent measurements.) Again the HF values are smaller by 9, 7, 5, 14, 10, and 7 percent from the observed ones, respectively, for the worst cases. This, coupled with the situation in the N = 5, 6, 7, 8, 9, 10 sequences indicates that agreement between the theory and observation is better for atoms and ions with less than half-filled outer p shells than for those with more than half-filled p shells. However, there is no general trend that can be related to the specific number of electrons. In the range of Z considered, these discrepancies between observation and HF results are mostly attributable to electron correlation.

Table 14[9]a. Differences in term energies (a.u.) for the np^2 configuration. The first entry is the observed and the second the Hartree-Fock value.

n	Element	$^1D - {}^3P$	$^1S - {}^3P$	$^1S - {}^1D$
2	C	.046306	.098501	.052195
		.057277	.139015	.081738
3	Si	.028018	.069460	.041442
		.03921	.09558	.05637
4	Ge	.02805	.07016	.04211
		.0383	—	—

Table 14[9]b. Differences in term energies (a.u.) for the np^3 configuration. The first entry is the observed and the second the Hartree-Fock value.

n	Element	$^2D - {}^4S$	$^2P - {}^4S$	$^2P - {}^2D$
2	N	.087609	.131401	.043792
		.104766	.172835	.068069
3	P	.051806	.085384	.033577
		.06994	.11549	.04555
4	As	.04914	.08426	.03512
		.0664	.1099	.0435

Table 14[9]c. Differences in term energies (a.u.) for the np^4 configuration. The first entry is the observed and the second the Hartree-Fock value.

n	Element	$^1D - {}^3P$	$^1S - {}^3P$	$^1S - {}^1D$
2	0	.071944	.153615	.081671
		.080147	.198392	.118245
3	S	.041202	.100167	.058965
		.05264	.13035	.07771
4	Se	.03933	.09797	.05864
		.0473	.1192	.0719

4[9]. Observed and HF term energies for the normal configurations of atoms

The total HF term energies for the normal configuration of certain atoms and ions have been calculated by Clementi [35[8]]. Because each energy represents an independent self-consistent field calculation, the

Table 15⁹. Values of $F^2(np, np)$ for the np^4 configuration, computed from the observed term values (first line), from the HF term values (second line), by a best fit to the experimental data (first line of the last column) and by direct calculation from the HF wave functions of the lowest terms (second line of the last column) (in a.u.).

n	Element	$^1D - {}^3P$	$^1S - {}^3P$	$^1S - {}^1D$	Best fit HF
2	0	.2998	.2560	.2269	.2997
		.3340	.3307	.3284	.3361
3	S	.1717	.1670	.1638	.1714
		.2193	.2173	.2159	.2201
4	Se	.1639	.1633	.1629	.1579
		.1971	.1987	.1997	.2039

differences in total energies for individual terms of the same configuration may be compared directly with the observed differences in term values. Examples of such comparisons are given in Tables 14⁹a, 14⁹b, and 14⁹c for np^2, np^3 and np^4 configurations with $n = 2, 3$, and 4, where the observed values [8⁴, 9⁴f] represent averages over J and have been converted to atomic units using R_∞.

The tables show that the Hartree-Fock calculations predict larger term splittings within the configurations considered than are observed. Although the agreement between observed and calculated values becomes somewhat better in those cases treated here for larger Z, the discrepancies are still large. This is at first surprising, because the spacings between term values may be expressed in terms of a single parameter $F^2(np, np)$ in each case considered by using the relations given in cases 3 and 9 of Appendix 3. This parameterization, which assumes that the core energy remains constant and that the integrals $F^2(np, np)$ are the same for each term in the configuration (approximations that are not made in the Hartree-Fock calculations), produces a reasonable fit to the observed term splittings. This is shown for the np^4 configuration in Table 15⁹ where an effective parameter $F^2(np, np)$ obtained from the observed term splittings, using the relations of Appendix 3, is given for each case. Also shown in Table 15⁹ are effective F^2 (np, np) values calculated using the HF term energies, values of $F^2(np, np)$ computed directly by evaluating the numerical integrals using wave functions for the lowest terms obtained by a HF calculation [44⁸a], and $F^2(np, np)$ values found by a best fit to the experimental data [14]. Note that the effective $F^2(np, np)$ values [1] obtained from the HF term energies are almost constant and close to the value obtained by direct evaluation but are systematically larger than those obtained from the observed data.

Table 16[9]a. $3d^2$ and $3d^8$ term energies (a.u.) relative to the lowest term of the configuration. First entry is observed, second is from term splittings obtained from Hartree-Fock calculations [35[8]], and the third is computed with $F^2(3d, 3d)$ and $F^4(3d, 3d)$ radial integrals found by employing HF wave functions for the lowest terms [44[8]a]

Element	${}^3P - {}^3F$	${}^1G - {}^3F$	${}^1D - {}^3F$	${}^1S - {}^3F$
Ti	.03793	.05420	.03204	
	.05757	.07377	.04752	.17518
	.05849	.07552	.04823	.18634
Ni	.06709	.09628	.05718	
	.0888	.1145	.0732	.2811
	.08888	.11472	.07325	.28303

Table 16[9]b. $3d^3$ and $3d^7$ term energies (a.u.) relative to the lowest term of the configuration. First entry is observed, the second is from term splittings obtained from Hartree-Fock calculations [35[8]], and the third is computed with $F^2(3d, 3d)$ and $F^4(3d, 3d)$ radial integrals found by employing HF wave functions for the lowest terms [44[8]a].

Element	${}^4_3P - {}^4_3F$	${}^2_3H - {}^4_3F$	${}^2_3G - {}^4_3F$	${}^2_3F - {}^4_3F$	${}^2_1D - {}^4_3F$	${}^2_3D - {}^4_3F$	${}^2_3P - {}^4_3F$
V	.04281	.06776	.04870			.06477	.06144
	.06412	.08538	.06429	.14787			.08538
	.06468	.08653	.06497	.15121	.23753	.09402	.08653
CO	.06007	.09707	.07297				.09088
	.0842	.1123	.0844	.1960			.1123
	.08438	.11278	.08466	.19716	.30967	.12252	.11278

Table 17[9]. Values of $F^2(3d, 3d)$ and $F^4(3d, 3d)$. The first entry is obtained from HF wave functions [44[8]a] and the second from observed data (in a.u.).

Z	Element	$F^2(3d, 3d)$	$F^4(3d, 3d)$
22	Ti	.2916	.1810
		.1837	.1056
23	V	.3226	.2004
		.2015	.1305
27	Co	.4207	.2611
		.3398	.2256
28	Ni	.4430	.2748
		.3691	.2492

The simple example given above is typical of results obtained by HF calculations for other configurations. In cases where the formulas of Appendix 3 can be used to obtain a good fit of observed term differences by treating the Slater F and G integrals as free parameters the values of these parameters are found to be systematically smaller than those obtained directly from HF calculations.

Another example of this type is shown in Tables 16⁹a and 16⁹b where term splittings for $3d^2$, $3d^8$, $3d^3$, $3d^7$ configurations are presented. The theoretical term splittings shown in the tables are computed both directly from the energies of individual terms as given by Clementi and by using the formulas of Appendix 3 and the F^2 and F^4 values obtained from Froese's HF calculation for the lowest term. The tables show that the Hartree-Fock values of F^2 and F^4 are larger than those obtained directly from the experimental data. Direct comparison of the F^2 and F^4 values is shown in Table 17⁹.

The reason that the Hartree-Fock method fails to produce term splittings that agree with the experiment is that it ignores the effects of interactions with other configurations. In spite of this defect, calculations by the Hartree-Fock method serve a useful purpose in providing estimates of electrostatic interaction parameters; although the values of these parameters are generally too large, their variation along periods or isoelectronic sequences as well as their ratio to values obtained directly from observation is uniform. Values obtained from Hartree-Fock calculations suitably scaled may therefore be used as estimates.

5⁹. Isoelectronic sequences; observed and calculated term energies

An isoelectronic sequence starts with a neutral atom and contains all ions with increasing Z with the same number of electrons. The Z dependence of term values for isoelectronic sequences may be studied using the concept of effective screening. Equation 1⁹1 may be rewritten as in 3⁴4

$$T_{nl} = -R\,\frac{(Z - S_{nl})^2}{n^2} \tag{1}$$

where $Z - S_{nl}$ is the effective screening charge. The parameter S_{nl} thus represents the effect of screening by core electrons. If the effective screening were complete, (1) would reduce to the Rydberg formula for an ion of charge $C = Z - N + 1$ with $S_{nl} = N - 1$. Therefore, it is useful to rewrite (1) as

$$T_{nl} = -R\,\frac{(C + P_{nl})^2}{n^2} \tag{2}$$

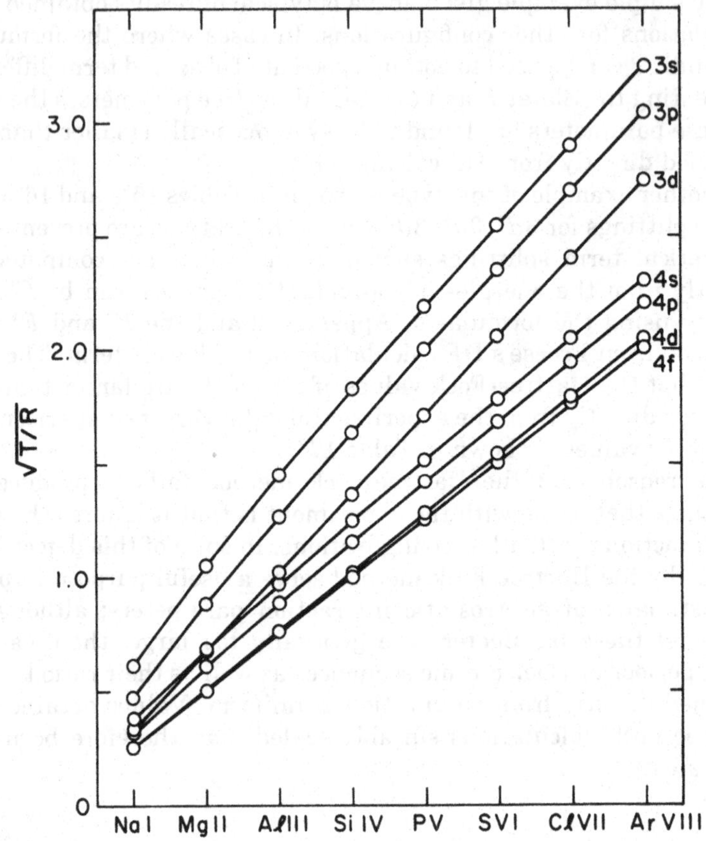

Figure 18[9]. Observed values of $\sqrt{T/R}$ in the $N = 11$ isoelectronic sequence.

where P_{nl} represents the deviation of the effective nuclear charge from complete screening by the core electrons and thus is a measure of the core penetration of a single electron.

The use of (1) and (2), which characterize term values by effective screening or penetration numbers, may be considered as an alternative to 1^91.

The relationships between S_{nl}, P_{nl} and the quantum defects of 1^91 are

$$S_{nl} = Z\left(1 - \frac{n}{n - \Delta_{nl}}\right) + \frac{n}{n - \Delta_{nl}}(N - 1) \tag{3}$$

and

$$P_{nl} = \frac{(Z - N + 1)\Delta_{nl}}{(n - \Delta_{nl})} \tag{4}$$

Table 18^9. Some observed values of penetration parameters P_{nl} in the $N = 11$ isoelectronic sequence, as computed with term values given by C.E. Moore [8^4, 9^4f].

nl	Na I	Mg II	Al III	Si IV	P V	S VI	Cl VII	Ar VIII
$3s$	0.84377	1.15372	1.33799	1.46455	1.55849	1.63191	1.69139	1.74157
$3p$	0.41699	0.64867	0.79509	0.89789	0.97492	1.03518	1.08386	1.12483
$3d$	0.00344	0.02025	0.05088	0.08753	0.12505	0.16056	0.19295	0.22279
$4s$	0.51344	0.73924	0.88064	0.98028	1.05560	1.11500	1.16332	1.20528
$4p$	0.27668	0.43384	0.53616	0.60936	0.66512	0.70912	0.74492	0.77552
$4d$	0.00308	0.01892	0.04664	0.07848	0.10960	0.13792	0.16284	0.18572
$4f$	0.00016	0.00124	0.00260	0.00416	0.00592	0.00760	0.00928	0.01184

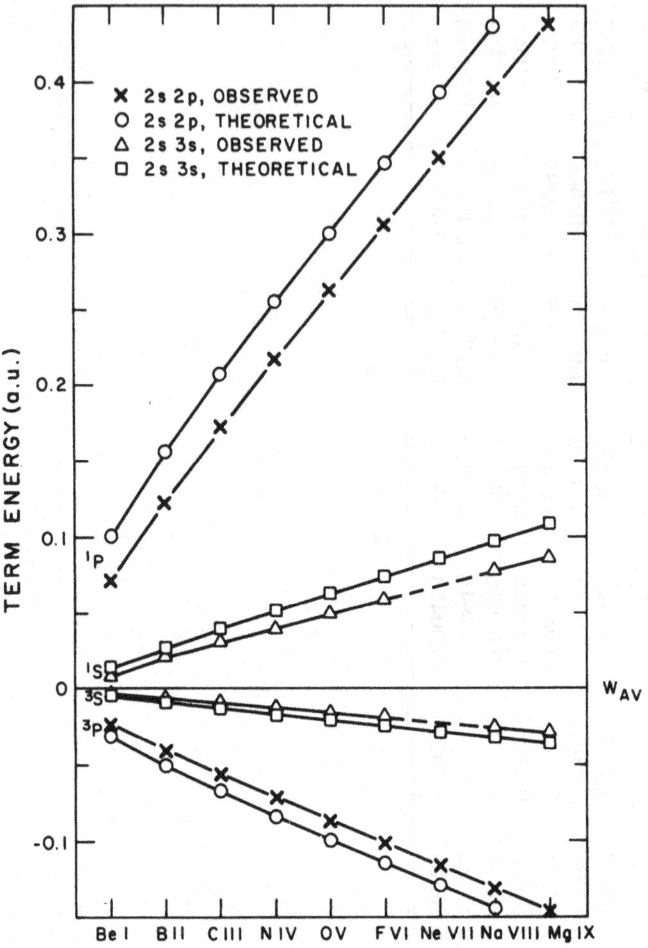

Figure 19[9]. Comparison of the observed [8[4], 9[4]] and the HFS $(2s2p)^3P$, ^1P and $(2s3s)^3S$, ^1S terms in $N = 4$ (beryllium) isoelectronic sequence.

(3) shows that for small values of Δ_{nl} (or for large n) S_{nl} is independent of Z. To the extent that this is so, plots of $\sqrt{T/R}$ should be linear when plotted as a function of Z (or $Z - N + 1$) for given n and l. The linearity of such plots is illustrated in Figure 18[9], which shows the observed values of $\sqrt{T/R}$ for the $n = 3$ and 4, $l = 0, 1, 2, 3$ terms of the $N = 11$ isoelectronic sequences. Penetration parameters obtained from the observed data are presented in Table 18[9].

Graphs similar to Figure 18[9] may be used to study the regularities of term structure along isoelectronic sequences. Edlén [1 pp. 150–201] gives many examples of this. The graphical methods used also clearly

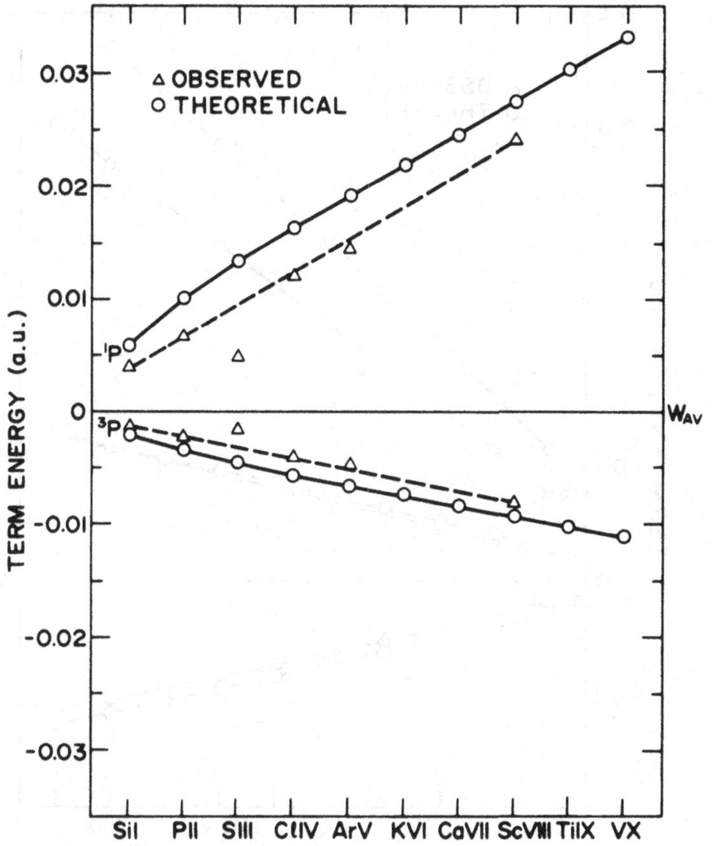

Figure 20⁹. Comparison of the observed [8⁴, 9⁴] and the HFS $(3p4s)^3P$, ^1P terms in the $N = 14$ (silicon) isoelectronic sequence.

show in specific cases evidence of term perturbation and provide estimates of the approximate magnitudes of such perturbations.

In comparing observed energies along isoelectronic sequences with calculated values, a somewhat different procedure than is used for absolute term values is appropriate. Because term energies relative to the configuration average may always be represented by a linear combination of Coulomb integrals as given in Appendix 3 (provided that magnetic interactions are ignored), such energies will have the same Z dependence as the Coulomb integrals. The Coulomb integrals however, represent the radial parts of the basic Coulomb interaction integral given by 9⁴2. That is,

$$\left\langle a, b \left| \frac{e^2}{r_{12}} \right| c, d \right\rangle = \iint \rho(a, c; 1) \frac{1}{|\mathbf{r}_1 - \mathbf{r}_2|} \rho(b, d; 2)\, d\tau_1\, d\tau_2 \quad (5)$$

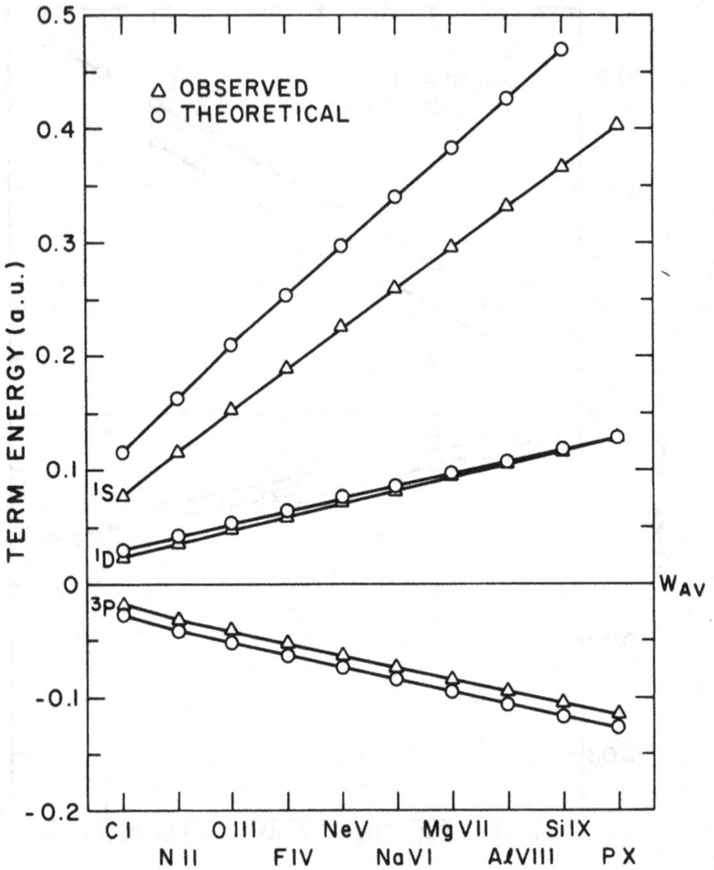

Figure 21⁹. Comparison of the observed [8⁴, 9⁴] and the HFS $(2s^22p^2)^3P$, ^1D, ^1S terms in the $N = 6$ (carbon) isoelectronic sequence.

If (5) is evaluated with hydrogenic wave functions as in Table 9⁴ a simple scaling law exists since $\rho(a, c, 1)$ and $\rho(b, d, 2)$ represent normalized charge distributions. Thus the only Z-dependent term in the integral will be $1/|\mathbf{r}_1 - \mathbf{r}_2|$, which by 1⁴8 can be represented as $Z/|\boldsymbol{\rho}_1 - \boldsymbol{\rho}_2|$ for all Z. The Coulomb integrals given in Table 9⁴ evaluated using hydrogenic wave functions for $Z = 1$ therefore can be used to obtain estimates of Coulomb integrals for arbitrary Z simply by multiplying the values given by Z.

Although the actual values of individual Coulomb integrals may deviate from those given using hydrogenic wave functions, their dependence on Z is expected to be similar. Therefore, it is reasonable to plot term energy differences against Z. This procedure will be followed in comparing observed and calculated term energies.

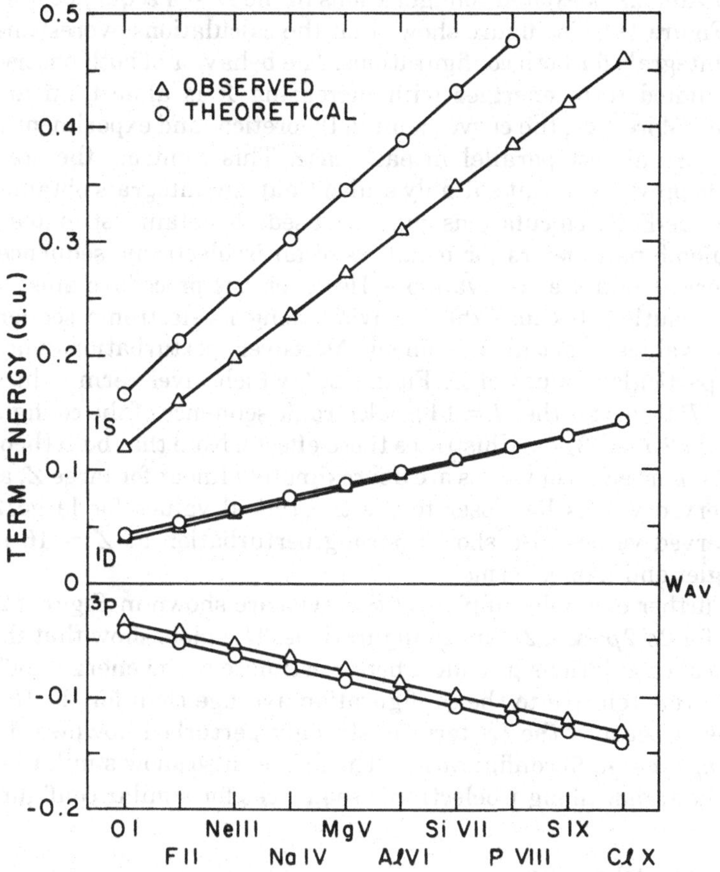

Figure 22[9]. Comparison of the observed [8[4], 9[4]] and the HFS $(2s^22p^4)^3P$, ^1D, ^1S terms in the $N = 8$ (oxygen) isoelectronic sequence.

Because relatively few HF calculations are available for term values of ions [35[8] and 46[8]], the discussion here will be limited to calculations made using the HFS procedure [6a]. The method used in these computations was to compute central-field wave functions for each case and to use these wave functions to obtain the electrostatic interaction integrals by numerical evaluation. The formulas of Appendix 3 were then used to compute theoretical term energies relative to the configuration average.

The simplest two-electron configurations resulting in more than one term are the $(ns, n'l)\ ^{1,3}L$ configurations where the separation between singlets and triplets as well as the deviation from the configuration average, is given by a single exchange integral $G^l(ns, n'l)$ (case 2 of App. 3). Observed [8[4], 9[4]] and theoretical [6a] term energies for the

$2s2p$ and $2s3s$ excited configurations in the $N = 4$ sequence are shown in Figure 19[9]. The figure shows that the calculations overestimate the G^l integrals for both configurations. The behavior of both observed and calculated term energies with increasing Z is almost linear in all cases. Moreover, the curves joining theoretical and experimental energies are almost parallel in each case. This confirms the statement made previously that suitably scaled Coulomb integrals obtained from Hartree-Fock calculations may be used to obtain estimates of the Coulomb parameters for members of an isoelectronic sequence when observed data are not available. However, the procedure must be used with caution, because the behavior along isoelectronic sequences of term values is not always linear. Moreover, perturbations may occur for particular values of Z. Figure 20[9], which gives term values of ^1P and ^3P terms for the $N = 14$ isoelectronic sequence of the configuration $(1s^22s^22p^63s^2)3p4s$, illustrates these effects. Note that both theoretical and observed term values are approximately linear for large Z, and the observed values lie closer to the theoretical values for large Z. The observed values also show a strong perturbation at $Z = 16$ in both singlet and triplet terms.

Further examples of plots of this type are shown in Figures 21[9] and 22[9] for $2s^22p^2$ and $2s^22p^4$ configurations. The plots show that the theoretical calculations provide a better estimate of the energies of ^3P and ^1D terms relative to the configuration average than for the ^1S terms. This is because the ^1S term is strongly perturbed downward by the $ns^0np^q(q = 4, 6)$ configuration. The figures also show similar behavior in the trends along isoelectronic sequences for similar configurations.

6[9]. Parental mixing

For configurations containing more than two open-shell electrons a term cannot be completely characterized by its configuration and SL values, because such a configuration may contain several terms of the same kind. This situation requires the introduction of the concept of parentage (5[4]) to complete the labeling of terms.

Even in cases where pure LS (Russell-Saunders) coupling is a good approximation, the classification of terms according to parentage requires an additional approximation. This is because the expressions for term energies based on a particular parent scheme contain off-diagonal elements. The actual term energies correspond to the proper-values of the energy matrix, so the assignment of parentage for a particular term means that the off-diagonal elements connecting that particular term with terms of the same L and S values but with different parentage must be small. As mentioned in Section 5[4], it often happens that this is the case. An example of parental mixing has

Table 19⁹. HFS values (a.u.) of diagonal elements of the energy submatrix (App. 3, case 11) for the $(1s^22s^22p^63s^23p^24p)$ ^2D, ^2P terms in the $N = 15$ sequence. First entry is before and second is after diagonalization.

Z, Element	$(^3P)\,^2D$	$(^1D)\,^2D$	$(^3P)\,^2P$	$(^1D)\,^2P$	$(^1S)\,^2P$
15, P	−.02253	.02512	−.01648	.03328	.10882
	−.02355	.02613	−.01756	.03405	.10913
16, S	−.02126	.02890	−.00818	.04526	.12993
	−.02614	.03377	−.01346	.04944	.13103
17, Cl	−.02048	.03251	−.00106	.05632	.15009
	−.03016	.04219	−.01166	.06488	.15214
18, Ar	−.02002	.03600	.00528	.06679	.16961
	−.03469	.05068	−.01089	.07992	.17265
19, K	−.01975	.03942	.01113	.07688	.18866
	−.03941	.05909	−.01061	.09458	.19271
20, Ca	−.01959	.04280	.01667	.08670	.20739
	−.04420	.06741	−.01061	.10892	.21244
21, Sc	−.01951	.04615	.02198	.09633	.22586
	−.04902	.07566	−.01077	.12303	.23192
22, Ti	−.01948	.04949	.02714	.10582	.24416
	−.05384	.08385	−.01105	.13695	.25121
23, V	−.01948	.05281	.03218	.11520	.26230
	−.05866	.09198	−.01141	.15073	.27036
24, Cr	−.01951	.05612	.03713	.12449	.28033
	−.06347	.10008	−.01182	.16438	.28938
25, Mn	−.01956	.05942	.04200	.13371	.29826
	−.06828	.10814	−.01227	.17794	.30831
26, Fe	−.01962	.06272	.04683	.14288	.31612
	−.07308	.11617	−.01275	.19142	.32715
27, Co	−.01970	.06601	.05160	.15199	.33391
	−.07787	.12418	−.01326	.20483	.34593
28, Ni	−.01979	.06930	.05633	.16107	.35164
	−.08265	.13217	−.01379	.21819	.36465

already been given in Figure 1⁹, which shows energy levels of terms converging to the ^4S, ^2D, and ^2P ionization limits of O II. The parentage assignments are rather obvious in this case owing to the large spacing between the ^4S, ^2D, and ^2P limits relative to the binding energies of the lowest terms and the existence of well-defined Rydberg series converging to each of the limits.

A single electron far outside of an open shell is well aware of its own parentage, because the off-diagonal elements of the energy matrix

Table 20[9]. HFS values (a.u.) of diagonal elements of the energy submatrix (App. 3, case 12) for the $(1s^22s^22p^63s^23p^23d)$ ^2F, ^2P, ^2D terms in the $N = 15$ sequence. First entry is before and second is after diagonalization.

Z, Element	$(^3P)\,^2F$	$(^1D)\,^2F$	$(^3P)\,^2P$	$(^1D)\,^2P$	$(^3P)\,^2D$	$(^1D)\,^2D$	$(^1S)\,^2D$
15, P	−.00640	.02254	−.03076	.04139	.01103	.03176	.10689
	−.02218	.03832	−.03412	.04475	.00742	.03271	.10955
16, S	.04102	.02316	−.03880	.07577	.09428	.05074	.12449
	.12407	−.05989	−.06124	.09821	.07257	.03319	.16376
17, Cl	.07295	.02555	−.04535	.10105	.15146	.06565	.14378
	.18816	−.08966	−.08323	.13894	.22584	.04002	.09504
18, Ar	.09471	.02781	−.05182	.12125	.19266	.07738	.16319
	.23412	−.11159	−.10069	.17011	.27568	.04491	.11264
19, K	.11149	.02983	−.05832	.13898	.22610	.08744	.18234
	.27096	−.12965	−.11578	.19644	.31779	.04913	.12896
20, Ca	.12553	.03164	−.06484	.15534	.25527	.09653	.20120
	.30274	−.14557	−.12953	.22003	.35533	.05300	.14467
21, Sc	.13791	.03332	−.07136	.17086	.28187	.10502	.21982
	.33143	−.16020	−.14245	.24195	.39004	.05664	.16002
22, Ti	.14922	.03489	−.07788	.18582	.30678	.11312	.23824
	.35810	−.17399	−.15483	.26277	.42287	.06012	.17514
23, V	.15979	.03639	−.08440	.20039	.33053	.12093	.25650
	.38339	−.18720	−.16684	.28283	.45439	.06347	.19009
24, Cr	.16984	.03784	−.09091	.21469	.35346	.12853	.27463
	.40768	−.20000	−.17857	.30234	.48497	.06674	.20491
25, Mn	.17950	.03924	−.09741	.22876	.37576	.13599	.29265
	.43124	−.21250	−.19009	.32144	.51485	.06994	.21962
26, Fe	.18887	.04061	−.10391	.24268	.39759	.14333	.31059
	.45424	−.22476	−.20146	.34023	.54419	.07307	.23424
27, Co	.19802	.04195	−.11040	.25647	.41907	.15058	.32846
	.47681	−.23684	−.21271	.35878	.57313	.07617	.24880
28, Ni	.20698	.04327	−.11688	.27014	.44025	.15775	.34625
	.49903	−.24878	−.22387	.37713	.60173	.07922	.26329

Table 21⁹. Coefficients of parental mixing for the ^2D terms in the $3p^24p$-excited configuration of the $N = 15$ sequence

Z, Element	a_{11}	a_{12}
15, P	.98971	.14311
16, S	.95847	.28519
17, Cl	.93068	.36583
18, Ar	.91002	.41458
19, K	.89464	.44679
20, Ca	.88291	.46955
21, Sc	.87370	.48647
22, Ti	.86629	.49955
23, V	.86020	.50995
24, Cr	.85511	.51844
25, Mn	.85080	.52548
26, Fe	.84709	.53146
27, Co	.84387	.53654
28, Ni	.84105	.54096

$$\Psi(1, {}^2D) = a_{11}\Phi[({}^3P)\,{}^2D] + a_{12}\Phi[({}^1D)\,{}^2D]$$
$$\Psi(2, {}^2D) = a_{21}\Phi[({}^3P)\,{}^2D] + a_{22}\Phi[({}^1D)\,{}^2D]$$
$$a_{11} = a_{22};\ a_{12} = -a_{21}$$

always contain Coulomb integrals involving wave functions of both the electron outside the open shell and open-shell electrons. Because the overlap of these wave functions is small, the off-diagonal elements of the energy matrix will be small compared with the diagonal terms. The same situation does not prevail when the off-diagonal matrix elements involve wave functions of penetrating electrons. In such cases the actual term values are linear combinations of terms of different parents, and parental mixing occurs.

Direct diagonalization of the energy matrices may be used to determine the degree of parental mixing. If the wave functions corresponding to a given parentage scheme are represented by $\Phi[({}^{2S'+1}L')^{2S+1}L]$, where L and S refer to individual terms and L' and S' to individual parents, the wave functions in a mixed parent scheme correspond to

$$\Psi(i, {}^{2S+1}L) = \sum_j a_{ij}\Phi[({}^{2S'+1}L')\ {}^{2S+1}L] \tag{1}$$

Table 22[9]. Coefficients of parental mixing for the ^2P terms in the $3p^24p$-excited configuration of the $N = 15$ sequence

Z, Element	a_{11}	a_{12}	a_{13}	a_{21}	a_{22}	a_{23}	a_{31}	a_{32}	a_{33}
15, P	.99010	−.13582	.03543	.13734	.98951	−.04484	−.02897	.04927	.99837
16, S	.95932	−.27195	.07578	.27651	.95926	−.05803	−.05692	.07662	.99543
17, Cl	.93156	−.34878	.10269	.35585	.93257	−.06073	−.07459	.09311	.99286
18, Ar	.91086	−.39452	.12122	.40351	.91295	−.06079	−.08669	.10428	.99076
19, K	.89546	−.42427	.13474	.43475	.89855	−.05997	−.09563	.11228	.98907
20, Ca	.88371	−.44500	.14505	.45666	.88770	−.05880	−.10260	.11820	.98768
21, Sc	.87449	−.46022	.15320	.47282	.87928	−.05750	−.10825	.12272	.98652
22, Ti	.86707	−.47186	.15981	.48522	.87259	−.05613	−.11297	.12621	.98555
23, V	.86097	−.48105	.16530	.49502	.86715	−.05477	−.11699	.12898	.98472
24, Cr	.85587	−.48848	.16992	.50296	.86266	−.05343	−.12049	.13119	.98401
25, Mn	.85155	−.49461	.17387	.50951	.85888	−.05212	−.12356	.13298	.98339
26, Fe	.84781	−.49978	.17731	.51503	.85566	−.05085	−.12630	.13443	.98284
27, Co	.84459	−.50415	.18029	.51971	.85290	−.04964	−.12875	.13562	.98236
28, Ni	.84174	−.50793	.18293	.52374	.85050	−.04847	−.13096	.13661	.98193

$$\Psi(1, {}^2P) = a_{11}\Phi[({}^3P)\,{}^2P] + a_{12}\Phi[({}^1D)\,{}^2P] + a_{13}\Phi[({}^1S)\,{}^2P]$$
$$\Psi(2, {}^2P) = a_{21}\Phi[({}^3P)\,{}^2P] + a_{22}\Phi[({}^1D)\,{}^2P] + a_{23}\Phi[({}^1S)\,{}^2P]$$
$$\Psi(3, {}^2P) = a_{31}\Phi[({}^3P)\,{}^2P] + a_{32}\Phi[({}^1D)\,{}^2P] + a_{33}\Phi[({}^1S)\,{}^2P]$$

Table 23^9. Coefficients of parental mixing for the ^2F and ^2P terms in the $3p^23d$-excited configuration of the $N = 15$ sequence

Z, Element	^2F		^2P	
	a_{11}	a_{12}	a_{11}	a_{12}
15, P	.85976	.51070	.97842	.20663
16, S	.74065	−.67190	.92696	.37516
17, Cl	.76505	−.64397	.91076	.41294
18, Ar	.77250	−.63502	.90529	.42479
19, K	.77584	−.63093	.90331	.42898
20, Ca	.77763	−.62872	.90274	.43019
21, Sc	.77870	−.62740	.90281	.43005
22, Ti	.77938	−.62655	.90318	.42927
23, V	.77983	−.62599	.90370	.42817
24, Cr	.78014	−.62561	.90428	.42694
25, Mn	.78015	−.62534	.90489	.42565
26, Fe	.78050	−.62516	.90549	.42437
27, Co	.78061	−.62503	.90607	.42312
28, Ni	.78068	−.62493	.90663	.42192

$$\Psi(1, {}^2F) = a_{11}\Phi[({}^3P)\,{}^2F] + a_{12}\Phi[({}^1D)\,{}^2F],$$
$$\Psi(2, {}^2F) = a_{21}\Phi[({}^3P)\,{}^2F] + a_{22}\Phi[({}^1D)\,{}^2F]$$
$$\Psi(1, {}^2P) = a_{11}\Phi[({}^3P)\,{}^2P] + a_{12}\Phi[({}^1D)\,{}^2P],$$
$$\Psi(2, {}^2P) = a_{21}\Phi[({}^3P)\,{}^2P] + a_{22}\Phi[({}^1D)\,{}^2P]$$
$$a_{11} = a_{22};\ a_{12} = -a_{21}$$

where j indicates alternative values of L' and S'. The wave functions in the mixed scheme are thus obtained from those in the pure parent scheme by a unitary transformation. The coefficients a_{ij}, which are the matrix elements of A, the unitary matrix that diagonalizes the energy matrix expressed in the pure parentage scheme, provide a measure of the degree of parental mixing. If no mixing occurred, A would be a unit matrix.

One might expect a partial breakdown of parentage assignments with increasing Z along isoelectronic sequences owing to increased core penetration. Theoretical claculations confirm this. The simplest examples are provided by the $np^2n'd$ and $np^2n'p(n \neq n')$ configurations. Diagonal elements of the energy submatrices for $(3s^23p^24p)^2$D, ^{2}P, and $(3s^23p^23d)^2$F, ^{2}P, ^{2}D terms both before and after diagonalization are given in Tables 19^9 and 20^9. The corresponding coefficients of transformation from pure parentage to mixed parentage are given in

Table 24[9]. Coefficients of parental mixing for the ^2D terms in the $3p^23d$-excited configuration of the $N = 15$ sequence

Z, Element	a_{11}	a_{12}	a_{13}	a_{21}	a_{22}	a_{23}	a_{31}	a_{32}	a_{33}
15, P	.96207	−.22111	.15979	.22796	.97333	−.02568	−.14985	.06113	.98682
16, S	.58862	.66716	.45653	−.53863	.74478	−.39393	−.60283	−.01402	.79775
17, Cl	.73009	.05715	−.68096	−.49159	.73611	−.46528	.47467	.67445	.56552
18, Ar	.76488	.06791	−.64059	−.47629	.72916	−.49140	.43372	.68096	.59007
19, K	.77875	.07040	−.62337	−.46656	.72927	−.50049	.41937	.68059	.60077
20, Ca	.78533	.07002	−.61511	−.45913	.73238	−.50281	.41528	.67729	.60731
21, Sc	.78865	.06846	−.61102	−.45309	.73550	−.50228	.41562	.67297	.61186
22, Ti	.79030	.06639	−.60912	−.44801	.74077	−.50054	.41799	.66847	.61518
23, V	.79100	.06411	−.60845	−.44368	.74487	−.49831	.42127	.66412	.61764
24, Cr	.79113	.06177	−.60852	−.43993	.74867	−.49595	.42495	.66006	.61947
25, Mn	.79091	.05946	−.60904	−.43665	.75212	−.49362	.42872	.65634	.62082
26, Fe	.79047	.05721	−.60982	−.43375	.75524	−.49140	.43245	.65294	.62181
27, Co	.78990	.05504	−.61077	−.43117	.75807	−.48931	.43607	.64985	.62253
28, Ni	.78923	.05298	−.61181	−.42886	.76062	−.48738	.43953	.64703	.62302

$$\Psi(1, {}^2D) = a_{11}\Phi[({}^3P)\,{}^2D] + a_{12}\Phi[({}^1D)\,{}^2D] + a_{13}\Phi[({}^1S)\,{}^2D]$$
$$\Psi(2, {}^2D) = a_{21}\Phi[({}^3P)\,{}^2D] + a_{22}\Phi[({}^1D)\,{}^2D] + a_{23}\Phi[({}^1S)\,{}^2D]$$
$$\Psi(3, {}^2D) = a_{31}\Phi[({}^3P)\,{}^2D] + a_{32}\Phi[({}^1D)\,{}^2D] + a_{33}\Phi[({}^1S)\,{}^2D]$$

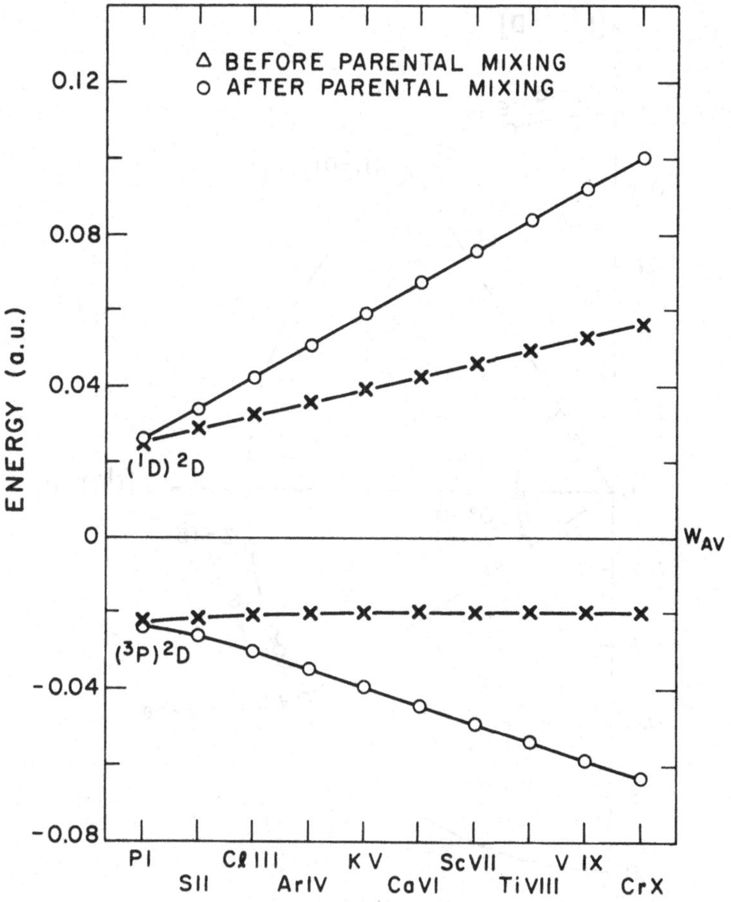

Figure 23⁹. The effect of nondiagonal elements in the energy submatrix for $(3s^23p^24p)^2D$ terms in the $N = 15$ (phosphorus) sequence.

Tables 21⁹, 22⁹, 23⁹, and 24⁹. The tables show that with the exception of the ^2F terms all terms for neutral atoms are characterized by almost pure parentage, but that parental mixing becomes important for higher terms in the isoelectronic sequence. Diagonal elements of the 2×2 submatrix for the $(3s^23p^24p)\ ^2D$ terms before and after diagonalization in the $N = 15$ sequence are plotted in Figure 23⁹, which illustrates the additional repulsion of the ^2D terms due to the effects of parental mixing.

The mixing coefficients can be considered as projections of rotating orthonormal vectors representing the mixed-wave functions in an

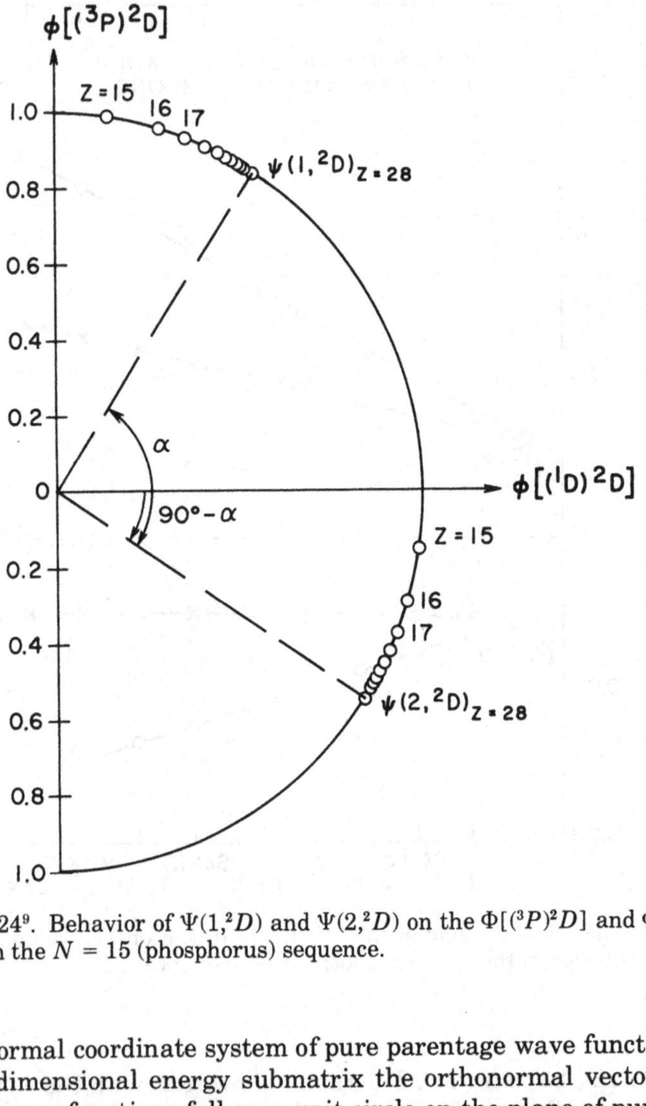

Figure 24^9. Behavior of $\Psi(1,{}^2D)$ and $\Psi(2,{}^2D)$ on the $\Phi[({}^3P)^2D]$ and $\Phi[({}^1D)^2D]$ plane in the $N = 15$ (phosphorus) sequence.

orthonormal coordinate system of pure parentage wave functions. For a two-dimensional energy submatrix the orthonormal vectors of the mixed-wave functions fall on a unit circle on the plane of pure parentage wave functions. In this case the mixing coefficients can be represented in terms of a single rotation, because the unitary transformation matrix may be written as $a_{11} = a_{22} = \cos\alpha$, $a_{21} = \sin\alpha$, $a_{12} = -a_{21}$. This is demonstrated in Figure 24^9 for the $(3s^23p^24p)\ {}^2D$ terms of the $N = 15$ sequence. In a three-dimensional case the vectors would fall on the surface of a unit sphere and three angles would be necessary to represent the transformation.

Appendix 1. Atomic magnitudes

> Changing the units does not affect the velocity of light. Whether you say light travels 186,000 miles a second or whether you say it is so many inches an hour makes no difference to the velocity. An algebraic symbol ought to represent the thing itself, not a mere number of units. Altering the numerical specification – which is what you do by altering units – means no difference to the thing itself."
>
> Sir Oliver Lodge, *Monthly Notices,* Royal Astronomical Society, 80, 107 (1919).

A physical quantity Q is specified [1] by giving its numerical magnitude (Q/Q_0) relative to another quantity Q_0 of like kind, and the basic relation is

$$Q = (Q/Q_0)Q_0 \tag{1}$$

in which Q_0 is usually designated by writing the name or an abbreviation of the name of Q_0. The rule of transformation from (Q/Q_0) to the magnitude of Q in terms of Q_1 as unit is

$$(Q/Q_1) = (Q/Q_0)(Q_0/Q_1) \tag{2}$$

Although it is not usual to indicate the unit explicity in this way, it is convenient to do so when otherwise confusion might arise.

Derived quantities that are defined in the usual way in terms of mass (M), length (L), and time (T) may be conveniently represented [2] in a space having three mutually orthogonal axes carrying logarithmic scales for M, L, and T. A quantity whose dimensional formula is

$$Q = M^{\mu} L^{\lambda} T^{\tau}$$

is represented in this space by the plane whose formula is

$$\log (Q/Q_0) = \mu \log (M/M_0) + \lambda \log (L/L_0) + \tau \log (T/T_0) \tag{3}$$

The quantity Q^n is then represented by this same plane: Q^n and Q are alike physically in that if one is known, the other can be found without making additional physical observations. The nature of Q is thus specified solely by the ratios of $\mu{:}\lambda{:}\tau$, that is, by the orientation of the plane. Various particular Q's are represented by parallel planes whose separation is simply related to the log of the ratio.

In particular the mass M_0 is represented by a plane orthogonal to the M-axis and similarly for the length L_0 and time T_0. Thus each point in the space corresponds to a choice of units (M_0, L_0, T_0). The two-space of points on the plane (3) representing Q then represents all possible systems of units in which Q has the value unity. The one-space of points on the line of intersection of the planes for Q and R, where these are of unlike kind, represents all possible systems of units in which Q and R have unit magnitude. Likewise, if Q, R, and S are unlike, their planes intersect in a point which determines a unique system of units in which each of them has magnitude unity.

There is nothing fundamental about the usual custom of expressing quantities in terms of exactly three basic units. For electromagnetic units more than three are convenient.

The numerical values assigned to basic quantities change because of changes (1) in definitions of the fundamental units, (2) in definitions of derived units, (3) in new measurements that relate a particular Q to a well-defined basis.

The basic magnitudes such as m, $\hbar$, e, c, . . . are usually inferred indirectly from measurements of various Q, R, S . . . , which are related to the m, $\hbar$, e, c . . . by theoretical expressions. Improved accuracy of measurement may reveal the need for revision of the theoretical relations, which gives rise to a fourth reason for inconsistency of values published in the literature.

It usually happens that the number of Q, R, S . . . that can be measured is more than the number of independent parameters sought so the results are over-determined. Compromise and judgment then enter into the task of choosing a "best" set of basic magnitudes from inconsistent measurements. Critical studies of the data were initiated by R. T. Birge [3] and continued by E. R. Cohen and J. W. M. DuMond [4] whose latest full reduction [5] was carried out in 1963. A later study was carried out by B. N. Taylor, W. H. Parker and D. N. Langenberg [6]. These latter values are adopted in this book.

Système Internationale (SI)

The form of the metric system now in use by international agreement is called the Système Internationale. It was initially adopted in 1960 by the Eleventh General Conference on Weights and Measures, the atomic definition of time having been added in 1967 by the Thirteenth General Conference.

Unit mass is the *kilogram* (kg), which is the mass of a platinum standard kept at the International Bureau of Weights and Measures, Sèvres, France.

Unit length is the *meter* (m), defined as a multiple of λ_{Kr}, the vacuum wavelength of the $2p_{10} - 5d_5$ transition in ^{86}Kr under specified conditions,

$$1 \text{ m} = 1{,}650{,}763.73\lambda_{Kr}$$

The value adopted for (m/λ_{Kr}) is known [7] to agree, to an imprecision of 0.2 ppm, with the meter as previously defined by the 1889 standard platinum-iridium bar. Imprecision of a number is defined as the ratio of its standard deviation to its mean value expressed in parts per million.

Unit time is the *second* (s), defined in 1967 as a multiple of t_{Cs}, which is the duration of one cycle of the radiation in the (4, 0)–(3, 0) hyperfine transition in the normal state of ^{133}Cs,

$$1 \text{ s} = 9{,}192{,}631{,}770 t_{Cs}$$

This definition supersedes the one adopted in 1955 by the International Astronomical Union, and in 1960 by the Eleventh General Conference,

$$1 \text{ tropical year (1900)} = 31{,}556{,}925.9747 \text{ s}$$

Unit temperature is the *degree* Kelvin (K) defined on the absolute thermodynamic scale by the relation

$$\text{triple point of water} = 273.16 \text{ K}$$

This was adopted in 1954 at the Tenth General Conference. At normal pressure

$$\text{melting ice water} = 273.15 \pm 0.0001 \text{ K}$$
$$\text{boiling water–steam} = 373.1464 \pm 0.0001 \text{ K}$$

Temperature expressed on the Celsius scale (°C) is equal to that on the Kelvin scale less 273.15.

From 1964 nations that use the pound (lb) and the yard have agreed that

$$1 \text{ lb} = 0.45359237 \text{ kg}$$
$$1 \text{ yd} = 0.9144 \text{ m}$$

This latter choice makes 1 in. = 25.4 mm, superseding the nearly equivalent relation, 1 m = 39.37 in., formerly in common use.

The following standard prefixes designate decimal power multiples of SI units:

10^{-18}	atto-	10^{18}	nebu-
10^{-15}	femto-	10^{15}	astra-
10^{-12}	pico-	10^{12}	tera-
10^{-9}	nano-	10^{9}	giga-
10^{-6}	micro	10^{6}	mega
10^{-3}	milli-	10^{3}	kilo-
10^{-2}	centi-	10^{2}	hekto-
10^{-1}	deci-	10^{1}	deka-

Hartree atomic units

Hartree [8] recommended for use in atomic physics a system of units in which m, $\hbar$, and e have unit value. Other choices have been advocated, but this choice is the one most widely used.

We write a for the length unit, which is the radius of the first Bohr orbit (7^13) in hydrogen of infinite nuclear mass, and τ for the time required to traverse one radian in this orbit, $2\pi\tau$ being therefore the time required for one cycle.

Accordingly, the Hartree basis is

$$\begin{aligned} &\text{mass: mass of the electron, } m \\ &\text{length: Bohr radius, } a = \hbar^2/me^2 \\ &\text{time: one radian in Bohr orbit, } \tau = \hbar^3/me^4 \end{aligned} \tag{1}$$

The speed in the first Bohr orbit is the unit of velocity,

$$a\tau^{-1} = e^2/\hbar$$

and the velocity of light c is given in this unit by the reciprocal of the fine structure constant (7^111),

$$c = \alpha^{-1}e^2/\hbar \tag{2}$$

Various powers of α enter in combination with the length and energy units to give auxiliary quantities of physical interest as shown in the table of conversion factors below.

Nuclear, atomic and ionic masses are expressed (Sec. 2^1) in terms of the unified mass units

$$u = M(^{12}\text{C})/12 = u_m m \tag{3}$$

where u_m is written for the value of this unit expressed in electron masses.

In a mass $\mathfrak{M}$ of material whose individual molecules have mass M, the number of molecules is $(\mathfrak{M}/M)$. The quantity

$$N = (g/u) \tag{4}$$

is the number of unified mass units contained in one gram, and is commonly called the *Avogadro number*. We can write

$$\frac{\mathfrak{M}}{M} = \left(\frac{\mathfrak{M}/g}{M/u}\right) N \tag{5}$$

The factor in brackets is the mass in grams of material divided by the molecular mass with u as unit. It is called the number of (gram) moles in the mass $\mathfrak{M}$. In (4) and (5) any other mass unit than g can be used with a corresponding change in the value of the Avogadro number and the definition of the mole.

The parameter u_m is useful also in defining auxiliary atomic units. Thus $u_m ma^2$, rather than ma^2, is a natural unit for molecular moment of inertia. Molecular electronic energy as a function of internuclear distance is naturally expressed in the atomic unit, $ma^2\tau^{-2} = e^2/a$, so the harmonic law force constant k in $U = (1/2)k(r - r_0)^2$ is naturally expressed in the atomic unit $m\tau^{-2}$. But the molecular reduced mass $M = M_1M_2/(M_1 + M_2)$ is of order u so the natural atomic unit for vibration frequency is

$$\nu_v = (2\pi\tau)^{-1}u_m^{1/2} \tag{6}$$

Similarly, the vibrational amplitude is naturally expressed in a unit of length b defined by

$$\frac{1}{2}\,kb^2 = \frac{1}{2}\,h\nu_v$$

That is,

$$b = a u_m^{-1/4} \tag{7}$$

Mechanical conversion factors

Values given here are based on the 1969 reduction of Taylor, Parker, and Langenberg, and the 1973 reduction of Cohen and Taylor [6].

$\alpha = 7.2973506 \times 10^{-3}$ (0.82)
$u_m = 1822.88743$ (4.0)
$\hbar = h/2\pi = 1.0545887 \times 10^{-34}$ joule·second (5.4)

The imprecision is given in parts per million (ppm).

Length (L)	m	(ppm)
Fermi	10^{-15}	
Nuclear radius is given as $r_n A^{1/3}$, a rough empirical relation with r_n	1.37×10^{-15}	def. (approx.)
Electron radius, $e^2/mc^2 = \alpha^2 a$	$2.8179380 \times 10^{-15}$	(2.5)
$\hbar/mc = \alpha a$	$3.8615905 \times 10^{-13}$	(1.7)
Muon Bohr radius, $a_\mu = a(m/m_\mu)$	$2.5592712 \times 10^{-13}$	(2.5)
Compton wavelength, $2\pi\alpha a$	$24.263089 \times 10^{-13}$	(1.6)
X unit*	$1.0020772 \times 10^{-13}$	(5.3)
Vibration amplitude, $b = a u_m^{-1/4}$	$0.80986205 \times 10^{-11}$	(1.5)
Bohr radius, $a = \hbar^2/me^2$	$5.2917706 \times 10^{-11}$	(0.82)
Lyman limit, $\lambda_L = 4\pi\alpha^{-1}a(1 + m/M_p)$	9.1176335×10^{-8}	(0.1)
Krypton standard λ_{Kr}	$6.057802106 \times 10^{-7}$	(def.)

* (Defined as true grating space of calcite at 18°C is $d^1 = 3029.049$ X unit)

Wave number (L^{-1})	cycle/m	(ppm)
$\alpha^2 R_\infty$	584.3657655	(1.6)
λ_{Kr}^{-1}	1650763.73	(def.)
R_H	10967758.54	(0.08)
R_D	10970742.86	(0.08)
R_{He}	10972227.55	(0.08)
$(\alpha/4\pi)a^{-1} = R_\infty$	10973731.77	(0.075)

Area (L^2)	m^2	(ppm)
Thomson cross section, $(8\pi/3)(\alpha^2 a)^2$	6.652448×10^{-29}	(4.9)
a^2	$2.8002836 \times 10^{-21}$	(1.6)
πa^2	$8.7973504 \times 10^{-21}$	(1.6)

Volume (L^3)	m^3	(ppm)
Nuclear $(4\pi/3)r_n^3$	1.077×10^{-44}	(approx.)
Atomic a^3	$1.4818458 \times 10^{-31}$	(2.4)
Perfect gas, volume per molecule at standard conditions	$3.7219632 \times 10^{-26}$	(32)
Molar volume of perfect gas at s.t.p.	22.41383×10^{-3}	(31)

Time (T)	s	(ppm)
$\tau = \hbar^3/me^4 = (4\pi c R_\infty)^{-1}$	$2.4188843 \times 10^{-17}$	(0.08)
$2\pi\tau$	$1.51982982 \times 10^{-16}$	(0.08)
t_{Cs}	$1.08782775 \times 10^{-10}$	(def.)
Day	86400.	(def.)
Tropical year (1900)	31556925.9747	(old def.)

Velocity (LT^{-1})	ms^{-1}	(ppm)
First Bohr orbit, $a\tau^{-1} = c\alpha$	2.18769066×10^6	(0.82)
Light, c	2.99792458×10^8	(0.004)
α^{-1}	137.03604	(0.82)
$(\alpha = e^2/\hbar c = 7.2973506 \times 10^{-3})$		

Acceleration (LT^{-2})	ms^{-2}	(ppm)
$a\tau^{-2} = e^2/ma^2$	9.0442137×10^{22}	(0.82)
Normal gravity g_n	9.80665	(def.)

Diffusivity (L^2T^{-1})	m^2s^{-1}	(ppm)
$a^2\tau^{-1} = \alpha^2 c/4\pi R_\infty$	$1.15767571 \times 10^{-4}$	(1.6)

Mass (M)	m	(ppm)	kg	(ppm)
Electron, m,	1.	(def.)	9.109534×10^{-31}	(5.1)
Proton, M_p	1836.15152	(0.38)	$1.6726485 \times 10^{-27}$	(5.1)
Neutron defect Δ_n	1.531189		1.394846×10^{-30}	
Deuterium defect $\Delta(^2H)$	4.353318		3.965680×10^{-30}	
Muon	206.76865	(2.3)	1.883566×10^{-28}	(5.6)
$u = M(^{12}C)/12$	1822.88743	(4.0)	$1.6605655 \times 10^{-27}$	(5.1)
$u_0 = M(^{16}O)/16$	1822.30803	(4.0)	$1.6600377 \times 10^{-27}$	(5.1)

	u	(ppm)	kg	(ppm)
Proton M_p	1.007276470	(0.011)	$1.6726485 \times 10^{-27}$	(5.1)
Hydrogen $M(^1H)$	1.007825036	(0.011)	$1.6735595 \times 10^{-27}$	(5.1)
Neutron M_n	1.008665012	(0.037)	$1.6749543 \times 10^{-27}$	(5.1)
Deuterium $M(^2H)$	2.014101795	(0.010)	$3.3445479 \times 10^{-27}$	(5.1)
Carbon $M(^{12}C)$	12.	(def.)	$1.9926786 \times 10^{-26}$	(5.1)

Avogadro number, $N = u^{-1}$

$= 6.022045 \times 10^{26}$ molecules/(kg mole) (5.1)

Mass density (ML^{-3})	kg m^{-3}	(ppm)
Nuclear, $u(4\pi r_n^3/3)^{-1}$	1.5417×10^{17}	(approx.)
Electron structure, ma^{-3}	6.1474235	(5.8)
1 g cm^{-3}	10^3	(def.)

Force constant (MT^{-2})	kg s^{-2}	(ppm)
$m\tau^{-2}$	1.5569188×10^3	(5.2)

Force (MLT^{-2})	N (newton)	(ppm)
$ma\tau^{-2}$	8.2388572×10^{-8}	(5.2)
10^5 dyne	1	(def.)

Energy (ML^2T^{-2})	J (joule)	(ppm)
$\alpha^2(e^2/a) = 4\pi\alpha e^2 R_\infty$	$2.3216589 \times 10^{-22}$	(6.2)
Electron-volt	$1.6021892 \times 10^{-19}$	(2.9)
$(e^2/a) = 4\pi\alpha^{-1}e^2R_\infty$	$4.3598142 \times 10^{-18}$	(5.5)
$mc^2 = \alpha^{-2}(e^2/a)$	8.187241×10^{-14}	(5.1)
uc^2	$1.4924418 \times 10^{-10}$	(5.1)
Erg	10^{-7}	(def.)

Power (ML^2T^{-3})	watt	(ppm)
$ma^2\tau^{-3}$	1.8024071×10^{-1}	(5.4)

Pressure ($ML^{-1}T^{-2}$)	newton m^{-2}	(ppm)
$ma^{-1}\tau^{-2}$	2.942151×10^{13}	(5.6)
Dyne cm^{-2}	10^{-1}	(def.)
Millibar	10^2	(def.)
Torr (1 mm Hg)	1.33322×10^2	
Inch Hg	3.38639×10^3	
Bar	10^5	(def.)
Normal atmosphere (760 mm Hg)	1.01325×10^5	(def.)

Gravitation ($M^{-1}L^3T^{-2}$)	N m^2 kg^{-2}	(ppm)
G	6.6720×10^{-11}	(615)

The force of gravitational attraction between masses M and M′ at distance r is GMM'/r^2.

The ratio of the Coulomb repulsion between two electrons to their gravitational attraction is

$$e^2/Gm^2 = 4.166973 \times 10^{42}$$

Planck [9] suggested a system of natural units making h, G, and c equal to unity,

$$\text{Mass} \quad (hc/G)^{1/2} = 5.456496 \times 10^{-8}\ \text{kg}$$
$$\text{Length} \quad (Gh/c^3)^{1/2} = 4.050685 \times 10^{-35}\ \text{m}$$
$$\text{Time} \quad (Gh/c^5)^{1/2} = 1.351163 \times 10^{-43}\ \text{s}$$

but this has found very little use in physics.

Temperature

Temperature (T) occurs in fundamental relations multiplied by the Boltzmann constant (k) to give an energy (kT) characterizing the statistical distribution of energy in an equilibrium ensemble,

	J deg^{-1}	(ppm)
Boltzmann constant $k = R/N$	1.380662×10^{-23}	(32)
	J deg^{-1} × $kilomole^{-1}$	
Molal gas constant R	8.31441×10^3	(31)

In the Planck radiation law (4¹5) the combination hc/k is called the second radiation constant, c_2.

	m deg	(ppm)
$c_2 = hc/k$	1.438786×10^{-2}	(31)

The temperature T at which $kT = W$ for various energies is

W	T deg	(ppm)
$\alpha^2(e^2/a)$	1.6815549×10^1	(33)
e^2/au_m	1.7322907×10^2	(33)
$e^2/au_m^{1/2}$	7.396064×10^3	(33)
e^2/a	3.157771×10^5	(33)
mc^2	5.929939×10^9	(33)
uc^2	1.080961×10^{13}	(33)

Electromagnetic magnitudes

Units for electromagnetic magnitudes depend (1) on the choice of (M, L, T), (2) on whether the relation to mechanical units is electric or magnetic, and (3) on where 4π is placed in the relations.

The three CGS systems are based on $(M, L, T) = $ (g, cm, s): CGSE (electric), CGSM (magnetic), and Gaussian, a mixed system in which electric quantities are expressed in electric units and magnetic quantities in magnetic units. The Hartree system is the Gaussian system based on $(M, L, T) = (m, a, \tau)$. The MKSA is based on $(M, L, T) = $ (kg, m, s) and a definition of the ampere.

The energy W of interaction between charges q and q' separated by distance r in each system is

$$W = \frac{qq'}{4\pi\epsilon_0 r} \tag{8}$$

Names of the units are: statcoulomb (stC) in CGSE and Gaussian, abcoulomb (abC) in CGSM, electron (e) in Hartree, and coulomb (C) in MKSA. The values of $4\pi\epsilon_0$ and the relations of these units are as follows.

Charge ($Q = M^{1/2}L^{3/2}T^{-1}\epsilon_0^{1/2}$)

	electron Hartree	statcoulomb CGSE, Gaussian	abcoulomb CGSM	coulomb MKSA	(ppm)
$4\pi\epsilon_0$	1	1	$1/c^2$	$10^7/c^2$	(def.)
		c	1	10	(def.)
e	1	4.803242×10^{-10}	$1.6021892 \times 10^{-19}$		(2.9)

Charge/mass ratio (QM^{-1})		(ppm)
electron, e/m	$= 5.272764 \times 10^{17}$ stC g^{-1}	
	$= 1.7588047 \times 10^{11}$ C kg^{-1}	(2.8)
proton, e/M_p	$= 2.871639 \times 10^{14}$ stC g^{-1}	
	$= 9.578756 \times 10^{7}$ C kg^{-1}	(3.0)

Potential V and electric vector **E** in each system are measured in units such that the energy W, of charge q and the force F on charge q are

$$W = qV, \qquad \mathbf{F} = e\mathbf{E} \tag{9}$$

The potential at **r**, caused by a distribution of charges q', and surface distributions $\sigma(\mathbf{r}')$ and volume distributions $\rho(\mathbf{r}')$ is given by

$$V(\mathbf{r}) = (4\pi\epsilon_0)^{-1}\left[\sum_{q'} \frac{q'}{|\mathbf{r}-\mathbf{r}'|} + \iint \frac{\sigma(\mathbf{r}')\,dS'}{|\mathbf{r}-\mathbf{r}'|} + \iiint \frac{\rho(\mathbf{r}')\,dv'}{|\mathbf{r}-\mathbf{r}'|}\right] \tag{9a}$$

and the resulting **E** is

$$\mathbf{E}(\mathbf{r}) = -\mathbf{grad}\ V(\mathbf{r}) \tag{9b}$$

In (9a) the first two terms may be put in the form of the third by admitting appropriate singularities in $\rho(\mathbf{r}')$.

Charge density (σ, surface, and ρ, volume)		(ppm)
surface, ea^{-2}	$= 1.7152698 \times 10^{7}$ stC cm^{-2}	
	$= 5.1422500 \times 10^{11}$ C m^{-2}	(2.7)
volume, ea^{-3}	$= 3.2413912 \times 10^{15}$ stC cm^{-3}	
	$= 9.7174472 \times 10^{21}$ C m^{-3}	(2.9)

(ea^{-3} is of the order of volume charge density of electrons in atoms, $\alpha^{-6}ea^{-3}$ is of the order of volume charge density of protons in atomic nuclei.)

Potential ($QL^{-1}\epsilon_0^{-1}$)

Hartree	statvolt CGSE, Gaussian	abvolt CGSM	volt MKSA	(ppm)
	1	c	$c \times 10^{-8}$	
e/a	9.076814×10^{-2}	27.2116076×10^{8}	27.2116076	(2.6)

The corresponding relations for **E** are

Electric vector $(QL^{-2}\epsilon_0^{-1})$

Hartree	statvolt cm^{-1} CGSE, Gaussian	abvolt cm^{-1} CGSM	volt m^{-1} MKSA	(ppm)
e/a^2	1.7152698×10^7	5.1422500×10^{17}	5.1422500×10^{11}	(7)
	1	c	$c \times 10^{-6}$	

For $|\mathbf{r}|$ large compared with the largest $|\mathbf{r}'|$ of a charge distribution of finite extent, $V(\mathbf{r})$ can be developed in a series of spherical harmonics.

$$V(\mathbf{r}) = (4\pi\epsilon_0)^{-1} \sum_{k=0}^{\infty} \frac{1}{r^{k+1}} \iiint r'_k \rho(\mathbf{r}')P_k(\cos\omega)\, dv' \qquad (9c)$$

in which P_k is a Legendre polynomial and ω is the angle between $\mathbf{r}'$ and $\mathbf{r}$. The integrals are spherical tensors of the direction of $\mathbf{r}$. For $k = 0$ it is Q, the total charge. For $k = 1$ the integral is $\mathbf{p}\cdot\mathbf{r}$ in which $\mathbf{p}$ is the electric dipole moment of the distribution. For $k = 2$ it is related to the quadrupole moment.

Electric dipole and quadrupole moments

Hartree	CGSE, Gaussian	MKSA	(ppm)
	1 debye = 10^{-18} stC cm		
ea	2.5417655 debye	$8.4784177 \times 10^{-30}$ C m	(3.3)
ea^2	$1.3450440 \times 10^{-26}$ stC cm^2	$4.4865842 \times 10^{-40}$ C m^2	(3.9)

Current in a linear conductor is the time rate at which positive charge passes a point in the circuit, in electric measure. The CGSE unit is the statampere (stA), whereas the Hartree unit is $e\tau^{-1}$, the current in the first Bohr orbit being $e/2\pi\tau$. The magnetic measure of current is defined on the basis of Ampere's law, expressing the force between two circuits in vacuum carrying currents I and I'. The total force, $\mathbf{F}$, is

$$\mathbf{F} = \frac{\mu_0}{4\pi} II' \,\mathbf{grad}\, N \qquad (10)$$

in which

$$N = \oint\oint \frac{\mathbf{dr}\cdot\mathbf{dr}'}{|\mathbf{r} - \mathbf{r}'|}$$

and **grad** N is the rate of change of N with rigid parallel translation of the first circuit. The CGSM unit is called the abampere (abA) and the MKSA unit the ampere (A).

Current QT^{-1} or $(M^{1/2}L^{1/2}T^{-1}\mu_0^{-1/2})$

Hartree	statampere CGSE	abampere CGSM	ampere MKSA	(ppm)
$\mu_0/4\pi$		1	10^{-7}	
	c	1	10	
$e\tau^{-1}$	1.9857262×10^{7}	6.6236703×10^{-4}	6.6236703×10^{-3}	(3.0)
$e(2\pi\tau)^{-1}$	3.1603815×10^{6}	1.0541899×10^{-4}	1.0541899×10^{-3}	(3.0)

The definition of the ampere agrees with that adopted by the Ninth General Conference: "The ampere is the strength of an unvarying current sustained in two parallel rectilinear conductors of infinite length and negligibly small cross section separated by a distance of one meter in vacuum, producing a force between these conductors equal to 2×10^{-7} newtons per meter length."

Current density, **J**, is defined so that **J dS** is the time rate of transfer of positive charge toward the positive side of **dS**, where **dS** is an element of area. All known electromagnetic phenomena are consistent with the law of conservation of charge so that **J** and ρ are related by the equation of continuity

$$\mathbf{div\,J} + \dot{\rho} = 0 \tag{11}$$

where **J** and ρ are both in eletric or both in magnetic measure. From Gauss' theorem, $\mathbf{div\,E} = \rho/\epsilon_0$, with both **E** and ρ in electric measure. With **J** in magnetic measure and **E** in electric, we have

$$\mathbf{div}\,(\mathbf{J} + \epsilon\dot{\mathbf{E}}_0) = 0 \tag{12}$$

Current density

Hartree	stA cm^{-2} CGSE	abA cm^{-2} CGSM	A m^{-2} MKSA	(ppm)
	$c \times 10^{-5}$	10^{-5}	1	(def.)
$ea^{-2}\tau^{-1}$	7.0911612×10^{23}	2.365357×10^{13}	2.365357×10^{18}	(2.6)

The magnetic field, **B(r)**, is defined by the relation for the resultant force **F** acting on a rigid volume distribution of current density **J(r)**,

$$\mathbf{F} = \iiint \mathbf{J(r)} \times \mathbf{B}(r)\, dv \tag{13}$$

in which **J(r)** is in magnetic measure. In particular, if **J(r)** arises solely from a charge q_m (magnetic measure) moving with velocity **v**, the magnetic force on it is

$$\mathbf{F} = q_m\mathbf{v} \times \mathbf{B} = (q/c)\mathbf{v} \times \mathbf{B} \tag{14}$$

Thus the complete form of (9) is

$$\mathbf{F} = q(\mathbf{E} + k\mathbf{v} \times \mathbf{B}) \tag{15}$$

in which $k = 1/c$ for Gaussian and $k = 1$ for MKSA units. For the magnetic field the CGSM unit is the gauss, the MKSA unit the tesla.

Magnetic vector

Hartree	gauss CGSM	tesla MKSA	(ppm)
	10^4	1	(def.)
$e/ac\tau$	1.2516925×10^5	12.516925	(2.6)
$\mu_B a^{-3}$	0.6258463×10^5	6.258463	(2.6)

where $(e/ac\tau)$ is the magnetic field at the center of the first Bohr orbit.

The magnetic field caused by a steady volume distribution of current $\mathbf{J}(\mathbf{r}')$ is

$$\mathbf{B}(\mathbf{r}) = \frac{\mu_0}{4\pi} \iiint \frac{J(\mathbf{r}') \times (\mathbf{r} - \mathbf{r}')}{|\mathbf{r} - \mathbf{r}'|^3}\, dv' \tag{16}$$

In addition to the part of $\mathbf{E}$ given by (9) for which $\mathbf{curl\,E} = 0$, there is another part of $\mathbf{E}$ associated with changing magnetic field that is given by Faraday's law of electromagnetic induction,

$$\mathcal{E} = \oint \mathbf{E} \cdot \mathbf{dr} = -k\dot{\Phi}, \quad \text{where} \quad \Phi = \iint \mathbf{B} \cdot \mathbf{ds} \tag{17}$$

Here Φ is the magnetic flux through the circuit Φ and the surface integral is over any surface bounded by the path on which the line integral for $\mathbf{E}$ is calculated. Unit Φ is called the maxwell or abV·s in CGSM, and weber or V·s in MKSA.

Magnetic flux

Hartree	maxwell CGS, Gaussian	weber MKSA	(ppm)
	10^8	1	(def.)
flux quantum, $hc/2e$	2.0678506×10^{-7}	$2.0678506 \times 10^{-15}$	(2.6)

The field equations for $\mathbf{E}$ from (9) and (17) are

$$\begin{aligned} \mathbf{div\,E} &= \rho/\epsilon_0 \\ \mathbf{curl\,E} + k\dot{\mathbf{B}} &= 0 \end{aligned} \tag{18}$$

The field equations for $\mathbf{B}$ follow from the fact that (1) Φ is independent of the surface over which it is calculated and (2) Ampere's law must be formulated with the divergence-free vector in (12),

$$\begin{aligned} \mathbf{div\,B} &= 0 \\ \mathbf{curl\,B} - k\epsilon_0\mu_0\dot{\mathbf{E}} &= \mu_0\mathbf{J} \end{aligned} \tag{19}$$

From the two curl equations we find the wave equations for **E** and **B**.

$$\begin{aligned} \mathbf{curl\,curl\,E} + k^2\epsilon_0\mu_0\ddot{\mathbf{E}} &= -k\mu_0\dot{\mathbf{J}} \\ \mathbf{curl\,curl\,B} + k^2\epsilon_0\mu_0\ddot{\mathbf{B}} &= k\mu_0\,\mathbf{curl\,J} \end{aligned}$$

It follows from these that in all systems

$$k^2\epsilon_0\mu_0 = c^{-2} \tag{20}$$

In the Gaussian system $k = c^{-1}$ so $\epsilon_0\mu_0 = 1$, whereas in the MKSA system $k = 1$ so $\epsilon_0\mu_0 = c^{-2}$.

The magnetic dipole moment of current I flowing in a plane circuit of area A is IA abA·cm² or A·m². The magnetic moment for an electron in the first Bohr orbit is

$$\mu_B = (e/2\pi\tau c)\cdot\pi a^2 = e\hbar/2mc = 9.274078\times10^{-21}\ \text{erg gauss}^{-1} \tag{3.9}$$
$$= 9.274078\times10^{-24}\ \text{J tesla}^{-1} \tag{3.9}$$

This atomic magnitude is called the Bohr magneton. The nuclear magneton, μ_N, is defined as

$$\mu_N = e\hbar/2M_pc = (m/M_p)\mu_B = 5.050824\times10^{-27}\ \text{J tesla}^{-1} \tag{3.9}$$

The measured magnetic moment of the proton is

$$\mu_p = 1.4106171\times10^{-26}\ \text{J tesla}^{-1} \tag{3.9}$$

Appendix 2. The nj symbols

Some properties of the $3j$, $6j$, and $9j$ symbols are introduced in Chapter 3. Here we give additional ones. Recursion relations and formulas for special cases are presented which are useful in evaluating the nj symbols. Most of the material in this appendix is taken from A. R. Edmonds [1³c]. Other sources are indicated.

The $3j$ symbol

1. Orthogonality properties

See equations 9³11a, 9³11b.

2. Recursion relations

Here J stands for the sum of j_1, j_2, and j_3. Two of the j values may be increased by 1/2 by the relation

$$[(J+1)(J-2j_1)]^{1/2}\begin{pmatrix} j_1 & j_2 & j_3 \\ m_1 & m_2 & m_3 \end{pmatrix}$$
$$= [(j_2+m_2)(j_3-m_3)]^{1/2}\begin{pmatrix} j_1 & j_2-1/2 & j_3-1/2 \\ m_1 & m_2-1/2 & m_3+1/2 \end{pmatrix}$$
$$- [(j_2-m_2)(j_3+m_3)]^{1/2}\begin{pmatrix} j_1 & j_2-1/2 & j_3-1/2 \\ m_1 & m_2+1/2 & m_3-1/2 \end{pmatrix}$$

Louck [1] gives a similar formula:

$$(2j_3+1)(j_2+m_2)^{1/2}\begin{pmatrix} j_1 & j_2 & j_3 \\ m_2-m_3 & -m_2 & m_3 \end{pmatrix}$$
$$= -[(J-2j_1)(J+1)(j_3+m_3)]^{1/2}\begin{pmatrix} j_1 & j_2-1/2 & j_3-1/2 \\ m_2-m_3 & -m_2+1/2 & m_3-1/2 \end{pmatrix}$$
$$-[(J-2j_3)(J-2j_2+1)(j_3-m_3+1)]^{1/2}\begin{pmatrix} j_1 & j_2-1/2 & j_3+1/2 \\ m_2-m_3 & -m_2+1/2 & m_3-1/2 \end{pmatrix}$$

which is not valid for $m_2 = -j_2$.

One of the j values may be increased by unity with

$$[(J+1)(J-2j_1)(J-2j_2)(J-2j_3+1)]^{1/2}\begin{pmatrix} j_1 & j_2 & j_3 \\ m_1 & m_2 & m_3 \end{pmatrix}$$

$$= [(j_2-m_2)(j_2+m_2+1)(j_3+m_3)(j_3+m_3-1)]^{1/2}\begin{pmatrix} j_1 & j_2 & j_3-1 \\ m_1 & m_2+1 & m_3-1 \end{pmatrix}$$

$$- 2m_2[(j_3+m_3)(j_3-m_3)]^{1/2}\begin{pmatrix} j_1 & j_2 & j_3-1 \\ m_1 & m_2 & m_3 \end{pmatrix}$$

$$- [(j_2+m_2)(j_2-m_2+1)(j_3-m_3)(j_3-m_3-1)]^{1/2}\begin{pmatrix} j_1 & j_2 & j_3-1 \\ m_1 & m_2-1 & m_3+1 \end{pmatrix}$$

The recursion relation among the magnetic quantum numbers (TAS, $14^3 3$) in terms of $3j$ symbols is

$$-[(j_3+m_1+m_2+1)(j_3-m_1-m_2)]^{1/2}\begin{pmatrix} j_1 & j_2 & j_3 \\ m_1 & m_2 & -m_3+1 \end{pmatrix}$$

$$= [(j_1+m_1+1)(j_1-m_1)]^{1/2}\begin{pmatrix} j_1 & j_2 & j_3 \\ m_1 & m_2+1 & -m_3 \end{pmatrix}$$

$$+ [(j_2+m_2+1)(j_2-m_2)]^{1/2}\begin{pmatrix} j_1 & j_2 & j_3 \\ m_1+1 & m_2 & -m_3 \end{pmatrix}$$

3. Special formulas

$$\begin{pmatrix} j_1 & j_2 & j_3 \\ 0 & 0 & 0 \end{pmatrix} = (-1)^{J'}\left[\frac{(J-2j_1)!(J-2j_2)!(J-2j_3)!}{(J+1)!}\right]^{1/2}\frac{J'!}{(J'-j_1)!(J'-j_2)!(J'-j_3)!}$$

where $J' = J/2$. The $3j$ symbol is zero for J odd.

$$\begin{pmatrix} j_1 & j_2 & j_3 \\ 0 & 1 & -1 \end{pmatrix} = \frac{1}{2}\left[\frac{(J+2)(J-2j_1+1)(J-2j_2+1)(J-2j_3)}{j_2(j_2+1)j_3(j_3+1)}\right]^{1/2}\begin{pmatrix} j_1 & j_2 & j_3+1 \\ 0 & 0 & 0 \end{pmatrix}$$

in which J must be odd.

$$\begin{pmatrix} j_1 & j_2 & j_3 \\ 0 & -1/2 & 1/2 \end{pmatrix} = \frac{1}{2}\left[\frac{(J+2)(J-2j_1+1)}{(j_2+1/2)\,(j_3+1/2)}\right]^{1/2}\begin{pmatrix} j_1 & j_2+1/2 & j_3+1/2 \\ 0 & 0 & 0 \end{pmatrix}$$

which is not valid for even J.

$$\begin{pmatrix} j & j & 0 \\ m & -m & 0 \end{pmatrix} = (-1)^{j-m}(2j+1)^{-1/2}$$

$$\begin{pmatrix} j & j & 1 \\ m & -m & 0 \end{pmatrix} = (-1)^{j-m}\frac{m}{[j(j+1)(2j+1)]^{1/2}}$$

$$\begin{pmatrix} j & j & 1 \\ m & -m-1 & 1 \end{pmatrix} = (-1)^{j-m}\left[\frac{(j-m)(j+m+1)}{j(2j+1)(2j+2)}\right]^{1/2}$$

$$\begin{pmatrix} j+1 & j & 1 \\ m & -m & 0 \end{pmatrix} = (-1)^{j-m-1}\left[\frac{(j+m+1)(j-m+1)}{(j+1)(2j+1)(2j+3)}\right]^{1/2}$$

$$\begin{pmatrix} j+1 & j & 1 \\ m & -m-1 & 1 \end{pmatrix} = (-1)^{j-m-1}\left[\frac{(j-m)(j-m+1)}{(2j+1)(2j+2)(2j+3)}\right]^{1/2}$$

$$\begin{pmatrix} j+1/2 & j & 1/2 \\ m & -m-1/2 & 1/2 \end{pmatrix} = (-1)^{j-m-1/2}\left[\frac{j-m+1/2}{(2j+1)(2j+2)}\right]^{1/2}$$

$$\begin{pmatrix} j+1/2 & j & 3/2 \\ m & -m-1/2 & 1/2 \end{pmatrix} = (-1)^{j-m-1/2}\frac{1}{2}(j+3m+3/2)\left[\frac{j-m+1/2}{j(j+1)(2j+1)(2j+3)}\right]^{1/2}$$

$$\begin{pmatrix} j+1/2 & j & 3/2 \\ m & -m-3/2 & 3/2 \end{pmatrix} = (-1)^{j-m-1/2}\frac{\sqrt{3}}{2}\left[\frac{(j-m-1/2)(j-m+1/2)(j+m+3/2)}{j(j+1)(2j+1)(2j+3)}\right]^{1/2}$$

$$\begin{pmatrix} j+3/2 & j & 3/2 \\ m & -m-1/2 & 1/2 \end{pmatrix} = (-1)^{j-m+1/2}\frac{\sqrt{3}}{2}\left[\frac{(j-m+1/2)(j-m+3/2)(j+m+3/2)}{(j+1)(j+2)(2j+1)(2j+3)}\right]^{1/2}$$

$$\begin{pmatrix} j+3/2 & j & 3/2 \\ m & -m-3/2 & 3/2 \end{pmatrix} = (-1)^{j-m+1/2}\frac{1}{2}\left[\frac{(j-m-1/2)(j-m+1/2)(j-m+3/2)}{(j+1)(j+2)(2j+1)(2j+3)}\right]^{1/2}$$

$$\begin{pmatrix} j & j & 2 \\ m & -m & 0 \end{pmatrix} = (-1)^{j-m}\frac{3m^2-j(j+1)}{[(2j-1)j(j+1)(2j+1)(2j+3)]^{1/2}}$$

$$\begin{pmatrix} j & j & 2 \\ m & -m-1 & 1 \end{pmatrix} = (-1)^{j-m}(1+2m)\left[\frac{3(j+m+1)(j-m)}{2(2j-1)j(j+1)(2j+1)(2j+3)}\right]^{1/2}$$

$$\begin{pmatrix} j & j & 2 \\ m & -m-2 & 2 \end{pmatrix} = (-1)^{j-m}\left[\frac{3(j-m-1)(j-m)(j+m+1)(j+m+2)}{2(2j-1)j(j+1)(2j+1)(2j+3)}\right]^{1/2}$$

$$\begin{pmatrix} j+1 & j & 2 \\ m & -m & 0 \end{pmatrix} = (1)^{j-m+1}\, m\left[\frac{3(j-m+1)(j+m+1)}{j(j+1)(j+2)(2j+1)(2j+3)}\right]^{1/2}$$

$$\begin{pmatrix} j+1 & j & 2 \\ m & -m-1 & 1 \end{pmatrix} = (-1)^{j-m+1}(j+2m+2)\left[\frac{(j-m)(j-m+1)}{2j(j+1)(j+2)(2j+1)(2j+3)}\right]^{1/2}$$

$$\begin{pmatrix} j+1 & j & 2 \\ m & -m-2 & 2 \end{pmatrix} = (-1)^{j-m+1}\left[\frac{(j-m-1)(j-m)(j-m+1)(j+m+2)}{2j(j+1)(j+2)(2j+1)(2j+3)}\right]^{1/2}$$

$$\begin{pmatrix} j+2 & j & 2 \\ m & -m & 0 \end{pmatrix} = (-1)^{j-m}\left[\frac{3(j-m+1)(j-m+2)(j+m+1)(j+m+2)}{2(j+1)(j+2)(2j+1)(2j+3)(2j+5)}\right]^{1/2}$$

$$\begin{pmatrix} j+2 & j & 2 \\ m & -m-1 & 1 \end{pmatrix} = (-1)^{j-m}\left[\frac{(j-m)(j-m+1)(j-m+2)(j+m+2)}{(j+1)(j+2)(2j+1)(2j+3)(2j+5)}\right]^{1/2}$$

$$\begin{pmatrix} j+2 & j & 2 \\ m & -m-2 & 2 \end{pmatrix} = (-1)^{j-m}\frac{1}{2}\left[\frac{(j-m-1)(j-m)(j-m+1)(j-m+2)}{(j+1)(j+2)(2j+1)(2j+3)(2j+5)}\right]^{1/2}$$

The 6*j symbol*

1. Orthogonality properties

In addition to $9^3 12$, an incomplete orthogonality relation exists:

$$\sum_{l_3} 2(2l_3+1)(2j_3''+1)\begin{Bmatrix} j_1 & j_2 & j_3' \\ l_1 & l_2 & l_3 \end{Bmatrix}\begin{Bmatrix} j_1 & j_2 & j_3'' \\ l_1 & l_2 & l_3 \end{Bmatrix}$$

$$= \delta(j_3', j_3'') \pm (-1)^{j_1+l_2+j_3'+j_3''}(2j_3''+1)\begin{Bmatrix} j_1 & j_2 & j_3' \\ l_2 & l_1 & j_3'' \end{Bmatrix}$$

The summation runs over either odd or even values of l_3.

2. Sum rules.

Equations 9³13a and 9³13b give two of the summation properties. Schwartz and de-Shalit [2] give the following two:

$$\sum_k (2k+1)\begin{Bmatrix} j & j & k \\ j & j & l \end{Bmatrix} = 1;$$

$$\sum_k (-1)^{k+2j}\frac{(2k+1)}{(2j+1)}\begin{Bmatrix} j & j & k \\ j & j & l \end{Bmatrix} = \delta(l, 0)$$

3. Recursion relations

$$(-1)^{\alpha+\beta-a-b-c}\begin{Bmatrix} a & b & c \\ d & e & f \end{Bmatrix}\begin{Bmatrix} b & f & d \\ 1/2 & d+\alpha & f+\beta \end{Bmatrix}$$

$$= -(2e+1)\begin{Bmatrix} a & b & c \\ d+\alpha & e+1/2 & f+\beta \end{Bmatrix}\begin{Bmatrix} a & f & e \\ 1/2 & e+1/2 & f+\beta \end{Bmatrix}$$

$$\times\begin{Bmatrix} c & d & e \\ 1/2 & e+1/2 & d+\alpha \end{Bmatrix} + 2e\begin{Bmatrix} a & b & c \\ d+\alpha & e-1/2 & f+\beta \end{Bmatrix}$$

$$\times\begin{Bmatrix} a & f & e \\ 1/2 & e-1/2 & f+\beta \end{Bmatrix}\begin{Bmatrix} c & d & e \\ 1/2 & e-1/2 & d+\alpha \end{Bmatrix}$$

in which α and β take the value $+1/2$ or $-1/2$ independently.

4. Special formulas.

Here $S = a + b + c$.

$$\begin{Bmatrix} a & b & c \\ 0 & c & b \end{Bmatrix} = (-1)^S[(2b+1)(2c+1)]^{-1/2} = (-1)^S\left[\frac{1}{(2b+1)(2c+1)}\right]^{1/2}$$

$$\begin{Bmatrix} a & b & c \\ 1/2 & c+1/2 & b+1/2 \end{Bmatrix} = (-1)^{S+1}\left[\frac{(S+2)(S-2a+1)}{(2b+1)(2b+2)(2c+1)(2c+2)}\right]^{1/2}$$

$$\begin{Bmatrix} a & b & c \\ 1/2 & c+1/2 & b-1/2 \end{Bmatrix} = (-1)^S\left[\frac{(S-2c)(S-2b+1)}{2b(2b+1)(2c+1)(2c+2)}\right]^{1/2}$$

$$\begin{Bmatrix} a & b & c \\ 1/2 & c-1/2 & b+1/2 \end{Bmatrix} = (-1)^S\left[\frac{(S-2b)(S-2c+1)}{(2b+1)(2b+2)2c(2c+1)}\right]^{1/2}$$

$$\begin{Bmatrix} a & b & c \\ 1/2 & c-1/2 & b-1/2 \end{Bmatrix} = (-1)^S\left[\frac{(S+1)(S-2a)}{2b(2b+1)2c(2c+1)}\right]^{1/2}$$

$$\begin{Bmatrix} a & b & c \\ 1 & c & b \end{Bmatrix} = (-1)^{S+1}\frac{b(b+1)+c(c+1)-a(a+1)}{2\sqrt{b(b+1)(2b+1)c(c+1)(2c+1)}}$$

$$\begin{Bmatrix} a & b & c \\ 1 & c-1 & b \end{Bmatrix} = (-1)^S \frac{1}{2}\left[\frac{(S+1)(S-2a)(S-2b)(S-2c+1)}{b(b+1)(2b+1)c(2c-1)(2c+1)}\right]^{1/2}$$

$$\begin{Bmatrix} a & b & c \\ 1 & c-1 & b+1 \end{Bmatrix}$$

$$= (-1)^S \frac{1}{2}\left[\frac{(S-2b-1)(S-2b)(S-2c+1)(S-2c+2)}{(b+1)(2b+1)(2b+3)c(2c-1)(2c+1)}\right]^{1/2}$$

$$\begin{Bmatrix} a & b & c \\ 1 & c-1 & b-1 \end{Bmatrix} = (-1)^S \frac{1}{2}\left[\frac{S(S+1)(S-2a-1)(S-2a)}{b(2b-1)(2b+1)c(2c-1)(2c+1)}\right]^{1/2}$$

$$\begin{Bmatrix} a & b & c \\ 3/2 & c-1/2 & b+1/2 \end{Bmatrix}$$

$$= (-1)^S \frac{[(S-2b-1)(S-2c)-2(S+2)(S-2a)][(S-2b)(S-2c+1)]^{1/2}}{4[b(b+1)(2b+1)(2b+3)c(c+1)(2c-1)(2c+1)]^{1/2}}$$

$$\begin{Bmatrix} a & b & c \\ 3/2 & c-1/2 & b-1/2 \end{Bmatrix}$$

$$= (-1)^S \frac{[2(S-2b)(S-2c)-(S+2)(S-2a-1)][(S+1)(S-2a)]^{1/2}}{4[b(b+1)(2b-1)(2b+1)c(c+1)(2c-1)(2c+1)]^{1/2}}$$

$$\begin{Bmatrix} a & b & c \\ 3/2 & c-3/2 & b+1/2 \end{Bmatrix}$$

$$= (-1)^S \left[\frac{3(S+1)(S-2a)(S-2b-1)(S-2b)(S-2c+1)(S-2c+2)}{2b(2b+1)(2b+2)(2b+3)(2c-2)(2c-1)2c(2c+1)}\right]^{1/2}$$

$$\begin{Bmatrix} a & b & c \\ 3/2 & c-3/2 & b-1/2 \end{Bmatrix}$$

$$= (-1)^S \left[\frac{3S(S+1)(S-2a-1)(S-2a)(S-2b)(S-2c+1)}{(2b-1)2b(2b+1)(2b+2)(2c-2)(2c-1)2c(2c+1)}\right]^{1/2}$$

$$\begin{Bmatrix} a & b & c \\ 3/2 & c-3/2 & b+3/2 \end{Bmatrix}$$

$$= (-1)^S \left[\frac{(S-2b-2)(S-2b-1)(S-2b)(S-2c+1)(S-2c+2)(S-2c+3)}{(2b+1)(2b+2)(2b+3)(2b+4)(2c-2)(2c-1)2c(2c+1)}\right]^{1/2}$$

$$\begin{Bmatrix} a & b & c \\ 3/2 & c-3/2 & b-3/2 \end{Bmatrix}$$

$$= (-1)^S \left[\frac{(S-1)S(S+1)(S-2a-2)(S-2a-1)(S-2a)}{(2b-2)(2b-1)2b(2b+1)(2c-2)(2c-1)2c(2c+1)}\right]^{1/2}$$

$$\begin{Bmatrix} a & b & c \\ 2 & c & b \end{Bmatrix}$$

$$= (-1)^S \frac{2[3A(A-1)-4b(b+1)c(c+1)]}{[(2b-1)2b(2b+1)(2b+2)(2b+3)(2c-1)2c(2c+1)(2c+2)(2c+3)]^{1/2}}$$

in which $A = b(b+1) + c(c+1) - a(a+1)$

$$\begin{Bmatrix} a & b & c \\ 2 & c-1 & b \end{Bmatrix} = (-1)^S \frac{[(a+b+1)(a-b)-c^2+1][3(S+1)(S-2a)(S-2b)(S-2c+1)]^{1/2}}{2[b(b+1)(2b-1)(2b+1)(2b+3)c(c-1)(c+1)(2c-1)(2c+1)]^{1/2}}$$

$$\begin{Bmatrix} a & b & c \\ 2 & c-1 & b+1 \end{Bmatrix} = (-1)^S \frac{[(a+b+2)(a-b-1)-(c-1)(b+c+2)][(S-2b-1)(S-2b)(S-2c+1)(S-2c+2)]^{1/2}}{2[b(b+1)(b+2)(2b+1)(2b+3)c(c-1)(c+1)(2c-1)(2c+1)]^{11/2}}$$

$$\begin{Bmatrix} a & b & c \\ 2 & c-1 & b-1 \end{Bmatrix} = (-1)^S \frac{[(a+b)(a-b+1)-(c-1)(c-b+1)][S(S+1)(S-2a-1)(S-2a)]^{1/2}}{2[b(b-1)(b+1)(2b-1)(2b+1)c(c-1)(c+1)(2c-1)(2c+1)]^{1/2}}$$

$$\begin{Bmatrix} a & b & c \\ 2 & c-2 & b \end{Bmatrix} = (-1)^S \left[\frac{6S(S+1)(S-2a-1)(S-2a)(S-2b-1)(S-2b)(S-2c+1)(S-2c+2)}{(2b-1)2b(2b+1)(2b+2)(2b+3)(2c-3)(2c-2)(2c-1)2c(2c+1)}\right]^{1/2}$$

$$\begin{Bmatrix} a & b & c \\ 2 & c-2 & b+1 \end{Bmatrix} = (-1)^S \left[\frac{(S+1)(S-2a)(S-2b-2)(S-2b-1)(S-2b)(S-2c+1)(S-2c+2)(S-2c+3)}{2b(2b+1)(2b+2)(2b+3)(2b+4)c(c-1)(2c-3)(2c-1)(2c+1)}\right]^{1/2}$$

$$\begin{Bmatrix} a & b & c \\ 2 & c-2 & b-1 \end{Bmatrix} = (-1)^S \left[\frac{(S-1)S(S+1)(S-2a-2)(S-2a-1)(S-2a)(S-2b)(S-2c+1)}{(2b-2)(2b-1)2b(2b+1)(2b+2)c(c-1)(2c-3)(2c-1)(2c+1)}\right]^{1/2}$$

$$\begin{Bmatrix} a & b & c \\ 2 & c-2 & b+2 \end{Bmatrix} = (-1)^S \left[\frac{(S-2b-3)(S-2b-2)(S-2b-1)(S-2b)(S-2c+1)(S-2c+2)(S-2c+3)(S-2c+4)}{(2b+1)(2b+2)(2b+3)(2b+4)(2b+5)(2c-3)(2c-2)(2c-1)2c(2c+1)}\right]^{1/2}$$

$$\begin{Bmatrix} a & b & c \\ 2 & c-2 & b-2 \end{Bmatrix} = (-1)^S \left[\frac{(S-2)(S-1)S(S+1)(S-2a-3)(S-2a-2)(S-2a-1)(S-2a)}{(2b-3)(2b-2)(2b-1)2b(2b+1)(2c-3)(2c-2)(2c-1)2c(2c+1)}\right]^{1/2}$$

5. Relations between 3j and 6j symbols

The 6j symbol can be expressed as a sum of 3j symbols as [9³a, p. 17]

$$\begin{Bmatrix} j_1 & j_2 & j_3 \\ l_1 & l_2 & l_3 \end{Bmatrix} = \sum_{m,n} (-1)^S \begin{pmatrix} j_1 & j_2 & j_3 \\ m_1 & m_2 & m_3 \end{pmatrix} \begin{pmatrix} j_1 & l_2 & l_3 \\ m_1 & n_2 & -n_3 \end{pmatrix} \times \begin{pmatrix} l_1 & j_2 & l_3 \\ -n_1 & m_2 & n_3 \end{pmatrix} \begin{pmatrix} l_1 & l_2 & j_3 \\ n_1 & -n_2 & m_3 \end{pmatrix}$$

where the summation runs over all m and n values; moreover,

$$S = l_1 + l_2 + l_3 + n_1 + n_2 + n_3$$

In addition to 9³8, there is another relation of this kind [9³a, p. 17]:

$$\sum_{m_3} \begin{pmatrix} j_1 & j_2 & j_3 \\ m_1 & m_2 & m_3 \end{pmatrix} \begin{pmatrix} l_1 & l_2 & j_3 \\ n_1 & n_2 & -m_3 \end{pmatrix} = \sum_{l_3, n_3} (-1)^{m_1+n_1+j_3+l_3} (2l_3+1) \begin{Bmatrix} j_1 & j_2 & j_3 \\ l_1 & l_2 & l_3 \end{Bmatrix} \times \begin{pmatrix} l_1 & j_2 & l_3 \\ n_1 & m_2 & n_3 \end{pmatrix} \begin{pmatrix} j_1 & l_2 & l_3 \\ m_1 & n_2 & -n_3 \end{pmatrix}$$

The 9j *symbol*

1. Orthogonality properties (See equation 9³23.)

2. Sum rules

$$\begin{Bmatrix} j_{11} & j_{12} & j_{13} \\ j_{21} & j_{22} & j_{23} \\ j_{31} & j_{32} & j_{33} \end{Bmatrix}$$

$$= \sum_{j,j'} (-1)^{2j_{12}+j_{32}+j'-j_{23}} (2j+1)(2j'+1) \begin{Bmatrix} j_{11} & j_{12} & j_{13} \\ j_{22} & j_{21} & j_{23} \\ j & j' & j_{33} \end{Bmatrix} \begin{Bmatrix} j_{11} & j_{22} & j \\ j_{21} & j_{12} & j' \\ j_{31} & j_{32} & j_{33} \end{Bmatrix}$$

$$\sum_{j_{31},j_{32},j_{33}} (2j_{31}+1)(2j_{32}+1)(2j_{33}+1) \begin{Bmatrix} j_{11} & j_{12} & j_{13} \\ j_{11} & j_{12} & j_{13} \\ j_{31} & j_{32} & j_{33} \end{Bmatrix}$$

$$= \frac{1}{4}\left[1 + \frac{(-1)^{2j_{11}}}{2j_{11}+1} + \frac{(-1)^{2j_{12}}}{2j_{12}+1} + \frac{(-1)^{2j_{13}}}{2j_{13}+1}\right]$$

3. Recursion relation.

One that relates two special cases of the vector coupling of 9³20 is

$$\begin{Bmatrix} j_1 & j_1 & 1 \\ j_3 & j_4 & j \\ j_{13} & j_{14} & j \end{Bmatrix} = \frac{j_3(j_3+1) + j_{14}(j_{14}+1) - j_{13}(j_{13}+1) - j_4(j_4+1)}{2[j_1(j_1+1)(2j_1+1)j(j+1)(2j+1)]^{1/2}} \begin{Bmatrix} j_1 & j_1 & 0 \\ j_3 & j_4 & j \\ j_{13} & j_{14} & j \end{Bmatrix}$$

in which the 9j symbol on the right can be evaluated with the help of

$$\begin{Bmatrix} f & f & 0 \\ d & c & e \\ b & a & e \end{Bmatrix} = \begin{Bmatrix} a & b & e \\ c & d & e \\ f & f & 0 \end{Bmatrix} = \begin{Bmatrix} 0 & e & e \\ f & d & b \\ f & c & a \end{Bmatrix} = \begin{Bmatrix} e & 0 & e \\ c & f & a \\ d & f & b \end{Bmatrix}$$

$$= \begin{Bmatrix} f & b & d \\ 0 & e & e \\ f & a & c \end{Bmatrix} = \begin{Bmatrix} a & f & c \\ e & 0 & e \\ b & f & d \end{Bmatrix} = \begin{Bmatrix} b & a & e \\ f & f & 0 \\ d & c & e \end{Bmatrix} = \begin{Bmatrix} e & d & c \\ e & b & a \\ 0 & f & f \end{Bmatrix}$$

$$= \begin{Bmatrix} c & e & d \\ a & e & b \\ f & 0 & f \end{Bmatrix} = \frac{(-1)^{b+c+e+f}}{\sqrt{(2e+1)(2f+1)}} \begin{Bmatrix} a & b & e \\ d & c & f \end{Bmatrix}$$

4. Relations among 3j, 6j, and 9j symbols

Another special case of the vector coupling is obtained when $j_1 = j_2 = 1/2$ and $j_{12} = j_1 + j_2 = 1$ in 9³20. The resulting expression is [9³a, p. 22]

$$\begin{Bmatrix} 1/2 & 1/2 & 1 \\ j_3 & j_4 & j_{34} \\ j_{13} & j_{24} & j \end{Bmatrix} = (-1)^K \frac{1}{3}\left[(-1)^{2k} \begin{Bmatrix} j_{13} & j_{24} & j \\ 1/2 & k & j_4 \end{Bmatrix} \begin{Bmatrix} j_4 & j_3 & j_{34} \\ 1/2 & k & j_{13} \end{Bmatrix} \right.$$

$$\left. + (-1)^{j_3+j_{24}-k+1} \frac{\delta(j_{34}, j)}{2(2j+1)} \begin{Bmatrix} j & j_4 & j_3 \\ 1/2 & j_{13} & j_{24} \end{Bmatrix} \right] \begin{Bmatrix} j & j_{34} & 1 \\ 1/2 & 1/2 & k \end{Bmatrix}^{-1}$$

in which $K = j_3 + j_4 + j_{34} + j_{13} + j_{24} + j$ and

$$k = \frac{1}{2}(j_{34} + j) \quad \text{if} \quad j_{34} \neq j$$

$$k = j_3 + \frac{1}{2} \quad \text{if} \quad j_{34} = j$$

De-Shalit [3] gives the following relation between $3j$ and $9j$ symbols

$$\sum_{\gamma x} (2x+1) \begin{pmatrix} a & b & x \\ \alpha & \beta & \gamma \end{pmatrix} \begin{pmatrix} x & c & d \\ \gamma & \epsilon & \zeta \end{pmatrix} \begin{Bmatrix} a & b & x \\ e & f & c \\ g & h & d \end{Bmatrix}$$

$$= \sum_{\text{primes}} \begin{pmatrix} a & e & g \\ \alpha & \mu' & \nu' \end{pmatrix} \begin{pmatrix} b & f & h \\ \beta & \rho' & \sigma' \end{pmatrix} \begin{pmatrix} e & f & c \\ \mu' & \rho' & \epsilon \end{pmatrix} \begin{pmatrix} g & h & d \\ \nu' & \sigma' & \zeta \end{pmatrix}$$

A similar relation to $9^3 24$ between $6j$ and $9j$ symbols is [9^3a, p. 24]

$$\begin{Bmatrix} a & b & c \\ a' & b' & c' \\ d & e & f \end{Bmatrix} \begin{Bmatrix} d & e & f \\ h & i & j \end{Bmatrix} = \sum_{\lambda\mu} (-1)^{X+\lambda}(2\lambda+1)(2\mu+1) \begin{Bmatrix} c & c' & f \\ h & i & \lambda \end{Bmatrix}$$

$$\times \begin{Bmatrix} a' & b' & c' \\ h & \lambda & \mu \end{Bmatrix} \begin{Bmatrix} b & b' & e \\ i & j & \mu \end{Bmatrix} \begin{Bmatrix} a & b & c \\ a' & \mu & \lambda \\ d & j & i \end{Bmatrix}$$

in which $X = b + c - a' - c' - d - e + 2h + j$.

5. Special formulas.

De-Shalit [3] gives the following:

$$\begin{Bmatrix} 0 & b & c \\ d & e & f \\ d & e & f \end{Bmatrix} = (-1)^{b+d+e+f} \frac{\delta(b, c)}{[2(2b+1)]^{1/2}} \begin{Bmatrix} b & f & f \\ d & e & e \end{Bmatrix}$$

Kennedy, Sears, and Sharp [4] tabulated some of the X coefficients that are precisely $9j$ symbols.

The $12j$ symbol

The definition of the $12j$ symbol, $9^3 25$a and $9^3 25$b, which are given by Jahn and Hope [5] and by Ord-Smith [6] are not unique [7]. Other definitions and properties of the $12j$ symbols are given by Sharp [8]. These are briefly discussed by Rotenberg et al [9^3a, pp. 25–7].

Appendix 3. Russell-Saunders term energies

The term energies given here are found by the method of Chapter 5. Two electron cases are covered by $2^5 6$ and $2^5 7$. Three electron cases are given by $4^5 4$ and $4^5 4'$ for the configurations that contain nonequivalent electrons; by $5^5 2$ and $5^5 2'$ for the configurations where two of the electrons are equivalent; by the equations of Sections 9^5 and 10^5 for equivalent-electron configurations. The $l^n l'$ configurations are covered by $11^5 3'$ or, alternatively, by $13^5 6$. The term energies for the $l^n s$ and $l^n l's$ configurations are given by $12^5 2$ and $12^5 3$-5, respectively. One-hole and one-electron cases are covered by $14^5 11$. Two-equivalent holes and one-electron cases are given by $14^5 14$. The rules given in Section 14^5 for the correlated and conjugated configurations generalize the following results to the configurations that contain more than half-filled shells.

Notations used in the following tables are:

> ${}^{2S+1}L$–energy of this term, relative to the configuration average, when the labels L and S are sufficient to specify it
>
> $({}^{2S'+1}L')\,{}^{2S+1}_{\nu}L$–diagonal element of the Coulomb energy matrix, relative to the configuration average. If, in specifying a term, either the seniority number ν [14] or the parentage labels $L'S'$ are not needed, they are omitted.
>
> $({}^{2S'+1}L')\,{}^{2S+1}_{\nu}L{:}({}^{2S''+1}L'')\,{}^{2S+1}_{\nu'}L$–off-diagonal element of the Coulomb energy matrix
>
> E_{Av}–configuration average
>
> $E(cs)$–written for interactions of the pairs of electrons in closed shells, plus their $I(nl)$ one-electron integrals
>
> $E(cs, nl)$–interaction of the nl electron with those in closed shells
>
> $E(nl)$–sum of the $I(nl)$ and pairwise interactions for electrons in open shells
>
> $E(nl^n, {}^{2S+1}L)$–energy of the parent term relative to the configuration average of the nl^n core

In the classification of the Russell-Saunders terms of the $f^n (n \geqslant 3)$ configurations additional labels are needed. These are (Chap. 7) the $SO(7)$ and G_2 group representation symbols [Racah's $W \equiv (w_1, w_2, w_3)$ and $U \equiv (u_1, u_2)$ labels, respectively [15]].

To simplify the tabulation of the off-diagonal energy matrix elements of the $f^n (n \geqslant 3)$ configurations we assign a sequential index to each term with the same L, S and ν values and write it as a right-subscript to the L designation. The correspondence of these indices to the W and U labels are found from the diagonal elements where they are first introduced.

Off-diagonal energy matrix elements that vanish but appear formally are omitted.

Additional configurations involving g electrons have been given by Shortley and Fried [11].

0. Closed shells only

$$^1S = E_{\text{Av}} = E(cs)$$

1. Closed shells plus one (nl) electron

$$E_{\text{Av}} = E(cs) + E(cs, nl) + I(nl)$$
$$^2L = E_{\text{Av}}$$

2. Closed shells plus (ns, $n'l$) [1, 2, 4]

$$E_{\text{Av}} = E(cs) + E(cs, ns) + E(cs, n'l) + I(ns) + I(n'l) + F^0(ns, n'l) - \frac{1}{2(2l+1)} G^l(ns, n'l)$$

$$^1L = \frac{3}{2(2l+1)} G^l(ns, n'l)$$
$$^3L = -\frac{1}{2(2l+1)} G^l(ns, n'l)$$

3. Closed shells plus $(np)^2$[1, 2, 4, 6, 7, 24]

$$E_{\text{Av}} = E(cs) + 2E(cs, np) + 2I(np) + F^0(np, np) - \frac{2}{25} F^2(np, np)$$

$$^1S = \frac{12}{25} F^2(np, np)$$
$$^3P = -\frac{3}{25} F^2(np, np)$$
$$^1D = \frac{3}{25} F^2(np, np)$$

4. Closed shells plus (np, $n'l$) [1, 2, 3, 4, 5, 6, 11]

$$E_{\text{Av}} = E(cs) + E(cs, np) + E(cs, n'l) + I(np) + I(n'l) + E(n'l)$$
$$E(n'p) = F^0(np, n'p) - \frac{1}{6} G^0(np, n'p) - \frac{1}{15} G^2(np, n'p)$$

$$E(n'd) = F^0(np, n'd) - \frac{1}{15}G^1(np, n'd) - \frac{3}{70}G^3(np, n'd)$$

$$E(n'f) = F^0(np, n'f) - \frac{3}{70}G^2(np, n'f) - \frac{2}{63}G^4(np, n'f)$$

The singlet or triplet terms for $L = l - 1, l, l + 1$ are called a *triad:*

$l = 1:(np, n'p)$

	$F^2/25$	$G^0/6$	$G^2/75$
^1S	10	7	35
^3S	10	−5	−25
^1P	−5	−5	20
^3P	−5	7	−10
^1D	1	7	8
^3D	1	−5	2

$l = 2:(np, n'd)$

	$F^2/35$	$G^1/15$	$G^3/490$
^1P	7	2	147
^3P	7		−105
^1D	−7	−2	63
^3D	−7	4	−21
^1F	2	7	27
^3F	2	−5	15

$l = 3:(np, n'f)$

	$F^2/75$	$G^2/350$	$G^4/189$
^1D	12	21	42
^3D	12	9	−30
^1F	−15	−15	15
^3F	−15	45	−3
^1G	5	105	7
^3G	5	−75	5

5. Closed shells plus $(nd)^2$ [1, 2, 4, 6, 13, 24]

$$E_{\text{Av}} = E(cs) + 2E(cs, nd) + 2I(nd) + F^0(nd, nd) - \frac{2}{63}F^2(nd, nd) - \frac{2}{63}F^4(nd, nd)$$

	$F^2/441$	$F^4/441$
^1_0S	140	140
^3_2P	77	−70
^1_2D	−13	50
^3_2F	−58	5
^1_2G	50	15

6. Closed shells plus $(nd, n'l)$ [2, 3, 4, 6, 11]

$$E_{\mathrm{Av}} = E(cs) + E(cs, nd) + E(cs, n'l) + I(nd) + I(n'l) + E(n'l)$$

$$E(n'd) = F^0(nd, n'd) - \frac{1}{10} G^0(nd, n'd) - \frac{1}{35} G^2(nd, n'd) - \frac{1}{35} G^4(nd, n'd)$$

$$E(n'f) = F^0(nd, n'f) - \frac{3}{70} G'(nd, n'f) - \frac{2}{105} G^3(nd, n'f) - \frac{5}{231} G^5(nd, n'f)$$

The singlet or triplet terms for $L = l - 2, l - 1, l, l + 1, l + 2$ are called a *pentad:*

$l = 2{:}(nd, n'd)$

	$F^2/49$	$F^4/441$	$G^0/10$	$G^2/245$	$G^4/2205$
^1S	14	126	11	77	693
^3S	14	126	−9	−63	−567
^1P	7	−84	−9	−28	483
^3P	7	−84	11	42	−357
^1D	−3	36	11	−8	243
^3D	−3	36	−9	22	−117
^1F	−8	−9	−9	47	108
^3F	−8	−9	11	−33	18
^1G	4	1	11	27	68
^3G	4	1	−9	−13	58

$l = 3{:}(nd, n'f)$

	$F^2/105$	$F^4/693$	$G^1/70$	$G^3/315$	$G^5/7623$
^1P	24	66	5	30	1815
^3P	24	66	1	−18	−1485
^1D	6	−99	−3	−36	990
^3D	6	−99	9	48	−660
^1F	−11	66	15	25	440
^3F	−11	66	−9	−13	−110
^1G	−15	−22	−17	41	220
^3G	−15	−22	23	−29	110
^1H	10	3	33	16	170
^3H	10	3	−27	−4	160

7. Closed shells plus $(nf)^2$ [3, 4, 24]

$$E_{\mathrm{Av}} = E(cs) + 2E(cs, nf) + 2I(nf) + F^0(nf, nf) - \frac{4}{195} F^2(nf, nf) - \frac{2}{143} F^4(nf, nf) - \frac{100}{5577} F^6(nf, nf)$$

	$F^2/2925$	$F^4/14157$	$F^6/184041$
${}^1_0S$	840	2772	46200
${}^3_2P$	645	627	−28875
${}^1_2D$	307	−1089	21175
${}^3_2F$	−70	−231	−3850
${}^1_2G$	−330	1459	5250
${}^3_2H$	−265	−465	2975
${}^1_1I$	385	315	3325

8. Closed shells plus $(nf, n'f)$ [3, 4]

$$E_{Av} = E(cs) + E(cs, nf) + E(cs, n'f) + I(nf) + I(n'f) + F^0(nf, n'f) - \frac{1}{14}G^0(nf, n'f) - \frac{2}{105}G^2(nf, nf') - \frac{1}{77}G^4(nf, n'f) - \frac{50}{3003}G^6(nf, n'f)$$

The singlet or triplet terms for $L = l - 3, l - 2, \ldots, l + 2, l + 3$ are called a *heptad*.

	$F^2/225$	$F^4/1089$	$F^6/184041$	$G^0/14$	$G^2/1575$	$G^4/7623$	$G^6/1288287$
${}^1S$	60	198	42900	15	450	1485	321750
${}^3S$	60	198	42900	−13	−390	−1287	−278850
${}^1P$	45	33	−32175	−13	−285	−132	246675
${}^3P$	45	33	−32175	15	345	330	−203775
${}^1D$	19	−99	17875	15	163	−594	146575
${}^3D$	19	−99	17875	−13	−103	792	−103675
${}^1F$	−10	−33	−7150	−13	100	330	71500
${}^3F$	−10	−33	−7150	15	−40	−132	−28600
${}^1G$	−30	97	1950	15	−180	778	35100
${}^3G$	−30	97	1950	−13	240	−580	7800
${}^1H$	−25	−51	−325	−13	205	456	23725
${}^3H$	−25	−51	−325	15	−145	−258	19175
${}^1I$	25	9	25	15	205	162	21625
${}^3I$	25	9	25	−13	−145	36	21275

9. Closed shells plus $(np)^3$ [1, 2, 4, 5, 7, 24]

$$E_{Av} = E(cs) + 3E(cs, np) + 3I(np) + 3F^0(np, np) - \frac{6}{25}F^2(np, np)$$

${}^4S$: $-9F^2/25$
${}^2P$: $6F^2/25$
${}^2D$: 0

10. Closed shells plus p^2s [2, 3, 4, 7]

$$E_{\mathrm{Av}} = E(cs) + 2E(cs, np) + E(cs, n's) + 2I(np) + I(\mathrm{n}'s) + 2F^0(np, n's) - \frac{2}{6}G^1(np, n's) + F^0(np, np) - \frac{2}{25}F^2(np, np)$$

$$^2S = \frac{12}{25}F^2(np, np)$$

$$^2P = -\frac{3}{25}F^2(np, np) + \frac{2}{3}G^1(np, n's)$$

$$^4P = -\frac{3}{25}F^2(np, np) - \frac{1}{3}G^1(np, n's)$$

$$^2D = \frac{3}{25}F^2(np, np)$$

11. Closed shells plus p^2p' [2, 12, 17]

$$E_{\mathrm{Av}} = E(cs) + 2E(cs, np) + E(cs, n'p) + 2I(np) + I(n'p) + 2F^0(np, n'p) - \frac{2}{6}G^0(np, n'p) - \frac{2}{15}G^2(np, n'p) + F^0(np, np) - \frac{2}{25}F^2(np, np)$$

	$F^2(np, np)/25$	$F^2(np, n'p)/25$	$G^0(np, n'p)/6$	$G^2(np, n'p)/150$
$(^3P)\,^2S$	−3	−10	−4	50
$(^3P)\,^4S$	−3	−10	14	−40
$(^1S)\,^2P$	12			
$(^3P)\,^2P$	−3	5	5	50
$(^1D)\,^2P$	3	7	1	−26
$(^1S)\,^2P$:$(^3P)\,^2P$			−6	30
$(^1S)\,^2P$:$(^1D)\,^2P$		$4\sqrt{5}$	$-2\sqrt{5}$	$-2\sqrt{5}$
$(^3P)\,^2P$:$(^1D)\,^2P$			$3\sqrt{5}$	$12\sqrt{5}$
$(^3P)\,^4P$	−3	5	−4	−40
$(^3P)\,^2D$	−3	−1	5	32
$(^1D)\,^2D$	3	−7	5	−4
$(^3P)\,^2D$:$(^1D)\,^2D$			−9	18
$(^3P)\,^4D$	−3	−1	−4	−4
$(^1D)\,^2F$	3	2	−4	14

12. Closed shells plus p^2d [2, 12, 17]

$$E_{\mathrm{Av}} = E(cs) + 2E(cs, np) + E(cs, n'd) + 2I(np) + I(n'd) + 2F^0(np, n'd) - \frac{2}{15}G^1(np, n'd) - \frac{6}{70}G^3(np, n'd) + F^0(np, np) - \frac{2}{25}F^2(np, np)$$

	$F^2(np, np)/25$	$F^2(np, n'd)/35$	$G^1(np, n'd)/30$	$G^3(np, n'd)/490$
$(^1D)\,^2S$	3	14	2	−84
$(^3P)\,^2P$	−3	−7		105
$(^1D)\,^2P$	3	7	4	−63
$(^3P)\,^2P{:}(^1D)\,^2P$			6	63
$(^3P)\,^4P$	−3	−7	12	−84
$(^1S)\,^2D$	12			
$(^3P)\,^2D$	−3	7	10	105
$(^1D)\,^2D$	3	−3	6	−27
$(^1S)\,^2D{:}(^3P)\,^2D$			$-6\sqrt{3}$	$42\sqrt{3}$
$(^1S)\,^2D{:}(^1D)\,^2D$		$4\sqrt{7}$	$-2\sqrt{7}$	$-6\sqrt{7}$
$(^3P)\,^2D{:}(^1D)\,^2D$				$15\sqrt{21}$
$(^3P)\,^4D$	−3	7	−8	−84
$(^3P)\,^2F$	−3	−2	10	60
$(^1D)\,^2F$	3	−8	4	12
$(^3P)\,^2F{:}(^1D)\,^2F$			$-6\sqrt{6}$	$12\sqrt{6}$
$(^3P)\,^4F$	−3	−2	−8	6
$(^1D)\,^2G$	3	4	−8	36

13. Closed shells plus p^2f [2, 12, 17]

$$E_{\mathrm{Av}} = E(cs) + 2E(cs, np) + E(cs, n'f) + 2I(np) + I(n'f) + 2F^0(np, n'f) - \frac{6}{70}G^2(np, n'f) - \frac{4}{63}G^4(np, n'f) + F^0(np, np) - \frac{2}{25}F^2(np, np)$$

	$F^2(np, np)/25$	$F^2(np, n'f)/75$	$G^2(np, n'f)/350$	$G^4(np, n'f)/189$
$(^1D)\,^2P$	3	24	24	−24
$(^3P)\,^2D$	−3	−12	12	30
$(^1D)\,^2D$	3	6	36	−15
$(^3P)\,^2D{:}(^1D)\,^2D$			$12\sqrt{6}$	$9\sqrt{6}$
$(^3P)\,^4D$	−3	−12	66	−24
$(^1S)\,^2F$	12			
$(^3P)\,^2F$	−3	15	75	30
$(^1D)\,^2F$	3	−11	39	−4
$(^1S)\,^2F{:}(^3P)\,^2F$			$-30\sqrt{6}$	$9\sqrt{6}$
$(^1S)\,^2F{:}(^1D)\,^2F$		$4\sqrt{30}$	$-6\sqrt{30}$	$-\sqrt{30}$
$(^3P)\,^2F{:}(^1D)\,^2F$			$-9\sqrt{5}$	$9\sqrt{5}$
$(^3P)\,^4F$	−3	15	−60	−24
$(^3P)\,^2G$	−3	−5	75	16
$(^1D)\,^2G$	3	−15	15	6
$(^3P)\,^2G{:}(^1D)\,^2G$			$-45\sqrt{5}$	$3\sqrt{5}$
$(^3P)\,^4G$	−3	−5	−60	4
$(^1D)\,^2H$	3	10	−60	11

14. Closed shells plus d^3 [2, 3, 4, 9, 13, 24]

$$E_{\mathrm{Av}} = E(cs) + 3E(cs, nd) + 3I(nd) + 3F^0(nd, nd) - \frac{6}{63} F^2(nd, nd) - \frac{6}{63} F^4(nd, nd)$$

	$F^2/441$	$F^4/441$
${}^2_3P$	−12	30
${}^4_3P$	42	−105
${}^2_1D$	105	105
${}^2_3D$	69	−15
${}^2_1D:{}^2_3D$	$27\sqrt{21}$	$-15\sqrt{21}$
${}^2_3F$	123	−45
${}^4_3F$	−93	−30
${}^2_3G$	−57	55
${}^2_3H$	−12	30

15. Closed shells plus d^2s [2, 3, 4]

$$E_{\mathrm{Av}} = E(cs) + 2E(cs, nd) + E(cs, n's) + 2I(nd) + I(n's) + F^0(nd, nd) - \frac{2}{63} F^2(nd, nd) - \frac{2}{63} F^4(nd, nd) + 2F^0(nd, n's) - \frac{2}{10} G^2(nd, n's)$$

$${}^2L = E(d^2, {}^1L) \qquad \text{for even } L$$

$${}^2L = E(d^2, {}^3L) + \frac{2}{5} G^2(nd, n's) \qquad \text{for odd } L$$

$${}^4L = E(d^2, {}^3L) - \frac{1}{5} G^2(nd, n's) \qquad \text{for odd } L$$

16. Closed shells plus d^2p [2, 13]

$$E_{\mathrm{Av}} = E(cs) + 2E(cs, nd) + E(cs, n'p) + 2I(nd) + I'(np) + F^0(nd, nd) - \frac{2}{63} F^2(nd, nd) - \frac{2}{63} F^4(nd, nd) + 2F^0(nd, n'p) - \frac{2}{15} G^1(nd, n'p) - \frac{6}{70} G^3(nd, n'p)$$

	$E(d^2, {}^{2s+1}L)$	$F^2(nd, n'p)/1225$	$G^1(nd, n'p)/1050$	$G^3(nd, n'p)/17150$
$({}^3P)\,{}^2S$	$E(d^2, {}^3P)$	−490	−70	2940
$({}^3P)\,{}^4S$	$E(d^2, {}^3P)$	−490	560	−1470
$({}^1S)\,{}^2P$	$E(d^2, {}^1S)$			
$({}^3P)\,{}^2P$	$E(d^2, {}^3P)$	245	350	3675
$({}^1D)\,{}^2P$	$E(d^2, {}^1D)$	−105	210	−945
$({}^1S)\,{}^2P{:}({}^3P)\,{}^2P$			$-210\sqrt{3}$	$1470\sqrt{3}$
$({}^1S)\,{}^2P{:}({}^1D)\,{}^2P$		$140\sqrt{7}$	$-70\sqrt{7}$	$-210\sqrt{7}$
$({}^3P)\,{}^2P{:}({}^1D)\,{}^2P$				$525\sqrt{21}$
$({}^3P)\,{}^4P$	$E(d^2, {}^3P)$	245	−280	−2940
$({}^3P)\,{}^2D$	$E(d^2, {}^3P)$	−49	308	2499
$({}^1D)\,{}^2D$	$E(d^2, {}^1D)$	105		−525
$({}^3F)\,{}^2D$	$E(d^2, {}^3F)$	84	112	4536
$({}^3P)\,{}^2D{:}({}^1D)\,{}^2D$			$-42\sqrt{105}$	$189\sqrt{105}$
$({}^3P)\,{}^2D{:}({}^3F)\,{}^2D$		$84\sqrt{14}$	$42\sqrt{14}$	$126\sqrt{14}$
$({}^1D)\,{}^2D{:}({}^3F)\,{}^2D$			$14\sqrt{30}$	$462\sqrt{30}$
$({}^3P)\,{}^4D$	$E(d^2, {}^3P)$	−49	−196	−588
$({}^3F)\,{}^4D$	$E(d^2, {}^3F)$	84	196	−4662
$({}^3P)\,{}^4D{:}({}^3F)\,{}^4D$		$84\sqrt{14}$	$-84\sqrt{14}$	$-252\sqrt{14}$
$({}^1D)\,{}^2F$	$E(\mathrm{d}^2, {}^1D)$	−30	−90	780
$({}^3F)\,{}^2F$	$E(d^2, {}^3F)$	−105	175	3150
$({}^1G)\,{}^2F$	$E(d^2, {}^1G)$	275	125	−2250
$({}^1D)\,{}^2F{:}({}^3F)\,{}^2F$			$-100\sqrt{21}$	$300\sqrt{21}$
$({}^1D)\,{}^2F{:}({}^1G)\,{}^2F$		$120\sqrt{3}$	$-60\sqrt{3}$	$-180\sqrt{3}$
$({}^3F)\,{}^2F{:}({}^1G)\,{}^2F$			$45\sqrt{7}$	$810\sqrt{7}$
$({}^3F)\,{}^4F$	$E(d^2, {}^3F)$	−105	70	−1890
$({}^3F)\,{}^2G$	$E(d^2, {}^3F)$	35	455	1890
$({}^1G)\,{}^2G$	$E(d^2, {}^1G)$	−385	245	210
$({}^3F)\,{}^2G{:}({}^1G)\,{}^2G$			$-105\sqrt{15}$	$210\sqrt{15}$
$({}^3F)\,{}^4G$	$E(d^2, {}^3F)$	35	−490	630
$({}^1G)\,{}^2H$	$E(d^2, {}^1G)$	140	−280	1260

17. Closed shells plus d^2d'

$$\begin{aligned} E_{\mathrm{Av}} = {} & E(cs) + 2E(cs, nd) + E(cs, n'd) + 2I(nd) + I(n'd) \\ & + F^0(nd, nd) - \frac{2}{63}F^2(nd, nd) - \frac{2}{63}F^4(nd, nd) + 2F^0(nd, n'd) \\ & - \frac{2}{10}G^0(nd, n'd) - \frac{2}{35}G^2(nd, n'd) - \frac{2}{35}G^4(nd, n'd) \end{aligned}$$

	$E(d^2, {}^{2s+1}L)$	$F^2(d, d')/1715$	$F^4(d, d')/3087$	$G^0(d, d')/70$	$G^2(d, d')/3430$	$G^4(d, d')/30870$
$({}^1D)\,{}^2S$	$E(d^2, {}^1D)$	−210	504	−56	406	−756
$({}^3P)\,{}^2P$	$E(d^2, {}^3P)$	−343		−7	392	3528
$({}^1D)\,{}^2P$	$E(d^2, {}^1D)$	−105	−336	49	196	−1176
$({}^3F)\,{}^2P$	$E(d^2, {}^3F)$	168	−462		−140	6090
$({}^3P)\,{}^2P{:}({}^1D)\,{}^2P$				$7\sqrt{105}$	$-70\sqrt{105}$	$252\sqrt{105}$
$({}^3P)\,{}^2P{:}({}^3F)\,{}^2P$		$28\sqrt{14}$	$-126\sqrt{14}$	$-14\sqrt{14}$	$14\sqrt{14}$	$126\sqrt{14}$
$({}^1D)\,{}^2P{:}({}^3F)\,{}^2P$				$-14\sqrt{30}$	$-70\sqrt{30}$	$546\sqrt{30}$
$({}^3P)\,{}^4P$	$E(d^2, {}^3P)$	−343		56	−196	−1764
$({}^3F)\,{}^4P$	$E(d^2, {}^3F)$	168	−462	42	868	−6888
$({}^3P)\,{}^4P{:}({}^3F)\,{}^4P$		$28\sqrt{14}$	$-126\sqrt{14}$	$28\sqrt{14}$	$-28\sqrt{14}$	$-252\sqrt{14}$
$({}^1S)\,{}^2D$	$E(d^2, {}^1S)$					
$({}^3P)\,{}^2D$	$E(d^2, {}^3P)$	343		49	294	4704
$({}^1D)\,{}^2D$	$E(d^2, {}^1D)$	45	144	29	−214	−1296
$({}^3F)\,{}^2D$	$E(d^2, {}^3F)$	42	693	49	518	5544
$({}^1G)\,{}^2D$	$E(d^2, {}^1G)$	550	143	13	102	−4612
$({}^1S)\,{}^2D{:}({}^3P)\,{}^2D$				−42	−294	3528
$({}^1S)\,{}^2D{:}({}^1D)\,{}^2D$		$196\sqrt{5}$		$-14\sqrt{5}$	$42\sqrt{5}$	$-504\sqrt{5}$
$({}^1S)\,{}^2D{:}({}^3F)\,{}^2D$				$-14\sqrt{21}$	$112\sqrt{21}$	$126\sqrt{21}$
$({}^1S)\,{}^2D{:}({}^1G)\,{}^2D$			588	−42	−168	−42
$({}^3P)\,{}^2D{:}({}^1D)\,{}^2D$				$-21\sqrt{5}$		$1764\sqrt{5}$
$({}^3P)\,{}^2D{:}({}^3F)\,{}^2D$		$56\sqrt{21}$	$42\sqrt{21}$		$56\sqrt{21}$	$210\sqrt{21}$
$({}^3P)\,{}^2D{:}({}^1G)\,{}^2D$				84	588	294
$({}^1D)\,{}^2D{:}({}^3F)\,{}^2D$				$8\sqrt{105}$		$378\sqrt{105}$
$({}^1D)\,{}^2D{:}({}^1G)\,{}^2D$		$48\sqrt{5}$	$330\sqrt{5}$	$-12\sqrt{5}$	$-204\sqrt{5}$	$-282\sqrt{5}$
$({}^3F)\,{}^2D{:}({}^1G)\,{}^2D$				$3\sqrt{21}$	$126\sqrt{21}$	$798\sqrt{21}$
$({}^3P)\,{}^4D$	$E(d^2, {}^3P)$	343		−56		−4116
$({}^3F)\,{}^4D$	$E(d^2, {}^3F)$	42	693	−56	−448	−5796
$({}^3P)\,{}^4D{:}({}^3F)\,{}^4D$		$56\sqrt{21}$	$42\sqrt{21}$		$-112\sqrt{21}$	$-420\sqrt{21}$
$({}^3P)\,{}^2F$	$E(d^2, {}^3P)$	−98		28	462	2688
$({}^1D)\,{}^2F$	$E(d^2, {}^1D)$	120	−36	−26	196	−126
$({}^3F)\,{}^2F$	$E(d^2, {}^3F)$	−77	−462	−35	315	4305
$({}^1G)\,{}^2F$	$E(d^2, {}^1G)$	55	−286	19	441	−1981
$({}^3P)\,{}^2F{:}({}^1D)\,{}^2F$				$-6\sqrt{70}$	$-30\sqrt{70}$	$234\sqrt{70}$
$({}^3P)\,{}^2F{:}({}^3F)\,{}^2F$		336	−42	42	−252	672
$({}^3P)\,{}^2F{:}({}^1G)\,{}^2F$				$-30\sqrt{7}$	$60\sqrt{7}$	$120\sqrt{7}$
$({}^1D)\,{}^2F{:}({}^3F)\,{}^2F$				$-3\sqrt{70}$	$-60\sqrt{70}$	$342\sqrt{70}$
$({}^1D)\,{}^2F{:}({}^1G)\,{}^2F$		$60\sqrt{10}$	$-165\sqrt{10}$	$15\sqrt{10}$		$-210\sqrt{10}$
$({}^3F)\,{}^2F{:}({}^1G)\,{}^2F$				$-15\sqrt{7}$	$-195\sqrt{7}$	$1185\sqrt{7}$
$({}^3P)\,{}^4F$	$E(d^2, {}^3P)$	−98		−14	−336	−84
$({}^3F)\,{}^4F$	$E(d^2, {}^3F)$	−77	−462	112	−42	−3318
$({}^3P)\,{}^4F{:}({}^3F)\,{}^4F$		336	−42	−84	504	−1344
$({}^1D)\,{}^2G$	$E(d^2, {}^1D)$	−60	4	−6	−144	1294
$({}^3F)\,{}^2G$	$E(d^2, {}^3F)$	−105	154	49	371	2849
$({}^1G)\,{}^2G$	$E(d^2, {}^1G)$	−325	234	−1	81	99
$({}^1D)\,{}^2G{:}({}^3F)\,{}^2G$				$-5\sqrt{210}$		$70\sqrt{210}$
$({}^1D)\,{}^2G{:}({}^1G)\,{}^2G$		$20\sqrt{110}$	$15\sqrt{110}$	$-5\sqrt{110}$	$20\sqrt{110}$	$-30\sqrt{110}$
$({}^3F)\,{}^2G{:}({}^1G)\,{}^2G$				$5\sqrt{231}$	$-35\sqrt{231}$	$105\sqrt{231}$
$({}^3F)\,{}^4G$	$E(d^2, {}^3F)$	−105	154	−56	−154	−406
$({}^3F)\,{}^2H$	$E(d^2, {}^3F)$	70	−21	49	546	1974
$({}^1G)\,{}^2H$	$E(d^2, {}^1G)$	−350	−91	49	−294	1274
$({}^3F)\,{}^2H{:}({}^1G)\,{}^2H$				105	210	420
$({}^3F)\,{}^4H$	$E(d^2, {}^3F)$	70	−21	−56	−504	1344
$({}^1G)\,{}^2I$	$E(d^2, {}^1G)$	280	14	−56	−84	1694

18. Closed shells plus f^3 [13, 24]

$$E_{\mathrm{Av}} = E(cs) + 3E(cs, nf) + 3I(nf) + 3F^0(nf, nf) - \frac{12}{195}F^2(nf, nf) - \frac{6}{143}F^4(nf, nf) - \frac{300}{5577}F^6(nf, nf)$$

W	U	${}^{2s+1}_{v}L$	$F^2/184275$	$F^4/297297$	$F^6/11594583$
(111)	(00)	${}^{4}_{3}S$	−13230	−14553	−727650
(210)	(11)	${}^{2}_{3}P$	−9135	462	848925
(210)	(20)	${}^{2}_{3}D_1$	−11709	15048	398475
(210)	(21)	${}^{2}_{3}D_2$	22923	−7299	439425
		${}^{2}_{3}D_1 : {}^{2}_{3}D_2$	$-1404\sqrt{33}$	$7956\sqrt{33}$	$-245700\sqrt{33}$
(111)	(20)	${}^{4}_{3}D$	31815	3465	−2304225
(100)	(10)	${}^{2}_{1}F$	44100	48510	2425500
(210)	(21)	${}^{2}_{3}F$	68670	6741	−1014300
		${}^{2}_{1}F : {}^{2}_{3}F$	$-8190\sqrt{22}$	$-3276\sqrt{22}$	$286650\sqrt{22}$
(111)	(10)	${}^{4}_{3}F$	−13230	−14553	−727650
(210)	(20)	${}^{2}_{3}G_1$	13680	−4608	541800
(210)	(21)	${}^{2}_{3}G_2$	20466	42933	−1274400
		${}^{2}_{3}G_1 : {}^{2}_{3}G_2$	$702\sqrt{4290}$	$-156\sqrt{4290}$	$-9450\sqrt{4290}$
(111)	(20)	${}^{4}_{3}G$	3150	−8001	−1300950
(210)	(11)	${}^{2}_{3}H_1$	9975	8106	180075
(210)	(21)	${}^{2}_{3}H_2$	−24969	−8547	1495725
		${}^{2}_{3}H_1 : {}^{2}_{3}H_2$	$1092\sqrt{455}$	$-1092\sqrt{455}$	$14700\sqrt{455}$
(210)	(20)	${}^{2}_{3}I$	7245	10836	143325
(111)	(20)	${}^{4}_{3}I$	−41895	−26019	275625
(210)	(21)	${}^{2}_{3}K$	−21420	12747	683550
(210)	(21)	${}^{2}_{3}L$	11340	−4725	595350

19. Closed shells plus f^2s [2]

$$E_{\mathrm{Av}} = E(cs) + 2E(cs, nf) + E(cs, n's) + 2I(nf) + I(n's) + F^0(nf, nf) - \frac{4}{195}F^2(nf, nf) - \frac{2}{143}F^4(nf, nf) - \frac{100}{5577}F^6(nf, nf) + 2F^0(nf, n's) - \frac{2}{14}G^3(nf, n's)$$

$$({}^1L)\ {}^2L = E(f^2, {}^1L) \qquad \text{for even } L$$

$$({}^3L)\ {}^2L = E(f^2, {}^3L) + \frac{2}{7}G^3(nf, n's) \qquad \text{for odd } L$$

$$({}^3L)\ {}^4L = E(f^2, {}^3L) - \frac{1}{7}G^3(nf, n's) \qquad \text{for odd } L$$

20. Closed shells plus f^2p

$$E_{\rm Av} = E(cs) + 2E(cs, nf) + E(cs, n'p) + 2I(nf) + I(n'p)$$
$$+ F^0(nf, nf) - \frac{4}{195}F^2(nf, nf) - \frac{2}{143}F^4(nf, nf) - \frac{100}{5577}F^6(nf, nf)$$
$$+ 2F^0(nf, n'p) - \frac{6}{70}G^2(nf, n'p) - \frac{4}{63}G^4(nf, n'p)$$

	$E(f^2, {}^{2s+1}L)$	$F^2(f, p)/17325$	$G^2(f, p)/26950$	$G^4(f, p)/87318$
$({}^3P)\,{}^2S$	$E(f^2, {}^3P)$	-6930		9702
$({}^3P)\,{}^4S$	$E(f^2, {}^3P)$	-6930	6930	-2772
$({}^1S)\,{}^2P$	$E(f^2, {}^1S)$			
$({}^3P)\,{}^2P$	$E(f^2, {}^3P)$	3465	5775	13860
$({}^1D)\,{}^2P$	$E(f^2, {}^1D)$	-2541	3003	-1848
$({}^1S)\,{}^2P{:}({}^3P)\,{}^2P$			$-2310\sqrt{6}$	$4158\sqrt{6}$
$({}^1S)\,{}^2P{:}({}^1D)\,{}^2P$		$924\sqrt{30}$	$-462\sqrt{30}$	$-462\sqrt{30}$
$({}^3P)\,{}^2P{:}({}^1D)\,{}^2P$			$-693\sqrt{5}$	$4158\sqrt{5}$
$({}^3P)\,{}^4P$	$E(f^2, {}^3P)$	3465	-4620	-11088
$({}^3P)\,{}^2D$	$E(f^2, {}^3P)$	-693	4851	9702
$({}^1D)\,{}^2D$	$E(f^2, {}^1D)$	2541	-693	-2310
$({}^3F)\,{}^2D$	$E(f^2, {}^3F)$	-924	1848	14784
$({}^3P)\,{}^2D{:}({}^1D)\,{}^2D$			-6237	8316
$({}^3P)\,{}^2D{:}({}^3F)\,{}^2D$		$2772\sqrt{3}$	$1386\sqrt{3}$	$1386\sqrt{3}$
$({}^1D)\,{}^2D{:}({}^3F)\,{}^2D$			$-462\sqrt{3}$	$6006\sqrt{3}$
$({}^3P)\,{}^4D$	$E(f^2, {}^3P)$	-693	-2772	-2772
$({}^3F)\,{}^4D$	$E(f^2, {}^3F)$	-924	3234	-12936
$({}^3P)\,{}^4D{:}({}^3F)\,{}^4D$		$2772\sqrt{3}$	$-2772\sqrt{3}$	$-2772\sqrt{3}$
$({}^1D)\,{}^2F$	$E(f^2, {}^1D)$	-726	-792	2442
$({}^3F)\,{}^2F$	$E(f^2, {}^3F)$	1155	4620	12705
$({}^1G)\,{}^2F$	$E(f^2, {}^1G)$	495	2640	-5445
$({}^1D)\,{}^2F{:}({}^3F)\,{}^2F$			$-462\sqrt{210}$	$462\sqrt{210}$
$({}^1D)\,{}^2F{:}({}^1G)\,{}^2F$		$396\sqrt{110}$	$-198\sqrt{110}$	$-198\sqrt{110}$
$({}^3F)\,{}^2F{:}({}^1G)\,{}^2F$				$693\sqrt{231}$
$({}^3F)\,{}^4F$	$E(f^2, {}^3F)$	1155	-2310	-8778
$({}^3F)\,{}^2G$	$E(f^2, {}^3F)$	-385	6160	7777
$({}^1G)\,{}^2G$	$E(f^2, {}^1G)$	-693	924	-693
$({}^3H)\,{}^2G$	$E(f^2, {}^3H)$	2002	2156	18326
$({}^3F)\,{}^2G{:}({}^1G)\,{}^2G$			$-924\sqrt{55}$	$693\sqrt{55}$
$({}^3F)\,{}^2G{:}({}^3H)\,{}^2G$		$308\sqrt{110}$	$154\sqrt{110}$	$154\sqrt{110}$
$({}^1G)\,{}^2G{:}({}^3H)\,{}^2G$			$462\sqrt{2}$	$6930\sqrt{2}$
$({}^3F)\,{}^4G$	$E(f^2, {}^3F)$	-385	-5390	1078
$({}^3H)\,{}^4G$	$E(f^2, {}^3H)$	2002	2618	-20020
$({}^3F)\,{}^4G{:}({}^3H)\,{}^4G$		$308\sqrt{110}$	$-308\sqrt{110}$	$-308\sqrt{110}$
$({}^1G)\,{}^2H$	$E(f^2, {}^1G)$	252	-2436	4032

	$E(f^2, {}^{2s+1}L)$	$F^2(f,p)/17325$	$G^2(f,p)/26950$	$G^4(f,p)/87318$
$({}^3H)\,{}^2H$	$E(f^2, {}^3H)$	-3003	2541	10626
$({}^1I)\,{}^2H$	$E(f^2, {}^1I)$	3675	2205	-9114
$({}^1G)\,{}^2H{:}({}^3H)\,{}^2H$			$-1134\sqrt{33}$	$630\sqrt{33}$
$({}^1G)\,{}^2H{:}({}^1I)\,{}^2H$		$252\sqrt{65}$	$-126\sqrt{65}$	$-126\sqrt{65}$
$({}^3H)\,{}^2H{:}({}^1I)\,{}^2H$			$21\sqrt{2145}$	$168\sqrt{2145}$
$({}^3H)\,{}^4H$	$E(f^2, {}^3H)$	-3003	1848	-4620
$({}^3H)\,{}^2I$	$E(f^2, {}^3H)$	1155	8085	6468
$({}^1I)\,{}^2I$	$E(f^2, {}^1I)$	-5775	3465	1848
$({}^3H)\,{}^2I{:}({}^1I)\,{}^2I$			$-1155\sqrt{21}$	$462\sqrt{21}$
$({}^3H)\,{}^4I$	$E(f^2, {}^3H)$	1155	-9240	3696
$({}^1I)\,{}^2K$	$E(f^2, {}^1I)$	2310	-4620	5082

21. Closed shells plus p^3s [2, 4, 7, 17]

$$E_{\mathrm{Av}} = E(cs) + 3E(cs, np) + E(cs, n's) + 3I(np) + I(n's) + 3F^0(np, np) - \frac{6}{25}F^2(np, np) + 3F^0(np, n's) - \frac{3}{6}G^1(np, n's)$$

$${}^3S = E(p^3, {}^4S) + \frac{5}{6}G^1(np, n's)$$

$${}^5S = E(p^3, {}^4S) - \frac{1}{2}G^1(np, n's)$$

$${}^1L = E(p^3, {}^2L) + \frac{1}{2}G^1(np, n's) \quad \text{for } L = 1, 2$$

$${}^3L = E(p^3, {}^2L) - \frac{1}{6}G^1(np, n's) \quad \text{for } L = 1, 2$$

22. Closed shells plus p^3p' [2, 17]

$$E_{\mathrm{Av}} = E(cs) + 3E(cs, np) + E(cs, n'p) + 3I(np) + I(n'p) + 3F^0(np, np) - \frac{6}{25}F^2(np, np) + 3F^0(np, n'p) - \frac{3}{6}G^0(np, n'p) - \frac{3}{15}G^2(np, n'p)$$

	$E(p^3, {}^{2s+1}L)$	$F^2(np,n'p)/25$	$G^0(np,n'p)/6$	$G^2(np,n'p)/150$
$({}^2P)\,{}^1S$	$E(p^3, {}^2P)$		21	30
$({}^2P)\,{}^3S$	$E(p^3, {}^2P)$		-3	-30
$({}^2P)\,{}^1P$	$E(p^3, {}^2P)$		-3	15
$({}^2D)\,{}^1P$	$E(p^3, {}^2D)$		-3	-3
$({}^2P)\,{}^1P{:}({}^2D)\,{}^1P$		$3\sqrt{15}$		$-9\sqrt{15}$
$({}^4S)\,{}^3P$	$E(p^3, {}^4S)$		5	50
$({}^2P)\,{}^3P$	$E(p^3, {}^2P)$		3	-15
$({}^2D)\,{}^3P$	$E(p^3, {}^2D)$		7	-29

	$E(p^3, {}^{2s+1}L)$	$F^2(np, n'p)/25$	$G^0(np, n'p)/6$	$G^2(np, n'p)/150$
$({}^4S)\,{}^3P{:}({}^2P)\,{}^3P$			$4\sqrt{3}$	$-20\sqrt{3}$
$({}^4S)\,{}^3P{:}({}^2D)\,{}^3P$			$-4\sqrt{5}$	$-4\sqrt{5}$
$({}^2P)\,{}^3P{:}({}^2D)\,{}^3P$		$3\sqrt{15}$	$-2\sqrt{15}$	$\sqrt{15}$
$({}^4S)\,{}^5P$	$E(p^3, {}^4S)$		-3	-30
$({}^2P)\,{}^1D$	$E(p^3, {}^2P)$		3	39
$({}^2D)\,{}^1D$	$E(p^3, {}^2D)$		15	33
$({}^2P)\,{}^1D{:}({}^2D)\,{}^1D$		$-3\sqrt{3}$	$6\sqrt{3}$	$-21\sqrt{3}$
$({}^2P)\,{}^3D$	$E(p^3, {}^2P)$		-3	-3
$({}^2D)\,{}^3D$	$E(p^3, {}^2D)$		-3	-21
$({}^2P)\,{}^3D{:}({}^2D)\,{}^3D$		$-3\sqrt{3}$		$9\sqrt{3}$
$({}^2D)\,{}^1F$	$E(p^3, {}^2D)$		-3	42
$({}^2D)\,{}^3F$	$E(p^3, {}^2D)$		-3	6

23. Closed shells plus p^3d [2, 17]

$$E_{\text{Av}} = E(cs) + 3E(cs, np) + E(cs, n'd) + 3I(np) + I(n'd) + 3F^0(np, np) - \frac{6}{25}F^2(np, np) + 3F^0(np, n'd) - \frac{3}{15}G^1(np, n'd) - \frac{9}{70}G^3(np, n'd)$$

	$E(p^3, {}^{2s+1}L)$	$F^2(np, n'd)/35$	$G^1(np, n'd)/30$	$G^3(np, n'd)/490$
$({}^2D)\,{}^1S$	$E(p^3, {}^2D)$		-6	-63
$({}^2D)\,{}^3S$	$E(p^3, {}^2D)$		14	-63
$({}^2P)\,{}^1P$	$E(p^3, {}^2P)$		21	63
$({}^2D)\,{}^1P$	$E(p^3, {}^2D)$		3	-21
$({}^2P)\,{}^1P{:}({}^2D)\,{}^1P$		$7\sqrt{3}$	$\sqrt{3}$	$-42\sqrt{3}$
$({}^2P)\,{}^3P$	$E(p^3, {}^2P)$		-1	-63
$({}^2D)\,{}^3P$	$E(p^3, {}^2D)$		9	-63
$({}^2P)\,{}^3P{:}({}^2D)\,{}^3P$		$7\sqrt{3}$	$-5\sqrt{3}$	
$({}^2P)\,{}^1D$	$E(p^3, {}^2P)$		-3	21
$({}^2D)\,{}^1D$	$E(p^3, {}^2D)$		15	45
$({}^2P)\,{}^1D{:}({}^2D)\,{}^1D$		$3\sqrt{7}$	$3\sqrt{7}$	$-36\sqrt{7}$
$({}^4S)\,{}^3D$	$E(p^3, {}^4S)$		10	105
$({}^2P)\,{}^3D$	$E(p^3, {}^2P)$		3	-21
$({}^2D)\,{}^3D$	$E(p^3, {}^2D)$		1	-57
$({}^4S)\,{}^3D{:}({}^2P)\,{}^3D$			12	-84
$({}^4S)\,{}^3D{:}({}^2D)\,{}^3D$			$-4\sqrt{7}$	$-12\sqrt{7}$
$({}^2P)\,{}^3D{:}({}^2D)\,{}^3D$		$3\sqrt{7}$	$-3\sqrt{7}$	$6\sqrt{7}$
$({}^4S)\,{}^5D$	$E(p^3, {}^4S)$		-6	-63
$({}^2P)\,{}^1F$	$E(p^3, {}^2P)$		6	93
$({}^2D)\,{}^1F$	$E(p^3, {}^2D)$		18	99

	$E(p^3, {}^{2s+1}L)$	$F^2(np, n'd)/35$	$G^1(np, n'd)/30$	$G^3(np, n'd)/490$
$({}^2P)\ {}^1F{:}({}^2D)\ {}^1F$		$-6\sqrt{2}$	$12\sqrt{2}$	$-54\sqrt{2}$
$({}^2P)\ {}^3F$	$E(p^3, {}^2P)$		-6	-3
$({}^2D)\ {}^3F$	$E(p^3, {}^2D)$		-6	-33
$({}^2P)\ {}^3F{:}({}^2D)\ {}^3F$		$-6\sqrt{2}$		$30\sqrt{2}$
$({}^2D)\ {}^1G$	$E(p^3, {}^2D)$		-6	87
$({}^2D)\ {}^3G$	$E(p^3, {}^2D)$		-6	27

24. Closed shells plus p^3f [2, 17]

$$E_{\mathrm{Av}} = E(cs) + 3E(cs, np) + E(cs, n'f) + 3I(np) + I(n'f) + 3F^0(np, np) - \frac{6}{25} F^2 np, np) + 3F^0(np, n'f) - \frac{9}{70} G^2(np, n'f) - \frac{6}{63} G^4(np, n'f)$$

	$E(p^3, {}^{2s+1}L)$	$F^2(np, n'f)/75$	$G^2(np, n'f)/350$	$G^4(np, n'f)/378$
$({}^2D)\ {}^1P$	$E(p^3, {}^2D)$		-3	-36
$({}^2D)\ {}^3P$	$E(P^3, {}^2D)$		81	-36
$({}^2P)\ {}^1D$	$E(p^3, {}^2P)$		129	36
$({}^2D)\ {}^1D$	$E(p^3, {}^2D)$		63	
$({}^2P)\ {}^1D{:}({}^2D)\ {}^1D$		$18\sqrt{2}$	$18\sqrt{2}$	$-36\sqrt{2}$
$({}^2P)\ {}^3D$	$E(p^3, {}^2P)$		-3	-36
$({}^2D)\ {}^3D$	$E(p^3, {}^2D)$		39	-36
$({}^2P)\ {}^3D{:}({}^2D)\ {}^3D$		$18\sqrt{2}$	$-42\sqrt{2}$	
$({}^2P)\ {}^1F$	$E(p^3, {}^2P)$		-15	9
$({}^2D)\ {}^1F$	$E(p^3, {}^2D)$		117	39
$({}^2P)\ {}^1F{:}({}^2D)\ {}^1F$		$3\sqrt{15}$	$18\sqrt{15}$	$-15\sqrt{15}$
$({}^4S)\ {}^3F$	$E(p^3, {}^4S)$		75	60
$({}^2P)\ {}^3F$	$E(p^3, {}^2P)$		15	-9
$({}^2D)\ {}^3F$	$E(p^3, {}^2D)$		-9	-31
$({}^4S)\ {}^3F{:}({}^2P)\ {}^3F$			$60\sqrt{2}$	$-36\sqrt{2}$
$({}^4S)\ {}^3F{:}({}^2D)\ {}^3F$			$-12\sqrt{30}$	$-4\sqrt{30}$
$({}^2P)\ {}^3F{:}({}^2D)\ {}^3F$		$3\sqrt{15}$	$-12\sqrt{15}$	$3\sqrt{15}$
$({}^4S)\ {}^5F$	$E(p^3, {}^4S)$		-45	-36
$({}^2P)\ {}^1G$	$E(p^3, {}^2P)$		45	57
$({}^2D)\ {}^1G$	$E(p^3, {}^2D)$		105	63
$({}^2P)\ {}^1G{:}({}^2D)\ {}^1G$		$-5\sqrt{15}$	$30\sqrt{15}$	$-11\sqrt{15}$
$({}^2P)\ {}^3G$	$E(p^3, {}^2P)$		-45	-1
$({}^2D)\ {}^3G$	$E(p^3, {}^2D)$		-45	-15
$({}^2P)\ {}^3G{:}({}^2D)\ {}^3G$		$-5\sqrt{15}$		$7\sqrt{15}$
$({}^2D)\ {}^1H$	$E(p^3, {}^2D)$		-45	48
$({}^2D)\ {}^3H$	$E(p^3, {}^2D)$		-45	20

25. Closed shells plus d^4 [2, 10, 13, 24]

$$E_{\mathrm{Av}} = E(cs) + 4E(cs, nd) + 4I(nd) + 6F^0(nd, nd) - \frac{12}{63}F^2(nd, nd) - \frac{12}{63}F^4(nd, nd)$$

	$F^2/441$	$F^4/441$
${}^1_0S$	210	210
${}^1_4S$	138	-30
${}^1_0S\colon{}^1_4S$	$54\sqrt{21}$	$-30\sqrt{21}$
${}^3_2P$	21	70
${}^3_4P$	57	-55
${}^3_2P\colon{}^3_4P$	$36\sqrt{14}$	$-20\sqrt{14}$
${}^1_2D$	219	30
${}^1_4D$	111	-15
${}^1_2D\colon{}^1_4D$	$108\sqrt{2}$	$-60\sqrt{2}$
${}^3_4D$	39	-45
${}^5_4D$	-105	-105
${}^1_4F$	84	
${}^3_2F$	66	45
${}^3_4F$	12	-30
${}^3_2F\colon{}^3_4F$	108	-60
${}^1_2G$	30	135
${}^1_4G$	48	20
${}^1_2G\colon{}^1_4G$	$36\sqrt{11}$	$-20\sqrt{11}$
${}^3_4G$	-24	-10
${}^3_4H$	-69	15
${}^1_4I$	-51	75

26. Closed shells plus d^3s [2]

$$E_{\mathrm{Av}} = E(cs) + 3E(cs, nd) + E(cs, n's) + 3I(nd) + I(n's) + 3F^0(nd, nd) - \frac{6}{63}F^2(nd, nd) - \frac{6}{63}F^4(nd, nd) + 3F^0(nd, n's) - \frac{3}{10}G^2(nd, n's)$$

$$({}^4_vL)\,{}^5_vL = E(d^3, {}^4_vL) - \frac{3}{10}G^2(nd, n's) \quad \text{for} \quad L = 1, 3$$

$$({}^4_vL)\,{}^3_vL = E(d^3, {}^4_vL) + \frac{1}{2}G^2(nd, n's) \quad \text{for} \quad L = 1, 3$$

$$({}^2_vL)\,{}^3_vL = E(d^3, {}^2_vL) - \frac{1}{10}G^2(nd, n's) \quad \text{for} \quad L = 1, 2, 3, 4, 5$$

$$({}^2_vL)\ {}^1_vL = E(d^3, {}^2_vL) + \frac{3}{10} G^2(nd, n's) \quad \text{for} \quad L = 1, 2, 3, 4, 5$$

$$ {}^1_1D\!:{}^1_3D = {}^3_1D\!:{}^3_3D = \frac{\sqrt{21}}{147}[9F^2(nd, nd) - 5F^4(nd, nd)]$$

27. Closed shells plus d^3p [2, 18]

$$E_{Av} = E(cs) + 3E(cs, nd) + E(cs, n'p) + 3I(nd) + I(n'p) + 3F^0(nd, nd) - \frac{6}{63}F^2(nd, nd) - \frac{6}{63}F^4(nd, nd) + 3F^0(nd, n'p) - \frac{3}{15}G^1(nd, n'p) - \frac{9}{70}G^3(nd, n'p)$$

	$E(d^3, {}^{2s+1}_{\ \ v}L)$	$F^2(nd, n'p)/2450$	$G^1(nd, n'p)/6300$	$G^3(nd, n'p)/34300$
$({}^2_3P)\ {}^1S$	$E(d^3, {}^2_3P)$	-280	-2940	2310
$({}^2_3P)\ {}^3S$	$E(d^3, {}^2_3P)$	-280	2380	-2170
$({}^4_3P)\ {}^3S$	$E(d^3, {}^4_3P)$	980	2660	9310
$({}^2_3P)\ {}^3S\!:({}^4_3P)\ {}^3S$			$-560\sqrt{14}$	$2240\sqrt{14}$
$({}^4_3P)\ {}^5S$	$E(d^3, {}^4_3P)$	980	-2940	-10290
$({}^2_3P)\ {}^1P$	$E(d^3, {}^2_3P)$	140	2730	6930
$({}^2_1D)\ {}^1P$	$E(d^3, {}^2_1D)$	245	2625	7350
$({}^2_3D)\ {}^1P$	$E(d^3, {}^2_3D)$	105	2205	630
$({}^2_3P)\ {}^1P\!:({}^2_1D)\ {}^1P$		$105\sqrt{42}$	$-315\sqrt{42}$	$210\sqrt{42}$
$({}^2_3P)\ {}^1P\!:({}^2_3D)\ {}^1P$		$-105\sqrt{2}$	$315\sqrt{2}$	$-4410\sqrt{2}$
$({}^2_1D)\ {}^1P\!:({}^2_3D)\ {}^1P$		$105\sqrt{21}$	$315\sqrt{21}$	$-1260\sqrt{21}$
$({}^2_3P)\ {}^3P$	$E(d^3, {}^2_3P)$	140	-770	-3570
$({}^4_3P)\ {}^3P$	$E(d^3, {}^4_3P)$	-490	1400	7350
$({}^2_1D)\ {}^3P$	$E(d^3, {}^2_1D)$	245	-105	-5880
$({}^2_3D)\ {}^3P$	$E(d^3, {}^2_3D)$	105	315	-3360
$({}^2_3P)\ {}^3P\!:({}^4_3P)\ {}^3P$			$-560\sqrt{14}$	$840\sqrt{14}$
$({}^2_3P)\ {}^3P\!:({}^2_1D)\ {}^3P$		$105\sqrt{42}$	$-105\sqrt{42}$	$-280\sqrt{42}$
$({}^2_3P)\ {}^3P\!:({}^2_3D)\ {}^3P$		$-105\sqrt{2}$	$105\sqrt{2}$	$1680\sqrt{2}$
$({}^4_3P)\ {}^3P\!:({}^2_1D)\ {}^3P$			$-840\sqrt{3}$	$1960\sqrt{3}$
$({}^4_3P)\ {}^3P\!:({}^2_3D)\ {}^3P$			$840\sqrt{7}$	$840\sqrt{7}$
$({}^2_1D)\ {}^3P\!:({}^2_3D)\ {}^3P$		$105\sqrt{21}$	$-315\sqrt{21}$	$210\sqrt{21}$
$({}^4_3P)\ {}^5P$	$E(d^3, {}^4_3P)$	-490	840	-4410
$({}^2_3P)\ {}^1D$	$E(d^3, {}^2_3P)$	-28	1218	3318
$({}^2_1D)\ {}^1D$	$E(d^3, {}^2_1D)$	-245	-735	2940
$({}^2_3D)\ {}^1D$	$E(d^3, {}^2_3D)$	-105	2205	3780
$({}^2_3F)\ {}^1D$	$E(d^3, {}^2_3F)$	168	1512	-1638
$({}^2_3P)\ {}^1D\!:({}^2_1D)\ {}^1D$		$-21\sqrt{210}$	$-63\sqrt{210}$	$252\sqrt{210}$

	$E(d^3, {}^{2s+1}_{\ v}L)$	$F^2(nd, n'p)/2450$	$G^1(nd, n'p)/6300$	$G^3(nd, n'p)/34300$
$({}^2_3P)\,{}^1D{:}({}^2_3D)\,{}^1D$		$21\sqrt{10}$	$1071\sqrt{10}$	$-1764\sqrt{10}$
$({}^2_3P)\,{}^1D{:}({}^2_3F)\,{}^1D$		-336	-504	-504
$({}^2_1D)\,{}^1D{:}({}^2_3D)\,{}^1D$		$-105\sqrt{21}$	$315\sqrt{21}$	$-210\sqrt{21}$
$({}^2_1D)\,{}^1D{:}({}^2_3F)\,{}^1D$		$28\sqrt{210}$	$84\sqrt{210}$	$-336\sqrt{210}$
$({}^2_3D)\,{}^1D{:}({}^2_3F)\,{}^1D$		$252\sqrt{10}$	$252\sqrt{10}$	$-168\sqrt{10}$
$({}^2_3P)\,{}^3D$	$E(d^3, {}^2_3P)$	-28	-770	-70
$({}^4_3P)\,{}^3D$	$E(d^3, {}^4_3P)$	98	2408	6958
$({}^2_1D)\,{}^3D$	$E(d^3, {}^2_1D)$	-245	1155	-1470
$({}^2_3D)\,{}^3D$	$E(d^3, {}^2_3D)$	-105	-105	-2030
$({}^2_3F)\,{}^3D$	$E(d^3, {}^2_3F)$	168	840	-3710
$({}^4_3F)\,{}^3D$	$E(d^3, {}^4_3F)$	-168	1092	9142
$({}^2_3P)\,{}^3D{:}({}^4_3P)\,{}^3D$			$448\sqrt{14}$	$-952\sqrt{14}$
$({}^2_3P)\,{}^3D{:}({}^2_1D)\,{}^3D$		$-21\sqrt{210}$	$63\sqrt{210}$	$-42\sqrt{210}$
$({}^2_3P)\,{}^3D{:}({}^2_3D)\,{}^3D$		$21\sqrt{10}$	$-399\sqrt{10}$	$546\sqrt{10}$
$({}^2_3P)\,{}^3D{:}({}^2_3F)\,{}^3D$		-336	840	840
$({}^2_3P)\,{}^3D{:}({}^4_3F)\,{}^3D$			1848	1848
$({}^4_3P)\,{}^3D{:}({}^2_1D)\,{}^3D$			$-504\sqrt{15}$	$1176\sqrt{15}$
$({}^4_3P)\,{}^3D{:}({}^2_3D)\,{}^3D$			$-168\sqrt{35}$	$-168\sqrt{35}$
$({}^4_3P)\,{}^3D{:}({}^2_3F)\,{}^3D$			$336\sqrt{14}$	$336\sqrt{14}$
$({}^4_3P)\,{}^3D{:}({}^4_3F)\,{}^3D$		$-168\sqrt{14}$	$-168\sqrt{14}$	$-168\sqrt{14}$
$({}^2_1D)\,{}^3D{:}({}^2_3D)\,{}^3D$		$-105\sqrt{21}$	$105\sqrt{21}$	$280\sqrt{21}$
$({}^2_1D)\,{}^3D{:}({}^2_3F)\,{}^3D$		$28\sqrt{210}$	$-84\sqrt{210}$	$56\sqrt{210}$
$({}^2_1D)\,{}^3D{:}({}^4_3F)\,{}^3D$			$84\sqrt{210}$	$-196\sqrt{210}$
$({}^2_3D)\,{}^3D{:}({}^2_3F)\,{}^3D$		$252\sqrt{10}$	$-588\sqrt{10}$	$-448\sqrt{10}$
$({}^2_3D)\,{}^3D{:}({}^4_3F)\,{}^3D$			$-84\sqrt{10}$	$1316\sqrt{10}$
$({}^2_3F)\,{}^3D{:}({}^4_3F)\,{}^3D$			336	3136
$({}^4_3P)\,{}^5D$	$E(d^3, {}^4_3P)$	98	-2184	-3234
$({}^4_3F)\,{}^5D$	$E(d^3, {}^4_3F)$	-168	1764	-9786
$({}^4_3P)\,{}^5D{:}({}^4_3F)\,{}^5D$		$-168\sqrt{14}$	$504\sqrt{14}$	$504\sqrt{14}$
$({}^2_1D)\,{}^1F$	$E(d^3, {}^2_1D)$	70	2100	4200
$({}^2_3D)\,{}^1F$	$E(\mathrm{d}^3, {}^2_3D)$	30	180	6480
$({}^2_3F)\,{}^1F$	$E(d^3, {}^2_3F)$	-210	3780	5355
$({}^2_3G)\,{}^1F$	$E(d^3, {}^2_3G)$	110	1500	4125
$({}^2_1D)\,{}^1F{:}({}^2_3D)\,{}^1F$		$30\sqrt{21}$	$-360\sqrt{21}$	$690\sqrt{21}$
$({}^2_1D)\,{}^1F{:}({}^2_3F)\,{}^1F$		$-140\sqrt{3}$	$1470\sqrt{3}$	$-2730\sqrt{3}$
$({}^2_1D)\,{}^1F{:}({}^2_3G)\,{}^1F$		$-60\sqrt{105}$	$90\sqrt{105}$	$90\sqrt{105}$
$({}^2_3D)\,{}^1F{:}({}^2_3F)\,{}^1F$		$-180\sqrt{7}$	$90\sqrt{7}$	$-510\sqrt{7}$
$({}^2_3D)\,{}^1F{:}({}^2_3G)\,{}^1F$		$-60\sqrt{5}$	$-990\sqrt{5}$	$-990\sqrt{5}$
$({}^2_3F)\,{}^1F{:}({}^2_3G)\,{}^1F$		$-60\sqrt{35}$	$-180\sqrt{35}$	$495\sqrt{35}$
$({}^2_1D)\,{}^3F$	$E(d^3, {}^2_1D)$	70	-1680	420

	$E(d^3, {}^{2s+1}_{v}L)$	$F^2(nd, n'p)/2450$	$G^1(nd, n'p)/6300$	$G^3(nd, n'p)/34300$
$({}^2_3D)\,{}^3F$	$E(d^3, {}^2_3D)$	30	-960	-260
$({}^2_3F)\,{}^3F$	$E(d^3, {}^2_3F)$	-210	-420	-2345
$({}^4_3F)\,{}^3F$	$E(d^3, {}^4_3F)$	210	2100	8050
$({}^2_3G)\,{}^3F$	$E(d^3, {}^2_3G)$	110	1380	-6495
$({}^2_1D)\,{}^3F{:}({}^2_3D)\,{}^3F$		$30\sqrt{21}$	$60\sqrt{21}$	$-290\sqrt{21}$
$({}^2_1D)\,{}^3F{:}({}^2_3F)\,{}^3F$		$-140\sqrt{3}$	$-210\sqrt{3}$	$1190\sqrt{3}$
$({}^2_1D)\,{}^3F{:}({}^4_3F)\,{}^3F$			$840\sqrt{3}$	$-1960\sqrt{3}$
$({}^2_1D)\,{}^3F{:}({}^2_3G)\,{}^3F$		$-60\sqrt{105}$	$90\sqrt{105}$	$90\sqrt{105}$
$({}^2_3D)\,{}^3F{:}({}^2_3F)\,{}^3F$		$-180\sqrt{7}$	$330\sqrt{7}$	$530\sqrt{7}$
$({}^2_3D)\,{}^3F{:}({}^4_3F)\,{}^3F$			$-840\sqrt{7}$	$1160\sqrt{7}$
$({}^2_3D)\,{}^3F{:}({}^2_3G)\,{}^3F$		$-60\sqrt{5}$	$450\sqrt{5}$	$450\sqrt{5}$
$({}^2_3F)\,{}^3F{:}({}^4_3F)\,{}^3F$			-1680	-980
$({}^2_3F)\,{}^3F{:}({}^2_3G)\,{}^3F$		$-60\sqrt{35}$	$180\sqrt{35}$	$-45\sqrt{35}$
$({}^4_3F)\,{}^3F{:}({}^2_3G)\,{}^3F$				$900\sqrt{35}$
$({}^4_3F)\,{}^5F$	$E(d^3, {}^4_3F)$	210	-1260	-6510
$({}^2_3F)\,{}^1G$	$E(d^3, {}^2_3F)$	70	-840	7035
$({}^2_3G)\,{}^1G$	$E(d^3, {}^2_3G)$	-154	1176	4221
$({}^2_3H)\,{}^1G$	$E(d^3, {}^2_3H)$	364	924	1554
$({}^2_3F)\,{}^1G{:}({}^2_3G)\,{}^1G$		$140\sqrt{3}$	$-840\sqrt{3}$	$1785\sqrt{3}$
$({}^2_3F)\,{}^1G{:}({}^2_3H)\,{}^1G$		$-140\sqrt{22}$	$-210\sqrt{22}$	$-210\sqrt{22}$
$({}^2_3G)\,{}^1G{:}({}^2_3H)\,{}^1G$		$28\sqrt{66}$	$-42\sqrt{66}$	$-882\sqrt{66}$
$({}^2_3F)\,{}^3G$	$E(d^3, {}^2_3F)$	70	-1120	455
$({}^4_3F)\,{}^3G$	$E(d^3, {}^4_3F)$	-70	2660	5810
$({}^2_3G)\,{}^3G$	$E(d^3, {}^2_3G)$	-154	336	-2079
$({}^2_3H)\,{}^3G$	$E(d^3, {}^2_3H)$	364	1484	-7126
$({}^2_3F)\,{}^3G{:}({}^4_3F)\,{}^3G$			1120	-980
$({}^2_3F)\,{}^3G{:}({}^2_3G)\,{}^3G$		$140\sqrt{3}$		$-875\sqrt{3}$
$({}^2_3F)\,{}^3G{:}({}^2_3H)\,{}^3G$		$-140\sqrt{22}$	$350\sqrt{22}$	$350\sqrt{22}$
$({}^4_3F)\,{}^3G{:}({}^2_3G)\,{}^3G$			$-1680\sqrt{3}$	$1820\sqrt{3}$
$({}^4_3F)\,{}^3G{:}({}^2_3H)\,{}^3G$			$280\sqrt{22}$	$280\sqrt{22}$
$({}^2_3G)\,{}^3G{:}({}^2_3H)\,{}^3G$		$28\sqrt{66}$	$-42\sqrt{66}$	$-238\sqrt{66}$
$({}^4_3F)\,{}^5G$	$E(d^3, {}^4_3F)$	-70	-2940	210
$({}^2_3G)\,{}^1H$	$E(d^3, {}^2_3G)$	56	1176	4746
$({}^2_3H)\,{}^1H$	$E(d^3, {}^2_3H)$	-546	3024	5754
$({}^2_3G)\,{}^1H{:}({}^2_3H)\,{}^1H$		-168	4032	-3528
$({}^2_3G)\,{}^3H$	$E(d^3, {}^2_3G)$	56	-2184	2226
$({}^2_3H)\,{}^3H$	$E(d^3, {}^2_3H)$	-546	504	-1806
$({}^2_3G)\,{}^3H{:}({}^2_3H)\,{}^3H$		-168	-1008	1512
$({}^2_3H)\,{}^1I$	$E(d^3, {}^2_3H)$	210		5250
$({}^2_3H)\,{}^3I$	$E(d^3, {}^2_3H)$	210	-2520	2730

28. Closed shells plus f^4 [20, 24]

$$E_{\mathrm{Av}} = E(cs) + 4E(cs, nf) + 4I(nf) + 6F^0(nf, nf) - \frac{24}{195}F^2(nf, nf) - \frac{12}{143}F^4(nf, nf) - \frac{600}{5577}F^6(nf, nf)$$

W	U	${}^{2s+1}_{\ \nu}L$	$F^2/184275$	$F^4/297297$	$F^6/11594583$
(000)	(00)	${}^1_0S$	88200	97020	4851000
(220)	(22)	${}^1_4S$	137340	13482	2028600
		${}^1_0S{:}{}^1_4S$	$-16380\sqrt{22}$	$-6552\sqrt{22}$	$573300\sqrt{22}$
(111)	(00)	${}^5_4S$	−26460	−29106	−1455300
(110)	(11)	${}^3_2P$	39879	187803/5	1382535
(211)	(11)	${}^3_4P_1$	−14994	−33418/5	436590
(211)	(30)	${}^3_4P_2$	77644	17122	−2347100
		${}^3_2P{:}{}^3_4P_1$	−18018	−36036/5	630630
		${}^3_2P{:}{}^3_4P_2$	$8190\sqrt{11}$	$3276\sqrt{11}$	$-286650\sqrt{11}$
		${}^3_4P_1{:}{}^3_4P_2$	$-12285\sqrt{11}$	$11648\sqrt{11}$	$-143325\sqrt{11}$
(200)	(20)	${}^1_2D$	90657	67683/5	2013165
(220)	(20)	${}^1_4D_1$	25083/4	62631/10	3975615/4
(220)	(21)	${}^1_4D_2$	−18855	46827	121275
(220)	(22)	${}^1_4D_3$	274995/4	40509	−6052725/4
		${}^1_2D{:}{}^1_4D_1$	$-1287\sqrt{21}$	$-(2574/5)\sqrt{21}$	$45045\sqrt{21}$
		${}^1_2D{:}{}^1_4D_2$	$-936\sqrt{2310}$	$-(1872/5)\sqrt{2310}$	$32760\sqrt{2310}$
		${}^1_2D{:}{}^1_4D_3$	$-2457\sqrt{715}$	$-(4914/5)\sqrt{715}$	$85995\sqrt{715}$
		${}^1_4D_1{:}{}^1_4D_2$	$351\sqrt{110}$	$(11349/5)\sqrt{110}$	$-85995\sqrt{110}$
		${}^1_4D_1{:}{}^1_4D_3$	$-(1053/4)\sqrt{15015}$	$-(117/5)\sqrt{15015}$	$(25515/4)\sqrt{15015}$
		${}^1_4D_2{:}{}^1_4D_3$	$1755\sqrt{546}$	$-1755\sqrt{546}$	$23625\sqrt{546}$
(211)	(20)	${}^3_4D_1$	−13707	14652	−329175
(211)	(21)	${}^3_4D_2$	54153	−11439	−861525
		${}^3_4D_1{:}{}^3_4D_2$	$-5148\sqrt{66}$	$780\sqrt{66}$	$81900\sqrt{66}$
(111)	(20)	${}^5_4D$	18585	−11088	−3031875
(220)	(21)	${}^1_4F$	39060	48699	−1168650
(110)	(10)	${}^3_2F$	30870	33957	1697850
(211)	(10)	${}^3_4F_1$	−980	−1078	−53900
(211)	(21)	${}^3_4F_2$	17220	−2079	−404250
(211)	(30)	${}^3_4F_3$	−2254	−11270	325850
		${}^3_2F{:}{}^3_4F_2$	$-3276\sqrt{165}$	$-(6552/5)\sqrt{165}$	$114660\sqrt{165}$
		${}^3_4F_1{:}{}^3_4F_2$	$728\sqrt{165}$	$(1456/5)\sqrt{165}$	$-25480\sqrt{165}$
		${}^3_4F_1{:}{}^3_4F_3$	$-1820\sqrt{143}$	$1820\sqrt{143}$	$-24500\sqrt{143}$
		${}^3_4F_2{:}{}^3_4F_3$	$364\sqrt{195}$	$-2912\sqrt{195}$	$93100\sqrt{195}$
(111)	(10)	${}^5_4F$	−26460	−29106	−1455300
(200)	(20)	${}^1_2G$	27594	373443/5	1812510
(220)	(20)	${}^1_4G_1$	6066	71037/5	723240
(220)	(21)	${}^1_4G_2$	76968	−1221	557865
(220)	(22)	${}^1_4G_3$	1386	−9177	1631700
		${}^1_2G{:}{}^1_4G_1$	$-468\sqrt{21}$	$-(936/5)\sqrt{21}$	$16380\sqrt{21}$

W	U	${}^{2S+1}_{\nu}L$	$F^2/184275$	$F^4/297297$	$F^6/11594583$
		${}^{1}_{2}G{:}{}^{1}_{4}G_2$	$-780\sqrt{3003}$	$-312\sqrt{3003}$	$27300\sqrt{3003}$
		${}^{1}_{2}G{:}{}^{1}_{4}G_3$	$-5460\sqrt{39}$	$-2184\sqrt{39}$	$191100\sqrt{39}$
		${}^{1}_{4}G_1{:}{}^{1}_{4}G_2$	$2340\sqrt{143}$	$-156\sqrt{143}$	$-44100\sqrt{143}$
		${}^{1}_{4}G_1{:}{}^{1}_{4}G_3$	$2340\sqrt{91}$	$-4251\sqrt{91}$	$97650\sqrt{91}$
		${}^{1}_{4}G_2{:}{}^{1}_{4}G_3$	$468\sqrt{77}$	$-468\sqrt{77}$	$6300\sqrt{77}$
(211)	(20)	${}^{3}_{4}G_1$	13320	−14286	100800
(211)	(21)	${}^{3}_{4}G_2$	−7272	16225	−519750
(211)	(30)	${}^{3}_{4}G_3$	17766	19670	−1168650
		${}^{3}_{4}G_1{:}{}^{3}_{4}G_2$	$-312\sqrt{2145}$	$-416\sqrt{2145}$	$21000\sqrt{2145}$
		${}^{3}_{4}G_1{:}{}^{3}_{4}G_3$	$-1092\sqrt{15}$	$6188\sqrt{15}$	$-191100\sqrt{15}$
		${}^{3}_{4}G_2{:}{}^{3}_{4}G_3$	$3276\sqrt{143}$	$-728\sqrt{143}$	$-44100\sqrt{143}$
(111)	(20)	${}^{5}_{4}G$	−10080	−22554	−2028600
(220)	(21)	${}^{1}_{4}H_1$	−4347	−16821	2017575
(220)	(22)	${}^{1}_{4}H_2$	71001	16758	−738675
		${}^{1}_{4}H_1{:}{}^{1}_{4}H_2$	$-3276\sqrt{78}$	$3276\sqrt{78}$	$-44100\sqrt{78}$
(110)	(11)	${}^{3}_{2}H$	28413	164871/5	1783845
(211)	(11)	${}^{3}_{4}H_1$	2842	2254/5	−187670
(211)	(21)	${}^{3}_{4}H_2$	2933	25585	−1059625
(211)	(30)	${}^{3}_{4}H_3$	20496	−14910	−29400
		${}^{3}_{2}H{:}{}^{3}_{4}H_1$	4914	9828/5	−171990
		${}^{3}_{2}H{:}{}^{3}_{4}H_3$	$8190\sqrt{39}$	$3276\sqrt{39}$	$-286650\sqrt{39}$
		${}^{3}_{4}H_1{:}{}^{3}_{4}H_2$	$910\sqrt{182}$	$-910\sqrt{182}$	$12250\sqrt{182}$
		${}^{3}_{4}H_1{:}{}^{3}_{4}H_3$	$455\sqrt{39}$	$-1092\sqrt{39}$	$28175\sqrt{39}$
		${}^{3}_{4}H_2{:}{}^{3}_{4}H_3$	$5642\sqrt{42}$	$-546\sqrt{42}$	$-100450\sqrt{42}$
(200)	(20)	${}^{1}_{2}I$	36603	181251/5	2952495
(220)	(20)	${}^{1}_{4}I_1$	71883/4	72324/5	1849995/4
(220)	(22)	${}^{1}_{4}I_2$	50589/4	47880	−2326275/4
		${}^{1}_{2}I{:}{}^{1}_{4}I_1$	$819\sqrt{21}$	$(1638/5)\sqrt{21}$	$-28665\sqrt{21}$
		${}^{1}_{2}I{:}{}^{1}_{4}I_2$	$-4095\sqrt{51}$	$-1638\sqrt{51}$	$143325\sqrt{51}$
		${}^{1}_{4}I_1{:}{}^{1}_{4}I_2$	$-(12285/4)\sqrt{119}$	$1638\sqrt{119}$	$(33075/4)\sqrt{119}$
(211)	(20)	${}^{3}_{4}I_1$	−5985	2016	−55125
(211)	(30)	${}^{3}_{4}I_2$	26775	3654	−804825
		${}^{3}_{4}I_1{:}{}^{3}_{4}I_2$	$16380\sqrt{3}$	$-8736\sqrt{3}$	$-44100\sqrt{3}$
(111)	(20)	${}^{5}_{4}I$	−55125	−40572	−452025
(220)	(21)	${}^{1}_{4}K$	−14994	18669	1014300
(211)	(21)	${}^{3}_{4}K_1$	11214	−24101	485100
(211)	(30)	${}^{3}_{4}K_2$	−36288	8750	352800
		${}^{3}_{4}K_1{:}{}^{3}_{4}K_2$	$6552\sqrt{17}$	$3640\sqrt{17}$	$-264600\sqrt{17}$
(220)	(21)	${}^{1}_{4}L_1$	9030	9933	808500
(220)	(22)	${}^{1}_{4}L_2$	−34104	1470	2013900
		${}^{1}_{4}L_1{:}{}^{1}_{4}L_2$	$-2184\sqrt{95}$	$2184\sqrt{95}$	$-29400\sqrt{95}$
(211)	(21)	${}^{3}_{4}L$	−34650	−3717	749700
(211)	(30)	${}^{3}_{4}M$	−22365	−16002	915075
(220)	(22)	${}^{1}_{4}N$	2205	1470	1245825

29. Closed shells plus f^3s [2]

$$E_{Av} = E(cs) + 3E(cs, nf) + E(cs, n's) + 3I(nf) + I(n's) + 3F^0(nf, nf) - \frac{12}{195}F^2(nf, nf) - \frac{6}{143}F^4(nf, nf) - \frac{300}{5577}F^6(nf, nf) + 3F^0(nf, n's) - \frac{3}{14}G^3(nf, n's)$$

$$({}^2_\nu L)\,{}^1_\nu L = E({}^2_\nu L) + \frac{3}{14}G^3(nf, n's) \quad \text{for } L = 0, 2, 3, 4, 6$$

$$({}^2_\nu L)\,{}^3_\nu L = E({}^2_\nu L) - \frac{1}{14}G^3(nf, n's) \quad \text{for } L = 0, 2, 3, 4, 6$$

$$({}^4_\nu L)\,{}^3_\nu L = E({}^4_\nu L) + \frac{5}{14}G^3(nf, n's) \quad \text{for } L = 1, 2, 3, 4, 5, 6, 7, 8$$

$$({}^4_\nu L)\,{}^5_\nu L = E({}^4_\nu L) - \frac{3}{14}G^3(nf, n's) \quad \text{for } L = 1, 2, 3, 4, 5, 6, 7, 8$$

$$({}^{2S+1}_{\nu}L_n)\,{}^{2S}_{\nu}L_n{:}({}^{2S+1}_{\nu'}L_{n'})\,{}^{2S}_{\nu'}L_{n'} = ({}^{2S+1}_{\nu}L_n)\,{}^{2S+2}_{\nu}L_n{:}({}^{2S+1}_{\nu'}L_{n'})\,{}^{2S+2}_{\nu'}L_{n'} = {}^{2S+1}_{\nu}L_n{:}{}^{2S+1}_{\nu'}L_{n'}$$

30. Closed shells plus p^4p' [2, 17]

$$E_{Av} = E(cs) + 4E(cs, np) + E(cs, n'p) + 4I(np) + I(n'p) + 6F^0(np, np) - \frac{12}{25}F^2(np, np) + 4F^0(np, n'p) - \frac{4}{6}G^0(np, n'p) - \frac{4}{15}G^2(np, n'p)$$

	$E(p^4, {}^{2s+1}L)$	$F^2(np, n'p)/25$	$G^0(np, n'p)/6$	$G^2(np, n'p)/150$
$({}^3P)\,{}^2S$	$E(p^4, {}^3P)$	10	−2	−20
$({}^3P)\,{}^4S$	$E(p^4, {}^3P)$	10	−2	−20
$({}^1S)\,{}^2P$	$E(p^4, {}^1S)$			
$({}^3P)\,{}^2P$	$E(p^4, {}^3P)$	−5	16	25
$({}^1D)\,{}^2P$	$E(p^4, {}^1D)$	−7	8	−19
$({}^1S)\,{}^2P{:}({}^3P)\,{}^2P$			6	−30
$({}^1S)\,{}^2P{:}({}^1D)\,{}^2P$		$-4\sqrt{5}$	$2\sqrt{5}$	$2\sqrt{5}$
$({}^3P)\,{}^2P{:}({}^1D)\,{}^2P$			$6\sqrt{5}$	$-3\sqrt{5}$
$({}^3P)\,{}^4P$	$E(p^4, {}^3P)$	−5	−2	−20
$({}^3P)\,{}^2D$	$E(p^4, {}^3P)$	1	−2	61
$({}^1D)\,{}^2D$	$E(p^4, {}^1D)$	7	−2	−11
$({}^3P)\,{}^2D{:}({}^1D)\,{}^2D$				−27
$({}^3P)\,{}^4D$	$E(p^4, {}^3P)$	1	−2	−20
$({}^1D)\,{}^2F$	$E(p^4, {}^1D)$	−2	−2	16

31. Closed shells plus p^4d [2, 17]

$$E_{Av} = E(cs) + 4E(cs, np) + E(cs, n'd) + 4I(np) + I(n'd) + 6F^0(np, np) - \frac{12}{25}F^2(np, np) + 4F^0(np, n'd) - \frac{4}{15}G^1(np, n'd) - \frac{12}{70}G^3(np, n'd)$$

	$E(p^4, {}^{2s+1}L)$	$F^2(np, n'd)/35$	$G^1(np, n'd)/30$	$G^3(np, n'd)/245$
$({}^1D)\,{}^2S$	$E(p^4, {}^1D)$	-14	16	-21
$({}^3P)\,{}^2P$	$E(p^4, {}^3P)$	7	11	-21
$({}^1D)\,{}^2P$	$E(p^4, {}^1D)$	-7	11	-21
$({}^3P)\,{}^2P{:}({}^1D)\,{}^2P$			15	
$({}^3P)\,{}^4P$	$E(p^4, {}^3P)$	7	-4	-21
$({}^1S)\,{}^2D$	$E(p^4, {}^1S)$			
$({}^3P)\,{}^2D$	$E(p^4, {}^3P)$	-7	23	42
$({}^1D)\,{}^2D$	$E(p^4, {}^1D)$	3	3	-18
$({}^1S)\,{}^2D{:}({}^3P)\,{}^2D$			$6\sqrt{3}$	$-21\sqrt{3}$
$({}^1S)\,{}^2D{:}({}^1D)\,{}^2D$		$-4\sqrt{7}$	$2\sqrt{7}$	$3\sqrt{7}$
$({}^3P)\,{}^2D{:}({}^1D)\,{}^2D$			$3\sqrt{21}$	$-3\sqrt{21}$
$({}^3P)\,{}^4D$	$E(p^4, {}^3P)$	-7	-4	-21
$({}^3P)\,{}^2F$	$E(p^4, {}^3P)$	2	-4	69
$({}^1D)\,{}^2F$	$E(p^4, {}^1D)$	8	-4	-6
$({}^3P)\,{}^2F{:}({}^1D)\,{}^2F$				$-15\sqrt{6}$
$({}^3P)\,{}^4F$	$E(p^4, {}^3P)$	2	-4	-21
$({}^1D)\,{}^2G$	$E(p^4, {}^1D)$	-4	-4	24

32. Closed shells plus p^4f [2, 17]

$$E_{\mathrm{Av}} = E(cs) + 4E(cs, np) + E(cs, n'f) + 4I(np) + I(n'f) + 6F^0(np, np) - \frac{12}{25}F^2(np, np) + 4F^0(np, n'f) - \frac{12}{70}G^2(np, n'f) - \frac{8}{63}G^4(np, n'f)$$

	$E(p^4, {}^{2s+1}_{\ \ v}L)$	$F^2(np, n'f)/75$	$G^2(np, n'f)/175$	$G^4(np, n'f)/378$
$({}^1D)\,{}^2P$	$E(p^4, {}^1D)$	-24	48	-24
$({}^3P)\,{}^2D$	$E(p^4, {}^3P)$	12	48	-24
$({}^1D)\,{}^2D$	$E(p^4, {}^1D)$	-6	27	-24
$({}^3P)\,{}^2D{:}({}^1D)\,{}^2D$			$21\sqrt{6}$	
$({}^3P)\,{}^4D$	$E(p^4, {}^3P)$	12	-15	-24
$({}^1S)\,{}^2F$	$E(p^4, {}^1S)$			
$({}^3P)\,{}^2F$	$E(p^4, {}^3P)$	-15	75	57
$({}^1D)\,{}^2F$	$E(p^4, {}^1D)$	11	3	-19
$({}^1S)\,{}^2F{:}({}^3P)\,{}^2F$			$15\sqrt{6}$	$-18\sqrt{6}$
$({}^1S)\,{}^2F{:}({}^1D)\,{}^2F$		$-4\sqrt{30}$	$3\sqrt{30}$	$2\sqrt{30}$
$({}^3P)\,{}^2F{:}({}^1D)\,{}^2F$			$18\sqrt{5}$	$-9\sqrt{5}$
$({}^3P)\,{}^4F$	$E(p^4, {}^3P)$	-15	-15	-24
$({}^3P)\,{}^2G$	$E(p^4, {}^3P)$	5	-15	81
$({}^1D)\,{}^2G$	$E(p^4, {}^1D)$	15	-15	-3
$({}^3P)\,{}^2G{:}({}^1D)\,{}^2G$				$-21\sqrt{5}$
$({}^3P)\,{}^4G$	$E(p^4, {}^3P)$	5	-15	-24
$({}^1D)\,{}^2H$	$E(p^4, {}^1D)$	-10	-15	32

33. Closed shells plus d^5 [2, 13, 16, 24]

$$E_{\mathrm{Av}} = E(cs) + 5E(cs, nd) + 5I(nd) + 10F^0(nd, nd) - \frac{20}{63}F^2(nd, nd) - \frac{20}{63}F^4(nd, nd)$$

	$F^2/441$	$F^4/441$
${}^2_5S$	113	−55
${}^6_5S$	−175	−175
${}^2_3P$	320	−100
${}^4_3P$	−112	35
${}^2_1D$	140	140
${}^2_3D$	104	20
${}^2_5D$	86	−40
${}^2_1D{:}{}^2_5D$	$54\sqrt{14}$	$-30\sqrt{14}$
${}^4_5D$	−22	−85
${}^2_3F$	−85	125
${}^2_5F$	59	−25
${}^4_3F$	23	−40
${}^2_3G$	167	−15
${}^2_5G$	23	−5
${}^4_5G$	−85	−50
${}^2_3H$	−58	110
${}^2_5I$	−76	50

34. Closed shells plus d^4s [2]

$$E_{\mathrm{Av}} = E(cs) + 4E(cs, nd) + E(cs, n's) + 4I(nd) + I(n's) + 6F^0(nd, nd) - \frac{12}{63}F^2(nd, nd) - \frac{12}{63}F^4(nd, nd) + 4F^0(nd, n's) - \frac{4}{10}G^2(nd, n's)$$

$$({}^1_\nu L)\ {}^2_\nu L = E(d^4, {}^1_\nu L) \quad \text{for } L = 0, 2, 3, 4, 6$$

$$({}^3_\nu L)\ {}^2_\nu L = E(d^4, {}^3_\nu L) + \frac{2}{5}G^2(nd, n's) \quad \text{for } L = 1, 2, 3, 4, 5$$

$$({}^3_\nu L)\ {}^4_\nu L = E(d^4, {}^3_\nu L) - \frac{1}{5}G^2(nd, n's) \quad \text{for } L = 1, 2, 3, 4, 5$$

$$({}^5_4D)\ {}^4_4D = E(d^4, {}^5_4D) + \frac{3}{5}G^2(nd, n's)$$

$$({}^5_4D)\ {}^6_4D = E(d^4, {}^5_4D) - \frac{2}{5}G^2(nd, n's)$$

$$({}^{2s+1}_{\nu}L)\ {}^{2s}_{\nu}L{:}\ ({}^{2s+1}_{\nu'}L)\ {}^{2s}_{\nu'}L = ({}^{2s+1}_{\nu}L)\ {}^{2s+2}_{\nu}L{:}({}^{2s+1}_{\nu'}L)\ {}^{2s+2}_{\nu'}L = {}^{2s+1}_{\nu}L{:}{}^{2s'+1}_{\nu'}L$$

35. Closed shells plus d^4p [2, 19]

$$E_{\mathrm{Av}} = E(cs) + 4E(cs, nd) + E(cs, n'p) + 4I(nd) + I(n'p) + 6F^0(nd, nd) - \frac{12}{63}F^2(nd, nd) - \frac{12}{63}F^4(nd, nd) + 4F^0(nd, n'p) - \frac{4}{15}G^1(nd, n'p) - \frac{12}{70}G^3(nd, n'p)$$

For the Russell-Saunders term energies see Slater [2] or Ishidzu and Obi [19].

36. Closed shells plus f^5 [21, 24]

$$E_{\mathrm{Av}} = E(cs) + 5E(cs, nf) + 5I(nf) + 10F^0(nf, nf) - \frac{40}{195}F^2(nf, nf) - \frac{20}{143}F^4(nf, nf) - \frac{1000}{5577}F^6(nf, nf)$$

For the Russell-Saunders term energies see Wybourne [21] or Nielson and Koster [24].

37. Closed shells plus f^4s [2]

$$E_{\mathrm{Av}} = E(cs) + 4E(cs, nf) + E(cs, n's) + 4I(nf) + I(n's) + 6F^0(nf, nf) - \frac{24}{195}F^2(nf, nf) - \frac{12}{143}F^4(nf, nf) - \frac{600}{5577}F^6(nf, nf) + 4F^0(nf, n's) - \frac{4}{14}G^3(nf, n's)$$

$$({}^1_\nu L)\,{}^2_\nu L = E(f^4, {}^1_\nu L) \quad \text{for } L = 0, 2, 3, 4, 5, 6, 7, 8, 10$$

$$({}^3_\nu L)\,{}^2_\nu L = E(f^4, {}^3_\nu L) + \frac{2}{7}G^3(nf, n's) \quad \text{for } L = 1, 2, 3, 4, 5, 6, 7, 8, 9$$

$$({}^3_\nu L)\,{}^4_\nu L = E(f^4, {}^3_\nu L) - \frac{1}{7}G^3(nf, n's) \quad \text{for } L = 1, 2, 3, 4, 5, 6, 7, 8, 9$$

$$({}^5_\nu L)\,{}^4_\nu L = E(f^4, {}^5_\nu L) + \frac{3}{7}G^3(nf, n's) \quad \text{for } L = 0, 2, 3, 4, 6$$

$$({}^5_\nu L)\,{}^6_\nu L = E(f^4, {}^5_\nu L) - \frac{2}{7}G^3(nf, n's) \quad \text{for } L = 0, 2, 3, 4, 6$$

$$({}^{2S+1}_{\nu}\mathrm{L}_n)\,{}^{2S}_{\nu}L_n{:}({}^{2S+1}_{\nu'}L_{n'})\,{}^{2S}_{\nu'}L_{n'} = ({}^{2S+1}_{\nu}L_n)\,{}^{2S+2}_{\nu}L_n{:}({}^{2S+1}_{\nu'}L_{n'})\,{}^{2S+2}_{\nu'}L_{n'} = {}^{2S+1}_{\nu}L_n{:}\,{}^{2S+1}_{\nu'}L_{n'}$$

38. Closed shells plus p^5p' [1, 2, 4, 8, 17]

$$E_{Av} = E(cs) + 5E(cs, np) + E(cs, n'p) + 5I(np) + I(n'p) + 10F^0(np, np) - \frac{20}{25}F^2(np, np) + 5F^0(np, n'p) - \frac{5}{6}G^0(np, n'p) - \frac{5}{15}G^2(np, n'p)$$

	$F^2(np, n'p)/25$	$G^0(np, n'p)/6$	$G^2(np, n'p)/75$
^1S	−10	35	−5
^3S	−10	−1	−5
^1P	5	−1	−5
^3P	5	−1	−5
^1D	−1	−1	31
^3D	−1	−1	−5

39. Closed shells plus p^5d [2, 4, 17]

$$E_{Av} = E(cs) + 5E(cs, np) + E(cs, n'd) + 5I(np) + I(n'd) + 10F^0(np, np) - \frac{20}{25}F^2(np, np) + 5F^0(np, n'd) - \frac{5}{15}G^1(np, n'd) - \frac{15}{70}G^3(np, n'd)$$

	$F^2(np, n'd)/35$	$G^1(np, n'd)/15$	$G^3(np, n'd)/490$
^1P	−7	19	−21
^3P	−7	−1	−21
^1D	7	−1	−21
^3D	7	−1	−21
^1F	−2	−1	159
^3F	−2	−1	−21

40. Closed shells plus p^5f [2, 4, 17]

$$E_{Av} = E(cs) + 5E(cs, np) + E(cs, n'f) + 5I(np) + I(n'f) + 10F^0(np, np) - \frac{20}{25}F^2(np, np) + 5F^0(np, n'f) - \frac{15}{70}G^2(np, n'f) - \frac{10}{63}G^4(np, n'f)$$

	$F^2(np, n'f)/75$	$G^2(np, n'f)/350$	$G^4(np, n'f)/189$
^1D	−12	237	−6
^3D	−12	−15	−6
^1F	15	−15	−6
^3F	15	−15	−6
^1G	−5	−15	50
^3G	−5	−15	−6

41. Closed shells plus d^5s [2]

$$E_{\mathrm{Av}} = E(cs) + 5E(cs, nd) + E(cs, n's) + 5I(nd) + I(n's) + 10F^0(nd, nd) - \frac{20}{63}F^2(nd, nd) - \frac{20}{63}F^4(nd, nd) + 5F^0(nd, n's) - \frac{5}{10}G^2(nd, n's)$$

$$({}^6_5S)\,{}^5_5S = E(d^5, {}^6_5S) + \frac{7}{10}G^2(nd, n's)$$

$$({}^6_5S)\,{}^7_5S = E(d^5, {}^6_5S) - \frac{1}{2}G^2(nd, n's)$$

$$({}^2_vL)\,{}^1_vL = E(d^5, {}^2_vL) + \frac{3}{10}G^2(nd, n's) \quad \text{for } L = 0, 1, 2, 3, 4, 5, 6$$

$$({}^2_vL)\,{}^3_vL = E(d^5, {}^2_vL) - \frac{1}{10}G^2(nd, n's) \quad \text{for } L = 0, 1, 2, 3, 4, 5, 6$$

$$({}^4_vL)\,{}^3_vL = E(d^5, {}^4_vL) + \frac{1}{2}G^2(nd, n's) \quad \text{for } L = 1, 2, 3, 4$$

$$({}^4_vL)\,{}^5_vL = E(d^5, {}^4_vL) - \frac{3}{10}G^2(nd, n's) \quad \text{for } L = 1, 2, 3, 4$$

42. Closed shells plus d^5p [2, 19]

$$E_{\mathrm{Av}} = E(cs) + 5E(cs, nd) + E(cs, n'p) + 5I(nd) + I(n'p) + 10F^0(nd, nd) - \frac{20}{63}F^2(nd, nd) - \frac{20}{63}F^4(nd, nd) + 5F^0(nd, n'p) - \frac{5}{15}G^1(nd, n'p) - \frac{15}{70}G^3(nd, n'p)$$

For the Russell-Saunders term energies see Slater [2] or Ishidzu and Obi [19].

43. Closed shells plus f^6 [22, 24]

$$E_{\mathrm{Av}} = E(cs) + 6E(cs, nf) + 6I(nf) + 15F^0(nf, nf) - \frac{60}{195}F^2(nf, nf) - \frac{30}{143}F^4(nf, nf) - \frac{1500}{5577}F^6(nf, nf)$$

For the Russell-Saunders term energies see Wybourne [22] or Nielson and Koster [24].

44. Closed shells plus f^5s [2]

$$E_{\text{Av}} = E(cs) + 5E(cs, nf) + E(cs, n's) + 5I(nf) + I(n's) + 10F^0(nf, nf) - \frac{40}{195}F^2(nf, nf) - \frac{20}{143}F^4(nf, nf) - \frac{1000}{5577}F^6(nf, nf) + 5F^0(nf, n's) - \frac{5}{14}G^3(nf, n's)$$

$$({}^4_\nu L)\ {}^3_\nu L = E(f^5, {}^4_\nu L) + \frac{5}{14}G^3(nf, n's) \quad \text{for } L = 0, 1, 2, 3, 4, 5, 6, 7, 8, 9$$

$$({}^4_\nu L)\ {}^5_\nu L = E(f^5, {}^4_\nu L) - \frac{3}{14}G^3(nf, n's) \quad \text{for } L = 0, 1, 2, 3, 4, 5, 6, 7, 8, 9$$

$$({}^2_\nu L)\ {}^1_\nu L = E(f^5, {}^2_\nu L) + \frac{3}{14}G^3(nf, n's) \quad \text{for } L = 1, 2, 3, 4, 5, 6, 7, 8, 9, 10, 11$$

$$({}^2_\nu L)\ {}^3_\nu L = E(f^5, {}^2_\nu L) - \frac{1}{14}G^3(nf, n's) \quad \text{for } L = 1, 2, 3, 4, 5, 6, 7, 8, 9, 10, 11$$

$$({}^6_\nu L)\ {}^5_\nu L = E(f^5, {}^6_\nu L) + \frac{7}{14}G^3(nf, n's) \quad \text{for } L = 1, 3, 5$$

$$({}^6_\nu L)\ {}^7_\nu L = E(f^5, {}^6_\nu L) - \frac{5}{14}G^3(nf, n's) \quad \text{for } L = 1, 3, 5$$

$$({}^{2S+1}_{\nu}L_n)\ {}^{2S}_{\nu}L_n\!:({}^{2S+1}_{\nu'}L_{n'})\ {}^{2S}_{\nu'}L_{n'} = ({}^{2S+1}_{\nu}L_n)\ {}^{2S+2}_{\nu}L_n\!:({}^{2S+1}_{\nu'}L_{n'})\ {}^{2S+2}_{\nu'}L_{n'} = {}^{2S+1}_{\nu}L_n\!:\ {}^{2S+1}_{\nu'}L_{n'}$$

45. Closed shells plus d^6p [2, 19]

$$E_{\text{Av}} = E(cs) + 6E(cs, nd) + \text{E}\,(cs, n'p) + 6I(nd) + I(n'p) + 15F^0(nd, nd) - \frac{30}{63}F^2(nd, nd) - \frac{30}{63}F^4(nd, nd) + 6F^0(nd, n'p) - \frac{6}{15}G^1(nd, n'p) - \frac{18}{70}G^3(nd, n'p)$$

For the Russell-Saunders term energies see Slater [2] or Ishidzu and Obi [19].

46. Closed shells plus f^7 [23, 24]

$$E_{\text{Av}} = E(cs) + 7E(cs, nf) + 7I(nf) + 21F^0(nf, nf) - \frac{84}{195}F^2(nf, nf) - \frac{42}{143}F^4(nf, nf) - \frac{2100}{5577}F^6(nf, nf)$$

For the Russell-Saunders term energies see Runciman [23] or Nielson and Koster [24].

47. Closed shells plus f^6s [2]

$$E_{\mathrm{Av}} = E(cs) + 6E(cs, nf) + E(cs, n's) + 6I(nf) + I(n's) + 15F^0(nf, nf) - \frac{60}{195}F^2(nf, nf) - \frac{30}{143}F^4(nf, nf) - \frac{1500}{5577}F^6(nf, nf) + 6F^0(nf, n's) - \frac{6}{14}G^3(nf, n's)$$

$({}^{1}_{\nu}L)\,{}^{2}_{\nu}L = E(f^6, {}^{1}_{\nu}L)$ for $L = 0, 1, 2, 3, 4, 5, 6, 7, 8, 9, 10, 12$

$({}^{3}_{\nu}L)\,{}^{2}_{\nu}L = E(f^6, {}^{3}_{\nu}L) + \frac{2}{7}G^3(nf, n's)$ for $L = 1, 2, 3, 4, 5, 6, 7, 8, 9, 10, 11$

$({}^{3}_{\nu}L)\,{}^{4}_{\nu}L = E(f^6, {}^{3}_{\nu}L) - \frac{1}{7}G^3(nf, n's)$ for $L = 1, 2, 3, 4, 5, 6, 7, 8, 9, 10, 11$

$({}^{5}_{\nu}L)\,{}^{4}_{\nu}L = E(f^6, {}^{5}_{\nu}L) + \frac{3}{7}G^3(nf, n's)$ for $L = 0.1, 2, 3, 4, 5, 6, 7, 8$

$({}^{5}_{\nu}L)\,{}^{6}_{\nu}L = E(f^6, {}^{5}_{\nu}L) - \frac{2}{7}G^3(nf, n's)$ for $L = 0, 1, 2, 3, 4, 5, 6, 7, 8$

$({}^{7}_{6}F)\,{}^{6}_{6}F = E(f^6, {}^{7}_{6}F) + \frac{4}{7}G^3(nf, n's)$

$({}^{7}_{6}F)\,{}^{8}_{6}F = E(f^6, {}^{7}_{6}F) - \frac{3}{7}G^3(nf, n's)$

$$({}^{2S+1}_{\nu}L_n)\,{}^{2S}_{\nu}L_n:({}^{2S+1}_{\nu'}L_{n'})\,{}^{2S}_{\nu'}L_{n'} = ({}^{2S+1}_{\nu}L_n)\,{}^{2S+2}_{\nu}L_n:({}^{2S+1}_{\nu'}L_{n'})\,{}^{2S+2}_{\nu'}L_{n'} = {}^{2S+1}_{\nu}L_n\text{: }{}^{2S+1}_{\nu'}L_{n'}$$

48. Closed shells plus d^7p [2, 18]

$$E_{\mathrm{Av}} = E(cs) + 7E(cs, nd) + E(cs, n'p) + 7I(nd) + I(n'p) + 21F^0(nd, nd) - \frac{42}{63}F^2(nd, nd) - \frac{42}{63}F^4(nd, nd) + 7F^0(nd, n'p) - \frac{7}{15}G^1(nd, n'p) - \frac{21}{70}G^3(nd, n'p)$$

	$E(d^7\,{}^{2s+1}_{\nu}L)$	$F^2(nd, n'p)/2450$	$G^1(nd, n'p)/6300$	$G^3(nd, n'p)/17150$
$({}^2_3P)\,{}^1S$	$E(d^7, {}^2_3P)$	280	-1260	-2205
$({}^2_3P)\,{}^3S$	$E(d^7, {}^2_3P)$	280	700	-525
$({}^4_3P)\,{}^3S$	$E(d^7, {}^4_3P)$	-980	7700	3675
$({}^2_3P)\,{}^3S:({}^4_3P)\,{}^3S$			$1120\sqrt{14}$	$-840\sqrt{14}$
$({}^4_3P)\,{}^5S$	$E(d^7, {}^4_3P)$	-980	-1260	-2205
$({}^2_3P)\,{}^1P$	$E(d^7, {}^2_3P)$	-140	4410	2205
$({}^2_1D)\,{}^1P$	$E(d^7, {}^2_1D)$	-245	6195	735
$({}^2_3D)\,{}^1P$	$E(d^7, {}^2_3D)$	-105	1575	1575
$({}^2_3P)\,{}^1P:({}^2_1D)\,{}^1P$		$-105\sqrt{42}$	$315\sqrt{42}$	$-105\sqrt{42}$
$({}^2_3P)\,{}^1P:({}^2_3D)\,{}^1P$		$105\sqrt{2}$	$-2835\sqrt{2}$	$945\sqrt{2}$
$({}^2_1D)\,{}^1P:({}^2_3D)\,{}^1P$		$-105\sqrt{21}$	$-315\sqrt{21}$	$630\sqrt{21}$

	$E(d^{7}\,{}^{2s+1}_{v}L)$	$F^{2}(nd, n'p)/2450$	$G^{1}(nd, n'p)/6300$	$G^{3}(nd, n'p)/17150$
$({}^{2}_{3}P)\,{}^{3}P$	$E(d^{7}, {}^{2}_{3}P)$	-140	-770	-1085
$({}^{4}_{3}P)\,{}^{3}P$	$E(d^{7}, {}^{4}_{3}P)$	490	980	1715
$({}^{2}_{1}D)\,{}^{3}P$	$E(d^{7}, {}^{2}_{1}D)$	-245	-315	-1470
$({}^{2}_{3}D)\,{}^{3}P$	$E(d^{7}, {}^{2}_{3}D)$	-105	945	-1890
$({}^{2}_{3}P)\,{}^{3}P{:}({}^{4}_{3}P)\,{}^{3}P$			$280\sqrt{14}$	$-560\sqrt{14}$
$({}^{2}_{3}P)\,{}^{3}P{:}({}^{2}_{1}D)\,{}^{3}P$		$-105\sqrt{42}$	$105\sqrt{42}$	$140\sqrt{42}$
$({}^{2}_{3}P)\,{}^{3}P{:}({}^{2}_{3}D)\,{}^{3}P$		$105\sqrt{2}$	$735\sqrt{2}$	$-420\sqrt{2}$
$({}^{4}_{3}P)\,{}^{3}P{:}({}^{2}_{1}D)\,{}^{3}P$			$840\sqrt{3}$	$-980\sqrt{3}$
$({}^{4}_{3}P)\,{}^{3}P{:}({}^{2}_{3}D)\,{}^{3}P$			$840\sqrt{7}$	$420\sqrt{7}$
$({}^{2}_{1}D)\,{}^{3}P{:}({}^{2}_{3}D)\,{}^{3}P$		$-105\sqrt{21}$	$315\sqrt{21}$	$-105\sqrt{21}$
$({}^{4}_{3}P)\,{}^{5}P$	$E(d^{7}, {}^{4}_{3}P)$	490	-1260	-2205
$({}^{2}_{3}P)\,{}^{1}D$	$E(d^{7}, {}^{2}_{3}P)$	28	-126	3087
$({}^{2}_{1}D)\,{}^{1}D$	$E(d^{7}, {}^{2}_{1}D)$	245	-525	4410
$({}^{2}_{3}D)\,{}^{1}D$	$E(d^{7}, {}^{2}_{3}D)$	105	1575	2100
$({}^{2}_{3}F)\,{}^{1}D$	$E(d^{7}, {}^{2}_{3}F)$	-168	756	903
$({}^{2}_{3}P)\,{}^{1}D{:}({}^{2}_{1}D)\,{}^{1}D$		$21\sqrt{210}$	$63\sqrt{210}$	$-126\sqrt{210}$
$({}^{2}_{3}P)\,{}^{1}D{:}({}^{2}_{3}D)\,{}^{1}D$		$-21\sqrt{10}$	$-567\sqrt{10}$	$1134\sqrt{10}$
$({}^{2}_{3}P)\,{}^{1}D{:}({}^{2}_{3}F)\,{}^{1}D$		336	-1512	-756
$({}^{2}_{1}D)\,{}^{1}D{:}({}^{2}_{3}D)\,{}^{1}D$		$105\sqrt{21}$	$-315\sqrt{21}$	$105\sqrt{21}$
$({}^{2}_{1}D)\,{}^{1}D{:}({}^{2}_{3}F)\,{}^{1}D$		$-28\sqrt{210}$	$-84\sqrt{210}$	$168\sqrt{210}$
$({}^{2}_{3}D)\,{}^{1}D{:}({}^{2}_{3}F)\,{}^{1}D$		$-252\sqrt{10}$	$756\sqrt{10}$	$588\sqrt{10}$
$({}^{2}_{3}P)\,{}^{3}D$	$E(d^{7}, {}^{2}_{3}P)$	28	-434	-567
$({}^{4}_{3}P)\,{}^{3}D$	$E(d^{7}, {}^{4}_{3}P)$	-98	1652	4851
$({}^{2}_{1}D)\,{}^{3}D$	$E(d^{7}, {}^{2}_{1}D)$	245	105	-735
$({}^{2}_{3}D)\,{}^{3}D$	$E(d^{7}, {}^{2}_{3}D)$	105	-315	-1295
$({}^{2}_{3}F)\,{}^{3}D$	$E(d^{7}, {}^{2}_{3}F)$	-168	1764	-2093
$({}^{4}_{3}F)\,{}^{3}D$	$E(d^{7}, {}^{4}_{3}F)$	168	4788	-581
$({}^{2}_{3}P)\,{}^{3}D{:}({}^{4}_{3}P)\,{}^{3}D$			$-392\sqrt{14}$	$504\sqrt{14}$
$({}^{2}_{3}P)\,{}^{3}D{:}({}^{2}_{1}D)\,{}^{3}D$		$21\sqrt{210}$	$-63\sqrt{210}$	$21\sqrt{210}$
$({}^{2}_{3}P)\,{}^{3}D{:}({}^{2}_{3}D)\,{}^{3}D$		$-21\sqrt{10}$	$231\sqrt{10}$	$-357\sqrt{10}$
$({}^{2}_{3}P)\,{}^{3}D{:}({}^{2}_{3}F)\,{}^{3}D$		336	-168	-84
$({}^{2}_{3}P)\,{}^{3}D{:}({}^{4}_{3}F)\,{}^{3}D$			1848	924
$({}^{4}_{3}P)\,{}^{3}D{:}({}^{2}_{1}D)\,{}^{3}D$			$504\sqrt{15}$	$-588\sqrt{15}$
$({}^{4}_{3}P)\,{}^{3}D{:}({}^{2}_{3}D)\,{}^{3}D$			$-168\sqrt{35}$	$-84\sqrt{35}$
$({}^{4}_{3}P)\,{}^{3}D{:}({}^{2}_{3}F)\,{}^{3}D$			$336\sqrt{14}$	$168\sqrt{14}$
$({}^{4}_{3}P)\,{}^{3}D{:}({}^{4}_{3}F)\,{}^{3}D$		$168\sqrt{14}$	$-672\sqrt{14}$	$-336\sqrt{14}$
$({}^{2}_{1}D)\,{}^{3}D{:}({}^{2}_{3}D)\,{}^{3}D$		$105\sqrt{21}$	$-105\sqrt{21}$	$-140\sqrt{21}$
$({}^{2}_{1}D)\,{}^{3}D{:}({}^{2}_{3}F)\,{}^{3}D$		$-28\sqrt{210}$	$84\sqrt{210}$	$-28\sqrt{210}$
$({}^{2}_{1}D)\,{}^{3}D{:}({}^{4}_{3}F)\,{}^{3}D$			$-84\sqrt{210}$	$-98\sqrt{210}$
$({}^{2}_{3}D)\,{}^{3}D{:}({}^{2}_{3}F)\,{}^{3}D$		$-252\sqrt{10}$	$252\sqrt{10}$	$56\sqrt{10}$
$({}^{2}_{3}D)\,{}^{3}D{:}({}^{4}_{3}F)\,{}^{3}D$			$756\sqrt{10}$	$-322\sqrt{10}$
$({}^{2}_{3}F)\,{}^{3}D{:}({}^{4}_{3}F)\,{}^{3}D$			2016	-392
$({}^{4}_{3}P)\,{}^{5}D$	$E(d^{7}, {}^{4}_{3}P)$	-98	-1260	-2205
$({}^{4}_{3}F)\,{}^{5}D$	$E(d^{7}, {}^{4}_{3}F)$	168	-1260	-2205
$({}^{4}_{3}P)\,{}^{5}D{:}({}^{4}_{3}F)\,{}^{5}D$		$168\sqrt{14}$		
$({}^{2}_{1}D)\,{}^{1}F$	$E(d^{7}, {}^{2}_{1}D)$	-70	420	1470
$({}^{2}_{3}D)\,{}^{1}F$	$E(d^{7}, {}^{2}_{3}D)$	-30	900	2550
$({}^{2}_{3}F)\,{}^{1}F$	$E(d^{7}, {}^{2}_{3}F)$	210	2835	2730

	$E(d^7\,{}^{2s+1}_{v}L)$	$F^2(nd,n'p)/2450$	$G^1(nd,n'p)/6300$	$G^3(nd,n'p)/17150$
$(^2_3G)\,^1F$	$E(d^7,\,^2_3G)$	-110	4665	-1080
$(^2_1D)\,^1F{:}(^2_3D)\,^1F$		$-30\sqrt{21}$	$360\sqrt{21}$	$-345\sqrt{21}$
$(^2_1D)\,^1F{:}(^2_3F)\,^1F$		$140\sqrt{3}$	$-1470\sqrt{3}$	$1365\sqrt{3}$
$(^2_1D)\,^1F{:}(^2_3G)\,^1F$		$60\sqrt{105}$	$-90\sqrt{105}$	$-45\sqrt{105}$
$(^2_3D)\,^1F{:}(^2_3F)\,^1F$		$180\sqrt{7}$	$-810\sqrt{7}$	$-105\sqrt{7}$
$(^2_3D)\,^1F{:}(^2_3G)\,^1F$		$60\sqrt{5}$	$-1170\sqrt{5}$	$-585\sqrt{5}$
$(^2_3F)\,^1F{:}(^2_3G)\,^1F$		$60\sqrt{35}$	$45\sqrt{35}$	$-315\sqrt{35}$
$(^2_1D)\,^3F$	$E(d^7,\,^2_1D)$	-70	-840	-420
$(^2_3D)\,^3F$	$E(d^7,\,^2_3D)$	-30	-1080	160
$(^2_3F)\,^3F$	$E(d^7,\,^2_3F)$	210	-945	-1610
$(^4_3F)\,^3F$	$E(d^7,\,^4_3F)$	-210	3780	3115
$(^2_3G)\,^3F$	$E(d^7,\,^2_3G)$	-110	765	-1980
$(^2_1D)\,^3F{:}(^2_3D)\,^3F$		$-30\sqrt{21}$	$-60\sqrt{21}$	$145\sqrt{21}$
$(^2_1D)\,^3F{:}(^2_3F)\,^3F$		$140\sqrt{3}$	$210\sqrt{3}$	$-595\sqrt{3}$
$(^2_1D)\,^3F{:}(^4_3F)\,^3F$			$-840\sqrt{3}$	$980\sqrt{3}$
$(^2_1D)\,^3F{:}(^2_3G)\,^3F$		$60\sqrt{105}$	$-90\sqrt{105}$	$-45\sqrt{105}$
$(^2_3D)\,^3F{:}(^2_3F)\,^3F$		$180\sqrt{7}$	$-90\sqrt{7}$	$-145\sqrt{7}$
$(^2_3D)\,^3F{:}(^4_3F)\,^3F$			$360\sqrt{7}$	$-820\sqrt{7}$
$(^2_3D)\,^3F{:}(^2_3G)\,^3F$		$60\sqrt{5}$	$270\sqrt{5}$	$135\sqrt{5}$
$(^2_3F)\,^3F{:}(^4_3F)\,^3F$			-1260	-980
$(^2_3F)\,^3F{:}(^2_3G)\,^3F$		$60\sqrt{35}$	$-135\sqrt{35}$	$45\sqrt{35}$
$(^4_3F)\,^3F{:}(^2_3G)\,^3F$			$540\sqrt{35}$	$-180\sqrt{35}$
$(^4_3F)\,^5F$	$E(d^7,\,^4_3F)$	-210	-1260	-2205
$(^2_3F)\,^1G$	$E(d^7,\,^2_3F)$	-70	315	2520
$(^2_3G)\,^1G$	$E(d^7,\,^2_3G)$	154	1281	1218
$(^2_3H)\,^1G$	$E(d^7,\,^2_3H)$	-364	4284	-1323
$(^2_3F)\,^1G{:}(^2_3G)\,^1G$		$-140\sqrt{3}$	$1155\sqrt{3}$	$-735\sqrt{3}$
$(^2_3F)\,^1G{:}(^2_3H)\,^1G$		$140\sqrt{22}$	$-630\sqrt{22}$	$-315\sqrt{22}$
$(^2_3G)\,^1G{:}(^2_3H)\,^1G$		$-28\sqrt{66}$	$-462\sqrt{66}$	$189\sqrt{66}$
$(^2_3F)\,^3G$	$E(d^7,\,^2_3F)$	-70	-1225	700
$(^4_3F)\,^3G$	$E(d^7,\,^4_3F)$	70	-700	6475
$(^2_3G)\,^3G$	$E(d^7,\,^2_3G)$	154	-315	-1050
$(^2_3H)\,^3G$	$E(d^7,\,^2_3H)$	-364	1820	-2135
$(^2_3F)\,^3G{:}(^4_3F)\,^3G$			-140	980
$(^2_3F)\,^3G{:}(^2_3G)\,^3G$		$-140\sqrt{3}$	$-105\sqrt{3}$	$385\sqrt{3}$
$(^2_3F)\,^3G{:}(^2_3H)\,^3G$		$140\sqrt{22}$	$-70\sqrt{22}$	$-35\sqrt{22}$
$(^4_3F)\,^3G{:}(^2_3G)\,^3G$			$420\sqrt{3}$	$1540\sqrt{3}$
$(^4_3F)\,^3G{:}(^2_3H)\,^3G$			$280\sqrt{22}$	$140\sqrt{22}$
$(^2_3G)\,^3G{:}(^2_3H)\,^3G$		$-28\sqrt{66}$	$210\sqrt{66}$	$-35\sqrt{66}$
$(^4_3F)\,^5G$	$E(d^7,\,^4_3F)$	70	-1260	-2205
$(^2_3G)\,^1H$	$E(d^7,\,^2_3G)$	-56	-924	5103
$(^2_3H)\,^1H$	$E(d^7,\,^2_3H)$	546	1764	1617
$(^2_3G)\,^1H{:}(^2_3H)\,^1H$		168	-1008	3276
$(^2_3G)\,^3H$	$E(d^7,\,^2_3G)$	-56	-1260	315
$(^2_3H)\,^3H$	$E(d^7,\,^2_3H)$	546	-1260	-1575
$(^2_3G)\,^3H{:}(^2_3H)\,^3H$		168		-1260
$(^2_3H)\,^1I$	$E(\mathrm{d}^7,\,^2_3H)$	-210	-1260	5145
$(^2_3H)\,^3I$	$E(d^7,\,^2_3H)$	-210	-1260	945

49. Closed shells plus f^7s [2]

$$E_{\mathrm{Av}} = E(cs) + 7E(cs, nf) + E(cs, n's) + 7I(nf) + I(n's) + 21F^0(nf, nf) - \frac{84}{195}F^2(nf, nf) - \frac{42}{143}F^4(nf, nf) - \frac{2100}{5577}F^6(nf, nf) + 7F^0(nf, n's) - \frac{7}{14}G^3(nf, n's)$$

$$({}^8_7S)\ {}^7_7S = E(f^7, {}^8_7S) + \frac{9}{14}G^3(nf, n's)$$

$$({}^8_7S)\ {}^9_7S = E(f^7, {}^8_7S) - \frac{1}{2}G^3(nf, n's)$$

$$({}^2_\nu L)\ {}^1_\nu L = E(f^7, {}^2_\nu L) + \frac{3}{14}G^3(nf, n's) \quad \text{for } L = 0, 1, 2, 3, 4, 5, 6, 7, 8, 9, 10, 11, 12$$

$$({}^2_\nu L)\ {}^3_\nu L = E(f^7, {}^2_\nu L) - \frac{1}{14}G^3(nf, n's) \quad \text{for } L = 0, 1, 2, 3, 4, 5, 6, 7, 8, 9, 10, 11, 12$$

$$({}^4_\nu L)\ {}^3_\nu L = E(f^7, {}^4_\nu L) + \frac{5}{14}G^3(nf, n's) \quad \text{for } L = 0, 1, 2, 3, 4, 5, 6, 7, 8, 9, 10$$

$$({}^4_\nu L)\ {}^5_\nu L = E(f^7, {}^4_\nu L) - \frac{3}{14}G^3(nf, n's) \quad \text{for } L = 0, 1, 2, 3, 4, 5, 6, 7, 8, 9, 10$$

$$({}^6_\nu L)\ {}^5_\nu L = E(f^7, {}^6_\nu L) + \frac{1}{2}G^3(nf, n's) \quad \text{for } L = 1, 2, 3, 4, 5, 6$$

$$({}^6_\nu L)\ {}^7_\nu L = E(f^7, {}^6_\nu L) - \frac{5}{14}G^3(nf, n's) \quad \text{for } L = 1, 2, 3, 4, 5, 6$$

$$({}^{2S+1}_{\nu}L_n)\ {}^{2S}_{\nu}L_n{:}({}^{2S+1}_{\nu'}L_{n'})\ {}^{2S}_{\nu'}L_{n'} = ({}^{2S+1}_{\nu}L_n)\ {}^{2S+2}_{\nu}L_n{:}({}^{2S+1}_{\nu'}L_{n'})\ {}^{2S+2}_{\nu'}L_{n'} = {}^{2S+1}_{\nu}L_n{:}{}^{2S+1}_{\nu'}L_{n'}$$

50. Closed shells plus d^8p [2, 12]

$$E_{\mathrm{av}} = E(cs) + 8E(cs, nd) + E(cs, n'p) + 8I(nd) + I(n'p) + 28F^0(nd, nd) - \frac{56}{63}F^2(nd, nd) - \frac{56}{63}F^4(nd, nd) + 8F^0(nd, n'p) - \frac{8}{15}G^1(nd, n'p) - \frac{24}{70}G^3(nd, n'p)$$

	$E(d^8, {}^{2s+1}_{\ v}L)$	$F^2(nd, n'p)/1225$	$G^1(nd, n'p)/1050$	$G^3(nd, n'p)/17150$
$({}^3P)\,{}^2S$	$E(d^8, {}^3P)$	490	−140	−1470
$({}^3P)\,{}^4S$	$E(d^8, {}^3P)$	490	−140	−1470
$({}^1S)\,{}^2P$	$E(d^8, {}^1S)$			
$({}^3P)\,{}^2P$	$E(d^8, {}^3P)$	−245	805	2940
$({}^1D)\,{}^2P$	$E(d^8, {}^1D)$	105	105	−1260
$({}^1S)\,{}^2P{:}({}^3P)\,{}^2P$			$210\sqrt{3}$	$-1470\sqrt{3}$
$({}^1S)\,{}^2P{:}({}^1D)\,{}^2P$		$-140\sqrt{7}$	$70\sqrt{7}$	$210\sqrt{7}$
$({}^3P)\,{}^2P{:}({}^1D)\,{}^2P$			$105\sqrt{21}$	$-210\sqrt{21}$
$({}^3P)\,{}^4P$	$E(d^8, {}^3P)$	−245	−140	−1470
$({}^3P)\,{}^2D$	$E(d^8, {}^3P)$	49	49	3822
$({}^1D)\,{}^2D$	$E(d^8, {}^1D)$	−105	105	−210
$({}^3F)\,{}^2D$	$E(d^8, {}^3F)$	−84	1036	−1092
$({}^3P)\,{}^2D{:}({}^1D)\,{}^2D$			$21\sqrt{105}$	$-252\sqrt{105}$
$({}^3P)\,{}^2D{:}({}^3F)\,{}^2D$		$-84\sqrt{14}$	$126\sqrt{14}$	$378\sqrt{14}$
$({}^1D)\,{}^2D{:}({}^3F)\,{}^2D$			$98\sqrt{30}$	$-126\sqrt{30}$
$({}^3P)\,{}^4D$	$E(d^8, {}^3P)$	49	−140	−1470
$({}^3F)\,{}^4D$	$E(d^8, {}^3F)$	−84	−140	−1470
$({}^3P)\,{}^4D{:}({}^3F)\,{}^4D$		$-84\sqrt{14}$		
$({}^1D)\,{}^2F$	$E(d^8, {}^1D)$	30	−120	690
$({}^3F)\,{}^2F$	$E(d^8, {}^3F)$	105	280	1365
$({}^1G)\,{}^2F$	$E(d^8, {}^1G)$	−275	400	−1425
$({}^1D)\,{}^2F{:}({}^3F)\,{}^2F$			$20\sqrt{21}$	$-540\sqrt{21}$
$({}^1D)\,{}^2F{:}({}^1G)\,{}^2F$		$-120\sqrt{3}$	$60\sqrt{3}$	$180\sqrt{3}$
$({}^3F)\,{}^2F{:}({}^1G)\,{}^2F$			$180\sqrt{7}$	$-135\sqrt{7}$
$({}^3F)\,{}^4F$	$E(d^8, {}^3F)$	105	−140	−1470
$({}^3F)\,{}^2G$	$E(d^8, {}^3F)$	−35	−140	6405
$({}^1G)\,{}^2G$	$E(d^8, {}^1G)$	385	−140	−945
$({}^3F)\,{}^2G{:}({}^1G)\,{}^2G$				$-525\sqrt{15}$
$({}^3F)\,{}^4G$	$E(d^8, {}^3F)$	−35	−140	−1470
$({}^1G)\,{}^2H$	$E(d^8, {}^1G)$	−140	−140	1680

51. Closed shells plus d^8d'

$$E_{\mathrm{Av}} = E(cs) + 8E(cs, nd) + E(cs, n'd) + 8I(nd) + I(n'd)$$
$$+ 28F^0(nd, nd) - \frac{56}{63}F^2(nd, nd) - \frac{56}{63}F^4(nd, nd) + 8F^0(nd, n'd)$$
$$- \frac{8}{10}G^0(nd, n'd) - \frac{8}{35}G^2(nd, n'd) - \frac{8}{35}G^4(nd, n'd)$$

	$E(d^8, {}^{2s+1}L)$	$F^2(d, d')/1715$	$F^4(d, d')/3087$	$G^0(d, d')/70$	$G^2(d, d')/3430$	$G^4(d, d')/30870$
$({}^1D)\,{}^2S$	$E(d^8, {}^1D)$	210	−504	−14	784	−1764
$({}^3P)\,{}^2P$	$E(d^8, {}^3P)$	343		−14	833	−1764
$({}^1D)\,{}^2P$	$E(d^8, {}^1D)$	105	336	−14	49	−1764
$({}^3F)\,{}^2P$	$E(d^8, {}^3F)$	−168	462	−14	980	−1764
$({}^3P)\,{}^2P{:}({}^1D)\,{}^2P$					$49\sqrt{105}$	
$({}^3P)\,{}^2P{:}({}^3F)\,{}^2P$		$-28\sqrt{14}$	$126\sqrt{14}$		$-294\sqrt{14}$	
$({}^1D)\,{}^2P{:}({}^3F)\,{}^2P$					$-98\sqrt{30}$	
$({}^3P)\,{}^4P$	$E(d^8, {}^3P)$	343		−14	−196	−1764
$({}^3F)\,{}^4P$	$E(d^8, {}^3F)$	−168	462	−14	−196	−1764
$({}^3P)\,{}^4P{:}({}^3F)\,{}^4P$		$-28\sqrt{14}$	$126\sqrt{14}$			
$({}^1S)\,{}^2D$	$E(d^8, {}^1S)$					
$({}^3P)\,{}^2D$	$E(d^8, {}^3P)$	−343		112	245	5292
$({}^1D)\,{}^2D$	$E(d^8, {}^1D)$	−45	−144	56	−151	−1044
$({}^3F)\,{}^2D$	$E(d^8, {}^3F)$	−42	−693	280	1148	−1575
$({}^1G)\,{}^2D$	$E(d^8, {}^1G)$	−550	−143	112	−52	−1763
$({}^1S)\,{}^2D{:}({}^3P)\,{}^2D$				42	294	−3528
$({}^1S)\,{}^2D{:}({}^1D)\,{}^2D$		$-196\sqrt{5}$		$14\sqrt{5}$	$-42\sqrt{5}$	$504\sqrt{5}$
$({}^1S)\,{}^2D{:}({}^3F)\,{}^2D$				$14\sqrt{21}$	$-112\sqrt{21}$	$-126\sqrt{21}$
$({}^1S)\,{}^2D{:}({}^1G)\,{}^2D$			−588	42	168	42
$({}^3P)\,{}^2D{:}({}^1D)\,{}^2D$				$42\sqrt{5}$	$-63\sqrt{5}$	$-1008\sqrt{5}$
$({}^3P)\,{}^2D{:}({}^3F)\,{}^2D$		$-56\sqrt{21}$	$-42\sqrt{21}$	$42\sqrt{21}$	$-168\sqrt{21}$	$252\sqrt{21}$
$({}^3P)\,{}^2D{:}({}^1G)\,{}^2D$				126	252	−84
$({}^1D)\,{}^2D{:}({}^3F)\,{}^2D$				$14\sqrt{105}$	$24\sqrt{105}$	$-36\sqrt{105}$
$({}^1D)\,{}^2D{:}({}^1G)\,{}^2D$		$-48\sqrt{5}$	$-330\sqrt{5}$	$42\sqrt{5}$	$-36\sqrt{5}$	$12\sqrt{5}$
$({}^3F)\,{}^2D{:}({}^1G)\,{}^2D$				$42\sqrt{21}$	$-96\sqrt{21}$	$-3\sqrt{21}$
$({}^3P)\,{}^4D$	$E(d^8, {}^3P)$	−343		−14	−196	−1764
$({}^3F)\,{}^4D$	$E(d^8, {}^3F)$	−42	−693	−14	−196	−1764
$({}^3P)\,{}^4D{:}({}^3F)\,{}^4D$		$-56\sqrt{21}$	$-42\sqrt{21}$			
$({}^3P)\,{}^2F$	$E(d^8, {}^3P)$	98		−14	308	4535
$({}^1D)\,{}^2F$	$E(d^8, {}^1D)$	−120	36	−14	124	486
$({}^3F)\,{}^2F$	$E(d^8, {}^3F)$	77	462	−14	−70	−189
$({}^1G)\,{}^2F$	$E(d^8, {}^1G)$	−55	286	−14	254	−1739
$({}^3P)\,{}^2F{:}({}^1D)\,{}^2F$					$48\sqrt{70}$	$-450\sqrt{70}$
$({}^3P)\,{}^2F{:}({}^3F)\,{}^2F$		−336	42		252	3150
$({}^3P)\,{}^2F{:}({}^1G)\,{}^2F$					$-180\sqrt{7}$	$-150\sqrt{7}$
$({}^1D)\,{}^2F{:}({}^3F)\,{}^2F$					$24\sqrt{70}$	$-225\sqrt{70}$
$({}^1D)\,{}^2F{:}({}^1G)\,{}^2F$		$-60\sqrt{10}$	$165\sqrt{10}$		$-120\sqrt{10}$	$75\sqrt{10}$
$({}^3F)\,{}^2F{:}({}^1G)\,{}^2F$					$-90\sqrt{7}$	$-75\sqrt{7}$
$({}^3P)\,{}^4F$	$E(d^8, {}^3P)$	98		−14	−196	−1764
$({}^3F)\,{}^4F$	$E(d^8, {}^3F)$	77	462	−14	−196	−1764
$({}^3P)\,{}^4F{:}({}^3F)\,{}^4F$		−336	42			
$({}^1D)\,{}^2G$	$E(d^8, {}^1D)$	60	−4	−14	−116	986
$({}^3F)\,{}^2G$	$E(d^8, {}^3F)$	105	−154	−14	854	4011
$({}^1G)\,{}^2G$	$E(d^8, {}^1G)$	325	−234	−14	354	−1539
$({}^1D)\,{}^2G{:}({}^3F)\,{}^2G$					$20\sqrt{210}$	$-275\sqrt{210}$
$({}^1D)\,{}^2G{:}({}^1G)\,{}^2G$		$-20\sqrt{110}$	$-15\sqrt{110}$		$20\sqrt{110}$	$75\sqrt{110}$
$({}^3F)\,{}^2G{:}({}^1G)\,{}^2G$					$50\sqrt{231}$	$-75\sqrt{231}$
$({}^3F)\,{}^4G$	$E(d^8, {}^3F)$	105	−154	−14	−196	−1764
$({}^3F)\,{}^2H$	$E(d^8, {}^3F)$	−70	21	−14	−196	9261
$({}^1G)\,{}^2H$	$E(d^8, {}^1G)$	350	91	−14	−196	−539
$({}^3F)\,{}^2H{:}({}^1G)\,{}^2H$						−3675
$({}^3F)\,{}^4H$	$E(d^8, {}^3F)$	−70	21	−14	−196	−1764
$({}^1G)\,{}^2I$	$E(d^8, {}^1G)$	−280	−14	−14	−196	3136

52. Closed shells plus d^9p [2, 4, 8]

$$E_{\mathrm{Av}} = E(cs) + 9E(cs, nd) + E(cs, n'p) + 9I(nd) + I(n'p) + 36F^0(nd, nd) - \frac{72}{63}F^2(nd, nd) - \frac{72}{63}F^4(nd, nd) + 9F^0(nd, n'p) - \frac{9}{15}G^1(nd, n'p) - \frac{27}{70}G^3(nd, n'p)$$

The Russell-Saunders term energies are the same as of the p^5d configuration.

53. Closed shells plus d^9d' [2, 4, 8]

$$E_{\mathrm{Av}} = E(cs) + 9E(cs, nd) + E(cs, n'd) + 9I(nd) + I(n'd) + 36F^0(nd, nd) - \frac{72}{63}F^2(nd, nd) - \frac{72}{63}F^4(nd, nd) + 9F^0(nd, n'd) - \frac{9}{10}G^0(nd, n'd) - \frac{9}{35}G^2(nd, n'd) - \frac{9}{35}G^4(nd, n'd)$$

	$F^2(d, d')/49$	$F^4(d, d')/441$	$G^0(d, d')/10$	$G^2(d, d')/35$	$G^4(d, d')/315$
^1S	−14	−126	99	−1	−9
^3S	−14	−126	−1	−1	−9
^1P	−7	84	−1	−1	−9
^3P	−7	84	−1	−1	−9
^1D	3	−36	−1	19	−9
^3D	3	−36	−1	−1	−9
^1F	8	9	−1	−1	−9
^3F	8	9	−1	−1	−9
^1G	−4	−1	−1	−1	91
^3G	−4	−1	−1	−1	−9

54. Closed shells plus d^9f [4]

$$E_{\mathrm{Av}} = E(cs) + 9E(cs, nd) + E(cs, n'f) + 9I(nd) + I(n'f) + 36F^0(nd, nd) - \frac{72}{63}F^2(nd, nd) - \frac{72}{63}F^4(nd, nd) + 9F^0(nd, n'f) - \frac{27}{70}G^1(nd, n'f) - \frac{18}{105}G^3(nd, n'f) - \frac{45}{231}G^5(nd, n'f)$$

	$F^2(f,d)/105$	$F^4(f,d)/693$	$G^1(f,d)/70$	$G^3(f,d)/105$	$G^5(f,d)/2541$
^1P	−24	−66	137	−2	−55
^3P	−24	−66	−3	−2	−55
^1D	−6	99	−3	−2	−55
^3D	−6	99	−3	−2	−55
^1F	11	−66	−3	38	−55
^3F	11	−66	−3	−2	−55
^1G	15	22	−3	−2	−55
^3G	15	22	−3	−2	−55
^1H	−10	−3	−3	−2	645
^3H	−10	−3	−3	−2	−55

55. Closed shells plus $f^{12}p$

$$E_{\mathrm{Av}} = E(cs) + 12E(cs, nf) + E(cs, n'p) + 12I(nf) + I(n'p)$$
$$+ 66F^0(nf, nf) - \frac{264}{195}F^2(nf, nf) - \frac{132}{143}F^4(nf, nf) - \frac{6600}{5577}F^6(nf, nf)$$
$$+ 12F^0(nf, n'p) - \frac{36}{70}G^2(nf, n'p) - \frac{24}{63}G^4(nf, n'p)$$

	$E(f^{12}, {}^{2s+1}L)$	$F^2(f,p)/17325$	$G^2(f,p)/26950$	$G^4(f,p)/87318$
$(^3P)\,^2S$	$E(f^{12}, {}^3P)$	6930	−2310	−5544
$(^3P)\,^4S$	$E(f^{12}, {}^3P)$	6930	−2310	−5544
$(^1S)\,^2P$	$E(f^{12}, {}^1S)$			
$(^3P)\,^2P$	$E(f^{12}, {}^3P)$	−3465	11550	13167
$(^1D)\,^2P$	$E(f^{12}, {}^1D)$	2541	462	−4389
$(^1S)\,^2P{:}(^3P)\,^2P$			$2310\sqrt{6}$	$-4158\sqrt{6}$
$(^1S)\,^2P{:}(^1D)\,^2P$		$-924\sqrt{30}$	$462\sqrt{30}$	$462\sqrt{30}$
$(^3P)\,^2P{:}(^1D)\,^2P$			$2772\sqrt{5}$	$-2079\sqrt{5}$
$(^3P)\,^4P$	$E(f^{12}, {}^3P)$	−3465	−2310	−5544
$(^3P)\,^2D$	$E(f^{12}, {}^3P)$	693	1848	13167
$(^1D)\,^2D$	$E(f^{12}, {}^1D)$	−2541	1848	231
$(^3F)\,^2D$	$E(f^{12}, {}^3F)$	924	10164	−2772
$(^3P)\,^2D{:}(^1D)\,^2D$			4158	−10395
$(^3P)\,^2D{:}(^3F)\,^2D$		$-2772\sqrt{3}$	$4158\sqrt{3}$	$4158\sqrt{3}$
$(^1D)\,^2D{:}(^3F)\,^2D$			$4158\sqrt{3}$	$-2310\sqrt{3}$
$(^3P)\,^4D$	$E(f^{12}, {}^3P)$	693	−2310	−5544
$(^3F)\,^4D$	$E(f^{12}, {}^3F)$	924	−2310	−5544
$(^3P)\,^4D{:}(^3F)\,^4D$		$-2772\sqrt{3}$		
$(^1D)\,^2F$	$E(f^{12}, {}^1D)$	726	−1518	1716
$(^3F)\,^2F$	$E(f^{12}, {}^3F)$	−1155	8085	9702
$(^1G)\,^2F$	$E(f^{12}, {}^1G)$	−495	3135	−4950
$(^1D)\,^2F{:}(^3F)\,^2F$			$198\sqrt{210}$	$-726\sqrt{210}$
$(^1D)\,^2F{:}(^1G)\,^2F$		$-396\sqrt{110}$	$198\sqrt{110}$	$198\sqrt{110}$
$(^3F)\,^2F{:}(^1G)\,^2F$			$495\sqrt{231}$	$-198\sqrt{231}$

	$E(f^{12}, {}^{2s+1}L)$	$F^2(f,p)/17325$	$G^2(f,p)/26950$	$G^4(f,p)/87318$
$({}^3F)\,{}^4F$	$E(f^{12}, {}^3F)$	-1155	-2310	-5544
$({}^3F)\,{}^2G$	$E(f^{12}, {}^3F)$	385	-1155	19866
$({}^1G)\,{}^2G$	$E(f^{12}, {}^1G)$	693	231	-1386
$({}^3H)\,{}^2G$	$E(f^{12}, {}^3H)$	-2002	18018	-4620
$({}^3F)\,{}^2G{:}({}^1G)\,{}^2G$			$231\sqrt{55}$	$-1386\sqrt{55}$
$({}^3F)\,{}^2G{:}({}^3H)\,{}^2G$		$-308\sqrt{110}$	$462\sqrt{110}$	$462\sqrt{110}$
$({}^1G)\,{}^2G{:}({}^3H)\,{}^2G$			$5082\sqrt{2}$	$-1386\sqrt{2}$
$({}^3F)\,{}^4G$	$E(f^{12}, {}^3F)$	385	-2310	-5544
$({}^3H)\,{}^4G$	$E(f^{12}, {}^3H)$	-2002	-2310	-5544
$({}^3F)\,{}^4G{:}({}^3H)\,{}^4G$		$-308\sqrt{110}$		
$({}^1G)\,{}^2H$	$E(f^{12}, {}^1G)$	-252	-2184	4284
$({}^3H)\,{}^2H$	$E(f^{12}, {}^3H)$	3003	1848	3465
$({}^1I)\,{}^2H$	$E(f^{12}, {}^1I)$	-3675	5880	-5439
$({}^1G)\,{}^2H{:}({}^3H)\,{}^2H$			$126\sqrt{33}$	$-1638\sqrt{33}$
$({}^1G)\,{}^2H{:}({}^1I)\,{}^2H$		$252\sqrt{65}$	$126\sqrt{65}$	$126\sqrt{65}$
$({}^3H)\,{}^2H{:}({}^1I)\,{}^2H$			$126\sqrt{2145}$	$-21\sqrt{2145}$
$({}^3H)\,{}^4H$	$E(f^{12}, {}^3H)$	3003	-2310	-5544
$({}^3H)\,{}^2I$	$E(f^{12}, {}^3H)$	-1155	-2310	28413
$({}^1I)\,{}^2I$	$E(f^{12}, {}^1I)$	5775	-2310	-3927
$({}^3H)\,{}^2I{:}({}^1I)\,{}^2I$				$-1617\sqrt{21}$
$({}^3H)\,{}^4I$	$E(f^{12}, {}^3H)$	-1155	-2310	-5544
$({}^1I)\,{}^2K$	$E(f^{12}, {}^1I)$	-2310	-2310	7392

56. Closed shells plus $f^{13}p$ [4]

$$\begin{aligned} E_{\mathrm{Av}} = {} & E(cs) + 13E(cs, nf) + E(cs, n'p) + 13I(nf) + I(n'p) \\ & + 78F^0(nf, nf) - \frac{312}{195}F^2(nf, nf) - \frac{156}{143}F^4(nf, nf) - \frac{7800}{5577}F^6(nf, nf) \\ & + 13F^0(nf, n'p) - \frac{39}{70}G^2(nf, n'p) - \frac{26}{63}G^4(nf, n'p) \end{aligned}$$

The Russell-Saunders term energies are the same as of the p^5f configuration.

57. Closed shells plus $f^{13}d$ [4]

$$\begin{aligned} E_{\mathrm{Av}} = {} & E(cs) + 13E(cs, nf) + E(cs, n'd) + 13I(nf) + I(n'd) \\ & + 78F^0(nf, nf) - \frac{312}{195}F^2(nf, nf) - \frac{156}{143}F^4(nf, nf) - \frac{7800}{5577}F^6(nf, nf) \\ & + 13F^0(nf, n'd) - \frac{39}{70}G^1(nf, n'd) - \frac{26}{105}G^3(nf, n'd) - \frac{65}{231}G^5(nf, n'd) \end{aligned}$$

The Russell-Saunders term energies are the same as of the d^9f configuration.

58. Closed shells plus $f^{13}f'$

$$E_{\text{Av}} = E(cs) + 13E(cs, nf) + E(cs, n'f) + 13I(nf) + I(n'f) + 78F^0(nf, nf) - \frac{312}{195}F^2(nf, nf) - \frac{156}{143}F^4(nf, nf) - \frac{7800}{5577}F^6(nf, nf) + 13F^0(nf, n'f) - \frac{13}{14}G^0(nf, n'f) - \frac{26}{105}G^2(nf, n'f) - \frac{13}{77}G^4(nf, n'f) - \frac{650}{3003}G^6(nf, n'f)$$

	$F^2(f,f')/225$	$F^4(f,f')/1089$	$F^6(f,f')/184041$	$G^0(f,f')/14$	$G^2(f,f')/525$	$G^4(f,f')/693$	$G^6(f,f')/39039$
^1S	−60	−198	−42900	195	−10	−9	−650
^3S	−60	−198	−42900	−1	−10	−9	−650
^1P	−45	−33	32175	−1	−10	−9	−650
^3P	−45	−33	32175	−1	−10	−9	−650
^1D	−19	99	−17875	−1	382	−9	−650
^3D	−19	99	−17875	−1	−10	−9	−650
^1F	10	33	7150	−1	−10	−9	−650
^3F	10	33	7150	−1	−10	−9	−650
^1G	30	−97	−1950	−1	−10	187	−650
^3G	30	−97	−1950	−1	−10	−9	−650
^1H	25	51	325	−1	−10	−9	−650
^3H	25	51	325	−1	−10	−9	−650
^1I	−25	−9	−25	−1	−10	−9	9150
^3I	−25	−9	−25	−1	−10	−9	−650

References

Chapter 1. Pre-quantum mechanical developments

1 I. Newton, *Mathematical Principles of Natural Philosophy* (Cambridge, 1686), preface to the first edition.

2 (a) J. R. Partington, *A Short History of Chemistry* (London: Macmillan, 1937); also (New York: Harper & Row, 1960), pp. 167–74.
(b) D. S. L. Cardwell, ed., *John Dalton and the Progress of Science* (Manchester: Manchester University Press, 1968).
(c) Arnold W. Thackray, "The Emergence of Dalton's Chemical Atomic Theory, 1801–08," *Brit. J. Hist. Sci.* 3: 1 (1966).

3 Partington, *A Short History of Chemistry,* pp. 208–10, 256–8.

4 Hermann von Helmholtz, *J. Chem. Soc.* 39: 277 (1881).

5 S. G. Brush, "The Development of the Kinetic Theory of Gases," *Ann. Sci.* 13: 188, 273 (1957); 14: 185, 243 (1958); *Am. J. Phys.* 29: 593 (1961); 30: 269 (1962). See also his *Translator's Introduction* to his translation of L. Boltzmann, *Lectures on Gas Theory,* (Berkeley: University of California Press, 1964).

6 (a) Partington, *A Short History of Chemistry,* Chap. XV.
(b) F. P. Venable, *The Development of the Periodic Law,* (Easton, Pa.: Chemical Publishing Co., 1896).
(c) J. D. Main-Smith, *Chemistry and Atomic Structure,* (London: Ernest Benn, 1924).
(d) D. Mendeleev, *J. Russ. Phys.-Chem. Soc.* 1: 60 (1869); *Z. Chem.* 5: 405 (1869).
(e) K. Seulurt, ed., "Das natürliche System der Chemischen Elemente," in Ostwald, *Klassiker der exacten Naturwissenschaften* Vol. 68 (Leipzig: 1895).

7 (a) K. Lonsdale, *Crystals and X-rays,*" (New York: Van Nostrand, 1949), Chap. 1.
(b) W. C. Röntgen, *Sitzungsberichte der physikalisch-medizinischen Gesellschaft zu Würzburg* (Dec. 1895). [First announcement.]

8 (a) H. Becquerel, *Compt. Rend.* 122: 501 (1896).
(b) Pierre and Marie Curie, *Compt. Rend.* 127: 175 (1898) [polonium]; *Compt. Rend.* 127: 1215 (1898) [radium].
(c) Alfred Romer, ed., *The Discovery of Radioactivity and Transmutation,* (New York: Dover, 1964). [Reprints of basic papers.]
(d) Marie Curie, *Pierre Curie,* (New York: Dover, 1963).
(e) Eve Curie, *Madame Curie,* (New York: Doubleday, Doran, 1939).

9 David L. Anderson, *The Discovery of the Electron,* (Princeton, N.J.: Van Nostrand Reinhold, 1963).

10 (a) E. U. Condon, *Physics Today* 15: 37 (1962).
(b) G. N. Lewis, *Nature* 118: 874 (1926) [photon].
11 (a) J. B. Birks, ed., *Rutherford at Manchester,* (London, Heywood, 1962).
(b) E. Rutherford, *Phil. Mag.* 21: 669 (1911); 27: 488 (1914); *Nature* 92, 423 (1913); *Phil. Mag.* 26: 702 (1913).
12 (a) J. J. Thomson, *Proc. Roy. Soc. (London)* A89: 1 (1913).
(b) K. Fajans, *Physik. Z.* 14: 131, 136 (1913).
(c) F. Soddy, *Chem. News* 107: 97 (1913).
13 (a) C. G. Darwin, "The Discovery of Atomic Numbers," in *Niels Bohr and the Development of Physics,* (New York: McGraw-Hill, 1955).
(b) W. V. Farrar, "Nineteenth Century Speculations on the Complexity of the Chemical Elements," *Brit. J. Hist. Sci.* 2: 297 (1964).
14 (a) James Chadwick, *Proc. Roy. Soc. (London)* A136: 692 (1932).
(b) W. Heisenberg, *Z. Physik* 77: 1 (1932).
(c) D. Iwanenko, *Nature* 129: 798 (1932).
15 International Union of Pure and Applied Chemistry recommendation on notation.
16 (a) A. Einstein, *Ann. Physik* 17: 132 (1905).
(b) K. T. Bainbridge, *Phys. Rev.* 44: 123 (1933).
(c) E. B. Jordan and K. T. Bainbridge, *Phys. Rev.* 50: 98 (1936).
17 (a) E. Fermi, *Z. Physik.* 88: 161 (1934).
(b) C. S. Wu, "The Neutrino," in *Theoretical Physics in the Twentieth Century* (New York: Wiley, 1960), Pauli memorial volume.
(c) B. W. Ridley, "The Neutrino," *Prog. Nuc. Phys.* 5: 188 (1956).
(d) P. M. S. Blackett, *Proc. Roy. Soc.* (*London*) A251: 293 (1959).
18 *Phil. Trans. Roy. Soc. London,* no. 80, 19 February 1671/72; reproduced in *Isaac Newton's Papers and Letters on Natural Philosophy,* I. Bernard Cohen, ed. (Cambridge, Mass.: Harvard University Press, 1958).
19 (a) Ernst Mach, *The Principles of Physical Optics,* (London: Methuen, 1926).
(b) E. T. Whittaker, *A History of the Theories of Aether and Electricity* (London: Thomas Nelson, 1951, 1953), 2 vols.
(c) Thomas Young, *Phil. Trans. Roy. Soc. London* 92: 12, 387 (1802). [Basic paper on interference.]
(d) Herbert Dingle, "A Hundred Years of Spectroscopy," *Brit. Jour. Hist. Sci.* 1: 199 (1963).
20 (a) W. H. Wollaston, *Phil. Trans. Roy. Soc. London* 92: 365 (1802).
(b) Joseph von Fraunhofer, *Ann. Physik* 56, (1817). Later diffraction grating measurements of wavelengths in *Denkschr. Akad. München* 8: 1 (1821–22).
(c) M. von Rohr, *Trans. Opt. Soc. London* 27: 272 (1925). [Biography of Fraunhofer.]
21 W. Herschel, *Phil. Trans. Roy. Soc. London* 255: 284, 293, 437 (1800).
22 J. W. Ritter, *Ann. Physik* 7: 527 (1801); 12: 409 (1803).
23 W. Herschel, *Edinburgh Trans.* 9: 445 (1923); *On the Theory of Light* (London: 1828).
24 (a) D. Brewster, *Phil. Mag.* (3) 2: 360 (1833); 8: 384 (1836); *Edinburgh Trans.* 12: 519 (1834).

(b) C. Wheatstone, *Phil. Mag.* 7: 299 (1835); *Chem. News* 3: 198 (1861).
(c) H. F. Talbot, *Brewsters J. Sci.* 77: (1825); *Phil. Mag.* 4: 122 (1834); 7: 113 (1835); 9: 1 (1836).
25 D. Brewster and J. H. Gladstone, *Phil. Trans. Roy. Soc. London* 150: 149 (1860).
26 (a) E. Becquerel, *Bibl. Univ. de Genève,* 40: 341 (1842).
(b) J. W. Draper, *Phil. Mag.* (3) 21: 348 (1842); 22: 360 (1843).
27 (a) Léon Foucault, *L'Institut,* p. 45 (1849), reprinted in *Ann. Chim. Phys.* (3) 48: 476 (1960).
(b) J. Plücker, *Ann. Physik* 103: 88; 104: 113; 105: 67; 107: 497, 638 (1858, 1859).
(c) F. Klein, *Vorlesungen über die Entwicklung der Mathematik im 19 Jahrhundert* (Berlin: Springer, 1926), vol. 1, pp. 119–21. Reprinted by Chelsea, New York, 1956.
28 A. J. Ångstrom, *Öf Svensk Vet. Akad.,* p. 229 (1852); *Ann. Physik,* 94: 141 (1855).
29 Balfour Stewart, *Trans. Roy. Soc. Edinburgh* 22: 1 (1858). Comments by Lord Rayleigh, *Phil. Mag.* 1: 98 (1901).
30 (a) G. Kirchhoff, *Ber. Berlin Akad.,* pp. 662, 783 (1859); *Phil. Mag.* 19: 193 (1860); *Ann. Physik* 109: 148, 275 (1860).
(b) G. Kirchhoff and R. Bunsen, *Ann. Physik* 110: 160 (1860); 113: 337 (1861). [The Kirchhoff papers are collected in Ostwald, *Klassiker der exakten Naturwissenschaften,* Vol. 100 (1898).]
31 W. N. Hartley, *J. Chem. Soc. (London)* 43: 390 (1883).
32 G. D. Liveing and James Dewar, *Proc. Roy. Soc. (London)* 29: 398 (1879).
33 J. J. Balmer, *Basel Verhandl.* 7: 548, 750 (1885); *Ann. Physik* 25: 80 (1885).
34 J. R. Rydberg, *Koningl. Svenska Vetensk. Akad. Handl. (Stockholm)* 23: No. 11 (1890); *Phil. Mag.* 29: 331 (1890); *Ann. Physik* 58: 674 (1896); *Astrophys. J.* 4: 91 (1896). Historical remarks by N. Bohr and W. Pauli, *Proceedings of the Rydberg Centennial Conference on Atomic Spectroscopy* (Lund: C. W. K. Gleerup, 1954).
35 H. Kayser and C. Runge, *Abhandl. Berlin Akad.* (1890); *Ann. Physik* 41: 302 (1890).
36 (a) J. R. Rydberg, *Rapports presentes au congresse internationale,* Vol. 2 (Paris: 1900), p. 200.
(b) W. Ritz, *Physik. Z.* 9: 521 (1908); *Gesammelte Werke* (Paris: Gauthier-Villars, 1911), p. 141; *Astrophys. J* 28: 237 (1908).
37 P. Zeeman, *Phil. Mag.* 43: 226 (1896); 44: 55, 255 (1897); 45: 197 (1898).
38 H. A. Lorentz, *Phil. Mag.* 43: 232 (1897); *Ann. Physik* 63: 278 (1897).
39 (a) Thomas Preston, *Trans. Roy. Soc. Dublin* 6: 385 (1898).
(b) A. Cornu, *Compt. Rend.* 126: 181 (1898).
(c) C. Runge and F. Paschen, *Physik. Z.* 1: 480 (1900).
40 Max Planck, *Ann. Physik* 4: 553 (1901).
41 Max Planck, *Arch. Neerl.* 6: 55–66 (1902).
42 S. H. Burbury, *Phil. Mag.* 3: 225–40 (1902).
43 J. D. Van der Waals, Jr., *Akad. Wetenschap. Amsterdam* 11: 79–88, 243–50 (1902).

44 Max Planck, *Scientific Autobiography and Other Papers* (New York: Philosophical Library, 1949), pp. 34–43.
45 E. B. Wilson, *Sci. Monthly* 32: 211 (1931); L. P. Wheeler, *Josiah Willard Gibbs,* rev. ed. (New Haven: Yale University Press, 1952), p. 194.
46 J. W. Gibbs, *Elementary Principles in Statistical Mechanics* (New Haven: Yale University Press, 1902); also in *Collected Works* Vol. 2 (New York: Longmans, Green, 1928); reprinted by Dover Publications, New York.
47 (Lord) Rayleigh, *Phil. Mag.* 49: 539 (1900).
48 (Lord) Rayleigh, *Phil. Mag.* 11: 117 (1906).
49 (Lord) Rayleigh, *Phil. Mag.* 44: 362 (1897).
50 Sir G. G. Stokes, *Phil. Trans. Roy. Soc. London* 153: 463 (1852).
51 A. L. Hughes and L. A. DuBridge, *Photoelectric Phenomena* (New York: McGraw-Hill, 1932), pp. 15–26.
52 J. J. Thomson, *Conduction of Electricity Through Gases* (Cambridge: Cambridge University Press, 1904), p. 258.
53 J. J. Thomson, *Electricity and Matter,* (New Haven: Yale University Press, 1904), pp. 63–5.
54 A. Einstein, *Ann. Physik* 22: 180, 800 (1907).
55 (a) P. L. Dulong and A. T. Petit, *Ann. Chim. Phys.* 10: 395 (1819); *Phil. Mag.* 54: 267 (1819).
(b) Robert Fox, "The Background to the Discovery of Dulong and Petit's Law," *Brit. J. Hist. Sci.* 4: 1 (1968).
56 H. F. Weber, *Ann. Physik* 154: 575 (1875).
57 Lothar Meyer, *Moderne Theorien der Warme* (1884), p. 167.
58 F. E. Neumann, *Ann. Physik* 23: 32 (1831).
59 J. P. Joule, *Phil. Mag.* (3) 25: 334 (1844).
60 H. Kopp, *Liebigs Ann. Chem.* (Specialband) 3: 289 (1864–5).
61 A. Winkelmann, *Ann. Physik* 49: 401 (1893).
62 L. Boltzmann, *Sitzber. Koningl. Akad. Wiss. Wien* (2)63: 679 (1871).
63 Sir James Dewar, *Proc. Roy. Soc. (London)* A74: 122 (1904).
64 Peter Debye, *Ann. Physik* 39: 789 (1912).
65 M. Born and T. von Karman, *Physik. Z.* 13: 297 (1912); 14: 15, 65 (1913).
66 A. Eucken, *Sitzber. Preuss. Akad. Wiss.,* 141 (1912).
67 (a) A. Einstein, *Ann. Physik* 40: 551 (1913).
(b) O. Stern and P. Ehrenfest, *Collected Scientific Papers,* p. 333, [originally in *Verhandl. Deut. Phys. Ges.* 15: 451–7 (1913)].
68 D. M. Dennison, *Proc. Roy. Soc. (London)* A115: 483 (1927).
69 Niels Bohr, *Phil. Mag.* 26: 1, 476, 875 (1913); 27: 506 (1914); 29: 332 (1915); 30: 394 (1915).
70 F. Paschen, *Ann. Physik* 27: 565 (1908).
71 T. Lyman, *Phys. Rev.* 3: 504 (1914); *Phil. Mag.* 24: 284 (1915).
72 (a) F. S. Brackett, *Nature,* 109: 209 (1922); *Astrophys. J.* 56: 154 (1922).
(b) A. H. Pfund, *J. Opt. Soc. Am.*, 9: 193 (1924).
(c) C. J. Humphreys, *J. Res. Nat. Bur. Stands.,* 50: 1 (1953).
73 (a) E. C. Pickering, *Astrophys. J.* 4: 369 (1896); 5: 92 (1897); 8: 119 (1898); 13: 230 (1901).
(b) A. Fowler, *Monthly Not. Roy. Astron. Soc.* 73: 62 (1912); *Phil. Trans. Roy. Soc. London* A214: 225 (1914).

(c) E. J. Evans, *Phil. Mag.* (6) 29: 284 (1915).
(d) N. Bohr, *Nature* 92: 231 (1913); *Phil. Mag.* (6) 26: 1 (1913).
(e) H. H. Plaskett, *Nature* 108: 209 (1921); *Pubs. Dominion Astrophys. Observ.* 1: 325 (1922).
(f) W. Huggins, *Scientific Papers of Sir William Huggins* (London: Wesley, 1909), p. 112.
(g) N. Lockyer et al., *Phil. Trans. Roy. Soc. London* A197: 151 (1901).
(h) T. Lyman, *Astrophys. J.* 60: 1 (1924).
(i) H. D. Babcock and C. E. Moore, *Carnegie Inst. Washington* Pub. 579 (1947).
(j) R. Wilson, *Monthly Not. Roy. Astron. Soc.* 113: 557 (1953).
(k) L. H. Aller, I. S. Bowen, and R. Minkowski, *Astrophys. J.* 122: 62 (1955).
74 H. C. Urey, F. G. Brickwedde, and G. M. Murphy, *Phys. Rev.* 40: 1, 464 (1932); *J. Chem. Phys.* 1: 512 (1933).
75 W. Wilson, *Phil. Mag.* 29: 795 (1915); 31: 156 (1916).
76 A. Sommerfeld, *Munchener Akad.* (1915), pp. 425, 459; (1916) p. 131; *Ann. Physik* 51: 1 (1916).
77 A. Sommerfeld, *Physik. Z.* 17: 491 (1916); P. Debye, *Physik. Z.* 17: 507 (1916).
78 (a) O. Stern, *Z. Physik* 7: 249 (1922).
(b) W. Gerlach and O. Stern, *Z. Physik* 8: 110; 9: 349, 353 (1922); *Ann. Physik* 74: 673 (1924).
79 (a) J. Franck and G. Hertz, *Verhandl. Dent. Physik. Ges.,* 16: 512 (1914).
(b) J. Franck and P. Jordan, *Anregung von Quantensprüngen durch Stösse* (Berlin: Springer, 1926).
(c) P. D. Foote, W. F. Meggers, and F. L. Mohler, *Astrophys. J.* 55: 145 (1922).
(d) P. D. Foote and F. L. Mohler, *The Origin of Spectra,* (New York: Chemical Catalog Co., 1922).
(e) F. M. Penning and W. de Groot in *Handbuch der Physik,* Vol. 23, Pt. I, H. Geiger and K. Scheel, eds., 2d ed. (Berlin: Springer, 1933), pp. 23–184.
80 R. W. Wood studied, in 1904, the excitation of the yellow Na *D* lines in a bulb of sodium vapor excited by absorption of a beam of the yellow Na *D* light, calling it *resonance radiation.*
(a) R. W. Wood, *Physical Optics* (New York: Macmillan, 1934), Chap. 18.
(b) R. J. Strutt, *Proc. Roy. Soc. (London)* A98: 272 (1916). [When a bulb of sodium vapor is excited by absorbing λ3303, the second line of the principal series, the excited spectrum includes both that line and also the *D* lines.]
(c) A. C. G. Mitchell and M. W. Zemansky, *Resonance Radiation and Excited Atoms* (Cambridge: Cambridge University Press, 1934).
81 (a) G. Cario, *Z. Physik* 10: 185 (1922).
(b) G. Cario and J. Franck, *Z. Physik* 17: 202 (1923).
(c) S. Loria, *Phys. Rev.* 26: 573 (1925).
82 (a) W. Kossel and A. Sommerfeld, *Verhandl. deut. physik. Ges.* 21: 240 (1919).
(b) E. Schrödinger, *Z. Physik* 4: 347 (1921).
(c) E. Fues, *Z. Physik* (1922); 12: 1 (1923).

(d) A. Th. van Urk, *Z. Physik* 13: 268 (1923).
(e) D. R. Hartree, *Proc. Cambridge Phil. Soc.* 21: 625 (1923); 22: 409, 464 (1924).
(f) M. Born and W. Heisenberg, *Z. Physik* 23: 388 (1924).
(g) R. B. Lindsay, *J. Opt. Soc. Am.* 11: 17 (1925).
83 (a) A. Fowler, *Proc. Roy. Soc. (London)* A103: 413 (1923); *Phil. Trans. Roy. Soc. London* 225: 1 (1925).
(b) F. Paschen, *Ann. Physik* 71: 142 (1923).
84 R. A. Millikan and I. S. Bowen, *Phys. Rev.* 23: 1 (1924); 24: 223, 289 (1924); 25: 591, 600 (1925); 26: 150, 310 (1925).
85 G. E. Uhlenbeck and S. A. Goudsmit, *Naturwiss.* 13: 593 (1925); *Nature* 117: 264 (1926).
86 H. G. J. Moseley, *Phil. Mag.* 26: 1024 (1913); 27: 703 (1914).
87 C. J. Barkla, *Phil. Mag.* 16: 550 (1908).
88 (a) W. Kossel, *Verhandl. deut. physik. Ges.* 16: 898, 953 (1914); 18: 339, 396 (1916); *Physik. Z.* 18: 240 (1917); *Z. Physik* 1: 119 (1920); 2: 470 (1920).
(b) A. Sommerfeld, *Ann. Physik* 51: 125 (1916).
(c) A. Smekal, *Z. Physik* 5: 91 (1921).
(d) G. Wentzel, *Z. Physik* 6: 84 (1921).
(e) D. Coster, *Phil. Mag.* 43: 1070 (1922); 44: 546 (1922).
(f) N. Bohr and D. Coster, *Z. Physik* 12: 342 (1923).
89 M. de Broglie, *Compt. Rend.* 158: 1493 (1913); 163: 87, 352 (1916); *J. Phys.* 6: 161 (1916).
90 (a) D. Mendeleev, *J. Russ. Phys.-Chem. Soc.* 1: 60 (1869).
(b) T. Moeller, *Inorganic Chemistry* (New York: Wiley, 1952), Chap. 4. [Historical summary.]
91 N. Bohr, *Z. Physik* 9: 1 (1922); *Ann. Physik* 71: 228 (1923).
92 W. Pauli, Jr., *Z. Physik* 31: 765 (1925).
93 (a) A. Rubinowicz, *Physik. Z.* 19: 441, 465 (1918).
(b) N. Bohr, "On the Quantum Theory of Line Spectra," *Det Koningl. Danske Vidensk. Selskab,* Nat. Math. Afd. Series 4, No. 1 (1918).
(c) N. Bohr, *Z. Physik* 2: 423 (1920).
94 A. Einstein, *Physik. Z.* 18: 121 (1917).
95 (a) L. Lorenz, *Phil. Mag.* 39: 287 (1867). [Retarded potentials.]
(b) G. F. FitzGerald, *Trans. Roy. Soc. Dublin* 3: 57 (1883). [Radiative solution for an oscillating magnetic dipole.]
(c) Heinrich Hertz, *Ann. Physik* 34: 373 (1887); 34: 551 (1888); *Ann. Physik* 36: 1 (1889). [Latter is about experimental demonstration of radiation from a radiating electric dipole, and solution for this case.]
(d) J. H. Poynting, *Phil. Trans. Roy. Soc. London* 175: 343 (1884); Oliver Heaviside, *Electrician,* 14: 178, 306 (1885). [Discovery of Poynting vector.]
(e) Joseph Larmor, *Phil. Mag.* 44: 503 (1897).
96 (a) M. Born, *Vorlesungen über Atommechanik* (Berlin: Springer, 1925); English translation, *The Mechanics of the Atom* (London: G. Bell, 1927).
(b) J. H. Van Vleck, *Quantum Principles and Line Spectra,* Bulletin 54, National Research Council, Washington, D.C., (1926).
(c) W. Pauli, Jr., "*Quantentheorie,* in *Handbuch der Physik,* Vol. 23, H. Geiger and K. Scheel, eds. (Berlin: Springer, 1926).

97 H. A. Kramers, *Det Koningl. Danske Vidensk. Selskab,* series 8, vol. 3, No. 3 (1919).
98 N. Bohr, *Z. Physik* 13: 117 (1923). [Bohr's later views on the role of the correspondence principle in quantum theory.]
99 J. H. Van Vleck, *Phys. Rev.* 24: 330, 362 (1924).
100 R. W. Wood, *Phil. Mag.* (6) 8: 293 (1904).
101 (a) R. Ladenburg, *Z. Physik* 4: 451 (1921).
(b) R. Ladenburg and F. Reiche, *Naturwiss.* 11: 584 (1923).
102 (a) H. A. Kramers, *Nature* 113: 673 (1924); 114: 310 (1924).
(b) H. A. Kramers and W. Heisenberg, *Z. Physik* 31: 681 (1924).
103 R. Ladenburg, *Rev. Mod. Phys.* 5: 243 (1933).
104 (a) J. J. Thomson, *Recent Researches in Electricity and Magnetism,* (Oxford: Clarendon, 1893), p. 13. [Momentum transfer in the electromagnetic field.]
(b) H. Poincaré, *Arch. Neerl.* (2) 5: 252 (1900).
(c) M. Abraham, *Nachrichten Wissenschaftliche Gesselschaft Göttingen,* Math. Phys. Klasse, 20 (1902).
(d) Max Planck, *Verhandl. Deut. Physik. Ges.* 10: 728 (1908); *Physik. Z.* 9: 828 (1908).
(e) A. Einstein, *Physik. Z.* 10: 185, 817 (1909). [Momentum concept applied to quanta.]
(f) P. Lebedeff, *Arch. Sci. Phys. Nat.* (4) 8: 184 (1899); and *Ann. Physik* 6: 433 (1901). [Pressure of light.]
(g) E. F. Nichols and G. F. Hull, *Phys. Rev.* 13: 293 (1901); *Astrophys. J.* 17: 315 (1903).
105 A. H. Compton and S. K. Allison, *X-rays in Theory and Experiment* (Princeton, N. J.: Van Nostrand Reinhold, 1935).
106 (a) A. H. Compton, Bulletin 20, National Research Council, Washington, D.C. (1922); *Phys. Rev.* 21: 483 (1923); 22: 409 (1923).
(b) P. Debye, *Physik. Z.* 24: 161 (1923).
107 (a) W. Bothe and H. Geiger, *Z. Physik* 26: 44 (1924); 32: 639 (1925).
(b) A. H. Compton and A. W. Simon, *Phys. Rev.* 26: 289 (1925).
108 (a) W. Duane, *Proc. Nat. Acad. Sci.,* 9: 159 (1923).
(b) A. H. Compton, *Proc. Nat. Acad. Sci.,* 9: 359 (1923).
(c) P. S. Epstein and P. Ehrenfest, *Proc. Nat. Acad. Sci.,* 10: 133 (1924).
109 (a) G. I. Taylor, *Proc. Cambridge Phil. Soc.* 15: 114 (1909).
(b) A. J. Dempster and H. F. Batho, *Phys. Rev.* 30: 644 (1927).
110 E. O. Lawrence and J. W. Beams. *Proc. Nat. Acad. Sci.* 13: 207 (1927).
111 J. H. Van Vleck, *Quantum Principles and Line Spectra,* Chap. VII. Ref. 97.
112 M. Born, *Z. Physik* 26: 379 (1924). [First use of the term *quantum mechanics.*]
113 Louis de Broglie, *Compt. Rend.,* 177: 507, 548, 630 (1923); 179: 39, 676, 1039 (1924); *Phil. Mag.* 47: 446 (1924); *Ann. Phys.* (10) 3: 22 (1925).
114 (a) E. Schrödinger, "Quantisierung als Eigenwertproblem," I, *Ann. Physik* 79: 361 (1926); II, *Ann. Physik* 79: 489 (1926); III, *Ann. Physik* 80: 437 (1926); IV, *Ann. Physik* 81: 109 (1926). "Der stetige Übergang von der Mikro – zur Makromechanik," *Naturwiss.* 14: 664 (1926); "Uber das Verhältnis der Heisenberg – Born – Jordansche Quantenmechanik zu der meinen," *Ann. Physik* (4) 79: 734 (1926). These papers were repub-

lished as a group under the title *Abhandlungen zur Wellenmechanik* (Leipzig: Barth, 1927); English transl., *Collected Papers on Wave Mechanics,* (London: Blackie & Son, 1928).
(b) C. Eckart, *Phys. Rev.* 28: 711 (1926).
115 (a) W. Heisenberg, *Z. Physik* 33: 879 (1925).
(b) M. Born and P. Jordan, *Z. Physik* 34: 858 (1925).
(c) M. Born, W. Heisenberg, and P. Jordan, *Z. Physik* 35: 557 (1926).
(d) M. Born and N. Wiener, *J. Math. Phys.* 5: 84 (1926).
116 M. Born, *Z. Physik* 38: 803 (1926).
117 (a) W. Heisenberg, *Z. Physik* 43: 172 (1927).
(b) W. Heisenberg, *The Physical Principles of the Quantum Theory* (Chicago: University of Chicago Press, 1930).
Related papers present specific examples:
(c) C. G. Darwin, *Proc. Roy. Soc. (London)* A117: 258 (1927).
(d) A. E. Ruark, *Phys. Rev.* 31: 311, 709 (1928).
(e) E. H. Kennard, *Phys. Rev.* 31: 344 (1928).
(f) E. U. Condon, *Science* 69: 573 (1929).
(g) H. P. Robertson, *Phys. Rev.* 34: 163 (1929); 35: 667 (1930); 46: 794 (1934).
(h) E. Schrödinger, *Berliner Berichte* (1930), p. 296.
(i) R. W. Ditchburn, *Proc. Irish Acad.* 39: 58 (1930).
(j) W. Heisenberg, *The Development of Physics* (New York: McGraw-Hill, 1955).
118 N. Bohr, *Nature* 121: 78, 580 (1928); *Naturwiss.* 16: 245 (1928); 17: 483 (1929); 18: 73 (1930); *J. Chem. Soc. (London)* (1932) p. 349; Faraday lecture, *Atomic Theory and the Description of Nature* (Cambridge: Cambridge University Press, 1934); "Discussion with Einstein on Epistemological Problems in Atomic Physics," in *Albert Einstein, Philosopher-Scientist,* Paul A. Schilpp, ed., [this volume is published as Vol. 7 of the Library of Living Philosophers], (Evanston, Ill.: 1949) pp. 199–242; and Einstein's comments, in ibid., pp. 663–88. This paper also appears in the collection of N. Bohr, *Atomic Physics and Human Knowledge* (London: Chapman & Hall, 1958).
119 (a) C. J. Davisson and C. H. Kunsman, *Phys. Rev.* 22: 242 (1922).
(b) W. Elsasser, *Naturwiss.* 13: 711 (1925).
(c) C. J. Davisson and L. H. Germer, *Nature* 119: 558 (1927); *Phys. Rev.* 30: 707 (1927); *Proc. Nat. Acad. Sci.* 14: 801 (1929).
(d) G. P. Thomson and A. Reid, *Nature* 119: 890 (1927).
(e) G. P. Thomson, *Proc. Roy. Soc. (London)* A117: 600 (1928); A128: 641 (1930).
(f) I. Estermann and O. Stern, *Z. Physik* 61: 95 (1930).
(g) Z. G. Pinsker, *Electron Diffraction* (London: Butterworth, 1953).

Chapter 2. Principles of quantum mechanics

1 Early books of this class are
(a) E. U. Condon and P. M. Morse *Quantum Mechanics* (New York: McGraw-Hill, 1929); paperback reprint, 1963.

(b) N. F. Mott, *An Outline of Wave Mechanics* (Cambridge: Cambridge University Press, 1930).
(c) A. Sommerfeld, *Wave Mechanics,* trans. H. L. Brose (London: Methuen, 1930).
(d) J. Frenkel, *Wave Mechanics, Elementary Theory* (Oxford: Oxford University Press, 1932); *Wave Mechanics, Advanced General Theory,* (Oxford: Oxford University Press, 1934).
(e) L. Pauling and E. B. Wilson, *Introduction to Quantum Mechanics* (New York: McGraw-Hill, 1935).
(f) E. C. Kemble, *The Fundamental Principles of Quantum Mechanics* (New York: McGraw-Hill, 1937).

2 P. A. M. Dirac, *The Principles of Quantum Mechanics* (London and New York: Oxford University Press, 1st ed., 1930; 2nd ed., 1935; 3rd ed., 1947; 4th ed., 1958).

3 J. Von Neumann, *Mathematische Grundlagen der Quantenmechanik* (Berlin: Springer, 1932); trans. R. T. Beyer, *Mathematical Foundations of Quantum Mechanics,* (Princeton, N.J.: Princeton University Press, 1955).

4 (a) H. Weyl, *Group Theory and Quantum Mechanics,* trans. H. P. Robertson (London: Methuen, 1931); reprinted by Dover, New York.
(b) E. P. Wigner, *Gruppentheorie und ihre Anwendung auf die Quantenmechanik der Atomspektren* (Braunschweig: Vieweg, 1931); trans. J. J. Griffin, *Group Theory and Its Application to the Quantum Mechanics of Atomic Spectra* (New York and London: Academic Press, 1959).
(c) M. Hamermesh, *Group Theory and Its Application to Physical Problems* (Reading, Mass.: Addison-Wesley, 1962).
(d) B. L. Van der Waerden, *Die Gruppentheoretische Methode in der Quantenmechanik* (Berlin: Springer, 1932).
(e) M. Tinkham, *Group Theory and Quantum Mechanics* (New York: McGraw-Hill, 1964).
(f) V. Heine, *Group Theory of Quantum Mechanics* (New York: Pergamon, 1960).
(g) H. Boerner, *Representations of Groups, with Special Consideration for the Needs of Modern Physics* (Amsterdam: North Holland, 1963).
(h) G. Ya. Lyubarskii, *The Application of Group Theory in Physics,* trans. S. Dedijer, (New York: Pergamon, 1960).

5 M. Born and P. Jordan, *Elementare Quantenmechanik* (Berlin: Springer, 1930).

6 (a) V. Rojansky, *Introductory Quantum Mechanics* (Englewood Cliffs, N.J.: Prentice-Hall, 1938).
(b) L. I. Schiff, *Quantum Mechanics,* 3rd ed. (New York: McGraw-Hill, 1968).
(c) N. F. Mott and I. N. Sneddon, *Wave Mechanics and Its Applications* (London: Oxford University Press, 1948).
(d) H. A. Kramers, *Quantum Mechanics*, trans. D. Ter Harr (Amsterdam: North Holland, 1958); reprinted by Dover, New York, 1964.
(e) D. Bohm, *Quantum Theory* (Englewood Cliffs, N.J.: Prentice-Hall, 1951).
(f) W. V. Houston, *Principles of Quantum Mechanics* (New York: McGraw-Hill, 1951).

(g) A. Messiah, *Quantum Mechanics,* trans. G. M. Temmer and J. Potter (Amsterdam: North Holland, and New York: Wiley, 1961), 2 vols.
(h) B. Kurşunoğlu, *Modern Quantum Theory* (San Francisco and London: Freeman, 1962).
(i) G. L. Trigg, *Quantum Mechanics* (Princeton, N.J.: Van Nostrand Reinhold, 1964).
(j) F. A. Kaempffer, *Concepts in Quantum Mechanics* (New York: Academic Press, 1965).
(k) E. Merzbacher, *Quantum Mechanics,* 2nd ed. (New York: Wiley, 1970).
(l) L. D. Landau and E. M. Lifshitz, *Quantum Mechanics, Non-relativistic Theory* (Oxford: Pergamon, 1958).
(m) W. Pauli, "Die allgemeinen Principien der Wellenmechanik" *Encyclopedia of Physics* (Berlin: Springer, 1958), vol. 5, part 1.
(n) P. Roman, *Advanced Quantum Theory* (Reading, Mass.: Addison-Wesley, 1965).
(o) A. S. Davydow, *Quantum Mechanics* (Reading, Mass.: Addison-Wesley, 1965).
(p) Kurt Gottfried, *Quantum Mechanics* (New York: Benjamin, 1966), Vol. 1.
(q) E. M. Corson, *Perturbation Methods in the Quantum Mechanics of n-Electron Systems* (London: Blackie & Son, 1951).
(r) M. Mizushima, *Quantum Mechanics of Atomic Spectra and Atomic Structure* (New York: Benjamin, 1970).
(s) H. A. Bethe and R. W. Jackiw, *Intermediate Quantum Mechanics,* 2nd ed. (New York: Benjamin, 1968).

7 Many books on linear algebra have been published in recent years. Particularly recommended are
(a) L. Mirsky, *An Introduction to Linear Algebra* (London and New York: Oxford University Press, 1955).
(b) H. L. Hamburger and M. E. Grimshaw, *Linear Transformations in N-Dimensional Vector Space* (Cambridge: Cambridge University Press, 1951).
(c) A. I. Mal'cev, *Foundations of Linear Algebra,* trans. T. C. Brown (San Francisco and London: Freeman, 1963).
(d) W. G. Greub, *Linear Algebra,* 2nd ed. (New York: Academic Press, 1963).
(e) G. E. Shilov, *An Introduction to the Theory of Linear Spaces* (Englewood Cliffs, N.J.: Prentice-Hall, 1961).
(f) V. I. Smirnov, *Linear Algebra and Group Theory,* trans. R. A. Silverman (New York: McGraw-Hill, 1961).
(g) Paul R. Halmos, *Finite-Dimensional Vector Spaces* (Princeton, N.J.: Princeton University Press, 1948).
(h) Hans Schwerdtfeger, *Introduction to Linear Algebra and the Theory of Matrices* (Groningen: P. Noordhoff, 1950).

8 (a) Von Neumann *Mathematische . . . ,* chaps. 1 and 2.
(b) R. G. Cooke, *Linear Operators* (London: Macmillan, 1953).
(c) C. Lanczos, *Linear Differential Operators* (London: Van Nostrand Reinhold, 1961).

(d) I. N. Sneddon, "Functional Analysis," in *Encyclopedia of Physics,* Vol. 2 (Berlin: Springer, 1955).
(e) A. E. Taylor, *Introduction to Functional Analysis* (New York: Wiley, 1958).
(f) A. C. Zaanen, *Linear Analysis* (Amsterdam: North Holland, 1956).
(g) M. H. Stone, *Linear Transformations in Hilbert Space* (New York: American Mathematical Society, 1932).
(h) Paul R. Halmos, *Introduction to Hilbert Space,* 2d ed. (New York: Chelsea, 1957).

9 (a) E. U. Condon, *Science* 69: 573 (1929).
(b) H. P. Robertson, *Phys. Rev.* 34: 163 (1929).
(c) E. Schrödinger, *Berliner Berichte,* 296 (1930).
(d) H. Weyl, *Group Theory and Quantum Mechanics,* Appendix 1.

10 (a) L. Landau, *Z. Physik* 45: 430 (1927).
(b) J. Von Neumann, *Nachrichten Wissenschaftliche Gesellschaft Göttingen,* Math. Phys. Klasse, 245, 273 (1927).
(c) P. A. M. Dirac, *Proc. Cambridge Phil. Soc.,* 25: 62 (1929); 26: 376 (1930); 27: 240 (1930).
(d) R. C. Tolman, *The Principles of Statistical Mechanics* (New York and London: Oxford University Press, 1938), Chap. 9.
(e) K. Husimi, *Proc. Phys. Math. Soc. Japan* 22: 264 (1940).
(f) U. Fano, *Rev. Mod. Phys.* 29: 74 (1957).

11 (a) Norbert Wiener, *Cybernetics* (New York: Wiley, 1948).
(b) C. E. Shannon and W. Weaver, *The Mathematical Theory of Communication* (Urbana, Illinois: University of Illinois Press, 1949).
(c) L. Brillouin, *Science and Information Theory,* 2d ed. (New York: Academic Press, 1962).
(d) A. I. Khinchin, *Mathematical Foundations of Information Theory* (New York: Dover, 1957).

12 R. Courant and D. Hilbert, *Methods of Mathematical Physics,* Vol. 1 (New York: Interscience, 1953), p. 80.

13 (a) P. A. M. Dirac, *Proc. Roy. Soc. (London),* A112: 661 (1926).
(b) W. Heisenberg, *Z. Physik* 40: 501 (1926).
(c) E. P. Wigner, *Z. Physik* 40: 883 (1927).
(d) J. C. Slater, *Phys. Rev.* 34: 1293, (1929).

14 (a) O. Klein and P. Jordan, *Z. Physik* 45: 751 (1927).
(b) P. Jordan and E. P. Wigner, *Z. Physik* 47: 631 (1928).
(c) V. Fock, *Z. Physik* 49, 339 (1928); 75: 622 (1932); 81: 195 (1933); *Physik Z. Soviet Union* 6: 425 (1934).
(d) P. A. M. Dirac, *Ann. de l'Institut Henri Poincaré,* 1: part 4 (1929); 11: part 7 (1949).

15 The variation method was devised in connection with classical problems of continuum mechanics first by (a) Lord Rayleigh, *Proc. London Math. Soc.* 4: 357 (1873) (*Scientific Papers,* Cambridge University Press, Vol. 1, p. 170), and then by (b) Walther Ritz, *Journal fur reine und angewandte Mathematik* 135: 1 (1907); *Gesammelte Werke* (Paris: Gauthier-Villars, 1911), p. 192. It was first used in quantum mechanics by (c) E. Hylleraas, *Z. Physik* 48: (1928); 54: 347 (1929); 65: 209 (1930). (d) J. K. L. Mac Donald, *Phys. Rev.* 43: 830 (1933).

16 (a) Lord Rayleigh, *Theory of Sound,* Vol. 1 (1894), p. 113.
(b) E. Schrödinger, *Ann. Physik* 80: 437 (1926).
(c) J. E. Lennard-Jones, *Proc. Roy. Soc. (London),* A129: 604 (1930).
(d) L. Brillouin, *J. de Phys.* 4: 1 (1933).
(e) E. P. Wigner, *Math. naturwiss. Anz. der ungarischen Akad. Wiss.,* 53: 477 (1935).
(f) K. O. Friedrichs, *Perturbation of Spectra in Hillbert Space* (Providence, R.I.: American Mathematical Society, 1965).

17 (a) P. A. M. Dirac, *Proc. Roy. Soc. (London)* A112: 661 (1926); A114: 243 (1927).
(b) E. Schrödinger, *Ann. Physik* 81: 109 (1926).
(c) J. C. Slater, *Proc. Nat. Acad. Sci.* 13: 7 (1927).

18 H. A. Lorentz, *Proc. Amsterdam Acad.* 14: 518, 577 (1905); 18: 134 (1915); *Theory of Electrons,* 2d ed.; reprinted by Dover, New York, 1952.

19 (a) G. Wentzel, *Z. Physik* 38: 518 (1926).
(b) L. Brillouin, *Compt. Rend.,* 183: 24 (1926).
(c) H. A. Kramers, *Z. Physik* 39: 828 (1926).
(d) H. Jeffreys, *Proc. London Math. Soc.* (2)23: 428 (1923).

20 J. Heading, *An Introduction to Phase-Integral Methods,* (New York: Wiley, 1962).

21 (a) Harold Jeffreys and Bertha Swirles Jeffreys, *Methods of Mathematical Physics,* 3d ed. (Cambridge: Cambridge University Press, 1956) pp. 508–11.
(b) G. N. Watson, *A Treatise on the Theory of Bessel Functions,* 2d ed. (Cambridge: Cambridge University Press, 1944), pp. 188–90.
(c) J. C. P. Miller, *The Airy Integral* British Association Tables, Vol. B, 1946. [Tables of $Ai(x)$ and $Bi(x)$.]
(d) A. D. Smirnov, *Tables of Airy Functions and Special Confluent Hypergeometric Functions* (New York: Pergamon, 1960).
(e) M. Abramowitz and I. A. Stegun, *Handbook of Mathematical Functions,* Nat. Bur. Stand. Applied Mathematics Series 55, (Washington, D.C.: Government Printing Office, 1964); later reprinted by Dover, New York.

Chapter 3. Angular momentum

1 Historically the development of the theory of angular momentum matrices began with M. Born and P. Jordan, *Z. Physik* 34: 858 (1925); M. Born, W. Heisenberg, and P. Jordan, *Z. Physik* 35: 557 (1926); W. Heisenberg and P. Jordan, *Z. Physik* 37: 263 (1926). Several books devoted entirely to this subject are available. Of great convenience is a collection of reprints and original papers, here cited as *QTAM:* L. C. Biedenharn and H. Van Dam, *Quantum Theory of Angular Momentum* (New York: Academic Press, 1965).
Other books are
(a) M. E. Rose, *Elementary Theory of Angular Momentum* (New York: Wiley, 1957).
(b) M. E. Rose, *Multipole Fields* (New York: Wiley, 1955).

(c) A. R. Edmonds, *Angular Momentum in Quantum Mechanics* (Princeton, N.J.: Princeton University Press, 1957).
(d) Eugene Feenberg and George Edward Pake, *Notes on the Quantum Theory of Angular Momentum,* (Reading, Mass.: Addison-Wesley, 1953); reissued by Stanford University Press, 1959.
(e) D. M. Brink and G. R. Satchler, *Angular Momentum,* 2d ed. (London and New York: Oxford University Press, 1968).
(f) A. P. Yutsis, I. B. Levinson, and V. V. Vanagas, *Mathematical Apparatus of the Theory of Angular Momentum* (Jerusalem: Israel Program for Scientific Translations, 1962).

2 (a) G. Racah, "Theory of Complex Spectra," I, *Phys. Rev.* 61: 186 (1942); II, *Phys. Rev.* 62: 438 (1942); III, *Phys. Rev.* 63: 367 (1943); IV, *Phys. Rev.* 76: 1352 (1949); *QTAM,* pp. 134–200.
(b) U. Fano and G. Racah, *Irreducible Tensorial Sets* (New York: Academic Press, 1959).

3 E. P. Wigner, *On the Matrices Which Reduce the Kronecker Products of Representations of S.R. Groups,* unpublished, 1940; *QTAM,* pp. 89–133.

4 J. Schwinger, *On Angular Momentum,* unpublished, 1952; *QTAM,* pp. 229–79.

5 L. D. Landau and E. M. Lifshitz, *Quantum Mechanics* (Reading, Mass.: Addison-Wesley, 1958), p. 357.

6 T. Regge, *Nuovo Cimento,* 10: 544 (1958); *QTAM,* pp. 296–7.

7 E. U. Condon and G. H. Shortley, *The Theory of Atomic Spectra* (Cambridge: Cambridge University Press, 1935), pp. 76–8.

8 (a) R. Saito and M. Morita, *Prog. Theoret. Phys.* (Kyoto) 13: 540 (1955). [Tables for $j_2 = 5/2$.]
(b) M. A. Melvin and N. V. V. J. Swamy, *Phys. Rev.* 107: 186 (1957). [Tables for $j_2 = 5/2$.]
(c) D. L. Falkoff, C. S. Calladay, and R. E. Sells, *Canad. J. Phys.* 30: 253 (1952). [Tables for $j_2 = 3$.]

9 (a) M. Rotenberg, R. Bivins, N. Metropolis, and J. K. Wooten, Jr., *The 3j and 6j symbols* (Cambridge, Mass.: Technology Press, 1959).
(b) A. Simon, *Numerical Tables of the Clebsch-Gordan Coefficients.* Oak Ridge National Laboratory Report 1718 (1954). [Decimal values up to $(j_1, j_2) \leqslant 9/2$.]
(c) B. J. Sears and M. G. Radtke, *Algebraic Values of the Clebsch-Gordan Coefficients,* Chalk River Report TPI-75 (1954).
(d) E. R. Cohen, *Tables of Clebsch-Gordan Coefficients,* AEC Report NAA-SR-2123 (1958). [Values expressed as square roots of rational fractions for $(j_1 + j_2) \leqslant 4$.]
(e) T. Shimpuku, "General Theory and Numerical Tables of Clebsch-Gordan Coefficients," *Prog. Theoret. Phys.* (Kyoto) Supplement 13: 1–135 (1960).
(f) R. S. Caswell and L. C. Maximon, *Fortran Programs for the Calculations of Wigner 3j, 6j and 9j Coefficients for Angular Momenta <80,* Nat. Bur. Stand. Tech. Note No. 409, (Washington, D.C.: Government Printing Office, 1966).
(g) K. W. H. Stevens, *Proc. Phys. Soc.* A65: 209 (1952).

10 The mathematical study of spherical harmonics has a large literature:
(a) T. M. MacRobert, *Spherical Harmonics* (Glasgow: Blackie, and New York: Dutton, 1927); reprinted by Dover, New York, 1948.
(b) E. W. Hobson, *The Theory of Spherical and Ellipsoidal Harmonics* (Cambridge: Cambridge University Press, 1931).
For formulas and numerical tables:
(c) E. Jahnke, F. Emde, and F. Lösch, *Tables of Higher Functions* (New York: McGraw-Hill, 1960).
(d) *Handbook of Mathematical Functions,* Nat. Bur Stand. Applied Mathematics Series 55 (Washington, D.C.: Government Printing Office, 1964).
11 (a) E. Merzbacher, *Am. J. Phys.* 30: 237 (1962).
(b) H. Buchdahl, *Am. J. Phys.* 30: 289 (1962); 31: 329 (1963).
12 E. U. Condon, *Science* 69: 573 (1929).
13 A. Ünsold, *Ann. Physik* 82: 355 (1927).
14 H. A. Kramers, *Quantum Mechanics* (Amsterdam: North Holland, 1958); reprinted by Dover, New York, 1964, pp. 171–8.
15 (a) W. Pauli, *Z. Physik* 43: 601 (1927); *QTAM,* pp. 14–36.
(b) C. G. Darwin, *Proc. Roy. Soc.* (*London*) A116: 227 (1927)
16. *Theoretical Physics in the Twentieth Century,* a memorial to Wolfgang Pauli (New York: Interscience, 1960), pp. 209-16, 236.
17. (a) B. L. Van der Waerden, *Nachrichten Wissenschaftliche Gesellschaft Göttingen,* Math. Phys. Klasse, 100, (1929).
(b) G. E. Uhlenbeck and O. Laporte, *Phys. Rev.* 37: 1380, 1552 (1931).
(c) W. L. Bade and and H. Jehle, *Rev. Mod. Phys.* 25: 714 (1953).
(d) C. Chevalley, *The Algebraic Theory of Spinors* (New York: Columbia University Press, 1954).
(e) Elie Cartan, *The Theory of Spinors* (Cambridge, Mass.: MIT Press, 1966).
18. (a) Felix Klein, *Math. Annal.* 9: 183 (1875).
(b) A. Cayley, *Math. Annal.* 15: 238 (1879).
(c) E. T. Whittaker, *A Treatise on the Analytical Dynamics of Particles and Rigid Bodies,* 3d. ed. (Cambridge: Cambridge University Press, 1927).
(d) Herbert Goldstein, *Classical Mechanics* (Reading, Mass.: Addison-Wesley, 1959), p. 109.
19 (a) G. Szegö, "Orthogonal Polynomials," *Am. Math. Soc.,* Colloquium Publications, Vol. 23 (1939).
(b) C. G. J. Jacobi, *Journal für die reine und angewandte Mathematik,* 56: 140–165 (1859); *Werke,* Vol. 6 (Berlin: 1891), pp. 184–202.
(c) Bateman Manuscript Project, *Higher Transcendental Functions,* Vol. 2 (New York: McGraw-Hill, 1953), p. 168.
20 (a) F. Reiche and H. Rademacher, *Z. Physik* 39: 444 (1926); 41: 453 (1927).
(b) D. M. Dennison, *Phys. Rev.* 28: 318 (1926); *Rev. Mod. Phys.* 3: 280 (1931).
(c) R. de L. Kronig and I. I. Rabi, *Phys. Rev.* 29: 262 (1927).
(d) E. U. Condon, *Phys. Rev.* 30: 781 (1927).
(e) D. M. Brink and G. R. Satchler, *Angular Momentum*, 2nd ed. (Oxford: Clarendon Press, 1968); p. 25.

21 (a) C. Eckart, *Rev. Mod. Phys.* 2: 305 (1930).
(b) E. P. Wigner, *Gruppentheorie* (Braunschweig: Vieweg, 1931), pp. 262–5.
22 B. R. Judd, *Operator Techniques in Atomic Spectroscopy* (New York: McGraw-Hill, 1963).

Chapter 4. Central-field approximation

1 W. Heisenberg, *Z. Physik* 38: 411 (1926); 39: 499 (1926); 41: 239 (1927).
2 P. A. M. Dirac, *Proc. Roy. Soc. (London)* A112: 661 (1926).
3 (a) J. C. Slater, *Phys. Rev.* 34: 1293 (1929).
(b) E. U. Condon, *Phys. Rev.* 36: 1121 (1930).
4 (a) H. N. Russell and F. A. Saunders, *Astrophys. J.* 61: 38 (1925).
(b) H. N. Russell, A. G. Shenstone, and L. A. Turner, *Phys. Rev.* 33: 900 (1929).
5 D. Layzér, *Ann. Phys.* 8: 271 (1959); *Monthly Not. Roy. Astron. Soc.* 114: 692 (1954).
6 (a) D. R. Hartree, *Proc. Cambridge Phil. Soc.* 24: 87, 111, 426 (1928).
(b) D. R. Hartree, *The Calculation of Atomic Structures* (New York: Wiley, 1957).
7 V. Fock, *Z. Physik* 61: 126 (1930).
8 C. E. Moore, *Atomic Energy Levels as Derived from the Analyses of Optical Spectra,* Circ. 467, Nat. Bur. Stand. (Washington, D.C.: Government Printing Office, Vol. 1, 1949; Vol. 2, 1952; Vol. 3, 1958); reissued as NSRDS-NBS 35 (1971). A volume covering the rare earths is in preparation, with publication expected during 1978: "Atomic Energy Levels – The Rare-Earth Elements," by W. C. Martin, R. Zalubas, and L. Hagan.
9 (a) R. F. Bacher and S. A. Goudsmit, *Atomic Energy States* (New York: McGraw-Hill, 1932).
(b) W. Grotrian, *Graphische Darstellungen der Spektren* 2 vols. (Berlin: Springer, 1928).
(c) Revised tables of atomic energy levels are being published by C. E. Moore, *Selected Tables of Atomic Spectra,* Nat. Stand. Ref. Data Series – Nat. Bur. Stand. NSRDS-NBS 3, (Washington, D.C.: Government Printing Office.) This is being issued in sections: Sec. 1 (1965), Sec. 2 (1967), Sec. 3 (1970), Sec. 4 (1971), Sec. 5 (1975), Sec. 6 (1972) and Sec. 7 (1976) have appeared up to mid-1976.
(d) C. E. Moore, *Bibliography on the Analysis of Optical Atomic Spectra,* Nat. Bur. Stand. Spec. Publ. 306, 1968; 306-2, 1969; 306-3, 1969; 306-4, 1969. (Washington, D.C.: Government Printing Office).
(e) L. Hagan and W. C. Martin, *Bibliography on Atomic Energy Levels and Spectra July 1968 through June 1971,* Nat. Bur. Stand. Spec. Publ. 363 (Washington, D.C.: Government Printing Office, 1972). A supplement (N.B.S. Spec. Pub. 363, Supp. 1) will appear in late 1976 and will cover July 1971–1975.
(f) C. E. Moore, *Ionization Potentials and Ionization Limits Derived from the Analyses of Optical Spectra,* NSRDS-NBS 34, (Washington, D.C.: Government Printing Office, 1970).
(g) W. C. Martin, L. Hagan, J. Reader, and J. Sugar, "Ground Levels and

Ionization Potentials for Lanthanide and Actinide Atoms and Ions", *J. Phys. Chem. Ref. Data,* 3: 771 (1974).
(h) Among other compilations we mention W. C. Martin, *J. Phys. Chem. Ref. Data,* 2: 257 (1973) on He I; and J. Reader and J. Sugar, *J. Phys. Chem. Ref. Data,* 4: 353 (1975) who cover all stages (I to XXVI) of ionization of iron. A similar compilation for Cr is in press as part of a series for the iron-group elements.
(i) An excellent bibliography on the term analysis of atomic spectra of atoms and ions in all stages of ionization, for hydrogen to nickel, was given by B. Edlén, in *Beam-Foil Spectroscopy.* Vol. 1: *Atomic Structure and Lifetimes,* ed., I. A. Sellin and D. J. Pegg (New York and London: Plenum, 1976), pp. 1–27.
(j) Grotrian diagrams, giving graphical representations of atomic energy levels and associated transitions, have been given for important spectra by C. E. Moore and P. W. Merrill, NSRDS-NBS 23, (Washington, D.C.: Government Printing Office, 1968). A much larger compilation of Grotrian diagrams is in publication: S. Bashkin and J. O. Stoner, *Atomic Energy Levels and Grotrian Diagrams.* Vol. I; *Hydrogen I–Phosphorus XV* (Amsterdam and Oxford: North Holland, 1976); other volumes will follow.
(k) Bibliographies of recent work on atomic spectra appear in the triennial reports of Commission 14 of the International Astronomical Union; the latest of these is *Trans. I.A.U.,* 16A: Part 1, 31 (1976).

10 (a) E. Laguerre, *Bull. Soc. Math. France* 7: 72 (1879); (Paris: Oeuvres, 1898), Vol. 1, p. 428.
(b) E. Schrödinger, *Ann. Physik* 79: 361 (1926).
(c) W. Gordon, *Ann. Physik* 2: 1031 (1929).
(d) A. Sommerfeld and G. Schur, *Ann. Physik* 4: 409 (1930).
(e) I. Waller, *Z. Physik* 38: 635 (1926).

11 (a) E. T. Whittaker and G. N. Watson, *A Course of Modern Analysis,* 4th ed. (Cambridge: Cambridge University Press, 1927), Chap. 16.
(b) Abramowitz and Stegun, *Handbook of Mathematical Functions,* Chap. 13.

12 H. A. Bethe and E. E. Salpeter, *Encyclopedia of Physics,* Vol. 35 (Berlin: Springer, 1957), p. 99; *Quantum Mechanics of One- and Two-Electron Atoms* (Berlin: Springer, 1957), pp. 13–15.

13 B. Podolsky and L. Pauling, *Phys. Rev.* 34: 108 (1928).

14 (a) E. Hylleraas, *Z. Physik* 74: 216 (1932).
(b) H. Weyl, *Z. Physik* 46: 1 (1928).
(c) V. Fock, *Z. Physik* 98: 145 (1935).

15 (a) G. H. Shortley, *Phys. Rev.* 50: 1072 (1936).
(b) J. C. Slater, *Quantum Theory of Atomic Structure,* Vol. 1 (New York: McGraw-Hill, 1960), p. 316.

16 N. M. Gray and L. A. Wills, *Phys. Rev.* 38: 248 (1931).

17 C. W. Ufford and G. H. Shortley, *Phys. Rev.* 42: 167 (1932).

Chapter 5. Racah methods

1 (a) G. Racah, *Phys. Rev.* 61: 186 (1942).
(b) G. Racah, *Phys. Rev.* 62: 438 (1942).

(c) G. Racah, *Phys. Rev.* 63: 367 (1943).
(d) G. Racah, *Phys. Rev.* 76: 1352 (1949).
2 C. W. Nielson and George F. Koster, *Spectroscopic Coefficients for the p^n, d^n and f^n Configurations* (Cambridge, Mass.: MIT Press, 1963).
3 J. H. Van Vleck, *Phys. Rev.* 45: 405 (1934).
4 P. A. M. Dirac, *Proc. Roy. Soc. (London)* A123: 714 (1929).
5 R. Serber, *Phys. Rev.* 45: 461 (1934).
6 F. Rohrlich, *Phys. Rev.* 74: 1372 (1948).
7 G. H. Shortley, *Phys. Rev.* 40: 185 (1932).
8 M. H. Johnson, *Phys. Rev.* 43: 632 (1933).
9 N. H. Möller, *Ark. Fysik* 18: 135 (1960).

Chapter 6. Group-theoretical methods

The standard books on group theory in relation to quantum mechanics are given in reference 4 of Chapter 2.
1 Books on finite groups, mainly as a topic in mathematics, include
(a) L. E. Dickson, *Linear Groups* (Leipzig: Teubner, 1901).
(b) H. Hilton, *Introduction to the Theory of Groups of Finite Order,* 2nd ed. (New York: Oxford University Press, 1908).
(c) W. Burnside, *Theory of Groups of Finite Order* (Cambridge: Cambridge University Press, 1911).
(d) G. A. Miller, H. F. Blichfeldt, and L. E. Dickson, *Theory and Application of Finite Groups* (Brooklyn, N.Y.: Scientific Press, 1916); reprinted by Dover, New York, 1961.
(e) H. F. Blichfeldt, *Finite Collineation Groups* (Chicago: University of Chicago Press, 1917).
(f) A. Speiser, *Die Theorie der Gruppen von endlicher Ordnung* (Berlin: Springer, 1927).
(g) W. Ledermann, *Introduction to the Theory of Finite Groups* (New York: Wiley, 1949).
(h) G. Gorenstein, *Finite Groups* (New York: Harper & Row, 1968).
(i) A. G. Kurosh, *The Theory of Groups,* 2nd ed., translated and edited by K. A. Hirsch (New York: Chelsea, 1960).
(j) H. J. Zassenhaus, *The Theory of Groups,* 2nd ed. (New York: Chelsea, 1958).
(k) J. S. Lomont, *Applications of Finite Groups* (New York: Academic Press, 1959).
(l) M. Hall, *The Theory of Groups* (New York: Macmillan, 1959).
(m) W. Magnus, A. Karrass, and D. Solitar, *Combinatorial Group Theory* (New York: Wiley, 1966).
(n) I. D. Macdonald, *The Theory of Groups* (London and New York: Oxford University Press, 1968).
2 Books on continuous groups, mainly as a topic in mathematics, include
(a) S. Lie and G. Scheffers, *Vorlesungen über continuierliche Gruppen* (Leipzig: Teubner, 1938).
(b) H. Weyl, *The Classical Groups, Their Invariants and Representations* 2nd ed. (Princeton, N.J.: Princeton University Press, 1946).
(c) C. Chevalley, *Theory of Lie Groups* (Princeton, N.J.: Princeton University Press, 1946).

(d) I. M. Gel'fand, R. A. Minlos, and Z. Ya. Shapiro, *Representations of the Rotation and Lorentz Groups and Their Applications* (Oxford: Pergamon, 1963).
(e) H. J. Lipkin, *Lie Groups for Pedestrians* (Amsterdam: North Holland, 1965).
(f) R. Hermann, *Lie Groups for Physicists* (New York: Benjamin, 1966).
(g) T. Kahan, *Theory of Groups in Classical and Quantum Physics* ed., trans. A. R. Edmonds (New York: Elsevier, 1965).
(h) P. M. Cohen, *Lie Groups* (Cambridge: Cambridge University Press, 1957).
(i) M. Hausner and J. T. Schwartz, *Lie Groups; Lie Algebras* (New York: Gordon and Breach, 1968).
(j) R. Gilmore *Lie Groups, Lie Algebras, and Some of Their Applications* (New York: Wiley, 1974).

3 Most general works on group theory deal with the theory of representations. In addition there are
(a) F. D. Murnaghan, *The Theory of Group Representations* (Baltimore: Johns Hopkins Press, 1938); reprinted by Dover, New York, 1963.
(b) G. de B. Robinson, *Representation Theory of the Symmetric Group* (Edinburgh: Edinburgh University Press, 1961).
(c) C. W. Curtis and I. Reiner, *Representation Theory of Finite Groups and Associative Algebras* (New York: Wiley, 1962).
(d) B. L. Van der Waerden, *Gruppen von linearen Transformationen* (New York: Chelsea, 1948).
(e) Henri Bacry, *Leçons sur la Théorie des Groupes et les Symétries des Particules Elémentaires* (New York: Gordon and Breach, 1967).
(f) L. Fonda and G. C. Ghirardi, *Symmetry Principles in Quantum Physics* (New York: Dekker, 1970).
(g) B. G. Wybourne, *Classical Groups for Physicists* (New York: Wiley, 1974).
(h) A. O. Barut, R. Raczka, *Theory of Group Representations and Applications* (Warsaw: Polish Scientific Publisher, 1977).

4 E. Cartan, *Sur la structure des groups de transformations finis et continus,* thesis. (Paris: Nony, 1894; 2nd ed. Paris: Vuibert, 1933).

5 S. Helgason, *Differential Geometry and Symmetric Spaces* (New York: Academic Press, 1962), Chap. IX, Sec. 4.

6 D. E. Littlewood, *The Theory of Group Characters and Matrix Representations of Groups,* 2nd ed., (London and New York: Oxford University Press, 1958).
Most general works on group theory develop the idea of characters of a group representation.

7 (a) B. G. Wybourne, *Symmetry Principles and Atomic Spectroscopy,* (New York: Wiley-Interscience, 1970).
(b) D. E. Littlewood and A. R. Richardson, *Phil. Trans. Roy. Soc. London* A233: 99 (1934).
(c) M. Zia-ud-Din, *Proc. London Math. Soc.* 39: 200 (1935); 42: 340 (1937).
(d) K. Kondo, *Proc. Phys. Math. Soc. Japan* 22: 585 (1940).

8 R. L. Bivins, N. Metropolis, P. R. Stein, and M. B. Wells, *Math. Tables and Other Aids to Computation* 8: 212 (1954).

9 J. S. Frame, G. de B. Robinson, and R. M. Thrall, *Canad. J. Math.* 6: 316 (1954).
10 (a) F. D. Murnaghan, *Am. J. Math* 60: 761 (1938).
(b) F. D. Murnaghan, *Proc. Nat. Acad. Sci.* 41: 515 (1955).
(c) A. Gamba and L. A. Radicati, *Atti Accad. naz. Lincei Rend.* 14: 632 (1953).
(d) F. D. Murnaghan, *Proc. Nat. Acad. Sci.* 42: 93 (1956).
(e) G. de B. Robinson and O. E. Taulbee, *Proc. Nat. Acad. Sci.* 40: 723 (1954).
(f) D. E. Littlewood, *J. London Math. Soc.* 31: 89 (1956); 32: 18 (1957).
(g) D. E. Littlewood, *Canad. J. Math.* 10: 1 (1958); 10: 17 (1958).
11 (a) H. Weyl, *The Theory of Groups and Quantum Mechanics* (New York: Dutton, 1932), p. 123; reprinted by Dover, New York.
(b) B. L. Van der Waerden, *Die gruppentheoretische Methode in der Quantenmechanik,* (Berlin: Springer, 1932), p. 70.
(c) A. Clebsch, *Theorie der binären algebraischen Formen,* (Leipzig: Teubner, 1872).
(d) Paul Gordan, *Math. Annal.* 5: 95–122 (1872).
(e) F. Klein, *Vorlesungen über die Entwicklung der Mathematik im 19 Jahrhundert;* reprinted by Chelsea, New York, 1956.
12 N. Jacobson, *Lie Algebras* Interscience Tracts in Pure and Applied Mathematics 10, (New York: Wiley, 1962).
13 E. Cartan, *Ann. Sci. Ecole Norm. Sup.* 31: 263 (1914).
14 (a) E. Dynkin, *Uspekhi Mat. Nauk* (N.S.), 2: 59 (1947); *Am. Math. Soc.*, trans. 17 (1950).
(b) E. Dynkin, *Mat. Sbornik,* 30: 349 (1952); *Am. Math. Soc.* trans. Ser. 2, 6: 111 (1957).
15 (a) B. L. Van der Waerden, *Math. Z.*, 37: 446 (1933).
(b) G. Racah, *Ergeb. Exakt. Naturwiss.* 37: 28 (1965).
16 H. Casimir, *Proc. Koninkl. Akad. Amsterdam* 34: 844 (1931).

Chapter 7. Applications of group theory

1 (a) B. R. Judd, *Second Quantization and Atomic Spectroscopy* (Baltimore: Johns Hopkins Press, 1967).
(b) B. R. Judd, in *Group Theory and Its Applications,* Vol. 1, ed. E. M. Loebl, (New York: Academic Press, 1968), pp. 183–220.
2 M. Moshinsky, *Group Theory and the Many-Body Problem* (New York: Gordon and Breach, 1968).
3 G. Racah, in *Group Theoretical Concepts and Methods in Elementary Particle Physics,* ed. F. Gürsey (New York: Gordon and Breach, 1964), pp. 1–36.
4 W. Feneuille, *J. de Phys.* 28: 61, 315, 497, 701, (1967).
5 (a) D. E. Littlewood, *J. London Math. Soc.* 11: 370 (1936).
(b) J. P. Elliott, *Proc. Roy. Soc. (London)* A245: 128 (1958).
(c) M. Kretzschmar, *Z. Physik* 158: 284 (1960).
6 B. G. Wybourne, *Spectroscopic Properties of Rare Earths* (New York: Wiley, 1965).

7 A. P. Stone, *Proc. Cambridge Phil. Soc.* 57: 460 (1961).
8 A. G. McLellan, *Proc. Phys. Soc. (London)* 76: 419 (1960).
9 B. G. Wybourne, *J. Chem. Phys.* 36: 2295 (1962).
10 (a) L. Armstrong, Jr., and B. R. Judd, *Proc. Roy. Soc. (London)* A315: 27 (1970).
(b) L. Armstrong, Jr., and B. R. Judd, *Proc. Roy. Soc. (London)* A315: 39 (1970).
11 C. L. B. Shudeman, *J. Franklin Inst.* 224: 501 (1937).
12 (a) B. G. Wybourne and P. H. Butler, *J. de Phys.* 30: 181 (1969).
(b) M. J. Cunningham and B. G. Wybourne, *J. Math. Phys.* 10: 2149 (1969); *J. Math. Phys.* 11: 1288 (1970).
(c) S. Feneuille, *J. de Phys.* 30: 923 (1969).
13 B. R. Judd, in *Topics in Atomic and Nuclear Theory,* lectures given at the University of Canterbury, New Zealand (Christchurch: Caxton Press, 1971), pp. 1–60.
14 (a) A. M. Lane, *Nuclear Theory* (New York: Benjamin, 1964).
(b) J. R. Schrieffer, *Theory of Superconductivity* (New York: Benjamin, 1964).
15 (a) V. Fock, *Z. Physik* 98: 145 (1935).
(b) V. Bargmann, *Z. Physik* 99: 576 (1936).
16 L. C. Biedenharn, *J. Math. Phys.* 2: 433 (1961).
17 W. Pauli, Jr., *Z. Physik* 36: 336 (1926).
18 (a) C. Runge, *Vektoranalysis* (Leipzig: S. Hirzel, 1919), p. 68; trans. (London: Methuen, 1923).
(b) W. Lenz, *Z. Physik* 24: 197 (1924).
19 N. V. V. J. Swamy, R. G. Kalkarni, and L. C. Biedenharn, *J. Math. Phys.* 11: 1165 (1970).
20 S. Pasternack and R. M. Sternheimer, *J. Math. Phys.* 3: 1280 (1962).
21 (a) A. O. Barut in *Symmetry Principles of High Energy,* ed. B. Kursunoğlu and A. Perlmutter (San Francisco, Freeman, 1964), pp. 81–92.
(b) A. O. Barut, *Phys. Rev.* 135B: 839 (1964).
(c) A. O. Barut and A. Böhm, *Phys. Rev.* 139B: 1107 (1965).
22 Y. Dothan, M. Gell-mann, and Y. Ne'eman, *Phys. Letters* 17: 148 (1965).
23 E. C. G. Sudarshan, N. Mukunda, and L. O'Raifeartaigh, *Phys. Letters* 19: 322 (1965).
24 (a) A. O. Barut in *High Energy Physics and Elementary Particles* (Vienna: IAEA, 1965), pp. 679–94, 707–21.
(b) A. O. Barut in *Non-Compact Groups in Particle Physics,* ed. Y. Chow (New York: Benjamin, 1966), pp. 1–23.
(c) A. O. Barut in *Symmetry Principles and Fundamental Particles,* ed. B. Kursunoğlu and A. Perlmutter (San Francisco: Freeman, 1967), pp. 92–125.
(d) Y. Nambu in *Proceedings of the 1967 International Conference on Particles and Fields,* ed. C. R. Hagen, G. Guralnik, and W. S. Mathur (New York: Wiley, 1967), pp. 383–93.
(e) A. Böhm in *Lectures in Theoretical Physics,* ed. W. E. Brittin and A. O. Barut (New York: Gordon and Breach, 1967), Vol. IXB, pp. 327–36.
(f) A. O. Barut in *Lectures in Theoretical Physics,* ed. W. E. Brittin and A. O. Barut (New York: Gordon and Breach, 1967), Vol. XB, pp. 377–426.

(g) H. Kleinert in *Lectures in Theoretical Physics,* ed. W. E. Brittin and A. O. Barut (New York: Gordon and Breach, 1967) Vol. XB, pp. 427–82.
(h) A. O. Barut and S. Malin, *Rev. Mod. Phys.* 40: 632 (1968).
25 (a) V. Bargmann, *Ann. Math.* 48: 568 (1947).
(b) A. O. Barut and C. Fronsdal, *Proc. Roy. Soc. (London)* A287: 532 (1965).
(c) A. O. Barut in *Lectures in Theoretical Physics,* ed. W. E. Brittin, A. O. Barut, and M. Guenin (New York: Gordon and Breach, 1967), Vol. IXA, pp. 124–71.
26 A. O. Barut and G. L. Bornzin, *J. Math. Phys.* 12: 841 (1971).
27 (a) A. O. Barut, D. Corrigan, and H. Kleinert, *Phys. Rev. Letters* 20: 167 (1968); *Phys. Rev.* 167: 1527 (1968).
(b) A. O. Barut and A. Baiquni, *Phys. Rev.* 184: 1342 (1969).
28 A. O. Barut and H. Kleinert, *Phys. Rev.* 160: 1149 (1967); *Phys. Rev.* 161: 1464 (1967).
29 A. O. Barut and W. Rasmussen, *Phys. Rev.* 3D: 956 (1971).
30 C. E. Wulfman in *Group Theory and Its Applications,* Vol. 2, ed. E. M. Loebl (New York: Academic Press, 1971), pp. 145–97.
31 L. Armstrong, Jr., *Phys. Rev.* A3: 1546 (1971).
32 L. Armstrong, Jr., *J. de Phys.* 31: Suppl. C4: 17 (1970).
33 B. R. Judd in *Advances in Atomic and Molecular Physics,* Vol. 7, ed. D. R. Bates and I. Esterman (New York: Academic Press, 1971), pp. 251–86.

Chapter 8. Self-consistent fields: Hartree-Fock theory; Thomas-Fermi model

1 L. H. Thomas, *Proc. Cambridge Phil. Soc.* 23: 542 (1927).
2 E. Fermi, *Atti Accad. nazl. Lincei* 6: 602 (1927); 7: 342 (1928); *Z. Physik* 48: 73; 49: 550 (1928).
3 (a) E. B. Baker, *Phys. Rev.* 36: 630 (1930).
(b) E. Guth and R. Peierls, *Phys. Rev.* 37: 217 (1931).
4 (a) P. Gombás, *Die Statistische Theorie des Atoms und ihre Anwendungen* (Vienna: Springer-Verlag, 1949).
(b) P. Gombás, "Statistische Behandlung de Atoms," in *Encyclopedia of Physics* Vol. 36 (Berlin: Springer-Verlag, 1956).
5 (a) E. Fermi, *Z. Physik* 36: 902 (1926); *Collected Papers of Enrico Fermi* Vol. 1 (Chicago: University of Chicago Press, 1962), p. 186.
(b) P. A. M. Dirac, *Proc. Roy. Soc. (London)* A112: 661 (1926).
(c) E. Fermi and E. Amaldi, *Mem. Accad. Ital.* 6: 117 (1934); *Collected Papers of Enrico Fermi,* Vol. 1 (Chicago: University of Chicago Press, 1962), p. 592.
6 (a) V. Bush and S. H. Caldwell, *Phys. Rev.* 38: 1898 (1931).
(b) C. Miranda, *Mem. Accad. Ital.* 5: 283 (1934).
(c) J. C. Slater and H. M. Krutter, *Phys. Rev.* 47: 559 (1935).
(d) S. Kobayashi, T. Matsukuma, S. Nagai, and K. Umeda, *J. Phys. Soc. Japan* 10: 759 (1955).
(e) C. A. Coulson and H. M. March, *Proc. Phys. Soc.* (London) A63: 367 (1950); correction, *Proc. Cambridge Phil. Soc.* 48: 665 (1952).
(f) S. Kobayashi, *J. Phys. Soc. Japan* 10: 824 (1955).

7 R. Latter, *Phys. Rev.* 99: 510 (1955).
8 A. Sommerfeld, *Z. Physik* 78: 283 (1932).
9 (a) P. A. M. Dirac, *Proc. Cambridge Phil. Soc.* 26: 376 (1930); 27: 240 (1931).
(b) H. Jensen, *Z. Physik* 89: 713 (1934); 93: 232 (1935).
10 (a) J. C. Slater and H. M. Krutter, *Phys. Rev.* 47: 559 (1935).
(b) H. Jensen, G. Meyer-Gossler, and H. Rohde, *Z. Physik* 110: 277 (1938). [Their results are also published in Gombás, *Die Statistiche.*]
(c) R. P. Feynman, N. Metropolis, and E. Teller, *Phys. Rev.* 75: 1561 (1949).
(d) N. Metropolis and J. R. Reetz, *J. Chem. Phys.* 19: 555 (1951).
(e) P. Gombás and R. Gáspár, *Acta Phys. Acad. Sci. Hungary* 1: 66 (1951). Calculations of K. Umeda, *Phys. Rev.* 58: 92 (1940); *J. Fac. Sci. Hokkaido Univ.* (2) 3: 171 (1942) for $Z = 1$ to 92 are unusable because they are based on an incorrect boundary value for n_0 (See Gombás, reference 4*b*, page 141).
(f) L. H. Thomas, *J. Chem. Phys.* 22: 1758 (1954).
(g) Maria G. Mayer, *Phys. Rev.* 60: 184 (1941).
11 (a) P. Gombás, *Acta Phys. Acad. Sci. Hungary* 13: 233 (1961); 14: 83 (1962); *Z. Physik* 121: 523 (1943).
(b) H. W. Lewis, *Phys. Rev.* 111: 1554 (1958).
12 C. F. von Weizsäcker, *Z. Physik* 96: 431 (1935).
13 K. Yonei and Y. Tomishima, *J. Phys. Soc. Japan* 20: 1051 (1965).
14 (a) D. A. Kirzhnits, *Zh. Eksperim. i Teor. Fiz.* 32: 115 (1957); trans. *Soviet Phys. – JETP* 5: 64 (1957).
(b) S. Golden, *Phys. Rev.* 105: 604 (1957); 107: 1283 (1957).
(c) G. A. Baraff and S. Borowitz, *Phys. Rev.* 121: 1704 (1961).
15 H. W. Schey and J. L. Schwartz, *Phys. Rev.* 137A: 709 (1965).
16 (a) J. S. Plaskett, *Proc. Phys. Soc.* (*London*) 66A: 178 (1953).
(b) N. H. March and J. S. Plaskett, *Proc. Roy. Soc. (London)* A235: 419 (1956).
17 N. A. Ashby and M. A. Holzman, *Phys. Rev.* A1: 764 (1970).
18 (a) P. Gombás and K. Ladányi, *Acta Phys. Acad. Sci. Hungary* 5: 313 (1955); 7: 255 (1957); 7: 263 (1957); 8: 301 (1958); *Z. Physik* 158: 261 (1960).
(b) P. Gombás and T. Szondy, *Acta Phys. Acad. Sci. Hungary* 14: 335 (1962).
19 P. Gombás, *Z. Physik* 118: 164 (1941); *Acta Phys. Acad. Sci. Hungary* 1: 285 (1952).
20 N. A. Ashby and M. A. Holzman, *Topics in Modern Physics – A Tribute to E. U. Condon,* ed. W. E. Brittin and H. Odabasi (Boulder: Colorado Associated University Press, 1971), pp. 1–14.
21 (a) D. R. Hartree, *Proc. Cambridge Phil. Soc.* 24: 89, 111, 426 (1927).
(b) D. R. Hartree, W. Hartree, and B. Swirles, *Phil. Trans. Roy. Soc. London* A238: 229 (1939).
(c) D. R. Hartree, *Rep. Progr. Phys.* 11: 113 (1948).
(d) D. R. Hartree, *The Calculation of Atomic Structure* (New York: Wiley, 1957).
22 V. Fock, *Z. Physik* 61: 126; 62: 795 (1930).
23 J. C. Slater, *Phys. Rev.* 35: 210 (1930).

24 (a) J. C. Slater, *Quantum Theory of Atomic Structure,* Vol. 2 (New York: McGraw-Hill, 1960).
(b) H. A. Bethe and R. Jackiw, *Intermediate Quantum Mechanics,* 2nd ed. (New York: Benjamin, 1968).
25 A. E. Taylor, *Advanced Calculus* (New York: Ginn, 1955).
26 M. Delbrück, *Proc. Roy. Soc. (London)* A129: 686 (1930).
27 C. C. J. Roothaan, *Rev. Mod. Phys.* 23: 69 (1951).
28 (a) J. C. Slater, *Phys. Rev.* 82: 538 (1951).
(b) R. Lefebvre, *Compt. Rend.* 237: 1158 (1953).
(c) G. Berthier, *J. Chim. Phys.* 51: 363 (1954).
(d) J. A. Pople and R. K. Nesbet, *J. Chem. Phys.* 22: 571 (1954).
(e) R. K. Nesbet, *Proc. Roy. Soc. (London)* A230: 312 (1955); *Rev. Mod. Phys.* 33: 28 (1961).
(f) P. O. Löwdin, *Phys. Rev.* 97: 1509 (1955); *Rev. Mod. Phys.* 32: 328 (1960).
(g) G. W. Pratt, Jr., *Phys. Rev.* 102: 1303 (1956).
(h) V. Heine, *Phys. Rev.* 107: 1002 (1957).
(i) J. H. Wood and G. W. Pratt, Jr., *Phys. Rev.* 107: 995 (1957).
(j) R. E. Watson and A. J. Freeman, *Phys. Rev.* 120: 1125 (1960).
(k) R. K. Nesbet and R. E. Watson, *Ann. Phys.* 9: 260 (1960).
(l) D. A. Goodings, *Phys. Rev.* 123: 1706 (1961).
(m) N. Bessis, H. Lefebvre-Brion, and C. M. Moser, *Phys. Rev.* 128: 213 (1962).
(n) R. K. Nesbet in *Quantum Theory of Atoms, Molecules, and the Solid State – A Tribute to J. S. Slater,* ed. P. O. Löwdin (New York: Academic Press, 1966), pp. 157–65.
29 (a) J. J. Vizbaraite, K. K. Eringis, and A. P. Jucys, *Doklady AN SSR* 135: 809 (1960).
(b) K. K. Eringis, J. J. Vizbaraite, and A. P. Jucys, *Liet. TSR MA Darbai, Trudy AN Lit. SSR,* (26) B3: 67 (1961).
(c) A. P. Jucys, V. K. Shugurow, J. J. Vizbaraite, and K. K. Eringis, *Liet. TSR MA Darbai, Trudy AN Lit. SSR,* (26) B3: 81 (1961).
(d) A. P. Jucys, *Int. J. Quant. Chem.* 1: 311 (1967); 1s: 37 (1967).
30 (a) P. O. Löwdin, *Phys. Rev.* 97: 1474, 1509 (1955); *J. Math. Phys. 3,* 1171 (1962); in *Quantum Theory of Atoms, Molecules, and the Solid State – A Tribute to J. C. Slater,* ed. P. O. Löwdin (New York: Academic Press, 1966), pp. 601–23.
(b) U. Kaldor and F. E. Harris, *Phys. Rev.* 183: 1 (1969) and references therein.
31 (a) C. C. J. Roothaan, *Rev. Mod. Phys.* 32: 179 (1960).
(b) S. Huzinaga, *Phys. Rev.* 120: 866 (1960); 122: 131 (1962).
(c) C. C. J. Roothaan and P. S. Bagus in *Methods in Computational Physics,* Vol. 2, ed. B. Alder, S. Fernbach, and M. Rotenberg (New York: Academic Press, 1963), p. 47.
32 (a) J. C. Slater, J. B. Mann, T. M. Wilson, and J. H. Wood, *Phys. Rev.* 184: 672 (1969).
(b) J. C. Slater, *Int. J. Quant. Chem.* S3: 727 (1970).
33 (a) C. Froese, *J. Chem. Phys.* 45: 1417 (1966).
(b) J. B. Mann, *Atomic Structure Calculations,* Los Alamos Scientific Lab. Rep., LA-3690 (1967); LA-3691 (1968).

34 (a) T. H. Koopmans, *Physica* 1: 104 (1933).
(b) S. T. Epstein, *The Variation Method in Quantum Chemistry* (New York: Academic Press, 1974), p. 46.
35 (a) E. Clementi, *Tables of Atomic Functions* (San Jose, Cal.: International Business Machines Corporation, 1965).
(b) E. Clementi and C. Roetti, *Atomic Data and Nuclear Data Tables* 14: 177 (1974).
36 (a) P. M. Morse, L. A. Young, and E. S. Haurwitz, *Phys. Rev.* 48: 948 (1935).
(b) L. Goldberg and A. M. Clogston, *Phys. Rev.* 56: 696 (1939).
(c) W. E. Duncanson and C. A. Coulson, *Proc. Roy. Soc. Edinburgh* 62: 37 (1944).
37 (a) A. M. Karo and L. C. Allen, *J. Chem. Phys.* 31: 968 (1959).
(b) L. C. Allen, *J. Chem. Phys.* 34: 1156 (1961).
(c) C. C. J. Roothaan, L. M. Sachs, and A. W. Weiss, *Rev. Mod. Phys.* 32: 186 (1960); 124: 1117 (1961).
(d) R. E. Watson, *Phys. Rev.* 118: 1036 (1960); 119: 1934 (1960).
(e) R. E. Watson and A. J. Freeman, *Phys. Rev.* 123: 521 (1961).
(f) E. Clementi, C. C. J. Roothaan, and M. Yoshimine, *Phys. Rev.* 127: 1618 (1962).
(g) J. W. Richardson, W. C. Nieuwpoort, R. R. Powell, and W. F. Edgell, *J. Chem. Phys.* 36: 1057 (1962).
(h) A. J. Freeman and R. E. Watson, *Phys. Rev.* 127: 2058 (1962).
(i) C. C. J. Roothaan and P. S. Kelly, *Phys. Rev.* 131: 1177 (1963).
(j) J. W. Richardson, R. R. Powell, and W. C. Nieuwpoort, *J. Chem. Phys.* 38: 796 (1963).
(k) E. Clementi, *J. Chem. Phys.* 38: 996, 1001 (1963); 41: 295, 303 (1964); *IBM J. Res. Develop.* 9: 2 (1965).
(l) C. C. J. Roothaan and M. Synek, *Phys. Rev.* 133A: 1263 (1964).
(m) M. Synek, *Phys. Rev.* 131: 1572 (1963); 133A: 961 (1964); 139A: 1049 (1965).
(n) M. Synek and G. E. Stungis, *J. Chem. Phys.* 41: 971 (1964); *Phys. Rev.* 136A: 112 (1964).
(o) M. Synek, A. E. Rainis, and C. C. J. Roothaan, *Phys. Rev.* 141: 174 (1966).
(p) G. L. Malli, *Canad. J. Phys.* 44: 3121 (1966).
(r) M. Synek, A. E. Rainis, and E. A. Peterson, *J. Chem. Phys.* 46: 2039 (1967).
(s) M. Synek, P. Grossgut, and F. Schmitz, *Phys. Rev.* 178: 73 (1969).
(t) M. Synek and W. Timmons, *Phys. Rev.* 185: 38 (1969).
(u) M. Synek and P. Grossgut, *Phys. Rev.* A1: 1 (1970).
38 D. R. Hartree and M. H. Black, *Proc. Roy. Soc. (London)* A139: 311 (1933).
39 D. R. Hartree and W. Hartree, *Proc. Roy. Soc. (London)* A150: 9 (1935).
40 E. C. Ridley, *Proc. Cambridge Phil. Soc.* 51: 702 (1955).
41 W. W. Piper, *Trans. AIEE* 75: 152 (1956).
42 C. Froese, *Proc. Cambridge Phil. Soc.* 53: 206 (1957).
43 B. H. Worsley, *Canad. J. Phys.* 36: 289 (1958).
44 (a) C. Froese, *Canad. J. Phys.* 41: 1895 (1963); 46: 2336 (1968).

(b) C. Froese Fischer, *Comp. Phys. Comm.* 1: 151 (1969).
45 D. F. Mayers and F. O'Brien, *J. Phys.* 1B: 145 (1968).
46 S. Fraga, J. Karwowski, and K. M. S. Saxena, *Handbook of Atomic Data* (New York: Elsevier, 1976).
47 D. C. Griffin, R. D. Cowan, and K. L. Andrew, *Phys. Rev.* 3A: 1233 (1971).
48 C. Froese Fisher, *J. Comp. Phys.* 10: 211 (1972).
49 J. C. Slater, *Phys. Rev.* 81: 385 (1951).
50 R. Gáspár, *Acta. Phys. Acad. Sci. Hungary* 3: 263 (1954).
51 W. Kohn and L. J. Sham, *Phys. Rev.* 140A: 1133 (1965).
52 R. D. Cowan, *Phys, Rev.* 163: 54 (1967).
53 F. Herman and S. Skillman, *Atomic Structure Calculations* (Englewood Cliffs, N.J.: Prentice Hall, 1963).
54 E. C. Snow, J. M. Canfield, and J. T. Waber, *Phys. Rev.* 135A: 969 (1964).
55 I. Lindgren, *Arkiv Fysik* 31: 59 (1966).
56 (a) R. D. Cowan, A. C. Larson, D. Liberman, J. B. Mann, and J. Waber, *Phys. Rev.* 144: 5 (1966).
(b) A. Rosén and I. Lindgren, *Phys. Rev.* 176: 114 (1968).
57 (a) J. A. Bearden and A. F. Burr, *Atomic Energy Levels,* Rept. No. NYO-2543-1, USAEC, Div. Tech. Info., Oak Ridge, Tenn., 1965.
(b) C. E. Moore, *Atomic Energy Levels,* Vol. I, *Nat. Bur. Stand. Circ.* No. 467 (Washington, D.C.: Government Printing Office, 1949).
58 (a) M. Ross, *Phys. Rev.* 179: 612 (1969).
(b) W. E. Rudge, *Phys. Rev.* 181: 1024, 1033 (1969).
59 M. Berrondo and O. Goscinski, *Univ. of Uppsala Rept.* No. 198, 1967; *Phys. Rev.* 184: 10 (1969).
60 J. H. Wood, *Int. J. Quant. Chem.* S3: 747 (1970).
61 F. Herman, J. P. Van Dyke, and I. B. Ortenburger, *Phys. Rev. Letters* 22: 807 (1969).
62 E. A. Kmetko, *Phys. Rev.* A1: 37 (1970).
63 (a) J. Yamashita, M. Fukuchi, and S. Wakoh, *J. Phys. Soc. Japan* 18: 999 (1963).
(b) J. Yamashita, *J. Phys. Soc. Japan,* 18: 1010 (1963).
(c) S. Wakoh and J. Yamashita, *J. Phys. Soc. Japan,* 19: 1342 (1964).
(d) S. Wakoh, *J. Phys. Soc. Japan* 20: 1894 (1965).
64 J. C. Slater and J. H. Wood, *Int. J. Quant. Chem.* S4: 3 (1971).
65 J. C. Slater in *Topics in Modern Physics – A Tribute to E. U. Condon,* ed. W. E. Brittin and H. Odabasi (Boulder: Colorado Associated University Press, 1971), pp. 297–303.
66 (a) A. P. Jucys and J. J. Vizbaraite, *Liet. TSR MA Darbai, Trudy AN Lit. SSR,* (24) B 1: 65 (1961).
(b) R. Karazija, P. Bogdanovicius, and A. P. Jucys, *Acta Phys. Acad. Sci. Hungary* 27: 467 (1969).
67 C. Froese, *Phys. Rev.* 137A: 1644 (1965).

Chapter 9. Configuration structure

1 B. Edlén in *Encyclopedia of Physics* Vol. 27 (Berlin: Springer, 1964).
2 W. Ritz, *Physik Z.* 4: 406 (1903).
3 H. Odabasi, *J. Opt. Soc. Am.* 57: 1459 (1967).

4 R. F. Bacher and S. Goudsmit, *Phys. Rev.* 46: 948 (1934).
5 (a) S. Meshkov and C. W. Ufford, *Phys. Rev.* 94: 75 (1954).
(b) R. E. Trees, *J. Res. Nat. Bur. Stand.* 53: 35 (1954); *J. Opt. Soc. Am.* 48: 293 (1958).
6 (a) E. U. Condon and H. Odabasi, *Joint Institute for Laboratory Astrophysics (JILA) Report* No. 95 (1968).
(b) H. Odabasi, *JILA Report* No. 97 (1968).
7 E. Madelung, *Mathematishe Hilfsmittel des Physikers,* 3rd ed. (Berlin: Springer, 1936), p. 359.
8 C. Froese Fischer, *Canad. J. Phys.* 49: 1205 (1971).
9 (a) D. Neubert, *Z. Naturf.* 25A: 210 (1970).
(b) P.-O. Löwdin, *Int. J. Quant. Chem.* S3: 331 (1969).
(c) O. Novaro and M. Berrondo, *J. Phys.* 5B: 1104 (1972).
10 A. O. Barut in *The Structure of Matter,* Rutherford Centennial Symposium, 1971, ed. B. G. Wybourne (Christchurch: University of Canterbury Publications, 1972), pp. 126–36.
11 A. O. Barut and A. Böhm, *J. Math. Phys.* 11: 2938 (1970).
12 H. Odabasi, *Int. J. Quant. Chem.* S7: 23 (1973).
13 B. Edlén in *Topics in Modern Physics – A Tribute to E. U. Condon,* ed. W. E. Brittin and H. Odabasi (Boulder: Colorado Associated University Press, 1971), p. 133.
14 (a) H. H. Marvin, *Phys. Rev.* 47: 521 (1935).
(b) M. A. Catalan, F. Rohrlich, and A. G. Shenstone, *Proc. Phys. Soc. (London)* A221: 421 (1954).

Appendix 1

1 (a) J. C. Maxwell, *A Treatise on Electricity and Magnetism,* 3rd. ed. (1891); reprinted by Dover, New York, 1954, pp. 1–6.
(b) Alfred O'Rahilley, *Electromagnetics, a Discussion of Fundamentals* (Cork: Cork University Press, 1938); later reprinted by Dover, New York. [Chapters XIV and XV present a spirited critique of many of the confusions in the treatment of physical "dimensions" in the literature.]
2 E. U. Condon, *Am. J. Phys.* 29: 487 (1961).
3 R. T. Birge, *Rev. Mod. Phys.* 1: 1 (1929); *Phys. Rev.* 40: 228 (1932); *Phys. Rev.* 43: 211 (1933).
4 E. R. Cohen and J. W. M. DuMond, *Encyclopedia of Physics,* Vol. 34 (Berlin: Springer, 1957), pp. 1–87; *Rev. Mod. Phys.* 37: 537 (1965).
5 *Nat. Bur. Stand. Tech. News Bull.,* 175, October 1963; *Phys. Today* 17: 48 (1964).
6 (a) B. N. Taylor, W. H. Parker, and D. N. Langenberg, *The Fundamental Constants and Quantum Electrodynamics* (New York: Academic Press, 1969); *Rev. Mod. Phys.* 41: 375 (1969).
(b) E. R. Cohen and B. N. Taylor, *J. Phys. Chem. Ref. Data* 2: 663 (1973).
7 H. Barrell, "The Metre," *Contemporary Physics* 3: 415 (1962).
8 D. R. Hartree, *Proc. Cambridge Phil. Soc.* 24: 89 (1926).
9 Max Planck, *The Theory of Heat Radiation* (Philadelphia: Blakiston's, 1914), p. 173.

Appendix 2

1 J. D. Louck, *Phys. Rev.* 110: 815 (1958).
2 C. Schwartz and A. de-Shalit, *Phys. Rev.* 94: 1257 (1954).
3 A. de-Shalit, *Phys. Rev.* 91: 1479 (1953).
4 J. M. Kennedy, B. J. Sears, and W. T. Sharp, *Tables of* X *Coefficients,* CRT 569, AECL No. 106, Atomic Energy of Canada Limited, Chalk River, Ontario (1954).
5 H. A. Jahn and J. Hope, *Phys. Rev.* 93: 318 (1954).
6 R. J. Ord-Smith, *Phys. Rev.* 94: 1227 (1954).
7 J. P. Elliott and B. H. Flowers, *Proc. Roy. Soc. (London)* A229: 536 (1955).
8 W. T. Sharp, *Some Formal Properties of the 12*j*-symbol,* TPI-81, Atomic Energy of Canada Limited, Chalk River, Ontario (1955).

Appendix 3

1 J. C. Slater, *Phys. Rev.* 34: 1293 (1929).
2 J. C. Slater, *Quantum Theory of Atomic Structure,* Vol. II (New York: McGraw-Hill, 1960), App. 21.
3 E. U. Condon and G. H. Shortley, *Phys. Rev.* 37: 1025 (1931).
4 E. U. Condon and G. H. Shortley, *The Theory of Atomic Spectra* (Cambridge: Cambridge University Press, 1935), pp. 197–207.
5 N. M. Gray and L. A. Wills, *Phys. Rev.* 38: 248 (1931).
6 M. H. Johnson, Jr., *Phys. Rev.* 38: 1628 (1931).
7 M. H. Johnson, Jr., *Phys. Rev.* 39: 197 (1932).
8 M. H. Johnson, Jr., *Phys. Rev.* 43: 632 (1933).
9 C. W. Ufford and G. H. Shortley, *Phys. Rev.* 42: 167 (1932).
10 M. Ostrofsky, *Phys. Rev.* 46: 604 (1934).
11 G. H. Shortley and B. Fried, *Phys. Rev.* 54: 739 (1938).
12 G. Racah, *Phys. Rev.* 61: 186 (1941); *Quantum Theory of Angular Momentum (QTAM)*, ed. L. C. Biedenharn and H. Van Dam, (New York: Academic Press, 1965), pp. 134–45.
13 G. Racah, *Phys. Rev.* 62: 438 (1942); *QTAM,* pp. 146–70.
14 G. Racah, *Phys. Rev.* 63: 367 (1943); *QTAM,* pp. 171–86.
15 G. Racah, *Phys. Rev.* 76: 1352 (1949); *QTAM,* pp. 187–200.
16 O. Laporte, *Phys. Rev.* 61: 302 (1942).
17 T. Yamanouchi and A. Amemiya, *J. Phys. Soc. Japan* 1: 18 (1946).
18 F. Rohrlich, *Phys. Rev.* 74: 1372 (1948).
19 T. Ishidzu and S. Obi, *J. Phys. Soc. Japan* 5: 124 (1950).
20 E. F. Reilly, *Phys. Rev.* 91: 876 (1953).
21 B. G. Wybourne, *J. Chem. Phys.* 35: 340 (1961).
22 B. G. Wybourne, *J. Chem. Phys.* 37: 450 (1962).
23 W. A. Runciman, *J. Chem. Phys.* 36: 1481 (1962).
24 C. W. Nielson and G. F. Koster, *Spectroscopic Coefficients for the p^n, d^n and f^n Configuration* (Cambridge, Mass: MIT Press, 1963).

Name index

Subject index